Archimedes

NEW STUDIES IN THE HISTORY AND PHILOSOPHY OF SCIENCE AND TECHNOLOGY

VOLUME 55

Archimedes has three fundamental goals; to further the integration of the histories of science and technology with one another: to investigate the technical, social and practical histories of specific developments in science and technology; and finally, where possible and desirable, to bring the histories of science and technology into closer contact with the philosophy of science. To these ends, each volume will have its own theme and title and will be planned by one or more members of the Advisory Board in consultation with the editor. Although the volumes have specific themes, the series itself will not be limited to one or even to a few particular areas. Its subjects include any of the sciences, ranging from biology through physics, all aspects of technology, broadly construed, as well as historically-engaged philosophy of science or technology. Taken as a whole, *Archimedes* will be of interest to historians, philosophers, and scientists, as well as to those in business and industry who seek to understand how science and industry have come to be so strongly linked.

More information about this series at http://www.springer.com/series/5644

H Darrel Rutkin

Sapientia Astrologica: Astrology, Magic and Natural Knowledge, ca. 1250–1800

I. Medieval Structures (1250-1500): Conceptual, Institutional, Socio-Political, Theologico-Religious and Cultural

Springer

H Darrel Rutkin
Dipartimento di Filosofia e Beni Culturali
Università Ca' Foscari Venezia
Dorsoduro, Venezia, Italy

ISSN 1385-0180 ISSN 2215-0064 (electronic)
Archimedes
ISBN 978-3-030-10778-9 ISBN 978-3-030-10779-6 (eBook)
https://doi.org/10.1007/978-3-030-10779-6

Library of Congress Control Number: 2018968012

This Springer imprint is published by the registered company Springer Nature Switzerland AG.
The registered company address is: Gewerbestrasse 11, 6330 Cham, Switzerland

Preface

Daily newspaper horoscopes are like a text message from Athena.

—Sophia Joy Rutkin, age 13.

Astrology currently plays a variety of complex and often controversial roles in the contemporary early twenty-first-century's increasingly globalized world—East and West, including Europe and America[1]—sometimes functioning even in the highest corridors of power. Nancy Reagan, the wife of Ronald Reagan (1911–2004), the 40th president of the United States (1981–1989), famously used an astrologer from San Francisco, Joan Quigley, to time a broad range of official activities and announcements to her husband's best political advantage. To do so, she used a traditional astrological practice called "elections," namely, the choosing of astrologically propitious times for beginning any sort of venture, many earlier examples of which will be seen in what follows.[2] Apparently, François Mitterrand (1916–1996), the 21st President of France (1981–1995), also regularly sought astrological advice.[3]

Despite the virtually ubiquitous appearance of the daily newspaper sun-sign "horoscope," an extremely watered down, twentieth-century expression of

[1] I avoid the East almost entirely in what follows, due mainly to issues of competence and space, not lack of interest or widespread expression in either historical or contemporary India and the Islamic world, China (with its own indigenous ancient system), or other Eastern societies. The *New York Times* recently reported (6 January 2015) about astrology in contemporary Sri Lankan politics: "As Vote Nears, Astrologer for Sri Lanka's President Faces Ultimate Test of his Skills," by Ellen Barry.

[2] Joan Quigley, "*What Does Joan Say?*": *My Seven Years as White House Astrologer to Nancy and Ronald Reagan*, New York: Birch Lane Press, 1990. Reagan was famously known as the "teflon president," perhaps in part due to Quigley's astrological interventions.

[3] Elizabeth Teissier was his astrologer between 1989 and 1994; see Jon Henley, "How Mitterand Sought Advice from Astrologer," *The Observer*, 25 June 2000 (http://www.theguardian.com/world/2000/jun/25/jonhenley.theobserver2). Controversially, Teissier also earned a PhD in sociology at the Sorbonne with a thesis entitled, "Situation épistémologique de l'astrologie à travers l'ambivalence fascination / rejet dans les sociétés postmodernes," in 2001. My thanks to Marco Gentile for bringing Teissier to my attention.

astrology's central structure[4]—a figure of the heavens coordinated for time and place, to be extensively discussed in what follows—and its periodic recrudescences in popular and/or esoteric psychological and religious contexts,[5] astrology still inspires controversy today due to its persistently broad range of societal roles, even though today, of course, astrology has no purchase among scholars, scientific or humanistic, and is relegated to aficionados of the art and to believers.

Surprisingly (or perhaps not!), astrology still provokes outcries of outrage, indignation and intolerance. In 1955, for example, at the time of the McCarthy investigations (often described as "witchhunts"), the dean of American historians of science and the founder of *Isis: An International Review Devoted to the History of Science and its Cultural Influences*, George Sarton (1884–1956) of Harvard University insisted that astrologers should be arrested and jailed for their beliefs and practices, due to his profound suspicion that they were preying on society.[6] Similarly, in 1972, Wayne Shumaker (1910–1999),[7] a distinguished professor of English at the University of California, Berkeley, devoted the first extensive chapter of his influential *The Occult Sciences in the Renaissance: A Study in Intellectual Patterns* to debunking the science, art, and/or practice of astrology, due explicitly to his alarm at its perceived deleterious effects on his impressionable undergraduates in the Berkeley of the late 1960s and early 1970s.[8]

Finally, in its September/October 1975 issue, *The Humanist*, an organ of secular humanism in America, published an attack on astrology that included an oath rejecting astrology signed by 186 scientists. This provoked a characteristically harsh critical firebombing (Feyer-bombing!) of their entire enterprise by Paul Feyerabend (1924–1994), a distinguished and controversial philosopher of science also at UC Berkeley.[9] Feyerabend's arguments are very interesting indeed and strike a suggestive tone for the study that follows, where far too much ideologically laden and mostly unconscious assumption, assertion, prejudice and wishful thinking has passed for critical historical scholarship.

*

[4] For newspaper astrology in the England of the 1930s, see Ellic Howe's illuminating, *Astrology: A Recent History including the Untold Story of its Role in World War II*, New York: Walker and Company, 1968, 66 ff.

[5] See, e.g., Wouter J. Hanegraaff, *New Age Religion and Western Culture: Esotericism in the Mirror of Secular Thought*, Leiden: Brill, 1996.

[6] Eugenio Garin provides the reference in his *Lo zodiaco della vita: la polemica sull'astrologia dal Trecento al Cinquecento*, Rome: Laterza, 1976, 129 (n. 2) to George Sarton, *Ancient Science and Modern Civilization*, New York: Harper, 1959, 61 ff.

[7] My thanks to John Heilbron for tracking down Shumaker's dates.

[8] UC Berkeley was famously satirized for precisely this sort of thing as the State University of Euphoria at Plotinus (colloquially known as Euphoric State) in David Lodge's splendid novel, *Changing Places: A Tale of Two Campuses*, London: Secker and Warburg, 1975.

[9] *Science in a Free Society*, London: NLB, 1978, 91–96. The examples offered here are merely illustrative. A "thicker" study of these and other responses to astrology in the twentieth and twenty-first centuries would be very interesting, I am sure.

Given its extraordinary adaptive flexibility and striking cultural centrality—in stark contrast to its current marginality—in numerous societies from Hellenistic Babylon to the Imperial Rome of Augustus and Tiberius, and from Sasanian Persia, Abbasid Baghdad and the Byzantine Empire to Renaissance Rome, Early Modern England, and the rest of Europe,[10] the history of astrology as an alternative discourse of knowledge and praxis, with its alternating periods of legitimacy and otherwise, can help us construct a revealing mirror (like Filippo Brunelleschi's perspective mirror) for capturing and fixing a broad range of insights valuable for understanding crucial and often profoundly misunderstood features of medieval, Renaissance, and early modern European intellectual, cultural and political history. The language of optics (geometrical and otherwise) is instructive, and resonates deeply with astrology's history, as we will see. To employ another comparison, astrology seems to provide a uniquely revealing backbone to history—cultural and political, religious and scientific—due to its previous centrality and subsequent marginalization over a 2000-year span of Western history from Augustus's rule to the present day.

*

Thinking about astrology's history can also inspire reflection on historical astrology, that is, the study of astrological patterns in order to discern the deeper structures of history.[11] Contemporary astrologers point to recent outer planet conjunctions and configurations as astrological indicators of profound changes in human nature and society, as we segue more or less gracefully into yet another phase of human existence.[12] They argue that the nature of these changes is indicated by two recent sets of "great conjunctions": of Uranus and Pluto in 1965, and of Uranus and Neptune in 1992.[13] Discovered in 1781, 1846, and 1930, respectively, Uranus, Neptune, and

---

[10] For Hellenistic Babylon, see Francesca Rochberg, *The Heavenly Writing: Divination, Horoscopy, and Astronomy in Mesopotamian Culture*, Cambridge: Cambridge University Press, 2004, and her *Before Nature: Cuneiform Knowledge and the History of Science*, Chicago: University of Chicago Press, 2016; for Ancient Rome, see, e.g., Tamsyn S. Barton, *Knowledge and Power: Astrology, Physiognomics, and Medicine under the Roman Empire*, Ann Arbor: University of Michigan Press, 1994; for Sasanian Persia, see David Pingree, *From Astral Omens to Astrology: From Babylon to Binaker*, Rome: Istituto Italiano per l'Africa et l'Oriente, 1997; for Abbasid Baghdad, see Dmitri Gutas, *Greek Thought, Arabic Culture: The Graeco-Arabic Translation Movement in Baghdad and Early Abbasid Society (2nd-4th/8th–10th Centuries)*, London: Routledge, 1998; and for Byzantium, see Paul Magdalino, *L'Orthodoxie des astrologues: La science entre le dogme et la divination à Byzance (VIIe-XIVe siècle)*, Paris: Lethielleux, 2006. The others will be treated in what follows.

[11] This also relates to themes arising from astrology's relationship to divine providence and fate, as we will see in Part II.

[12] When I initially wrote this section in 2011, there was an increasing popular interest in some major changes said to be about to occur in 2012 due to a major cycle ending in the Mayan Calendar. See, for example, Daniel Pinchbeck, *2012: The Return of Quetzalcoatl*, New York: Tarcher-Penguin, 2006.

[13] Richard Tarnas interestingly treats these and other issues of what he calls "archetypal astrology" in his *Cosmos and Psyche: Intimations of a New World View*, New York: Viking, 2006, and his earlier *Prometheus the Awakener: An Essay on the Archetypal Meaning of the Planet Uranus*, Woodstock, CN: Spring Publications, 1995. In 2015, astrologers were concerned with the slow

Pluto are the new "outer" planets, although Pluto's status as a planet has recently come under attack.[14] The discovery of three "new" planets in the modern era also raises a central issue concerning astronomy and astrology's dialectical dynamics throughout history. Only time will tell what ultimate effect Pluto's demotion from full planetary to dwarf planet status will have (if any) on astrological theory and/or practice.

Regardless, the use of astrology to gain insight into the broader patterns of history also has its own significant premodern history that, while in many respects continuous with the modern situation, also has significant transformations and/or ruptures, including that the relevant outer planets under discussion up to and including the seventeenth century were primarily Jupiter and Saturn and sometimes Mars. This framework of historical astrology was taken very seriously indeed for well over a millennium, from its origins in Sasanian Persia through its rise in Abbasid Islam and full flowering in Western Europe from the fourteenth through seventeenth centuries. The formerly broad cultural acceptance of such a view (to be developed below) also points to a striking historical transformation.[15]

*

In this study, I focus on the construction of astrology's centrality and its later marginalization over the extensive period from ca. 1250 CE to ca. 1800 in conceptual, institutional, socio-political, theologico-religious, and cultural respects. Throughout, my focus will primarily be on astrology's configurations within natural knowledge (particularly within the disciplines of natural philosophy, mathematics and medicine), but I will also treat astrology's often illuminating relationships to both theology/religion and magic. Although my principal focus will be on conceptual structures, I will also devote significant attention to astrology's characteristic disciplinary configurations and institutional foundations as well as to its socio-political and cultural expressions.

To ground this analysis, I will offer an interpretive framework—constructed flexibly enough, I hope, to be subjected easily to refinement and revision—that organizes a vast amount of evidence and highlights a rich problematic for further

moving multiple squaring of Uranus in Aries with Pluto in Capricorn, especially since (strikingly) Pluto's entry (or "ingress") into Capricorn in 2008 corresponded closely to the global financial meltdown and subsequent great recession, whereas Uranus's ingress into Aries in Spring 2011 corresponded (once again strikingly) with the revolutionary development of the Arab Spring and subsequent dramatic transformations in the region.

[14] For Pluto's current status, see the article by Mason Inman, "Pluto Not a Planet, Astronomers Rule," 26 August 2006, on the National Geographic website: http://news.nationalgeographic.com/news/2006/08/060824-pluto-planet.html

[15] Some of the themes in this preface and several others may be found in Anthony Grafton's marvelous and insightful, "Starry Messengers: Recent Work in the History of Western Astrology, *Perspectives on Science* 8 (2000): 70–83. See also his and William R. Newman's "Introduction: The Problematic Status of Astrology and Alchemy in Premodern Europe" to *Secrets of Nature: Astrology and Alchemy in Early Modern Europe*, Newman and Grafton (eds.), Cambridge, MA: MIT Press, 2001, 1–37.

scholarly research. There are many significant and more-or-less understudied areas to explore. Fortunately, there is no lack of sources in many languages and in many cultural and historical periods. Only when these sources have all been catalogued, edited, analyzed and integrated will we be able to tell astrology's history fully.[16] In the meantime, I hope to show that it still well repays the effort to organize what we now know and to place it into an explicitly constructed integrated framework in order to sharpen our understanding and orient future research.[17]

I have repeatedly found that astrology—at least during the period under discussion here—benefits greatly from being studied over a long duration. In this way, the larger patterns are more easily and effectively discerned by being cast into sharper relief, much like the aerial surveying of an archeologically rich district in the late afternoon. At that time, the angle of the sun causes the resultant extended shadows to more fully reveal underlying deeper patterns that are often hidden to a normal viewing perspective. With this framework and orientation, we can then dig more knowledgably, as it were, and thus more effectively explore a broad range of deeper archeological structures.[18]

My understanding has developed as a series of approximations, with different types of evidence regularly inspiring new questions and thus further refining my interpretations. I have no doubt but that the articulations—fragments—soundings in this study will themselves be subject to further refinement and revision as collective scholarly endeavors both expand our knowledge and bring it into sharper focus.[19]

---

[16] Following in the finest scholarly traditions of Franz Cumont's CCAG and the Warburg Institute, the indefatigable David Juste is doing his best to single handedly rectify this situation, with, among other things, the recent publication of the first two volumes of his survey of extant Latin astrological manuscripts in the *Catalogus Codicum Astrologorum Latinorum* (CCAL). The first volume is entitled, *Les manuscrits astrologiques latins conservés à la Bayerische Staatsbibliothek de Munich* (Paris: CNRS Éditions, 2011) as number 81 in the series *Documents, Études et Répertoires*. The second volume is *Les manuscrits astrologiques latins conservés à la Bibliothèque nationale de France à Paris* (2015, no. 84). Although there is much excellent scholarship in many languages over its entire history, there does not yet exist a fully satisfactory overall history of astrology in any language. For examples of reasonably useful recent attempts, see S. Jim Tester, *A History of Western Astrology*, Woodbridge, Suffolk: Boydell Press, 1987; Peter Whitfield, *Astrology: A History*, New York: Abrams, 2001 (with my review: *Journal for the History of Astronomy* 34 [2003]: 335–36); Kocku von Stuckrad, *Geschichte der Astrologie: Von den Anfängen bis zur Gegenwart*, Munich: Beck, 2003; and Nicholas Campion, *History of Western Astrology*, 2 vols., London: Continuum, 2009. Much more of the scholarship on the history of astrology will be explored in what follows.

[17] The lack of such a framework accounts for many of the weaknesses in Robert S. Westman's *Copernican Question: Prognostication, Skepticism, and Celestial Order* (Berkeley: University of California Press, 2011) as I will discuss further below. Although my monograph was not originally designed or conceived as a response to Westman's study, it will partially serve as one de facto, both implicitly and explicitly.

[18] I hereby gesture, of course, to Michel Foucault's suggestive concept of the archeology of knowledge.

[19] For example, the recent groundbreaking study of Babylonian astrology by Rochberg, *Heavenly Writing*, and now Dag Nikolaus Hasse's equally groundbreaking *Success and Suppression: Arabic Science and Philosophy in the Renaissance* (Cambridge, MA: Harvard University Press, 2016), which treats astrology valuably along with philosophy and medicine.

My study hopes to contribute to this worthy endeavor *inter alia* by providing a viable, sound and flexible interpretive framework.

*

There have now been three main stages in the development of this project. In the first phase, from soon before the beginning of my graduate studies at Indiana University (1996–2002) in the History and Philosophy of Science (ca. 1993) until ca. 2005, I strove to focus as clearly and exclusively as possible on astrology *qua* natural knowledge—especially its configurations with mathematics, natural philosophy and medicine—while keeping its relationship to magic and the other so-called occult sciences strictly to one side. I focused primarily on astrology's relation to mathematics and natural philosophy, but I also treated medicine at times where it seemed most relevant. Likewise, I paid some attention to the theological dimension, especially concerning astrology's status as legitimate knowledge and practice in relation to theologico-religious (primarily Roman Catholic) orthodoxy.

The second phase began in Spring 2006 when Irven M. Resnick kindly asked me to write the essay "Astrology and Magic" in the work of Albertus Magnus for a volume of invited essays,[20] which finally forced me to deal at least to some extent with astrology's relationship to magic. It quickly became obvious that the most interesting and illuminating way to approach this subject for my interests would be to focus on astrological images (*imagines astronomicae*) or talismans. Fortunately, a brilliant work of scholarship had recently appeared on precisely this subject, greatly facilitating my work: Nicolas Weill-Parot's penetrating and massive, *Les "images astrologiques" au Moyen Age et à la Renaissance: Spéculations intellectuelles et pratiques magiques (XIIe-XVe siècle)*.[21] Furthermore, my originally one-volume study split into two volumes while I was a Dibner fellow at the Huntington Library in 2008–2009, and then again into three in Summer 2014 while at the IKGF in Erlangen.[22] The third phase on astrology and theology/religion began in earnest in the last couple of years and is just now beginning to come to fruition, with the first fruits appearing primarily in Chap. 5.

*

I have consciously chosen to err (if it be such) on the side of providing too many primary sources—in both the original languages (Latin or vernacular) and in English translation—than too few, primarily because many have been overlooked for far too long, even if they have long been published in modern editions, and many have

[20] *A Companion to Albert the Great: Theology, Philosophy, and the Sciences*, Irven M. Resnick (ed.), Leiden: Brill, 2013, 451–505. There is much of interest in this volume overall for the material reconstructed in my study, especially in Parts I and II.

[21] Paris: Champion, 2002.

[22] This explains why Part IV of volume I, originally intended to bridge the two halves of one volume, seems somewhat out of place at the end of volume I here. I must beg my readers' indulgence for this structural inelegance.

never before been turned into English. Some that have been studied and/or translated, moreover, have required more or less significant reinterpretation and revision. I have repeatedly found that astrological texts more than those of many (perhaps most) other subjects are readily misinterpreted due primarily to reading modernist and postmodern assumptions back into the premodern material. This is at least in part because much of the terminology and their correlative conceptual structures—including of Aristotelian natural philosophy—are no longer part of the mental furniture of most twenty-first-century readers.

For all of these reasons (and others to be discussed below), I have presented a great deal of the primary evidence as well as offered extensive argument, analysis and interpretation. I have consistently found that it is always best to verify and thus confirm, or to criticize and revise what other scholars say about astrology (and magic) based on the primary sources they have used to make their interpretations (no matter how fine a scholar they may otherwise be), especially when it comes to making larger claims and generalizations. Such is the lot of pre-paradigmatic reconstructions of ideologically overburdened subjects.

In this way, I also hope to make my own interpretations more easily subject to *both* verification and acceptance *and* criticism and revision, as I attempt to establish or refine at least some of the foundations required for further study. Please rest assured that what you hold in your hands (or are perusing in a digital format) is not "a" let alone "the" history of astrology, but rather a contribution to that history by presenting and interpreting a broad range of textual and other sources, and by reconstructing their meanings and contexts within a consciously and explicitly articulated interpretive framework. Much of the story remains to be understood.

Venice, Italy H Darrel Rutkin

Acknowledgments

Fata sua habent libelli! This volume—the first part of a significant development and expansion of my 2002 Indiana PhD thesis, "Astrology, Natural Philosophy and the History of Science, c. 1250–1700: Studies Toward an Interpretation of Giovanni Pico della Mirandola's *Disputationes adversus astrologiam divinatricem*"—has taken a very long time to write, but I could not have undertaken it under more favorable educational and institutional circumstances. The intellectual and educational odyssey that ultimately led to this book effectively began in March 1983 when my boss Philip Sansone's girlfriend (Clio!), at the fondly remembered but no longer extant Grok Books in Austin, Texas, fatefully placed Frances Yates's *Giordano Bruno and the Hermetic Tradition*[1] into my eager hands at the impressionable age of 23. This inspired me to study classics and ancient philosophy, primarily Plato and Plotinus, in order to explore and hopefully verify Yates's lively, imaginative and fruitful, but ultimately unreliable account.

Over the subsequent years, I have been inspired by an extraordinary range of teachers and colleagues. As an undergraduate in the departments of classics and philosophy at the University of Texas at Austin, I was fortunate to study (i.a.) with David Armstrong, Paul Woodruff and Alex Mourelatos. In graduate school for the first time in classics at Stanford, I was broadly inspired by Jack Winkler, who introduced me to Apuleius, and I was properly introduced to Aristotle by Alan Code.

Since I switched from studying ancient philosophy per se to the history of astrology as a part of the history of science on reading Otto Neugebauer's *Exact Sciences in Antiquity*[2] with my newborn daughter Sophia on my lap in October/November 1992, I have had the great good fortune to study with and learn from Bill Newman, Nico Bertoloni Meli and Michael Friedman in the Department of History and Philosophy of Science at Indiana University, and to spend a valuable year studying with Noel Swerdlow at the University of Chicago. I also benefitted immensely from spending a year working informally but fruitfully with Edith Sylla and Edward Cranz while in exile in North Carolina.

[1] Chicago: University of Chicago Press, 1964.

[2] Second ed., corrected, New York: Dover Publications, 1969.

I have been no less fortunate with a splendid collection of pre- and postdoctoral fellowships at the American Academy in Rome, the late-lamented Dibner Institute of the History and Philosophy of Science and Technology at MIT, Villa I Tatti (the Harvard University Center for Italian Renaissance Studies) in Florence, the Huntington Library and Gardens in San Marino, CA, NYU's Institute for the Study of the Ancient World (ISAW), the Italian Academy for Advanced Studies in America of Columbia University, and most recently at the Internationales Kolleg für Geisteswissenschaftliche Forschung (IKGF) of the Friedrich-Alexander Universität, Erlangen-Nürnberg.

These periods of research fellowships have been happily punctuated by teaching and research in the Department of the History of Science at the University of Oklahoma, the Department of History at the University of Nevada, Reno, the Unit (now School) for the History and Philosophy of Science at the University of Sydney, and now in the Dipartimento di Filosofia e Beni Culturali at Università Ca' Foscari Venezia, and by periodically becoming a visiting scholar in the History Department and in the Program for the History and Philosophy of Science and Technology at Stanford University, where I must particularly thank Paula Findlen and Michael Friedman for their continued interest and generosity.

Other friends and colleagues who have provided encouragement, friendship, stimulating conversations and/or helpful readings of chapters or sections include (in alphabetical order) Michael J.B. Allen, Josefina Rodriguez Arribas, Ana Cecilia Avalos, Monica Azzolini, Craig Berent, Jean-Patrice Boudet, Steven Vanden Broecke, Charles Burnett, Thony Christie, Brian Copenhaver, Geoffrey Cornelius, Patrick Curry, Benno Van Dalen, Germana Ernst, Ofer Gal, Paolo Galluzzi, Owen Gingerich, Dorian Greenbaum, Dag Nikolaus Hasse, Stephan Heilen, Hiro Hirai, Alex Jones, David Juste, Michael Lackner, Craig Martin, Maria Mavroudi, Barry and Zeke Mazur, Tayra Lanuza Navarro, Katy Park, Larry Principe, Mary Quinlan-McGrath, Irv Resnick, Nancy Siraisi, Daniel Stolzenberg, Laura Sumrall, Noel Swerdlow, Rick Tarnas, Eric Travis, Rienk Vermij and Glen Worthey. I hope I have not inadvertently left anyone out. I am particularly grateful to David Juste, Katy Park and an anonymous reader for the press for reading a prepublication version of the entire volume and offering many helpful suggestions. My special thanks to Jed Buchwald for so patiently waiting for me to deliver the manuscript. I hope it is worth the wait! Pride of place, however, must go to Anthony Grafton, whose scholarship, scholarly guidance and example have contributed immeasurably to this project. Whatever faults remain are surely my own. I hope there are not too many.

*

I would also like to acknowledge that I have completed the proofs under the auspices of a project that has received funding from the European Union's Horizon 2020 Research and Innovation Programme (GA n. 725883 Early Modern Cosmology), and as an Honorary Associate in History of Science, School for the History and Philosophy of Science, University of Sydney.

*

I dedicate this work to the memory of Norman K. Rutkin and Jacob Alexander, my father and stepfather, who did not live to see its publication, but would have heartily rejoiced to hold it in their hands. To Sondra Alexander and Judith Rutkin, my mother and stepmother, who have happily lived long enough to see that day. And to my extraordinary children, Isaac Flanagan and Sophia Rutkin, *sine quibus non*.

H Darrel Rutkin.
1 July 2017, Bondi Beach, Sydney, Australia, overlooking the Pacific.
16 May 2018, San Francisco.
8 Feb 2019, Venice, Italy.

Overall Introduction to I–III

Constructing the Problem

Astrology was finally removed from the domain of legitimate natural knowledge during the seventeenth and eighteenth centuries, but the precise contours of this story remain obscure. Although there is ample evidence, most people—historians included—do not realize that astrology was taught from at least the beginning of the fourteenth century as a regular part of the arts and sciences curriculum at the finest medieval and Renaissance universities, including Padua, Bologna, and Paris. In fact, astrology and its mathematical and natural philosophical foundations were taught in different respects within three distinct scientific disciplines: (1) in the mathematics course (the quadrivium), as astronomy's practical sister "science of the stars"; (2) in the natural philosophy course, in relation to core texts of Aristotle's natural philosophy; and (3) in the medical course, primarily in relation to Galen's *De diebus decretoriis* (*On Critical Days*).

Furthermore, astrology served to integrate the highly developed mathematical sciences of antiquity—mathematical astronomy and geography, and geometrical optics (*perspectiva*)—with natural philosophy. Drawing deeply from Al-Kindi's *De radiis stellarum* (*On Stellar Rays*), Robert Grosseteste, Albertus Magnus, and Roger Bacon in the thirteenth century integrated these mathematical disciplines with Aristotelian natural philosophy by developing a geometrical-optical model of celestial influences within the broader, essentially Ptolemaic cosmographic framework provided by mathematical astronomy and geography. This deeply mathematicized natural philosophy provided fundamental patterns of interpretation and analysis in premodern (i.e., pre-Newtonian) science. It also provided the scientific foundations for the various branches of practical astrology.

Within this conceptual and institutional context (among others), Giovanni Pico della Mirandola (1463–1494)—author of the famous humanist manifesto, the so-called *Oration on the Dignity of Man*—wrote his extensive attack on astrology, the *Disputationes adversus astrologiam divinatricem* (*Disputations against Divinatory Astrology*) in 1493–1494; it was published posthumously in 1496. It has been claimed that Pico's work was profoundly influential in subsequently dislodging astrology from its formerly central scientific, cultural and socio-political locations. In this regard, Ernst Cassirer claimed in no uncertain terms—and, ironically, using an astrological metaphor—that Pico "with one blow [...] destroys the sphere of influence of astrology."[1] This judgment is belied, however, by the fact that astrology continued to be taught for another century and more at the distinguished Italian universities mentioned above, as well as in the newly formed (and reformed) Lutheran universities under Philip Melanchthon's well-articulated and deeply astrological pedagogical influence.

It has also been claimed that Nicolaus Copernicus's *De revolutionibus orbium coelestium* (*On the Revolutions of the Celestial Orbs*, 1543) completed the process of astrology's undoing. But this too is belied by the fact that two of the most important warriors for the Copernican cause, Galileo Galilei and Johannes Kepler, were both practicing astrologers into the seventeenth century. In fact, Kepler was profoundly influenced by Pico's *Disputations*, but, instead of rejecting astrology *tout court* (as Pico would have liked), Kepler set out to reform astrology on a sounder natural philosophical and mathematical foundation. The rejection of astrology from the scientific world view is fully apparent, finally, in Isaac Newton's *Philosophiae naturalis principia mathematica* (*Mathematical Principles of Natural Philosophy*, 1687), with its revisionary mathematico-physical approach and radically new causal heuristic, which fully rejected Aristotle's causal structures and thus ultimately undermined astrology's natural philosophical foundations.

I contend that the history of astrology—in particular, the story of the protracted criticism and ultimate removal of astrology from the realm of legitimate knowledge—is crucial for fully understanding the transition from premodern Aristotelian-Ptolemaic natural philosophy to modern Newtonian science. This removal, I argue, was neither obvious nor unproblematic. Astrology was not some sort of magical nebulous hodgepodge of beliefs. Rather, astrology emerged in the thirteenth century as a richly mathematical system that served to integrate astronomy and natural philosophy, precisely the aim of the "New Science" of the seventeenth century. As such, it becomes a fundamentally important historical question to determine why this promising astrological synthesis should have been rejected in favor of a rather different mathematical natural philosophy—and one with a very different causal structure than Aristotle's.

When we look at four of the central figures of the New Science—Johannes Kepler, Galileo Galilei, Francis Bacon and Isaac Newton—we can begin to discern the spectrum of opinion. Kepler was fully committed to an astrologizing natural philosophy, albeit a reformed one, which he significantly helped to develop; he was also a practicing astrologer, despite also proposing significant reforms for practical astrology. Galileo, while practicing astrology over most of his career, worked hard to undermine its Aristotelian causal foundations. For Bacon, on the other hand, we have no evidence of practice whatsoever, although elements of his proposals for

---

[1] *The Individual and the Cosmos in Renaissance Philosophy*, Mario Domandi (tr), New York: Harper and Row, 1964 (originally published 1927), 115.

astrological reform at the end of his career point in that direction. Newton, finally, although an alchemist, was no astrologer nor does his integrated system of celestial and terrestrial mechanics embrace astrology.

How and why, then, did astrology come to lose its previous centrality in the domain of legitimate natural knowledge both conceptually and institutionally—and in relation more broadly to theology/religion, politics and culture—during the seventeenth and eighteenth centuries? These are the main questions I address in what follows by analyzing and integrating a broad range of sources.

Interpretive Framework

Although there has been a good measure of excellent scholarship within the history of science, philosophy, and culture more generally, there is still much research to be undertaken on the history of astrology, both in discovering and publishing further source material and in interpreting what is already known. Fundamental misconceptions remain to be rectified and false dichotomies exploded, even regarding such basic issues as astrology's relationship to Aristotelian natural philosophy and humanism, to Catholic and Protestant theology, and to magic and the other so-called occult sciences.[2] In this historiographic circumstance, discussing astrology in a historically and conceptually sound manner is fraught with interpretive difficulties. The studies that collectively comprise this monograph are intended to function as an integrated set of interpretive lenses, in which related conceptual and historical issues are treated. The more accurate perspectives to emerge may collectively, I hope, set the study of astrology's history on more solid scholarly foundations.

The disciplinary and curricular patterns characteristic of premodern Aristotelian-Ptolemaic-Galenic natural knowledge as taught at European universities ca. 1250–1500 provide the primary conceptual and institutional structures for developing sharper, more accurate analytic tools. With these tools, we may then construct an interpretive framework to use in more sharply focusing and ultimately reframing the larger historical problem of how astrology came to be removed from its previously central conceptual and institutional locations in premodern natural knowledge during the seventeenth and eighteenth centuries.

In approaching this question, we must first accurately understand astrology's status and locations in the premodern map of knowledge (ca. 1250–1500), which came to be so greatly transformed. In this study, I focus on two scientific disciplines in particular: mathematics and natural philosophy, although I also take the third, medicine, into account. Reconstructing characteristic features of these two bodies of knowledge—in both of which astrology was significantly configured—is a major component of my interpretive framework.

[2] All of these themes would need to be taken into account to provide a comprehensive interpretive framework.

In the mathematics curriculum, astrology as a theoretical and practical doctrine was taught as a central feature of the "science of the stars" along with mathematical astronomy, but only after the propaedeutic study of arithmetic and geometry. In this context, Johannes Müller von Königsberg, better known as Regiomontanus (1434–1474)—the first great figure in the "Renaissance of mathematics" and the first head of a scientific publishing firm—praised astrology as the queen of the mathematical disciplines in the inaugural lecture for his 1464 course in mathematical astronomy at the University of Padua. This integral configuration of astrology with astronomy in the mathematics curriculum differs greatly, of course, from our modern disciplinary structures, where astrology is no longer considered a legitimate part of mathematics, nor is it taught as such in modern universities.

Astrology's foundations in nature were taught in the natural philosophy course by reading core texts of Aristotle, most notably the *De caelo (On the Heavens)*, *De generatione et corruptione (On Generation and Corruption)*, and *Meteorologica (Meteorology)*. To establish this study's interpretive framework, I explore in detail the characteristic structures of these conceptual and curricular patterns by considering their articulation in the thirteenth century, and their further development and institutionalization during the fourteenth and fifteenth centuries, with a focus on those Northern Italian universities where Giovanni Pico della Mirandola, Nicolaus Copernicus and many others studied, namely, Padua, Bologna, and Ferrara.

Establishing and properly historicizing this framework serves two fundamental and interrelated purposes. First, it articulates astrology's primary conceptual and institutional structures within the premodern map of knowledge. Second, it provides a means to analyze in detail astrology's gradual removal from this central position. For example, in volume III, I discuss two extremely influential sixteenth-century educational reformers: Philip Melanchthon (1497–1560) and Christoph Clavius (1538–1612). One of Martin Luther's major allies, Melanchthon reformed the Lutheran universities (inter alia) by increasing the emphasis on mathematical instruction within the broader context of an explicitly astrological and deeply religious natural philosophy. As is well known, Tycho Brahe and Johannes Kepler, major participants in the astronomical revolution and both practicing astrologers, were educated within this framework.

Likewise, Christoph Clavius firmly advocated the importance of mathematics in Jesuit education. By using the interpretive framework constructed in volumes I and II, some of the innovative features of Clavius's reform can be detected. In his influential commentary on that best-selling medieval textbook, the *Sphere* of Sacrobosco (originally written ca. 1220; Clavius's first edition, 1570), Clavius explained his reasons for removing the teaching of astrology from its hitherto normal position in the mathematics curriculum, and he cited Pico's *Disputations Against Divinatory Astrology* to support this removal. Yet he also made it clear that astrology still had a place within natural philosophy and medicine, its two other main disciplinary locations.

This is an important moment, I contend, in astrology's weakening position within the premodern curricular framework—and a significant move towards the modern configuration of the mathematical disciplines—precisely because Clavius's textbooks were taught in mathematics courses throughout the burgeoning international Jesuit educational network. The interpretive framework presented here is intended to orient future research in this utterly central, yet incompletely understood part of the history of science. It will also be used to grind more accurate conceptual lenses to examine the premodern material in greater depth and in sharper detail.

Periodization

To set the historical frame more crisply, I embrace (and slightly modify) Charles B. Schmitt's suggestion in the introduction to his *John Case and Aristotelianism in Renaissance England*[3]:

> Western history is conventionally divided into medieval and early modern—the term "Renaissance" often being used for the later period, especially when cultural and intellectual history are at the centre of focus. An alternative periodization, but one seldom employed, might consider the centuries from the twelfth to the seventeenth as a unit. Such a division of history has its attractions, especially when philosophy and science are seen as central driving forces. Aristotelianism in the West lasted as a key intellectual force for precisely that period. From the time of the cultural ferment of the twelfth century renaissance down to the scientific revolution of the seventeenth century there was a historical continuity that should be considered as a single unit and not broken down into medieval and early modern. The force of the Aristotelian tradition did not end with the thirteenth or fourteenth century but persisted as a backdrop against which intellectual achievement had to be evaluated for several more centuries. Only with Bacon, Descartes, and Galileo—feeding on the critical tradition of Copernicus, Ramus, Telesio, Patrizi, and Bruno, among others—do we see that Aristotelianism failed to revive itself as a viable general philosophy, though even then several branches of the *Peripatetic* system displayed continuing stamina (7–8).[4]

This broader historical framework will be employed in what follows; I often refer to it simply as "premodern." Much of what I discuss in this study will apply during the entire period in which the Aristotelian, Ptolemaic, and Galenic traditions of natural knowledge held sway. Printing's importance as an epoch-making invention that bore its first rich harvest toward the end of the fifteenth century should also be duly noted.[5]

*

[3] Kingston: McGill-Queen's University Press, 1983.

[4] Edward Grant for similar reasons holds similar views; *Planets, Stars, and Orbs: The Medieval Cosmos, 1200–1687,* Cambridge: Cambridge University Press, 1994, 9–10. Schmitt developed his groundbreaking analysis of the Aristotelian tradition's continuing viability in the sixteenth and seventeenth centuries in much greater detail in his *Aristotle and the Renaissance*, Cambridge, MA: Harvard University Press, 1983. My work may be viewed as supplementing Schmitt's by reconstructing another style of "Aristotelianism," what I call an "astrologizing Aristotelianism," that remained in force during the period just described.

[5] See Elizabeth L. Eisenstein, *The Printing Press as an Agent of Change: Communications and Cultural Transformations in Early-Modern Europe*, Cambridge: Cambridge University Press, 1979, with which one should read Anthony Grafton's incisive review essay, "The Importance of Being Printed," *Journal of Interdisciplinary History* 11 (1980), 265–286. See also Adrian Johns, *The Nature of the Book: Print and Knowledge in the Making*, Chicago: University of Chicago Press, 1998, which further problematizes Eisenstein's claims, with extensive bibliography. Joseph A. Dane takes to task all such talk of a stabilizing "print culture" in his *The Myth of Print Culture: Essays on Evidence, Textuality, and Bibliographical Method*, Toronto: University of Toronto Press, 2003.

Within this broader framework, I would now like to further refine the chronological dimension of my interpretive framework by sharpening our understanding of the transition from the medieval Aristotelian-Ptolemaic and Roman Catholic map of knowledge to a characteristically early modern one. In my view, the transition from medieval (1.0) to early modern (2.0) was precipitated in three major steps by the cumulatively epoch-making alternatives offered by what we usually call the Renaissance (1.1), the Reformation (1.2) and Copernicanism (1.3), including Copernicus's own profound contributions to mathematical astronomy and cosmology. To fine-tune our understanding, I will now articulate the structural chronology underpinning my interpretive framework, which is designed to non-arbitrarily inform and thereby further structure my analysis.

[1.0] Medieval: I begin with the fundamental medieval structures, which are Aristotelian-Ptolemaic in natural knowledge and Roman Catholic in religion and theology. These structures provided astrology's properly *medieval* scientific and theological foundations, composed primarily of Aristotelian natural philosophy and Ptolemaic mathematical astronomy, as integrated with Roman Catholic views of divine providence and fate. These conceptual structures were articulated fully from the mid-thirteenth century onwards, and were taught at the finest universities as well as in the *studia* of the major religious orders. Thus were provided both the essential conceptual patterns passed down from generation to generation as "normal science" and what became normative theology (especially after the Council of Trent in the sixteenth century), and also their main institutional foundations, as I reconstruct in detail below.

Essentially, every educated European thinker between ca. 1250 and 1482 could properly be characterized as "medieval," meaning Aristotelian-Ptolemaic in natural knowledge/science and Roman Catholic in religion/theology. Then, profound changes started to take place in three major steps that individually and collectively led from the medieval to the early modern map of knowledge and culture via the Renaissance, the Reformation, and Copernicanism, which I will define straightforwardly and unambiguously here, at least in their strict senses with respect to my story. This historical structure takes as significant culture-historical and ultimately world-historical turning points the full-blown and well-articulated appearances of the first reasonably comprehensive and systematic alternatives to Aristotle, Ptolemy and Roman Catholicism. Each of these alternative systems introduce at a particular moment—1482/84, 1517 and 1543—the beginnings of the next major step in the transformation/development of the map of European knowledge and culture from medieval to early modern, and each of them develops both in itself and in relation to the others in complex, manifold and unexpected ways.

I should also reiterate that each of these three fundamental and transformative contributions took place within the significant technological context of the printing press with movable type. The printing press itself began a new culture-historical stage and, in so doing, provided an essential communications technology for supporting and sustaining the next levels of development articulated here, as well as the many that came after. I will now describe the three transitional stages that collectively and cumulatively may be characterized as early modern.

[1.1] Renaissance: The first stage is indicated by characteristic Renaissance contributions that properly introduced Plato and the Neoplatonists into the solidly Aristotelian medieval philosophical (and theological) scene.[6] In the second half of the fifteenth century and especially from the 1480s on with the publication of Marsilio Ficino's philosophical magnum opus, the *Platonic Theology* (1482), and his translations into Latin of Plato (1484) and the major Neoplatonists (1492) from their recently acquired Greek manuscripts, Platonism became an increasingly accessible philosophical alternative to Aristotelianism.[7]

Although Aristotle still reigned supreme in natural philosophy (and in university education), Plato became a viable alternative in cosmology and metaphysics (and also was taught at some universities), just as for the late-antique Neoplatonists, on whom Ficino seems to have successfully modeled himself.[8] Also, with respect to humanistic contributions to shifting views of Aristotle himself, the medieval translations of Aristotle and their use in scholastic theology became subject to increasing critical scrutiny. This derived in large measure from the new trend of reading Aristotle directly in Greek with the recently acquired broader knowledge of Greek, and as facilitated by the epoch-making Aldine *Editio princeps* (1495–1498) of the *Corpus Aristotelicum* in Greek, as well as from new humanistic translations directly from the Greek.[9] All of these developments, and especially the fully formed appearance in Latin of Plato and the Neoplatonists, had complex relations to and effects upon astrology, cosmology and theology (including views of providence) that are worth sorting out, as I attempt to do in my volume II.

[1.2] Reformation: With respect to religion and theology, the second transitional stage begins with the Reformation, when Lutheran Protestantism became an increasingly viable alternative to Catholicism, from 1517 on, with the initial provocation of Martin Luther's 95 theses. This was followed by an explosion of theological and polemical writings that took full advantage of and richly exploited the revolutionary potentials of the printing press, and within a culture-historical context where theology mixed increasingly explosively with politics, in part as incited by pamphlet

---

[6] This is not, of course, to imply that Aristotelianism between ca. 1250 and 1482 and beyond was some sort of homogenous monolithic structure, as one can see by both the plethora of medieval and Renaissance commentaries (for which, see Charles H. Lohr's foundational catalogues of medieval and Renaissance Aristotle commentaries), and further by major developments in the fifteenth century (for which, see, e.g., Schmitt, *Aristotle and the Renaissance*).

[7] See, e.g., James Hankins, *Plato in the Italian Renaissance*, 2 vols., Leiden: Brill, 1991.

[8] For a splendid introduction to Neoplatonism, see Richard T. Wallis's brilliant *Neoplatonism*, second edition with a foreword and updated bibliography by Lloyd P. Gerson (London, 1995 [originally published, 1972]). Peter Adamson brings the bibliography further up to date with his "Neoplatonism: The Last Ten Years" in *The International Journal of the Platonic Tradition* 9 (2015): 205–20. And for a insightful up-to-date treatment of Ficino's self-fashioning of a "humanist rhetorical persona as a Platonic philosopher," see now Denis J.-J. Robichaud, *Plato's Persona: Marsilio Ficino, Renaissance Humanism, and Platonic Traditions*, Philadelphia: University of Pennsylvania Press, 2018.

[9] See, once again, Schmitt's *Aristotle and the Renaissance*.

wars.[10] Calvinism arose soon after with Jean Calvin's influential theological treatises, including famously his *Institutes of the Christian Religion* (1536–1559/1560). Nevertheless, God's providence remained a central theme, even in the new confessional context, and one that can valuably be studied in relation to astrology, embracing both pro and contra positions both within and between the Catholic and Protestant confessional domains. There appear to be no simple dichotomies in this story. On the contrary, complexity reigns, as we can easily see with the strikingly different perspectives on astrology's relation to providence provided by Luther, Melanchthon and Calvin, to name but three major cases within the Protestant confessional camp.[11]

[1.3] Copernicus: With respect to mathematical astronomy and cosmology, the third and final transitional stage between the medieval and early modern maps of knowledge—as informed in various ways by both Renaissance and Reformation developments—begins with the publication of Copernicus's epoch-making and Janus-faced *De revolutionibus orbium coelestium* in 1543. With *De rev*, incipient Copernicanism became a fully functioning mathematical-astronomical alternative to Ptolemy (and vis-à-vis both Aristotelian and Platonic cosmologies) from 1543 on, and increasingly so (and in a very complex manner) in the 70 years of its initial reception, as Pietro Daniel Omodeo rightly emphasizes.[12] I treat the Reformation (1.2) and Copernicanism (1.3) in my volume III on the period ca. 1500–1800.

[2.0] Early Modern: All three movements—Renaissance, Reformation and Copernican—each added a profoundly new and transformative dimension to both conceptual and institutional realms. Thereby, they individually and especially collectively catalyzed the development of significantly expanded and transformed conceptual and ideological spaces, in which to reconfigure and locate both continuities and transformations concerning astrology, cosmology and providence within their relevant conceptual, institutional, confessional, socio-political and cultural contexts. Thus, they should all be used, first, to characterize someone's position in relation to astrology, cosmology and theology, and then, on that basis, to make proper comparisons within and between the relevant structures, as I will demonstrate in all three volumes. Furthermore, these new structures collectively set the stage for further developments in the astronomical and/or scientific revolutions, whatever these phrases mean precisely.

Thus, every thinker who lived and worked between ca. 1250 and 1482 can reasonably be characterized as "medieval," meaning precisely: Aristotelian in natural philosophy and cosmology, Ptolemaic in mathematical astronomy, and Roman Catholic in religion/theology. This statement comes with the significant caveat, however, that each of these three larger structures—both in themselves and in relation to their Renaissance, Reformation and Copernican alternatives—are anything

[10] See, e.g., David Bagchi, "Printing, Propaganda, and Public Opinion in the Age of Martin Luther," *Oxford Research Encyclopedia of Religion*, Aug. 2016 (online publication date), 1–31.

[11] For a valuable recent study, see Robin B. Barnes, *Astrology and Reformation*, New York: Oxford University Press, 2016.

[12] *Copernicus in the Cultural Debates of the Renaissance*, Leiden: Brill, 2014.

but monolithic configurations, since each has their own characteristic local and temporal variations within a broad range of contexts. In what follows in volumes II and III, I will focus, among other things, on the complex astrological dynamics within these rich and both cosmologically and providentially informed conceptual, institutional and confessional contexts. First, however, in volume I, I will reconstruct the fundamental medieval structures against which the later continuities and transformations should all be measured.

The cumulative Renaissance, Reformation and Copernican contributions in all their complexity then characterize the early modern map of knowledge, both in themselves and in relation to the complex and well-established medieval Aristotelian, Ptolemaic and Roman Catholic conceptual and institutional structures, which certainly did not disappear by 1543, and themselves continued to develop both internally and in relation to the new alternatives. Thus, by the mid-sixteenth century, all three domains—(1) natural philosophy, (2) mathematical astronomy, and (3) theology—have all received significant and accessible epoch-making alternative possibilities to Aristotle, Ptolemy and Roman Catholicism. Collectively, these factors are all extremely significant for characterizing a thinker's positions vis-à-vis astrology, cosmology and providence—as well as many other topics—thus evoking a fuller, more historically nuanced and structurally sound picture of each thinker's intellectual and cultural physiognomy. As we will amply see in what follows, there are no simple dichotomies either within or between positions concerning astrology, cosmology, or providence. By the seventeenth century, of course, even more alternative possibilities had arisen with, e.g., Descartes and Cartesianism (from 1644), Newton and Newtonianism (from 1687), and a host of splintering Protestant sects. All of these factors will be taken into account more or less fully in what follows.

Methodological Preliminaries

Deconstruction

These methodological preliminaries are intended to help us solidly but flexibly construct the broader interpretive framework—with its clearly articulated fundamental conceptual and disciplinary structures—that may then be used to further sharpen the picture. To accurately grasp astrology in its proper historical perspective, we should first identify and correct for two broader distorting modern biases drawn from a fundamentally anachronistic yet still virtually ubiquitous interpretation of astrology's complex range of places within the premodern map of knowledge. Since the focuses of conceptual lenses seem mainly to be ground, as it were, on the basis of fundamental distinctions and disciplinary configurations, I will endeavor to replace these outdated historiographical lenses with new more accurate ones, ground in accordance with three fundamental premodern conceptual and disciplinary structures.

Before we can see more clearly, however, we must first remove the distorting older spectacles. To know both which lenses distort and how to properly grind new ones, we must be keenly aware of when we are using actors' categories and when we are imposing modern distinctions on the past. Getting the right focus is particularly difficult when past disciplinary configurations resonate strongly with modern assumptions and/or prejudices, which we then tend—usually unconsciously—to read back into the past. I hope that the analysis adumbrated here and developed in detail in what follows will provide a new more accurate prescription for use in future investigations, allowing us to see in sharper focus both the broader patterns and the many specific details of past conceptual and disciplinary structures related to the history of astrology. Therefore, close attention to the range of relevant terminology and its respective conceptual referents will be a central concern in what follows.

*

I will first simply—and proscriptively—identify some of what I consider the more problematic conceptual and disciplinary structures pervading the historiography that should be removed at the very beginning:[13] first, a fundamentally anachronistic disciplinary configuration. In modern scholarship, astrology is almost always closely associated with the other so-called occult sciences, especially magic, alchemy and the kabbalah, as we find it in numerous influential studies by (among others) Wayne Shumaker, Brian Vickers and Brian Copenhaver.[14] Although not entirely mistaken, this presumed configuration with the occult sciences or occult philosophy is deeply problematic conceptually—unless skillfully nuanced—when applied to astrology ca. 1250–1800.[15]

[13] There is no doubt that this is my particular (and perhaps peculiar) take on the subject, based on my reading and understanding over the past 20 some-odd years of engagement with the history of astrology. Other scholars have different interpretations. Respectful criticism and further refinements are heartily encouraged!

[14] Some of the relevant studies are Yates, *Giordano Bruno and the Hermetic Tradition*; Keith Thomas, *Religion and the Decline of Magic*, New York: Scribner, 1971; Wayne Shumaker, *The Occult Sciences in the Renaissance*; and *Occult and Scientific Mentalities in the Renaissance*, Brian Vickers (ed.), Cambridge: Cambridge University Press, 1984. As I discuss more fully in the introduction to volume II, Brian Copenhaver also centrally uses the categories of "the occult sciences" and "the occult tradition" in numerous authoritative studies, including his erudite chapters in the *Cambridge History of Renaissance Philosophy* and the *Cambridge History of 17th Century Philosophy*. He cites Heinrich Cornelius Agrippa's *De occulta philosophia* to justify his usage. I discuss other non-anglophone studies in due course. The most historically and conceptually sound usage of the term "occult" I know of is that discussed in the introduction to Daniel Stolzenberg's recent monograph, *Egyptian Oedipus: Athanasius Kircher and the Secrets of Antiquity*, Chicago: The University of Chicago Press, 2013, 24–29. Liana Saif's explicit statement on the subject also sets the correct tone (n. 14); *The Arabic Influences on Early Modern Occult Philosophy*, Basingstoke: Palgrave Macmillan, 2015.

[15] For a penetrating analysis of these issues, especially concerning the relationship between astrology and alchemy, see Newman and Grafton, "Introduction," 17 ff.

Whether called the hermetic, occult or esoteric traditions (essentially progressive variations on a theme), the same strictures apply. Accurately historicizing these overarching terms is required but very difficult to achieve, especially if we are dealing with a broader audience or one of nonspecialists, that is, most readers. At this point, my suggestion is to reject these terms altogether, at least for the present, except perhaps as signposts.[16] I will take up these issues in greater detail in the historiographic introduction to volume II.

In my view, our understanding is fundamentally compromised and ultimately distorted by predisposing it from the very beginning in a way that deeply influences and orients a broad range of further assumptions and the related "natural" questions to ask, especially when some sort of deeper unity—if not a fundamental identity or even a "family resemblance"—among these disciplines is also assumed.[17] This is particularly problematic—as we will see repeatedly—where modern scholars come out of the gate assuming (as in Keith Thomas's classic *Religion and the Decline of Magic*) that astrology is somehow a part or subset of magic.[18]

For now, we will simply not make any assumptions along these lines, highlighting astrology's configuration with these and other disciplines as a question to ask and historicize, rather than an assumption to make a priori based on our modern (or postmodern) map of knowledge, with its characteristic conceptual and disciplinary structures. We shall verify, refine or entirely reject this configuration of astrology with the occult sciences based on both its accuracy, and its related utility for characterizing the premodern map of knowledge.

The first steps in reconstruction must be taken carefully, since the basic patterns established early on become a foundation for all that follows. To shift the metaphor again: with conceptual lenses, each distortion tends to compound the next, as in a telescope. To further clarify these issues, I will explicitly discuss astrology's relationship to divination and magic in Parts II and III by focusing on two paradigmatic

---

[16] For related suggestions concerning what he calls "tainted terminology," which include the terms "magic," "occult," and "superstition," see Wouter Hanegraaff, *Esotericism and the Academy: Rejected Knowledge in Western Culture*, Cambridge: Cambridge University Press, 2012.

[17] To my lights, it is best to treat such phrases as "the occult (or "esoteric") tradition" as signposts for modern scholars pointing to a set of specific disciplines in the past, which should then be properly historicized both in themselves and in relation to each other.

[18] For other examples of such lumping of astrology (and alchemy) within magic, see Lynn Thorndike's vastly influential *History of Magic and Experimental Science*, 8 vols., New York: Columbia University Press, 1923–58; the more recent but also very influential *Magic in the Middle Ages* by Richard Kieckhefer (Cambridge: Cambridge University Press, 1989); and Karen Jolly, "Medieval Magic: Definitions, Beliefs, Practices," in *Witchcraft and Magic in Europe: The Middle Ages*, Bengt Ankarloo and Stuart Clark (eds.), Philadelphia: University of Pennsylvania Press, 2002, 3–71. For the marvelous historiographical distinction between "lumpers" and "splitters," see J. H. Hexter, "The Historical Method of Christopher Hill," in his *On Historians: Reappraisals of Some of the Masters of Modern History*, Cambridge, MA: Harvard University Press, 1979, 227–51 (originally published as a book review of Christopher Hill's *Change and Continuity in 17th-Century England* in the *Times Literary Supplement*, October 24, 1975). I discuss my understanding of astrology's relationship to magic in Part III and in the overall conclusion.

practices in particular: (1) predicting the future (in relation to divination), and (2) making talismans (in relation to magic).

*

The second conceptual structure to be set aside and brought up for review is the equally pervasive historiographic distinction between "natural" and "judicial" astrology. Although this may end up being a useful distinction, we must first clarify what it actually means and trace its existence up to and beyond its modern use in Ephraim Chambers's *Cyclopaedia* article, "Astrology" (1728). By contrast, Francis Bacon did *not* use this distinction in his 1623 proposals for astrological reform. We find it in a recognizable form, however, in the *Index of Prohibited Books* from 1564 on.[19]

In normal premodern usage, all astrological predictions—whether relating to a person's nativity or annual revolution, the weather, medicine, or political events—were called astrological *judgments* (*iudicia*) and, thus, in some real sense, may be called "judicial astrology." We can see this as early as the 1260s in the *Speculum astronomiae* (to be discussed more fully below), but also in the work of Placido Titi (1601–1668), professor of astrology at the University of Pavia, who makes this very point (among many others) in the mid-seventeenth century.[20] Likewise, astrology's causal "naturalness" vis-à-vis its legitimacy (or otherwise) will be discussed extensively in what follows. In the meantime, we will set this distinction aside along with astrology's configuration among the hermetic, occult and/or esoteric sciences and traditions, removing them both (at least for the time being) from our interpretive apparatus in order to assess their value. They will not be missed, nor their consequent built-in confusions. By the end of this study, we will have a much greater sense of their utility or otherwise.

Finally, if we begin approaching the history of astrology by importing a typical view of modern-day astrological thought and practice, we will also have started off on the wrong foot, introducing at the outset significant conceptual distortions, especially if we have in mind the sorts of low-level practices found in daily newspaper "horoscopes," a twentieth-century innovation.

---

[19] To further historicize this issue, in volume III, I will use a very interesting discussion published recently by Ugo Baldini and Leen Spruit, which draws on and develops Thomas Aquinas's influential discussion in *Summa theologiae* II.II.92–95, to be discussed more fully in Chap. 5; *Catholic Church and Modern Science: Documents from the Archives of the Roman Congregation of the Holy Office and the Index,* U. Baldini and L. Spruit (eds.); Vol. 1: 16th Century Documents (4 vols.), Rome: Libreria Editrice Vaticana, 2009.

[20] Placido Titi, *Tocco di paragone*, Giuseppe Bezza (ed.), Milan: Nuovi Orizzonti, 1992, ch. 6, "Il titolo di giudiciaria si conviene ad ogni scienza," 50–54. The *Tocco di paragone* was originally published in Pavia in 1666. I discuss Titi in volume III.

Reconstruction

Having removed the old distorting spectacles and set them aside, the first step of reconstruction will begin by identifying and grinding the basic framing structures for new interpretive lenses as deeply informed by the three following fundamental distinctions and configurations. Not superimposed on the historical material by questionable modern (mis)understandings, these structures are derived, rather, from within the patterns of premodern practices. More accurately reflecting our premodern actors' conceptual and disciplinary categories, we may thus perceive them more accurately. This principle is at the heart of my historicizing methodology and will permit, I hope, a more accurate "thick description" of the material in question.[21]

The first conceptual structure involves the most basic terminology, namely, the term "astrology" itself, and concerns the utterly fundamental distinction between what *we* call "astronomy" and "astrology." This fundamental conceptual distinction is found in what came to be its classic formulation in the first chapter of Ptolemy's foundational text for this entire tradition of "scientific astrology," namely, the "Apotelesmatika (Gk)," *Tetrabiblos* or *Quadripartitum*, that was composed in the middle of the second century CE, and has had an extraordinarily influential international Nachleben. In brief (and to be refined below), "astronomy" (i.e., mathematical astronomy, as opposed to physical astronomy) is concerned primarily with analyzing and predicting the *motions* of the luminaries (the sun and moon), the planets and the fixed stars, whereas "astrology" treats their *influences* or *effects* on the earth itself, its atmosphere and inhabitants.

Ptolemy used the same overarching descriptive phrase for both, namely, "foreknowledge through the science of the stars."[22] In fact, both of the terms *astronomia* and *astrologia*—in Latin, Greek, and numerous vernaculars—were normally used *interchangeably* throughout the entire premodern period to refer to *both* of the intimately related, but conceptually distinct parts of the overarching category, "the science of the stars."[23]

We distinguish them today (and have for some time now) as "astrology" and "astronomy," employing a distinctive terminology. This does not mean, however,

[21] For a splendid discussion of the anthropological term "thick description," see its classic formulation in Clifford Geertz, "Thick Description: Toward an Interpretive Theory of Culture," in his *The Interpretation of Cultures: Selected Essays*, New York: Basic Books, 1973, 3–30. In a nutshell, my goal is to reveal Chomskyan-style deep structures by means of Geertzian-style thick descriptions.

[22] The Greek for this phrase is 'to di' astronomias prognostikon telos' (Gk). The Latin is: '*prognosticabilis scientiae stellarum profectio*'. The Latin is from Plato of Tivoli's twelfth-century translation in David Juste's unpublished transcription from Erhard Ratdolt's 1484 Venice edition. My thanks for permission to cite it here. On this configuration in Ptolemy, see now Jacqueline Feke, *Ptolemy's Philosophy: Mathematics as a Way of Life*, Princeton: Princeton University Press, 2018, chap. 8.

[23] For this and other astrological terminology in the Middle Ages, see Charles Burnett, "Astrology," in *Medieval Latin: An Introduction and Bibliographical Guide*, F.A.C. Mantello and A.G. Rigg (eds.), Washington, D.C.: Catholic University of America Press, 1996, 369–82, and Appendix I, "An Introduction to Medieval Astrology (334–94)," and Appendix II, "English and Latin Glossary of Astrological Terms (395–439)" in Robert S. Hand, "The Use of Military Astrology in Late Medieval Italy: The Textual Evidence," PhD thesis, The Catholic University of America, 2014.

that the premodern actors confused the disciplines because they used the same terms to refer individually and collectively to both major parts, as is sometimes claimed in the scholarship. From Ptolemy on, the disciplinary distinction both conceptually and in practice was well understood. What is meant in detail by both "astrology" and "astronomy" is a major subject of what follows.[24]

*

The two other essential framing structures for our new conceptual spectacles derive from two disciplinary configurations, one of which situates astrology within the broader realm of natural knowledge. The other differentiates astrology's practical dimension. In addition to revealing astrology's normal locations within the premodern map of knowledge, the first disciplinary configuration also serves to situate astrology within one of its most important institutional locations—the premodern university—where it was studied, taught and passed down as "normal science" from generation to generation for roughly 500 years.[25] As already noted, astrology was integrally configured within mathematics, natural philosophy and medicine, and it was studied and taught within these disciplines at the finest premodern universities, as we can see clearly from the University of Bologna's 1405 statutes, and from much other corroborating evidence to be discussed in detail in Part IV.

The third and final fundamental structure is the four main types of astrological praxis: general astrology or revolutions, nativities, elections and interrogations. Revolutions were concerned with large-scale changes, primarily in the weather, but also with state affairs. This was a major feature of the annual prognostications found later in many printed almanacs, and included the doctrine of great conjunctions.[26] Nativities, on the other hand, involved the astrological configuration at a person's birth. Interrogations entertained questions on a broad range of topics, including personal, medical and business affairs. Finally, elections determined the most propitious moment to begin an enterprise or perform an activity, such as crowning a ruler,

[24] Historicizing the terminological distinction is also a goal of this study. Whenever I use the terms "astrology" and "astronomy" in English, I always employ our modern distinction. When I use either *astrologia* or *astronomia* in Latin, Greek or a vernacular, it always embodies the premodern usage, and usually that of a particular premodern thinker. I hope that my usage adds to both clarity and understanding, and minimizes confusion and/or distortion, another goal of my historicizing methodology.

[25] I discuss some of the Kuhnian dimensions of my story below.

[26] According to Bernard Capp, *English Almanacs, 1500–1800: Astrology and the Popular Press*, Ithaca: Cornell University Press, 1979, 25: "An almanac was, technically, a table of the astronomical and astrological events of the coming year: the movements and conjunctions of the planets and stars in the zodiac, and details of eclipses. As such it can be traced back, like astrology, to antiquity. The manuscript almanacs of the Middle Ages often combined a kalendar, which supplied ecclesiastical information, notably the dates of festivals of the Church. [...] Following the invention of printing, almanacs (supplying astrological data) and prognostications (predictions derived thence) were among the earliest works to be published, at first separately." During the medieval period, however, prognostications usually stood on their own. I will discuss these matters further in the excursus to this introduction on practical astrology and in Chap. 10.

passing the baton of command to a general, or laying the cornerstone of an important building, including Saint Peter's in Rome or the Fortezza da Basso in Florence. Elections also included the controversial practice of making astrological images or talismans. These practices all required the construction and interpretation of horoscopes, and will be discussed more fully later in this introduction in an excursus and exemplified (inter alia) in Chap. 11.[27]

*

Finally, I would like to reiterate the fundamental distinction between [1] practical astrology (with its four canonical branches) and [2] astrology's scientific foundations—both conceptual and institutional—in mathematics, natural philosophy and medicine. I will also argue in Part II that, in addition to its scientific foundations, astrology also had theological foundations in highly influential thirteenth-century views of divine providence and fate in the works of Albertus Magnus and Thomas Aquinas. Furthermore, these theological foundations seem to have surrounded, embraced and further supported its scientific foundations, resulting in a beautifully wrought, deeply integrated architectonic structure, the details of which much await the analysis in Part II.

*

I should also say something about why I begin my story in the thirteenth and not in the twelfth century, when many of the fundamental natural philosophical, astrological, medical and magical texts were translated, primarily from Arabic, but also from Greek into Latin.[28] To be sure, the decision is somewhat arbitrary, but in my view, it was only in the thirteenth century in the Latin West that astrology became

[27] I follow the deeply influential and deliberately anonymous *Speculum astronomiae* (1260s) on these matters, which will be discussed further below. See also Burnett, "Astrology," and Hand, "Use of Military Astrology." For further methodological reflections, see my "Understanding the History of Astrology Accurately: Methodological Reflections on Terminology and Anachronism," *Philosophical Readings* 7 (2015): 42–54 (a special issue on astrology edited by Donato Verardi).

[28] There is much valuable scholarship on astrology in the twelfth century. One should begin with Jean-Patrice Boudet, *Entre science et nigromance: astrologie, divination et magie dans l'Occident médiéval, XIIe-XVe siècle*, Paris: Publications de la Sorbonne, 2006, especially Chap 1: "Le renouveau de la science des astres (XIIe-XIIIe siècle)," 35–87, and Joshua Lipton, "The Rational Evaluation of Astrology in the Period of the Arabo-Latin Translations, ca. 1126–1187," PhD thesis, UCLA, 1978. Much of Charles Burnett's scholarship is very useful, including on the translation movement itself. Several of these studies are collected in his *Arabic into Latin in the Middle Ages: The Translators and their Intellectual and Social Context*, Farnham, England: Ashgate/Variorum, 2009. Marie-Thérèse d'Alverny's synthetic article is also very useful: "Translations and Translators," in *Renaissance and Renewal in the 12th Century*, R.L. Benson and G. Constable with C. Lanham (eds.), Cambridge, MA: Harvard University Press, 1982, 421–62. Charles H. Haskins, *Studies in the History of Mediaeval Science*, second ed., Cambridge, MA: Harvard University Press, 1927 is still valuable, primarily for the texts that it published. Maria Mavroudi offers a valuable revisionist study; "Translations from Greek into Latin and Arabic during the Middle Ages: Searching for the Classical Tradition," *Speculum* 90 (2015): 28–59.

fully (if not seamlessly) integrated with Aristotelian, Ptolemaic and Galenic natural, mathematical and medical knowledge—all having been rediscovered and translated into Latin in the twelfth century—which were then further integrated within the theological structures of Roman Catholicism.[29]

For me, Albertus Magnus's edition of the complete works of Aristotle (including his supplements to the canonical *Corpus Aristotelicum*) provided a solid conceptual structure of Aristotelian natural knowledge, within which astrology was significantly integrated for the first time in the Latin West, as I argue in Part I. At the same time, the universities also arose in Europe for the first time, thus providing solid institutional structures that still continue today, but without Aristotle, Ptolemy, Galen, or astrology, except as historical studies. Together, these conceptual and institutional structures provide essential foundations for my reconstruction. I should also note that I do not discuss Arabic authors in any depth except insofar as they were used by later Latin authors.[30] Finally, I will put off discussing the critics of astrology until volume II.[31] Volume I will thus be devoted almost entirely to astrology's defenders, promoters and practitioners, who have received relatively less scholarly attention than its critics.

---

[29] In fact, as I will argue in what follows, three of these disciplines—practical astrology, Aristotelian natural philosophy, and Roman Catholic theology—seem to have mutually informed each other in their respective developments as a medieval synthesis was forged in the thirteenth century, primarily in the works of Albertus Magnus and Thomas Aquinas.

[30] Although this would be a valuable and some might argue essential undertaking, this volume is already long enough and would thereby become much longer. Inter alia, see Richard J. Lemay's groundbreaking if somewhat exaggerated, *Abu Ma'shar and Latin Aristotelianism in the 12th Century: The Recovery of Aristotle's Natural Philosophy through Arabic Astrology*, Beirut: Publications of the Faculty of Arts and Sciences, 1962; Dmitri Gutas, *Greek Thought, Arabic Culture: The Graeco-Arabic Translation Movement in Baghdad and Early 'Abbasid Society (2nd-4th/8th–tenth Centuries*; and now Hasse, *Success and Suppression*. All three of these valuable works draw on the earlier work (i.a.) of Haskins, Carmody and Sezgin. Hasse's work in particular will be discussed further below. My profound thanks go to Prof. Dr. Hasse for sharing his brilliant and insightful study with me in a prepublication version. I want to emphasize that I in no way intend to minimize the importance of the Arabic contribution to European astrology or to European learning in general. Au contraire! In this light, it is worth quoting the conclusion to Hasse's chapter on astrology: "In the final analysis the question may arise whether the label "Arabic" can be sensibly employed in the field of astrology, given that so much of the allegedly Arabic doctrines ultimately are of Greek, Indian, or Persian origin. This is true to some extent. But in another sense, the label "Arabic" is perfectly appropriate. Hardly any other science in medieval and Renaissance culture was based as massively on Arabic sources. It is the Arabic version of interrogations, elections, anniversary horoscopes, and so forth that the reformers had to confront. In addition, Arabic astrologers clearly added to and transformed astrological theory also in the centuries relevant for the Latin West, that is, in the period between Messahalah in the eighth century and Haly Abenragel in the eleventh century. But modern scholarship on this development is still in its infancy. It is to be expected from the coming decades of research that the contributions of Greek, Indian, Persian, and Arabic scholars to astrological theory will be much clarified and will thus also offer a better basis for understanding the development of Renaissance astrology (292)."

[31] For a recent valuable study, see Philipp Nothaft, "*Vanitas vanitatum et super omnia vanitas*: The Astronomer Heinrich Selder and a Newly Discovered 14th-Century Critique of Astrology," *Erudition and the Republic of Letters* 1 (2016), 261–304.

Further methodological issues will be treated as they arise. Likewise, relevant scholarship will be reviewed in its proper context. To treat all of this as front matter would be excessively burdensome. Furthermore, the reader should know that this study is meant to be a contribution to a vast and central, but still far from fully understood topic. Expecting exhaustive or definitive coverage will only result in disappointment. My main hope is that this study will be read as a series of preliminary investigations, soundings and/or case studies that collectively provide a more soundly historicized interpretive framework, based on fundamental conceptual and institutional structures, for understanding both astrology's central places in the premodern Western map of knowledge and the complex patterns of its removal. Some questions will be settled; many more will be raised. The curious reader has now been properly forewarned!

Structure of the Book

To properly understand astrology's marginalization within and ultimate removal from the domains of legitimate natural and theological knowledge during the seventeenth and eighteenth centuries and the epistemological significance of this epoch-making transformation, one must first accurately understand its previous centrality. Therefore, volume I is devoted to tracing in detail the basic structures in conceptual, institutional, socio-political, theologico-religious and cultural respects. The structure of the book ends up being essentially chronological. Primarily treating astrology's conceptual structures within the premodern Aristotelian-Ptolemaic-Galenic and Roman Catholic map of knowledge, volume I, Parts I–III, fall in the thirteenth century, with most of the action focusing around 1250–80. Volume I, Part IV, which mainly concerns astrology's institutional, socio-political and cultural configurations, spans the entire fourteenth and fifteenth centuries as it shifts the geographical focus from Northern Europe to Northern Italy. Returning primarily to conceptual patterns, volume II pertains to the late-fifteenth century, focusing particularly on the 1480s and 1490s, also in Italy. Volume III treats ca. 1500–1800 as it expands to embrace a broader European panorama.

Thus, volumes I and II (1250–1500) and III (1500–1800) reflect modern historiographical divisions between the Middle Ages, the Renaissance and the Early Modern period. Although they can each profitably be read separately, when they are read together, I hope they will encourage the bridging of this unfortunate, mostly arbitrary, and still institutionally perpetuated historiographic divide. This breach has caused particularly acute problems for understanding astrology and magic, as we will amply see in what follows, especially when medieval forms of knowledge and practice are sharply contrasted with so-called "Renaissance" forms without a clear understanding of the medieval structures. I attempt to bridge this gap in particular in volumes I and II.

Volume I

Medieval Structures (1250–1500): Conceptual, Institutional, Socio-Political, Theologico-Religious and Cultural

Part I (Chapters 1, 2 and 3) reconstructs fundamental and deeply influential astrologizing patterns in premodern natural knowledge, as fully articulated in the thirteenth century—and taught in some of its universities—by arguably the finest natural philosophers of the day: Robert Grosseteste (ca. 1168–1253), Albertus Magnus (ca. 1200–80), Roger Bacon (ca. 1214/20-ca. 1292), and Thomas Aquinas (ca. 1225–74). Chapter 1 focuses primarily on recovering the natural philosophical foundations of what I have come to call an astrologizing Aristotelianism by focusing primarily on Albertus Magnus's translation-commentaries on core texts of Aristotle's natural philosophy, especially the *De generatione et corruptione*.[32] This is a detailed study that contributes to the history of philosophy as well as to the histories of science and of Western esotericism.[33] Chapter 1 also confronts an utterly central question for understanding premodern natural knowledge, namely, the relationship between astrology and scholastic Aristotelian natural philosophy.

In Chap. 2, this astrologizing Aristotelian natural philosophical model is further articulated by adding its specifically mathematical elements. To do so, I reconstruct a geometrical-optical model of planetary influences within a fundamentally Ptolemaic cosmographic framework that integrated mathematical astronomy and geography. In this system, rays of light understood both physically and geometrically play the key mediating role between the heavens and the earth in articulating how celestial influences work. Charting the mathematical dimensions of this astrologizing Aristotelian natural philosophy also provides telling evidence for reevaluating the intellectual relationship between Albertus Magnus and Roger Bacon, the two main authors treated in Chap. 2. Their views on mathematics (including astronomy and geometrical optics [*perspectiva*]) in relation to astrology are investigated in detail.

Part I, Chaps. 1 and 2 thus reconstruct a richly mathematicized astrologizing Aristotelianism, in which a fundamentally Ptolemaic cosmographic framework of mathematical astronomy calibrated with mathematical geography is fitted out with a geometrical-optical model of celestial influences, thus integrating the realm of the heavens with the natural world on earth. This reconstruction is meant to contribute to our understanding of medieval natural philosophy and cosmology, refining, supplementing, and in part replacing Edward Grant's analysis in his magnum opus,

[32] I prefer "translation-commentary" to the more normal "paraphrase-commentary" because these works are more enhanced translations than proper paraphrases, and even though Albert himself did not make the translations, but drew on earlier ones.

[33] One of the main virtues of my study for students of Western esoterica is that my research has not been unduly influenced by the questions usually asked or the approaches usually taken in such studies. I have tried as much as I am able to historicize the subject without too many additional assumptions.

Planets, Stars, and Orbs: The Medieval Cosmos, 1200–1687,[34] where Albertus Magnus, Roger Bacon, and others sharing similar views are not treated as fully as they deserve. As we will see, a rather different picture emerges based on the different texts considered.

Chapter 3 further develops our understanding of astrology's natural philosophical foundations by focusing on Albert's articulation of these foundations in relation to the astrological practice of revolutions or general astrology, including great conjunctions. He does so in his commentary on the pseudo-Aristotelian *De causis proprietatum elementorum (On the Causes of the Properties of the Elements).* By discussing these foundations for revolutions, Albert complements his articulation of the natural philosophical foundations for nativities analyzed in Chap. 1.

My reconstruction in Part I also further develops Charles B. Schmitt's pioneering research into the different medieval, Renaissance and early modern Aristotelianisms. Part I also provides the central natural philosophical patterns missing in both Steven vanden Broecke's and Robert Westman's more or less recent treatments of Renaissance and early modern astrology.[35] Primarily, though, Part I articulates the premodern Aristotelian and Ptolemaic natural philosophical and mathematical foundations that underlay astrological theory and practice ca. 1250 to 1500, that is, the structures of knowledge that were later subject to criticism, reform, partial appropriation, marginalization, and ultimately rejection from the map of legitimate knowledge during the sixteenth through eighteenth centuries (as explored in volume III).

*

Chapter 4 is an intermezzo devoted to how astrology could be and actually was defended, where I focus on Roger Bacon and the *Speculum astronomiae*. Many of the themes explored here—including the defense of human free will—will be developed further in what follows.

*

Parts II (Chaps. 5 and 6) and III (Chaps. 7 and 8) shift the focus and seek to relate this configuration of astrologizing Aristotelian natural knowledge to two other major realms of knowledge and practice—what we call "religion" (primarily theology) and "magic" (primarily talismans)—by focusing particularly on how the boundaries of legitimate thought and practice were demarcated, defended, and at times transgressed. In Part II, Chap. 5 with theology/religion, I focus on astrology's surprising relationship to divination—as well as to fate and providence—in Albertus Magnus and Thomas Aquinas. An essential feature of Chap. 5 is to reconstruct and establish what I have come to call astrology's theological foundations in Thomas Aquinas's influential doctrine of divine providence in the *Summa contra gentiles*. We will find that the astrologizing Aristotelian natural philosophy reconstructed in

[34] Cambridge: Cambridge University Press, 1994.

[35] Vanden Broecke, *The Limits of Influence: Pico, Louvain, and the Crisis of Renaissance Astrology*, Louvain: Brill, 2003; Westman, *The Copernican Question*. I discuss these works further below.

Part I provides the central conceptual structures for analyzing fate, divination and providence. I then contrast the views of Albert and Thomas with Roger Bacon's very different approach to astrology and religion in Chap. 6, in which (among other things) Roger embraces a much more practical-astrological approach to the subject.

With magic in Part III, I analyze in detail the controversial issue of talismans (*imagines astronomicae*) in Chap. 7. The deliberately anonymous *Speculum astronomiae* orients the discussion at first. This influential analysis is then compared with authentic texts by Albertus Magnus, especially the *De mineralibus*. These views are then compared and contrasted with Thomas Aquinas's strikingly different and much stricter position, at least in his later works, including the *Summa theologiae*. In them, he rejected talismans altogether, in sharp contrast to his earlier *Summa contra gentiles*, in which he seems to have provided a gaping loophole for legitimate all-natural talismans of the kind defended by Albert and the Magister Speculi. I also draw conclusions in Chap. 7 concerning Albert's purported authorship of the *Speculum astronomiae*.

In Chap. 8, I explore Roger Bacon's much more expansive position on what *we* call magic, but which *he* most assuredly did not, a position articulated in a much more apocalyptically inflected idiom. Not only does he accept talismans, but he also promotes using powerful words to complement and amplify the power of his talismans—thus embracing both of what he calls the *opera et verba sapientiae*—by linking the soul's powers in speech with those of the celestial bodies in a talisman. Unlike with natural knowledge and theology, we will discover that magic does not provide any foundations for astrology whatsoever. Rather, astrology and its natural philosophical foundations seem to have provided magic's foundations for both its practice and its legitimacy, such as it was, at least regarding talismans.

*

Part IV (Chaps. 9, 10 and 11) turns away from conceptual structures in themselves to explore how this astrologizing Aristotelian natural knowledge—in both theoretical and practical respects—was taught in the top-tier premodern universities and practiced at university, in cities and at court, as we move from astrology's thirteenth-century foundations to its more established state in the fourteenth and fifteenth centuries. In these chapters, I also shift gears and relocate the geographical focus from Northern Europe in the thirteenth century to Northern Italy in the fourteenth and fifteenth.

Using Regiomontanus (1436–1476) as an initial focus for Chaps 9 and 10, I primarily examine the disciplinary and curricular patterns of astrological study and teaching at a range of universities, focusing on the University of Bologna, whose 1405 statutes show clearly that astrology and its natural philosophical foundations were taught in three distinct disciplinary locations: in the mathematics course with mathematical astronomy; in the natural philosophy course in relation to fundamental Aristotelian texts, and in the medical course. Using these disciplinary and curricular patterns, one may more accurately characterize both the premodern map of knowledge and its educational institutionalization, by means of which this knowledge was passed down as "normal science" from generation to generation. Chapter 10 also develops the socio-professional profile of the university astrologer in some detail. Part IV also treats another significant historiographical issue, namely, the relationship between astrology and humanism, which has been construed—falsely, I argue—as a diametric opposition.

Having recovered central features of its conceptual and institutional foundations, Chap. 11 locates astrology within the socio-political and cultural landscape of Quattrocento Italy. In both secular and ecclesiastical realms, astrology ramified into many social, cultural and political arenas from art to public health. Drawing primarily on published scholarship and the primary sources (archival and literary) that it mined, Chap. 11 offers a panoramic view of astrology's broad spectrum of practices in the richly dynamic culture of the Northern Italian courts. I evoke several of astrology's influential roles in these courts—for example, at Ferrara, Milan, Florence and Rome—which have been discussed in numerous studies dispersed throughout various disciplinary literatures, including history proper, history of science and art history. A series of synthetic studies by experts in these various fields would be of great service in integrating our current knowledge. Volume I thus provides the fundamental structures whose continuities and transformations will be explored in volumes II and III.

Volume II

Renaissance Structures (1450–1500): Continuities and Transformations

Volume II concentrates on two dominant intellectuals in Renaissance Florence and in our historiography: Marsilio Ficino (1433–1499) and Giovanni Pico della Mirandola (1463–1494). In these chapters, among many other things, I challenge and revise Frances Yates's still influential genealogy of Renaissance magic, a view in which Pico simply added kabbalah to Ficino's astrologically grounded and medically oriented natural magic. Asking how their strikingly different magics relate to astrology (and to kabbalah for Pico) and then to each other—as well as to the patterns reconstructed in volume I—are key heuristic features of volume II. I hope thereby to provide a more solid conceptual and historiographic foundation for more accurately relating medieval and Renaissance astrology and magic.

Part I primarily reconstructs Pico's early views of astrology in relation to magic in key texts of 1486–1487, namely, the so-called *Oration on the Dignity of Man* (*Oratio de hominis dignitate*), the *900 Conclusiones*, and the decidedly ineffectual *Apologia* therefor. By reading these works closely (primarily the magical and kabbalistic conclusions and related passages), I show that Yates profoundly misinterpreted Pico's thought by assuming a fundamental similarity between Ficino's and Pico's ideas, whereby Pico as student naturally adopted many of his teacher Ficino's views, based in large measure on similarities in terminology.

The problems with this interpretation (and all its corollaries) have become painfully clear in the best recent scholarship (primarily in Michael J.B. Allen's work), which points to a much more conflicted—and at times downright antagonistic—interpersonal and intellectual dynamic. Instead, I argue that Pico offered a radical revision of traditional magical theory and practice. In fact, in my reading, Pico's magic had no normal astrological underpinnings whatsoever. Rather, he attempted to substitute the kabbalistic sephirot for the planets, in a much more religiously oriented magic, as Yates rightly pointed out. Thus, Pico attempted to radically subvert magic's otherwise normal astrological foundations, as reconstructed in volume I. I also explore some of Pico's views in the transitional *Heptaplus* of 1489.

Part II primarily explores Marsilio Ficino's rich and influential *De vita libri tres* (*Three Books on Life*, 1489), and especially Book III, the famous *De vita coelitus comparanda* (*On Deriving Life from the Heavens*), after a preliminary chapter on his earlier views. To highlight the differences between Ficino's deeply astrological medically oriented magic in his mature *De vita* and Pico's non-astrological and non-medical but deeply religious kabbalistic magic in his early works, I analyze and present in detail *De vita's* numerous astrological dimensions, and I relate them to—in part by contrasting them with—the astrologizing Aristotelianism reconstructed in volume I. There is much to learn by this comparison about both continuities and transformations.

I focus in particular on what amounts to a mini-treatise on the theory and practice of making and using talismans, which Ficino also calls *imagines astronomicae*, in *De vita* III, Chaps. 13–23. This also makes a striking point of contrast with Pico's magic, from which talismans are conspicuously absent. I also relate Ficino's position on talismans to their proper medieval roots as reconstructed in volume I. Celestial influences understood as rays are utterly central to both. Recovering the integral thought of both Ficino and Pico allows a proper comparison that fundamentally revises and richly complicates Yates's picture. I also offer arguments as to why it is appropriate to call Ficino's *De vita* a properly "Renaissance" expression of astrology and magic, in which I highlight Ficino's own radical and subversive transgressions of many well-established medieval safeguards, including those for protecting human free will and for warding off the use of demons, as also reconstructed in volume I.

Part III focuses on Pico's late *Disputations against Divinatory Astrology*, which was composed in 1493–1494 and published posthumously in 1496. In particular, I provide a close reading of Book III, Chaps. 1–6, which offer an excellent point of entry into this complex and difficult text. Roughly a quarter of the whole, Book III presents central features of Pico's attack on astrology's natural philosophical foundations as reconstructed in volume I. In addition to both its intrinsic interest and its profound neglect in the scholarship, I use Pico's treatment of astrology in Book III of the *Disputations* to argue for a major shift in his own views on astrology over the course of his short but passionate life, from an essentially neutral early position to a profoundly negative later view. This chapter also provides a much-needed corrective and complement to Westman's analysis of Pico in his *Copernican Question*.

Volume II, Part III also explores the complexly related broader world views in Ficino's *De vita* and in Pico's rabidly anti-astrological *Disputations*. I focus here on an essential feature of Pico's anti-astrological analysis, namely, his use of *spiritus*, which offers a useful focus of comparison for grasping further his intellectual relationship to Ficino, who also used *spiritus*, but as a central feature of his *pro*-astrological *De vita*. Furthermore, this theme of *spiritus*—the primary energetic link between human beings and the heavens—also permits discussion of some of astrology's physiological and medical dimensions while also raising themes that recur in volume III.

In the overall conclusion to volume II—in addition to summarizing the results so far attained and thus setting the stage for volume III—I also discuss Jerome Torella's *Important Work on Astrological Talismans* (*Opus praeclarum de imaginibus astronomicis)*, which appeared in the *Annus Mirabilis* of 1496, the same year in which Pico's *Disputations* was published in Bologna and Copernicus arrived there to pursue the next phase of his studies with Domenico Maria da Novara. Torrella had studied in Italy for the previous 20 years and knew Ficino and Pico personally. As we will see, Torrella's *Opus praeclarum* used the comparisons and contrasts between (1) Albertus Magnus's (and the *Speculum astronomiae*'s) and (2) Thomas Aquinas's analysis and evaluation of talismans to structure his entire work. In addition, he incorporated significant features of Ficino's *De vita* III into his *Opus praeclarum*, but oddly under cover, representing these often verbatim passages as by his former teacher, Pierleone da Spoleto.

*

A larger point I hope to establish in volumes I and II (and in the study overall) concerns the relationship between Platonic and Aristotelian cosmology in relation to astrology. In the thirteenth-century texts analyzed in volume I, astrology is configured within a mathematicized but essentially Aristotelian cosmology and natural philosophy. With early Pico and Ficino in volume II, on the other hand, we find their magic (and for Ficino, his astrology) configured within an essentially Neoplatonic overarching framework, where the universe is alive and the cosmos ensouled in both and, in Pico, where kabbalah also provides essential cosmological structures. In Pico's later *Disputations*, however, the world is no longer animated as he both rejects (or neglects) kabbalah and shifts from a Platonic living to an Aristotelian non-ensouled cosmos.

That astrology could exist in either a Platonic animated or an Aristotelian non-ensouled universe points to astrology's profound cosmological neutrality.[36] As we

[36]This is entirely unclear in the historiography, at least from Walker on, who begins his reconstruction with Ficino's Platonizing system, which is followed by Yates, Garin, Copenhaver, etc., as I discuss in detail in the overall introduction to volume II.

will see in volume III, this neutrality is further apparent in those astrologers (or astrologizing natural philosophers) who used either the older Ptolemaic geocentric or the newer Copernican heliocentric (or, more precisely, heliostatic) cosmology, as it was for those who adopted Tycho Brahe's geo-heliocentric compromise.[37] We can also see this today, where, although astrology no longer has any toehold whatsoever in the domains of legitimate knowledge—nor is it taught, except as history, in contemporary universities—it still continues to exist unabated in both the East and the West, thus exhibiting both its cosmological and more broadly scientific neutrality as well as its extraordinary adaptive flexibility.

Volume III

Early Modern Structures (1500–1800): Continuities and Transformations

In its turn, volume III analyzes significant patterns of astrology's continuities and transformations from the sixteenth through the eighteenth centuries within the interpretive framework constructed in volumes I and II. I provide significant landmarks in a vast terrain as well as instructive examples of how to employ the interpretive framework for analyzing this crucial yet underexplored area of scientific transformation, where modern science and philosophy emerged from their astrologizing premodern matrix.

These chapters have somewhat different textures, with some figures treated in less detail, especially where extensive research has already been undertaken, for example, with Kepler. Others are treated more extensively and in greater depth, for example, Galileo's astrological practice within the context of his education and career, and Francis Bacon's proposals for astrological reform. With the framework and foundations provided in volumes I and II—and neither neglecting nor disregarding the numerous and long-lasting deep-structural and more superficial continuities (some extending well into the eighteenth century and sometimes even beyond)—volume III more broadly reinvestigates the crucially important problems of (1) astrology's relationship to the New Science and related philosophical developments of the sixteenth through eighteenth centuries, and [2] its marginalization and ultimate removal from legitimate natural knowledge during the Scientific Revolution and the Enlightenment, perhaps the most significant desiderata in the history of science.[38] Exploring the deeper historical structures and the epistemologi-

[37] For much relevant analysis that also takes astrology into account, see Omodeo, *Copernicus in the Cultural Debates of the Renaissance*.

[38] The basic outlines of an earlier phase of this argument appear in my article, "Astrology," in *The Cambridge History of Science, Vol. 3: Early Modern Science*, Lorraine Daston and Katharine Park (eds.), Cambridge: Cambridge University Press, 2006, 541–61. An updated and expanded version

cal significance of this story's various dimensions will be of primary concern in what follows.[39]

*

Overall, I intend to trace through all three volumes the complex and manifold relationships between practical astrology, on the one hand, and (1) natural philosophy/science, (2) theology/religion and (3) magic, on the other. I have given a preliminary indication so far of how I will approach the relationship between astrology and natural knowledge. I would like to offer an indication here of how I will likewise trace the relationship of astrology to theology/religion—both continuities and transformations—through all three volumes.[40] In volume I, I primarily treat the themes of fate, divination and providence in the works of Albertus Magnus and Thomas Aquinas in a predominantly Aristotelian natural philosophical and Roman Catholic theological context in order to establish astrology's theological foundations. In volume II, I will treat astrology and theology in the works of Marsilio Ficino and Giovanni Pico della Mirandola, but in the revised context of a richly (and newly) articulated Platonic cosmology and theology, including firsthand knowledge of all of Plato's works, and of numerous influential Neoplatonic texts, most of which were translated from Greek into Latin by Ficino himself.

Finally, in volume III, I will examine (i.a.) how the astrologizing features of Thomas Aquinas's increasingly authoritative theological writings were read in the sixteenth century, especially in printed editions of the *Summa theologiae* and *Summa contra gentiles* with the richly astrological commentaries by Tommaso de Vio (Cardinal Caietanus) and Francesco Silvestri Ferrariensis respectively, both of whom became generals of the Dominican Order. These will first be explored in themselves and then compared with influential and informative texts by Philip Melanchthon for the Lutheran Reformers, including his richly astrological and theological natural philosophy textbook for the reformed German universities, the *Initia doctrinae physicae* (1549).

As it happened, the Council of Trent decreed that Thomas's texts with Caietanus's and Sylvestri's commentaries were to become the officially endorsed and authoritative versions. Finally (for the sixteenth century), I will also explore John Calvin's profoundly anti-astrological *Admonitio adversus astrologiam divinatricem* (also

---

has recently appeared in *Early Science and Medicine* 23 (2018):217–43, "How to Accurately Account for Astrology's Marginalization in the History of Science and Culture: The Central Importance of an Interpretive Framework."

[39] These three volumes thus represent a significant development of my PhD thesis, "Astrology, Natural Philosophy and the History of Science, c. 1250–1700: Studies Toward an Interpretation of Giovanni Pico della Mirandola's *Disputationes adversus astrologiam divinatricem*," Indiana University, 2002. Material from its Chaps. 1, 2 and 3 will be found amplified and reworked in volume I, from its Chaps. 4, 5 and 6 in volume II, and from its Chap. 7 in volume III. The interested reader may rest assured that he or she will find much new evidence and argumentation in each volume.

[40] I will briefly indicate how I will trace out the magical themes through all three volumes in the overall introduction to Part III, and more fully in the overall introduction to volume II.

1549). Thus, I will explore how astrology related to Catholic Aristotelian and Platonic as well as to Reformed Lutheran and Calvinist theological views in at least some of their profound complexity, with a particular focus on their respective views of divine providence vis-à-vis astrology. These themes will then be further traced throughout the seventeenth and eighteenth centuries. I will thus explore and I hope to explain what is becoming an increasingly central question in my research, namely, how astrology came to be marginalized, dislodged, and ultimately removed from its theological foundations during this period.

Astrology's Conceptual Structure: A First Approximation

We should now return to astrology *qua* natural knowledge in order to zoom in a little closer and sharpen our focus, asking first where astrology fits into the medieval map of knowledge.[41] To begin to answer this question, we should briefly review our initial distinctions in the broader conceptual field indicated by the term "astrology," and thereby place our premodern conceptual spectacles back on, but this time with their newly ground lenses. We may now begin to look more deeply with our clearer and sharper vision. First is the fundamental distinction between what we call "astronomy" and "astrology," namely, the study of the *motions* of the luminaries, planets and fixed stars (astronomy) and their *influences and effects* (astrology).[42]

With regard to astrology proper, one should then distinguish between theory and practice.[43] The practical side is straightforward and concerns the actual construction and interpretation of horoscopes—making judgments or interpretations of various sorts—including on nativities (birth horoscopes) and for annual prognostications.[44]

[41] I will also now start layering in more of the scholarship.

[42] Astrology and astronomy are clearly differentiated conceptually in, e.g., Ptolemy, Albertus Magnus and Roger Bacon, even though the same term is used for both; they are both called *astronomia* in Albert, which in this sense should be translated neutrally as "the science of the stars." This is well known and requires no further discussion here; see (eg) Richard Lemay, "The True Place of Astrology in Medieval Science and Philosophy: Towards a Definition," in *Astrology, Science and Society: Historical Essays,* Patrick Curry (ed.), Woodbridge: Boydell, 1987, 57–73, 65.

[43] Westman devotes much attention to this distinction in *The Copernican Question*, but I develop it rather differently than he does.

[44] The terminology is somewhat ambiguous. In normal contemporary English parlance, "horoscope" refers to the astrological figure (normally square in shape during the premodern period), whether a natal (birth) horoscope, or one for an election, interrogation or revolution. *Horoscopus* in Latin, however, refers to the rising point of the astrological figure, the degree of the ecliptic (and zodiac) rising on the eastern horizon at the time for which the figure is cast at a given location. When I use the term "horoscope" in English, it will thus refer to an astrological figure; when I use the Latin term *horoscopus*, it will refer to the rising (ascending) point, but I will normally call it the "ascendant," "rising sign," or "rising point" in English. See John D. North's masterly study, *Horoscopes and History*, London: Warburg Institute, 1986, 1, for a discussion of terminology; he also published numerous historical examples of astrological figures. I discuss horoscopes further in the excursus to this introduction just below.

In the *Speculum astronomiae* (ca. 1260s), the deliberately anonymous author[45] usefully discusses the different types of astrological praxis: revolutions (also often called general or universal astrology), nativities, interrogations and elections, in what became an authoritative description of practical astrology's basic structures.[46] I discuss the basic features of practical astrology just below in an excursus to this introduction.

Astrology's theoretical side, however, reveals a more complex structure, requiring a further distinction which turns on how the theorizing is oriented. It can be directed toward practical applications, for example, answering such questions as what is the nature of each planet, sign and house. Thus, when a practitioner would need to interpret what, say, Jupiter in Scorpio in the fifth house signifies in a given astrological figure, he or she[47] would have some theoretical basis for the interpretation. I call this "theoretical astrology" proper; it is the sort of thing found in most astrology textbooks.

But theorizing about astrology can also be oriented, although in fact it often is not, toward astrology's natural philosophical foundations, asking more fundamental scientific questions concerning how astrology works, for example, what is the nature of planetary and more broadly celestial influences on the terrestrial world, and how do they function. This side of theorizing, which I will normally call "astrologizing natural philosophy," corresponds to the natural philosophical side of David Lindberg's distinction between practical astrology and astrology as a branch of nat-

[45] I no longer accept Paola Zambelli's arguments for Albertus Magnus as the author, but I also do not have a compelling alternative to suggest, whether Roger Bacon, Campanus of Novara or Richard de Fournival. In other words, the case is still open. The main recent studies treating this issue are, first, Paola Zambelli, *The* Speculum Astronomiae *and its Enigma: Astrology, Theology and Science in Albertus Magnus and his Contemporaries*, Dordrecht: Kluwer, 1992. She discusses the attribution throughout much of her extensive introductory study, especially Chaps. 1–7. See also Agostino Paravicini Bagliani, *Le Speculum Astronomiae, Une enigme? Enquete sur les manuscrits*, Florence: Galluzzo, 2001, and Bruno Roy, "Richard de Fournival, auteur du *Speculum astronomiae*?" *Archives d'Histoire Doctrinale et Littéraire du Moyen Age* 67 (2000): 159–80. I discuss this issue in more detail in Part I.

[46] 4, 2–8, and, for each in turn, Chaps. 7–10. The *Speculum astronomiae* provides an excellent introduction to the structure and literature of medieval astrology for historians of premodern thought, without being itself a practical textbook. Pietro d'Abano (c. 1250-c. 1315) provides the same schema in the same order; *Conciliator*, Diff. 10, fol. 14r-v. See Siraisi, *Arts and Sciences at Padua: The Studium of Padua before 1350*, Toronto: Pontifical Institute of Medieval Studies, 1973, 86. I discuss Pietro d'Abano more fully in Part IV.

[47] Although there were some women with astrological knowledge in premodern Europe (e.g., Christine de Pizan, 1364–1430), the vast majority were men. Therefore, I will use the masculine pronoun as normative. For Christine, whose father was the physician and astrologer, Tommaso di Benvenuto da Pizzano, see, e.g., Charity C. Willard, *Christine de Pizan: Her Life and Works*, New York: Cannon, 1984. See also Joan Cadden, "Charles V, Nicole Oresme, and Christine de Pizan: Unities and Uses of Knowledge in 14th-Century France," in *Texts and Contexts in Ancient and Medieval Science*, Edith Sylla and Michael McVaugh (eds.), Leiden: Brill, 1997, 208–44.

ural philosophy[48] and to what Richard Lemay calls theoretical, learned astrology.[49] This is also the realm of astrology discussed by John North,[50] Edward Grant,[51] Steven Vanden Broecke,[52] and now Robert Westman[53] and Mary Quinlan-McGrath in their different ways, as discussed in more detail below.[54] Learned astrologizing natural philosophy will be the primary focus of this study, but astrological theory and practice will also be treated where relevant.[55]

---

[48] David C. Lindberg, *The Beginnings of Western Science: The European Scientific Tradition in Philosophical, Religious, and Institutional Context, 600 B.C. to A.D. 1450*, Chicago: University of Chicago Press, 1992, 74–80, especially 74: "It will help if we begin by distinguishing between (1) astrology as a set of beliefs about physical influences within the cosmos and (2) astrology as the art of casting horoscopes, determining propitious moments, and the like. The former was a respectable branch of natural philosophy, the conclusions of which were rarely called into question." See also his salutary remarks about how modern scholars should approach astrology: "But medieval astrology also had a serious scholarly side, and we must not allow our attitude toward it to be colored by the low regard in which astrology is held today. Medieval scholars judged astrological theory and practice by *medieval* criteria of rationality and by the *contemporary* evidence to which they had access; and it is only as we do the same that we can hope to understand the importance and the changing fortunes of astrology during the Middle Ages (74, his emphases)." He restates this position in the second edition as well (2007, 270–77).

[49] Lemay, "True Place of Astrology," 60–63.

[50] John D. North, "Celestial Influence—The Major Premiss of Astrology," in "Astrologi Hallucinati"*: Stars and the End of the World in Luther's Time*, Paola Zambelli (ed.), Berlin: de Gruyter, 1986, 45–100, now conveniently reprinted in his *Stars, Minds and Fate: Essays in Ancient and Medieval Cosmology*, London: Hambledon, 1989, 243–98.

[51] *Planets, Stars, and Orbs*, especially Chap. 19, "The Influence of the Celestial Region on the Terrestrial," 569–615. Grant notes that he drew heavily for this chapter on his "Medieval and Renaissance Scholastic Conceptions of the Influence of the Celestial Region on the Terrestrial," *Journal of Medieval and Renaissance Studies* 17 (1987): 1–23. This article adds nothing of substance to the treatment in *Planets, Stars, and Orbs* with regard to our concerns.

[52] *Limits of Influence.* Vanden Broecke interprets astrology's conceptual field differently than I do here. From my perspective, he conflates two components that I distinguish and unnecessarily distinguishes another. In my dissertation, I used the phrase "scientific astrology" to describe what I here call "astrologizing natural philosophy." I am grateful to David Juste for persuading me that my previous usage of "scientific astrology" was ambiguous, primarily because the natural philosophical foundations of astrological practice are not themselves a form of astrology, but rather they provide its broader framework and deeper scientific foundations, as Albertus Magnus indicates clearly in his discussion of Aristotle's *De caelo* II.3.5, which I discuss in Part I.

[53] In significant ways, the present volume provides the proper background—including conceptual, institutional, and socio-political foundations—for Westman's study, as we will see in what follows.

[54] *Influences: Art, Optics and Astrology in the Italian Renaissance,* Chicago: University of Chicago Press, 2013. Among many other things, Quinlan McGrath usefully reconstructs astrology's geometrical-optical model of celestial influences. I discuss her book more fully in volume II. For some of my reflections on her valuable study, see my essay review, "'See, Reflect, Be Changed': Marsilio Ficino, *Imagines astronomicae* and a New Perspective on Renaissance Art and Science," *Early Science and Medicine* 19 (2014): 584–93.

[55] In her *History, Prophecy, and the Stars: The Christian Astrology of Pierre D'Ailly, 1350–1420* (Princeton: Princeton University Press, 1994), Laura Smoller usefully describes the theoretical and practical levels in her introductory chapter on the state of astrology at Pierre's time (10–24). She touches only very briefly on astrologizing natural philosophy, particularly in her discussion of

*

The broader conceptual field that includes astrology's natural philosophical foundations warrants further discussion. Astrology's overall structure as just delineated has three distinctive levels: (1) a practical level where a practitioner applies general information derived from (2) a theoretical understanding of astrology as a more or less systematic doctrine, gained from teachers and/or textbooks, and oriented toward practical concerns. This theoretical astrology (2) is then grounded ultimately in and justified by (3) a certain natural philosophical and cosmological understanding of how the world works, supporting the view that celestial bodies have this putative relationship with the terrestrial realm, and that it is at least partially knowable by human beings.[56] Properly speaking, however, (3) is not itself a part of astrological doctrine but, rather, provides astrology's natural philosophical foundations. There are thus three main levels: practical, theoretical and natural philosophical/scientific. This appears to be the same structure we discover in another of the most important and best studied scientific disciplines of the premodern period, namely, medicine.

Ptolemy's *Tetrabiblos* is mainly a theoretical treatise oriented toward astrology's practical application.[57] There is very little on astrology's natural philosophical foundations, for which one would have to turn, ultimately, to Aristotle.[58] Galen's medical

---

Thomas Aquinas (31–32). I discuss her work further below. I should also note that there is a major gap in my representation of this system between the theoretical and the natural philosophical levels. Planetary influences may well work by means of rays, as we will see, but why should the planets or signs symbolize what they do? This symbolic dimension is provided by a long-standing and strikingly consistent mythologizing with roots reaching back ultimately to ancient Babylonian and Greco-Roman religious structures. Ptolemy tries more or less successfully to philosophically justify some of these traditions, but for the very most part, they are simply assumed as given, presumably because they were considered either empirically or symbolically to "work" or be "true" on some level. However interesting, this gap (the symbolic turn)—which should be explored by historians and anthropologizing theoreticians of myth, religion and folklore—will not be treated in this study.

[56] The epistemological dimension concerned with astrology's "knowability" provides the crux of skeptical attacks on astrology from Antiquity through the eighteenth century, from Cicero and Sextus Empiricus through Nicole Oresme and Giovanni Pico della Mirandola to Pierre Bayle, among others. I will treat astrology's increasingly vocal critics more fully in volumes II and III.

[57] Ptolemy discussed the physical (Aristotelian) foundations very briefly and in little depth in Book I, Chapters 4 ff. It was left to his medieval followers to provide a much fuller articulation, as we will see, although it seems that Ptolemy may have provided the germ for the geometrical-optical model of celestial influences developed later by al-Kindi. This is according to Jacqueline Feke's interesting but somewhat speculative reconstruction; *Ptolemy's Philosophy*, 176 ff. Vanden Broecke discusses Ptolemy's natural philosophical foundations, what he calls "astrological physics," which is grounded on the four primary Aristotelian qualities—hot and cold, dry and moist—and their natures as active or passive. This is the extent of Vanden Broecke's analysis of astrological physics and thus of what he thinks Pico was attacking in the *Disputations*. As the present study will argue in detail, I think that there is much more to this story. Unfortunately, Westman implicitly embraces Vanden Broecke's interpretation, which he then uses as the profoundly inadequate basis of his broader argument concerning astrology in relation to natural philosophy and Pico's attack thereon, as we will see below.

[58] See Lemay, "True Place," 57: "[T]he focus of historians who set out to understand and explain the place of astrology in ancient and medieval science was directed the wrong way. Stimulated by the Renaissance favoritism toward Greek classics, historians have stressed the capital role of

writings provide an instructive contemporary parallel to Ptolemy.[59] Galen's extensive medical corpus comprises practical, theoretical and natural philosophical works in our sense.[60] The practical part is straightforward and requires no discussion here. The theoretical part, on the other hand, is easily identifiable as medicine and, although quite different from our own, is still predominantly concerned with diagnosing and curing patients, describing and understanding disease processes, and designing and prescribing medicaments. All of this occurs within the framework of its own theoretical patterns, in which the body is composed of parts that each relate to the four humors and their relative balance, resulting, ultimately, in a person's complexion or constitution.[61]

But Galen hardly stops there. He often relates his theoretical and practical discussions directly to the scientific foundations of his medical theory, which fall deeply, as one would expect, within the fundamental patterns of Aristotle's natural philosophy: not only his biology and embryology, but ultimately the deepest structures of Aristotle's natural philosophy, including his views on the soul, the four elements and qualities, the four causes, and the four types of change (*motus* or "kinesis" [Gl]).[62] The patterns

---

Ptolemy's *Tetrabiblos*, which was advanced by nearly all of them as a substitute, on the one hand for the more basic and dominant works of Aristotle in the Greek tradition, and on the other for the role of Abu Ma'shar in the Arab tradition, in providing the 'scientific' background of ancient astrology throughout the Middle Ages and the Renaissance." This criticism may be applied, e.g., to Mary Ellen Bowden, "Scientific Revolution in Astrology: The English Reformers, 1558–1686," PhD thesis, Yale University, 1974; Vanden Broecke, *Limits of Influence*, and Westman, *Copernican Question*. Albumasar's role in introducing Aristotle to the West by melding astrology with Aristotelian natural philosophy was significant, even if somewhat exaggerated by Lemay; "True Place," 58–60, and *Abu Mashar and Latin Aristotelianism*, especially chapter II, "Aristotle's Natural Philosophy in the *Introductorium*," 41–132. For the primary texts, see his nine-volume edition of Albumasar's Arabic text and its Latin translations, with an extremely long and characteristically acerbic introduction: *Liber introductorii majoris ad scientiam iudiciorum astrorum*, Naples: Istituto Universitario Orientale, 1995. I mention Albumasar from time to time in what follows.

[59] On the importance of studying Ptolemy and Galen for getting a proper understanding of ancient science, particularly with regard to counterbalancing the overemphasis on Plato, see Otto Neugebauer, *The Exact Sciences in Antiquity*, second ed., New York: Dover, 1969 (originally published, 1957), 152.

[60] For excellent orientations, see Vivian Nutton, *Ancient Medicine*, second ed., Oxford: Routledge, 2013, and Lawrence I. Conrad, Michael Neve, Vivian Nutton, Roy Porter and Andrew Wear, *The Western Medical Tradition: 800 B.C.-1800 A.D.*, Cambridge: Cambridge University Press, 1995.

[61] For a clear and concise discussion of fundamental theoretical patterns of medieval Galenic medicine, see Nancy G. Siraisi, *Medieval and Early Renaissance Medicine: An Introduction to Knowledge and Practice*, Chicago: University of Chicago Press, 1990, and Michael McVaugh's admirably learned and clear introduction to *Arnaldi de Villanova, Aphorismi De gradibus*, Granada: Seminarium Historiae Medicae Granatensis, 1975, esp. 4 and 20 ff. These theoretical patterns are also directly relevant for astrological medicine.

[62] Roger French puts a fine point on this in his description of a medieval medical practitioner trying to impress his patients and so attract more of them: "[H]e could, as indeed Galen himself had done, not only persuade the patient and pupil that he knew by experience and reading about this or that disease, but that he could *explain* it on the basis of a chain of argument that reached from the patient's symptoms back to the very fundamentals of the world picture." "Astrology in Medical

of thought at this natural philosophical or scientific level appear to be the same for both astrology and medicine,[63] as we might have expected.

*

There are essentially three distinguishable but by no means independent realms to this cosmological and natural philosophical picture.[64] The three main physical realms are (1) the elemental, comprising earth, air, fire, and water as yet uncombined, and their transformations into each other; (2) the corporeal, comprising mixed bodies, plants, people, etc., that is, *res naturales* (natural things) in general, "substances" in Aristotle's sense[65]; and (3) the celestial, including the sun, moon, and planets, with their motions and influences, and the sphere of the fixed stars. Part I of volume I is primarily concerned with mapping out the basic structures of (3), and its most significant modes of interaction with the other two realms, especially (2).[66]

One of the main practical results to be gained from articulating this medieval natural philosophical/scientific model in its basic structures and in sufficient detail in the works to be analyzed here is that we will then have a proper basis for comparing this astrologizing Aristotelian natural knowledge with the particularities of other natural philosophical systems, including those developed during the Renaissance and the Scientific Revolution. As we will see repeatedly below, without properly articulating the characteristic features of these scientific structures and their dynamic

---

Practice," in *Practical Medicine from Salerno to the Black Death,* L. Garcia-Ballestar, R. French, J. Arrizabalaga, and A. Cunningham (eds.), Cambridge: Cambridge University Press, 1994, 30–59, 33, his emphasis; and see the bibliography cited there. I discuss the deeper structures of Aristotelian natural philosophy primarily in Chap. 1.

[63] For the philosophical foundations of medieval medicine, see Nancy G. Siraisi, *Taddeo Alderotti and His Pupils: Two Generations of Italian Medical Learning*, Princeton: Princeton University Press, 1981, *passim*, and especially Chap. 6, "The Uses of Philosophy: Reconciling the Philosophers and the Physicians," and McVaugh's introduction, *Aphorismi de gradibus*, *passim*, and especially Chap. 5, "The *Aphorismi De Gradibus* as Natural Philosophy," 89–122. Both authors argue strongly and persuasively that the study of medicine was very important to the study of natural philosophy (and vice versa) during the medieval period. See also Luis Garcia-Ballestar's introduction to *Practical Medicine from Salerno to the Black Death*, 1–29, esp. 22, and *passim* for other relevant classic and recent bibliography, including the fundamental studies by Paul O. Kristeller and Charles B. Schmitt cited just below.

[64] To be more precise, a fourth nonphysical realm is also associated with separated substances and God, as we will see. This metaphysico-theological realm is only touched upon where necessary in what follows.

[65] As discussed in the *Categories* and other works.

[66] The fourth theological or metaphysical realm from God to the celestial bodies encircles, embraces and supports the natural realm from the celestial bodies to the earth, its atmosphere and inhabitants. In fact, this divine realm stretches from God (= the first cause) on high via the angels (= separated intellectual substances) to the very same celestial realm, including the planets and luminaries in the zodiac. Except for the celestial dimension, I ignore this realm almost entirely in what follows, except in Chap. 5, where I treat astrology's relationship to theology/religion, focusing in particular on Thomas Aquinas's and Albertus Magnus's analyses of fate in relation to divine providence.

processes in depth and detail, it is simply impossible to compare two thinkers' views in an accurate and systematic manner, let alone to construct a historically sound thematic survey over extended periods of time.

*

How, then, should we best describe a system of natural knowledge at this deep structural level, especially a system no longer in use,[67] that is, one no longer the subject of intensive educational preparation?[68] It is well known that a broad range of Aristotle's works, with a plethora of commentaries,[69] solidly grounded this educational preparation from circa 1200, with the initial foundation of the medieval universities,[70] up to and including Newton's time.[71] I have adopted a compromise

---

[67] See Lemay, "True Place," 60: "Conceptually, the effect of the scientific revolution was to render nearly impossible for historians henceforward to grasp the different scope of ancient science, of its epistemology and orientation. Yet a close study of earlier paradigms remains theoretically possible, through patient and open-minded examination of texts and their context. This approach alone can reveal the key concepts and tools that buttressed ancient science, as well as the structure of thought elicited through them." This, of course, refers to all premodern science, not only ancient.

[68] Kuhn, *Structure of Scientific Revolutions*, 4–5, 10–11, etc. Kuhn is quite interesting on this. The point seems to hold also for the premodern period, even with the significant differences in the social organization of "science" then: "The study of paradigms, including many that are far more specialized than those named illustratively above (e.g., Aristotle's *Physica*, Ptolemy's *Almagest*, Newton's *Principia*, etc.) is mainly what prepares the student for membership in the particular scientific community with which he will later practice (10–11)."

[69] Charles H. Lohr's fundamental biobibliographical studies make this point very effectively for the entire period 1200–1700; see, e.g., *Latin Aristotle Commentaries*, Florence: Olschki, 1988.

[70] For translations of Aristotle's works into Latin from Arabic and Greek as the foundation of medieval university curricula, see, e.g., Gordon Leff, *Paris and Oxford in the 13th and 14th Centuries*, New York: Wiley, 1968, 116–184; John Marenbon, *Later Medieval Philosophy (1150–1350): An Introduction*, London: Routledge, 1987, 7–82; and various essays in *The Cambridge History of Later Medieval Philosophy*, N. Kretzmann, A. Kenny, and J. Pinborg (eds.), Cambridge: Cambridge University Press, 1982, and now in the *Cambridge History of Medieval Philosophy*, Robert Pasnau (ed.), 2 vols., Cambridge: Cambridge University Press, 2010. It is extremely important, however, not to focus only or primarily on the Northern, predominantly theologically oriented universities, but to pay equal attention also to the legally and medically oriented Italian universities: Salerno and Bologna in the earlier period, Padua and Naples soon after. See the ground-breaking work of Paul Oskar Kristeller on Salerno, "The School of Salerno: Its Development and Its Contribution to the History of Learning," in his *Studies in Renaissance Thought and Letters*, Rome: Edizioni di Storia e Letteratura, 1956, 495–551; Garcia-Ballestar, "Introduction," in *Practical Medicine from Salerno to the Black Death*, 13–29; and Charles B. Schmitt, "Aristotle among the Physicians," in *The Medical Renaissance of the 16th Century*, A. Wear, R.K. French, and I.M. Lonie (eds.), Cambridge: Cambridge University Press, 1985, 1–15, with historical background on 2–4. Kristeller, "Thomism and Italian Thought of the Renaissance," in *Medieval Aspects of Renaissance Learning: Three Essays by Paul Oskar Kristeller*, E.P. Mahoney (ed. and tr.), second ed., New York: Columbia University Press, 1992 (originally published, 1974), 27–91, makes the point about not focusing overmuch on Northern theologically oriented universities very strongly at 43 ff. On the Italian universities, see Paul F. Grendler, *The Universities of the Italian Renaissance*, Baltimore: Johns Hopkins University Press, 2002. I discuss the institutional and some social structures of medieval university education in Part IV.

[71] See particularly J.E. McGuire and Martin Tamny's introduction to *Certain Philosophical Questions: Newton's Trinity Notebook*, Cambridge: Cambridge University Press, 1983, esp. 15–20,

solution to be described in the overall introduction to Part I. Before we arrive there, however, we should begin to explore some of the scholarship on the history of astrology.

Review of the Scholarship

Most broader studies of astrology (and magic) in premodern Europe[72] either focus almost entirely on (1) the medieval period ca. 1250–1500 (or even earlier, ca. 1100–1500, thus including the translation of Arabic and Greek texts into Latin [e.g., Boudet and Weill-Parot]),[73] or on (2) the Renaissance and Early Modern periods, which almost always begin with either Marsilio Ficino or Giovanni Pico della Mirandola. This depends primarily on whether their main concern is (a) the history of astrology and/or magic in itself (e.g., Walker, Yates and Copenhaver) that begin with Ficino,[74] or more properly assessing (b) astrology's role in the Scientific Revolution (e.g., Westman, Vanden Broecke and Bowden) that begin with Pico.[75]

---

on Newton's Greek and Latin notes in the extant manuscript of the notes he took during his undergraduate days at Trinity College, Cambridge, beginning in 1661. McGuire and Tamny are insightful on the impact of Aristotle in Newton's education, certainly a major part of the scientific tradition he was schooled in: "[With regard to Newton's studies at Cambridge] […] Aristotelianism, in its variety of forms, continued to provide a coherent and powerful account of human experience. As a systematic metaphysical system, it had few serious pedagogic rivals throughout the seventeenth century. […] [I]t would be a mistake to think that Newton's intellectual pilgrimage began only after he had rejected the traditional curriculum in favor of the 'mechanical philosophy' (19)." That Aristotle was greatly studied in the Renaissance—even more so than in the medieval period—has now been well established, especially in the works of F. Edward Cranz, Charles Lohr and Charles B. Schmitt; for a magisterial survey, see Schmitt, *Aristotle and the Renaissance*, *passim* and the rich bibliography there.

[72] I am not speaking here of focused studies on particular texts, people or places, but rather on more synthetic studies. I offer many examples of more focused studies just below.

[73] *Entre science et nigromance*; *Les "images astrologiques*." There is also much valuable information in Graziella Federici Vescovini's *Medioevo Magico. La magia tra religione e scienza nei secoli XIII e XIV*, Torino: UTET, 2008. In his textbook on medieval science, Lindberg treats astrology seriously, but all too briefly (as cited above); *Beginnings of Western Science*, 1992, 2007². John D. North's treatment in the *Cambridge History of Science* is even briefer; "Astronomy and Astrology," in *The Cambridge History of Science, Vol 2: Medieval Science*, David C. Lindberg and Michael H. Shank (eds.), Cambridge: Cambridge University Press, 2013, 456–84, 473–78.

[74] Respectively, *Spiritual and Demonic Magic from Ficino to Campanella* (London: Warburg Institute, 1958), *Giordano Bruno and the Hermetic Tradition*, and various works by Copenhaver. These studies are primarily inflected toward the history of philosophy.

[75] Respectively, *Copernican Question*, *Limits of Influence,* and "Scientific Revolution of Astrology." These studies are primarily inflected toward the history of science. Furthermore, most analyses of the Scientific Revolution proper still marginalize astrology (e.g., Wootton, Dear, Osler, Principe) or ignore it entirely: David Wootton, *The Invention of Science: A New History of the Scientific Revolution*, New York: HarperCollins, 2015; Peter Dear, *Revolutionizing the Sciences: European Knowledge and its Ambitions, 1500–1700*, second ed., Princeton: Princeton University Press, 2009; Margaret J. Osler, *Reconfiguring the World: Nature, God, and Human Understanding from*

The main problem with both (2a) and (2b), however, is that they almost never take the studies in (1) as foundational, one of my main criticisms of Westman's *Copernican Question*.

This is not true, of course, for Lynn Thorndike's monumental, groundbreaking, and still extremely valuable eight-volume *History of Magic and Experimental Science* (New York: Columbia University Press, 1923–1958), which treated both astrology and magic (and many other more-or-less related disciplines) throughout the entire period. But, as rich and valuable as it is, Thorndike's primarily manuscript-based spade work lacks a viable interpretive framework. He did, however, offer a 5-page synthetic study entitled "The True Place of Astrology in the History of Science" (1955),[76] which strongly asserts the more general case I am trying to make here, but several orders of magnitude too briefly. In his valuable *Planets, Stars, and Orbs*, Edward Grant too links these normally divided historical periods, but he does not see astrology's place in either as significant for the history of science, as I will discuss more fully in Chap. 1.

*

On the basis of many particular studies (some of which are mentioned just below) and a penetrating reading of a range of primary sources, there have recently appeared extremely valuable interpretive syntheses of what we now know about medieval astrology and magic, especially two in French by Nicolas Weill-Parot (on astrological images or talismans within all of their relevant intellectual and institutional contexts [2002]) and Jean-Patrice Boudet (on astrology, magic and divination [2006]).[77] Both fundamentally treat the medieval period ca. 1100–1500, and both provide foundations for both broader interpretations and further research. Needless to say, I have learned a great deal from both of them.

Building on these foundations, one of my main innovations is to integrate the Medieval with the Renaissance and Early Modern periods into one extended but clearly differentiated time frame ca. 1250–1800.[78] I begin by establishing the medieval foundations (in volume I) and then show, first, how they were transformed and Platonized during the Renaissance (in volume II) and then (in volume III) how both

the Middle Ages to Early Modern Europe, Baltimore: Johns Hopkins University Press, 2010; Lawrence Principe, *The Scientific Revolution: A Very Short Introduction*, Oxford: Oxford University Press, 2011. Of these four examples, Principe treats astrology most fully.

[76] *Isis* 46, 273–78.

[77] *Entre science et nigromance; Les "images astrologiques."*

[78] Frank Klaassen also bridges this historiographic divide for magic in *The Transformations of Magic: Illicit Learned Magic in the Later Middle Ages and Renaissance*, University Park, PA: The Pennsylvania State University Press, 2013. Dag Nikolaus Hasse does this as well for the study of Arabic authors in Latin translation and their teaching in medieval and Renaissance universities—for astrology as well as for philosophy and medicine—in relation to both their manuscript production and printing history through 1700; *Success and Suppression*. He points out both continuities and transformations. I am certain that this groundbreaking study will bear rich fruit in subsequent research.

were transformed during the Reformation, Scientific Revolution and Enlightenment, when astrology was ultimately downgraded from legitimate to illegitimate knowledge. In this investigation, I focus much more on astrology's natural philosophical or scientific foundations (both conceptual and institutional) than on either its theological foundations or its various practices. Allowing the significant continuities and transformations over this *longue durée* to be more accurately traced, my integrated framework may be used to more fully comprehend both the broader patterns and the particular contours of astrology's rich and eventful history.

*

Why, then, has there been a recent turn toward a fuller and less ideologically distorted understanding of astrology's many roles in premodern science, theology and culture? With respect to the history of science, one main factor seems to be the shift away from focusing on Galileian and other precursors of modern physics (e.g., Duhem, Koyré, Maier) in research on the history of medieval Latin science during the 1940s and 1950s (and before)[79] to exploring more accurately the integral patterns of medieval natural knowledge in its own terms, including bodies of knowledge that are no longer considered scientific, such as alchemy and astrology. Significantly, this shift took place in the broader historiographic context of a major increase of interest in the study of the so-called hermetic or occult sciences, primarily in the Renaissance and Early Modern periods. This followed in the wake of Frances Yates's phenomenally influential *Giordano Bruno and the Hermetic Tradition* (1964) and Keith Thomas's *Religion and the Decline of Magic* (1970) as well as against the broader background of Thorndike's recently completed *History of Magic and Experimental Science* (1923–1958) and Thomas Kuhn's *Structure of Scientific Revolutions* (first ed., 1962).

In this shift of historiographic focus, studies of alchemy have led the way by greatly expanding the domains of legitimate research in the study of medieval science on previously unsuspected topics that have now convincingly been shown to provide valuable historical insight into conceptual structures and experimental practices that were further developed during the Scientific Revolution. Scholarly research into the history of alchemy in the 1980s and more recently[80] has thereby opened up

---

[79] For Pierre Duhem (with bibliography), see sect. 3, "History of Science," in the article s.v. in the Stanford Encyclopedia of Philosophy (https://plato.stanford.edu/entries/duhem/). For Alexandre Koyré, see, e.g., *Galileo Studies*, J. Mepham (tr), Atlantic Highlands, N.J.: Humanities Press, 1978 (originally published 1939). For Anneliese Maier, see, e.g., her *Die Vorläufer Galileis im 14. Jahrhundert: Studien zur Naturphilosophie der Spätscholastik*, Rome: Edizioni di Storia e Letteratura, 1949 (second ed. 1966; 1977), and her *Zwei Grundprobleme der scholastischen Naturphilosophie. Das Problem der intensiven Grösse. Die Impetustheorie*, Rome: Edizioni di Storia e Letteratura, 1951 (second ed.; third ed. 1968). For all of these scholars in their proper historiographic contexts, see John E. Murdoch, "Pierre Duhem and the History of Late Medieval Science and Philosophy in the Latin West," in *Gli studi di filosofia medievale fra Otto e Novecento*, R. Imbach and A. Maierù (eds.), Rome: Edizioni di Storia e Letteratura, 1991, 253–302.

[80] See, e.g., William R. Newman, *The* Summa Perfectionis *of Pseudo-Geber: A Critical Edition, Translation and Study*, Leiden: Brill, 1991 (based on his 1986 Harvard PhD thesis), and many

significant conceptual space for the scholarly study of both astrology and magic (and other associated topics), allowing them to be pursued more comprehensively, and without being as heavily burdened by so many of the explicitly negative judgments and associated ideologically laden baggage that had heretofore normally plagued such studies. Another important factor in our increasing understanding is more fully appreciating the medieval Arabic contributions to—and influences on—medieval Latin science, astrological, alchemical, medical, magical, and otherwise.[81] Confident in this propitious historiographic moment, my study hopes to contribute to this trend.

*

One of the major challenges to fully understanding astrology in the premodern period (ca. 1250–1800) is that astrology ramified so broadly and deeply throughout European culture—North and South, East and West—during this time: from art history to political history and thence to the history of medicine and science (primarily natural philosophy, astronomy and mathematics) and theology. My contributions to this study in volume I are to focus intensively on the works of three influential thirteenth-century thinkers—Albertus Magnus, Roger Bacon and Thomas Aquinas—in order to provide a solid foundation for the study of astrology in the Middle Ages and beyond. I do this within a more integrated periodization, treating the relevant structures over a *longue* but integrated *durée* that will ultimately encompass all three volumes. I also integrate conceptual and institutional patterns as well as compare and contrast astrology and magic with each other, and I strive to do so (as much as possible) within premodern conceptual patterns, not ours.[82]

*

Within this broader historiographic framework, five basic types of studies currently hold the field in recent scholarship on the history of astrology[83]:

---

other more recent groundbreaking and synthetic studies by him and Lawrence Principe. For earlier pioneering works on the influence of alchemy on Isaac Newton's science, see, e.g., Betty Jo Teeter Dobbs, *The Foundations of Newton's Alchemy or 'The Hunting of the Greene Lyon'*, Cambridge: Cambridge University Press, 1975, and Richard S. Westfall, "Newton and Alchemy," in *Occult and Scientific Mentalities*, 315–35 (1984). See now Rob Iliffe, *Priest of Nature: The Religious Worlds of Isaac Newton*, Oxford: Oxford University Press, 2017, and William R. Newman, *Newton the Alchemist: Science, Enigma, and the Quest for Nature's "Secret Fire*," Princeton: Princeton University Press, 2019. See also important studies by Chiara Crisciani, Sylvain Matton, Didier Kahn and Karen Figala.

[81] For example, see Dmitri Gutas, *Greek Thought, Arabic Culture*, many studies by Charles Burnett, and now Hasse, *Success and Suppression*. These studies are all founded on the penetrating earlier researches of Charles Homer Haskins, Francis Carmody, Fuat Sezgin, and others.

[82] The interested reader would also be well advised to read Anthony Grafton's magisterial review of the historiography of astrology in his "Starry Messengers" and his and William R. Newman's "Introduction" to their *Secrets of Nature*.

[83] This brief sketch is admittedly overly schematic. There are other important types of scholarship on astrology as well, textual, bibliographic, etc., some of which I will discuss in due course. For the full range, see David Juste's virtually comprehensive web bibliography for the history of astrology on the Warburg Institute's website (http://warburg.sas.ac.uk/mnemosyne/orientation/astronolinks.htm), updated regularly.

1. First is studies of astrology's different roles (including teaching and practice) in a specific locale, for example, Michael H. Shank, Helmuth Grössing and Darin Hayton on Vienna[84]; Steven vanden Broecke on Louvain[85]; Stefano Caroti on Italy overall;[86] Monica Azzolini on Milan;[87] Rodolfo Signorini on Mantua[88]; Jean-Patrice Boudet on medieval France;[89] Herve Drevillon on France in the seventeenth century[90]; Gerd Mentgen on medieval Germany[91]; Robin Barnes on sixteenth-century Germany[92]; Tayra Lanuza Navarro for early modern Spain[93]; Ana Cecilia Avalos for New Spain[94]; Luis Miguel Nunes Carolino for Portugal[95]; Jane Ridder Patrick on early modern Scotland[96]; Hilary M. Carey on medieval England; and Mary Ellen Bowden, Patrick Curry, and Ann Geneva on early modern England.[97]

---

[84] Shank, "Academic Consulting in 15th-Century Vienna: The Case of Astrology," in *Texts and Contexts in Ancient and Medieval Science*, 245–70; Grössing, *Humanistische Naturwissenschaft: Zur Geschichte der Wiener Mathematischen Schulen des 15. und 16. Jahrhunderts*, Baden-Baden: Valentin Koerner, 1983, and Hayton, *The Crown and the Cosmos: Astrology and the Politics of Maximilian I*, Pittsburgh: University of Pittsburgh Press, 2015.

[85] *Limits of Influence.*

[86] *L'astrologia in Italia: profezie, oroscopi e segreti celesti, dagli zodiaci romani alla tradizione islamica, dalle corti rinascimentali alle scuole moderne: storia, documenti, personaggi*, Rome: Newton Compton, 1983.

[87] *The Duke and the Stars: Astrology and Politics in Renaissance Milan*, Cambridge, MA: Harvard University Press, 2013.

[88] *Fortuna dell'astrologia a Mantova: arte, letteratura, carte d'archivio*, Mantua: Sometti, 2007. This book is richly and beautifully illustrated.

[89] *Recueil des plus celebres astrologues de Simon de Phares*, J. P. Boudet (ed.), 2 vols., Paris: Champion, 1997–99.

[90] *Lire et ecrire l'avenir: l'astrologie dans la France du Grand Siecle, 1610–1715*, Seyssel: Champ Vallon, 1996.

[91] *Astrologie und Öffentlichkeit im Mittelalter*, Stuttgart: Hiersmann, 2005.

[92] *Astrology and Reformation.*

[93] "Astrologia, Ciencia y Sociedad en la Espana de los Austrias," Phd thesis, Universitat de Valencia, 2005. Michael A. Ryan does so much less successfully for fourteenth-century Aragon; *A Kingdom of Stargazers: Astrology and Authority in the Late Medieval Crown of Aragon*, Ithaca: Cornell University Press, 2011.

[94] "As Above, So Below: Astrology and the Inquisition in 17th-Century New Spain," PhD thesis, European University Institute, 2007.

[95] "*Agant corpora coelestia in sublunarem mundum an non?* Ciencia, Astrologia e Sociedade em Portugal (1593–1755)," PhD thesis, Universidade de Évora, 2000.

[96] "Astrology in Early Modern Scotland ca. 1560–1726," PhD thesis, University of Edinburgh, 2012. Ridder-Patrick, Lanuza Navarro, and Carolino have also all made valuable contributions to a recent volume of *Early Science and Medicine* (22, 2017) on the marginalization of astrology edited by Rienk Vermij and Hiro Hirai.

[97] Carey, *Courting Disaster: Astrology at the English Court and University in the Later Middle Ages*, London: Macmillan, 1992; Bowden, "Scientific Revolution in Astrology"; Curry, *Prophecy and Power: Astrology in Early Modern England*, Cambridge: Polity, 1989; Geneva, *Astrology and the 17th Century Mind: William Lilly and the Language of the Stars*, Manchester: Manchester University Press, 1995.

2. Second is the treatment of certain astrologically significant events, such as the famous predictions for a great flood in 1524,[98] the eclipse of 1654,[99] and analyses surrounding the Black Death[100] and the Great Pox.[101]
3. Third is the more or less loosely structured thematic survey over time, which John D. North[102] and Edward Grant[103] exemplify with respect to astrology's natural philosophical foundations.
4. Fourth is the essay-length pastiche of texts from many different works of an author, e.g., Betsey Barker Price on Albertus Magnus.[104]
5. Fifth is the intensive book-length study of an author's work(s) in their historical context, for example, Thomas Litt's book on Thomas Aquinas's astronomy and astrology,[105] Paola Zambelli's historical and historiographical study of Albertus Magnus and the *Speculum astronomiae*,[106] Laura A. Smoller's on Pierre d'Ailly,[107] Graziella Federici Vescovini's on Biagio Pelacani da Parma,[108] Claudia Brosseder's on Caspar Peucer,[109] Pierre Brind'Amour's on Nostradamus,[110] Günther Oestmann's on Heinrich Rantzau,[111] Tommaso Duranti's on Girolamo

[98] *"Astrologi Hallucinati": Stars and the End of the World in Luther's Time*, Paola Zambelli (ed.), Berlin: de Gruyter, 1986, and Heike Talkenberger, *Sintflut: Prophetie und Zeitgeschehen in Texten und Holzschnitten astrologischer Flugschriften*, Tübingen: Niemeyer, 1990.

[99] Elisabeth Labrousse, *L'entrée de Saturne au Lion (L'eclipse de soleil du 12 Aout 1654)*, The Hague: Nijhoff, 1974.

[100] Anna M. Campbell, *The Black Death and Men of Learning*, New York: Columbia University Press, 1931, and Melissa P. Chase, "Fevers, Poisons and Apostemes: Authority and Experience in Montpellier Plague Treatises," *Annals of the New York Academy of Sciences* 441 (1985): 153–170.

[101] Jon Arrizabalaga, John Henderson, and Roger French, *The Great Pox: The French Disease in Renaissance Europe*, New Haven: Yale University Press, 1997.

[102] "Celestial Influence." His detailed studies: *Horoscopes and History*; *Richard of Wallingford: An Edition of His Writings*, Oxford: Clarendon Press, 1976; and *Chaucer's Universe*, Oxford: Clarendon Press, 1988, do not share the faults of his survey to be discussed in the introduction to Part I.

[103] *Planets, Stars, and Orbs.*

[104] "The Physical Astronomy and Astrology of Albertus Magnus," in *Albertus Magnus and the Sciences: Commemorative Essays 1980,* James A. Weisheipl (ed.), Toronto: Pontifical Institute, 1980, 155–185, and Zambelli's relevant chapters in *The* Speculum Astronomiae *and its Enigma.*

[105] *Les corps celestes dans l'univers de Saint Thomas d'Aquin,* Louvain: Publications Universitaires, 1963.

[106] *The* Speculum Astronomiae *and Its Enigma.*

[107] *History, Prophecy and the Stars: The Christian Astrology of Pierre d'Ailly, 1350–1420.*

[108] *Astrologia e scienza: La crisi dell'aristotelismo sul cadere del Trecento e Biagio Pelacani da Parma*, Florence: Vallecchi, 1979.

[109] *Im Bann der Sterne: Caspar Peucer, Philipp Melanchthon und andere Wittenberger Astrologen*, Berlin: Akademie Verlag, 2004.

[110] *Nostradamus astrophile: Les astres et l'astrologie dans la vie et l'oeuvre de Nostradamus*, Ottawa: Presses de l'Universite d'Ottawa, 1993.

[111] *Heinrich Rantzau und die Astrologie: Ein Beitrag zur Kulturgeschichte des 16. Jahrhunderts*, Braunschweig: Braunschweig-druck GmbH, 2004.

Manfredi,[112] Angus G. Clarke's Warburg PhD thesis on Giovanni Antonio Magini,[113] and Anthony Grafton's study of Girolamo Cardano.[114]

At this stage in the historiography, the first and fifth types are by far the most successful, in that—methodologically speaking—one may thereby adequately reconstruct what must be considered the fundamental unit of intellectual history: a writer's work(s) in his or her cultural context, broadly construed. It is only then, on this basis of intensive study, that proper generalizations—and, only then, accurate thematic surveys[115]—can be made; otherwise, there is simply no sound foundation for such generalizing.[116]

*

Finally, I should register my profound disappointment with the recent appearance of Robert S. Westman's long-awaited *The Copernican Question: Prognostication, Skepticism and Celestial Order* (2011). Arguing that the astrological context of his education at the universities of Cracow and Bologna played a significant role in Copernicus's project for astronomical reform, Westman makes Giovanni Pico della Mirandola the fulcrum of his argument. Unfortunately *not* relying on the finest contemporary scholarship on the intersection between Pico and astrology,[117] Westman's argument lacks the necessary foundations to be persua-

---

[112] *Mai sotto Saturno: Girolamo Manfredi, medico e astrologo*, Bologna: CLUEB, 2008.

[113] "Giovanni Antonio Magini (1555–1617) and Late Renaissance Astrology," PhD thesis, University of London, 1985.

[114] *Cardano's Cosmos: The Worlds and Works of a Renaissance Astrologer*, Cambridge, MA: Harvard University Press, 1999, in which he also provides an elegant review of the literature (13–15).

[115] Including general histories of astrology, such as those mentioned above. I would also like to mention three valuable recent contributions to knowledge that do not fit easily under the other rubrics but should be singled out: Robert Hand's PhD thesis on military astrology (cited above); Andreas Lerch, Scientia astrologiae: *Der Diskurs über die Wissenschaftlichkeit der Astrologie und die lateinischen Lehrbücher 1470–1610*, Leipzig: Akademischen Verlagsanstalt, 2015; and Nothaft, "*Vanitas vanitatum et super omnia vanitas*."

[116] There are also several more or less valuable collections of essays on astrology, for example, *Astrology, Science and Society*; *Nella luce degli astri: l'astrologia nella cultura del Rinascimento*, Ornella Pompeo Faracovi (ed.), Sarzana, La Spezia: Agorà, 2004; *Horoscopes and Public Spheres: Essays on the History of Astrology*, Günther Oestmann, H Darrel Rutkin and Kocku von Stuckrad (eds.), Berlin: Walter de Gruyter, 2005; *Nello specchio del cielo: Giovanni Pico della Mirandola e le Disputationes contro l'astrologia divinatoria*, Marco Bertozzi (ed.), Florence: Olschki, 2008; the unfortunate *A Companion to Astrology in the Renaissance*, Brendan Dooley (ed.), Leiden: Brill, 2014; *Astrologers and their Clients in Medieval and Early Modern Europe*, Wiebke Deimann and David Juste (eds.), Cologne: Böhlau Verlag, 2015, and *From Masha'allah to Kepler: Theory and Practice in Medieval and Renaissance Astrology*, Charles Burnett and Dorian Gieseler Greenbaum (eds.), Ceredigion, Wales: Sophia Centre Press, 2015.

[117] These include but are hardly limited to my 2002 Indiana PhD thesis with its reconstruction of the contemporary astrologizing Aristotelian natural philosophy ("Astrology, Natural Philosophy and the History of Science, c. 1250–1700: Studies Toward an Interpretation of Giovanni Pico della Mirandola's *Disputationes adversus astrologiam divinatricem*") and my 2006 chapter, "Astrology"

sive.[118] My volumes I and II (along with the earlier work just mentioned) provide these foundations and should thus be used to more fully ground, shore up, and ultimately revise a broad range of his arguments.[119]

Some of the areas of greatest interest and also (and unfortunately) of the most egregious misprisions are his description of (1) astrology and Aristotelian-Ptolemaic natural knowledge during the Middle Ages, including the geometrical-optical model of celestial influences (which I reconstruct in my volume I), and of (2) Pico's relationship thereto (which I analyze in volume II).[120] (3) Misunderstanding both (1) and (2) necessitates a fundamental reconfiguration and deepening of Westman's overall argument concerning Pico's influence.[121] I treat relevant topics in all three volumes. For better or worse, I have been able to take Westman's book into account before issuing mine.

One Final Preliminary Historiographic Consideration

On further reflection upon completing my volume I in relation to writing a review essay on Mary Quinlan-McGrath's provocative and stimulating *Influences: Art, Optics and Astrology in the Italian Renaissance* (2013)—and in thinking more about Westman's book and the responses he has provoked (including my own)—I wonder if we are in a pre-paradigmatic historiographical situation with the history of astrology on analogy with the story of the blind people and the elephant. The history of astrology, of course, is the elephant, and we who are trying to understand it are the blind people who think we understand all of it because we have a feeling for

in *The Cambridge History of Science*, that sketches out the arc of my larger argument concerning astrology and the Scientific Revolution as well as the basic structures of my interpretive framework. He is also apparently unaware of most of the relevant recent scholarship in Italian, but I will cite those in my volume II.

[118] For incisive critical essay reviews, see Noel M. Swerdlow's and John L. Heilbron's in *Science in Context*, and Michael H. Shank's in *Isis*. These three reviews also elicited Westman's published responses. Two of the reviews are published in the same issue of the same journal: *Perspectives on Science* 20 (2012): Swerdlow, "Copernicus and Astrology, with an Appendix of Translations of Primary Sources," 353–78; Heilbron, "Robert Westman on Galileo and Related Matters," 379–388. Robert S. Westman, "The Copernican Question Revisited: A Reply to Noel Swerdlow and John Heilbron," *Perspectives on Science* 21 (2013): 100–36. Shank's essay review, Westman's reply and Shank's rejoinder are all in *Isis* 105 (2014): Shank's review, 167–76; Westman, "Reply to Michael Shank," 177–84; Shank's rejoinder, 185–87. I discuss some interpretive implications just below.

[119] From my perspective, the main problem with Westman's book is that unsuspecting nonexpert students and scholars will assume that Westman provides a solid foundation for further research (as Miguel Granada stated in his review in *Journal for the History of Astronomy*), which is sadly not the case; "Planetary Order in the Long 16th Century," *JHA* 43 (2012), 239–243 at 243.

[120] I also treated both in my 2002 PhD thesis.

[121] His explicit statement that Garin used the original manuscript of Pico's *Disputations* in preparing his national edition is particularly unfortunate; *Copernican Question*, n. 55 (p. 528). Would that it were still extant!

the parts we have actually focused on. But when someone else reads our work, it sometimes feels like they understand it very differently than we had intended, and thus we feel profoundly misunderstood.

In my understanding, astrology functions as a particularly large blind spot in the Western psyche (as it were), after having been very effectively erased from the domain of legitimate knowledge in the seventeenth and eighteenth centuries. In fact, we are all trying to understand the same broader field—namely, the history of astrology, both its natural philosophical (and theological) foundations and its practices—in a historically and conceptually sound manner, and we treat many of the same texts and figures, that is, we are all trying to describe the very same elephant. This is very difficult, especially in the current pre-paradigmatic state of our understanding, and within the context of still powerful and often contrasting, countervailing or downright contradictory ideological force fields. Quinlan-McGrath's is a noble and valiant effort as is Westman's, but in Westman's case, a book promised for 40 years has brought upon itself a vastly heavier weight of expectation, at the very least to be more or less up-to-date on the relevant scholarship.

*

This is a first pass at reviewing the historiography and briefly reflecting upon it. Much more will be discussed in the following study. Before diving into the deeper conceptual waters of Part I, however, we should first explore and outline the basic tools and structures of practical astrology, including how to erect and interpret a horoscope, to which we will now turn.

Excursus

Astrological Basics: Horoscopes and Practical Astrology[1]

Introduction

Before plunging into the conceptual depths of Part I, we should first equip ourselves with an elementary understanding of practical astrology and its basic artifact, object and/or instrument—the horoscope—namely, a schematic map of the heavens coordinated for time and place, and made for any sort of event, past, present or future, including someone's birth. There are two distinct moments in an "astrological act"[2]: The first is actually constructing or erecting the horoscope, which is all astronomical, geographical and mathematical, that is, graphically mapping the heavens—namely, the planets and luminaries within the zodiac—in relation to a particular time and place on earth. This was performed with tables (including ephemerides and tables of houses) and/or instruments (including astrolabes) in the past. Today it is usually done with computers, although some modern astrologers find it a valuable meditative practice to construct their own horoscopes with the older methods. The second part is interpreting the horoscope, which itself has two moments: studying the horoscope (i.e., determining planetary dignities and thereby the lord of the horoscope or figure) and then offering a judgment or interpretation. I will discuss them each in turn. I will use Roger Bacon and the *Speculum astronomiae* as our primary premodern guides.[3]

[1] My thanks to Josefina Rodriguez Arribas for reading through this excursus and making useful suggestions.

[2] Regarding the distinction in Roger Bacon between *astronomia iudicialis et operativa* to be more fully explored in Chap. 4, this discussion here only concerns *iudicialis*, namely, the making of astrological judgments or interpretations. I use the phrase "astrological act" on analogy with speech acts, for which see Aloysius P. Martinich, *The Philosophy of Language*, fifth ed., New York: Oxford University Press, 2008.

[3] Two valuable modern guides to medieval astrological practice and terminology are Burnett, "Astrology," and Hand, Appendix I ("An Introduction to Medieval Astrology," 334–94) and II ("English and Latin Glossary of Astrological Terms," 395–439) of his "Use of Military Astrology."

I find it useful to think of astrology as a symbolic language—an alternative discourse of knowledge and practice—with its own distinctive vocabulary/terminology, grammar and rhetoric. The main terms are [1] the seven premodern planets (including the two luminaries): sun, moon, Mercury, Venus, Mars, Jupiter and Saturn; [2] the twelve zodiacal signs (or "celestial houses"): Aries, Taurus, Gemini, Cancer, Leo, Virgo, Libra, Scorpio, Sagittarius, Capricorn, Aquarius and Pisces; and [3] the related but different twelve "terrestrial houses" or "places" that only come into existence in an erected horoscope. I will always refer to these terrestrial houses simply as "houses" in what follows. This is in contradistinction to the "celestial houses," which I will always refer to as "signs" or "zodiacal signs" in order to minimize confusion. The terminology is often confusing in both the original sources and in the scholarship.[4] I discuss it further below.

The basic grammar for a proper and meaningful astrological statement is [1] that planets are always and only located in zodiacal signs (e.g., the sun is in Scorpio), and then [2] that planets-in-signs are themselves located—in an actual horoscope—within terrestrial houses (e.g., the sun in Scorpio is in the tenth house), which house divisions or cusps are determined by both the particular house system being employed and the latitude of the relevant location.[5]

The luminaries, planets and fixed stars actually exist in the heavens.[6] The zodiacal signs, however, are a man-made 360° grid projected onto the heavens, among other reasons, for measuring and calculatory purposes. The zodiacal signs have a historical relationship to the constellations of actual stars in the heavens, but are no longer located in the same places as when the zodiac was first constructed by the Babylonians in the fifth century BCE. This is due to the precession of the equinoxes that Hipparchus is said to have discovered in the second century BCE.[7] Likewise, the terrestrial houses do not exist in nature. They only come into existence when a horoscope is erected.[8]

In a judgment or interpretation (I will use these terms interchangeably in what follows)—astrology's primary rhetorical structure[9]—these strictly astronomical and

[4] For valuable discussions, see Burnett, "Astrology," 372–73, and Hand, "Use of Military Astrology," 342–57.

[5] See North, *Horoscopes and History*, and Edward S. Kennedy, "The Astrological Houses as Defined by Islamic Astronomers," in his *Astronomy and Astrology in the Medieval Islamic World*, Ashgate: Aldershot, 1998, 535–78 (originally published 1996) for a technical history of the different house systems.

[6] From now on, whenever I refer to the planets in general, this will also include the luminaries as well. Context should make my usage clear.

[7] Rochberg is very helpful on this; *Heavenly Writing*, 126–33.

[8] In the *Disputations*, Pico attacked both the signs and houses as arbitrary man-made inventions with no celestial influences because they do not physically exist. Kepler famously followed him in these matters, as I will discuss in volume III.

[9] Instantiated beautifully for a nativity in Giuliano Ristori's interpretation of Cosimo I de' Medici's horoscope, and for an annual revolution (*revolutio anni*) in the recent edition of Domenico Maria da Novara's prognostications. For Ristori, see Raffaella Castagnola, "Un oroscopo per Cosimo I," *Rinascimento* 29 (1989): 125–89. For Domenico Maria, see *I pronostici di Domenico Maria da*

most basic astrological morphemes are then given an interpretation relevant to one of the four canonical types of astrological practice: general astrology or revolutions, nativities, interrogations and elections.[10] Preliminary to an actual interpretation, however, the lord of the horoscope or figure must be determined, based on a weighing of planetary dignities.[11] This introductory excursus aims to provide the essential information that the reader will need to understand the following study, much as Roger Bacon provided a brief, useful and mostly astrological introduction to his learned edition of the *Secretum secretorum*. In this excursus, I will mainly use, integrate, and supplement two contemporary treatises, one certainly from the late 1260s, the other likely so: (1) Roger Bacon's *Astrologia* section, and (2) the deliberately anonymous *Speculum astronomiae*.[12]

[1] *Introductoria*: Toward the "Basic Horoscope"

In Roger Bacon's informative sketch for a treatise on practical astrology that his modern editor, John H. Bridges, placed at the end of *Opus maius*, Book IV (but actually belongs to the *Opus minus*),[13] Roger describes the basic elements required for an astrological judgment or interpretation, namely, the planets and signs and their qualitative natures. In this, he closely follows Alcabitius's *Liber introductorius (Introduction to Astrology)*.[14] It is a detailed sketch for a treatise he hopes to write if

Novara, Fabrizio Bonolì, Giuseppe Bezza, Salvo de Meis and Cinzia Colavita (eds.), Florence: Olschki, 2012. I discuss the former in volume III and the latter in Part IV.

[10] I sketch these out below.

[11] As discussed in detail (i.a.) in Ptolemy's *Tetrabiblos* I.17–24, in Marsilio Ficino's *De vita* III.9–10, and in Roger Bacon's *Astrologia* section to be discussed below as well as in Alcabitius's *Liber Introductorius,* 1.14–22, on which Bacon mostly drew. This numeration is in Charles S.F. Burnett and Keiji Yamamoto's valuable edition and translation of both the Arabic and Latin texts of Alcabitius; *Al-Qabisi (Alcabitius): The Introduction to Astrology: Editions of the Arabic and Latin Texts and an English Translation*, London: Warburg Institute, 2004.

[12] Many of these structures will be developed in what follows. These studies also offer useful brief introductions to astrological theory and practice: Smoller, *History, Prophecy and the Stars*; Grafton, *Cardano's Cosmos*, Azzolini, *The Duke and the Stars*, and more fully in J. C. Eade, *The Forgotten Sky: A Guide to Astrology in English Literature*, Oxford: Clarendon, 1984.

[13] *The* Opus maius *of Roger Bacon*, 3 vols., Oxford: Clarendon, 1897–1900. I discuss relevant textual issues in Chap. 2.

[14] I use Bacon's description here instead of Alcabitius's, since the latter is much more schematic and Bacon more expository. Abu s-Saqr 'Abd al-'Aziz ibn 'Utman ibn 'Ali al-Qabisi was the author of the most widely read elementary introduction to astrology in the Middle Ages and the Renaissance. His exact dates are unknown, but he flourished in the tenth century CE, and he lived and worked in a close relationship with the Hamdanid court in Syria. His most popular work, the *Introductorius ad magisterium iudiciorum astrorum* (usually referred to simply as the *Liber introductorius*) was translated into Latin by John of Seville in the twelfth century, which began a reception in Europe far more successful than that in the Islamic world. In fact, it was the most often printed astrological work of Arabic provenance; Hasse, *Success and Suppression*, 328–30. For critical editions of the Arabic and Latin texts with an English translation of the Arabic, see *Al-Qabisi (Alcabitius): The Introduction to Astrology*.

he ever finds the leisure and support necessary to do so properly. I will treat the relevant features of his description fully here because Roger's discussion is clear and concise. He begins with the fixed stars and the constellations and the signs of the zodiac. The tradition is strikingly consistent from the time of Ptolemy up to the thirteenth century and beyond:

> It was said above that there are 1022 fixed stars—whose quantity can be apprehended by astronomical instruments[15]—which have powers (*virtutes*) varying in heat, cold, moistness and dryness, and all the other natural passions and qualitative alterations. Among which there are the principle stars of the twelve signs, through which everything else is qualitatively altered in particular ways (*per quas omnia alia specialiter alterantur*). The signs (*signa*) are Aries, Taurus, Gemini, Cancer, Leo, Virgo, Libra, Scorpio, Sagittarius, Capricorn, Aquarius and Pisces, which are named in this way because the stars in the heavens (*stellae in coelo*) have the disposition of the things thus named. And the signs Aries, Leo and Sagittarius are fiery in their efficient causality (*effective*); Taurus, Virgo and Capricorn are earthy; Gemini, Libra and Aquarius are airy; Cancer, Scorpio and Pisces are watery.[16]

Roger thus lists the zodiacal signs and their elemental triplicities, which we will see are also one of their planetary dignities.

Then Roger discusses the threefold modality of the zodiacal signs along another axis[17]:

> And [1] Aries, Cancer, Libra and Capricorn are mobile (*mobilia*), since the four principle complexions of all things—namely, hot and moist in Aries, hot and dry in Cancer, cold and dry in Libra, cold and moist in Capricorn[18]—are renewed (*renovantur*) in them. [2] Taurus and its followers [sc. Leo, Scorpio and Aquarius] are fixed signs (*signa fixa*), since the said complexions are fixed and perfected in them. [3] Gemini and its followers [sc. Virgo, Sagittarius and Pisces] are said to be common signs because the complexion which is

[15] See the extended discussion at *Opus maius* IV, 224–36.

[16] "Superius quidem dictum est quod sunt 1022 stellae fixae, quarum quantitas potest deprehendi per instrumenta astronomiae, quae habent virtutes varias in calore frigore humore et siccitate et omnibus aliis passionibus et alterationibus naturalibus. Inter quas sunt principales stellae duodecim signorum, per quas omnia alia specialiter alterantur. Signa vero sunt Aries, Taurus, Gemini, Cancer, Leo, Virgo, Libra, Scorpio, Sagitarrius, Capricornus, Aquarius, Pisces; quae sic nominantur, quia stellae in coelo habent dispositionem rerum sic nominatarum. Et haec signa Aries, Leo, et Sagitarrius sunt effective ignea; Taurus, Virgo, et Capricornus sunt terrea; Gemini, Libra, et Aquarius sunt aerea; Cancer, Scorpio et Pisces sunt aquatica (377, 8–20)." All translations are mine unless otherwise noted. There is an inadequate English translation; *The Opus Majus of Roger Bacon*, Robert Belle Burke (tr), 2 vols., Philadelphia: University of Pennsylvania Press, 1928, which does not provide the basis for my translation, but to which I infrequently turn when necessary.

[17] Ptolemy presents the same three sets of four signs but refers to them slightly differently at *Tetrabiblos* I.11. The signs in the first group are called "solstitial" (Cancer and Capricorn) and "equinoctial" (Aries and Libra); those in the second are called "solid" (sterea [Gk]) and those in the third "bicorporeal" (disoma [Gk]).

[18] According to well-established astrological traditions, Aries is a fire sign and is thus hot and dry; Cancer is a water sign and is thus cold and moist; Libra is an air sign and is thus hot and moist, and Capricorn is an earth sign and is thus cold and dry. With Cancer here, Bacon seems more naturally to be referring to the nature of the season that the sign is announcing—summer, which is hot and dry—than to the nature of the sign itself. My thanks go to the anonymous referee for this suggestion.

> renewed in the following sign now declines in them towards something new.[19] Moreover, these signs have many other variations and properties, which other treatises can determine.[20]

We now have our main initial distinctions concerning the zodiacal signs. There are twelve, and they are subdivided into four elements (which he will treat below as triplicities), and into three modalities (cardinal [= mobile], mutable [= common], and fixed). And we should note, as he said, that all the other stars—i.e., the luminaries and planets—are changed qualitatively (*alterantur*) when they find themselves in the different zodiacal signs. We will see more fully what Roger means by this below.

Planets

After briefly treating the zodiacal signs, Bacon moves on to the planets, enumerating in turn the elemental qualities of each, which are only the first differentiating feature (*diversitas*) of the planets[21]:

> The first differentiating feature of the planets is in their proper powers (*in virtutibus propriis*). For Saturn is cold and dry, and causes all indolence, death and destruction of things through the excess (*per egressum*) of dryness and cold. Mars is corruptive due to the excess of heat and dryness. And these two planets never do anything good except by accident, as when sometimes poison is good by accident, as scammony, which purges the matter of illness (*materia morbi*), but nevertheless in itself harms nature. And these planets are called unbalanced (*inaequales*), misfortunes (*infortunia*) and malevolent (*malevoli*).[22]

Roger begins with the two traditional great malefics, Saturn and Mars, and provides their basic qualitative natures following Ptolemy.

---

[19] In contemporary astrological terminology, the fixed signs remain fixed; Roger's mobile signs are called "cardinal" signs, and the common "mutable." I will normally use the modern terminology to describe these, but I will always provide the Latin. For further discussion, see Burnett, "Astrology," 372, and Hand, "Use of Military Astrology," 345–46.

[20] Et Aries, Cancer, Libra et Capricornus sunt mobilia, quia renovantur in eis quatuor complexiones principales omnium rerum, scilicet calida et humida in Ariete; calida et sicca in Cancro; frigida et sicca in Libra; frigida et humida in Capricorno. Taurus et sui sequaces sunt signa fixa, quia dictae complexiones figuntur in eis et perficiuntur. Gemini et sui sequaces dicuntur signa communia, quia iam declinat complexio in eis ad novam, quae renovantur in signo sequente. Habent autem signa haec quamplures alias varietates et proprietates, quas ceteri tractatus habent determinare (377, 20–30).

[21] He brings in others below.

[22] Planetarum prima diversitas est in virtutibus propriis. Nam Saturnus est frigidus et siccus, et omnis pigritiae et mortificationis et destructionis rerum causativus per egressum siccitatis et frigoris. Mars vero est corruptivus propter egressum caliditatis et siccitatis. Et isti duo planetae nunquam faciunt bonum nisi per accidens; sicut aliquando venenum est bonum per accidens, ut Scammonea, quae purgat materiam morbi, sed tamen per se laedit naturam. Et vocantur isti planetae inaequales, et infortunia, et malevoli (377, 31–378, 1).

Then he turns to the more positive planets, Jupiter and Venus, which were also traditionally known as the great benefics, followed by Mercury, the moon and the sun:

> Jupiter and Venus have heat and moisture, but Jupiter's is more and better; and these two planets are called balanced (*aequalis*), fortunate and benevolent. Mercury is in the middle (*mediocris*) between good and bad, and is of a changeable (*convertibilis*) nature. For with good planets it is good, and it is bad with bad planets. The moon is cold and moist. The sun has generative and vital heat (*generativum calorem et vitalem*), since it is the cause of life and generation in everything (*causa vitae et generationis in omnibus*), whence, although it is hot and dry, nevertheless, its heat is not corruptive but generative, and its dryness is not mortifying, and therefore [sc. its heat and dryness] exist differently than in Mars.[23]

Like Ptolemy, Roger analyzes the planets in terms of elemental qualities, precisely as he did with the signs. We will see more talk of the sun as *generans* in Chap. 1.

In the next steps of his informative introduction, Roger distinguishes the planetary powers/virtues from those of the signs, beginning to indicate how they work together. Here Bacon adds the next level of richness and complexity to the picture by bringing in what he calls special powers or *dignitates*—the planetary dignities—which are discovered by examining the particular location of each planet within the various signs:

> Moreover, planets have powers (*virtutes*) that are different from signs; for when they are in a hot sign, they have the power of heating (*virtus calefaciendi*), and thus about the others. But beyond these, they [sc. the planets] have special powers (*speciales* […] *virtutes*), about which we spoke above, which are house (*domus*), exaltation (*exaltatio*), triplicity (*triplicitas*), term (*terminus*) and face (*facies*), and in accordance with these dignities (*dignitates*), they [sc. the planets] have wondrous effects (*effectus mirabiles*).[24]

Ptolemy discusses these same dignities (and a few other minor ones) at *Tetrabiblos*, I.17–24. They will occur again and again in what follows. Roger discusses dignities more fully in Chap. 7 of his introduction to the *Secretum secretorum.* It is primarily by calculating the balance of planetary dignities that the lord of the figure is determined.

---

[23] Jupiter vero et Venus habent caliditatem et humiditatem; sed Jupiter magis et melius: et isti duo planetae dicuntur aequalis fortunae, et benevoli. Mercurius est mediocris inter bonum et malum, et convertibilis naturae. Nam cum bonis est bonus, et malus cum malis. Luna vero frigida est et humida. Sol habet generativum calorem et vitalem, quia est causa vitae et generationis in omnibus, unde licet sit calidus et siccus, suus tamen calor non est corruptivus, sed generativus, et sua siccitas non est mortificativa, et ideo aliter est quam in Marte (378, 1–10).

[24] Habent autem planetae virtutes alias a signis; nam quando sunt in signo calido, habent virtutem calefaciendi, et sic de aliis. Sed speciales praeter has habent virtutes, de quibus superius dictum est, quae sunt domus, exaltatio, triplicitas, terminus et facies, et secundum has dignitates habent effectus mirabiles (378, 11–16).

Then Roger discusses the so-called planetary "aspects," namely, the significant angular relationships between the planets:

> And again concerning aspects (*aspectus*), which are conjunction, opposition, etc. Planets are said to be conjoined when they are in the same sign. They are said to be opposed, when one is in the seventh [sc. zodiacal sign (counting inclusively)] from another. But a trine aspect is when they are distant by four signs. And a square aspect when they are distant by three signs. And a sextile, when by two. Both opposition and square aspects are bad by their nature (*mali ex sua natura*). Trine and sextile are good, and likewise conjunction. Moreover, these aspects are considered among the five planets, especially in relation to the moon and sun. Therefore, when a bad planet aspects [sc. another planet] in a bad aspect, it is detestable because then the malice is doubled (*duplicat malitiam*); but when a good planet aspects another planet in a bad aspect, it is tolerable. And if a bad planet aspects another planet in a good aspect, its malice is mitigated. But when a bad planet opposes or conjoins a bad planet, then there is a great evil (*magnum malum*).[25]

The main aspects are conjunction (0° angular separation), opposition (180°), trine (120°), square (90°), and sextile (60°), although the angle need not be precise, in that each planet has its own variable but limited "sphere of influence."[26] Then Roger shows how both good and bad planets are modified for better or worse by means of good or bad aspects. The aspects are essential to astrological syntax by linking the planets to each other in astrologically significant ways. They were normally indicated in the planetary tables of almanacs and ephemerides, both in manuscript and later in print (see Image 1).

*

[25] Et iterum penes aspectus, qui sunt coniunctio, oppositio, &c. Conjuncti dicuntur planetae, quando sunt in eodem signo; oppositi, quando unus est in septimo ab alio. Trinus vero aspectus est, quando per quatuor distant signa. Quartus, quando per tria. Et sextilis, quando per duo. Et oppositio et quartus aspectus sunt mali ex natura sua. Trinus et sextilis sunt boni, et coniunctio similiter. Aspectus autem isti considerantur in quinque planetis, praecipue respectu Lunae et Solis. Quando ergo malus planeta aspicit in malo aspectu, detestabile est, quia tunc duplicat malitiam; quando vero bonus planeta aspicit in malo aspectu, tolerabile est; et si malus in bono aspectu aspicit, mitigatur eius malitia. Quando vero malus opponitur aut coniungitur malo, tunc magnum malum est (378, 16–29).

[26] Hand usefully discusses these spheres of influence, which he calls "orbs" or "orbs of light"; "Use of Military Astrology," 363–69.

infirmitatē repellit. Stationari⁹ pluuiā ꝛ nutrimentū cōfert. ❡Mars in ♒ sub
radijs ariditatē. Uespertin⁹ absentiā ventoꝝ. Matutin⁹ motionē eoꝝ. Statio-
nari⁹ nocumentū significat in mari. ❡Uen⁹ in ♒ sub radijs nebulosa. Uesptina
calida ꝛ ventosa. Matutina caliginosa ꝛ pluuiosa. ❡Mercuri⁹ in ♒ sub radijs ē
niuosus. Uesptin⁹ gelid⁹ ꝛ turbulent⁹. Matutin⁹ pluuiosus. ❡Saturn⁹ in ♓ sub
radijs est niuosus. Uesptin⁹ pluuiosus. Matutin⁹ bonā cōplexionē. Stationari⁹
aeris inequalitatē significat. ❡Jupiter in ♓ sub radijs augmentū pisciū. Uesp-
tinus bonā conditionē. Matutin⁹ tranquillitatē in marib⁹ ꝛ fluminib⁹. Statio-
narius bonā ꝛ tpatam significat pluuiā. ❡Mars in ♓ raritatē pisciū. Uesptin⁹
ariditatē. Matutin⁹ fulgura ꝛ tonitrua. Stationarius inequalitatē significat in
corporib⁹. ❡Uen⁹ in ♓ sub radijs gelidū tps. Uespertina aquosum ꝛ ventosum
Matutina bonā cōplexionē ꝛ germia ꝛ fet⁹ nutriet. ❡Mercuri⁹ in ♓ sub radijs
tēpestuosus est. Uespertin⁹ bonā complexionē. Matutin⁹ tranquill⁹ est ꝛ bonā
conditionem ostendit. Explicitum est hoc opus anno xpi dn̄i. 1492.

Image 1 Planetary aspects (Bayerische Staatsbibliothek München, 4 Inc.c.a. 944, fol. a8r)

I would simply like to note here that the normal glyphs for the signs, planets, and aspects could also be known as "characters," as we see them printed in Erhard Ratdolt's 1492 edition of Regiomontanus's *Almanach* printed in Augsburg (see Image 2).[27] After a brief introduction explaining how to use any ephemeris, he prints

[27] *Almanach magistri Johannis de Monte Regio ad annos xv accuratissime calculata.* The astronomical tables from 1491 to 1506 are also called ephemerides on a2r in a brief two page description of how to use an ephemeris (*usum ephemeridis cuiuslibet breviter exponemus*). He also

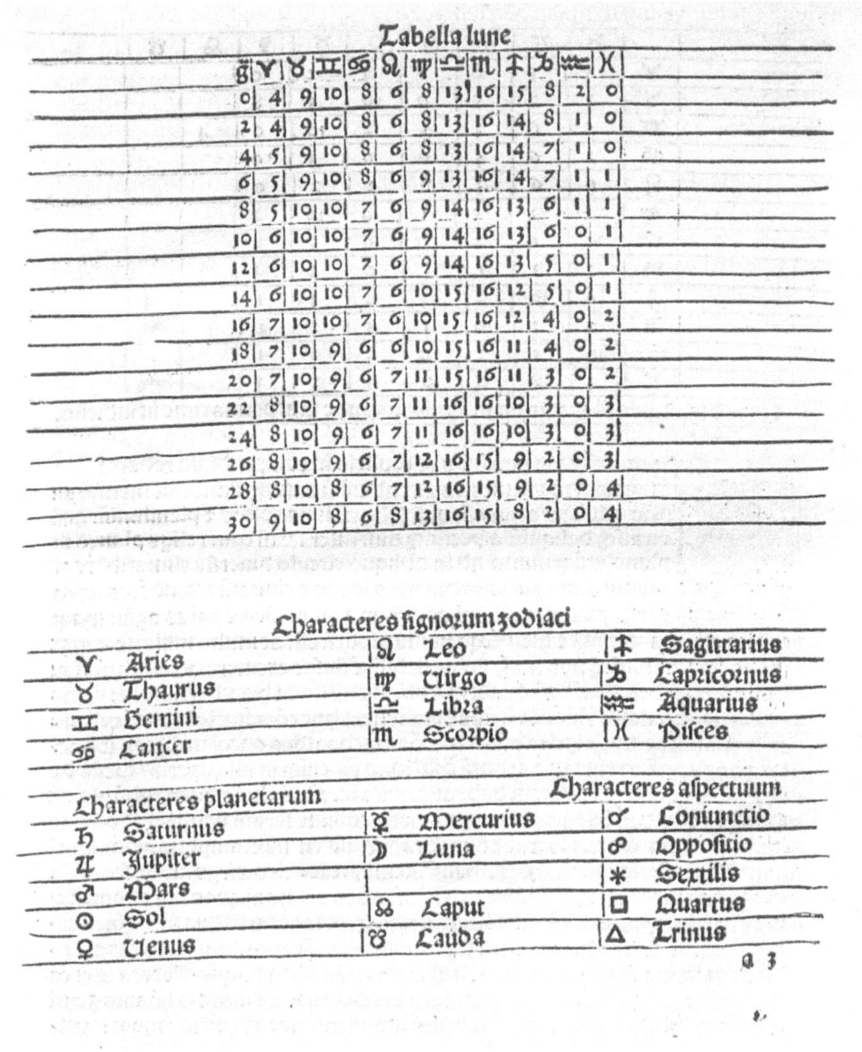

Tabella lune

☊	♈	♉	♊	♋	♌	♍	♎	♏	♐	♑	♒	♓
0	4	9	10	8	6	8	13	16	15	8	2	0
2	4	9	10	8	6	8	13	16	14	8	1	0
4	5	9	10	8	6	8	13	16	14	7	1	0
6	5	9	10	8	6	9	13	16	14	7	1	1
8	5	10	10	7	6	9	14	16	13	6	1	1
10	6	10	10	7	6	9	14	16	13	6	0	1
12	6	10	10	7	6	9	14	16	13	5	0	1
14	6	10	10	7	6	10	15	16	12	5	0	1
16	7	10	10	7	6	10	15	16	12	4	0	2
18	7	10	9	6	6	10	15	16	11	4	0	2
20	7	10	9	6	7	11	15	16	11	3	0	2
22	8	10	9	6	7	11	16	16	10	3	0	3
24	8	10	9	6	7	11	16	16	10	3	0	3
26	8	10	9	6	7	12	16	15	9	2	0	3
28	8	10	8	6	7	12	16	15	9	2	0	4
30	9	10	8	6	8	13	16	15	8	2	0	4

Characteres signorum zodiaci

♈	Aries	♌	Leo	♐	Sagittarius
♉	Thaurus	♍	Virgo	♑	Capricornus
♊	Gemini	♎	Libra	♒	Aquarius
♋	Cancer	♏	Scorpio	♓	Pisces

Characteres planetarum — Characteres aspectuum

♄	Saturnus	☿	Mercurius	☌	Coniunctio
♃	Jupiter	☽	Luna	☍	Oppositio
♂	Mars			✱	Sextilis
☉	Sol	☊	Caput	□	Quartus
♀	Venus	☋	Cauda	△	Trinus

a 3

Image 2 Characters/glyphs of planets, signs and aspects (Bayerische Staatsbibliothek München, 4 Inc.c.a. 944, fol. a3r)

a table at a3r with all of these glyphs, and he explicitly calls them *characteres signorum zodiaci*, *characteres planetarum* including the *caput* and *cauda draconis* (i.e., the north and south nodes of the moon) and *characteres aspectuum*. These and other "characters" will become more a part of our story in Part III, when I discuss

explains how to make the prognostications in a *tacuinum* (a3v-a4r), how to predict the weather (a4v-a5r), and how to make elections (a5r-a6v), including for medical and agricultural purposes. And on a6v-a8r, he discusses al-Kindi's treatise on weather prediction. I use an online version from the Bayerische StaatsBibliothek, from which I also take the images printed in this excursus.

astrology's relationship to magic. Needless to say, Regiomontanus does not mention or imply any explicitly magical uses here. We will now return to Bacon's exposition.

*

The next level of diversity arises from the locations of the planets in their cycles (378, 30–379, 16):

> Planets also have a great variety of actions (*actiones*) in relation to eccentrics and epicycles. For when they are in their higher parts (*in partibus superioribus istorum*), they make the strongest actions (*operationes fortissimas*), but when they are in the lower parts, they make weak actions. Because, when they are in the superior parts of their circles, which are called '*auges*', then they are moved around the world (*circa mundum*) with the daily motion in great circles, and then they are borne swiftly (*velociter*), and the speed of their motion (*velocitas motus*) induces a strength of action (*inducit fortitudinem actionis*) in things that were born to be moved (*in rebus quae natae sunt moveri*), of which sort stars are.[28] They also have different effects due to the parts of their revolutions, and in relation to their entire revolutions. For according to whether they transit (*transeunt*) a quarter of the heavens, a half, three quarters or the entire circle, they have perceptible variety in their effects (*habent sensibiles varietates in effectibus*). And not only thus, but when they are at any determinate revolution (*penes quamlibet revolutionum determinatarum*); just as Saturn, having completed its ten revolutions, induces a great transformation of the world (*inducit magnam mundi alterationem*), as was touched on before in treating religions (*in sectis*).[29]

Thus, regardless of its relation to any other planet, a planet's location in its own revolutionary motion through the zodiac or in the various locations of its own cycle affects the strength of its actions. I will discuss Bacon's account of the larger historical effect of ten 29-year revolutions of Saturn in Part II.

Then Roger briefly discusses the terrestrial houses or places (see Image 3). He will have more to say about them and their importance for an interpretation in texts I discuss in Chap. 6:

> Also the twelve houses, into which the entire heaven is divided, which were touched on above, are especially considered in relation to planetary powers. For the planets collect (*contrahunt*) different powers (*potestates*) from them, and the planets do different things in

[28] Galileo also indicates the speed of planetary motions in his astrological MS 81, as I will discuss in volume III. See now Germana Ernst's brilliant analysis and clear presentation (along with photographs of the entire manuscript) in the third appendix to Galileo's ongoing national edition; *Le opere di Galileo Galilei, Appendice Volume III, Testi*, Florence: Giunti, 2017, 103–93.

[29] "Habent etiam planetae magnam varietatem actionum penes eccentricos et epicyclos. Nam quando sunt in partibus superioribus istorum, faciunt operationes fortissimas, quando vero in inferioribus debiles; quia quando sunt in superioribus partibus suorum circulorum, qui vocantur auges, tunc moventur circa mundum moto diurno in circulis magnis, et tunc velociter feruntur, et velocitas motus inducit fortitudinem actionis in rebus quae natae sunt moveri, cuiusmodi sunt stellae. Habent autem effectus varios penes partes revolutionum suarum, et penes totas revolutiones. Nam secundum quod transeunt quartam coeli et medietatem et tres quartas, et totum circulum, habent sensibiles varietates in effectibus. Et non solum sic, sed penes quamlibet revolutionum determinatarum; sicut Saturnus, completis decem revolutionibus suis, inducit magnam mundi alternationem, sicut prius tactum est in sectis (378, 30–379, 7)." Roger treated Saturn's ten revolutions in discussing historical astrology, as we will see in Chap. 6.

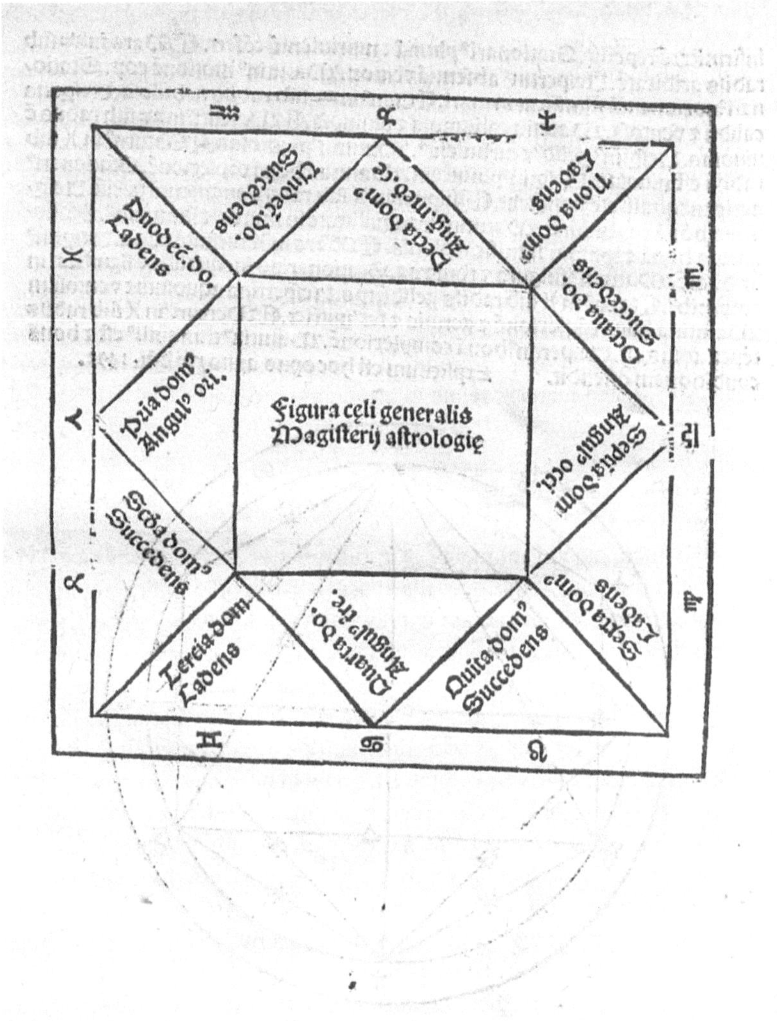

Image 3 The basic structure of a horoscope with the twelve houses (Bayerische Staatsbibliothek München, 4 Inc.c.a. 944, fol. a8v)

the world (*varia* […] *operantur planetae in hoc mundo*) through the powers of these houses (*per virtutes harum domorum*). And therefore, the astrologers (*mathematici*) teach especially to consider these houses. Therefore, these are the principle roots in the actions of the stars (*radices principales in actionibus stellarum*), which have infinite branches (*rami*), flowers (*flores*) and fruit (*fructus*). This, therefore, is the intention of a treatise about the powers and actions of the stars (*de virtutibus et actionibus stellarum*), which I have proposed to make in this work.[30]

[30] Domus etiam duodecim, in quas dividitur totum coelum, quae superius tactae sunt, maxime considerantur respectu virtutum planetarum. Nam varias potestates contrahunt planetae ex eis, et

Roger then goes on to consider how the places and things of the world are altered in their complexions by these just mentioned roots, but given that this is more concerned with natural philosophy than with astrology per se—and especially its elementary features—I will treat it elsewhere, namely, in Chap. 2.

Even so, we can begin to get a sense of astrology's interpretive complexity from Roger Bacon's sketch here of the introductory material for a proper treatise on practical astrology. So far, Roger has only discussed the four primary qualities (hot and cold, wet and dry) in relation to the planets, signs, houses and aspects, the basic elements of practical astrology. There has been no mention yet of stellar rays. We will see in Chap. 2, however, that there is a more specifically mathematical dimension to Roger's analysis that understands the communication of celestial influences in terms of a geometrical-optical model of planetary action. We will now see how these basic astrological features relate to an actual horoscope.

[2] Erecting a Horoscope

There are two fundamental moments of a fully fledged astrological act: [1] constructing (or erecting) the horoscope (the "basic horoscope") and [2] interpreting it, which both take place within and are expressed through the overarching context of the four canonical types of astrological practice (to be discussed in the fourth section of this excursus). Both of these processes are seen clearly in textbooks,[31] and in the remains of astrological practice, as we can see, for example, in the horoscopes Galileo drew up for his daughters and for Gianfrancesco Sagredo in MS Galileana 81, to be discussed more fully in volume III.[32] The material just discussed offers the basic building blocks to be used ultimately toward the interpretation. Now I will discuss how to erect a horoscope and what its main constituent parts are.

The first moment, constructing the basic horoscope, initially requires locating the main celestial bodies (luminaries and planets, sometimes also the lunar nodes and significant fixed stars) within the signs of the zodiac (their "celestial houses") for a particular date and time. During the entire period of this study (ca. 1250–1800), planetary locations for most practitioners were primarily derived from ephemerides, a type of planetary table that indicates the daily locations for every planet

---

varia per virtutes harum domorum operantur planetae in hoc mundo: et ideo has domos considerare maxime docent mathematici. Haec igitur sunt radices principales in actionibus stellarum, quae habent ramos et flores, et fructus infinitos. Haec igitur est intentio tractatus quam in hoc opere facere proposui de virtutibus et actionibus stellarum (379, 7–16).

[31] It should be noted, as David Juste informs me, that medieval textbooks teach one either how to cast horoscopes or how to interpret them, but not both together. Teaching both together only begins in the sixteenth century.

[32] I treat these to some extent in my "Galileo Astrologer: Astrology and Mathematical Practice in the Late-sixteenth and Early-seventeenth Centuries," *GALILAEANA: Journal of Galilean Studies* 2 (2005): 107–43. One may see this clearly now for Galileo's astrological practice in Ernst's national edition of Galileo's MS 81, in which, among other things, she has clearly transcribed all of the interpretations, many of which are exceedingly difficult to read in the manuscript.

month by month within the zodiac—usually for noon or midnight—over a given number of years for a particular location with its longitude and latitude, usually for a major city (e.g., Rome, see Images 4 and 5).[33] Locating six of the seven planets mainly requires finding the celestial longitude for that day (e.g., the sun at 23 degrees of Scorpio; Jupiter at 9 degrees of Sagittarius). Only the moon requires a more complex calculation, since it moves +/− 12 degrees during the course of a day. We see these simple calculations performed in Galileo's astrological MS 81.[34]

Once the planets and luminaries have been located, the next step in erecting a horoscope is to particularize this information for a given time in relation to a specific place on earth, usually to a city. This orientation takes place using a fixed mathematico-geographical grid of the horizon and meridian for each place on earth, with which one can then fix the two essential points of the terrestrial houses in a horoscope, namely, [1] the rising point, ascendant or *horoscopus*, where the horizon cuts the ecliptic at a particular degree of the zodiac at a particular time, and [2] the midheaven (or *medium coelum*), where the meridian cuts the ecliptic.[35] These main angles on a horoscope are often called *cardines*, and they too are significant for interpretations.

The zodiacal signs and the terrestrial houses offer two overlapping but related grids in a horoscope. [1] The planets (here, as often, shorthand for "the planets and the luminaries") in the heavens travel along the belt of the zodiac, an 8-degree band on either side of the sun's annual motion around the heavens, which marks the ecliptic proper. According to our most up-to-date understanding, the zodiac was constructed from the observations and sexagesimal calculations by Babylonian astronomer-astrologers in the fifth century BCE based on centuries of observation and analysis.[36] The zodiac is composed of 12 30-degree divisions called the signs of the zodiac or sometimes, and confusingly, celestial houses. In an almanac or ephemeris (pl. ephemerides), the planetary locations in the zodiac are particularized for a particular meridian—e.g., Rome—at a particular time, usually noon or midnight. The information in ephemerides can then be oriented more specifically to particular places on earth at particular times, especially when employing the tables of houses that were often contained—and later published—with ephemerides.

[2] With respect to the terrestrial houses, there are several different house systems, but they all employ either the ascendant or midheaven, and almost always both, although equal house systems sometimes require a *cor coeli* as well as a tenth

[33] This is also true for almanacs, which are made normally for one year, whereas ephemerides are normally made for several or many years. There is some confusion in the terminology, both premodern and modern.

[34] Ephemerides are derived by calculations from planetary tables, Alfonsine and others. See (i.a.) *Astronomy before the Telescope*, Christopher Walker (ed.), New York: St. Martins Press, 1997, and Jose Chabas and Bernard R. Goldstein, *A Survey of European Astronomical Tables in the Late Middle Ages*, Leiden: Brill, 2012.

[35] For a technical historical analysis of the horoscope and the different house systems, see North's now classic *Horoscopes and History*. His *Chaucer's Universe* is also very useful on this and related issues discussed in this excursus.

[36] For a lucid and accessible discussion, see Rochberg, *Heavenly Writing*.

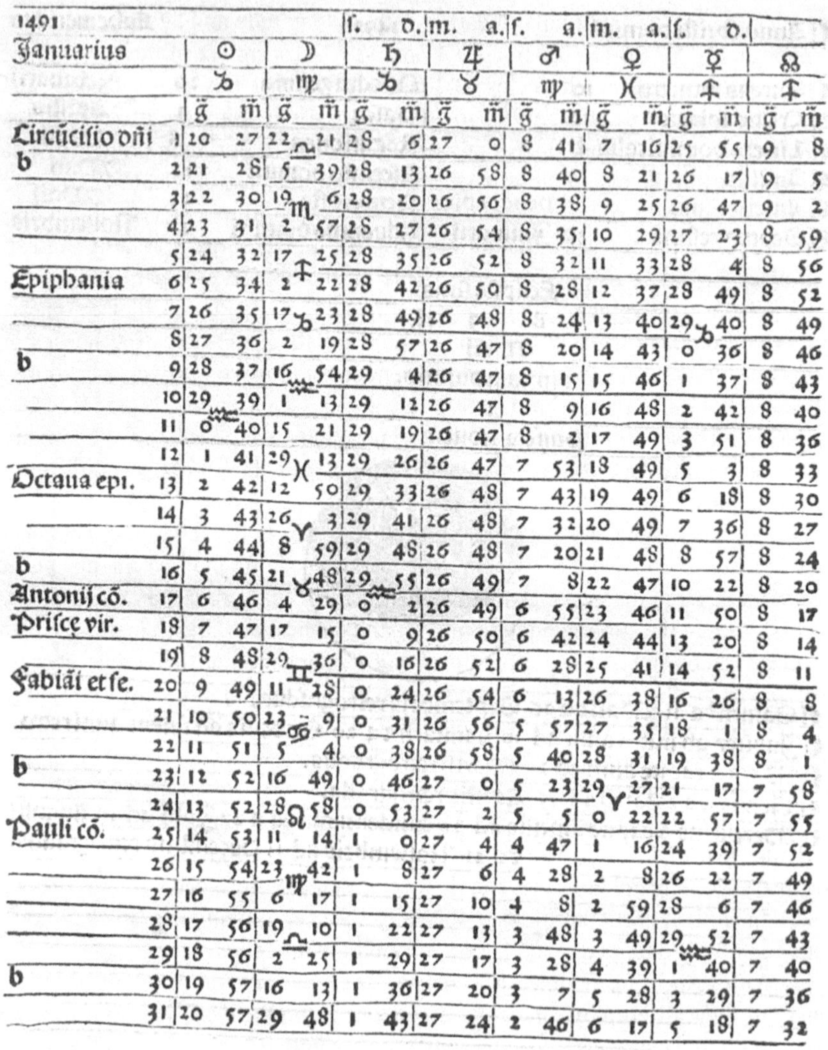

1491						s.	d.	m.	a.	s.	a.	m.	a.	s.	d.		
Januarius		☉		☽		♄		♃		♂		♀		☿		☊	
		♑		♍		♑		♉		♍		♓		♐		♐	
		g	m	g	m	g	m	g	m	g	m	g	m	g	m	g	m
Circūcisio dn̄i	1	20	27	22 ♎	21	28	6	27	0	8	41	7	16	25	55	9	8
b	2	21	28	5	39	28	13	26	58	8	40	8	21	26	17	9	5
	3	22	30	19 ♏	6	28	20	26	56	8	38	9	25	26	47	9	2
	4	23	31	2	57	28	27	26	54	8	35	10	29	27	23	8	59
	5	24	32	17 ♐	25	28	35	26	52	8	32	11	33	28	4	8	56
Epiphania	6	25	34	2	22	28	42	26	50	8	28	12	37	28	49	8	52
	7	26	35	17 ♑	23	28	49	26	48	8	24	13	40	29 ♑	40	8	49
	8	27	36	2	19	28	57	26	47	8	20	14	43	0	36	8	46
b	9	28	37	16 ♒	54	29	4	26	47	8	15	15	46	1	37	8	43
	10	29 ♒	39	1	13	29	12	26	47	8	9	16	48	2	42	8	40
	11	0	40	15	21	29	19	26	47	8	2	17	49	3	51	8	36
	12	1	41	29 ♓	13	29	26	26	47	7	53	18	49	5	3	8	33
Octaua epi.	13	2	42	12	50	29	33	26	48	7	43	19	49	6	18	8	30
	14	3	43	26 ♈	3	29	41	26	48	7	32	20	49	7	36	8	27
	15	4	44	8	59	29	48	26	48	7	20	21	48	8	57	8	24
b	16	5	45	21 ♉	48	29 ♒	55	26	49	7	8	22	47	10	22	8	20
Antonij cō.	17	6	46	4	29	0	2	26	49	6	55	23	46	11	50	8	17
Priscę vir.	18	7	47	17	15	0	9	26	50	6	42	24	44	13	20	8	14
	19	8	48	29 ♊	36	0	16	26	52	6	28	25	41	14	52	8	11
Fabiāi et se.	20	9	49	11	28	0	24	26	54	6	13	26	38	16	26	8	8
	21	10	50	23 ♋	9	0	31	26	56	5	57	27	35	18	1	8	4
	22	11	51	5	4	0	38	26	58	5	40	28	31	19	38	8	1
b	23	12	52	16	49	0	46	27	0	5	23	29 ♈	27	21	17	7	58
	24	13	52	28 ♌	58	0	53	27	2	5	5	0	22	22	57	7	55
Pauli cō.	25	14	53	11	14	1	0	27	4	4	47	1	16	24	39	7	52
	26	15	54	23 ♍	42	1	8	27	6	4	28	2	8	26	22	7	49
	27	16	55	6	17	1	15	27	10	4	8	2	59	28	6	7	46
	28	17	56	19 ♎	10	1	22	27	13	3	48	3	49	29 ♒	52	7	43
	29	18	56	2	25	1	29	27	17	3	28	4	39	1	40	7	40
b	30	19	57	16	13	1	36	27	20	3	7	5	28	3	29	7	36
	31	20	57	29	48	1	43	27	24	2	46	6	17	5	18	7	32

Image 4 Ephemeris for January 1491, Planetary locations (Bayerische Staatsbibliothek München, 4 Inc.c.a. 944, fol. [h]2v)

house cusp. The ascendant and midheaven provide the structural basis for the system of terrestrial houses. The ascendant provides the first house cusp and the midheaven the tenth house cusp (except in the equal house system, where each house is 30 degrees in length just as the zodiacal signs are). Since the horizon is a great circle, it also provides the western horizon or descendant, the seventh house cusp,

1491	Aſpectus lunę ad ſolē ⁊ planetas.						Solis ⁊ planetaꝝ inter ſe
Jan.	☉	♄	♃	♂	♀	☿	
		oc	oc	or	oc	or	
1		△ 11	△ 6			□ 7	
2							☍ ♂ ♀
3	□ 6	□ 17				✱ 14	
4				✱ 9	△ 13		
5	✱ 12	✱ 18	☍ 15				
6				□ 10	□ 18		
7						☌ 21	☽ ☊ △ ☉ ♃
8				△ 10	✱ 22		
9 ☌	20 52	☌ 20	△ 16				☌ ☉ ♄
10							
11			□ 20				
12				☍ 15		✱ 11	
13					☌ 14		
14	✱ 15	✱ 7	✱ 1				△ ♂ ☿
15						□ 0	
16		□ 15					
17	□ 4			△ 4		△ 16	
18			☌ 19		✱ 16		
19	△ 20	△ 1		□ 14			
20							☽ ☋ ✱ ♃ ♀
21					□ 9		
22				✱ 2			
23			✱ 20			☍ 9	
24		☍ 4			△ 3		
25 ☍	7 21						✱ ♄ ♀
26			□ 7	☌ 20			△ ♃ ☿
27							
28		△ 22	△ 15	1		△ 23	
29					☍ 4		☌ ♄ ☿
30	△ 7						
31		□ 4		✱ 5		□ 11	

Image 5 Facing Ephemeris Page for January 1491, Planetary Aspects (Bayerische Staatsbibliothek München, 4 Inc.c.a. 944, fol. [h]3r)

which is exactly opposite to (= 180° away from) the ascendant. Likewise, the meridian also provides the lower midheaven (*imum medium coelum*) and thus the fourth house cusp. The other eight houses are subdivided differently depending on the house system in use.[37] This basic structure of the horizon-meridian grid is fixed for

[37] Once again, North gives the mathematical and some of the cultural history of these techniques, *Horoscopes and History*.

every place on earth, based on a terrestrial location's longitude and latitude. Thus, mathematical geography is integrated with mathematical astronomy, as we will see in detail, especially in Chap. 2. This system was fully developed in Ptolemy's *Almagest*, which provided the astronomical foundations for the astrology articulated in the somewhat later *Tetrabiblos*, as Ptolemy himself says in the latter's opening chapter (I.1).

At any given place on earth, then, the circle of the zodiac (with the planets in their respective annual motions) is understood to move—for us, appears to move—in a 24 h daily motion within this fixed grid of the terrestrial houses. At any given time of the day, a particular degree of a zodiacal sign will be rising on a given place's Eastern horizon. If a horoscope is constructed for this moment, this degree of the sign will be the ascendant for that horoscope. The zodiac with the located luminaries, planets and nodes (etc.) may then be understood to turn regularly like a clockwise turning clock face within the fixed horizon-meridian grid for every given place on earth. Each sign of 30 degrees thus takes 2 h (more or less) to pass over the horizon, with each degree taking approximately 4 min. This represents the diurnal or daily motion of the heavens.[38] The Babylonians articulated these foundational structures using a sexagesimal mathematics that Ptolemy fully adopted (and adapted).[39] Nevertheless, the Babylonians themselves did not employ the ascendant in their "horoscopes," thus rendering the phrase "Babylonian horoscopes" somewhat of a misnomer.

Most horoscopes in the premodern period under study here were of a basic square structure, in which the circular motion of the zodiac was represented as a square (see Image 3). Numerous examples survive from the twelfth century on, as we can see (e.g.,) in the plates of North's *Horoscopes and History* and in *Adelard of Bath: An English Scientist and Arabist of the Early 12th Century*.[40] Changes to this square structure came in the sixteenth century with Tycho Brahe's round horoscope with 8 instead of the normal 12 terrestrial houses. Nevertheless, both round and square shapes convey precisely the same information. By far the most creative design of horoscopes that I know of is found toward the end of William Lilly's *Christian Astrology*.[41]

Astrolabes were the relevant analogue computers of the day for learning and practicing the science of the stars, just like slide rules were in the middle of the twentieth century for various scientific and engineering computations. Astrolabes

[38] Of course, the planets continue to move along the zodiac during the course of the year at different speeds with their respective annual motions, rendering a somewhat different celestial configuration for each day.

[39] Otto Neugebauer, *A History of Ancient Mathematical Astronomy*, Berlin: Springer, 1975.

[40] Charles Burnett (ed.), London: Warburg Institute, 1987.

[41] I use the 1659 online edition, 385–401 (https://archive.org/stream/ChristianAstrologyByWilliamLilly/Lilly_William-Christian_astrology#page/n1/mode/2up). As it turns out, these creative horoscope structures were designed by Elias Ashmole for Lilly as Vittoria Feola has discovered; "Antiquarianism, Astrology, and the Press in William Lilly's Network(?)," in *Antiquarianism and Science in Early Modern Urban Networks*, V. Feola (ed), Paris: Blanchard, 2014, 186–203, 195–98.

were very sophisticated observational and computational devises, which were used, among other things, for easily determining house cusps.[42] Today most astrologers use computer programs to construct horoscopes quickly and accurately, such as that found on www.astrodienst.com. These are the basic structures of the horoscope itself as a schematic map of the planets and luminaries at a specific time for a specific place. A properly erected horoscope is absolutely necessary for a fully fledged astrological interpretation.

[3] Toward an Interpretation: Finding the Lord of the Horoscope

For the second moment in an astrological act, namely, the interpretation or judgment (*iudicium*) itself, let us return to the planets. Each planet as the basic unit of astrological meaning (the basic vocabulary) has a particular but also polyvalent significance for interpretation. It has a qualitative nature composed of the four primary qualities (as we saw above) as well as a rich mythological overlay, with roots stretching down into Babylonian and Greco-Roman antiquity. The sign a planet is "in" on a particular date also fundamentally conditions its interpretation, as we also saw in Roger Bacon's description above. Likewise, the terrestrial "house" the "planet-in-a-sign" inhabits in the horoscope further conditions its interpretation for the native or event, depending on which type of practice is involved. The final basic features for the interpretation of a horoscope are the significant angular relations between the planets, the so-called astrological aspects, and the planetary dignities (see Image 6). We will see an extended example of an interpretation in Chap. 6, which Roger Bacon used to argue for astrology's utility for religion.

To get our next (and final for now) insight into how horoscopes were interpreted, we will return to Roger Bacon's *Astrologia* section, which he introduces thus:

> After these things the fourth follows, which is a consideration about judgment (*iudicium*) and the knowledge (*cognitio*) of past, present and future things. For if the cause of the complexions of things (*causa complexionum rerum*) is the celestial configuration (*coelestis constellatio*), the effect of this sort of thing can be known through this cause (*poterit huiusmodi effectus sciri per hanc causam*).[43]

Roger then supports this strong claim with religious and philosophical authorities. We will explore the causal nature of Bacon's analysis in Chap. 2.

Roger now addresses the fundamentals of interpretation:

> Moreover, the universal manner of making judgments by means of the planets (*iudicandi per planetas*) consists in this, that a human being (*homo*) can know by means of tables and instruments [1] how to calculate (*aequare*) the motions of the heavens and discover the

[42] For a splendid analysis of how astrolabes work, see John D. North, *Chaucer's Universe*, Chap. 2: "The Astrolabe: Bread and Milk for Children," 38–86. House cusps can also be calculated using tables of houses that were normally to be found later in all published ephemerides.

[43] Post haec sequitur quartum, quod est consideratio de iudicio et cognitione praeteritorum praesentium et futurorum. Nam si causa complexionum rerum est coelestis constellatio, poterit huiusmodi effectus sciri per hanc causam (388, 27–30).

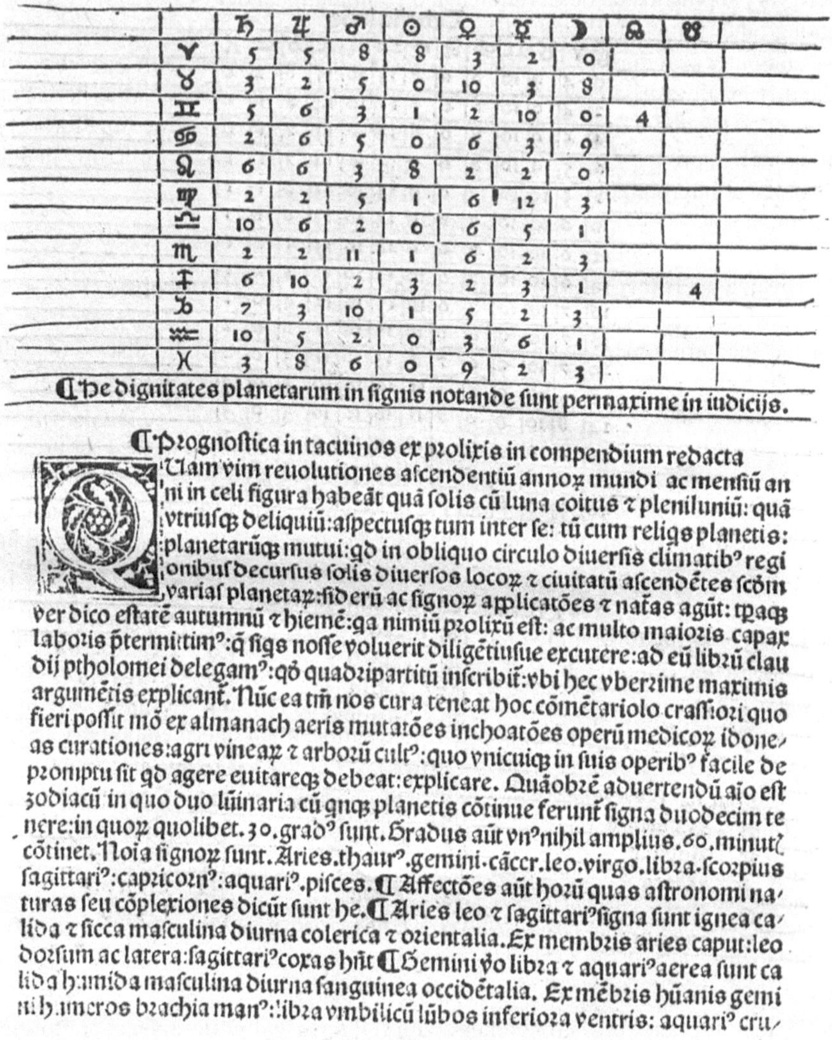

	♄	♃	♂	☉	♀	☿	☽	☊	☋
♈	5	5	8	8	3	2	0		
♉	3	2	5	0	10	3	8		
♊	5	6	3	1	2	10	0	4	
♋	2	6	5	0	6	3	9		
♌	6	6	3	8	2	2	0		
♍	2	2	5	1	6	12	3		
♎	10	6	2	0	6	5	1		
♏	2	2	11	1	6	2	3		
♐	6	10	2	3	2	3	1		4
♑	7	3	10	1	5	2	3		
♒	10	5	2	0	3	6	1		
♓	3	8	6	0	9	2	3		

¶ He dignitates planetarum in signis notande sunt permaxime in iudiciis.

¶ Prognostica in tacuinos ex prolixis in compendium redacta

Quam vim reuolutiones ascendentiũ annoꝝ mundi ac mensiũ anni in celi figura habeãt quã solis cũ luna coitus ꝛ pleniluniũ: quã vtriusq; deliquiũ: aspectusq; tum inter se: tũ cum reliqs planetis: planetarũq; mutui: qd in obliquo circulo diuersis climatib⁹ regionibus decursus solis diuersos locoꝝ ꝛ ciuitatũ ascendẽtes scdm varias planetaꝝ: siderũ ac signoꝝ applicatões ꝛ natas agũt: tpaq; ver dico estatẽ autumnũ ꝛ hiemẽ: qa nimiũ prolixũ est: ac multo maioris capax laboris ꝑternitim⁹: q̃ sigs nosse voluerit diligẽtiusue excutere: ad eũ librũ claudij ptholomei delegam⁹: qđ quadripartitũ inscribit̃: vbi hec vberrime maximis argumẽtis explicant̃. Nũc ea tm̃ nos cura teneat hoc cõmẽtariolo crassiori quo fieri possit mõ ex almanach aeris mutatões inchoatões operũ medicoꝝ idoneas curationes: agri vineaꝝ ꝛ arborũ cult⁹: quo vnicuiq; in suis operib⁹ facile de promptu sit qđ agere euitareq; debeat: explicare. Quãobrẽ aduertendũ aĩo est zodiacũ in quo duo lũinaria cũ gnq; planetis cõtinue ferunt̃ signa duodecim tenere: in quoꝝ quolibet. 30. grad⁹ sunt. Gradus aũt vn⁹ nihil amplius. 60. minut⁹ cõtinet. Noĩa signoꝝ sunt. Aries. thaur⁹. gemini. cãcer. leo. virgo. libra. scorpius sagittari⁹: capricorn⁹: aquari⁹. pisces. ¶ Affectões aũt horũ quas astronomi naturas seu cõplexiones dicũt sunt he. ¶ Aries leo ꝛ sagittari⁹ signa sunt ignea calida ꝛ sicca masculina diurna colerica ꝛ orientalia. Ex membris aries caput: leo dorsum ac latera: sagittari⁹ coxas hñt ¶ Gemini vo libra ꝛ aquari⁹ aerea sunt calida humida masculina diurna sanguinea occidẽtalia. Ex mẽbris hũanis gemini humeros brachia man⁹: libra vmbilicũ lũbos inferiora ventris: aquari⁹ cru-

Image 6 Table of Dignities (Bayerische Staatsbibliothek München, 4 Inc.c.a. 944, fol. a3v)

locations of the planets; and [2] to consider which of them has more and greater strengths (*fortitudines*) from its location in accordance with the five famous [sc. dignities], which are *domus* [= zodiacal sign], exaltation, triplicity, term and face; and similarly with this [3] to see the strengths which happen (*contingunt*) from [i] their aspects and [ii] from their motion in the *auges* of their circles, and [iii] from the twelve [sc. terrestrial] houses (*domus* also).[44]

[44] Modus autem universalis iudicandi per planetas consistit in hoc, quod homo sciat per tabulas et instrumenta aequare motus caelorum et invenire loca planetarum; et considerare quis eorum habeat plures et maiores fortitudines ex loco suo secundum illas quinque famosas, quae sunt domus, exal-

Here Roger enumerates the significant features for determining relative planetary strengths, including their 5 dignities, their aspects and their location in the 12 terrestrial houses.

Roger then turns to the lord of the horoscope, that is, identifying which planet or luminary "rules" the horoscope. This provides the focus in terms of which the horoscope should be interpreted:

> And when he has well examined and seen these things and what are joined to them, then, if the sun will have more witnesses (*testimonia*) and strengths of this sort, he ought to make his judgment in accordance with the solar complexion in things because it is dominant in them; and thus about the others. But the particular and special consideration in individual things (*in singulis*) has its own laws determined according to the condition of things.[45]

Without himself going into the details (some of which we will see in the ensuing study), Roger implies that after analyzing these relevant features, the astrologer will discover the horoscope's ruling planet. In fact, each dignity has a certain strength to which a number can be attributed. The planet with the highest resulting number is thereby denominated the lord of the geniture (or any other type of horoscope).[46] We will see how this works in detail in Chap. 6.

Then Roger expands the picture to indicate the broad range of astrology's potential value:

> And a person (*homo*) can revolve history (*revolvere historiam*) to times past, and consider the effects of the heavens from the beginning of the world (*a principio mundi*), as floods, earthquakes, pestilence, hunger, comets, monsters, and an infinite number of other things that have happened (*contigerunt*) both in human affairs (*in rebus humanis*) and in nature (*in naturalibus*). When these things have been compared, he should revolve the astronomical tables and canons and he will find that proper celestial configurations correspond to individual effects (*constellationes proprias singulis effectibus respondere*). Then, he should consider similar celestial configurations (*consimiles constellationes*) by means of tables at a future time as close or distant as he wishes. And he could then foretell their effects (*pronuntiare in effectibus*), which will be similar to what they were in the past, because when the cause is established, the effect is established (*quia posita causa ponitur effectus*).[47]

---

tatio, triplicitas, terminus et facies: et similiter cum hoc videre fortitudines quae contingunt ex aspectibus illorum et ex motu in augibus suorum circulorum, et ex domibus duodecim (388, 38–389, 8).

[45] "Et cum bene examinaverit et viderit haec et eis annexa, tunc si Sol habuerit plura testimonia et fortitudines huiusmodi, debet iudicare secundum complexionem solarem in rebus, eo quod illa dominatur in eis; et sic de aliis. Particularis vero consideratio et specialis in singulis habet suas leges determinatas secundum rerum conditionem (389, 8–13)." For more on witnesses, see Josefina Rodriguez Arribas, "*Testimonies* in Medieval Astrology: Finding Degrees of Certitude in Astrological Judgments," in *Doxa: Études sur les formes et la construction de la croyance*, Pascale Hummel (ed.), Paris: Philologicum, 2010, 115–133.

[46] Alexandre Tur discusses how to calculate the lord of the horoscope for annual prognostications, the *dominus anni*, in his immensely valuable Master's thesis on medieval Latin annual prognostications in manuscript; "À l'entrée du soleil en Bélier: Les prédictions astrologiques annuelles latines dans l'Europe du XVème siècle (1405–1484)," 3 vols., Thèse pour le diplome d'archiviste paléographe, École Nationale des Chartes, 2014, 64–67.

[47] Et potest homo revolvere historiam ad tempora praeterita, et considerare effectus coelorum a principio mundi, ut sunt diluvia, terrae motus, pestilentiae, fames, cometae, monstra, et alia

Analyzing astrological revolutions, an astrologer can thus discover causal structures in the past that he may then use to make predictions about future events. We will explore these causal structures more fully in the body of this study.[48]

*

Another interpretive feature that should be mentioned is that the basic horoscope can then be manipulated with various techniques to make predictions about the future, primarily by "revolutions" (or "transits"), since, although the horoscope is constructed for a particular time, the planets themselves keep moving throughout the zodiac over the course of time. The significant relationships of future planetary motions to those in the root horoscope (whether a nativity or otherwise) can signify future developments, as Roger just indicated.

This is also the case for another technique called "directions" (or "progressions" or "prorogations"), where a planet or a point in the horoscope (e.g., the ascendant degree) can be directed according to more or less complex geometrical and numerical techniques along the zodiac to another point thereon that represents a future time. For example, the sun can be directed by one degree per year and thus directed into benign or malignant aspects with other benign or malignant planets or with itself.[49]

The resultant interpretations derived from some or all of these features and/or techniques are usually called *iudicia* or judgments in the premodern texts examined here. Hence, in some sense, all astrological predictions may be called "judicial" or "judiciary" astrology because they are all judgments made by an astrologer. For this reason (among many others), I will not use the natural-judicial distinction in what follows unless the primary sources themselves use this terminology or a closely related underlying conceptual structure. Needless to say, these proper horoscopes (both historical and contemporary) should not be confused with the normal watered-down sun-sign "horoscope" found in most daily newspapers today.

Although horoscopes are used in slightly different ways and toward different ends in the four types of canonical practice—general astrology or revolutions, nativities, elections and interrogations—the basic normative horoscope as thus described provides the basic structural and interpretive unit for all astrological practices,

---

infinita, quae contigerunt tam in rebus humanis quam in naturalibus. Quibus comparatis, revolvat tabulas et canones astronomiae, et inveniet constellationes proprias singulis effectibus respondere. Deinde consideret per tabulas consimiles constellationes in futuro tempore propinquo vel remoto sicut vult; et poterit tunc pronuntiare in effectibus, qui consimiles erunt sicut fuerunt in praeterito, quia posita causa ponitur effectus (389, 13–23).

[48] Ptolemy makes similar claims in *Tetrabiblos* I.2, and Francis Bacon makes a similar argument for establishing astrological principles on the basis of historical study at the end of his proposal for astrological reform in his *De augmentis scientiarum* of 1623, as I have discussed in my "Astrology," 550–52, and which I will discuss more fully in volume III.

[49] Monica Azzolini discusses some of these techniques in practice in *The Duke and the Stars*, 107 ff, with further references. See also Otto Neugebauer and H.B. Van Hoesen, *Greek Horoscopes*, Philadelphia: American Philosophical Society, 1959, 12 ("starter"). Their discussion of astrological terminology at 2–13 is very useful.

where the astronomical, geographical, and mathematical foundations become astrologically meaningful and thus potentially valuable in many parts of life for both individuals and society. To these four basic types of astrological practice we will now turn.

[4] The Four Canonical Types of Astrological Praxis

In the following chapters (as well as in volumes II and III), examples of the different types of astrological praxis will be presented within their socio-political, theologico-religious and cultural contexts. To complete this introductory excursus, it will be useful here to outline the four canonical types of astrological praxis.[50]

As we will see, the *Speculum astronomiae* provides an excellent introduction to the study of astrology, its structure and problems, ca. the 1260s. It also had a rich and influential Nachleben, as we can tell from references to it, and its uses by, for example, Pietro d'Abano and Taddeo Alderotti in the fourteenth century, Pierre d'Ailly and Jean Gerson in the early fifteenth century, Regiomontanus and Marsilio Ficino in the later fifteenth century, and Andrea Argoli in the seventeenth.[51] The part of the *Speculum* that describes the four main types of astrological practice (Chaps. 7–10), however, is by far the poorest in organization and clarity of expression, sometimes reading more as rough notes than as polished prose. I will present a clarified (and somewhat simplified) version here, which is based closely on the *Speculum*'s structure and content.[52] This map will help orient us in the often strange terrain to follow. It will also give us a further taste of some of the technical features and terminology of astrological practice, all of which required the use of the central and iconic artifact of practical astrology, namely, the horoscope, figure or *thema*.[53]

[50] Medical astrology presents a special case which uses all four types of practical astrology toward a particular medical end: (e.g.) nativities in various circumstances (as we will see in the case of Andrea Argoli in volume III); general astrology or revolutions (for critical days); medical interrogations, and elections for determining times for bleeding, and, with Ficino in volume II, for compounding medicines to optimize their potency.

[51] Zambelli, Speculum astronomiae *and Its Enigma*, 116–20. This basic fourfold structure of astrological praxis continued well into the seventeenth century, as we can see in the relevant articles of Girolamo Vitali's *Lexicon mathematicum astronomicum geometricum* (1668) to be discussed more fully in Volume III.

[52] For the full text of each branch with translation, see Zambelli, *The* Speculum astronomiae *and its Enigma*.

[53] In his Chap. 6 on astrology, Hasse (*Success and Suppression*) has valuable discussions of interrogations, elections, anniversary horoscopes (*revolutiones annorum nati*) for individuals, the length of life (with the hyleg and alcocoden), lots, and annual revolutions (*revolutiones annorum mundi*) for the world. He also has an extensive discussion of great conjunctions. All of these topics are treated in relation to the antagonism between Greek and Arabic astrology during the Renaissance that is the overarching theme of his study. His discussions are primarily post-1500, but many of them also bridge the medieval-early modern gap as well, and all are relevant for the three volumes of my study.

General Astrology or Revolutions

The first type of practical astrology is general astrology or revolutions (*Speculum astronomiae*, Chap. 7), which deals with the individual and interrelated cycles of the sun, moon and planets over time. It is divided into three parts: The first treats the 120 different possible conjunctions of the planets and their eclipses. The second concerns the revolution of the years of the world (*revolutio annorum mundi*) or annual revolutions.[54] The third addresses the changes of times (*mutatio temporum*), that is, of the seasons, primarily with respect to the weather. Charles Burnett describes this first type of practical astrology as follows[55]:

> The first of these, according to the Magister Speculi, is general astrology (*de revolutionibus*), which pertains to whole nations and regions. This is divided into three parts: (1) conjunctions (*coniunctiones*), by which significant events happen when planets—in particular Saturn and Jupiter—are in conjunction; (2) revolutions of the years of the world (*revolutiones annorum mundi*), in which the events of the year are predicted from the planet that is most important in the astrological chart when the sun enters the first minute of Aries—this planet being the lord of the year (*dominus anni*); (3) astrometeorology, by which the weather can be forecast. Texts on the last subject are called "about rain" (*de imbribus, de pluviis*) or "on changes of weather" (*de mutationibus temporum*) and preserve a division of the zodiac into 28 parts according to the course of the moon (these are the lunar mansions: *astra*/MANSIONES *lune*).

Ptolemy treated this branch of practical astrology in book II of the *Tetrabiblos.*

According to the *Speculum astronomiae*, the first type of revolution consists in identifying and determining the influences of all the combinations of planetary conjunctions, from two planets to all seven, during the course of a year and over larger periods. It particularly considers the conjunctions of the three outer planets, Mars, Jupiter and Saturn, and thus includes historical astrology more generally and great conjunctions in particular under its purview. It also identifies and interprets the eclipses of all the planets with each other, but especially of the luminaries.

The second type—revolutions of the years of the world (*revolutio annorum mundi*) or annual revolutions—consists in interpreting a horoscope constructed for the time of the sun's entry (or "ingress") into Aries.[56] Knowledge of the planetary ruler of the year and the aspects of the planets to it, and also a number of details concerning the planets, can indicate to a skilled astrologer what God will do in that year with the stars as His instruments.[57] This concerns the rich and powerful men of

---

[54] As opposed to a *revolutio annorum nati*, which we will encounter under nativities.

[55] "Astrology," 375. I changed Burnett's attribution of this work to Albertus Magnus to the more neutral and anonymous Magister Speculi, following Weill-Parot's usage. I also make minor changes in the other descriptions. For an extended treatment, see Giuseppe Bezza, *Arcana mundi: Antologia del pensiero astrologico antico*, 2 vols, Milan: Rizzoli, 1995, ch. XI, "L'astrologia catolica," 549–669.

[56] Tur discusses these types of revolutions in depth in his "À l'entrée du soleil en Bélier." Among many other things, in Chaps. 2 and 3, he analyzes in detail the astrological techniques required to make these prognostications. I discuss these prognostications further in Chap. 10.

[57] "[...] indicatur quid operetur Deus gloriosus et sublimis in eodem anno per stellas sicut per instrumenta [...]." This instantiates the more general formulation in *Speculum astronomiae*,

each region and the populace in general, the harvest, war and peace, earthquakes and floods, falling stars and terrible prodigies, and the other things that happen in this world (*accidunt in hoc mundo*). These are a number of the themes found in annual prognostications as well as those we just saw reference to in Roger Bacon's discussion that astrological judgments could be made about past, present and future events. Annual revolutions also indicate what comes about from the actions (*opera*) of the fixed stars, and what the head and tail of the dragon (*caput* and *cauda draconis*, that is, the north and south nodes of the moon) and comets signify.

The third type of revolution, the change of times (*mutatio temporum*), relates particularly to the weather. It consists in knowing the accidents of the planets and their effect on the air as well as on the different seasons, including the moist and dry times of year. To make predictions, it uses in particular the 28 lunar mansions,[58] the direct and retrograde motions of the planets and their latitudes, and the 12 gates of the moon (especially their openings). This part also consists in the knowledge of the blowing of the winds and their parts.[59] Thus, these three types of revolution together make up the primary contents of an annual prognostication, as we will see.

Nativities

The second branch of practical astrology, nativities, concerns what can be learned about an individual's life by means of the natal (birth) horoscope. Burnett describes it thus:

> Next comes genethlialogy or nativities (*nativitates: genezia*), by which birth charts are drawn up and the course of life of the newborn child—or native (*natus*)—is predicted. One must determine the "hyleg" or "prorogator" (*dator vite: hileg, hylech, yleg* [Arabic *haylaj*], *dominus/significator vite*), which is a point on the ecliptic worked out by complicated means, and the planet with the most dignities in the point, which is called the lord of the geniture (*dominus geniture: alc*[*h*]ochoden, altothoden—Arabic *al-kadhkhudah*). The combined evidence of these two gives the length beyond which the life of the native cannot naturally extend, as well as indicating diseases and hardships in the course of the life. The events in different areas of the native's life can be discovered from looking at the situation of the planets in regard to the 12 places [= houses]. Moreover, each planet as chronocrator rules over a certain number of years of the native's life; these are called the planet's "firdaria" *(*:* afraadet [pl.*], alfardaria, alfirdariech, alfridaria*, FIRDARIA [Arabic

Chap. 3 (as discussed in Chap. 1), where human beings can gain insight through astrology into God's providential governance of the world, a theme to be explored more broadly in Chap. 5. This view ultimately provides astrology's theological foundations, as we will see in Part II. In this respect, Thomas Aquinas in *Summa contra gentiles* III.82–94 is fully in agreement with the *Speculum astronomiae*, as we will see in Chap. 5.

[58] The moon has 28 "mansions," each corresponding to the length of time the moon moves in 1 day of its cycle around the zodiac (roughly 13 degrees). For a list of all 28, see Gerolamo Vitali's rich article "Lunae mansiones, seu stationes," *Lexicon*, 270–74.

[59] Tur discusses astrometeorology in Chap. 6 of his "À l'entrée du soleil en Bélier."

> *fardar*].[60] It was a common practice to cast a horoscope when the sun returned to the same degree of the zodiac where it was at the time of birth, and to compare it to the astrological chart of the birth (*radix*). This subbranch of genethlialogy is called solar returns or anniversary horoscopes (*revolutiones nativitatum*, 375).[61]

According to the *Speculum astronomiae* (Chap. 8), nativities teach how to choose the place of the *hyleg* and the *alchochoden* in order to judge the length of the native's life[62]; not, that is, the length of time he ought to live by necessity, but that beyond which his life will not extend by nature.[63] It also teaches us how to prorogate (= direct) the degree of the ascendent and the degree of the moon for the events of the body with respect to sickness and health, and to prorogate the degree of the part of fortune for the acquisition of wealth. It also teaches us how to compute (*revolvere*) the years of the native for determining more important and lesser events. Today, natal astrology is probably the best known branch of astrological practice.

Interrogations

The third type of practical astrology is interrogations, namely, the posing of questions to an astrologer, of which medical interrogations are a significant subset. We will see a particularly colorful example in Part IV. Burnett describes it thus:

> The Magister Speculi next deals with interrogations (*interrogationes*), in which the astrological chart of the moment when the question is posed informs the outcome of what is asked. In this case, the state of mind of the inquirer—the radical intent (*intentio radicalis*)—is important. This branch of astrology is often confused (in medieval and modern

---

[60] Hand discusses "firdaria" in depth in his "Use of Military Astrology," 115–28, in a section entitled "The Firdaria—A Study in the Transmission of a Complex Astrological System."

[61] Anniversary horoscopes of nativities were also very popular. These *revolutiones annorum nati* should be contrasted with the *revolutio annorum mundi* or annual revolutions just encountered. We can see numerous historical examples in Johannes Kepler's personal astrological practices. See, for example, Friederike Boockmann, "Die Horoskopsammlung von Johannes Kepler," in *Miscellanea Kepleriana: Festschrift für Volker Bialas*, F. Boockmann, D. di Liscia and H. Kothmann (eds.), Augsburg: Rauner, 2005, 183–203. My thanks to Patrick Boner for this reference. Eade's treatment focuses solely on nativities; *Forgotten Sky*. See also Bezza, *Arcana Mundi*, II, 853–962. Ptolemy treats nativities in *Tetrabiblos*, Books III and IV. Hasse (*Success and Suppression*) discusses anniversary horoscopes at 260–62 and annual revolutions at 267–72.

[62] Azzolini shows how this works in the interpretation of Galeazzo Maria Sforza's nativity by Raffaele Vimercati; *The Duke and the Stars*, 108 ff. Hasse's description is also valuable; *Success and Suppression*, 262–65, as is Hand's "Use of Military Astrology," s.v. hyleg and alcocoden.

[63] These are akin to the periods of life discussed in Albertus Magnus's commentary on *De generatione et corruptione*, II.9–10 and *De fato*, as we will see in Chaps. 1 and 5. This issue is also discussed in relation to necessity in Chap. 13 of the *Speculum astronomiae* as we will see in Chap. 4. For the *hyleg* and *alcocoden* (including rules for their calculation), see also Eade, *Forgotten Sky*, II.30 and 32, and II.31, respectively. Giovanni Pico della Mirandola rejected the *hyleg* in his *Disputationes adversus astrologiam divinatricem*, but that did not stop Galileo from using it in the horoscopes he made in MS 81.

> sources) with “catarchic” astrology, the *Speculum astronomiae*’s next division, in which the best time for beginning an activity is determined. There is no evidence that interrogational astrology was known in classical times; it seems to have been an Indian invention that achieved great popularity among Islamic astrologers (375).[64]

According to the *Speculum astronomiae* (Chap. 9), interrogations teach us to judge a matter about which an interrogation has been made, concerning whether it will come to pass or not. If yes, it can reveal the cause and when it will be; if not, what prohibits it from coming into being. It can also show when something will appear which ought not to come into existence. We will see a broad range of the sorts of questions that might be posed, especially in Part IV.[65]

Elections

Burnett describes the fourth and final astrological practice, elections, thus:

> The medieval Latin term for catarchic astrology is *electiones* or *electiones horarum laudabilium.* In both interrogations and elections we find the first place [= house] assigned to the querent (*querens*) and the seventh to the thing asked about (*quesitum*), and the relationship of this chart to the birth chart (*radix*) can also be considered. Interrogational and electional chapters often occur together in works loosely entitled *iudicia*. This last branch includes the making of astrological talismans (*imagines astronomicae*), which is often accompanied by fumigation (*suffumigatio*), the invocation of angels or demons (*angeli, demones, spiritus*), and the inscription of magical signs (*characteres*, 376).[66]

According to the *Speculum astronomiae* (Chap. 10), elections teach us to choose (*eligere*) a praiseworthy time for beginning a project (*opus*) for him whose nativity is known, by means of the correspondence of the lord of the thing (*dominus rei*) with the signifier of his nativity. But if his nativity is unknown, take a most certain interrogation (*interrogatio certissima*)[67] for this reason that, when a man makes an interrogation, what his nativity signifies toward a good or a bad end already comes forth from his nativity. In place of the nativity, take this interrogation as a root (*radix*) for this reason that, although nativities are natural things (*res naturales*), interrogations are things similar to natural things.

Elections, then, are the astrological practice for choosing propitious times; they are closely related to both nativities and interrogations. We will see many examples in what follows: from choosing the proper moment for passing the baton of command to the captain general of the Florentine army to crowning a pope or king, or

---

[64] Bezza treats both elections and interrogations together under the broader rubric of catarchic astrology; *Arcana Mundi*, 519–48. Hasse (*Success and Suppression*) discusses interrogations at 256–58.

[65] Robert Hand discusses the astrological techniques for both interrogations and elections in his “Use of Military Astrology.” Although his focus is on their military uses, his analysis is more broadly applicable.

[66] Hasse (*Success and Suppression*) discusses elections at 259–60.

[67] In what the certainty resides, the anonymous author does not inform us; perhaps that the interrogation was made at a precisely known time.

for entering a city, or beginning a significant building project, such as St. Peter's in Rome or the Fortezza da Basso in Florence. As we will see in Part III, one of the more controversial elements of practical astrology, namely, astrological images (*imagines astronomicae*) or talismans—where astrology borders on magic—was subsumed under elections.

*

Equipped with our newly ground conceptual spectacles and a range of preliminary information—including a guide map to practical astrology's basic elements, structures and terrains—it is now time to engage the thirteenth-century material directly and in detail.

Contents

Part I
Conceptual Structures (1)
Astrology and Natural Philosophy/Science: Reconstructing a Thirteenth Century Astrologizing Aristotelianism

Overall Introduction: Astrology and Aristotelian Natural Knowledge

What was the relationship between astrology and Aristotelian natural philosophy in the thirteenth century? In his *magnum opus* on Aristotelian cosmology from the thirteenth through the seventeenth centuries, Edward Grant strongly claims that there is essentially none beyond the adoption by the astrologers of different combinations of pairs of Aristotle's four qualities to characterize the different planets. This claim will be tested in detail in Part I. The resulting reconstruction will then provide the basis for what follows in the rest of the study overall. For the purposes here of reconstructing the fundamental patterns of a thirteenth century astrologizing Aristotelian natural philosophy, I am required to compromise in the interests of space. Thus, I sharply delimit the texts discussed to those of a very select yet central group of thinkers whose works were still profoundly influential in the fifteenth century and beyond.

I choose these texts because they clearly express their underlying patterns of thought in their own terms; that is, they are not an assemblage of a medieval writer's views organized according to modern notions of what he was trying to do. Further, these texts are chosen for their continuous exposition, and because they are not overly burdened with detail. They may thus function as a framework in relation to which other views (with their variations) may profitably be compared and contrasted. These texts were all written by well known authors during a well defined and relatively well known culture-historical period, ca. 1250–1280. Most importantly, however, they articulate patterns of thought that remained vitally active well into the sixteenth and seventeenth centuries. They may thus be used to reconstruct the broader natural philosophical context in which astrology was established as a legitimate knowledge-making enterprise in thirteenth-century Europe.

*

The first set of primary sources for this reconstruction, discussed in detail in Chap. 1, are the *Speculum astronomiae* later attributed to Albertus Magnus,[1] his unquestionably authentic translation-commentary on Aristotle's *De generatione et corruptione*,[2] and his student Thomas Aquinas's letter, *De operationibus occultis naturae*.[3] Although these authors and texts are all very well known, they still suffer uneven treatment by modern commentators. I also examine central patterns of thought in Aristotle's *De generatione et corruptione*,[4] one of the three Aristotelian "charters of scientific astrology" mentioned by Richard Lemay along with the *De caelo* and *Meteorologica*.[5] John North also emphasized the *De generatione et corruptione* along with the *Meteorologica* in his influential survey of celestial influences.[6]

Edward Grant, on the other hand, focuses almost exclusively on the *De caelo* in his treatment of medieval cosmology—even though he cites North's very survey there—using only two *quaestiones* from commentaries on *De generatione et corruptione*;[7] he does not even mention Albert's commentary. For that matter neither does North. Furthermore, Grant uses Thomas's *De operationibus occultis natu-*

[1] The Latin text (established by S. Caroti, M. Pereira, S. Zamponi and P. Zambelli, with a facing English translation by C.S.F. Burnett, K. Lippincott, D. Pingree and P. Zambelli) is in Zambelli, Speculum astronomiae *and its Enigma*, 208–273. I discuss the contested attribution to Albert in Part I.

[2] Albertus Magnus, *Opera omnia* 5.2, *De generatione et corruptione*, Paul Hossfeld (ed.), Monasterii Westfalorum: Aschendorff, 1980, 107–219. This is the so-called Cologne edition (*Editio Coloniensis*).

[3] The critical text is in the Leonine edition; Thomas Aquinas, *Opera omnia* 43, Rome: De propaganda fide, 1976, 183–86. For Juan Luis Vives's text and an English translation with introduction and discussion, see Joseph Bernard McAllister, *The Letter of Saint Thomas Aquinas* De operationibus occultis naturae ad quemdam militem ultramontanum*: A Dissertation*, Washington, D.C: Catholic University of America Press, 1939, 191–97; McAlister's translation, 20–30. I discuss several other of Thomas's influential texts in Chaps. 5 and 7, including the *Summa theologiae* and *Summa contra gentiles*.

[4] Latin: Aristoteles Latinus IX.1, *De generatione et corruptione, Translatio Vetus*, Joanna Judycka (ed.), Leiden: Brill, 1986. Greek: *Aristotle On Coming-To-Be and Passing-Away* (De generatione et corruptione): *A Revised Text with Introduction and Commentary,* H. H. Joachim (ed.), Oxford: Clarendon Press, 1922. There is a helpful translation in the Loeb Classical Library series by E. S. Forster, London: Heinemann, 1955.

[5] Lemay, "True Place," 57. I also discuss *De caelo* in Part I.

[6] North, "Celestial Influence," 45–6.

[7] Grant, *Planets, Stars and Orbs*, *passim*, and esp. 571–72 for our purposes.

rae as one of only three texts to characterize the natural philosophical foundations for celestial influences in the medieval period, along with Aristotle's *De caelo* and Ptolemy's *Tetrabiblos*, but North does not mention it at all in his six-page treatment of Thomas's views on celestial influences.[8] This uneven treatment of central texts points to a pre-paradigmatic situation in the historiography of this part of medieval natural knowledge. I hope to establish a more solid foundation in what follows.

[8] North, "Celestial Influence," 75–81.

Chapter 1
Ligamentum naturalis philosophiae et metaphysicae: Astrology and Aristotelian Natural Philosophy

1.1 Introduction: *Speculum astronomiae*

We will begin to approach our central question of astrology's place within the premodern Aristotelian-Ptolemaic map of knowledge ca. 1250–1500 with what amounts to an explicit broadly-framing answer to just such a question. This is found in Chap. 3 of the extraordinarily influential and deliberately anonymous text that later came to be called the *Speculum astronomiae*, and was normally attributed to Albertus Magnus.[1] It seems to have been composed in the 1260s.[2] Among other

[1] Zambelli argues strongly for the attribution of this work to Albert; Speculum astronomiae *and its Enigma,* 25–41. Based on an intensive codicological investigation, however, Paravacini Bagliani argues convincingly that in the earliest manuscripts the text was both anonymous and not entitled *Speculum astronomiae.* This title only arose in the middle of the fourteenth century as did the attribution to Albert. In fact, the earliest manuscript with both an attribution to Albert and the title is Kues, Hospitalbibliothek, MS 209 (Ku), which is dated to "after 1339" and was formerly owned by Cardinal Nicholas of Cusa; *Le* Speculum astronomiae, 27–8. At the time of its composition, the text explicitly hid the identity of its author as *quidam vir zelator fidei et philosophiae*. Paravacini-Bagliani suggests that the original author may have been Campanus of Novara. Bruno Roy, on the other hand, argues for Richard of Fournival as the putative author; "Richard de Fournival, auteur du *Speculum astronomiae*?" I am now convinced that the anonymous author was not Albert, despite some strong similarities with his authentic works, as we will see. The main evidence for rejecting Albert is presented together in Chap. 7 below. Regardless, the identity of the author is not crucial for my argument, since the relevant ideas were extremely influential, and from the mid-fourteenth century it was increasingly attributed to Albert, as we will see. For convenience, I refer to the text as *Speculum astronomiae*, but I usually use a circumlocution for the author, including Magister Speculi, as Weill-Parot suggests.

[2] See Zambelli, Speculum astronomiae *and its Enigma,* 3, and also 3–23 for the arguments. The oldest datable manuscript (Firenze, Biblioteca Medicea Laurenziana, MS Plut. XXX.29 [Fi 2]) is dated to ca. 1280, thus providing a *terminus ante quem* for the composition. In this earliest manuscript, the text circulated without title or author; Paravacini-Bagliani, *Le* Speculum astronomiae, 22–3. Nicolas Weill-Parot gives 1271–72 as the *terminus ante quem* because Thomas Aquinas used the phrase '*imagines quas vocant "astronomicas*"' for talismans, which precise

H D. Rutkin, *Sapientia Astrologica: Astrology, Magic and Natural Knowledge, ca. 1250–1800*, Archimedes 55, https://doi.org/10.1007/978-3-030-10779-6_1

things, the *Speculum astronomiae* aimed to clearly indicate which astrological texts are legitimate and may thus be used, and which are not.

The *Speculum astronomiae* is comprised of two main sections: first, a descriptive and bibliographic prolegomenon to the study of "the science of the stars" (*astronomia*), which is divided, according to Ptolemy's standard division, into what we call astronomy and astrology (Chaps. 1–11). The second part (Chaps. 12–17) contains an informal series of *quaestiones*[3] that address the more problematic areas where astrology touches on theological, moral and magical issues, such as determinism in nature, human free will, and the making and use of talismans, as I will discuss in detail in Chap. 4.[4] Here we are concerned with Chap. 3, which immediately follows the rather long section on astronomy (Chaps. 1 and 2). It situates astrology clearly within the medieval map of knowledge (including theology) before turning to describe the four types of practical astrology that we just explored in the excursus.

Chapter 3 begins by immediately answering our question in general terms—with a whopper of a sentence—and then defining the answer more closely. First the author makes a bold claim: "The second great *sapientia*, which is likewise called *astronomia*, is the science of the judgments of the stars (*scientia iudiciorum astrorum*), which provides a link between natural philosophy and metaphysics (*ligamentum naturalis philosophiae et metaphysicae*)."[5] He then makes an argument to support this claim:

---

terminology was coined in the *Speculum astronomiae*, in his *Summa Theologiae* II.II.96.2, and this text can be securely dated to 1271–72. See Weill-Parot, *Les "images astrologiques,"* 38–40 for the arguments. The precise dating of the text also does not affect my argument. David Juste informs me that a firm *terminus post quem* is as early as 1217 because the *Speculum astronomiae* refers to Michael Scot's translation of Alpetragius's *De motibus celorum* (Zambelli, 214). For the purposes of my argument in this study—and without further particularizing evidence—I will hold to the 1260s as the decade of composition. We will also see significant similarities between it and Roger Bacon's *Opus maius*, which was certainly written ca. 1266–67, in my Chap. 4 below.

[3] These have a less formal structure than standard scholastic *quaestiones*, that is, their structure is not as formal as those, for example, in Albert's own commentaries on, e.g., [ps.-] Dionysius the Areopagite's *De divinis nominibus*, or in the *De fato* to be discussed in Chap. 5 below. The more formal argument structure normatively comprises (1) several numbered points to the contrary, (2) an authoritative *solutio*, and then (3) affirmative responses, on the basis of the *solutio*, to the negative points. See Martin Grabmann, *Die Geschichte der Scholastischen Methode: Nach den Gedruckten und Ungedruckten Quellen*, 2 vols, Graz: Akademische Druck- und Verlagsanstalt, 1957 (reprint of 1909 edition). Thomas Aquinas also uses this argument structure in the *Summa contra gentiles* and *Summa Theologiae* that I discuss in Chaps. 5 and 7.

[4] Without discussing it in detail here, one of the arguments (14, 40–48) claims that freedom of the will is quite compatable with astrology. In fact, disparaging astrology without actually using it where it can be helpful does more harm than good with respect to acting freely (*talia destruere plus esset contra liberum arbitrium quam pro eo* […] 44–5). I discuss these interesting arguments defending astrology in Chap. 4 below. For useful accounts of the relationship of astrology to arguments on free will during this period, see (e.g.) Smoller, *History, Prophecy and the Stars*, 25–32, and Zambelli, Speculum astronomiae, *passim*.

[5] "Secunda magna sapientia, quae similiter astronomia dicitur, est scientia iudiciorum astrorum, quae est ligamentum naturalis philosophiae et metaphysicae." I have found the English translation

(A) For if the highest God (*Deus altissimus*) ordered (*ordinavit*)[6] this world (*mundum istum*) by His great wisdom (*sua summa sapientia*)[7] in such a way that He who is a living God (*Deus vivus*)—but who is God of a heaven which is not alive (*Deus caeli non vivi*)[8]—wished to act (*operari*) within the realm of created things (*in rebus creatis*), which are found among these four lower elements (*in his quattuor elementis inferioribus*), by means of deaf and mute[9] stars as [His] instruments—

Especially when attributed to Albert, this densely packed clause also provides an authoritative foundation for articulating God's providential ordering of and action in the world by means of the stars, with its corollary of our ability to understand the same thereby.[10]

Still in the same sentence, he then briefly (and parenthetically) describes the two *scientiae* mentioned above, for which astrology is the *ligamentum*:

(B)—and we have (1) a metaphysical *scientia,* which teaches us to consider the causer of causes among the causes of things (*in rerum causis causatorem causarum considerare*), and (2) another, natural *scientia*, which teaches us to experience the creator of creatures among created things (*in rebus creatis creatorem creaturarum experiri*)—

The reader should note the striking parallelisms in this carefully constructed period.

Continuing the same sentence, having now presented the complementary subject matter and kinds of inquiries of these two *scientiae*, the author furnishes the point of his period as he completes it with a rhetorical question:

(C)—what then would be more desirable to a person of sense than to have a middle science (*media scientia*) that would teach us how (*qualiter*) a change (*mutatio*) of the terrestrial

---

by Burnett, Lippincott, Pingree and Zambelli to be extremely valuable, but I modify it where necessary; *The* Speculum astronomiae *and its Enigma.* I include a significant amount of Latin in my translation in order to introduce the reader to the technical philosophical vocabulary that mainly derives from the Latin translations of Aristotle. It is remarkably consistent over the entire premodern period, as we will see. See Schmitt's discussion in *Aristotle and the Renaissance,* and Minio-Paluello in *DSB*, s.v. Aristotle.

[6] This divinely structured *ordinatio* is central to this model of medieval natural knowledge, as we will see repeatedly in what follows.

[7] The order of the Latin more accurately reflects the order of creation than my translation does: "Sic ordinavit Deus altissimus sua summa sapientia mundum istum […] (3, 4–5).

[8] The crucial cosmological issue of the heavens (*caelum*) being alive or not will be discussed in much greater detail in volume II, where this factor is fundamental for distinguishing between Aristotelian and (Neo)Platonic cosmological patterns. I also discuss this in my forthcoming, "Were the Heavens Alive in the Renaissance?: Marsilio Ficino's and Giovanni Pico della Mirandola's Contrasting Views on the Animation of the Heavens."

[9] This issue of the stars' deafness and muteness is directly related to them not being ensouled and alive. They are thus unable to hear people's prayers and exhortations. Zambelli discusses this in her Speculum astronomiae *and its Enigma*, Chap. 8.

[10] I discuss the other relevant passages for the *Speculum astronomiae*'s views on divine providence from the section of *Quaestiones* (Chaps. 12–15) in Chap. 4 below. Thomas Aquinas specifically, emphatically and repeatedly says in *Summa contra gentiles* III.82–94 that God governs and rules the inferior world by means of the celestial bodies. I will discuss divine providence more fully below in relation to Thomas Aquinas in Chap. 5, and in volume III with respect (e.g.) to Philipp Melanchthon.

> world (*mundanorum*), with respect to this and that [sc. particular thing] (*ad hoc et ad illud*), came to be by a *mutatio* of the celestial bodies?[11]

The anonymous author thus locates astrology—the second great *sapientia* that comes under the umbrella term '*astronomia*', namely, the *scientia iudiciorum astrorum*—as the linking middle *scientia* between the two well-established bodies of knowledge: metaphysics (the realm of causes) and natural philosophy (the created realm of nature), quite an important place indeed. He also clearly indicates that this entire *scientia* is concerned with the judgments of the stars, thus authorizing that the term "judicial" be applied to the entire realm of practical astrology.

The Magister Speculi then goes on to offer a proof (*probatio*) for his claim:

> And indeed is this not one of the most powerful proofs, that if a *motus inferior* [sc. a movement[12] of the lower world] obeys (*oboedit*) a *motus superior* [sc. of the celestial bodies], it would not be the case unless there were one God—glorious and sublime—in the heavens and on earth (*in caelo et in terra*)? For, if there were divergent principles (*diversa principia* [that is, if there were a different *principium* governing each realm]), or He were to have an associate (*particeps*) in the heavens or on earth, so that the kingdom (*regnum*) of heaven and the kingdom of earth were different (*diversa*), it is not likely that this *oboedientia* would be fixed and permanent, persisting without deviation. But now, one is obviously convinced by this *scientia* [sc. astrology] that the aforesaid *oboedientia* is the case, and that it continues without change. Wherefore, it [sc. astrology] inspires humankind to love God more ardently to the extent that He is declared by it the leader and first principle (*princeps atque principium*) of everything.[13]

The *Speculum astronomiae* thus describes a divinely embracing unified field theory (as it were) that not only provides the basis for natural law,[14] namely, a coherently ordered and consistent structure of the world—the heavens *and* the earth—in one direction. In the other direction, it is also directly responsible, through the study of nature as God's creation, for inspiring the student to a greater love of and admiration for the Creator of such an extraordinarily structured system. Thus, the study of nature is invested with a deeply theologico-religious dimension.

---

[11] (A) Si enim sic ordinavit Deus altissimus sua summa sapientia mundum istum, ut ipse qui est Deus vivus, Deus caeli non vivi, velit operari in rebus creatis, quae inveniuntur in his quattuor elementis inferioribus, per stellas surdas et mutas sicut per instrumenta (B) (et nos habemus unam scientiam metaphysicam, quae docet nos in rerum causis causatorem causarum considerare, et aliam naturalem, quae docet nos in rebus creatis creatorem creaturarum experiri), (C) quid desideratius concionatori quam habere mediam scientiam, quae doceat nos qualiter mundanorum ad hoc et ad illud mutatio caelestium fiat corporum mutatione (3, 1–13)?

[12] *Motus* ([Gk] kinesis) is a central concept in Aristotelian natural philosophy to be explored more fully below.

[13] Numquid et haec una est ex praecipuis probationibus, quod non sit nisi unus Deus gloriosus et sublimis in caelo et in terra, si videlicet motus inferior motui superiori oboedit? Si enim essent diversa principia, aut haberet participem in caelo aut in terra, ut essent regnum caeli et regnum terrae diversa, non est verisimile quod esset haec oboedientia fixa permanens absque nutu. Nunc autem ex ista scientia convincitur evidenter, quod dicta oboedientia stet atque immutabiliter perseveret: quare tanto provocat hominem ad Deum ardentius deligendum, quanto per ipsam attingentius omnium princeps atque principium declaratur (3, 13–23).

[14] See Thorndike, "The True Place of Astrology in the History of Science."

The author then follows up on the theological side, in what is virtually a paean to the great value in studying astrology for cultivating one's passion for the divine:

> For, unknown (*incognitus*), He will not be loved, nor, since He is first (*primus*), will He be known through what is prior (*per prius*), nor through Himself (*per seipsum*), since He is incomprehensible [sc. unable to be known in Himself]. Therefore, it remains that He can be known through what comes after (*per posterius*), namely, through His glorious effects.[15] These are humankind (*homo*) and the arrangement (*ordinatio*) of the universe to it [sc. humankind],[16] namely, [sc. the order] of [1] supercelestial beings (*supercaelestium*)[17] to provide guidance to rational beings,[18] and of [2] the elements, in which the material aspects of rational beings are measured, the knowledge of which *ordinatio* of the whole, no human *scientia* attains as perfectly as the *scientia iudiciorum astrorum*.[19]

These are major claims indeed for scientific astrology! This last named *ordinatio* of the universe, the glorious effects of God's work, captures the basic structures of the medieval cosmos: celestial bodies, elements, and human beings (and mixed bodies in general). It is precisely this *ordinatio* that can best be known by means of astrology as well as its Creator. The anonymous author does not fill in the natural philosophical details of the picture here, nor in the rest of the *Speculum astronomiae*. He then "descends," as he puts it, to describe the practical parts of astrology within the limitations imposed by the existing but incomplete Latin translations, in order to instantiate what he has just said.[20]

---

[15] This refers both to God's creation of the world and also to his acting within it, thus picking up *operari in rebus creatis* from the first sentence.

[16] '*ad ipsum*' refers to humankind (*homo*); if it referred to God, it would be '*ad seipsum*', as he used just above.

[17] '*supercaelestium*' seems odd since the Magister Speculi is referring here to the observable effects of his creation. Supercelestials usually refer to the intelligences or separated substances above the planets, not to the planets themselves, as in the passage from Thomas Aquinas's *De operationibus occultis naturae* to be discussed in the next section. The translation in Zambelli et al. simply transliterates it as "supercelestial beings" but does not discuss it.

[18] Here we have an allusion to astrology's more practical side.

[19] Non enim diligetur incognitus, neque cum sit primus, cognoscetur per prius, neque per seipsum, cum sit incomprehensibilis. Restat ergo quod per posterius, per suos scilicet gloriosos effectus. Hi autem sunt homo et ordinatio universi ad ipsum, videlicet supercaelestium, ut praebeant ductum rationalibus, et elementorum, in quibus sumptus rationalium mensurentur, quam universi ordinationem nulla scientia humana perfecte attingit, sicut scientia iudiciorum astrorum (3, 23–30).

[20] "Quod ut liquidius appareat, descendam ad partes eius, commemorans quasi omnes libros laudabiles, quos de ea pauper latinitas ab aliarum linguarum divitiis per interpretes mendicavit (3, 31–33)." This trope on the paucity of scientific writings in Latin is found also in Roger Bacon; see Richard Lemay, "Roger Bacon's Attitude Toward the Latin Translations and Translators of the Twelfth and Thirteenth Centuries," in *Roger Bacon and the Sciences: Commemorative Essays*, J. Hackett (ed.), Leiden: Brill, 1997, 25–47. On the authentic Albert's knowledge of Greek—or rather the lack thereof—see Walter Berschin, *Greek Letters and the Latin Middle Ages: From Jerome to Nicholas of Cusa*, revised and expanded edition, J.C. Frakes (tr), Washington D.C: Catholic University of America Press, 1988, 260. For a magisterial survey with rich bibliography of the works translated into Latin from Greek and Arabic, see d'Alverny, "Translations and Translators," and numerous studies by Charles Burnett, a number of which are now conveniently collected in his *Arabic into Latin in the Middle Ages: The Translators and their Intellectual and Social Context*. For still valuable older treatments, see Charles Homer Haskins, *Studies in the*

Resisting the Magister Speculi's kind invitation to descend to the practical level, we shall remain at this broadly framing level of astrologizing natural philosophy, turning now to a work, Thomas Aquinas's *De operationibus occultis naturae*, which is not explicitly about astrology at all, but may profitably be used to articulate fundamental structures of the ordered universe adumbrated in the *Speculum astronomiae*.[21] We will focus here on one particular feature of Thomas's analysis, namely, the central role of the planets with respect to the most important regularizing feature in Aristotelian natural philosophy, namely, forms.[22] Since its language and presentation is systematic and clear, Thomas's contemporary *De operationibus occultis naturae* provides an illuminating window through which to examine the inner workings of the medieval cosmos.[23] Along with *Speculum astronomiae*, Chap. 3, it too provides a very useful framing structure for initially orienting my study.

1.2 Thomas Aquinas, *De operationibus occultis naturae*

Albertus Magnus's most famous student, Thomas Aquinas, wrote the *De operationibus occultis naturae* (*On the Hidden/Occult Operations of Nature*), a work of unquestioned authenticity, at some point after the *Speculum astronomiae* was composed, and quite likely very soon after.[24] In fact, the *Speculum astronomiae*

History of Medieval Science, and Francis J. Carmody, *Arabic Astronomical and Astrological Sciences in Latin Translation, A Critical Bibliography*, Berkeley: University of California Press, 1956. See now also Hasse, *Success and Suppression*.

[21] As noted, Grant uses the *De operationibus occultis naturae* for describing the philosophical foundations of the medieval view of celestial influences; *Planets, Stars and Orbs*, 573–4, and his "Medieval and Renaissance Scholastic Conceptions of the Influence of the Celestial Region on the Terrestrial," 3–5. He also emphasises the unique character of this work as an explicit discussion of how celestial influences work. In his "Scholastic Philosophy and Renaissance Magic in the *De Vita* of Marsilio Ficino," Brian P. Copenhaver also uses it to articulate underlying patterns of the magical world view; *Renaissance Quarterly*, 37 (1984): 523–54. To see in greater detail how the natural philosophical patterns articulated here relate to the broader patterns of Thomas's natural philosophy, see Litt's magisterial, *Les corps celestes dans l'univers de Saint Thomas d'Aquin*. For further discussion and analysis of Thomas's views on astrology, see especially Litt's ch. VI ("L'Étendue de la causalité des corps célestes"), section 2 ("La causalité des corps célestes sur les actions et mouvements des corps inférieurs irréductibles aux qualités des quatre éléments"), 113–129, where he treats (i.a.) the *De operationibus occultis naturae*, and ch. IX ("Répartition de la causalité entre les corps célestes. Saint Thomas et l'astrologie"), 220–241, for relevant comparative material. I discuss Thomas's *Summa Theologiae* II^{a} II^{ae} 92–96 and other relevant texts in Chaps. 5 and 7 below.

[22] I also discuss the *De operationibus occultis naturae* in Chap. 7 in relation to Thomas's views on talismans.

[23] This text is also highly pertinent for deciding whether we should characterize astrology as an "occult" science, as I will discuss primarily in volume II. I briefly mention this in the overall conclusion to Part III below.

[24] According to Jean-Pierre Torrell, Thomas wrote the *De operationibus occultis naturae* between 1268 and 1272 during his second sojourn in Paris; *Saint Thomas Aquinas*, Vol. 1, *The Person and*

and the *De operationibus occultis naturae* have structurally similar aims in that they are both concerned to distinguish between legitimate and illegitimate types of scientific endeavors: the *Speculum* with distinguishing astrology in both its epistemic and operative dimensions from necromancy, and the *De operationibus occultis naturae* with distinguishing natural—both elemental and occult—from demonic action. Both the necromantic and magical dimensions in both of these texts will be discussed at greater length in Chap. 7 below.

Unlike the *Speculum*, however, Thomas's text is not explicitly about astrology, let alone about defending legitimate from illegitimate varieties. It is, rather, a deep description of fundamental causal structures within the Aristotelian cosmos, which Thomas then uses as the basis for characterizing the essential differences between natural and supernatural (whether angelic or demonic) operations in nature. Our concern here is primarily with natural operations, the description of which reveals quite a full and clear picture of the underlying causal structures within the medieval cosmos. I focus here on two central features of Thomas's exposition: (1) the role of the planets in the overall economy of nature, and (2) certain features of their role in the generation of individual natural things (*res naturales*).

*

In *De operationibus occultis naturae*, Thomas develops the view that specific forms, the forms that account for each member of a species being a member of that species, are the regularizing structure in nature. They provide the internal structure by means of which natural things act in their characteristic ways and with their characteristic abilities. What role, then, do the planets play in relation to specific form? In Chap. 11, Thomas systematically ties together the components he had already constructed, integrating the form—the internal structural principle of every existing thing, every natural substance (as established in Chaps. 6, 7 and 8)—within the larger cosmological and metaphysical structures established in Chaps. 9 and 10:[25]

> Such forms [sc. substantial forms] proceed from separated substances as from first principles (*a substantiis separatis sicut a primis principiis*), which—by the mediating power (*virtus*) and motion (*motus*) of the celestial bodies—impress the forms, intellected in themselves,[26] into bodily matter (*imprimunt formas* [...] *in materiam corporalem*). And because we have shown that the actions (*actiones*) and powers (*virtutes*) of natural bodies (*naturalium corporum*) are caused from specific forms, it follows that they may be [sc. causally] led back (*reduci*) further, namely, to higher principles [sc. to the celestial bodies,

His Work, Robert Royal (tr), Washington, D.C.: Catholic University of America Press, 1996, 214 and 356.

[25] The chapter numbers in the translation are from McAlister's edition following Vives; the line numbers in the Latin quotations are from the critical text in the Leonine edition cited above.

[26] That is, in the separated substances themselves, as in the Neoplatonist's *nous* (Gk) or—in the Christianized version—in the mind of God.

> their power and motion]—and thence further to separated intellectual substances (*in substantias intellectuales separatas*, 11, 4–8).[27]

Thomas here lays out the basic patterns of his hierarchical cosmos, with the separated intellectual substances at the top:[28] they are form only, hence separated, that is, separated from matter. They intellect, that is, contemplate form within themselves, and they are completely incapable of motion or change (*immobiles*), and thus always exist in the same way.[29]

But these forms are also impressed on bodily matter by means of the mediating power (*virtus*) and motion (*motus*) of celestial bodies, the second level in his structure overall, and the first physical level. These celestial bodies are composed of form and matter and have a two-fold motion, but they are not themselves subject to generation and corruption:[30] "Therefore, it follows that the *principia* of the forms of such corruptible bodies are the celestial bodies (*corpora coelestia*),[31] which, existing in different ways (*diversemode*) in accordance with their coming towards (*accessus*) and moving away (*recessus*), cause generation and corruption in lower things (*in his inferioribus*, 10, 1–10)."[32]

The celestial level, then, is the first level of the natural world. It is the highest physical level because it is composed of matter and form, but at the same time it is not subject to generation and corruption. By means of the *virtus* and *motus* of these celestial bodies (the luminaries and planets), the separated substances somehow impress the forms that they have generated by their self-intellection on bodily matter,[33] the lowest level of the cosmic structure, in the generation of natural things. After thus describing the hierarchical structure from above to below, Thomas then shows how the principles of the actions and powers of natural bodies can readily be

---

[27] Procedunt tamen huiusmodi forme a substantiis separatis sicut a primis principiis, que mediante virtute et motu celestium corporum imprimunt formas apud se intellectas in materiam corporalem. Et quia actiones et virtutes naturalium corporum ex formis specificis causari ostendimus, consequens est quod ulterius reducantur sicut in altiora principia, et adhuc ulterius in substantias intellectuales separatas (161–69).

[28] But not at the very top; he mentions God later.

[29] Thomas describes these same structures in a different context in *Summa contra gentiles* III, as I discuss in Chap. 5 in exploring his views on divine providence.

[30] Although Thomas does not state this explicitly, it is a central tenet of Aristotelian cosmology, which I discuss more fully below in relation to Aristotle's *De generatione et corruptione* II.9–10.

[31] This conclusion seems to a modern like a startling leap for Thomas to have made, based on the stated premises. But there is an implied set of premises, which would have been deeply embedded in Thomas's world-view, namely, those from Aristotle's *De generatione et corruptione* II, 9–10, where Aristotle makes the sun one of the three necessary—and jointly sufficient—causes precisely of generation and corruption, along with form and matter, as discussed more fully below.

[32] [R]elinquitur igitur quod principia formarum huiusmodi corruptibilium corporum sunt celestia corpora, que diversimode se habentes secundum accessum et recessum ad generationem et corruptionem in hiis inferioribus causant (156–60).

[33] Thomas uses active verb forms here (*tales formae* [...] *imprimunt formas* [...] *in materiam corporalem*), but he provides little detail as to how this process actually works.

referred—in a causal sense—to higher principles, namely, the two just articulated: celestial bodies and separated substances.

Forms thus function on all three levels: they are generated on the first (the highest), transmitted by the second (the celestial bodies), and joined to matter in the third, thus providing a regularizing structural link between—and within—all levels of the system. This is obviously of central importance in understanding the inner structure of medieval natural knowledge: the luminaries and planets play an essential mediating (and integrating) role in this process.

*

Relying on these structures, Thomas makes some brief but highly significant remarks in Chap. 16 regarding the dynamics of his model, remarks which, whether he intended them to or not, indicate at least one major way his system could be put to astrological use. He begins by demarcating the areas where the particular planetary positions at an individual's generation (birth) do *not* have any import:

> Therefore, because such *virtutes* and *operationes* are derived from a specific form—which is common to every individual of the same species—it is impossible that a certain individual of a certain species could obtain some such *virtus* or *actio* beyond other individuals of the same species from this factor, namely, that it [sc. this or that individual] was generated under a specific placement of the heavenly bodies (16, 1–6).[34]

An individual cannot have a capability or activity fundamentally different from that of another individual of the same species on account of its generation under a particular planetary configuration because the specific form, the form of the species in question, provides for the characteristic capabilities and activities of that species, which every individual of the same species possesses by nature and, thus, by definition.

Thomas's next point, however, indicates precisely where the particular planetary configuration at an individual's generation *is* significant:

> Nevertheless, it is possible that in an individual of the same species, the *virtus* and *operatio*—which follows the species—can be found to be more remiss (*remissius*) or more intense (*intensius*) according to (1) the different disposition of the matter, and (2) the different placement of the heavenly bodies at the generation of this or that individual (16, 6–10).[35]

Thomas here accounts for the individual variation of each member of a given species with respect to its power (*virtus*) and action (*operatio*), but not by any difference in their essential (= formal) structures, which are by definition the same for

[34] Quia igitur huiusmodi virtutes et actiones a forma specifica derivantur, que est communis omnibus individuis eiusdem speciei, non est possibile quod aliquod individuum alicuius speciei aliquam talem virtutem vel actionem obtineat preter alia individua similis speciei, ex eo scilicet quod est sub determinato situ celestium corporum generatum (246–53).

[35] Possibile est tamen quod in uno individuo eiusdem speciei virtus et operatio consequens speciem vel intensius vel remissius inveniatur, secundum diversam dispositionem materie et diversum situm celestium corporum in generatione huius vel illius individui (253–58).

every member of that species. There is, however, a range—a latitude[36]—in the abilities and actions permitted to each individual within the parameters defined by their specific forms, which are, in any given case, either more or less able to fulfill their latent potentiality. This range is referred to by the terms *intensius* and *remissius*, which relate directly to one of the most characteristic features of medieval Aristotelian natural philosophy, namely, the intension and remission of forms and/or qualities.[37]

Thomas accounts for this variety among individuals in two different ways, based on (1) the peculiar disposition of each individual's matter, and (2) the particular planetary configuration at the time of its generation. In other words, as we shall see, by (1) the material cause and (2) the efficient cause, both of which admit variations. The other necessary cause, (3) the formal cause, however, must remain the same for each individual of the species in question because it provides the range within which there may be individual variation, but beyond which an entity would cease to be a member of its own species.[38]

There appear, then, to be two utterly central levels of planetary functioning which are both intimately connected with specific forms. The first concerns their regularizing aspect, in that the planets provide a mediating role necessary for impressing the regularizing principle, the specific form, on its appropriate matter in the generation of all natural things. The second concerns their individualizing aspect (along with matter), affecting the particular intensity that characterizes each individual's expression of the range of possibility allowed by their specific form.[39] Exactly how both operate is by no means fully articulated; the basic structure, at any rate, seems clear.

One need not look very closely to see that Thomas provides his readers here with a natural philosophical model in which the planets play a central (starring!) role, in both the formal structure of the cosmos overall as well as in its particular and varied instantiations, which could easily embrace an explicitly practical astrological view,

---

36 See McVaugh, *Aphorismi de gradibus*, 21–2, for Avicenna on latitudes, and esp. 92–94: "This term [*latitudo*] entered the Western medical vocabulary in the translations of Galen and Avicenna, as meaning a range of variability—specifically, the variability of complexions that, while differing slightly among themselves, remain characteristic of human health. By the mid-thirteenth century, these translations seem to have been well known to masters in arts and theology as well as in medicine, for they were beginning to enrich the natural-philosophical discussions of mid-century with observations, concepts and terminology (92–3)."

37 In addition to Anneliese Maier, "Das Problem der intensiven Größe," see Edith D. Sylla, "Medieval Quantification of Qualities: The 'Merton School'," *Archive for History of Exact Sciences* 8 (1971): 9–39. For relevant material, see also Sylla's "Medieval Concepts of the Latitude of Forms: The Oxford Calculators," *Archives d'Histoire Doctrinale et Litteraire du Moyen Age* 40 (1973): 223–83.

38 See McVaugh, *Aphorismi de gradibus*, 94. As we will see in Chap. 7 below, Thomas uses this type of argument in the slightly earlier *Summa contra gentiles* in relation to explaining how magicians might do their work.

39 For more on the question of individuation during this period, see *Individuation in Scholasticism: The Later Middle Ages and the Counter-Reformation (1150–1650)*, Jorge J.E. Gracia (ed.), Albany: State University of New York Press, 1994.

most obviously in the realm of nativities. In fact, one could hardly hope to find a better one-sentence authoritative foundation for justifying the practice of natal astrology than the last sentence of Chap. 16. But Thomas does not make this astrological turn.[40] Regardless, his treatise serves well for articulating the planets' fundamental roles within the deep structures of medieval natural knowledge.[41] After exploring these two broader framing texts, we will now turn to Thomas's master, Albertus Magnus, and to their ultimate master, Aristotle himself, to fill in the details of this structure.

1.3 Aristotle, *De Generatione et Corruptione*

Aristotle provided the basic conceptual structures that would ultimately link natural philosophy with astrology in his account of generation and corruption, and the closely related "periods of life." In this account, the sun's annual motion as efficient cause relates the heavenly motions integrally and necessarily to all earthly life and activity, indeed to all movement or change in every sense on earth.[42] Although the sun played utterly central roles in generation, Aristotle's natural philosophy was not strictly speaking astrological, primarily because astrology itself as a system of thought and practice—in particular, horoscopic astrology—did not yet exist in the Greek culture of the fourth century B.C.E.[43] Nevertheless, we will see how easily Aristotle's concepts were later sharpened and co-ordinated with the central structures of horoscopic astrology.

[40] For discussions of Thomas's views of astrology, see esp. Litt, *Les Corps Celestes*, 220–241. Litt was only able to find one astrological prediction in all of Thomas's works, and it is taken from Ptolemy's *Centiloquium* (233–34).

[41] We will see in volume II that Marsilio Ficino too offers a structure much like this in *De vita* III, but as part of an ultimately Platonizing system.

[42] In his translation-commentary on Aristotle's *Physica* 1.1.4 (lns. 44–46), Albertus Magnus states clearly (following Aristotle) that *corpus mobile* (body able-to-be-"moved," that is, to undergo the four types of change) is the ultimate subject of *scientia naturalis*, which may be translated as "natural knowledge"; *Opera omnia* 4.1, *Physica Pars I*, Paul Hossfeld (ed.), 1987, p. 6. For a fuller discussion of Aristotelian four-fold views of motion (or change), embracing (1) coming-into-being and passing away, that is, changes in the category "substance," (2) growth and diminution, changes in "quantity," (3) changes in "quality," and (4) local motion, see John Murdoch and Edith D. Sylla, "The Science of Motion," in *Science in the Middle Ages*, David C. Lindberg (ed.), Chicago, 1978, 206–64. For these views specifically in relation to Albert, see Steven Baldner's section in the chapter on "Albert's Physics," in *A Companion to Albert the Great*, 184–88. I discuss this further below.

[43] Recent research has revealed that a proto-version of horoscopic astrology was just being developed at this time within Babylonian culture. In fact, our earliest Babylonian "horoscope" can be dated to 410 B.C.E. See Rochberg, *The Heavenly Writing*, and her *Babylonian Horoscopes*, Philadelphia: American Philosophical Society, 1998. By contrast, the first monumental evidence for Greek horoscopic astrology does not appear until 62 B.C.E., and the first documentary horoscope, not until 10 B.C.E. See Neugebauer and Van Hoesen, *Greek Horoscopes*.

After treating generation and corruption in a variety of domains, including the generation and transmutation of the four elements, Aristotle argued at *De generatione et corruptione* II.10 that the sun in particular, as the universal efficient cause in its annual motion around the ecliptic, was ontologically prior to and thus necessary for generation and corruption, that is, coming-into-being and passing-away (which includes being born and dying for living things) in the sublunary world. Therefore, in biological reproductive processes of generation (human and otherwise), the sun as efficient cause was required along with the male, who provided the formal cause in the seed,[44] and the female, who provided the material cause in the womb.[45] *Inter alia*, Albertus Magnus (and others) added the rest of the planets to Aristotle's account, thus expanding the astrological possibilities and providing the natural philosophical foundations for nativities as well as for other astrological practices.[46] We should now examine these patterns in detail.

1.4 The Sun in Aristotle's *De generatione et corruptione*

Aristotle analyzed the sun's fundamental role in processes of generation on earth in *De generatione et corruptione*, a core text of his natural philosophy.[47] Both the centrality of these passages within Aristotle's system and their subsequently profound influence on medieval Aristotelianism—as well as their relative neglect in the scholarly literature—fully warrant their serious discussion in an analysis of astrology's place within medieval Aristotelian natural knowledge. In order to build

---

[44] This came to be called the "specific form," that is, the form of the species, as discussed in the previous section.

[45] For the basics of Aristotelian natural philosophy, see (e.g.) Friedrich Solmsen, *Aristotle's System of the Natural World: A Comparison with his Predecessors* (Ithaca, N.Y., 1960), and the rest of the discussion here in Part I.

[46] Gad Freudenthal treats these and related issues in "Providence, Astrology, and Celestial Influences on the Sublunar World in Shem-Tov ibn Falaquera's *De'ot ha-Filosofim,*" in *The Medieval Hebrew Encyclopedias of Science and Philosophy*, Steven Harvey (ed.), Dordrecht: Kluwer, 2000, 335–70, and "The Medieval Astrologization of Aristotle's Biology: Averroes on the Role of the Celestial Bodies in the Generation of Animate Beings," *Arabic Sciences and Philosophy* 12 (2002): 111–37. He too discusses what he calls "the *astrologization* of the Aristotelian doctrine in the Middle Ages" (2000, p. 343). In the conclusion to his valuable and insightful, "The Astrologization of the Aristotelian Cosmos: Celestial Influences on the Sublunar World in Aristotle, Alexander of Aphrodisias, and Averroes," he also claims that Averroes provided the final stage of astrologization within the Aristotelian tradition (*New Perspectives on Aristotle's* De caelo, Alan C. Bowen and Christian Wildberg (eds), Leiden: Brill, 2009, 239–81, 273). I beg to differe and offer this study as a thirteenth-century counterexample. Richard Lemay does so also for Albumasar in *Abu Ma'shar and Latin Aristotelianism in the Twelfth Century*. Neither Averroes nor Albumasar seems to be the main source for Albert's (or Roger Bacon's) astrologizing Aristotelianism. Rather, al-Kindi in the *De radiis stellarum* seems to be the essential source for the tradition reconstructed here with its geometrical-optical model of planetary action.

[47] John North ("Celestial Influences," 45) calls Aristotle's *De generatione et corruptione* II.10, "the fundamental text for those who try to justify astrology in the scholastic tradition." Lemay, as we will recall, calls it one of the three Aristotelian "charters of scientific astrology," along with *De Caelo* and *Meteorologica*.

up an analysis from its foundations in depth and detail, I begin by discussing the fundamental but brief passages in Aristotle, with reference both to his Greek text and to its medieval Latin translation; then I examine Albertus Magnus's translation-commentary on these passages.[48]

We should first compare Aristotle's text with *De operationibus occultis naturae* 16, where we just left off with Thomas's account of what individuates the different individual members of a species. The specific form provides the characteristic structures and abilities of the species, and is shared by every member of the species in question. Within this overall structure, there is a well-defined range of possibility allowed for by the specific form that cannot be transgressed, the variations in which account at least in part for the individuality of each different member of the species. Thomas accounts for this variation with two different types of explanation. The first concerns the disposition of the matter and its variable receptability; the second, the particular planetary configuration at the generation of each individual.

In addition to this causal structure providing a natural philosophical foundation for natal astrology, it also directly relates to one of the deepest conceptual structures in Aristotle's natural philosophy, namely, his view of the four causes, which were known in the scholastic period as material, formal, efficient, and final.[49] As we saw, Thomas explicitly used the formal and material causes, with the planets playing a crucial mediating role between the forms in themselves and their instantiation in substances. As it turns out, the planetary configuration plays the role of the efficient, that is, the moving, acting, making or doing cause. Before we leap too far ahead, however, we should first turn directly to Aristotle and see to what extent Thomas follows him closely or modifies his views.[50]

*

At the end of the section of *De generatione et corruptione* in which we are most interested, Aristotle is also concerned to account for a significant feature of the individuality within members of a species, in particular, the different lengths of their natural lives. He calls them "periods," which by nature—that is, with respect to the structure of their specific forms—are the same for any and every member of a given species. Aristotle first states this in general terms:

> Corruption and generation according to nature (*secundum naturam*) have an equal time period. Therefore both the time (*tempus*) and life (*vita*) of each thing have a number, and they are determined by this. There is an order of all things,[51] and every time and life are measured by a period (*omne tempus et vita mensuratur periodo*), but not all of them are the

[48] I will thus first review Aristotle's doctrine within the structure of his own argument, and then Albert's transformation of it, in the manner of Helen S. Lang's interpretation of Aristotle's *Physics* in her insightful *Aristotle's Physics and its Medieval Varieties*, Albany: SUNY Press, 1992.

[49] See (e.g.) William A. Wallace, *Causality and Scientific Explanation*, 2 vols., Ann Arbor: University of Michigan Press, 1972–74, I:13 ff.

[50] I only treat the directly relevant passages of *De generatione et corruptione* II.9–10, and, even at that, not as thoroughly as the material warrants. A full comparison would be very worthwhile.

[51] The theme of order was also important, we will recall, in the *Speculum astronomiae* and *De operationibus occultis naturae*.

> same: some are measured by a smaller period, and some by a greater. A year is the measure of the period for some, a longer period is the measure for some, a lesser period for others.
>
> These things appear to be so, moreover, and are affirmed by our sense experience (*secundum sensum*). For we see that when the sun comes closer (*adveniens* [sc. from winter to summer solstice]), there is generation, and when it recedes (*recedens* [from summer to winter solstice]), there is a diminishing, and both have an equal time, for there is an equal time for generation and corruption according to nature. But it often happens that something passes away in a shorter time because of its internal confusion [sc. the confusion of elements—and their qualities—in the mixture] (*ad invicem confusionem*).[52] For, since matter is unequal [sc. unbalanced] (*inaequalis*) and not everywhere the same, it is necessary that generations be unequal, that some be faster and some slower. Wherefore, it also comes about that the generation of some things generates corruption for others.[53]

As we can see, Aristotle uses the sun here—and its variation in coming closer and moving away during the course of the year[54]—as central to his analysis of generation and corruption. He uses the inequalities of matter, however, to account for the variety in the different actual periods of life for each individual member of a species, which would otherwise all be the same according to nature. Thus, the same three causal elements are present in both Thomas's and Aristotle's accounts, namely, formal, material, and efficient causes.

*

Aristotle begins *De generatione et corruptione* II.9 with the explicit intention of accounting for the principles of generation and corruption in general before proceeding to discuss it in individual cases, that is, the generation and corruption of actual natural things (*res naturales*), mainly in the biological works and

---

52 This seems to be the sense from the Greek: "pros allela sunkrasin." Joachim considers this phrase spurious, see his note *ad loc*. Regardless, it was considered genuine by medieval natural philosophers and appears in the Latin translation.

53 "Et in equali tempore et corruptio et generatio que secundum naturam. Ideoque et tempus et vita uniuscuiusque numerum habet, et hoc determinantur. Omnium est ordo, et omne tempus et vita mensuratur periodo, sed tamen non eadem omnes, sed hii quidem minori, hii autem maiore. His quidem enim annus, his autem maior, aliis autem minor periodos est mensura. Apparent autem et secundum sensum confessa his que a nobis dicta sunt. Videmus enim quoniam adveniente quidem sole generatio est, recedente autem diminutio, et in equali tempore alterutrum: equale enim tempus corruptionis et generationis que secundum naturam. Sed contingit multotiens in minori corrumpi propter ad invicem confusionem; inequali enim ente materia et non ubique eadem, necesse et generationes inequales esse et has quidem citiores, has autem tardiores; quocirca accidit propter horum generationem aliis generari corruptionem (75, 9–24; 336b10–23)." I cite the Latin translation of Aristotle from the Greek known as the *Translatio vetustior*; it is printed at the bottom of the page in Hossfeld's edition. I base my translation on it, although I turn to the Greek for clarification. The first set of line numbers are for the Latin in the Editio Coloniensis; the second for the Greek in Joachim's edition.

54 That is, moving closer to people in middle north latitudes on its annual path toward summer solstice, by definition the sun's northernmost point on the ecliptic during the course of the year; likewise, the opposite for winter solstice, when the sun moves furthest away. This orientation is, of course, flipped in the Southern Hemisphere.

De anima.[55] Aristotle immediately frames his discussion in terms of form and matter, and states that something else is required, since these two causal factors in themselves do not suffice to account for generation (*non enim sufficientia ad generandum duo*).[56] Something else is required, which does not itself exist among what is eternal and primary, where matter and form exist.

Aristotle then goes on to criticize those who hold both views, that forms alone are sufficient to account for generation, and that matter alone is sufficient (72, 15–74, 2; 335b7–336a12). Although Aristotle does not explicitly state that each is lacking an efficient cause, it is clear from the examples he uses in his criticism that this is in fact the case. Criticizing those who hold that form alone is sufficient, Aristotle cites the examples of health and learning:

> And further, in certain [sc. situations] we see that there is another cause (*alia causa*, "allo to aition" [Gk]); for a doctor makes (*facit*) health and a learned person, knowledge, even though both (1) health itself and knowledge, and (2) what is able to participate them both exist.[57]

(1) and (2) are the formal and material causes respectively, but without the *faciens* (= efficient cause), neither a healthy nor a learned person comes to be.

The other example comes from Aristotle's criticism of those who say that the material cause is sufficient:

> And further, they are doing the same thing as if someone were to attribute the cause of those things which are generated to a saw or any instrument. For it is necessary that there be a sawer for something to be divided, and a planer for something to be smoothed.[58]

Thus, although Aristotle mentioned only matter to account for the individuality of different periods of life in II.10, he quite clearly here in II.9, in his general discussion of the causes of generation, emphasises the necessity of the efficient cause.

To begin II.10, Aristotle immediately discusses motion in place (*allatio*, "phora" [Gk]) in relation to generation, in what appears to be a jarring non-sequitur. As we will see, this is not the case at all, and, although Aristotle is not as explicit as we might like in making the connection, his discussion of the three causes dovetails rather elegantly into his discussion, also in II.10, of another of his most characteristic doctrines—which is itself embedded in yet another—all three of which are central to medieval Aristotelian natural philosophy.

---

[55] This relates directly to the question, where in the system of Aristotle's works does *De generatione et corruptione* fall. Albert provides an influential response to this question at the very beginning of his commentary (109, 8–28).

[56] The language and content is very similar to *De operationibus occultis naturae* 9–10, where Thomas begins with forms and says: *Sed hoc principium non potest sufficere* (138). Thomas, however, does not explicitly criticize those who hold the view that matter itself is sufficient.

[57] Amplius autem in quibusdam videmus aliam causam entem: sanitatem enim medicus facit et doctrinam doctor, ente et sanitate et ipsa et doctrina et participalibus; similiter vero et in aliis secundum potentiam operatis (73, 4–7; 335b20–24).

[58] Amplius vero simile faciunt si quis serre et unicuique instrumentorum attribuat causam eorum que generantur. Necesse enim serrante dividi et incidente coequari, et in aliis similiter (73, 23–74, 1; 336a7–11).

Aristotle begins II.10 by stating that motion in place (*motio secundum allationem,* "he kata ten phoran kinesis" [Gk]) is perpetual, and that it is necessary for generation to be perpetual for the things that exist: "for *allatio*," he states, "will effect generation (*faciet generationem*) continuously because it provides the *generans*."[59] The language and content here make it clear that *allatio* is the efficient cause not explicitly discussed above.

Aristotle then demonstrates that the ontologically first of the changes in nature ([Gk]"metabole," *transmutatio*) is *allatio*, not *generatio*,[60] because it is more reasonable to assume that something that exists is ontologically prior to something that does not yet exist, and therefore that *allatio* is the cause ([Gk] "aitia," 336a26) of *generatio*. His next move is to discuss more fully the nature of the particular *allatio* which is the direct cause of generation and corruption. It is not the first movement, the [Gk] "prote phora," but rather that movement in accordance with the oblique circle ("kata ten loxon kuklon" [Gk], *circa obliquum circulum*) because it provides the two necessary motions (*motus*, "kineseis"[Gk]), namely, continuous and contrary.[61] This *circulus obliquus* will be discussed more fully below.

1.5 Deeper Patterns in Aristotle's Natural Philosophy: The Four Types of *Motus*/"kinesis" and the Ten Categories

To understand the deeper patterns of Aristotle's thought here—which to modern ways of thinking seem dubious at best, if not utterly bizarre—we need to place his analysis in its proper natural philosophical context. Although Aristotle does not explicitly refer to his *Physica* here, we find the proper background there for interpreting the otherwise curious statement at the beginning of this passage, that [Gk] "phora" (movement in place) is the first of the "kineseis" [Gk] or "metaboloi" [Gk] (motions, changes, transformations) in nature. In *Physics* III.1–3,[62] Aristotle

[59] "Amplius vero quoniam que secundum allationem motio demonstratum est quoniam perpetua, necesse his entibus et generationem continue: allatio enim faciet generationem continue, quia adducit generans (74, 5–8; 336a15–19)." The Greek is fuller as to how the *motio secundum allationem* effects generation because of the moving towards and moving away of the *generans*; [Gk] "to gennetikon, he gar phora poiesei ten genesin endelexon dia to prosagein kai apagein to gennetikon (336a16–18)."

[60] "[...] primam transmutationum allationem, sed non generationem dicendum" translates "to proten ten metabalon ten phoran alla me ten genesin eipein [Gk]."

[61] Ideoque non prima allatio causa generationis est et corruptionis, sed que circa obliquum circulum. In hac enim et continuum inest et moveri duos motus (74, 21–23 [336a31–34]).

[62] This topic is discussed further at *Physics* V.1-2, and at *Categories* 14. A proper understanding of these passages is essential for understanding the Aristotelian foundations of medieval views on the intension and remission of forms and qualities. In his translation-commentary on the *Physics*, Albertus Magnus has something of interest to say for the history of the subject explicitly concerning the intension and remission of qualities in V.1.8 (419, 51–86); *Opera omnia* 4.2, *Physica Pars*

discusses motion ("kinesis") in general.[63] This is utterly central to his view of nature overall because, as is well known, Aristotle defines the nature of a thing, its [Gk] "physis," as precisely its [Gk] "arche kineseos kai metaboles," its principle of motion and change.[64]

But, unlike what a modern would expect based on our understanding of the term "motion," "kinesis" does not refer only to motion in place, locomotion, which is the *allatio* ("phora" [Gk]) discussed above. It also refers to (1) generation and corruption ("genesis" [Gk] and "phthora [Gk]," *generatio* and *corruptio*), (2) growth and decay ("auxesis" [Gk] and "phthisis" [Gk]), and (3) change of quality ("alloiosis" [Gk]).[65] These refer respectively to change encountered in each of the following three Aristotelian categories respectively: (1) substance ("ousia [Gk]," *substantia*), (2) quantity ("to poson" [Gk]), and (3) quality ("to poion" [Gk]).[66] Together with "place," these are the only four of Aristotle's ten categories that admit of *motus*/"kinesis," as Aristotle discusses at *Physica* III.1 (200b33–201a3). But generation and corruption present a special case, in that their change is a movement from a state of non-being into a state of being (generation, coming-to-be) and vice versa, from a state of being to one of not-being (corruption, passing-away).[67] This is quite different from the other three, which do not express an ontological change of state, but rather a change of degree, whether in location, quantity, or quality.[68]

Thus, in *De generatione et corruptione* II.9–10, Aristotle is concerned to relate motion in place, that is, of the sun in particular (the *generans* par excellence, as we will see), to generation and corruption in general. He does not discuss the other two categories—quantity and quality—at all there. In fact, however, the physical instantiations of these last two categories (quantity and quality) depend for their very existence on what is generated, namely, what is called primary substance ("prote ousia" [Gk]) in *Categories* 4 and 5: primary substance in a physical sense, for example, an individual person or dog, not primary substance in a metaphysical sense, which are the forms themselves, as Aristotle discusses in the difficult passages

---

II, Paul Hossfeld (ed.), 1993. This fact corrects Anneliese Maier's bald statement to the contrary; "Das Problem," 20 (n. 12): "Die Kommentare Alberts zur Topik und zur Physik erörtern unser Problem nicht ausdrücklich."

[63] I sometimes translate it as *motion*, sometimes as *change*. This is also reflected in the Latin translations, here *transmutatio*, often *motus*. Here is one of the greatest and most characteristic conceptual differences between the older Aristotelian-Ptolemaic world-view and the modern, namely, their profoundly different conceptions of what each calls "motion."

[64] [Gk] "Epei d' he physis men estin arche kineseos kai metaboles, he de methodos hemin peri physeos esti, dei me lanthanein ti esti kinesis (200b12)."

[65] 201a11–16

[66] This doctrine of the ten categories is the third central pattern of Aristotelian natural philosophy referred to above, along with the four causes, and these four types of change.

[67] V.1.

[68] Once again—whatever appearances to the contrary—very little of this is simple or straightforward.

of *Metaphysics* Z and H.[69] But let us turn back now to the discussion at hand in *De generatione et corruptione.*

*

The confluence of the three causes sufficient for generation and corruption—form, matter and the efficient cause—thus jointly allows for the possibility of generation and corruption in general. This is, after all, precisely what Aristotle set out to account for at the beginning of II.9. But for the *actual* generation of, say, a human being, there needs to be an actually existing man, who provides the form in his seed (the formal cause); an actually existing woman, who provides the matter, her womb, her egg (the material cause);[70] and the sun in the world, without which there would be no life on earth at all, is the *generans*, the moving (efficient) cause which provides for the life of both parents, allowing both their coming-into-being in the first place and sustaining their lives while they live.

These three causes, then, may jointly cause something to come into being (*generatio*). This *generatio* is thus the second, ontologically derivative *motus* in the system after *allatio* ("phora" [Gk]), the efficient cause, that occurs with the necessary assistance of form and matter. This *motus* (*generatio*) also has a two-fold movement, coming-into-being and passing-away, which corresponds directly to the two-fold motion of the sun on the *circulus obliquus* (the inclined circle, that is, the ecliptic), moving north and then moving south, or "ascending" and "descending" as Albert calls it just below. This is as far as Aristotle takes the relationship between the movement of the efficient cause, the sun here, and those things that come-to-be and pass-away, the *res naturales*.

As we can now see, Thomas thoroughly embraced Aristotle's basic structures while, at the same time, modifying and developing them in *De operationibus occultis naturae* 16, where a generated thing's particular *virtus* and *actio* could be affected not only by a certain disposition of the matter—which he describes there in no further depth[71]—but also by the particular placement of the celestial bodies, now expanded from only the sun in Aristotle. Thomas also describes these modifications in more positive terms than Aristotle, who basically only says that something can pass away before its natural time (*in minore corrumpi*). By contrast, Thomas's analysis may be used to account for all variation among individuals, both positive and negative. Further, Thomas states his position quite simply, leaving many questions unanswered. This serves to leave his structure wide open, and allows it to

---

[69] The interested reader may find various works by Alan Code of interest in these matters.

[70] See Montgomery Furth, *Substance, Form and Psyche: An Aristotelian Metaphysics*, Cambridge: Cambridge University Press, 1988, 111 ff. and *passim* for an extremely interesting discussion. See also various essays in *Philosophical Issues in Aristotle's Biology,* Allan Gotthelf and James G. Lennox (eds.), Cambridge: Cambridge University Press, 1987, and *Biologie, Logique et Metaphysique Chez Aristote,* Daniel Devereux and Pierre Pellegrin (eds.), Paris: Editions du CNRS, 1990.

[71] Albert does, as we will see below.

be used in various ways as needed. Nevertheless, Thomas himself does not appear to have embraced the rather obvious astrological opportunities offered here.[72]

1.6 Albertus Magnus's Translation-Commentary on *De generatione et corruptione*

In what sense if any, then, can either of these works—Thomas Aquinas's *De operationibus occultis naturae* or Aristotle's *De generatione et corruptione*—be called astrological? At most, it seems, they may be considered as natural philosophical works which, by their placement of the sun and/or planets in such a central position, allow the basic tenet of astrology—celestial influences on the terrestrial world—to have a solid basis in the cosmic structure. This holds true even though neither of these works has anything to say directly about the practice of astrology.

Edward Grant's broader interpretation of the relationship of astrology to natural philosophy may thus be correct when he states strongly [1] that the astrologers used these natural philosophical models as the foundation of their belief in celestial influences, but without themselves contributing anything of value to their natural philosophical foundations. He also asserts, on the other hand, [2] that the natural philosophers did not engage with either astrological theory or practice, in our sense:

> If by astrology, however, we mean the prediction of natural events and human behavior on the basis of knowledge of alleged powers inherent in individual celestial bodies and their positions in the heavens, as well as their manifold configurations and interrelationships, then astrology plays very little role in scholastic natural philosophy. Commentaries and questions on Aristotelian treatises rarely contained specifically astrological discussions or predictions. Such discussions would have been deemed irrelevant, for which reason they will be inconspicuous in this account.[73]

For Grant, then, prediction is the key to describing something as being truly astrological.

To test this claim, we will examine Albertus Magnus's translation-commentary on Aristotle's *De generatione et corruptione*, since Albert is both a pre-eminent example of a scholastic natural philosopher and also someone who appears to have had a genuine interest in astrology. I will also use this opportunity to begin to compare the explicitly astrological views of the *Speculum astronomiae* with what Albert has to say here, in an undoubtedly authentic work, as a partial test of the *Speculum*'s authenticity.[74] Albert composed his translation-commentary on *De*

[72] In Chap. 5, however, we will see Thomas claim explicitly that astrology offers a legitimate mode of knowledge.

[73] Grant does not say much about astrology in his very large book on medieval cosmology, *Planets, Stars and Orbs*; see 35–36 (with n. 66) and 569, from which this quote is taken.

[74] On the authenticity of Albert's translation-commentary on *De generatione et corruptione*, see Hossfeld's prolegomena to his edition thereof, p. VI, ll. 61–75.

generatione et corruptione around a decade before the *Speculum* was most likely composed.[75] I will reconstruct Albert's position in four stages.[76]

In the first section (on Albert's chapter II.3.4), I will present Aristotle's ideas along with Albert's commentary. In this way, we can clearly see what Aristotle thought and how Albert did or did not transform it. In fact, the Latin text of Albert's commentary in the *Editio Coloniensis* is typeset to show the precise relationship between it and the Latin translation of Aristotle by italicizing the words in the commentary that Albert drew directly from the translation. To capture this in my translation, I put Albert's additions and explanations in brackets "{ }".[77]

1.6.1 Stage 1: The Sun's Role in Generation and Corruption[78]

In chapters II.3.1–5 of his translation-commentary, Albert explains Aristotle's analysis of the sun's essential role in generation from *De generatione et corruptione* II.10. In Chaps. 1–4, Albert presents Aristotle's text and explains what is obscure, often by filling out elliptical phrases in the original. By contrast, Chap. 5 is an explicit "digression," where Albert speaks in his own voice in explaining Aristotle's often challenging text.

In Chap. 4, "On the true opinion concerning the efficient cause of generation" (*De sententia vera de causa efficiente generationis*), Albert describes Aristotle's central conceptual structure, namely, the two-fold movement of the efficient or moving cause, and thus provides the material for the first stage in my reconstruction.

[75] Hossfeld dates it to 1251–4, while Albert was teaching at Cologne (p. V). It goes without saying that Albert's translation-commentary falls well within the boundaries of an Aristotelian commentary, Grant's primary subject.

[76] There is much of interest on Albert more generally in *A Companion to Albert the Great*, and much additional bibliography. For more on the astrological dimension of Albert's work, including his views on celestial influences, see Henryk Anzulewicz, "*Fatum*: Das Phänomen des Schicksals und die Freiheit des Menschen nach Albertus Magnus," *Miscellanea Mediaevalia* 28 (2001): 507–34, and his "Der Einfluss der Gestirne auf die sublunare Welt und die menschliche Willensfreiheit nach Albertus Magnus," in *Actes de la Vème Conférence Annuelle de la SEAC, Gdansk 1997*, T. Mikocki (ed.), Warsaw: Institute of Archaeology, 1999, 263–77.

[77] For Aristotle's Greek text, with commentary and introduction, I use *Aristotle On Coming-To-Be and Passing-Away*, whose numbering of the text I follow here. Joachim's commentary is very useful for elucidating the challenging features of this rich but difficult text and its relation to Aristotle's other works. I have found Forster's translation for the Loeb Classical Library series to be very helpful; Aristotle, *On Sophistical Refutations, On Coming-To-Be and Passing-Away, On the Cosmos*, whose translation I cite below. In addition to the Latin text of Albert's commentary, the *Editio Coloniensis* also publishes the Latin translation of Aristotle (the *translatio vetustior*) on which Albert primarily drew, namely, the Greco-Latin translation by Henricus Aristippus (d. 1162). Apparently Albert also knew the Arabo-Latin version by Gerard of Cremona (before 1187); *Albertus Magnus and the Sciences*, 567, and Albertus Magnus, *De generatione et corruptione*, p. x. In quoting Albert's text, I follow the practice of the *Editio Coloniensis*.

[78] My profound thanks to Katharine Park for helping me clarify the structure here. For some of this material, see my "Astrology and Magic," in *A Companion to Albert the Great*, 451–505.

Concerning heavenly motion, Aristotle distinguishes two complementary but different types (336a31–336b10). The first motion (*prima allatio*) is the first motion of the entire cosmos and accounts for the existence of all motions on earth, and for the fact that the cycles of generation and corruption are never-ending.

Being singular, however, the uniform *daily* motion cannot be the cause of both generation and corruption because contrary processes require contrary causes.[79] Rather, the sun's two-fold *annual* motion of "approaching" and "receding" along the inclined circle (the ecliptic) provides the two contrary efficient causes, as we also saw in Thomas's *De operationibus occultis naturae*.[80] We also learn that, by their nature, generation and corruption—both in general and for particular entitites—take place over the same period of time.

In his exposition (204, 54–205, 17), Albert follows Aristotle closely in describing this general structure, but then modifies it significantly as he explains the nature of the causality more precisely. The entire passage is Albert speaking in his own voice:

> Nevertheless, the motion of one part of a circle is not contrary to the different motion of another part of a circle per se,[81] but it is contrary with respect to the relationship of a ray to the place of generation (*sed secundum respectum radii ad locum generationis*)[82] because, when it approaches, it encounters the place of generation with a straighter ray (*rectior radius*) and therefore warms more (*plus calefacit*), and when it recedes, it encounters it with an oblique ray, and thence a killing cold (*frigus mortificans*) dominates.[83]

For Albert, generation and corruption take place not by the two-fold annual motions in themselves (*qua* local motion alone),[84] but by rays projected in the course of its different motions to the places of generation. The geometrical-optical model of celestial influences adumbrated here—and discussed more fully in my Chap. 2—is a significant addition to Aristotle's analysis.

Albert completes his explication of this passage in Aristotle by explaining why the time of generation and corruption are the same by nature. In so doing, Albert clearly identifies the general motions (and their implied rays) as being from the sun:

> {And because the motion of the sun in the oblique circle is the cause of generation and corruption}, therefore the time of generation is equal to the time of corruption {and vice

[79] We will see the daily motion's role more clearly below in discussing Albert's *De generatione et corruptione*, II.3.5.

[80] Albert discusses the two types of motion more fully at *De generatione et corruptione*, 205, 78–206, 29.

[81] That is, with respect to the motions themselves.

[82] The centrality of the "place" of generation will be discussed further below, especially in Chap. 2. See also Albert's discussion at *De generatione et corruptione*, 205, 49–58.

[83] "Non tamen est motus unius medietatis circuli contrarius alteri motui alterius medietatis per se, sed secundum respectum radii ad locum generationis, quia cum accedit, rectiori respicit radio locum generationis et ideo plus calefacit, et cum recedit, respicit obliquiori, et tunc dominatur frigus mortificans (205, 17–23)." Albert also points to this precise distinction in a very interesting digression at *Physica* 1.1.4 (ll. 64–74).

[84] Joachim (253–54, relying on Heath) explains the basics of Aristotle's cosmology, namely, the complex mechanical rotations of the system of homocentric spheres, where Aristotle followed Eudoxus and Callipus.

> versa, because the sun generates when ascending through the six signs which are from the beginning of Capricorn to the beginning of Cancer, and, in descending through the six signs, which are from the beginning of Cancer to the beginning of Capricorn, it corrupts; and these movements are between the two solstitial tropics} (205, 23–30).[85]

The two-fold nature of the sun's annual motion along the ecliptic is an essential cause of generation and corruption. Due to the obliquity of the ecliptic, in middle north latitudes the days do indeed get longer (and warmer) and the sun gets higher in the sky as it travels from winter to summer solstice, and thus increasingly approaches a place's meridian, and the days get shorter (and colder) and the sun gets increasingly lower in the sky as the sun moves back (recedes) from summer to winter solstice. For Aristotle, the approaching and increasing phase of the sun's annual motion corresponds to "generation," and the receding phase to "corruption."

Albert enhances this fundamentally empirical conception and makes it more precise by specifying the sun's location within the zodiac during both the approaching and receding phases, and by adding a geometrical-optical model of celestial influence. These passages are thus no more "astrological" than Thomas's and Aristotle's, although they do more explicitly use a significant part of astrology's language, namely, the signs of the zodiac. But this is no more astrological than astronomical, and would have been taught as such at an elementary level in every medieval undergraduate's university education, in particular, from Sacrobosco's *De sphera*, where such language is used to describe the structure of the world, the *machina mundi*, as Sacrobosco calls it.[86]

1.6.2 Stage 2: The Period of Life

In the second stage of my reconstruction, Aristotle clarifies the notion of generation and corruption by analyzing the period of something's life, that is, the time between its generation (birth) and ultimate corruption (death), including the time of growth and decay (336b10–15). Albert's interpretation develops the celestial dimension implicit in Aristotle, and thus moves us closer to an explicitly astrological view:

---

[85] Et quia motus solis in circulo obliquo causa est generationis et corruptionis, ideo tempus *generationis* est *aequale tempori corruptionis* et e converso, quia sol ascendendo per sex signa generat, quae sunt a principio Capricorni usque ad initium Cancri, et per sex signa descendendo corrumpit, quae sunt ab initio Cancri usque ad principium Capricorni, et isti motus sunt inter duos tropicos solstitiales (205, 23–30).

[86] For the teaching of astronomy as a part of the mathematics course, the quadrivium, see Part IV below. For the text and some commentaries on the *Sphere*, see Lynn Thorndike, *The Sphere of Sacrobosco and its Commentators,* Chicago: University of Chicago Press, 1949. Commentaries on the *Sphere*, furthermore, are another of the genres (along with Aristotelian commentaries) that Grant privileges, but he does not discuss their astrological content at all; *Planets, Stars and Orbs*, 33–4. This is particularly problematic for his account, since two of the *Sphere* commentaries he uses, Robertus Anglicus's and Cecco d'Ascoli's (both *passim*), have rather a high proportion of astrological content. Both texts are found in *The Sphere of Sacrobosco*; Robertus at 143–198 (Latin), 199–246 (Thorndike's English translation); Cecco, 344–411 (Latin only).

> {And because the generation and corruption of things below (inferiora) are caused in this way by a superior motion in the oblique circle}, therefore every time (*tempus*), {which is in a temporal thing}, and every life (*vita*) of each living thing has a number {from the celestial circle} and is determined {by the celestial circle itself, because from that circle is considered to what extent the power of the generating entity (virtus generantis) extends itself, in accordance with which it produces the existence of the thing (esse rei) before it produces the complete corruption of the thing, as is revealed from the science of the stars (ex scientia astronomica)}. For the order {of the causes} of all {lower things is dependent on the superior order} (*omnium enim* inferiorum rerum *est ordo* causarum pendens ex superiore ordine), and every time, {which is of a thing that arose in the duration of time}, and {every} life is measured by a period (205, 31–42).[87]

With this passage and its reminiscences of another astrologizing Aristotelian locus (namely, Aristotle's *Meteorologica* I.2), we find ourselves within an account of the cosmos where the causal structure of the lower world (*inferiora*) depends ontologically on that of the higher (*superiora*). Albert claims here that the science of the stars can play a central role in understanding the essential causal processes of generation and corruption, and the particular patterns and timing of each thing's existence. This is so because the power of the generating force to influence life on earth is derived from the celestial circle. We will return to a living thing's life (*vita*) and existence (*esse*), and the order (*ordo*) of its causes below.[88]

Albert then gestures to a crucial factor that was present but less fully articulated in Aristotle, namely, the role of matter in a full causal analysis. Here he explores why periods of life are not always of the same duration:

> Nevertheless, not everything is measured by the same period, but some have a larger, and some have a smaller one. The period of some is a year, but of others it is less, and of others it is more, {insofar as they are more susceptible and retentive of the power (*virtus*) which flowed into them from the celestial circle (secundum quod magis susceptibilia et retentiva sunt virtutis sibi ex circulo caelesti influxae)}.[89]

The variable receptivity of celestial influences thus significantly conditions the period of life in general, about which we will learn more below.

Although this interpretation is solidly grounded in his reading of Aristotle, Albert characteristically sharpens the astrologizing implications, as we can readily see.

[87] "Et quia ita inferiorum generatio et corruptio causantur a superiore allatione in circulo obliquo, *ideo* omne *tempus*, quod est in re temporali, *et* omnis *vita uniuscuiusque* vivi *habet numerum* ex circulo caelesti *et* ipso circulo caelesti *determinatur*, quia ex ipso circulo consideratur, usque quantum se extendit virtus generantis, secundum quod affert esse rei, antequam afferat perfectam rei corruptionem, sicut patet ex scientia astronomica. *Omnium enim* inferiorum rerum *est ordo* causarum pendens ex superiore ordine, *et omne tempus*, quod est rei exortae in tempore durativo, *et* omnis *vita periodo mensurantur* (205, 31–42)." This is precisely the argument Pico attacks as the first of the astrologers' strongest arguments in *Disputationes* book III, to be examined in volume II.

[88] Especially in Chap. 5, where I discuss Albert's *De fato*, in which he uses precisely this conceptual structure to explicate a fundamental component of his views on fate.

[89] *Sed tamen non omnia* mensurantur *eadem periodo*, *sed haec* habent *maiorem* et *haec* habent *minorem. Horum* enim periodus est *annus*, *horum autem minor et* horum *maior*, secundum quod magis susceptibilia et retentiva sunt virtutis sibi ex circulo caelesti influxae (205, 44–48).

Further, this passage, especially at the end, could be a gloss on what Thomas said in the *De operationibus occultis naturae* about the disposition of the matter, where he closely associated variations in the intention and remission of a person's *virtus* and *operatio* with the variable reception of planetary influences. With these passages, we still find ourselves in the same domain of inquiry with Aristotle and Thomas, namely, in the world of an astrologizing Aristotelian natural philosophy, but Albert has not yet taken an overtly astrological turn.

*

To support his present claims, Albert continues in the same chapter with evidence from the senses, as did Aristotle:

> Moreover, our sense perception bears witness that what we said is true because we see with our eyes that while the sun approaches the [sc. first] degree (punctum)[90] of Aries, when it touches our place (locus noster) with a straight ray (directo radio), generation begins to exist among what is born on the earth. But while the sun is receding, from the beginning of Libra, the diminution and corruption of things especially begins. And each time is equal because the time of generation, which is in accordance with nature, is equal to the time of corruption in accordance with nature because the ascent of the sun is just as great as its descent, as has been said (205, 49–58).[91]

This passage is quite similar to those in Thomas and Aristotle already discussed, except for the astrological "ray" physics,[92] and for the fact that Albert explicitly identified the *obliquus circulus,* on which the sun comes forward and recedes, as the wheel of the zodiac.

In accounting for the different actual periods of existing things, however, Albert emphasizes the role of the material cause. Here too Aristotle is quite explicit (336b20–23):

> It often happens that things pass away in too short a time owing to the commingling of things with one another (*ad invicem confusionem*). For, their matter being irregular (*inaequalis*) and not everywhere the same, their comings-to-be must also be irregular, sometimes too quick and sometimes too slow. Wherefore, it happens that due to the generation of these things corruption is generated for others.[93]

---

[90] We will see several instances of the Latin term '*punctum*' (or the Italian '*punto*') in practical astrological contexts in Part IV below.

[91] Quod *autem* vera sint, *quae dicta sunt*, testificatur *sensus*, quia nos *videmus* ad oculum, quod *sole adveniente* ad punctum Arietis, quando directo radio locum nostrum tangit, *generatio esse* incipit in terraenascentibus, *recedente autem sole* a principio Librae maxime incipit rerum *diminutio* et *corruptio. Et tempus utrumque* est *aequale*, quod *tempus generationis*, quod est *secundum naturam, aequale* est tempori corruptionis secundum naturam, quia tantus est ascensus solis, quantus est descensus, ut DICTUM EST (205, 49–58).

[92] This important feature of Albert's astrologizing natural philosophy—and of Roger Bacon's—is discussed in detail in Chap. 2.

[93] Sed contingit multotiens in minori corrumpi propter ad invicem confusionem; inaequali enim ente materia et non ubique eadem necesse est generationes inaequales esse, et has quidem citiores, has autem tardiores, quocirca accidit propter horum generationem aliis generari corruptionem (204, 71–73).

In interpreting this passage from Aristotle, Albert clarifies the astrologizing dimension of his own analysis:

> But if it so happens that at some time some thing becomes corrupted in less time, {that is, before the celestial generative force (vis generationis caelestis)—according to the structure of its period (ratio periodi)—recedes, this happens *per accidens*} because of the confusion {of its matter from some cause} because, when the matter is unbalanced (*inaequalis*) {and not able to receive or retain the *virtus caelestis*}, and does not exist in the same manner and in a balanced way {during its period}, then the generations and corruptions would necessarily be unequal; some will be faster and some slower.[94]

Albert here further articulates Aristotle's view that a confusion of the matter can limit the natural length of the period by describing what results from such a confusion: the matter is then unable to fully receive or retain the celestial force or power (*vis* or *virtus caelestis*).[95] We have seen so far that Albert consistently interprets Aristotle's more general formulations in a way that more fully astrologizes them by linking the causal structures of generation and corruption directly and precisely to the influences of the heavens. We are thus still well within the bounds of astrologizing natural philosophy without having yet crossed over into Grant's definition of astrology properly so-called.

*

We can more clearly see Albert's modification of Aristotle in a strongly astrological direction in *De generatione et corruptione* II.3.5—a digression added by Albert that is wholly in his own voice—where he explains more fully what Aristotle had just said about periods.[96] Albert's two main modifications provide the next two stages of my reconstruction, namely, he expands the role of the efficient cause from the sun alone to include all the planets, and he relates Aristotle's views more fully to fundamental astrological structures, in particular, the horoscope itself. This chapter is worthy of detailed study, both for its own intrinsic interest bearing on our subject, but also because of its virtually complete neglect in recent studies of celestial influences.[97]

To open the chapter, Albert begins with a brief but significant and deeply astrological definition of "period": "A period is a measure which is impressed on

[94] Si autem *contingit* aliquando *corrumpi* aliquam rem *in minori* tempore, hoc est, antequam vis generationis caelestis secundum rationem periodi recedat, hoc contingit per accidens *propter confusionem* materiae ex aliqua causa, quia cum *materia* est *inequalis* et non susceptibilis vel retentibilis virtutis caelestis *et non* est *eodem modo et aequaliter* se habens in periodo, tunc etiam *necesse est*, quod *generationes* et corruptiones sint *inaequales, et hae* erunt *citiores* et illae *tardiores* (205, 58–67).

[95] We will see this analysis also with the making of talismans in Part III below, and in volume II.

[96] Cap. 5. Et est digressio declarans ea quae sunt dicta de periodo (205, 70–71).

[97] There is complete neglect in Grant, *Planets, Stars and Orbs*; North, "Celestial Influence"; Lemay, "True Place"; Price, "Physical Astronomy and Astrology"; and Vanden Broecke, *Limits of Influence.* There is a brief mention in Zambelli, *The* Speculum astronomiae, 62, but with no discussion.

(*imprimitur*) or flows into (*influitur*) a thing from the celestial circle [sc. the zodiac], a thing caused by the [sc. celestial] circle in the lower realm (*rei causatae a circulo in inferioribus*)."[98] Thus, when anything is generated in the world, an essential part of its causal profile—namely, the efficient cause—derives from the celestial circle.[99] In this process of generation, its period—the measure of its life and/or existence—is also impressed upon it from the celestial circle. How this works in detail is discussed further below.

1.6.3 Stage 3: Expansion of the Efficient Cause

In the third stage, Albert further articulates the realm of the efficient cause, expanding it from the sun in Aristotle to include all the planets. So far, we have mainly discussed celestial influences in terms of motion; light will now receive more attention. Albert begins by distinguishing the two main types of motion within the celestial circles, beginning with a general statement:

> Moreover, although the celestial circle is multiplex, nevertheless, there are only two types of motions thereof (*duo motus ipsorum in genere*), namely, *motus planes* and *motus aplanes*, as Aristotle says in the 11th book of the *Metaphysics*, and as it is received in the proper philosophy of the *astronomi*.[100]

It is unclear whether Albert is referring here to astrologers or astronomers, but given the subject at hand, it is more likely to be to astronomers and their concern with celestial motions.

Albert then discusses the first motion:

> Moreover, the *motus aplanes* or *ordinatus* is the *motus* of the *primum mobile*,[101] which is called diurnal; [sc. the movement] is from east to west and is equidistant over the poles of

[98] Est autem periodus mensura, quae ex circulo caelesti imprimitur vel influitur rei causatae a circulo in inferioribus (205, 76–78).

[99] I will also explore the formal dimensions of this analysis in Chap. 5 in discussing Albert's *De fato*.

[100] "Cum autem multiplex sit caelestis circulus, non tamen sunt nisi duo motus ipsorum in genere, scilicet motus planes et motus aplanes, ut dicit ARISTOTELES in UNDECIMO PRIMAE PHILOSOPHIAE et sicut accipitur in propria philosophia ASTRONOMORUM (205, 78–82)." Hossfeld provides the following references for Aristotle, first, then the *astronomi*: 80–81 Arist., *Metaph.* 1.12 (11) c. 8 (1073b19.25); cf. Alb., *Metaph.* Ed. Colon. t. 16 p. 510 v. 74.76. Alb., *Metaph.* 1.11 tr. 2 c. 22 (l.c. p. 511 v. 94); *De caelo et mundo* 1.1 tr. 3 c. 5 (Ed. Colon. t. 5,1 p. 65 v. 57–61 cum nota 58–59), 1.2 tr. 3 c. 9 (p. 162 v. 64–65), — 82 Cf. Al-Bitruji, *De motibus celorum* I,4 sqq. (ed. F.J. Carmody p. 71–72). Johannes de Sacrobosco *Tractatus de spera* c. 2 (ed. L. Thorndike p. 86 v. 11 sqq.).

[101] Compare *The Sphere of Sacrobosco*, ch. II, 86, 4–16: "Inter circulos vero maiores primo dicendum est de equinoctialis circulus quidam dividens speram in duo equalia secundum quamlibet sui partem equedistans ab utroque polo. Et dicitur equinoctialis, quia, quando sol transit per illum, quod est bis in anno, scilicet in principio Arietis et in principio Libre, est equinoctium in universa terra. Unde appellatur equator diei et noctis, quia adequat diem artificialem nocti, et dicitur circulus primi motus. Unde sciendum quod primus motus dicitur motus primi mobilis, hoc est, spere none

> the earth (*mundus*) and over the circles, of which the greatest is the equinoctial circle. And this *motus* is properly called the *allatio* of that one which bears the *generans* (*affert generans*), that is, the sun; in the second place, the other planets (*alios planetas*); in the third place, all the stars (*omnes stellas*); and in the fourth, the degrees (*gradus*) of the entire orb from rise to setting, which *motus* is the cause of life and existence (*causa vitae et esse*) in certain things, to the extent that the existence and life of these things is not extended beyond the circle of this *motus*. And therefore, certain things are perfected in their existence (*perficiuntur in esse*) in a day, and their corruption is completed in a night, which is clear in the case of the fish (*in pisce*),[102] which is called "day" or "ephemeral," because on the same day it dies and is born, in accordance with its nature.[103]

Concerning this description, I only wish to point out the obvious, that we are here dealing with the same efficient cause, the *generans*, namely, the sun—the cause of life and existence—which Albert places at the first level. Here he differentiates four levels—including the second level with the other planets—which are discussed in more detail below.

Albert then discusses the second motion:

> The other *motus* is erratic (*erraticus*), which *motus* is called *planes*, and is the motion of the circle of the signs upon the poles of the zodiac, which stand apart from the poles of the earth by 23 degrees and 35 minutes.[104] Therefore, the sun and the other stars moved in it (*sol et aliae stellae motae in ipso*) are moved obliquely forward and vice versa, that is, from north to south and vice versa, as the sun is moved from Capricorn to Cancer and from Cancer to Capricorn. Moreover, in these two ways the elements are mixed at the same time toward generation (*simul commiscentur elementa ad generationem*), since while the sun is going toward Capricorn, the lower elements [sc. earth and water] win out over the superior elements, and when it comes towards Cancer, the superior [sc. air and fire] begin to win out over the inferior. In this way also it [sc. the sun] distinguishes the four seasons (*tempora*), which are spring, summer, autumn and winter. For at the two tropics it makes summer and winter, and at the two equinoctial points, namely, Aries and Libra, it makes spring and autumn.[105]

---

sive celi ultimi, qui est ab oriente per occidentem rediens iterum in orientem, qui etiam dicitur motus rationalis ad similitudinem motus rationis qui est in microcosmi, id est, in homine, scilicet quando fit consideratio a creatore per creaturas in creatorem ibi sistendo."

[102] Hossfeld gives a reference to Albert *De animalibus* 1.24 ch. 1 (ed. H. Stadler p. 1531 v. 18–21), 1.1, tr. 1 c. 7 (p. 36 v. 22–29).

[103] Motus autem aplanes sive ordinatus est motus primi mobilis, qui diurnus dicitur, et est ab oriente in occidentem super polos mundi et super circulos aequidistantes, quorum maximus est circulus aequinoctialis. Et hic motus dicitur allatio unius proprie, quod affert generans, hoc est solem, et secundo alios planetas et tertio omnes stellas et quarto gradus totius orbis ab ortu in occidentem, qui motus est causa vitae et esse in quibusdam intantum, quod quorundam esse et vita ultra circulum istius motus non extenditur. Et ideo quaedam perficiuntur in esse in die, et corruptio eorum completur in nocte, sicut patet in pisce, qui dicitur dies vel effimera, quia eodem die naturali moritur et oritur (205, 82–206, 14).

[104] This is Albert's measure of the obliquity of the ecliptic.

[105] Alius autem motus est erraticus, qui motus planes vocatur, et est circuli signorum super polos zodiaci, qui distant a polis mundi per 23 gradus et 35 minuta. Et ideo sol et aliae stellae motae in ipso oblique moventur ab ante in retro et e converso, hoc est ab aquilone in meridiem et e converso, sicut sol movetur a Capricorno in Cancrum et a Cancro in Capricornum. Istis autem duobus modis simul commiscentur elementa ad generationem, quoniam sole eunte versus Capricornum inferiora elementa vincunt super superiora, et quando venit versus Cancrum, incipiunt vincere superiora

So far the descriptions of both celestial motions have been straightforwardly cosmological, much like one finds in Sacrobosco's *De sphera*, with only a brief but important mention of the *motus* in relation to our terrestrial concerns of generation and corruption.[106] There is even a reference to one of the more obvious forms of celestial influence, namely meteorology, relating to the change of seasons.

Now Albert's description starts to heat up as he begins to move in a decidedly astrological direction, but still within the boundaries established by Grant:

> Moreover, it is to be observed that the sun is not the only generating cause (*causa generans*), although it is pre-eminently *generans*, unless it were said that generation comes to be by the light of the sun alone (*solo lumine solis fit generatio*), because, as the finest philosophers—Aristotle and Avicenna, Ptolemy and Messellach[107]—say, only the sun illuminates by its own light, and all the other planets and stars are illuminated by the sun, as also the moon is. But the light is received by them into their depths and imbues them with a different nature than it [sc. the light] has in the sun (*sed lumen recipitur ab eis in profundum sui et induit in eis aliam naturam, quam habeat in sole*).[108] And therefore, in the Moon it [sc. its light] is cold and moist intemperately; in Saturn, cold and dry intemperately; in Jupiter, hot and moist, temperately; and in Mars, hot and dry intemperately; and in Venus, cold and moist temperately. But in Mercury it is mixable (*commiscibile*) because it augments the property of each star it gazes upon. And in the sun, which moves all of them, it is hot and dry temperately. Moreover, all the other fixed stars follow the properties of some one of these. Since this is the case, it is to be noted that in no way does the access and recess of the sun alone make a period because otherwise no animal could be born in the winter, but rather

---

super inferiora. Sic etiam distinguit quattuor tempora, quae sunt ver, aestas, autumnus et hiems. In duobus enim tropicis facit aestatem et hiemem et in duobus aequinoctialibus punctis, scilicet Ariete et Libra, facit ver et autumnum (206,14–29).

[106] Compare *The Sphere of Sacrobosco*, ch. II, 87, 18, on the zodiac: "Est et alius circulus in spera qui intersecat equinoctialem et intersecatur ab eodem in duas partes equales, et una eius medietas declinat versus septentrionem, alia versus austrum. Et dicitur circulus iste zodiacus a *zoe*, quod est vita, quoniam secundum motum planetarum sub illo est omnis vita in rebus inferioribus. Vel dicitur a *zodias*, quod est animal, quia, cum dividatur in 12 partes equales, quelibet pars appellatur signum et nomen habet speciale a nomine alicuius animalis propter proprietatem aliquam convenientem tam ipsi quam animali, vel propter dispositionem stellarum fixarum in illis partibus ad modum huiusmodi animalium. Iste vero circulus latine dicitur signifer, quia fert signa vel quia dividitur in ea. Ab Aristotele vero *in libro de generatione et corruptione* dicitur circulus obliquus, ubi dicit quod secundum accessum et recessum solis in circulo obliquo fiunt generationes et corruptiones."

[107] Avicenna (Abu 'Ali al-Husayn ibn 'Abdallah ibn Sina, d. 1037) spent the greater part of his life serving the rulers of Persia. He is best known as a philosopher and physician. His extremely influential medical *Canon*, and several large parts of his philosophical magnum opus, *as-Sifa*, were translated into Latin in the twelfth and thirteenth centuries (357–64). Messahalah (Masa'allah, d. ca. 815), an early and influential Arabic astrologer, was a Jew from Basra. He was appointed by the caliph to determine the timing for the founding of Baghdad by an astrological election. Few of his numerous works survive in Arabic; they treated all parts of the science of astrology and drew on Sassanian, Greek and Indian theories. At least 15 treatises (or parts thereof) were translated into Latin in the twelfth and thriteenth centuries, some of which were later printed in the Renaissance (389–91). For more on these Arabic astrologers, see Hasse, *Success and Suppression*, who provides further references. The page references given here and elsewhere are to Hasse's valuable study.

[108] Albert discusses this fundamental idea further in *De caelo* II.3.6 and *De causis proprietatum elementorum* II.1.1. Fuller study of Albert's translation-commentary on *De caelo* II.3.1 ff. would be very valuable.

> everything would die, and the life of no living thing could be extended beyond a year, which is patently false.[109]

Albert here characterizes each planet's light in terms of the traditional astrological attribution of a set of primary qualities and their intensity to each planet—following Ptolemy himself and many others[110]—in order to more fully differentiate his natural philosophical picture. I should emphasize that Albert's argument, based on experience, requires the other planets also to have a share in efficient causality, otherwise everything would have the same annual period of generation and corruption by nature, which, as he says, is patently false.

In developing Aristotle, then, Albert removes the sun from its sole position as *generans*, and in the process differentiates the qualities of each planet's light, one of the primary modes of their influence, along with motion. This move is central to astrology's natural philosophical foundations.[111] Traditional astrology and Aristotelian natural philosophy thus appear to reinforce each other here, and so might appear to become dangerously intermingled. Regardless, they still remain on their proper sides of Grant's boundary, since he explicitly permits the attribution of qualities to planets in Aristotle commentaries.

---

[109] Est autem observandum, quod non solus sol est causa generans, licet ipse praecipue sit generans, nisi dicatur, quod solo lumine solis fit generatio, quia, sicut dicunt praecipui philosophi, ARISTOTELES et AVICENNA, PTOLOMEUS et MESSELLACH, solus sol lucet lumine proprio et omnes alii planetae et stellae illustrantur a sole quemadmodum et luna. Sed lumen recipitur ab eis in profundum sui et induit in eis aliam naturam, quam habeat in sole. Et ideo in luna est frigidum et umidum intemperate, in Saturno frigidum et siccum intemperate, in Iove calidum et umidum temperate, in Marte autem calidum et siccum intemperate et in Venere frigidum et umidum temperate. In Mercurio vero est commiscibile, quia auget proprietatem cuiuscumque stellae, quam respicit. Et in sole, qui movet omnes, est calidum et siccum temperate. Omnes autem aliae stellae fixae proprietates alicuius istarum sequuntur. Hoc habito notandum, quod nullatenus solus accessus solis et recessus facit periodum, quia aliter in hieme nullum animal deberet nasci, sed potius omnia mori, et nullius vita deberet extenti ultra annum, quod expresse falsum est (206, 29–51).

[110] See (e.g.) Ptolemy *Tetrabiblos*, 1.4.

[111] This view has a rich history. Among many others, Johannes Kepler embraced it in the seventeenth century in his *De fundamentis astrologiae certioribus* (1602). The text is published in Kepler's *Gesammelte Werke* [Munich, 1937-], vol. 4, 1941. For an English translation, see Judith V. Field, "A Lutheran Astrologer: Johannes Kepler," *Archive for History of Exact Sciences* 31 (1984): 189–272). Ephraim Chambers articulated and defended it in the early eighteenth century in the article, "Astrology," for his influential *Cyclopedia* (London, 1728), I: xxvii and pp. 162–3. We will see much more of this position in what follows. This is also where the limited discussion by Ptolemy of astrology's natural philosophical foundations becomes supplemented by and thereby much more fully articulated with the geometrical-optical model of celestial influences, thus filling out the picture described by Vanden Broecke and Westman. I develop this model in greater depth and detail in Chap. 2 below. The unique nature of each planet's light and thus influence was also one major focus of Pico's attack on astrology's natural philosophical foundations in *Disputations* Book III, as we will see in volume II.

1.6.4 Stage 4: The Period of Life and the Natal Horoscope

Albert now begins to move much more quickly in a properly astrological direction as he further develops his analysis of the period of life and links it explicitly to the structure of the natal horoscope:

> But what makes a period is the relation of the sign ascending over the horizon (*relatio ascendentis signi super horizontem*) to all the other signs of the circle [sc. the zodiac] (*ad omnia alia signa circuli*) with their stars and planets (*cum suis stellis et planetis*) in the hour of conception or birth of the thing below (*in hora conceptus vel nativitatis rei inferioris*), which is caused by or caused together with the celestial circle (*quae causatur vel concausatur a circulo caelesti*).[112]

This is a striking description of the situation reflected in an astrological figure constructed for the time and place of birth, a geniture or nativity (natal horoscope) with the rising sign or ascendent calculated.[113]

Albert continues, further articulating the celestial dynamics of the birth of things on earth:

> For in this manner, the measure of certain things is a year, and of certain others more or less, in accordance with the effects of the signs (*secundum effectus signorum*) and the strength of the stars [i.e. the sun, moon and planets] situated in the signs (*fortitudines stellarum, quae sitae sunt in signis*). And in this manner it is true that the time of the generation of a thing is equal to the time of its corruption because the growth of a thing (*profectus rei*) is calculated from the first sign ascending in the hour of the thing [i.e., of the thing's generation (birth)] up to the seventh sign of the same circle, and the winding down (*defectus*) is computed from the seventh sign to the first. And therefore, the seventh sign in astrology (*in astronomia*) is called the house of death (*domus mortis*), and the ascending sign is called the house of life.[114] And therefore the growth up to the full natural state is called the generation of a thing, and after the *status* to the decline and from the decline to death is called the period of corruption because they are equal according to nature, because from the first to the seventh is the same distance as from the seventh to the first by measuring through the other part of the circle. For example, from Aries through Taurus and Gemini and thence to Libra is the same distance as from Libra through Scorpio and Sagittarius and so thence to Aries. And therefore, if the period of growth (*periodus profectus*) of a human being is 35 or 40 years, as the physicians (*medici*) say, the *periodus defectus* will be the same, and the age of a person will be 70 or 80 years. Nevertheless, this can be impeded *per accidens*, by bad food or a violent death or in some such other manner. Aristotle calls this the inequality [or

---

[112] Sed periodum facit relatio ascendentis signi super horizontem ad omnia alia signa circuli cum suis stellis et planetis in hora conceptus vel nativitatis rei inferioris, quae causatur vel concausatur a circulo caelesti (206, 51–55).

[113] For details on the actual construction of astrological nativities in this period, and for information on the different methods of determining the house cusps, see North's masterly *Horoscopes and History*, and my more general discussion in the excursus to the overall introduction.

[114] The "houses" of life and death refer to the terrestrial houses, and are to be contrasted (not conflated) with the "celestial" houses, which are more commonly called the signs of the zodiac. The eighth (not the seventh) house is normally considered the house of death, being associated, as it is, with the eighth sign of the zodiac, Scorpio, that, among other things, rules death. This indicates the authentic Albert's somewhat loose grasp of practical astrology, which we will also see just below in discussing his commentary on Aristotle's *De caelo*.

> imbalance] of matter (*inaequalitas materiae*), namely, because by many accidents it is disposed otherwise than it would be by the celestial circle [sc. in itself].[115]

Albert thus appears to have grafted Aristotle's doctrine of generation and corruption here onto an overtly and explicitly astrological framework by means of Aristotle's concept of the period of life.

The astrological dimension can no longer be denied or explained away. The simple description of the cosmos is the same for both astronomy and astrology, but Albert's use of the particularities of the structure of a natal horoscope, and the discussion of the different strengths and effects of the planets and signs has pushed us fully into the astrological realm, or at least into a fully astrologizing natural philosophy, where the influences of the planets, not simply their motions (as in astronomy proper) are concerned. This also relates directly to the similar but not explicitly astrological talk in Thomas's *De operationibus occultis naturae*, especially in Chap. 16, where the differing configurations of the planets at different times of generation are used to account for the individual abilities and actions of different members of the same species.

In so doing, Albert has clearly inverted one of Grant's major propositions about the relationship between natural philosophers and astrologers—that astrologers use conclusions based on the philosophical structures and arguments of the natural philosophers, but that natural philosophers would never stoop so low as to employ characteristic structures derived from astrology to support their natural philosophical positions, at least not beyond a simple enumeration of the qualitative nature of the planets. Nevertheless, Albert is still safely within the bounds of Grant's final defensive bastion, that is, he has not yet mentioned prognostications.

Albert finally crashes headlong through even this last barrier by defending an explicitly prognosticatory astrological position:

> And therefore, human beings die differently, more quickly and more slowly, than they would die by nature, and likewise the other animals. Also, the ages of all things exist in this manner because the planets located in the periodic circle (*in circulo periodali*, sc. the zodiac), when they are stronger, give more years of life; and when they are weaker, they give fewer. And in this manner it comes to be known, since he who knows the strengths of

[115] Hoc enim modo mensura quorundam est annus et quorundam plus vel minus secundum effectus signorum et fortitudines stellarum, quae sitae sunt in signis. Et hoc modo verum est, quod aequale est tempus generationis rei tempori corruptionis, quia a primo signo ascendente in hora rei computatur profectus rei usque in septimum signum eiusdem circuli et a signo septimo usque in primum computatur defectus. Et ideo septimum signum in astronomia domus mortis vocatur, et ascendens vocatur domus vitae. Et ideo generatio rei vocatur profectus usque ad statum, et post statum usque ad declinationem et a declinatione usque ad mortem vocatur periodus corruptionis, quia aequalia sunt secundum naturam, quia a primo usque ad septimum tantum est, quantum est a septimo usque ad primum per aliam circuli partem mensurando. Verbi gratia ab Ariete per Taurum et Geminos et deinceps usque ad Libram tantum est, quantum a Libra per Scorpionem et Sagittarium et sic deinceps usque ad Arietem. Et ideo si periodus profectus hominis sint 35 vel 40 anni, ut dicunt MEDICI, periodus defectus erit tantundem, et erit aetas hominis 70 vel 80 anni. Potest tamen hoc impediri per accidens, per cibum malum vel mortem violentiam vel alio quocumque modo. Et hoc vocat ARISTOTELES materiae inaequalitatem, quia scilicet per accidentia multa aliter disponitur, quam moveatur a circulo (206, 55–81).

> the signs and of the stars placed among them in the periodic circle when some thing is born (*qui scirit vires signorum et stellarum in ipsis positarum in circulo periodali, dum nascitur res aliqua*) can prognosticate about the entire life of the generated thing (*ipse* [...] *prognosticari posset de tota vita rei generatae*), as much as there is from the influence of the heavens (*quantum est de influentia caeli*). Nevertheless, this does not postulate necessity (*hoc necessitatem non poneret*) because it could be impeded *per accidens*, as has been said.[116]

So ends the digression. Albert has thus transformed Aristotle's analysis in *De generatione et corruptione* to become nothing less than the natural philosophical/scientific foundation for prognosticatory astrology—true astrology in Grant's sense—and particularly for nativities. Thus, the stark dichotomy Grant postulated between Aristotelian natural philosophy and astrology appears, on closer examination, to reveal a rather different and far more integrated relationship, at least in this important case. Furthermore and finally, I wish to note that Albert here takes care to state explicitly that necessity is not implied (or required) by his conclusions, a theme to be developed below.

1.7 *De Caelo et Mundo*

Albert confirms and develops this interpretation in his contemporary translation-commentary on Aristotle's closely related *De caelo et mundo* II.3.5, also a digression, which offers a natural causal analysis for the effects of the stars.[117] In it, Albert even more clearly relates astrological practice to its natural philosophical foundations, informing us explicitly that astrologers, particularly those who specialize in elections and nativities, will use the causal analysis provided here as the foundation of their proper activities. We will recall that elections concern the astrological choosing of propitious times:

> Moreover, on the different effects of the stars (*de effectibus* [...] *stellarum diversis* [i.e. the planets, stars and luminaries]), two questions are asked in [sc. natural] philosophy, namely, (1) what (*quis*), and (2) when (*quando*) and where (*ubi*) the effect of any star is. Investigating this is the job of the practitioner of elections (*elector*) and of the diviner by means of the

[116] "Et ideo diversemode moriuntur homines citius et tardius, quam per naturam mortales sint, et similiter alia animalia. Hoc etiam modo aetates sunt omnium rerum, quia planetae in circulo periodali positi, quando sunt fortiores, dant plures annos vitae, et quando sunt debiliores, dant pauciores. Hoc etiam modo innotescit, quoniam qui scirit vires signorum et stellarum in ipsis positarum in circulo periodali, dum nascitur res aliqua, ipse, quantum est de influentia caeli, prognosticari posset de tota vita rei generatae. Sed tamen hoc necessitatem non poneret, quia posset impediri per accidens, ut DICTUM EST (206, 81–207, 7)." In *Tetrabiblos* 1.2, Ptolemy strongly argues against those who claim that celestial influences alone can provide a full causal analysis. Among other factors, Ptolemy cited a person's upbringing, their matter, and their place of generation as mitigating circumstances. Albert also discusses this in *De fato*, as we will see in Chap. 5 below.

[117] Albertus Magnus, *Opera Omnia,* 5.1, *De caelo et mundo,* Paul Hossfeld (ed.), Monasterii Westfalorum: Aschendorff, 1971.

> stars (*divinans per astra*),[118] whose job is to choose (*eligere*) and to know hours. In accordance with this, what come to be in the lower world [sc. things generated on earth] are referred [sc. causally] to the configurations of the stars (*secundum quas ad figuras astrorum referuntur ea quae fiunt in inferioribus*).[119]

What are the effects of the stars? When and where do they take place? These central questions of a scientific astrologizing Aristotelianism relate directly to what is generated on earth and in the sublunar realm more generally.

Practitioners of elections and nativities, which Albert relates closely to each other here, need to find systematic answers to these questions:

> And it is proper to leave this to the knowledge of the practitioners of elections (*scientia electorum*), who are called *geneatici* [sc. genethlialogists or nativity astrologers][120] by another name, for this reason in particular, because what they investigate by the configurations and effects of the stars (*quod inquirunt per stellarum figuras et effectus*) are the births of what is generated in the lower world (*sunt nativitates eorum quae generantur in inferioribus*) and the outcomes of [i.e., what happens to, the life experiences of] what is born (*eventus nascentium*).[121]

Here Albert closely associates something's birth, its nativity, with the beginning of a project or endeavor, which would also metaphorically signify its "birth." What is generated and what happens to it after it begins to exist are both considered by these astrologers. Albert hereby indicates the natural philosophical foundations for elections. He will develop these more fully in the *De mineralibus* to be discussed in Chap. 7.

Albert then evokes Ptolemy, the foremost ancient authority in the field:

> For Ptolemy authoritatively articulated this body of knowledge (*scientia*) in two volumes, of which one concerns great universal events (*de accidentibus magnis universalibus in mundo*), having eight distinctions. These events are great, such as changes of kingdoms from one people to another, and the movements (*translationes*) of sects [= religions][122] and new religious doctrines and such things.[123]

---

[118] As we will see in Chap. 5, Albert here is much looser in his use of the term "divination" than either Thomas Aquinas or Roger Bacon is.

[119] De effectibus autem stellarum diversis duo in philosophia quaeruntur, quis videlicet, et quando et ubi sit effectus cuiuslibet stellae. Et hoc inquirere est electoris et divinantis per astra, cuius est eligere et scire horas, secundum quas ad figuras astrorum referuntur ea quae fiunt in inferioribus (150, 49–54).

[120] '*Geneatici*' can also be used as a term to refer to astrologers dealing with human affairs in general, that is, to all four branches of astrology, not just nativities. My thanks to David Juste for this clarification.

[121] "Et hoc oportet relinquere scientiae electorum, qui alio nomine vocantur geneatici, eo quod principalius, quod inquirunt per stellarum figuras et effectus, sunt nativitates eorum quae generantur in inferioribus, et eventus nascentium (150, 54–58)." My apologies for the stilted English in the translation of *eventus* and its dependent genitives here and in what follows.

[122] As we will see in Chap. 6, Roger Bacon also regularly uses the term '*secta*' to refer to religions.

[123] Hanc enim scientiam in duobus voluminibus determinavit PTOLEMAEUS, quorum unum est de accidentibus magnis universalibus in mundo, habens octo distinctiones. Accidentia autem magna sunt, sicut mutationes regnorum de gente in gentem et translationes sectarum et doctrinae novarum religionum et huiusmodi (150, 58–64).

Although Ptolemy treated these general concerns in *Tetrabiblos*, Book II, the description seems to refer much more naturally to Albumasar's *De magnis coniunctionibus*, which contains eight *differentiae* or *tractatus*.[124] Albert refers here to revolutions in the astrological sense, including the effects of "great conjunctions," that is, of the outer planets, Jupiter and Saturn, and their world-historical implications.[125]

Albert then turns from astrological concerns with the big picture of society and the world at large to that of individuals: "Moreover, the other volume concerns small particular events (*de accidentibus parvis particularibus*), as are the outcomes of a person born under this configuration or that."[126] This last phrase, *eventus unius hominis nati in hac constellatione vel illa*, thus instantiates *eventus nascentium* just above. Ptolemy treated these small particular outcomes or events in discussing nativities in *Tetrabiblos*, Books 3 and 4.

The passage concludes by refering to astrology's causal structure:

> Moreover, concerning what is asked about the effects of the stars (*quod quaeritur de effectibus stellarum*), there is a natural cause (*naturalis causa*) because of which a star is said to have this or that effect. This is to be determined here (*hic determinandum est*) and presupposed by nativity or election astrologers (*a geneaticis sive electoribus supponendum*).[127]

Thus, Albert states explicitly that astrological practitioners should presuppose and ground their practice on this astrologizing Aristotelian causal framework, with its relevance for understanding the broader worlds of politics and society—and their transformations—as well as what happens to individuals born under particular

---

[124] My thanks again to David Juste for confirming this identification. Albumasar (Abu Ma'sar Ga'far ibn Muhammed ibn 'Umar al-Balhi, 787–886 AD) was the most influential and best known astrologer in both the Arabic and Latin worlds. He was from Balh in Persia, but lived most of his life in Baghdad. Albumasar composed general reference works and practical astrological manuals, which became standard accounts in the fields of anniversary horoscopes and great conjunctions. Translated twice into Latin in the twelfth century, his most important work is the *Introductorium maius in astronomiam*. His second major work available in Latin translation was *De magnis coniunctionibus*. He was also known for his *Flores Albumasaris*, a text on general astrology (Hasse, *Success and Suppression*, 326–28). For more on Albumasar, see Richard Lemay, *Abu Ma'shar and Latin Aristotelianism in the Twelfth Century*, and his critical edition of Albumasar's *Liber introductorii maioris ad scientiam judiciorum astrorum*.

[125] See John D. North, "Astrology and the Fortunes of Churches," *Centaurus* 24 (1980), 181–211, now conveniently reprinted in his *Stars, Minds and Fate*. For a critical edition of Albumasar's profoundly influential text, see *Abu Ma'shar on Hištorical Astrology: The Book of Religions and Dynasties (On the Great Conjunctions)*, Keiji Yamamoto and Charles Burnett (eds.), 2 vols. (Leiden, 2000). Only the germs of such a doctrine are found in Ptolemy's *Tetrabiblos*. I discuss Albert's views on the natural philosophical foundations of revolutions in general, and of great conjunctions in particular, in Chap. 3.

[126] Aliud autem est de accidentibus parvis particularibus, sicut sunt eventus unius hominis nati in hac constellatione vel illa (150, 64–66).

[127] Secundum autem quod quaeritur de effectibus stellarum, est naturalis causa, propter quam stella dicitur habere hunc vel illum effectum, et hoc hic determinandum est et a geneaticis sive electoribus supponendum (150, 67–71).

celestial configurations. Astrological prognostications at several levels are thus explicitly provided with their natural philosophical foundations. With Grant's strongly stated but erroneous conclusions sufficiently overturned, we may now turn to the mathematical dimensions of this system and expand our analysis to include Roger Bacon's *Opus Maius* and related works.

Chapter 2
Mathematizing the Picture: Mathematics, *Perspectiva* and Astrology in Roger Bacon and Albertus Magnus

I have now reconstructed the central role of the sun and planets within the Aristotelian analysis of nature in Albertus Magnus's translation-commentaries on two central natural philosophical texts by Aristotle—both with respect to life on earth in general, and to the generation of each and every individual natural thing—and related it to the structure of the horoscope. In this chapter, I will show how this astrologizing Aristotelian natural philosophy was mathematically articulated within a threefold essentially Ptolemaic cosmographic structure, comprising an integrated mathematical-astronomical and -geographical framework equipped with a geometrical-optical model of planetary action.

I will first discuss Roger Bacon's explicitly and emphatically mathematicized astrologizing Aristotelian natural philosophy in order to articulate the model in detail. Then I will show that Albertus Magnus also used a strikingly similar (although less emphatically described) mathematical analysis. We will first look at the broader patterns of Roger's views on mathematics in *Opus maius*, Book IV, which segue directly into a discussion of *perspectiva*, which he articulates there in terms of the multiplication of species. We will see that mathematics, *perspectiva* and astrology are central to Bacon's intricately mathematicized natural philosophy in his three *Opera* (*Maius, Minus* and *Tertium*), a trilogy intended for and at least in part sent to Pope Clement IV in 1266–67 in order to solicit his patronage.[1] In order to focus and further historicize these issues, however, I will begin this chapter by reopening the question concerning the identity of the "Unnamed Master" in Roger Bacon's *Opus tertium.*

[1] For the distinguished scientific circle at the Roman curia when Bacon's works were received there, see David C. Lindberg, "Lines of Influence in Thirteenth-Century Optics: Bacon, Witelo, and Pecham," *Speculum* 46 (1971): 66–83, and several of the chapters in Agostino Paravicini Bagliani, *Medicina e scienze della natura alla corte dei papi nel Duecento*, Spoleto: Centro Italiano di studi sull'alto Medioevo, 1991. For Bacon in general, see the essays collected in *Roger Bacon and the Sciences: Commemorative Essays*; and for his astrology in particular, David Juste, "Astrologie et philosophie naturelle chez Roger Bacon," *Villiers* 15 (2000): 10–18.

H D. Rutkin, *Sapientia Astrologica: Astrology, Magic and Natural Knowledge, ca. 1250–1800*, Archimedes 55, https://doi.org/10.1007/978-3-030-10779-6_2

2.1 The "Unnamed Master" Revisited

What is the intellectual relationship between two of the most important and influential natural philosophers of the thirteenth century: Albertus Magnus and Roger Bacon? More specifically, is Albert the unidentified object of Roger's spleen, especially in his *Opus tertium*? The *status quaestionis* currently leans heavily in the affirmative. The two most recent scholars to discuss this issue in any depth—Stewart Easton[2] and Jeremiah Hackett[3]—both conclude that Albert is, in fact, the unnamed master.[4] David Lindberg's not inconsiderable scholarly weight further pushes the scales in this direction,[5] as does Paola Zambelli's;[6] and now Amanda Power's recent biography, *Roger Bacon and the Defense of Christendom*, uncritically follows suit.[7]

Nevertheless—and however plausible the attribution—I wish to re-open the question here and take an approach that seems strangely neglected. Easton stated that he wanted to compare what Roger said about the unnamed master with the known facts of Albert's life and works.[8] Surely this is the soundest approach. But in practice, with regard to Albert's works, Easton focused almost exclusively on theological issues, specifically the relationship between theology and science (220–30), dismissing the evidence of Albert's scientific writings on his final page (230–31). Among other things, Easton states: "Albert had omitted optics, and was deficient in mathematics." Hackett, on the other hand, did not even deem it worthwhile to compare the content of ideas. He was content, rather, to show that the named attributions to Albert in

[2] Stewart C. Easton, *Roger Bacon and His Search for a Universal Science*, New York: Columbia University Press, 1952, 210–31; Appendix B: "Who was the Unnamed Master?"

[3] Jeremiah M.G. Hackett, "The Attitude of Roger Bacon to the *Scientia* of Albertus Magnus," in *Albertus Magnus and the Sciences*, 53–72 (1980). Hackett continues to maintain the identification in his essays in his 1997 collection, *Roger Bacon and the Sciences*.

[4] Both Easton (210–11) and Hackett (56–7) review some of the more significant earlier scholarship on this question.

[5] David C. Lindberg, "On the Applicability of Mathematics to Nature: Roger Bacon and his Predecessors," *The British Journal for the History of Science* 15 (1982): 3–25. Lindberg simply asserts this as the case: "It is clear, in any case, from Bacon's bitter attacks on Albert, that Bacon knew Albert's work and general philosophical position (14)." Lindberg references both (and only) Easton and Hackett as above, stating in addition that "[t]he best work on the relationship between Bacon and Albert" is Hackett's (n. 54). Lindberg's position becomes firmer and more explicit in his 1987 paper, "Roger Bacon and the Origins of *Perspectiva* in the West," in *Mathematics and its Applications to Science and Natural Philosophy in the Middle Ages: Essays in Honor of Marshall Clagett*, Edward Grant and John E. Murdoch (eds.), Cambridge: Cambridge University Press, 1987, 249–68, 258: "That the 'unnamed master' against whom Bacon directed his diatribes was Albert has been convincingly demonstrated by Jeremiah Hackett."

[6] Paola Zambelli, Speculum astronomiae *and its Enigma*, 50 and n. 18 (not 19, as in the sloppily edited text at p. 155). She only refers the reader to Hackett's essay.

[7] Cambridge: Cambridge University Press, 2013, 185–88. She does, however, hedge her bet with a "probably."

[8] *Roger Bacon and His Search*, 210: "And if we are to determine convincingly who the second master is (the first, Alexander of Hales, Bacon names), we must not only see whether the known details of his life and works correspond to Bacon's statements, but also whether Bacon's own teachings would conflict with those of this master in such a way as to account for his remarkable spleen against him."

Roger's works were similar to what was said of the unnamed master, and that, therefore, the unnamed master was most likely Albert.[9] Hackett undercuts his own argument on the final page of his essay, however, when he explicitly states that proper names in Roger's manuscripts are not nearly as reliable as one might like.[10]

Do we possess the textual means to reconstruct a firmer conceptual foundation, on the basis of which we might come to a more probable conclusion? We are fortunate that some of Roger's criticisms of the unnamed master ca. 1266–67 are quite detailed,[11] and that this detail is sufficient, on the one hand, for building up a fuller picture of Roger's views on the relevant subjects. On the other hand, we also possess quite a lot of relevant material by Albert. Thus, we may well have a sufficient basis for comparison. In fact, the content of Roger's criticism on which we shall focus—that the unnamed master was ignorant of *perspectiva*—provides us with an excellent basis for comparison, as I will show in some detail in the first part of this chapter.

My method is straightforward: first I will examine in detail the contents of Roger's criticism of the unnamed master. Then I will build up a picture of Roger's views on the subjects under discussion there, subjects utterly central to his natural philosophy, namely, mathematics, *perspectiva* and astrology. I will then compare Roger's views thus established with Albert's views on the same and similar subjects. We will then be in a much better position to answer our original questions. Or at least to approach such an answer, for I fear that the results of this investigation will rather complicate the issue than resolve it. So it appears must be the case with the present state of our evidence.

2.2 Roger Bacon on Mathematics, *Perspectiva* and the Unnamed Master

Roger mentions the "unnamed master"—actually, unnamed '*auctor*': writer, authority—most frequently in the short compass of his discussion in the *Opus tertium* (30–42).[12] Within an overall prolegomenon to his set of three works addressed and at

---

[9] Hackett, "Attitude of Roger Bacon," 57–8: "The procedure will be to examine the works of Bacon, beginning (in reverse order of composition) with the works wherein Albert is mentioned by name to the earlier works in which 'that man who has made himself an authority' is, in fact, unnamed. [Hackett then lists the works in question in order.] From this examination, not only should the identity of the 'unnamed master' be clear, but also the reason for Roger Bacon's objections against the 'science'." In my view, such an argument is better used in a supporting than in a primary role, especially for questions of intellectual history.

[10] Hackett, "Attitude of Roger Bacon," 72: "One has to grant that there are problems [which he does not enumerate] in regard to scribal changes of personal names in the works of Bacon."

[11] Hackett too makes this point, albeit a bit too strongly, "Attitude of Roger Bacon," 56: "Bacon's objections to the *scientia* of the master whom the whole world followed were, in fact, always very specific: that master was ignorant of ancient languages and mathematics, specifically perspective." Easton takes a middle position with which I agree, namely, that some of the criticisms are specific and some not; *Roger Bacon and His Search*, 215.

[12] The text of the *Opus tertium* is that edited by J. S. Brewer in *Fr. Rogeri Bacon Opera quaedam hactenus inedita*, vol. I in the series *Rerum Britannicarum Medii Aevi Scriptores*, London:

least in part sent to Pope Clement IV ca. 1266–68,[13] Roger is currently complaining about how difficult it is to accomplish the kinds of high-level scholarly and scientific work he has undertaken, and how important it is to have the high level of patronage required to actually complete what he has (sometimes several times) begun.

The passages where Roger's criticisms occur have been treated adequately in a general way by Easton (210–19) and Hackett (*passim*), so we need not be detained by them here. Let us turn directly to the central passage of criticism, which will provide the foundation for my argument: *Opus tertium*, Chap. 11 (pp. 34–38), where Roger discusses mathematics and *perspectiva*, and explicitly criticizes the unnamed master for his utter ignorance of *perspectiva*. Chapter 11 may be divided into two main parts: the first is on mathematics; the second, *perspectiva*. Roger specifically discusses the unnamed master, but only on two occasions in the second part. I will thus treat the second part first, and in some detail. I will not, however, neglect the first part on mathematics, for, as we will see, *perspectiva* and mathematics are intimately connected in Roger Bacon's extremely interesting scientific vision.

*

To complete the first part of Chap. 11 on mathematics, Roger discusses at some length the great expense of astronomical instruments. He then briefly mentions other instruments, and tables of practical geometry, arithmetic and music—thus all four traditional members of the *quadrivium*—and that they are of great utility, and are necessary. He then makes some more general points about the necessity of perspective:

> But much more than these, it is necessary to have men who know *perspectiva* and its instruments well, nay excellently. For this body of knowledge (*scientia*) is about vision, and it is through vision that we know everything (*per visum scimus omnia*). For a blind man knows nothing of this world (*nihil scit de hoc mundo*); for sight shows us the *differentiae* of things,

---

Longman, 1859. The text of the *Opus tertium* is incomplete, and that of the *Opus maius* is inadequate (as I will discuss more fully below). See Thorndike's bibliographical note 1 (*HMES* II, 617–18), and the more up-to-date bibliographies in Easton, *Roger Bacon and His Search*, 236 ff. (with descriptions); *Roger Bacon and the Sciences*, and especially in David C. Lindberg, *Roger Bacon's Philosophy of Nature: A Critical Edition with English Translation, Introduction and Notes of* De multiplicatione specierum *and* De speculis comburentibus, Oxford: Clarendon, 1983, and his *Roger Bacon and the Origins of* Perspectiva *in the Middle Ages: A Critical Edition and English Translation of Bacon's* Perspectiva, *with Introduction and Notes*, Oxford: Clarendon, 1996. A.G. Little's edited collection of essays with an extensive bibliography of primary sources is still fundamental; *Roger Bacon Essays*, A.G. Little (ed.), Oxford: Clarendon Press, 1914. I discuss various textual issues in Bacon's three *Opera* as they arise.

[13] For the dating of the works, see Thorndike, *HMES* II, 622–25. Lindberg, *Roger Bacon's Philosophy of Nature*, p. xxv suggests that the works were sent in late 1267 or early 1268. On their reception, Lindberg says: "Extant documents reveal nothing of the Papal reaction to Bacon's work; it can be demonstrated, however, that they arrived safely, since the Polish mathematician, Witelo, who was attached to the curia demonstrates knowledge of their content." He then refers to his demonstration of this in his important article, "Lines of Influence." Steven J. Williams's foundational article, "Roger Bacon and his Edition of the Pseudo-Aristotelian *Secretum secretorum*" (*Speculum* 69 [1994]: 57–73), is very useful for dating Bacon's works, among many other things.

> as Aristotle says, and as we know through experience (*per experientiam*). And this [sc. *perspectiva*] authenticates (*certificat*) mathematics and everything because the instruments of astronomy do not work except through vision in accordance with the laws of this science. It is small wonder that everything is known by mathematics (*omnia sciantur per mathematicam*), and everything by this [sc. *perspectiva*] (*omnia per hanc*), because all sciences are connected (*omnes scientiae sunt connexae*) (as I said above), although each together with this one has its own particular nature (*proprietas*). Whence, each has its strength in its own field, nor can one be known without the others, as was said before.[14]

Bacon here raises issues of the essential unity of the sciences in relation to our knowledge while emphasizing the importance of both mathematics and *perspectiva*.

Roger then goes to a deeper level of analysis in explaining why *perspectiva* is so important for understanding the world. This discussion is utterly central for reconstructing Roger's views, *inter alia*, because it is immediately followed by the first explicit criticisms of the unnamed master:

> And it is necessary that everything be known by this science [sc. *perspectiva*] (*necesse est omnia sciri per hanc scientiam*) because all the actions of things come to be in accordance with the multiplication of species and powers (*omnes actiones rerum fiunt secundum specierum et virtutum multiplicationem*), from the agents of this world [sc. acting] on receptive matter (*ab agentibus huius mundi in materias patientes*). And the laws of multiplications of this sort are not known except by *perspectiva* (and they are not treated anywhere else), since not only are actions common with regard to vision, but they are so for all the senses, and in the entire system of the world, both in the heavens and on earth (*in totam mundi machinam, et in coelestibus et in inferioribus*). But this science has not yet been taught in Paris, nor among the Latins, except twice at Oxford in England; and there are not three people who know its power. Therefore, he who made himself an authority (*ille, qui fecit se auctorem* [the first explicit criticism]), about whom I spoke above, knew nothing of the power of this science (*nihil novit de huius scientiae potestate*), as appears in his books, because he did not compose a work on this science—he would have done so had he known it—nor does he say anything on this science in his other works. Nevertheless, it is fitting that the use of this science has ramifications in all the others, and that all are known through its power, and, therefore, he is unable to know anything of the wisdom of philosophy (*de sapientia philosophiae*). But those who know it are few, as are those who know mathematics, nor are they had without great expense; likewise, neither are the instruments of this science, which are used with great difficulty and are of greater expense than the instruments of mathematics.[15]

[14] Sed longe magis quam haec oporteret homines haberi, qui bene, immo optime scirent perspectivam et instrumenta eius. Nam haec scientia est de visu vero, et per visum scimus omnia. Caecus enim nihil scit de hoc mundo; visus enim ostendit nobis rerum differentiae, ut Aristoteles dicit, et scimus per experientiam. Haec autem certificat mathematica et omnia, quia instrumenta astronomiae non vadunt, nisi per visionem, secundum leges istius scientiae. Nec mirum si omnia sciantur per mathematicam, et omnia per hanc, quia omnes scientiae sunt connexae (ut superius dixi), licet quaelibet simul cum hac habeat suam proprietatem. Unde quaelibet habet potestatem in aliam, nec potest una sciri sine alia, ut praedictum est (36, 30–37, 11).

[15] Et necesse est omnia sciri per hanc scientiam, quia omnes actiones rerum fiunt secundum specierum et virtutum multiplicationem ab agentibus huius mundi in materias patientes; et leges huiusmodi multiplicationum non sciuntur nisi a perspectiva, nec alibi sunt traditae adhuc; cum tamen non solum sint communes actioni in visum, sed in omnem sensum, et in totam mundi machinam, et in coelestibus et in inferioribus. Haec autem scientia non est adhuc lecta Parisius, nec

Roger's more general points (and his bile) are of no interest for our purposes.

Of great interest, first of all, is the reasonably detailed discussion of *perspectiva* encountered before the criticism (to be discussed in more detail below) and, secondly, the clearly stated claims that the master, whoever he is, knows nothing about the power of this science (sc. *perspectiva*), as we can tell from his books, because [1] he never wrote a treatise (*librum*) on *perspectiva*, [2] nor did he treat anything on that subject in his other works (*nec in libris aliis aliquid de hac scientia recitavit*). Since we do not in fact possess a separate treatise on *perspectiva* by Albert, the attribution is supported. Its accuracy, therefore, will stand or fall on the second point: if, in fact, there are no references at all to *perspectiva* as understood by Roger in Albert's extant works, then we will be able to lend further weight to Albert's attribution as the unnamed master.

*

Roger then provides references to his own fuller treatments of this central topic, some of which I will use in my reconstruction:

> What I have discussed about *perspectiva* lies open, plain to view, in my *Opus maius*, and in the treatise that I brought together on *perspectiva*, which is the fifth principle part of that work.[16] And at the same time, a large part of the fourth part of the whole work [sc. the *Opus maius*] is to be consulted, that is, where I determined on the multiplication of species and powers (*virtutes*) of agents, which, although I ascribe it there to geometry because of its *rationes* [= types of arguments], nevertheless, I took it from the science of *perspectiva*. But I am sending you a more complete treatment of this multiplication, as I mention later.[17] And for this reason, because it is the highest and principal root of wisdom (*radix sapientiae*), both for philosophy and for theology, there is in these multiplications also unbounded beauty; and neither is *perspectiva* nor any bit of philosophy able to be known without it. But he—[and here we have the second criticism]—who multiplied volumes, ignores these roots, for he touches on none of them.[18] And therefore it is certain that he is ignorant of *res naturales*, and everything concerning philosophy; and not only him, but the whole horde of philosophizers (*vulgus philosophantium*) who err because of him.[19]

---

apud Latinos, nisi bis Oxoniae in Anglia; et non sunt tres qui sciant eius potestatem: unde ille, qui fecit se auctorem, de quo superius dixi, nihil novit de huius scientiae potestate, sicut apparet in libris suis, quia nec fecit librum de hac scientia, et fecisset si scivisset, nec in libris aliis aliquid de hac scientia recitavit; cum tamen oportet quod usus istius scientiae cadat in omnibus aliis, et quod per eius virtutem sciantur omnia. Et ideo non potest scire aliquid de sapientia philosophiae. Sed hi, qui sciunt haec, sunt pauci, sicut qui mathematicam, nec habentur sine magnis expensis; similiter nec instrumenta huius scientiae, quae sunt multum difficilia, et maioris sumptus quam instrumenta mathematicae (37, 11–38, 3).

[16] For the critical edition of *Opus maius*, book V, with extensive valuable commentary, see Lindberg, *Roger Bacon and the Origins of* Perspectiva.

[17] See the critical edition with extensive introduction and commentary by Lindberg, *Roger Bacon's Philosophy of Nature*. This separate treatise provides a great deal of deeper information on the natural philosophical and metaphysical structure of the multiplication of species.

[18] Sed ille, qui multiplicavit volumina, ignorat has radices: nam nihil de eis tangit […].

[19] Et quae de perspectiva narravi modo patent manifeste ex Opere Maiori, et tractatu quem collegi de perspectiva, qui est pars quinta principalis illius operis; et simul cum ea consulenda est magna pars quartae partis totius operis, scilicet ubi de multiplicatione specierum et virtutum agentium determinavi, quam licet ascribo ibi geometriae propter rationes suas, tamen extraxi de scientia

Albertus Magnus certainly did write a lot, as any glance at the last page of any volume of the magisterial *Editio Coloniensis* amply reveals,[20] and so the comment about the unnamed master "multiplying volumes" could surely apply to Albert as well. Likewise could the comment about his great influence on students of philosophy, at least amongst Dominicans and those trained in its *studia*.[21] There is no doubt of the reasonableness, in general, of thinking that the unnamed master could be Albert, which is further supported (if not with much substance) by the fact that the scholarly community has not yet been able to discover a more suitable candidate.

Roger completes Chap. 11 with some further relevant comments in a challenge to the pope:

> For you ought to write to him [sc. the unnamed master, and ask him] what he has to say about these roots, and you will find him unable to perform this task. And you may be assured that I say this because I am gravely concerned for his ignorance and for that of the horde, for without these roots, they are unable to know anything (*sine his nihil sciri possunt*). Therefore, this is worth a hundred-fold more than whatever they know. But none of the authorities—neither ancient masters nor modern—wrote about these [sc. the multiplication of species]. But I have labored for ten years, and with however much free time I had from the time of your mandate, I discussed everything I was able to, putting it into writing.[22]

This last statement points to the specific interpretation that Roger's criticism refers more to the multiplication of species in particular than to *perspectiva* in general because many writings of the ancients on *perspectiva* were known to and used by

---

perspectivae. Sed completiorem tractatum mitto vobis de hac multiplicatione, ut facio postea mentionem; et hoc ideo, quia summa et principalis radix sapientiae; et pro philosophia et pro theologia, est in istis multiplicationibus, et infinita pulchritudo; et nec perspectiva, nec aliquid de philosophia sciri potest sine hoc. Sed ille, qui multiplicavit volumina, ignorat has radices: nam nihil de eis tangit; et ideo certum est ipsum ignorare res naturales, et omnia quae de philosophia sunt; et non solum ipse, sed totum vulgus philosophantium, quod errat per ipsum (38, 4–21).

[20] I count 78 works in 40 volumes. Irven Resnick's comment is apropos in the section entitled "Literary Production" of his chapter, "Albert the Great: Biographical Introduction," in *Companion to Albert the Great*, 9: "Albert's literary output is quite simply enormous. Meersseman identifies over 470 distinct titles (many of them sermons or homilies) attributed to Albert, including some 70 philosophical, scientific, and theological treatises comprising more than 20,000 pages in manuscript. Although Meersseman's calculation includes treatises that were subsequently found to be inauthentic, Fernand van Steenberghen does not exaggerate when he remarks that Albert was the most prolific author of the whole of the Middle Ages."

[21] A full study of Albert's influence up to and including the seventeenth century, including the transmission of his manuscripts and their printing history, would be of tremendous interest. In the meantime, there is much valuable information in Martin Grabmann, "Der Einfluss Alberts der Grossen auf das Mittelalterliche Geistesleben," in his *Mittelalterliches Geistesleben: Abhandlungen zur Geschichte der Scholastik und Mystik*, Munich: M. Hueber, 3 vols, 1926–56; vol 2 (1936), and Edward P. Mahoney, "Albert the Great and the *Studio Patavino* in the Late Fifteenth and Early Sixteenth Centuries," in *Albertus Magnus and the Sciences*, 537–63; and in the prolegomena to the separate volumes of the *Editio Coloniensis* and associated studies.

[22] Scribatis enim ei quod pertractet de his radicibus, et invenietis ipsum impossibilem ad eas. Et certe hoc dico quia doleo de eius et vulgi ignoratia: nam sine his nihil scire possunt. Et ideo hoc solum valet centies plus quam quicquid sciunt. Nullus vero de auctoribus, nec de magistris antiquis, nec de modernis, scripsit de his; sed laboravi per annos decem, quantumcunque potui vacare, et discussi omnia ut potui, redigens in scriptum a tempore mandati vestri (38, 21–30).

Roger himself at this time. These include Euclid's *De aspectibus*, as Roger himself points out in what Lindberg characterizes as the "prologue to the revised version of *De multiplicatione specierum*."[23] So much for our first encounter with *perspectiva.*

*

What about Roger's views on mathematics? I should first point out that Roger in no way criticizes the unnamed master for his knowledge of mathematics in this section of the *Opus tertium*, or anywhere else that I have found. In the first part of Chap. 11, Roger discusses the importance of mathematics for his overall programme:

> But without mathematics, nothing worthwhile in philosophy can be known (*nihil* […] *sciri potest*), as I show at great length and in a diffuse manner in the fourth part of the *Opus maius*. For I show at first in general (*in universali*) that no science can be known without it [sc. mathematics] (*nulla scientia potest sciri sine ea*). Secondly, I show in particular, that things of this world and the places of the world cannot be known (*res istius mundi et loca mundi sciri non possunt*), and therefore neither are the *scientiae* of these *res* known. I show this likewise in my *Opus minus*. For it is obvious that without mathematics, celestial things cannot be known (*sine mathematica non possunt sciri coelestia*). And *coelestia* are the causes of the things here below (*coelestia sunt causae rerum inferiorum*), and what has been caused cannot be known without their causes (*causata non possunt sciri sine causis suis*). And there are other particular pathways (*speciales viae*) through which it is shown that no thing (*res*) can be known without mathematics, as is made abundantly clear in the parts on mathematics in the *Opus maius*.[24]

These are strong claims indeed! Nothing worthwhile can be known without mathematics, in a phrase following Robert Grosseteste that we will see in Bacon time and time again: no science can be known without it, nor can the things (*res*) themselves be known. Bacon justifies these claims with a remarkable brief argument: [1] *Coelestia* are not known without mathematics. [2] *Coelestia* are the *causae inferiorum*, and [3] what has been caused (*causata*) cannot be known without their causes. I will examine this argument in greater depth just below.

*

To explore more fully the role of mathematics in Bacon's thought, we should now turn to the very beginning of *Opus maius*, Book IV, which is wholly concerned with mathematics. This will provide the first step in developing upon the material

[23] *Roger Bacon's Philosophy of Nature*, xxxiii; the text in question is Appendix a (alpha), 62–76 (p. 347).

[24] "Sine vero mathematica nihil dignum sciri potest in philosophia, ut ostendo in quarta parte Operis Maioris multum diffuse. Nam primo ostendo in universali quod nulla scientia potest sciri sine ea; secundo, in particulari magis scilicet, quod res istius modi [mundi?] et loca mundi sciri non possunt, et ideo nec scientiae de rebus; et in Opere Minori similiter hoc ostendo. Nam planum est quod sine mathematica non possunt sciri coelestia; et coelestia sunt causae rerum inferiorum, et causata non possunt sciri sine causis suis. Et aliae viae sunt speciales, per quas ostenditur quod nulla res sciri potest sine mathematica, ut abundanter patet in partibus mathematicae in Opere Maiori (35, 4–17)." Roger here exactly duplicates the same structure as found in the first major section of *Opus maius* IV, as we will see presently.

provided in *Opus tertium* XI, in which Roger explicitly mentioned the unnamed master. Roger begins the extensive Book IV with a very strong statement about the central importance of mathematics for knowledge and wisdom:

> Since it has been shown that many important roots of wisdom depend on the power of languages, through which [sc. in translations from Arabic and Greek] an introduction has been made into the wisdom of the Latins, I now wish to consider the fundamentals of this wisdom among the great sciences (*penes scientias magnas*), in which there is a special power (*specialis potestas*) with respect to the other sciences and the things of this world (*respectu caeterarum scientiarum et rerum huius mundi*). There are four *scientiae magnae*, without which the other sciences cannot be known, nor can knowledge of things be had [sc. without them] (*sine quibus caeterae scientiae sciri non possunt, nec rerum notitiam haberi*). When they [sc. the four] are known, every person can go forth in glory into the power of wisdom (*in sapientiae potestate*) without difficulty or labor, not only in human sciences (*in scientiis humanis*), but divine (*sed divina*). And the strength of each of these is touched not only with respect to wisdom absolutely, but with respect to the others already mentioned. And the gate and key of these sciences is mathematics (*harum scientiarum porta et clavis est mathematica*), which the saints discovered at the beginning of the world, as I will show, and which was always in the use of all the saints and wise men above all the other sciences.[25] The neglect of which, for thirty or forty years now, has destroyed the entire education of the Latins, since he who does not know this [sc. mathematics] cannot know the other sciences (*qui ignorat eam non potest scire caeteras scientias*), nor the things of this world (*nec res huius mundi*), as I will prove.
>
> And what is worse, those who do not know it do not perceive their own ignorance, and therefore do not seek a remedy. And on the contrary, the knowledge of this science prepares the soul and lifts it up to the confirmed understanding of everything (*ad omnium certificatam cognitionem*), as if he understood the roots of wisdom given concerning it (*ut si radices sapientiae datas circa illam cognoscat*). And [sc. if] he applied these roots properly to the understanding of the other sciences and things, then he can know without error or doubt, and easily and powerfully. For without these, neither what precedes nor what follows can be known. Whence, they perfect and regulate what comes before (*perficiunt priora et regulant*) just as the end (*finis*), concerning those things which are in relation to their end, and they arrange and open the way to what follows. I intend to support this by authority and by reason: first in the human sciences and the things of this world (*primo in scientiis humanis et rebus istius mundi*), then in divine matters (*deinde in divina*), finally in so far as the other three are compared to the church.[26]

[25] For more on this, see the beginning of the following section of Book IV, *Mathematicae in divinis utilitas*, 175 ff.

[26] "Manifestato quod multae praeclarae radices sapientiae dependent ex potestate linguarum, per quas est introitus in sapientiam Latinorum, nunc volo revolvere fundamenta eiusdem sapientiae penes scientias magnas, in quibus est specialis potestas respectu caeterarum scientiarum et rerum huius mundi. Et sunt quattuor scientiae magnae, sine quibus caeterae scientiae sciri non possunt, nec rerum notitia haberi: quibus scitis, potest quilibet gloriose proficere in sapientiae potestate sine difficultate et labore, non solum in scientiis humanis, sed divina. Et cuiuslibet istarum tangetur virtus non solum propter sapientiam absolute, sed respectu caeterorum praedictorum. Et harum scientiarum porta et clavis est mathematica, quam sancti a principio mundi invenerunt, ut ostendam, et quae semper fuit in usu omnium sanctorum et sapientum prae omnibus aliis scientiis. Cuius neglegentia iam per triginta vel quadriginta annos destruxit totum studium Latinorum. Quoniam qui ignorat eam non potest scire caeteras scientias nec res huius mundi, ut probabo. Et, quod peius est, homines eam ignorantes non percipiunt suam ignorantiam et ideo remedium non quaerunt. Ac per contrarium huius scientiae notitia praeparat animum et elevat ad omnium certificatam cognitionem, ut si radices sapientiae datas circa illam cognoscat, et eas radices recte applicet

Mathematics thus provides the gate and key to all knowledge, human and divine.

Now that we have a clearer sense of the nature of Roger's attack on the unnamed master with respect to his ignorance of and therefore complete disregard for *perspectiva*, and of the overall importance of mathematics in Roger's understanding of natural (and divine) knowledge, we will now zoom in more closely to get a sharper and more richly articulated perspective on his views. On this basis, finally, we will compare Roger's understanding of *perspectiva* and mathematics to the relevant texts in Albert's understanding of nature.

2.3 Roger Bacon on Mathematics

We shall now inquire more deeply into the role of mathematics in Roger Bacon's thought.[27] In the *Opus maius*, toward the beginning of book IV, Roger describes how he understands the place of mathematics within his broader map of knowledge, and some of its subdivisions. The first part of *Opus maius* IV, entitled "The Utility of Mathematics in Natural Philosophy" (*Mathematicae in physicis utilitas,* 97–174) is clearly structured. The passage of interest (109–11) directly follows a discussion *De scientiis* (98–108), where Bacon shows by authority (98–103) and reason (103–8) why every science requires mathematics. After this more general theoretical discussion, Bacon moves on to a discussion *De rebus*, which takes up the rest of the section *In physicis* (109–74). Our passage is its introduction. I discuss the structure of what follows more fully below. Roger makes the transition as follows: "What has just now been shown *de scientiis*, is able to be made manifest *de rebus*."[28]

Roger begins with a characteristically strong claim: "For it is impossible for the things of this world (*res huius mundi*) to be known, unless mathematics is known."[29]

---

ad caeterarum scientiarum et rerum cognitiones, tunc omnia sequentia poterit scire sine errore et sine dubitatione, ac de facili et potenter. Sine his enim nec praecedentia nec consequentia sciri possunt; unde perficiunt priora et regulant, sicut finis ea quae sunt ad finem, et disponunt et aperiunt viam ad sequentia. Ad quod nunc intendo innuere per auctoritatem et rationem; et primo in scientiis humanis et rebus istius mundi, deinde in divina, et ultimo prout ad Ecclesiam et caetera tria comparantur (97, 8–98, 10)." The textual references are to *The* Opus majus *of Roger Bacon*, John H. Bridges (ed.), 3 vols., London: Williams and Norgate, 1897–1900. For an inadequate English translation, see *The* Opus maius *of Roger Bacon*, Robert B. Burke (tr), Philadelphia: University of Pennsylvania Press, 1928, which is based on this inadequate edition of the Latin. David Lindberg describes the textual situation concisely in the introduction to his critical edition with translation, extensive introduction and commentary on *Opus maius*, Book V; *Roger Bacon and the Origins of* Perspectiva. His edition should become the model for similar editions of the other six books. It is high time for proper critical editions of all three of Bacon's *Opera* for Pope Clement to be undertaken.

[27] In general, see George Molland, "Roger Bacon's Knowledge of Mathematics," in *Roger Bacon and the Sciences*, 151–74. See also his "Roger Bacon's *De laudibus mathematicae*: A Preliminary Study," in *Texts and Contexts*, 68–83. Both have further bibliography. I treat Roger's fundamental distinction between true and false mathematics in my Chap. 4 below.

[28] Quod de scientiis iam ostensum est, potest de rebus manifestari (109, 7–8).

[29] Nam impossibile est res huius mundi sciri, nisi sciatur mathematica (109, 8–9).

He immediately begins justifying this universal claim by turning to the science of the stars:

> For the celestial bodies (*coelestia*) are certain to everyone because there are two great mathematical sciences (*duae scientiae magnae mathematicae*) concerning them, namely *astrologia speculativa* and *astrologia practica*. The first contemplates (A) the quantities of all things that are in the heavens, and everything that is reduced to quantity, both discrete and continuous quantity. For it certifies the number of the heavens and of the stars, whose quantity can be grasped by instruments, and all of their figures, and their magnitudes and altitudes from the earth, and their thickness and number, magnitude and smallness, the rising and setting of the signs of the stars, and the motion of both the heavens and the stars, and the quantities and varieties of eclipses.[30] Then, it descends to (B) the quantity and shape of the habitable world, and of all its large parts, which are called climates (*climata*), and it shows the diversity of horizons, and of days and nights corresponding to individual climates. These things, then, are determined here, and much is joined to them.[31]

In the first division of *astrologia speculativa* (A), Roger straighforwardly describes the content and aim of what we call mathematical astronomy; in (B) he describes the basic divisions of mathematical geography. Both are also found well-developed (albeit at an elementary level) and in much the same terms in Sacrobosco's *De sphaera*, which will be discussed more fully below.[32]

Roger then descends (his term)[33] to the practical level:

> But [*astrologia*] *practica* descends to this, that (A) for every hour we would know the location of the planets and stars, and their aspects and relative positions and everything which is periodic (*renovantur*) among celestial bodies, and it descends to (B) those things which come to be in the air, of which sort are comets and rainbows and other things *renovata* there, in order that we know their locations, altitudes, magnitudes and shapes, and many things which it ought to consider in these matters. And these all come to be through instruments fitted to these ends, and by tables, and by canons, that is, rules devised for confirming these things [...][34]

---

[30] Roger discusses this astronomical material in some depth in the next section of Book IV, *Mathematicae in divinis utilitas* (224–36).

[31] De coelestibus enim certum est omnibus, quia duae scientiae magnae mathematicae sunt de eis, scilicet astrologia speculativa, et astrologia practica. Prima speculatur quantitates omnium quae sunt in coelestibus, et omnia quae ad quantitatem reducuntur, tam discretam quam continuam quantitatem. Nam numerum coelorum et stellarum, quarum quantitas potest per instrumenta comprehendi, certificat, et figuras omnium, et magnitudines et altitudines a terra ac spissitudines et numerum ac magnitudinem ac parvitatem, ortum et occasum signorum stellarum, et motum tam coelorum quam stellarum, et quantitates et varietates eclipsium. Item descendit ad quantitatem et figuram habitabilis, et omnium partium eius magnarum, quae vocantur climata, et ostendit diversitatem horizontium et dierum et noctium secundum singula climata. Haec ergo determinantur hic, et multa eis annexa (109, 9–24).

[32] See *The Sphere of Sacrobosco*, 76–117 (Latin text), 118–142 (English translation).

[33] The same term is used in the same context also in *Speculum astronomiae*, Chap. 3, as we saw in Chap. 1.

[34] Practica vero descendit ad hoc, ut ad omnem horam sciamus loca planetarum et stellarum, et aspectus et compositiones earum et omnia quae in coelestibus renovantur, atque descendit ad ea quae fiunt in aere, cuiusmodi sunt cometae, et irides, et caetera ibi renovata, ut sciamus loca eorum, et altitudines, et magnitudines, et figuras, et multa quae oportet considerare in his. Et haec omnia fiunt per instrumenta ad haec idonea, et per tabulas, et per canones, id est, regulas ad haec certificanda inventas [...] (109, 24–32).

Let us stop here momentarily—in mid-sentence—to note that this first part of practical *astrologia* concerns the knowledge of where the planets and stars are at every hour of every day, and of certain meteorological phenomena, and that this knowledge is attained with the help of certain unspecified instruments.[35] We are thus still well within the realm of astronomy proper, but as particularized for specific times.

The second part of *astrologia practica* continues thus:

> […] so that the way is prepared [C] for judgments (*ad iudicia*), which can be made in accordance with the power of philosophy (*secundum potestatem philosophiae*),[36] not only in the realm of natural things (*in naturalibus*), but in those which take their inclination from nature, and freely follow the celestial disposition.[37] Not only is the way prepared for judgments (*iudicia*) of the present, past and future, but for amazing works (*ad opera miranda*), in order that everything favorable for this world may be promoted, and what is adverse repressed, usefully and magnificently. These things are not in doubt (*nec sunt haec dubia*). For the patriarchs and the prophets have confirmed these and other things from the beginning of the world. And Aristotle renewed the confirmation of the ancients, and brought it into the light. And all wise men agree in this in great matters, and experience teaches it. There will be an exposition of these things in their place.[38]

Nec sunt haec dubia, Bacon proclaims, adducing the support of both authority—religious and secular: patriarchs and prophets, Aristotle and all wise men—and experience (*et experientia docet*).

These are strong claims indeed! But let us hear him out before we jump in with our criticisms. He does not stop at mere claims, but spends the next 64 pages (in Bridges's edition) justifying these claims *in physicis*, and he spends much of the following 228 pages toward the same end. We should also note that Roger makes a clear distinction here between two distinctly different practices: [1] making judgments (*iudicia*) about the past, present and future, that is, normal astrological knowledge-based interpretations, but also [2] making wondrous works (*opera miranda*), which will ultimately be identified as talismans, as I will discuss further in Chap. 8 below. This reflects the distinction stated explicitly in Roger's important edition of the *Secretum secretorum* between *astronomia iudiciaria et operativa* (3.1-3) that I will discuss in Chap. 4.

*

[35] For much relevant material on slightly later astronomical (and astrological) practice in England, see North, *Chaucer's Universe*, part I.

[36] Bacon thus places himself within the Ptolemaic tradition of philosophical/scientific astrology; see *Tetrabiblos* I.1–3.

[37] Roger discusses this much more fully in the section entitled *Iudicia astronomiae,* which I discuss more fully in Chap. 4. It is touched on briefly below.

[38] […] quatenus via paretur ad iudicia, quae fieri possunt secundum potestatem philosophiae, non solum in naturalibus, sed in his quae sumunt inclinationem ex natura, et gratis sequuntur coelestem dispositionem; et non solum ad iudicia praesentium praeteritorum et futurorum, sed ad opera miranda, ut omnia prospera huius mundi promoveantur, et adversa reprimantur, utiliter et magnifice. Nec sunt haec dubia. Nam patriarchae et prophetae a principio mundi certificaverunt haec, sicut caetera. Et Aristoteles renovavit certificationem antiquorum, et produxit in lucem. Et omnes sapientes in rebus magnis in hoc concordant, et experientia docet. Sed de his expositio fiet suo loco (109, 32–110, 12).

Having just laid out the basic patterns of these *duae scientiae magnae mathematicae*, what we would call astronomy and astrology, Roger now turns to some actual arguments to substantiate his claims. He begins the first argument by stating what he will take as two givens: "It is clear, therefore, that (1) *coelestia* are known by mathematics, and that (2) the way to things below (*via ad haec inferiora*) is prepared by it (*per eam*, namely, *mathematica*)."[39] He now begins the heart of his argument:

> But that *haec inferiora* cannot be understood without mathematics (*non possunt cognosci sine mathematica*) is clear first by this, that (1) we do not know *res* except by causes, if *scientia* is understood properly, as Aristotle says. But (2) *coelestia* are the causes of *inferiora*. Therefore (3) these *inferiora* will not be known unless the *coelestia* be known, and these cannot be known without mathematics. Therefore (4) the knowledge (*scientia*) of these *inferiora* depends on this same thing [sc. mathematics].[40]

With the first conclusion (3), Roger is concerned with the knowledge of *haec inferiora* themselves, that is, the *res* that can only be known *per causas*, which causes he here identifies as the celestial bodies. On this basis he draws the second conclusion (4): that the *scientia horum inferiorum* must also depend on mathematics (as just articulated), albeit at one remove. This passage points quite clearly to the force of Bacon's distinction between *res* and *scientia*: one learns about some aspect of nature (the *res*) directly,[41] and then, by a process of conceptual generalization, constructs a broader understanding (*scientia*).[42] This argument also expresses more fully the one adumbrated in the *Opus tertium* discussed above, and is thus in close conceptual proximity to Roger's attack on the unnamed master.

*

In his second argument, Bacon begins by implying the obvious, that the first argument made general claims about the causal relations obtaining between the heavens and the earth, and their importance for our scientific knowledge. On the other hand, the second argument, which turns toward particular problems, will rely on some of the central patterns of Bacon's thought, and will be spelled out in a series of physical examples to follow. We shall now begin our descent—with Roger Bacon himself as our guide—into the inner-workings of his particularly mathematicized brand of Aristotelian natural philosophy:

---

[39] Planum ergo est, quod coelestia sciuntur per mathematicam, et quod praeparatur per eam via ad haec inferiora (110, 13–14).

[40] Quod autem haec inferiora non possunt cognosci sine mathematica, patet primo per hoc, quod non scimus res nisi per causas, si proprie accipiatur scientia, sicut Aristoteles dicit. Sed coelestia sunt causae inferiorum. Ergo non scientur haec inferiora, nisi sciantur coelestia, et illa sine mathematica sciri non possunt. Ergo horum inferiorum scientia dependet ex eadem (110, 14–21).

[41] Bacon refers to the *notitia rerum* at *Opus maius*, 97.

[42] This passage provides the proper response to Lindberg's criticism; "On the Applications of Mathematics to Nature," 20: "Thus far [through p. 108] Bacon has argued that mathematics is required for an understanding of the sciences; he now undertakes briefly[!] to defend the proposition that mathematics is also required for an understanding of things." There is then a footnote (n. 76): "The distinction is obscure, since presumably the sciences treat things."

> Secondly, we are able to see from particular examples (*ex propriis*) that nothing of these (1) *inferiora* or (2) *superiora* can be known without the power of mathematics. For (*ad* 1) every *res naturalis* is brought forth into existence (*producitur in esse*) by an *efficiens* and the matter (*materia*) in which it acts; for these two come together at first. For the *agens* [= *efficiens*], by its power (*virtus*), moves and transforms (*movet et transmutat*) the matter, in order that a *res* comes to be. But the power of the *efficiens* and of the matter cannot be known without the great power of mathematics, nor likewise the produced effects. There are, therefore, these three: *efficiens*, *materia* and *effectus*. And (*ad* 2) among the celestial bodies (*in coelestibus*) there comes to be a mutual influence of powers (*mutua influentia virtutum*), as of light and of others, and there is in them [sc. *in coelestibus*] *alteratio*, although not toward corruption.[43] And thus it can be shown that nothing among *res* [that is, both *inferiora* and *superiora*, as in the first premise] can be known without the power of geometry.[44]

Roger here squarely articulates basic patterns of Aristotelian natural philosophy, as we saw in discussing Albert in Chap. 1, and as Roger himself will develop in his own particular way in what follows. Here Roger strongly claims the necessity of mathematics for a proper and complete understanding of nature both in the heavens and on earth.

How Roger demonstrates geometry's fundamental importance for natural philosophy is central for our purposes. He begins straightaway with the efficient cause, where we will now follow him, and thereby engage with the multiplication of species, which we first encountered in discussing *perspectiva* in the *Opus tertium*:

> For every *efficiens* acts (*agit*) by its power (*virtus*), which works in the underlying[45] matter, as the light (*lux*) of the sun works its power in the air, which is light (*lumen*) diffused through the entire world (*per totum mundum*) by solar light. And this *virtus* is called likeness (*similitudo*), and image (*imago*), and species (*species*) and with many names, and both substance and accident do this, and spiritual moreso than corporeal [sc. substances or accidents]. And this *species* makes every activity of this world; for it acts on the senses, on the intellect, and on all the matter of the world through the generation of *res*[46] because one and the same thing comes to be by a natural agent in whatsoever it acts because it does not have [sc. the power of] deliberation. And, therefore, whatever it encounters, it does the same thing. But if it acts (*agat*) on the senses and intellect, a *species* [sc. an image, a likeness] comes to be, as everyone knows [from Aristotle's *De anima*, Books 2 and 3]. Therefore, on the other hand, *species* also comes to be in matter. And in those things which have reason (*ratio*) and intellect, although they do many things in accordance with the deliberation and choice of the will (*secundum deliberationem et electionem voluntatis*), nevertheless, this activity (*operatio*), which is the generation of species (*generatio speciei*), is natural in them just as in the others [sc. which have no deliberation]. Wherefore, the substance of the soul (*substantia animae*) multiplies its power (*virtus*) in the body and beyond the body, and each body makes its power beyond itself, and the angels move the world by these sorts of powers.

---

[43] Roger refers here to change in place.

[44] Secundo possumus videre ex propriis, quod nihil horum inferiorum nec superiorum sciri potest sine mathematicae potestate. Nam omnis res naturalis producitur in esse per efficiens et materiam in quam operatur, nam haec duo concurrunt primo. Agens enim per suam virtutem movet et transmutat materiam, ut fiat res. Sed virtus efficientis et materiae sciri non potest sine magna mathematicae potestate, sicut nec ipsi effectus producti. Sunt ergo haec tria, efficiens, materia et effectus. Et in coelestibus fit mutua influentia virtutum, ut lucis et aliarum, et est in eis alteratio, licet non ad corruptionem. Et sic potest ostendi, quod nihil in rebus sciri potest sine geometriae potestate (110, 21–32).

[45] *Subjectam* is the literal Latin rendering of the Greek "hypokeimenon [Gk]."

[46] Et haec species facit omnem operationem huius mundi; nam operatur in sensum, in intellectum, et in totam mundi materiam per rerum generationem […].

> But God makes powers (*virtutes*) from nothing (*de nihilo*), which multiply in things; created agents (*agentia creata*) do not, but in another manner, about which we are not concerned at present.[47]

In this dense passage, Roger lays out the conceptual groundwork that we will see developed in what follows. Although it is not yet obvious how, these structures provide the basis for Roger's geometrical analysis.

After this generalized analysis of agents and their powers (*virtutes*)—both how they work and the range of substances and accidents on which they act—Bacon concludes:

> Powers of agents of this sort, therefore, make every activity (*operatio*) in this world. But these things may be attended to in two ways: one is this multiplication of *species* and *virtus* from the place of its generation;[48] and the other is the various workings in the world (*operatio varia in hoc mundo*) because of the generation and corruption of things (*propter rerum generationem et corruptionem*). The second cannot be known (*sciri non potest*) without the first; therefore, it is fitting that this multiplication be described first.[49]

This last phrase sets out *in nuce* the structure for the rest of the section *in physicis*: first (112–27) Bacon sets out the basic patterns of his doctrine of the multiplication of species. It is basically a fleshed out, better organized version of the doctrine presented in Robert Grosseteste's *De lineis, angulis et figuris* (60, 30–65, 24).[50]

Roger then turns to a series of physical problems (127–174), which he uses to exemplify how the doctrine of multiplication of species can help to clarify and explain difficult physical problems. In this overall structure, as well as in many

---

[47] Omne enim efficiens agit per suam virtutem quam facit in materiam subiectam, ut lux solis facit suam virtutem in aere, quae est lumen diffusum per totum mundum a luce solari. Et haec virtus vocatur similitudo, et imago, et species et multis nominibus, et hanc facit tam substantia quam accidens, et tam spiritualis quam corporalis. Et substantia plus quam accidens, et spiritualis plus quam corporalis. Et haec species facit omnem operationem huius mundi; nam operatur in sensum, in intellectum, et in totam mundi materiam per rerum generationem, quia unum et idem fit ab agente naturali in quodcunque operetur, quia non habet deliberationem; et ideo quicquid ei occurrat facit idem. Sed si in sensum et intellectum agat, fit species, ut omnes sciunt. Ergo in contrarium, et in materiam fit species. Et in his quae habent rationem et intellectum, licet multa faciant secundum deliberationem et electionem voluntatis, tamen haec operatio, quae est generatio speciei, est naturalis in eis sicut in aliis. Unde substantia animae multiplicat suam virtutem in corpore et extra corpus, et quodlibet corpus extra se facit suam virtutem, et angeli movent mundum per huiusmodi virtutes. Sed Deus facit virtutes de nihilo, quas multiplicat in rebus; agentia creata non sic, sed alio modo de quo non est ad praesens curandum (111, 1–23).

[48] [...] unum est ipsa multiplicatio speciei et virtutis a loco suae generationis [...].

[49] Huiusmodi ergo virtutes agentium in hoc mundo faciunt omnem operationem. Sed duo sunt modo attendenda circa ista: unum est ipsa multiplicatio speciei et virtutis a loco suae generationis; et aliud est operatio varia in hoc mundo propter rerum generationem et corruptionem. Secundum sciri non potest sine primo. Et ideo oportet primo ipsam multiplicationem describi (111, 23–29).

[50] For the text, see Ludwig Baur, "Die Philosophischen Werke des Robert Grosseteste," *Beiträge zur Geschichte der Philosophie des Mittelalters*, ix (1912). For a penetrating analysis of Roger's views on the multiplication of species, see Lindberg, *Roger Bacon's Philosophy of Nature*, and his *Roger Bacon and the Origins of* Perspectiva. See now also Mary Quinlan-McGrath, *Influences*, who uses and develops Lindberg's analysis primarily to interpret Marsilio Ficino's *De vita libri tres*, as I discuss more fully in volume II. She also usefully discusses Roger Bacon, Albertus Magnus and Thomas Aquinas. I discuss Grosseteste further just below.

points of detail, Bacon follows the general structure of Grosseteste's treatment in the two closely connected works *De lineis* and *De natura loci*. In these analyses, the geometrical dimension will be revealed. We should also note that Roger's discussion of action here in terms of the multiplication of species is precisely the same as that referred to as *perspectiva* in his criticism of the unnamed master at *Opus tertium* XI, the profound ignorance of which Roger very strongly claimed for him.

2.4 Bacon's Natural Philosophy *De rebus*

2.4.1 De natura loci

Now that we have a firmer grasp of the basic structure of Roger's views on mathematics and *perspectiva* (including the multiplication of species), we should turn to the detailed analysis of a specific problem, "the nature of place," which, we will find, is essential for understanding Roger's views of generation.[51] After outlining the basic structures of his geometrical-optical model of action (to which we will return), Roger discusses a series of physical problems that he uses to illustrate the power of his natural philosophical model, and thereby establish the power of geometry for analysis *de rebus*:

> Having presented these sorts of principles by the pathways of geometry (*per vias geometriae*), a person can verify every action of nature because every truth regarding the operation of an agent in a medium, whether in generable matter (*in materiam generabilem*), or in the celestial bodies (*in coelestia*), indeed, in the entire system of the world (*in totam mundi machinam*), takes its origin either mediately or directly from what has already been said, and from certain similar things. Because I could not put everything in this *persuasio* which the *Opus maius* requires, I wish to clarify what I am saying by means of some examples among diverse things (*in diversis rebus*) in the world. I will begin from above (*a superioribus*).[52]

The next 47 pages in Bridges's edition are then devoted to this series of examples. I will focus on one of them in the remainder of this chapter: Bacon's analysis of the generation and complexion of things as fundamentally influenced by their place, the problem *De natura loci* (IV, 137–9). I also discuss supporting material from elsewhere in *Opus maius* IV.

*

[51] On Bacon's view of place, see P. Gautier Dalché, "Vers une *perfecta locorum doctrina*: Lieu et espace géographique selon Roger Bacon," in *Représentations et conceptions de l'espace dans la culture médiévale. Repräsentationsformen und Konzeptionen des Raums in der Kultur des Mittelalters*, T. Suarez-Nani and M. Rohde (eds.), Berlin: De Gruyter, 2011, 9–43. My thanks to David Juste for this reference.

[52] His principiis et huiusmodi datis per vias geometriae, potest homo verificare omnem actionem naturae, quia omnis veritas circa operationem agentis in medium, vel in materiam generabilem, vel in coelestia, et in totam mundi machinam, sumit ortum mediate vel immediate ex jam dictis, et quibusdam similibus, quia non potui omnia in hac persuasione ponere, quae opus maius requirit. Et quod dico manifestare volo per aliqua exempla in diversis rebus mundi, et incipiam a superioribus (127, 14–22).

I will first situate Bacon within his historical context by briefly comparing what we have just examined in the *Opus maius* with Robert Grosseteste's *De lineis, angulis et figuris seu de fractionibus et reflexionibus radiorum* (*"On Lines, Angles and Figures," or, "On the Refractions and Reflections of Rays"*) and *De natura locorum (On the Nature of Places)* that were both composed ca. 1230.[53] It is well known that Bacon was deeply influenced by Grosseteste in both content and method, and this will be immediately apparent in what follows.[54] Bacon also refers to Grosseteste explicitly in his works.[55] As we will see, much of the structure and a good deal of the content of *Opus maius* IV, *In physicis*, closely follows Grosseteste's aforementioned works. In fact, *De lineis* provided Bacon's model for the brief treatment of the multiplication of species (IV, 112–127), and its introduction provided much of the material for Bacon's programmatic passages on mathematics and *perspectiva* just examined (109–11).[56]

After presenting his geometrical-optical model of action in terms of the multiplication of species in *De lineis*, Grosseteste turns to his *De natura locorum*, one in a series of examples meant to illustrate the preceding principles. It begins with a passage strikingly similar, in both content and language, to that in which Bacon began his section of examples:

> When these rules and roots and foundations have been given from the power of geometry, the serious investigator into natural things (*res naturales*) can give the causes of all natural effects by this path.[57] And it will be impossible otherwise, as is already clear in general, since every natural act is varied, with respect to strength and weakness, by means of the variety of lines, angles and figures. But this same thing is more obvious in particulars, and, first, in a natural action made on matter and afterward on the senses, as the truth of geometry will make completely clear. Moreover, the first and greatest variety of nature is in the places of the world (*prima autem et maxima variatio naturae est in locis mundi*), and this is

[53] According to Richard W. Southern, *Robert Grosseteste: The Growth of an English Mind in Medieval Europe,* 2nd ed., Oxford: Clarendon Press, 1992, 120.

[54] As pointed out long ago by Ludwig Baur in the introduction to his edition of Grosseteste's philosophical writings (cited above), which Southern characterized as "epoch making." Much of the material in Southern's book is relevant to the intellectual and social context of Roger's work. Lindberg also noted Grosseteste's influence on Bacon; "Roger Bacon and the Origins of *Perspectiva*," 258: "Bacon's own theoretical position was an extension, or development, of Grosseteste's, enriched by a much more thorough immersion in the new Greek and Arabic sources. Bacon was surely influenced by Grosseteste's writings, as well as the sources available to Grosseteste (Plato, Euclid, Tideus, al-Kindi, Aristotle, and Avicenna), but he also had at his disposal the works of Ptolemy and Alhazen, where the promise of geometrical optics had been more completely fulfilled."

[55] For example, in the final part of the first section (*De scientiis*) of book IV: "Ad omnem autem confirmationem potest ratio ultima sumi ex experientia sapientium; nam omnes sapientes antiqui laborarunt in mathematica, ut omnia scirent, sicut nos vidimus de aliquibus nostri temporis, et audivimus de aliis, qui per mathematicam, quam bene sciverunt, omnem scientiam cognoverunt. Inventi enim sunt viri famosissimi, ut Episcopus Robertus Lincolniensis [i.e. Grosseteste], et Frater Adam de Marisco, et multi alii, qui per potestatem mathematicae sciverunt causas omnium explicare, et tam humana quam divina sufficienter exponere (108, 6–14)."

[56] This is obvious on even a cursory examination. This material deserves much fuller analysis than I can devote to it here.

[57] It is obvious that Grosseteste also provides Bacon with his model for making bold claims.

> especially to be considered by the natural philosopher (*maxime consideranda est a naturali philosopho*), since the variety of things located (*varietas locatorum*) corresponds to the different natures of places (*secundum varias naturas locorum*).[58]

Grosseteste then jumps abruptly into his first problem on the different heating of different places. Bacon does not closely follow Grosseteste either in the problems treated or in their order of presentation.[59] In fact, Bacon's treatment is by far the more orderly, proceding as it does from the heavens by steps down to matter, thus reflecting the hierarchical structure of the world, whereas Grosseteste's presentation is far more abrupt, and with an order not only not explicitly set out, as Bacon's is, but with an order often difficult to discern.[60]

*

Bacon's first physical example, on the complexion of places, explains why this knowledge is crucial to a natural philosopher. We will attend closely to his patterns of exposition. The relevant passages are brief and worth treating in detail.[61] "And therefore I am returning to what was proposed," he begins, about to make one of his characteristically bold claims:

> saying that (1) complexions of the places of the world (*complexiones locorum mundi*) cannot be discovered unless a person knows the aforementioned laws of multiplications [sc. of species], since one will neither avoid what is false, nor be able to confirm what is true. But (2) place is a principle of generation (*locus est principium generationis*), as also a father is, as Porphyry says (a). And we see (b) that everything is varied in accordance with the diverse places of the world (*omnia variantur secundum loca mundi diversa*), not only among natural things (*non solum in naturalibus*), but also people in their customs (*sed homines in moribus*), since Ethiopians have different customs, Spaniards other, Romans other, and Gauls other. For the Picards, who are neighbors to the true Gauls, have such a diversity in customs and language that one cannot help but wonder why there exists such a diversity in nearby places.[62]

---

[58] "His igitur regulis et radicibus et fundamentis datis ex potestate geometriae, diligens inspector in rebus naturalibus potest dare causas omnium effectuum naturalium per hanc viam. Et impossibile erit aliter, sicut iam manifestum est in universali, quando variatur omnis actio naturalis penes fortitudinem et debilitatem per varietatem linearum, angulorum et figurarum. Sed in particulari magis est manifestum istud idem, et primo in actione naturali facta in materiam et postea in sensum, ut complete pateat veritas geometriae. Prima autem et maxima variatio naturae est in locis mundi, et maxime consideranda est a naturali philosopho, quoniam secundum varias naturas locorum est varietas locatorum (65, 27–66, 9)." The text here comes from Baur's edition cited above.

[59] A close investigation of this relationship would also be worthwhile, but is out of place here.

[60] Grosseteste's organization here seems much more worthy of Lindberg's epithet "helter-skelter" than Bacon's is; Lindberg, "On the Application of Mathematics to Nature," 19 (n. 72).

[61] Although Roger treats this problem briefly here, it may be significantly supplemented by other passages from *Opus maius* IV, as we will see below.

[62] Et ideo redeo ad propositum dicens quod complexiones locorum mundi inveniri non possunt, nisi homo sciat leges multiplicationum praesignatas, quoniam nec falsum vitabit, nec verum poterit confirmare. Sed locus est principium generationis, quemadmodum et pater, ut dicit Porphyrius. Et nos videmus, quod omnia variantur secundum loca mundi diversa non solum in naturalibus, sed homines in moribus; quoniam alios mores habent Aethiopes, alios Hispani, alios Romani, et alios Gallici. Nam et Picardi, qui sunt veris Gallicis vicini, habent tantam diversitatem in moribus et in

Here is Roger's big claim: (1) the complexion of the places of the world simply cannot be discovered without the laws of multiplications. This is crucially significant for natural philosophy—as we will see in more detail here, but in much more below—because, as he goes on to say, (2) place (*locus*) is a principle of generation. He confirms this with (a) the citation of an authority (Porphyry), and then (b) an argument based on experience (*nos videmus*) in the realms of both nature and culture.

Roger then sharpens the picture by providing the causal basis of his theory. Here we can begin to discern its mathematical character:

> But since the things of this world (*res huius mundi*) which have been established in diverse places, however close they are, receive cones of diverse pyramids coming from the entire heavens opposite them (*recipiunt conos diversarum pyramidum venientium a toto coelo objecto eis*), therefore unbounded diversity (*infinita diversitas*) arises. For the cones of individual pyramids (*coni pyramidum singularum*) come to single points of the earth (*ad singula puncta terrae veniunt*), and each point [sc. of the earth] is the center of a new horizon (*quilibet punctus est centrum unius Horizontis novi*).[63]

Roger thus begins to develop the analysis only adumbrated above in his general programmatic claims regarding mathematics, *perspectiva* and astrology (*Opus maius*, IV, 110): (1) that *res* cannot be known except by their causes, and (2) that these causes are the heavens, which have now been placed into the geometrical-optical framework described at pp. 112–127 (to be discussed more fully below). Roger states here that each point on earth is geometrically unique based on its unique angular relationship to the celestial situation. We will come to see this as a central feature of his system.

Roger then illustrates the profound implications of his analysis:

> And therefore we see that two plants born at the same time from the earth without anything in between them (*sine medio*), and therefore two twins in the womb of their mother, are allotted a difference of complexion (*sortiuntur diversitatem complexionis*), so that afterward they would have diverse customs, and they would pursue different arts and different occupations throughout their entire lives. And therefore the powers of the heavens and the stars (*virtutes coelorum et stellarum*) produce everywhere things diverse in properties and natures (*producunt ubique diversas res in proprietatibus et naturis*), that is, in *res* generated by propagation.[64]

---

lingua ut non sine admiratione possit esse unde sit tanta diversitas locorum propinquorum (137, 29–138, 10).

[63] Sed quoniam res huius mundi in diversis locis constitutae, quantumcunque propinquae sunt, recipiunt conos diversarum pyramidum venientium a toto coelo objecto eis, ideo accidit infinita diversitas. Nam ad singula puncta terrae veniunt coni pyramidum singularum, et quilibet punctus est centrum unius Horizontis novi (138, 11–16).

[64] "Et ideo videmus, quod duae herbae simul nascuntur de terra sine medio, et ideo duo gemelli in ventre matris sortiuntur diversitatem complexionis, ut postea mores habeant diversos, et sequantur artes diversas, et occupationes difformes per totam vitam. Et ideo virtutes coelorum et stellarum producunt ubique diversas res in proprietatibus et naturis, et in rebus generatis secundum propagationem (138, 16–23)." John Pecham has something quite similar in his *Perspectiva communis*; edited—with an introduction, English translation, and critical notes by David C. Lindberg in *John Pecham and the Science of Optics:* Perspectiva Communis, Madison: University of Wisconsin Press, 1970, proposition 6 (I, 44–50).

We learn here not only that celestial powers must be taken into account to understand a natural thing's complexion, but also its place of generation. Without going any further into the complexities of the medieval doctrine of complexions, the centrality of a thing's or a person's complexion should be noted, that is, the particular character of each individual's unique mix of qualities within the overarching structure of their specific form. One's complexion has deep ramifications in every aspect of life, as discussed in more detail just below.

Roger then permits us to see how a thirteenth-century natural philosopher understood the complex of forces at play in conception and birth:

> Not only the multiplication of the *virtus coelestis* is operative (*operatur*), but also the *virtus* of the father and mother, since the *virtutes* of the father are passed off in the semen, as the physicians (*medici*) teach. And especially from the soul of the mother is the multiplication of *virtus* and *species* on the fetus continued until the completion of gestation, and parturition. And when a child at birth (*in nativitate*) is exposed to a new air (*aeri novo*) as to another world, then he receives cones of celestial pyramids with respect to his individual parts (*recipit conos pyramidum coelestium secundum singulas partes*), and thus he receives new impressions which never leave (*recipit impressiones novas, quas nunquam dimittit*), because "what a new [= young] head grasps, it knows forever" (*quod nova testa capit, inveterata sapit*).[65] And then the root complexion (*complexio radicalis*) is confirmed, which always remains until the end of life, although the current complexion (*complexio currens*) changes with every day. And it is this root complexion which the inclinations to customs and to sciences and to languages follow, and to whatever handiworks and occupations, and to every difference that we see in everyone (*ad omnem diversitatem quam videmus in omnibus*).[66]

Roger then goes on to nuance this picture: if the heavens are disposed well or ill during the gestation or parturition of a child, then his complexion will be alternately helped or harmed,[67] and what follows from that, but we need not follow him. Roger thus provides here his version of the natural philosophical foundations for nativities.

In this section, Roger has shown the necessity of his doctrine of the multiplication of species for understanding the complexions of things in the world, a central factor in our understanding of *res naturales*. And he has shown that celestial powers or virtues are not the only necessary causal factors, but that there are other relevant powers/virtues as well, namely, those from the parents and the place of generation. The astrological component—celestial influences—is central, however, and directly

---

[65] A. G. Little's edition of a new fragment of the *Opus tertium* provides the reference for this tag from Horace, *Epistles*, i.2, 69–70; *Part of the* Opus Tertium *of Roger Bacon*, Aberdeen: The University Press, 1912, 5 (n. 6).

[66] Et non solum multiplicatio virtutis coelestis operatur, sed patris et matris, quoniam discinduntur virtutes in seminibus, ut docent medici. Et praecipue ab anima matris continuatur multiplicatio virtutis et speciei super foetum usque in complementum generationis et nativitatis. Et cum puer in nativitate exponitur aeri novo tanquam alteri mundo, tunc recipit conos pyramidum coelestium secundum singulas partes, et sic recipit impressiones novas, quas nunquam dimittit, quia quod nova testa capit, inveterata sapit. Et tunc confirmatur complexio radicalis quae semper manet usque ad finem vitae, licet complexio currens mutetur tota die. Et ad hanc radicalem complexionem sequuntur inclinationes ad mores et ad scientias et ad linguas, et ad quaecunque artificia et negotia, et ad omnem diversitatem quam videmus in omnibus (138, 23–37).

[67] Et si coeli dispositio sit mala in conceptione et nativitate pueri, tunc coni pyramidum laedunt complexionem, [etc] (138, 37–139, 1).

effects both levels of a complexion, the root and the current, as well as the nature of place, which, as we learn here, is a principle of generation.

Like Thomas in *De operationibus occultis naturae* 16, Roger used the celestial configuration to account for the variety among individuals. Of course, Roger here goes into a much more detailed causal analysis, but one which coheres well with Thomas's framework. Nevertheless—and whatever appearances to the contrary notwithstanding—Roger also makes it quite explicit that none of this involves determinism, for even those born with a poor or mediocre complexion "can change themselves by the freedom of the will, and by the grace of God, and by the temptation of the devil, and by good or bad counsel, especially in childhood."[68]

2.4.2 *Geographia*

In order to more fully understand Roger's ideas on the nature of place, we should turn now to another section of *Opus maius* IV, entitled *Geographia* (pp. 286–374). It breaks down into two well-defined parts: (1) a theoretical introductory section (286–304), and (2) a practical descriptive geography (304–74). I will only treat a small part of the introductory section here, which further develops the material just discussed.[69]

After an introductory paragraph on the necessity of mathematics for directing the republic of the faithful, Roger turns directly to our subject:

> And although—when we treated the defense of mathematics, and before that, the comparison of celestial powers (*virtutes*) to things below—the understanding of the places of the world and of the things generable in them by the celestial bodies were touched on. Nevertheless, I will now set this out more fully in passing over to medicine because of the human body, whose understanding is more necessary to men than any other thing in this world. Not only do I show how things (*res*) in diverse places of the world are understood, but also how they are caused in the same places through the different passages of time. But an effect is not understood except by its cause, as is certain to all; but the celestial bodies (*coelestia*) are the causes of the *inferiora*, whence it is fitting that they know what is generable (*generabilia*) by means of what is ungenerable (*ingenerabilia*), which are the celestial bodies (*coelestia*).[70]

---

[68] "[...] licet poterit se mutare per libertatem arbitrii, et per gratiam Dei, et per tentationem diaboli et per bonum aut malum consilium, maxime a iuventute (139, 7–9)." Roger treats the subject of determinism more fully at the beginning of the section entitled *Iudicia astronomiae*, which I treat in detail in Chap. 4.

[69] There is an English translation of the parts of the geography section "of interest to the history of geographical thought," by Herbert M. Howe, 1996, and published online (http://www.geography.wisc.edu/histcart/bacon.html). Here are the author's caveats about his translation: "Although it is an improvement over the only other English translation, by Robert Belle Burke (Philadelphia: University of Pennsylvania Press, 1928), it is a working document which includes queries and notes in square brackets. It is not intended as a fully annotated and polished translation and should not be quoted as such. It is posted here for the convenience of researchers." It came to my attention after making my translation. I have compared it with mine and made alterations accordingly.

[70] Et licet, ubi actum est de excusatione mathematicae, atque superius de comparatione virtutum coelestium ad haec inferiora, sit tactum de cognitione locorum mundi et rerum generabilium in eis per coelestia, tamen nunc uberius hoc exponam transeundo ad medicinam propter corpus huma-

This is all pretty straightforward, and is in some measure a repetition of material already covered, as Roger himself acknowledges. We should also note that Roger explicitly introduces the all-important element of time into his picture here, as well as indicating its importance for medicine. He continues with a discussion of *coelestia* as not only universal, but also as particular causes,[71] and not only of inanimate, but also of animate beings. Although it is of much interest with regard to Roger's views on astrological action in the world, it is not directly germane to our current problem, and so must be passed over here.

We pick up the trail of our interests just below:

> And if we were to descend further, we can investigate more closely the causes of things below by the *coelestia*. But first is this article, that each point of the earth is the cone of one pyramid of celestial power (*quilibet punctus terrae est conus unius pyramidis virtuosae coeli*). In order that what we are intending be seen more certainly and plainly, it is necessary to consider what the diversity of the regions of the world is, and how the same region is varied at different times, and how diverse *res* of the same region receive diverse *passiones* [sc. have different experiences (i.e. receive different celestial influences)] at the same time. But these cannot be known unless we distinguish the quantity and shape of the habitable earth and its climates.[72]

Roger then proceeds to provide the mathematical-geographical substructure without which, in his view, the higher-level natural-philosophical understanding is simply impossible.[73] He provides a 13-page "essentials of mathematical geography" (288–300), very much in the tradition and language of Sacrobosco's *De sphaera*. There is no need to follow him in the details of this exposition.[74]

Bacon then discusses why mathematical geography is essential for his natural philosophical concerns:

> (1) Therefore, since the utility of understanding the places of the world is the greatest (*locorum mundi cognitionis maxima utilitas est*), for this reason it is fitting to offer another description. For the things of the world cannot be known (*res mundi sciri non possunt*)

num, cuius cognitio magis est homini necessaria quam alicuius rei alterius in hoc mundo. Et non solum manifesto quomodo res in diversis locis mundi cognoscuntur, sed quomodo in eisdem per temporum diversa curricula causantur. Effectus vero non cognoscitur nisi per suam causam, ut certum est apud omnes; sed coelestia sunt causae inferiorum, unde oportet quod sciant haec generabilia per ingenerabilia quae sunt coelestia (286, 29–287, 10).

[71] This issue of the celestial bodies as universal or general vs. particular causes becomes very important for Pico's critique of astrology's natural philosophical foundations, as we will see in volume II.

[72] Et si ulterius descendamus, possumus causas rerum inferiorum magis prope investigare per coelestia. Primo vero articulus hic est, quod quilibet punctus terrae est conus unius pyramidis virtuosae coeli. Ut hoc autem certius planiusque videatur quod intendimus, necesse est considerare quae sit diversitas regionum mundi, et quomodo eadem regio in diversis temporibus variatur, et quomodo res diversae ejusdem regionis diversas recipiunt passiones in eodem tempore. Sed haec sciri non possunt, nisi quantitatem et figuram habitabilis terrae et climata eius distinguamus (288, 7–17).

[73] Quatenus vero ad haec deveniamus, oportet nos supponere mundum esse sphaericae figurae, [etc] (288, 17–18).

[74] For more on Bacon and geography, see David Woodward with Herbert M. Howe, "Roger Bacon on Geography and Cartography," in *Roger Bacon and the Sciences*, 199–222.

> except through knowledge of the places in which they are contained (*nisi per notitiam locorum in quibus continentur*). For place is a principle of the generation of things (*locus enim est principium generationis rerum*), as Porphyry says, because the diversity of things corresponds to the diversity of places, and not only of natural things, but also of moral and scientific, as we see among men that, in accordance with the diversity of regions, they have diverse customs, and they occupy themselves in diverse arts and sciences. Therefore, because philosophy introjects itself into the things of the world, it has been greatly lacking to the Latins up to this point because[75] it has no verification of the places of the world. (2) But this verification is based on understanding the longitude and latitude of any place. For then we know under which stars each place is (*sciremus sub quibus stellis est quilibet locus*), and how far it is from the path of the sun and planets [sc. the ecliptic and zodiac] (*quantum a via solis et planetarum*), and of which planets and signs the places receive dominion (*quorum planetarum et signorum loca recipiant dominium*), which all make the diverse complexions of places (*quae omnia faciunt diversas complexiones locorum*). If they were known, one could know the complexions of every *res* in the world, and the natures (*naturae*) and properties (*proprietates*) which they draw from the *virtus* of a place (*quas a virtute loci contrahunt*).[76]

The first part of this passage is basically a recapitulation, reemphasizing what went before.

The second part, however, adds significant new information. First, Roger explicitly states that the reason a sound knowledge of mathematical geography is essential is that, without it, the natural philosopher cannot accurately co-ordinate the equally necessary information he requires from mathematical astronomy. It is with the information gleaned from both, then, that the essential mathematical-astronomical and -geographical framework can be constructed, within which—and on the basis of which—one may then move on to the more properly natural philosophical pursuit of understanding the nature of places in the world.

A proper knowledge of place is essential to the natural philosopher, as Roger repeatedly emphasizes, because place is a principle of generation. Thus, if a natural philosopher wishes to fully understand the nature of generated things, a proper understanding of their place of generation is required, whose power (*virtus*), as he just stated, is imparted to what is generated there. Sacrobosco provides the calibrated mathematical-astronomical and -geographical framework in *De sphera*, sometimes pointing towards, but rarely sojourning into the more properly natural philosophical realms with which Roger is so profoundly concerned.[77]

---

[75] *postquam* is printed; should it be emended to *propter*?

[76] Quoniam igitur locorum mundi cognitionis maxima utilitas est, ideo aliam descriptionem oportet afferri. Nam res mundi sciri non possunt nisi per notitiam locorum in quibus continentur. Locus enim est principium generationis rerum, ut dicit Porphyrius; quia secundum diversitatem locorum est diversitas rerum; et non solum naturalium, sed moralium et scientialium, ut videmus in hominibus quod secundum diversitatem regionum habent mores diversos et occupant se in artibus et scientiis diversis. Quia igitur philosophia intromittit se rebus mundi, multum ei deest adhuc apud Latinos, [postquam?] <*propter*? > non habet certificationem locorum mundi. Sed haec certificatio stat in cognitione longitudinis et latitudinis cujuslibet loci; tunc enim sciremus sub quibus stellis est quilibet locus, et quantum a via solis et planetarum, et quorum planetarum et signorum loca recipiant dominium, quae omnia faciunt diversas complexiones locorum: quae si scirentur, possit homo scire complexiones omnium rerum mundi et naturas et proprietates quas a virtute loci contrahunt (300, 33–301, 16).

[77] These little sojourns seem to me not unimportant; see especially, *The Sphere of Sacrobosco*, 77–9 and 87–8.

This integrated cosmographic framework then receives Roger's other major natural philosophical structure, namely, the multiplication of species, with its mathematicized apparatus derived from geometrical optics (*perspectiva*). The general notion that each point on earth is geometrically unique is only interesting to a limited extent in itself. But when each particular point is the unique geometrical terminus of a given celestial configuration, each planetary element of which provides its own unique qualities with a range of intensities by means of rays, which individual intensities are determined in part by their angles of incidence, and which all together contribute to the particular qualitative complexion of that place at that moment, we then find ourselves with a richly mathematical system of great sophistication and interest. That this mathematicized natural philosophy is also deeply astrological provides a further level of interest.[78]

*

We will now turn our attention away from the broader cosmographic framework and zoom in to examine more closely the central feature of Bacon's causal analysis: pyramids of force. He states it clearly in three passages recently examined, and in much the same language:

1. But since the things of this world (*res huius mundi*) which have been established in diverse places, however close they are, receive cones of diverse pyramids coming from the entire heavens opposite them (*recipiunt conos diversarum pyramidum venientium a toto coelo objecto eis*), therefore unbounded diversity (*infinita diversitas*) arises. For the cones of individual pyramids (*coni pyramidum singularum*) come to single points of the earth (*ad singula puncta terrae veniunt*), and each point [sc. of the earth] is the center of a new horizon (*quilibet punctus est centrum unius Horizontis novi*, 138, 11–16).
2. And when a child at birth (*in nativitate*) is exposed to a new air (*aeri novo*) as to another world (*tamquam alteri mundo*), then he receives cones of celestial pyramids with respect to his individual parts (*recipit conos pyramidum coelestium secundum singulas partes*), and thus he receives new impressions which never leave (*recipit impressiones novas, quas numquam dimittit*, 138, 28–31)[.]

[78] Clulee ("Astrology, Magic and Optics: Facets of John Dee's Early Natural Philosophy," *Renaissance Quarterly* 30 [1977]: 632–80), and Shumaker and Heilbron (*John Dee on Astronomy:* Propaedeumata Aphoristica *(1558 &1668), Latin and English*, Berkeley: University of California Press, 1978) note and discuss Roger's astrologizing natural philosophy in some depth in their valuable works, primarily as background for understanding John Dee's *Propaedeumata aphoristica* (1558, 1568). Grant, unfortunately, omits this tradition—including Grosseteste, whose cosmology he explicitly excludes from consideration as "idiosyncratic" (*Planets, Stars, and Orbs*, 42–4). My Chap. 2 may thus be used to fill this gap in Grant's account, and thus begin to articulate more accurately the relationship between astrology and Aristotelian natural philosophy in the thirteenth century.

3. [W]e can investigate more closely the causes of the things below by the *coelestia* (*causas rerum inferiorum magis prope investigare per coelestia*). But first is this article, that each point of the earth is the cone of one pyramid of heavenly power (*quilibet punctus terrae est conus unius pyramidis virtuosae coeli*, 288, 7–10).

In order to understand what Bacon is doing here, I will reconstruct his causal analysis in two stages, which correspond directly to the patterns we found in the programmatic passage at *Opus maius* IV, 109–11—and thus in the passages of the *Opus tertium* referring and relating to the unnamed master—in terms of (1) mathematics and (2) *perspectiva*. First I treat *perspectiva*, namely, his views about pyramids of force as articulated in the geometrical-optical analysis at *Opus maius* IV, 112–27. Bacon describes there, in Grosseteste's manner, his abstract geometrical-optical model of action in terms of lines, angles and figures in order to provide a universal mathematical analysis for every kind of action. Secondly, I discuss how the abstract geometrical-optical model is fitted into the integrated mathematical-astronomical and -geographical cosmographic framework, as briefly outlined at *Opus maius* IV, 109–10, and as more fully articulated, as we just saw, at *Opus maius* IV, 288–300.

2.4.3 Roger Bacon's Geometrical-Optical Model of Action

As Roger indicated so clearly at *Opus maius* 111, all actions work as *species*, whether "on the senses, on the mind, or on the entire matter of the world through the generation of things."[79] In other places, he uses the geometrical-optical model to explain vision, in particular, in greater detail.[80] In fact, vision and the type of action with which we are concerned here—on places and the generation of people and things—both rely on the same kind of punctiform analysis which al-Kindi seems to have been the first to develop, and which Alhazen seems to have been the first to apply to the case of vision.[81] Bacon is not concerned here to apply his geometrical-optical model

[79] Et haec species facit omnem operationem hujus mundi; nam operatur in sensum, in intellectum, et in totam mundi materiam per rerum generationem (111, 7–10).

[80] See David C. Lindberg, *Theories of Vision from al-Kindi to Kepler*, Chicago: University of Chicago Press, 1976, 107–16. See also the first chapter of Katherine H. Tachau, *Vision and Certitude in the Age of Ockham: Optics, Epistemology, and the Foundations of Semantics, 1250–1345*, Leiden: Brill, 1988.

[81] See Lindberg's lucid discussion, *Theories of Vision*, 71 ff. Born in Basra in Iraq in 965 CE, Alhazen (Al-Hasan ibn al-Hasan ibn al-Haytam, d. 1040 in Cairo) was one of the foremost scientists in Arabic culture, especially for his contributions to mathematical and experimental optics, and for his cosmological interpretation of Ptolemy's mathematical astronomy (Hasse, *Success and Suppression*, 334–36). Al-Kindi (Ya'qub ibn Ishaq al-Kindi, d. 861–66 AD) was the first great philosopher and scientist of Arabic culture. He was the leading figure of a circle of translators from Greek into Arabic in Baghdad, and he wrote prolifically on almost all of the sciences in the Greco-Arabic tradition, except alchemy (336–39). I provide more references to scholarship on al-Kindi just below.

to vision,[82] but rather to showing how celestial influences act in relation to places in the world. One of the central explanatory features of his analysis is its account of the diversity of places, which then becomes—in the next step of his theory—a central feature for explaining how the diversity in *res naturales* comes into existence because, as Porphyry says, *locus* is a *principium* of generation.

The central feature of Roger's model of action is what Lindberg calls "pyramids of force." Treating the model abstractly, Bacon is quite emphatic that this pyramidal figure is of particular importance:

> And this is the figure which nature especially chooses in every multiplication and action, and not any pyramid, but that whose base is the surface of the *agens*, and whose cone falls on some point on the *patiens*, because thus the species can come from the entire surface of the *agens* (*potest a tota superficie agentis species venire*) to the individual points of the *patiens*[83] (*ad singula puncta patientis*) through an infinite number of individual pyramids (*per singulas pyramides et infinitas*), as is shown in the figure.[84] For an infinite number of rays come-to-be from each point of the *patiens* (*a quolibet puncto patientis fiunt radii infiniti*), and therefore they can be combined in an infinite number of ways, so that there come-to-be an infinite number of round pyramids, for all of which there is one base, namely, the surface of the entire *agens*; and to each part of the *patiens* comes one cone of one pyramid (*ad quamlibet partem patientis venit unus conus unius pyramidis*), so that the *virtus* comes from the entire *agens* to each point of the *patiens* (*ut virtus veniat a toto agente ad quodlibet punctum patientis*), and not from some determinate part [sc. of the *agens*] […].[85]

An *agens* thus acts on a *patiens* by these pyramids of force, which come from the entire surface of the *agens* (whatever it may be) to each of the infinite number of individual points on the surface of the *patiens* (whatever it too may be), thus generating an infinite number of such pyramids.

Bacon also describes structurally how the *agens* acts:

> It is fitting that the multiplication comes to be spherically (*multiplicatio fiat sphaerice*). For the *agens* multiplies itself equally on every side (*agens multiplicat se aequabiliter in omnem partem*), and in accordance with every diameter and every difference of position, which are up, down, fore, aft, right and left. Therefore, lines come forth from everywhere on all sides from an *agens* (*undique exeunt lineae in omnem partem ab agente*), just as from the center […].[86]

---

[82] He does this, rather, in *Opus maius* V, the following book; see Lindberg, *Roger Bacon and the Origins of* Perspectiva.

[83] I.e. any thing/*res* in the world.

[84] References to figures in the text begin here; several follow. Bridges does not print them.

[85] Et haec est figura, quam specialiter elegit natura in omni multiplicatione et actione, et non quamcunque pyramidem, sed illam cujus basis est superficies agentis, et cujus conus cadit in aliquod punctum patientis, quia sic potest a tota superficie agentis species venire ad singula puncta patientis per singulas pyramides et infinitas, ut patet in figura. Nam a quolibet puncto patientis fiunt radii infiniti, et ideo possunt combinari infinities, ut fiant pyramides rotundae infinitae, quarum omnium est una basis, scilicet, superficies totius agentis; et ad quamlibet partem patientis venit unus conus unius pyramidis, ut virtus veniat a toto agente ad quodlibet punctum patientis, et non ab aliqua parte determinata […] (119, 6–24).

[86] Et oportet quod multiplicatio fiat sphaerice. Nam agens multiplicat se aequabiliter in omnem partem, et secundum omnes diametros, et omnes differentias positionis, quae sunt sursum, deorsum, ante, retro, dextrorsum, sinistrorsum. Ergo undique exeunt lineae in omnem partem ab agente tanquam a centro […] (117, 25–30).

This omnidirectional radiation, the multiplication of an agent's species, is analyzed in terms of the pyramids described above for every point where its action is received. This is the central—and universal—geometrical expression of action in Bacon's model.

But not all pyramids are created equal, nor is every part of a given pyramid the same. Bacon further articulates his model as follows, and in the process begins to discuss its use in two specific contexts, namely, astronomy and *perspectiva*:

> One should also know that the rays (*radii*) that fall in the center of the spherical body from which they come are those by means of which we judge the stars through the apertures of instruments. Whence the astronomer (*astronomus*) and the perspectivist (*perspectivus*), who are experienced in such things, use these rays because they are perceptible and strong. For although from each portion of a spherical body opposite to a recipient thing comes a pyramid having infinite rays, which come to be from the individual points of that portion [sc. of the spherical body], and which all run together into the cone of the pyramid with the perpendicular ray, nevertheless, there is only one perpendicular in any one pyramid, and that perpendicular ray dominates in strength and is the axis of the pyramid, and the entire pyramid is named after it among the *experimentatores*, and it is called the ray of the acting body (*radius corporis agentis*), as is clear in the figure.[87]

Roger then provides a geometrical demonstration, whither we shall not follow him. The central perpendicular ray thus characterizes the entire pyramid, and it is to certain features of this perpendicular ray that we can turn to fully characterize the force of any given pyramid. In his discussion of the different strengths of different lines, Bacon stated at the outset that perpendicular lines are the strongest in nature: "And this is the cause, that the perpendicular is stronger and shorter, and, therefore, nature acts (*operatur*) in a better way through it,[88] as geometrical demonstrations teach [...]."[89]

Roger further develops this view in discussing the different strengths of actions in relation to the different sizes of their pyramids of force:

> And since a pyramid, as has been said, is required for an action of nature, one should consider that the cone of a shorter pyramid acts more strongly, both because it is less distant from the *agens*, and because the rays which terminate together around the cone of a shorter pyramid are closer together, and the closeness of rays and their gathering together acts more strongly (*vicinia radiorum ac congregatio fortius operatur*); and this is evident in the figure.[90]

---

[87] Est etiam sciendum, quod radii qui cadunt in centrum corporis sphaerici a quo veniunt, sunt illi per quos judicamus stellas per foramina instrumentorum. Unde astronomus et perspectivus, qui experiuntur hujusmodi, utuntur istis radiis quia sensuales sunt et fortes. Quamvis enim a portione aliqua corporis sphaerici objecta rei patienti veniat una pyramis habens radios infinitos, qui fiunt a singulis punctis illius portionis, et omnes concurrunt in conum pyramidis cum radio perpendiculari, tamen unus solus est perpendicularis in una pyramide, et ille perpendicularis dominatur in fortitudine, et est axis pyramidis et tota pyramis ab eo nominatur apud experimentatores, et vocatur radius corporis agentis, ut patet in figura (126, 14–26).

[88] Should the text read just *per* here instead of *super* as printed?

[89] Et haec causa est, quia perpendicularis fortior est et brevior, et ideo natura operatur meliori modo [su]per eam, sicut docent geometricae demonstrationes [...] (112, 18–21).

[90] Et cum pyramis, ut dictum est, requiratur ad actionem naturae, considerandum est quod conus brevioris pyramidis fortius operatur, tum quia minus distat ab agente, tum quia radii conterminales circa conum pyramidis brevioris magis vicinantur, et vicinia radiorum ac congregatio fortius operatur; et hoc patet in figura (123, 18–24).

This has been a brief sketch of Bacon's geometrical-optical analysis of action—in the abstract—in terms of pyramids of force. There are further nuances to be sure, but this will suffice for our purposes.[91]

*

Once these pyramids of force are understood in the abstract, and specifically that the perpendicular ray, the axis, is the strongest ray in the pyramid and thus characterizes it, then we can show how this model applies to a particular physical phenomenon, in our case, celestial influences; this is the second stage of my reconstruction. As it turns out, celestial influences are one of the primary examples Bacon uses in describing his geometrical model.[92] He also discusses celestial influences in the three passages cited at the beginning of this section. In the first passage, Roger does not say that the *res* receive cones of abstract pyramids, but the cones of pyramids coming from the heavens; likewise in the second and third passages. We must first, however, come to terms with the mathematical-astronomical side of the picture, which we will then integrate with the mathematical-geographical just discussed. Once this integrated cosmographic framework has been established, then we will turn to the more explicitly astrological side of the question.

First, the astronomical. In order to accurately analyze the pyramid of force resulting from any given celestial configuration, each of the planets must first be accurately located. Once a working theory of each planet's motion has been devised (and, hopefully, verified by observation), then, for convenience, tables may be constructed for their location at every hour of every day.[93] This is critical because every planet as an *agens* in Bacon's model continuously emits an infinite number of rays in every direction (as we just saw). If each planet can be accurately located for every time (past, present and future), then its unique angular relationship to the center of the earth can be known, and thereby the axis of the resultant pyramid of force, with its cone focused on the center of the earth. With this same information for all the planets, their angular relationship to each other can also be calculated, and then their collective relationship to earth.

Once this astronomical foundation has been established, one must then be able to localize these planetary patterns of influence for particular places on earth at any given time in order to fully understand the particular planetary influences operative at a given place at a given time, or over a period of time, or at different places at the same time. As Bacon informed us in the *Geographia* section, the necessary mathematical foundation for this is provided by mathematical geography, by first

---

[91] For a much fuller analysis, see Lindberg, *Roger Bacon's Philosophy of* Nature, and his *Roger Bacon and the Origins of* Perspectiva.

[92] For example, this description of a *fractio radii et speciei*: "[…] si corpus secundum est densius, prout accidit descendendo a coelo in haec inferiora, tunc omnes virtutes stellarum quae non cadunt perpendiculariter super globum elementorum, franguntur inter incessum rectum et perpendicularem ducendam a loco fractionis (112, 22–27)."

[93] Bacon described *astrologia speculativa* and the first part of *practica* in this manner, as we saw, at 109–10.

accurately mapping the longitudes and latitudes of the earth, and then calibrating them with the patterns established by mathematical astronomy.

Having established the basic structures of this cosmographic framework, there must then be developed a means to integrate them for both a particular place and a particular time. For this we have the mathematico-geographic concept of the horizon, which is defined precisely—and uniquely for each place—by the relation of any given point on the sphere of the earth to the heavenly sphere. We shall use the definition of "horizon" from Sacrobosco's *Sphere*, where it is presented with another concept particularized for every place on earth, namely, the meridian. As we saw in the excursus, the horizon and meridian are transposed into two basic structures of the horoscope, namely, the ascendant and midheaven respectively:

> The meridian is a circle passing through the poles and through our zenith, and it is called "meridian" because, wherever a person (*homo*) may be and at whatever time of year, when the sun with the movement of the firmament reaches his meridian, it is noon for him.[94]

Sacrobosco defines "horizon" as follows (as with the meridian, I only quote the relevant parts): "The horizon is a circle dividing the lower hemisphere from the upper, whence it is called the horizon, that is, 'limiter of vision.' [...] Moreover, the zenith of our head (*zenith capitis*) is always the pole of the horizon."[95] Sacrobosco defined '*zenith capitis*' on the previous page: "[...][and the sun cannot approach further] toward our zenith (*ad zenith capitis nostri*). The zenith is a point in the firmament directly above our heads."[96]

The pole of the horizon for each and every person is thus unique to their particular location on the sphere of the earth; that is, if you were to draw a straight line from the center of the earth, through any individual, straight out to the sphere of the firmament, that would be their *zenith capitis*, and that unique point on the celestial sphere would also be the pole of their particular and unique horizon. Roger confirms this interpretation in a passage that will be treated in greater detail just below:

> [...] to every point of the earth falls the cone of one pyramid made virtuous[97] by the entire heaven (*ad omne punctum terrae incidit conus unius pyramidis virtuosae a toto coelo*). And these cones are diverse in nature (*coni isti sunt diversae in natura*), and the pyramids likewise (*et pyramides similiter*), because they have different bases because of the differences of their horizons, since each point of the earth is the center of its own horizon (*quilibet punctus terrae est centrum proprii horizontis*).[98]

---

[94] Est autem meridianus circulus quidam transiens per polos mundi et per zenith capitis nostri. Et dicitur meridianus quia, ubicumque sit homo et in quocumque tempore anni, quando sol raptu firmamenti pervenit ad suum meridianum, est illi meridies (91; Thorndike's translation, which I use here, 126).

[95] Orizon vero est circulus dividens inferius emisperium a superiori, unde appelatur orizon, id est terminator visus. [...] Zenith autem capitis nostri semper est polus orizontis (91; tr. 126–7).

[96] [...] ad zenith capitis nostri. Est autem zenith punctus in firmamento directe suprapositus capitibus nostris (90; tr. 126).

[97] "Virtuous" in the sense of full of celestial *virtutes* or powers.

[98] "[...] ad omne punctum terrae incidit conus unius pyramidis virtuosae a toto coelo. Et coni isti sunt diversae in natura, et pyramides similiter, quia diversas habent bases propter diversitates horizontum, quoniam quilibet punctus terrae est centrum proprii horizontis (250)." He makes the same point also in passage (1) above.

It is only at this point—when the abstract pyramids of force can be instantiated for a particular planetary configuration at a particular time for a particular place—that the astrologizing natural philosopher can approach the properly natural philosophical side of his problem. This simply cannot happen in Roger's view without the mathematical framework just articulated.

We now have a much better understanding of Roger's two mathematical models and how they are integrated into a coherent and powerful system: (1) the geometrical-optical model of action, planetary and otherwise, utilizing rays and their pyramids of force, and (2) the integrated mathematical-astronomical and -geographical cosmographic framework. Fitted together, they provide the necessary mathematical foundations for understanding utterly central features of Roger Bacon's deeply astrologizing natural philosophy. This is so because, as we will recall, celestial bodies are the causes of things below, and the celestial bodies themselves are not understood without mathematics.[99] Places also—a *principium* of generation, and thus central to an Aristotelian analysis of nature—are not understood (1) in their locational aspect without mathematical geography as calibrated with mathematical astronomy, and (2) in their qualitative (complexional) aspect without an understanding of how the pyramids of celestial force uniquely influence each and every point on the sphere of the earth. These are the basics of Bacon's system.

2.4.4 *Astrologia*

To complete this first approach to Roger Bacon's astrologizing Aristotelian natural philosophy,[100] we must also come to terms with his properly astrological views as found in a section entitled '*Astrologia*' printed at the end of *Opus maius* IV, which he describes there as merely a sketch.[101] It treats the more theoretical astrological

---

[99] Sed coelestia sunt causae inferiorum. Ergo non scientur haec inferiora, nisi sciantur coelestia, et illa sine mathematica sciri non possunt (*Opus maius* IV, 110, 17–20).

[100] I discuss Bacon's astrological and astrologizing views further in Chaps. 4, 6 and 8, and in the excursus. I find Hackett's treatment of the astrological dimension of Bacon's thought to be unsatisfactory, in part because of the inadequate nature of the studies on which he bases his view. See in particular "Roger Bacon on Astronomy-Astrology: The Sources of the *Scientia Experimentalis*," in *Roger Bacon and the Sciences*, 175–98, and his "Aristotle, *Astrologia,* and Controversy at the University of Paris (1266–1274)," in *Learning Institutionalized: Teaching in the Medieval University*, John van Engen (ed.), Notre Dame: University of Notre Dame Press, 2000, 69–110. Although Lindberg describes the mathematical models in rich detail, he basically ignores the astrological dimension. Ironically, in order to explicate John Dee's thought, Clulee, and Heilbron and Shumaker, give fuller treatments of the astrological dimensions of Bacon's astrologizing natural philosophy than Hackett or Lindberg do (as cited above). See now also *Filosofia, scienza, teologia dall'*Opus maius, Valeria Sorge and Fabio Sellar (eds. and trs.), Rome: Armando, 2010, who provide an insightful introduction, and a useful Italian translation of some of the material treated here, including all of the '*Astrologia*' section.

[101] We will recall that I discussed some of this section in the excursus. I discuss the textual situation just below.

questions (as defined in my overall introduction): for example, what is the qualitative nature of each planet and thus of its influence. Some of the details will be described more fully below.[102]

First, however, it will be useful to briefly set out the basic fourfold structure of planetary influences which correspond to these questions: (1) What is the qualitative nature of each planet? (2) What is the qualitative nature of each planet as modified by each sign of the zodiac? (3) What, then, is the qualitative nature of the state of the heavens at any given time, that is, the qualitative relationship of each planet in each of their respective signs at a particular time *in relation to each other* (thus including the planetary "aspects")? And, finally, (4) what is the qualitative nature of each planet (1) in each of their signs (2) all in relation to each other (3) in their unique angular relationship to each particular point on the sphere of the earth? This last level also includes the essential features of a horoscope with the ascendant (= horizon) and the midheaven (= meridian) determined, which thus allows the astrological "houses" or "places" to be calculated. One can easily see that the complexity and richness of the analysis is compounded at every level.

The fourth question relates directly to Roger's concern with place and thus fits squarely into the causal framework just reconstructed. The ever-changing state of the heavens at every point in time is related to every place (*locus*) in the world here below, irradiating it with its particular mix of planetary qualities at that time, and thus influencing it by informing its complexion. Then, since *locus* is a principle of generation, the entire planetary configuration focused on a particular place at a particular time affects the complexion of every natural thing (*res naturalis*) generated there. This irradiating celestial configuration is described by a complex pattern of planetary pyramids of force, which collectively have a unique angular relationship to every point on earth. This last angular relationship is significant because whether the angles of incidence are perpendicular or oblique further conditions their degree of power.

And of course the planets keep moving—and thus the planetary radiations in their changing patterns continue to flow down into the center of the cosmos, the *inferiora*, the realm of generation and corruption, namely, our earth and its atmosphere. We can now see that all of these patterns of celestial radiation in relation to particular places at particular times all converge to affect—and also to signify to the natural philosopher learned in such matters—the individual complexions of the places of this world, and thereby the *res naturales* generated there, as Roger had claimed he would show in his programmatic statements earlier. In this way, we can see much more fully how the *coelestia* are the causes of things below, as he had repeatedly claimed there.

---

[102] Roger also discusses his explicitly astrological views at *Opus maius* IV, 238–69, in a section entitled *Excusatio mathematicae* (Defense of Astrology), which I discuss in detail in Chap. 4 below. He supports mathematics there by defending its most maligned member, namely, *iudicia astronomiae* ("the judgments of the science of the stars" or "judicial astrology"). As we will see, he argues (among many other things) that astrology's domain may legitimately extend to the actions of human beings without implying necessity in nature or undermining human free will.

Furthermore, both places and the natural things generated there (including human beings) receive a twofold complexion: (1) the root complexion, which is deeply imprinted at the original formation of a place—and at the birth of each *res naturalis* generated there—and always remains; and (2) the current or running (*currens*) complexion, which changes continually over the course of time as the planets move in relation to the deeply imprinted patterns in (1). Roger does not explicitly inform us about how the two work together—at least as far as I have been able to find—but one may assume, I should think, that this interrelationship is also complex, and that it provides (among other things) the natural philosophical foundations for both types of astrological revolutions (i.e. *anni* and *nati*).[103]

2.5 Al-Kindi, *De radiis stellarum*

Before we return to Bacon's *Astrologia* section, however, we should first fill in the final source for these patterns of thought, namely, al-Kindi's *De radiis stellarum*.[104] We no longer possess the Arabic original, only the Latin translation. We also do not know the identity of the translator, nor the date of translation, but it seems to have been made around the turn of the thirteenth century.[105] We will see immediately that Grosseteste also seems to have been deeply influenced by it, as was Roger.[106] As it turns out, eleven of the twenty manuscripts described by d'Alverny and Hudry are of English provenance.[107]

---

[103] I will explore Albertus Magnus's analysis of the natural philosophical foundations for revolutions in Chap. 3.

[104] Edited, with an important introduction, by Marie-Thérèse d'Alverny and Françoise Hudry, "Al-Kindi: *De Radiis*," *Archives d'Histoire Doctrinale et Litteraire du Moyen Age* 41 (1975): 139–259. I discuss other features of al-Kindi's text in relation to Roger Bacon in Chap. 8 below, and in relation to Marsilio Ficino in volume II. I am grateful to Peter Adamson for allowing me to see his partial translation of the *De radiis* before it was published; see now Adamson and Peter E. Porman, *The Philosophical Works of Al-Kindi*, Oxford: Oxford University Press, 2012, 217–34. My translation is independent, but takes theirs into account. Charles Burnett translated the section on the power of words in his "The Theory and Practice of Powerful Words in Medieval Magical Texts," in *The Word in Medieval Logic, Theology and Philosophy*, Tetsuro Shimizu and Charles Burnett (eds.), Turnhout: Brepols, 2009, which I will discuss in Chap. 8 below. For a valuable up-to-date general treatment of al-Kindi and his philosophy, see Peter Adamson, *Al-Kindi*, in the Great Medieval Thinkers series; Oxford: Oxford University Press, 2007.

[105] *De radiis*, 169–72.

[106] This is the very same tradition to which Clulee relates the central patterns of John Dee's natural philosophy; "Astrology, Magic and Optics," 664 ff: "The optical works that Dee collected and studied in 1556 are almost entirely medieval Latin or Latin translations of Arabic treatises, and in many cases they are by authors, such as al-Kindi, Robert Grosseteste, and Roger Bacon, for whom optics was not just a science of ray geometry but also an integral part of a natural philosophy (664)."

[107] Further, some of the manuscripts in which the *De radiis* is found also contain works by Grosseteste and Bacon, and also Albert; see *De radiis*, 173–211. For other approaches to al-Kindi's work, see Hillary Suzanne Wiesner, "The Cosmology of al-Kindi," Phd thesis, Harvard University,

I will only discuss here a brief section of *De radiis* at the end of Chap. 1 and the beginning of Chap. 2, where al-Kindi begins to describe his own deeply astrologizing, intricately mathematized natural philosophy. After an introductory section, he turns to the stars, first sketching out some general patterns:

> [I.12] They [sc. wise men in the past] achieved an undoubting belief, by means of their senses, that the disposition of the stars disposes the world of the elements (*stellarum dispositio mundum elementorum disponit*) and everything composed from them in it (*omnia que ex ipsis composita in ipso*), in whatever place and at whatever time they are contained (*quocumque loco, quocumque tempore continentur*), such that no substance and no accident exists here which was not figured in its manner in the heavens (*nulla substantia, nullum accidens hic subsistit quod in celo suo modo non sit figuratum*), and it is not doubted that this comes forth from their rays sent into the world (*ex earumdem radiis in mundum missis*, 222).[108]

The astrologizing natural philosophical structure is immediately apparent, as well as the explicit language of rays.

Al-Kindi then further articulates his model at the beginning of Chap 2:

> [II.1] For every star has its own proper nature and condition (*omnis enim stella suam habet propriam naturam et condicionem*), in which the projection of rays is contained with other things (*in qua radiorum proiectio cum aliis continetur*). And just as each [sc. star] has its own proper nature (*unaqueque suam habet propriam naturam*), which happens to be found completely in no other (*quam totaliter in nulla alia contingit reperire*), in which the emission of rays (*emissio radiorum*) is contained, thus these rays in diverse stars are of a diverse nature (*sic ipsi radii in diversis stellis sunt diverse nature*), as also these stars are diverse in nature (*sicut et ipse stelle sunt in natura diverse*, 222).[109]

Thus, each of the stars—including the luminaries and planets—has its own unique and characteristic nature. And, since each different (and thus diverse) star contains rays among other things, the projectable rays of each different star will also be of a different nature. We notice here the emphasis on both rays and diversity. The unique nature of each individual star and its radiation established here is utterly essential to astrology's natural philosophical foundations. This central feature of al-Kindi's and Roger Bacon's (and others') astrologizing Aristotelian system was pointedly

---

1993, and Pinella Travaglia, *Magic, Causality and Intentionality: The Doctrine of Rays in al-Kindi*, Florence: SISMEL—Galluzzo, 1999.

[108] "[Nam sursum aspicientes, stellarum plurium condiciones quasdam perceperunt inter quas septem planetarum proprietates propensius ceteris investigare conati sunt et scire, utpote quos rerum mundialium precipuos dispensatores experiencia longa probaverunt.] Indubitatam ergo fidem per sensum sunt adepti quod stellarum dispositio mundum elementorum disponit et omnia que ex ipsis composita in ipso, quocumque loco, quocumque tempore continentur, adeo quod nulla substantia, nullum accidens hic subsistit quod in celo suo modo non sit figuratum, et hoc quidem ex earumdem radiis in mundum missis non dubitatur provenire (218, 19–28)." The lines, which I provide, are unnumbered in d'Alverny and Hudry's edition. The number in the translation is the page number in Adamson and Porman, who also provide the paragraph numbers.

[109] Omnis enim stella suam habet propriam naturam et condicionem in qua radiorum proiectio cum aliis continetur. Et sicut unaqueque suam habet propriam naturam quam totaliter in nulla alia contingit reperire, in qua radiorum emissio continetur, sic ipsi radii in diversis stellis sunt diverse nature, sicut et ipse stelle sunt in natura diverse (219, 1–5).

attacked by Giovanni Pico della Mirandola in his *Disputations* Book III, as we will see in volume II.

*

Although al-Kindi makes frequent allusion to other properties of the stars, the only one he actually discusses here are these rays and the effects they produce, based both on their different natures as coming from different stars, and as modified by their different relations to each other (*respectus*):

> [II.2-4] Moreover, every star has its place (*situs*) in the cosmic system (*in mundana machina*) different from all the others. Whence it comes about necessarily that every star is allotted a different relation (*respectus*) from the others to all the other stars and to all the *res* and places contained in the world (*ad* [...] *res omnes et loca in mundo contenta*). (3) Moreover, a different *respectus* varies the effect of the rays (*respectus autem varius variat effectum radiorum*) just as, also, the various other properties of these stars do. Whence it comes about that every star does something different and in a different manner on diverse places and things (*omnis stella aliud et aliter operatur in locis et rebus diversis*), even if they vary only a little. For, the entire stellar action (*operatio*) proceeds by means of rays (*tota stellaris operatio per radios procedat*) which are varied in themselves with each different aspect (*qui in omni aspectu vario in se ipsis variantur*). (4) For a ray which descends from the center of a star to the center of the earth is proved to be strongest in the activity of its species (*radius qui a centro stelle ad centrum terre descendit fortissimus esse probatur in operationibus sue specie*). Moreover, what strikes obliquely from the center of the earth are weakened in their effect in accordance with the proportion of the obliquity, except insofar as they are strengthened by the rays of other stars converging on the same places (222).[110]

Not only do we find al-Kindi interested in precisely the same issues as Grosseteste and Bacon in accounting astrologically for the diversity of places and things, but we find him doing so in precisely the same manner, using stellar rays in a geometrical-optical analysis, and in much the same language, albeit not as fully or as formally spelled out. Nevertheless, all the basic structures are there.

Al-Kindi continues:

> [II.5] For every star pours forth rays onto every place (*in omnem enim locum omnis stella radios effundit*) because of which the diversity of rays—conflated, as it were, into one (*propter quod radiorum diversitas quasi in unum conflata*)—varies the contents of every place (*variat omnium contenta locorum*), since in every diverse place, the tenor of the rays, which is derived from the entire harmony of the stars, is different (*cum in omni loco diverso diversus sit tenor radiorum qui a totali stellarum armonia derivatur*). In addition to these things, because it is changed continually by the continuous motion with respect to place

[110] Omnis autem stella suum habet situm in mundana machina, alium ab omnibus aliis. Unde provenit necessario quod omnis stella alium sortitur respectum quam alia ad omnes alias et res omnes et loca in mundo contenta. Respectus autem varius variat effectum radiorum sicut et alie proprietates ipsarum stellarum varie. Unde fit quod omnis stella aliud et aliter operatur in locis et rebus diversis, quantumcumque modicis et quam modice differentibus, cum tota stellaris operatio per radios procedat qui in omni aspectu vario in se ipsis variantur. Nam radius qui a centro stelle ad centrum terre descendit fortissimus esse probatur in operationibus sue specie. Qui autem a centro terre obliquantur secundum proportionem obliquationis in effectu debilitantur, nisi in quantum aliarum stellarum radiis concurrentibus in eisdem locis confortantur (219, 6–19).

> (*secundum locum*) of the planets and the other stars,[111] it moves the world of the elements and all their contents continually into diverse conditions, coming forth into actuality in accordance with the exigency of the harmony of the same time, even though to human senses the other things of the world seem to be permanent.[112] [...] Wherefore, one is convinced by reason that the rays of all the stars have diverse actions on the things of the world (*omnium stellarum radii diversas habent in rebus mundi operationes*) in accordance with the diverse properties of these very things, since all things arise and exist by means of rays (*omnes res per radios oriantur et extant*).[113]

It is interesting to recall in this context that it was apparently al-Kindi who originally devised a punctiform analysis for geometrical optics, even though he was not the first to apply it to the case of vision. Perhaps it was for just this sort of astrologizing place-oriented natural philosophy.[114]

2.6 Roger Bacon's *Astrologia* Section

Now that Roger's ideas have been situated in their proper historical contexts, we should turn to what is printed in Bridges's edition as the final section of *Opus maius* IV, entitled '*Astrologia*', which immediately follows the section '*Geographia*'. In fact, however, as A.G. Little showed over 100 years ago, it is really a part of the *Opus minus*, written soon after Roger sent the *Opus maius* to the pope.[115] Nevertheless, this section follows thematically on what we just examined at the end

---

[111] This refers to the efficient causality of the sun and planets in their local motion with respect to generation on earth, as discussed in Chap. 1.

[112] Al-Kindi discusses the epistemological dimensions of this knowledge in Chap. 4 of *De radiis*.

[113] "In omnem enim locum omnis stella radios effundit, propter quod radiorum diversitas quasi in unum conflata variat omnium contenta locorum, cum in omni loco diverso diversus sit tenor radiorum qui a totali stellarum armonia derivatur. Hec insuper, quia continue mutatur per planetarum et aliarum stellarum continuum motum[,] secundum locum<,> mundum elementorum et eorum contenta omnia continue movet in diversas condiciones, exeuntes in actum secundum exigenciam eiusdem temporis armonie, licet humanis sensibus res alie mundi videantur permanere. [...] Quocirca convincitur ratione quod omnium stellarum radii diversas habent in rebus mundi operationes secundum earundem rerum proprietates diversas, cum omnes res per radios oriantur et extant (219, 20–220, 21)." I revise the punctuation slightly.

[114] In "Powerful Words" (n. 9), Charles Burnett notes in passing that rays were not normally a part of Aristotelian natural philosophy. Al-Kindi seems here to have made this contribution, which fell on fertile ground in Roger Bacon, and (as we will see just below) in Albertus Magnus as well, thus inaugurating a new and influential phase of medieval Aristotelian natural philosophy.

[115] Little discusses this in the introduction to his edition of a new fragment of the *Opus tertium*; *Part of the* Opus Tertium *of Roger Bacon*, xvii: "It is clear [...] that the treatise on Astrology which follows the *Geographia* was not part of the *Opus Maius*, but has been inserted and adapted from the *Opus Minus*." The passage in the text indicating this is: "Hec autem que iam de locis mundi et alterationibus locorum et rerum per celestia et de iudiciis et operibus secretis tetigi, non posui omnia in Maiori Opere, sed de locis tantum. Alia posui in Minore Opere, quando veni ad declarandam intentionem istius partis Operis Maioris. Non enim proposui tunc plura ibi in Opere Maiori tractare, volens festinare propter Vestre Sanctitatis mandatum (18, 6–10)."

of *Opus maius* IV. Roger's section on astrology here is no more than a sketch as he explicitly informs us. Nevertheless, there is much of interest for our purposes.[116]

Roger introduces the section thus:

> After the description of places (1), four other parts ought to follow, namely, a verification of the natures of (2) the fixed stars and the planets, in order that by these a more certain certitude (*certior* [...] *certitudo*) would be had of (3) the complexions of all places and located things (*complexionum omnium locorum et rerum locatarum*), so that (4) [sc. astrological] judgments (*iudicia*) about present, past and future could be made, in order that, fifth and finally (5), works (*opera*) would come to be that would forward everything useful to a republic and exclude what is harmful.[117] But since I was unable—because of impediments—to complete the description of places in a figure in accordance with the ways of natural things, and these four follow, it is fitting that I desist from a treatment of these things. Nevertheless, I wish to treat here in summary the intent of a treatise for these four, as if it had been done as I have done for the others that I treated, in order that Your Wisdom (*Vestra Sapientia*) would see what is required for procuring a republic's utility, and what sort of writings and works of wisdom (*opera sapientiae*) you should seek from whomever.[118]

Roger thus alludes here to his treatment in *Opus maius* IV, including places and natural things (*res naturales*) in the world, and to astrological judgments and the works of wisdom, thus to both *astronomia iudiciaria et operativa*, with its implied words of wisdom as well.

Roger then launches immediately into the sketch proper with the fixed stars and the zodiacal signs and their qualitative natures, then the planets and their qualitative natures, followed by the planetary dignities and later the terrestrial (or mundane) houses that I discussed already in the introductory excursus. He completes his breathless survey as follows: "These then are the primary roots among the actions of the stars (*radices principales in actionibus stellarum*), which have branches and flowers and infinite fruit. This then is the intent of the treatise which I have proposed to put in this work on the powers (*virtutes*) and actions of the stars."[119]

*

[116] I will discuss how Roger instantiates this system in my Chap. 4 below.

[117] This sentiment recurs frequently in Roger's works, as we will see. I discuss the material specifically on *opera* in Chap. 8 below.

[118] Post locorum descriptionem deberent sequi alia quatuor, scilicet certificatio naturarum stellarum fixarum et planetarum, ut certior haberetur per haec certitudo complexionum omnium locorum et rerum locatarum, quatenus posset fieri iudicium de praesentibus, praeteritis, et futuris, ut tandem fierent, quinto, opera quae promoverent omnia rei publicae utilia et nociva excluderent. Sed cum non potui propter impedimenta perficere descriptionem locorum in figura secundum vias naturalium, et haec quatuor sequuntur, oportuit ab horum tractatu cessare; volo tamen hic in summa tangere intentionem tractatus horum quatuor, ac si factus esset, sicut de aliis feci de quibus tractavi; quatenus videat Vestra Sapientia, quid requirendum sit pro utilitate rei publicae procuranda, et qualiter scripturas et opera sapientiae petatis a quocunque (376, 21–377, 7).

[119] Haec igitur sunt radices principales in actionibus stellarum, quae habent ramos et flores, et fructus infinitos. Haec igitur est intentio tractatus quem in hoc opere facere proposui de virtutibus et actionibus stellarum (379, 14–16).

After this introductory material, Roger turns to a discussion more directly germane to our interests here:[120]

> Thirdly, one must consider how the places of the world and the *res* are altered in their complexions by these aforementioned roots.[121] Moreover, certain general statements were made about this matter before, as was fitting, but in this part of the work many things ought to be verified in particular. It is certain from Aristotle that the heavens are not only a universal cause, but also a particular cause of all the things of this world (*omnium rerum inferiorum*). For Aristotle says in *De generatione* II that the elements act more weakly with respect to the heavens than tools with respect to an artisan, as an axe and a hatchet with respect to a carpenter. Therefore, although everything here below comes to be by means of elemental qualities, nevertheless, they will not exist in relation to the heavens except as instruments in relation to an artisan (*artifex*). And, therefore, if the *artifex* is the primary universal and particular agent in relation to a work, much more so are the heavens in relation to a thing to be generated (*multo magis coelum respectu rei generandae*).[122]

This analysis should be quite familiar by now, although Roger here adds more interesting material concerning the instrumental nature of the heavens and universal vs. particular causality.

Roger then discusses the case of generation in much the same way as he did in the passage at *Opus maius* 137–9, where he treated the relation of the heavenly powers/virtues to that of the mother and father in generation. I will only quote the final part of the discussion here as a clearer statement of his position than he had made there:

> Averroes says that the sun does more than a man in the production of a thing. For the *virtus* of the sun continues in the seed from the beginning of the generation to the end, but not the *virtus* of the father. That happens once, namely, only in the dropping off of the semen, and, therefore, would do nothing unless the *virtus* of the heavens were continually multiplied and infused, regulating the entire generation.[123]

---

[120] There is much more of interest in the rest of this section (379–88), but I cannot treat it here.

[121] [...] quomodo loca mundi et res alterantur in complexionibus per iam dictas radices (379, 17–18).

[122] Tertio considerandum est quomodo loca mundi et res alterantur in complexionibus per iam dictas radices. Quidam autem universales sermones facti sunt de hac materia in prioribus, ut oportuit, sed in hac parte operis deberent plura certificari in particulari. Et certum est per Aristotelem, quod coelum non solum est causa universalis, sed particularis, omnium rerum inferiorum. Nam Aristoteles dicit secundo de Generatione quod elementa debilius agunt respectu coeli quam organa respectu artificis, ut securis et dolabrum respectu carpentatoris; et ideo licet per qualitates elementares fiant omnia hic inferius, non erunt tamen respectu coeli nisi sicut instrumenta respectu artificis. Et ideo si artifex est agens principale universale et particulare respectu operis, multo magis coelum respectu rei generandae (379, 17–30).

[123] "Sed Averroes dicit quod sol plus facit quam homo ad productionem rei. Nam virtus solis continuatur in semine a principio generationis usque ad finem, virtus autem patris non, sed fit semel, scilicet solum in seminis decisionem, et ideo nihil faceret nisi esset virtus coeli continue multiplicata et infusa, regulans totam generationem (380, 3–9)." Averroes (Abu l-Walid Muhammad ibn Rusd), was born in Cordoba in 1126 and died in Marrakesh in 1198. He had many facets, writing as a Muslim jurist and theologian, and as a physician and Aristotelian commentator, for which he was best know (often controversially) in the West. His commentaries on Aristotle began to be translated into Latin merely 20 years after his death, and had an extremely influential Nachleben throughout the sixteenth century and beyond (Hasse, *Success and Suppression*, 341–57).

Thus, the sun is required as a universal efficient cause along with the formal cause in the father's seed and the material cause in the mother's womb, although the latter is not explicitly mentioned here.

Roger then turns to our particular subject—the influence of place on generation—and clearly and concisely articulates its complexity:

> By means of the heavens, therefore, the complexions of all things are had (*per coelum ergo complexiones omnium rerum habentur*), and not only are regions diversified by means of the heavens, but the things of the same region, and the parts of the same thing; and not only in normal generation, but also monstrous, that is, the mistakes and errors of nature. For it is clear that everything is varied by means of the differences of horizons (*per diversitates horizontum* […] *patet omnia variari*), in accordance with which individual points of the earth are centers in new horizons (*secundum quod singula puncta terrae sunt centra in horizontibus novis*), as was noted above concerning the plants of diverse species, which are born, as it were, on the same point of the earth, and concerning the diversity of twins in the same womb. It happens for this reason, that the cones of diverse pyramids containing the powers of the stars and of the parts of the heavens (*coni diversarum pyramidum continentium virtutes stellarum et partium coeli*) over the heads of those living there (*super capita habitantium*) come to individual points of the earth (*veniunt ad singula puncta terrae*), in order that a full diversity comes about in things (*ut cadat diversitas plena in rebus*).[124]

This material and Roger's mode of analysis should be quite familiar to us by now as he reiterates and sharpens his earlier analysis in *Opus maius* IV.

Roger then goes on to complicate matters even further:

> But there is another diversity by means of the elongation from the poles and from the middle of the earth. And the cause of this is twofold. One is the universal cause, namely the distance or closeness of the sun, in accordance with what was recalled above concerning the places of the world. The other is the particular cause, namely the diversity of fixed stars over the heads of those living there (*diversitas stellarum fixarum super capita habitantium*). For *res naturales* in diverse regions are varied by means of these in particular ways, and men, not only in their natural features (*in naturalibus*), but in customs, in sciences, in arts, in languages, and in everything. A third cause is from the predominating power (*virtus*) of the twelve signs. For different signs rule in different regions (*diversa signa dominantur in diversis regionibus*), either because at the beginning of the world (*in mundi principio*) they came-to-be in directing those creatures receiving the first powers (*virtutes primas*), and because '*quod nova testa capit, inveterata sapit*'; or because they are similar in nature to the stars that revolve over the heads of those living there. And the fourth cause is by means of the planets (*per planetas*). For the planets are assigned to different regions by means of dominion [sc. their rulership], as the signs are, and this is the case from the twofold cause already stated. Now about the signs, how the signs and planets are assigned or in what regions is difficult to certify, for the authorities differ. Nevertheless, one needs to compose a treatise on these matters here.[125]

---

[124] Per coelum ergo complexiones omnium rerum habentur, et non solum regiones diversificantur per coelum, sed res ejusdem regionis et partes ejusdem rei, et non solum in generatione recta sed monstruosa, et peccatis ac erroribus naturae. Nam per diversitates horizontum, secundum quod singula puncta terrae sunt centra in horizontibus novis, patet omnia variari, ut superius notatum est de herbis diversarum specierum, quae nascuntur quasi in eodem puncto terrae, et de diversitate gemellorum in eadem matrice; propter hoc quod coni diversarum pyramidum continentium virtutes stellarum et partium coeli super capita habitantium veniunt ad singula puncta terrae, ut cadat diversitas plena in rebus (380, 10–21).

[125] Sed alia diversitas est per elongationem a polis et a medio mundi. Et hujus causa duplex est. Una est causa universalis, scilicet solis distantia vel propinquitas secundum quod superius in locis

The complications do not end here; they only begin. But we will not go any further into detail, as worthwhile as this would be in a fuller treatment of the natural philosophical implications of Roger's astrology.[126]

*

As we have now seen in sufficient detail, Roger Bacon is remarkably consistent in his analysis of our central theme: the variety of places in the world and their fundamental importance for understanding the complexions of natural things generated on earth. Although Roger repeatedly emphasizes the same themes, he provides us with more information each time he circles around. To be sure, one could have hoped for a more centralized and systematic treatment of this fundamental topic, instead of treating it piecemeal throughout a 300-page text. Nevertheless, the deeply astrologizing natural philosophical structures, which explicitly exemplify his views of mathematics and *perspectiva*, are—as I hope to have shown—quite clear and consistent. They shall thus provide the basis for our comparison with Albert, to which we will now turn.

I have gone into such detail because I want there to be no doubt just how central these astrologizing patterns are to Roger's natural philosophy. He emphasizes them repeatedly throughout the entirety of *Opus maius*, Book IV, his extended treatment of mathematics. He uses the same causal analysis throughout, namely, the multiplication of species—a geometrical-optical model of action, celestial and otherwise—within a mathematical-astronomically and -geographically described cosmos, that is, the very themes which inform his detailed criticism of the unnamed master.[127] Finally, it should also be clear by now that Roger's astrologizing Aristotelian natural philosophy reconstructed here provides more solid evidence to modify and correct Grant's claim about astrology's purported disjunction from the fundamental patterns of Aristotelian natural philosophy. Clearly Roger Bacon's astrological knowledge deeply informed his natural philosophy.

---

mundi est memoratum. Alia est causa particularis, scilicet diversitas stellarum fixarum super capita habitantium. Nam specialiter per has variantur res naturales in diversis regionibus, et homines, non solum in naturalibus, sed in moribus, in scientiis, in artibus, in linguis, et in omnibus. Tertia est ex praedominanti virtute duodecim signorum. Nam diversa signa dominantur in diversis regionibus, aut quia in mundi principio fuerunt in directo illarum creaturarum recipientium virtutes primas, et quod nova testa capit inveterata sapit; aut quia sunt similes in natura cum stellis quae super capita habitantium revolvuntur. Et quarta causa est per planetas. Nam planetae assignantur diversis regionibus per dominium sicut signa, et hoc ex duplici causa dicta. Nunc de signis, quomodo aut in quibus regionibus assignentur signa et planetae, difficile est certificare, nam auctores diversificantur. Tractatus tamen de his habet hic expedire (380, 20–381, 6).

[126] These further complications take up fully the next eight pages of Roger's text, which is followed by a brief algorithm, as it were, a strategy for how a natural philosopher should read the patterns in the heavens (388–89).

[127] I have also gone into such detail for two further reasons: (1) In order, more generally, to provide significant material for better understanding the astrologizing Aristotelianism of the premodern period, but also (2), more particularly, to provide a fuller picture of Bacon's views.

2.7 Albertus Magnus on the Science of Places

We are now ready to compare what has just been reconstructed with some further patterns of thought in Albert's understanding of nature. This will allow us both to more fully articulate his astrologizing Aristotelian natural philosophy as analyzed in Chap. 1 as well as to more accurately evaluate the likelihood that Roger Bacon had Albertus Magnus in mind when criticizing the unnamed master. There is an obvious work with which to begin, a treatise entitled *De natura locorum* (*On the Nature of Places*) which falls within the series of Albert's Aristotelian commentaries, but is a work that sets out to fill a lacuna in Aristotle's system where no authentic work exists.[128] Albert locates it just after the *Physics, De generatione and corruptione* and *De caelo*, thus among Aristotle's most fundamental works of natural philosophy.[129] There is no doubt that Albert wrote the text, which is entitled, in its simplest form, *De natura loci*.[130] We have the autograph manuscript and, thus, a rock-solid textual tradition,[131] as well as the information just refered to from Albert's *Physica* I.1.4. Hossfeld dates *De natura loci* to between 1251 and 1254, while Albert was teaching at Cologne; this is confirmed with internal evidence from the text.[132]

---

[128] Albert describes his *modus operandi* at *Physica* I.1.1, 23–41*; Opera* omnia 4.1 (Books 1–4), P. Hossfeld (ed.), Monasterii Westfalorum: Aschendorff, 1987: "Erit autem modus noster in hoc opere Aristotelis ordinem et sententiam sequi et dicere ad explanationem eius et ad probationem eius, quaecumque necessaria esse videbuntur, ita tamen, quod textus eius nulla fiat mentio. Et praeter hoc digressiones faciemus declarantes dubia suborientia et supplentes, quaecumque minus dicta in sententia Philosophi obscuritatem quibusdam attulerunt. Distinguemus autem totum hoc opus per titulos capitulorum, et ubi titulus simpliciter ostendit materiam capituli, significatur hoc capitulum esse de serie librorum Aristotelis, ubicumque autem in titulo praesignificatur, quod digressio fit, ibi additum est ex nobis ad suppletionem vel probationem inductum. Taliter autem procedendo libros perficiemus eodem numero et nominibus, quibus fecit libros suos Aristoteles. Et addemus etiam alicubi partes librorum imperfectas et alicubi libros intermissos vel omissos, quos vel Aristoteles non fecit vel forte si fecit, ad nos non pervenerunt."

[129] *Physica,* I.1.4 (p. 6, ll. 39–79). These are also the basic texts for university courses in philosophy, as discussed in Part 4 below. *De natura locorum* is published in the same volume of the *Editio Coloniensis* with *De generatione et corruptione*, and the also deeply astrological *De causis proprietatum elementorum*, which I discuss in Chap. 3 below.

[130] See the prolegomena to Hossfeld's edition (p.VI, ll. 76 ff) on its various titles, which begins thus: "Non semper eodem modo Albertus titulos operum suorum refert." I derive most of the information here from Hossfeld's informative discussion.

[131] There are 43 complete manuscripts, including the autograph; the 19 manuscripts with attributions are all to Albert (V, 29–38; VI, 44–53).

[132] "Praeterea ex his operibus discimus auctorem observationes Coloniae factas collegisse, quin etiam eum ibidem opus DE NATURA LOCI scripsisse; hic enim notavit […]: 'Agrippinam, quae nunc Colonia vocatur, in qua et istud volumen compilatum est'. His verbis auctorem Coloniae degisse ostenditur; nec quisquam hic esse potest nisi Albertus, qui certo primis annis sexti decenni illius saeculi ibi muneribus docti magistri atque etiam sequestri pacis in controversiis publicis dirimendis functus est (VI, 28–38)." For the dating of Albert's work, see also Weisheipl's clear account in *Albertus Magnus and the Sciences*, 565–77. Although he refers to but does not explore *De natura loci*, Henryk Anzulewicz does treat even earlier works by Albertus Magnus that reflect his knowledge of geometrical optics/*perspectiva*; "Perspektive und Raumvorstellung in den

Should we consider adding the *De natura loci* to the corpus pseudo-Aristotelicum? Schmitt and Knox do not include it in their inventory,[133] nor is it mentioned there at all. They do, however, include the closely associated *De causis proprietatum elementorum*.[134] But Albert states quite explicitly that Aristotle and Plato both wrote treatises *De natura locorum*. They had only come down to his time, however, as fragments.[135] Hossfeld disputes this, arguing that the explicit attribution to Aristotle actually applies only to the *De causis proprietatum elementorum*, and not to the *De natura loci* (VII, 70 ff.). At any rate, we have a genuine work of Albert's that he places among the central works of Aristotelian natural philosophy, representing it as originally of genuine Aristotelian provenance.[136] The only source for *De natura loci* that Hossfeld mentions is Pseudo-Ethicus's *Cosmographia*, but this is only for Tractatus III.[137] He does not discuss any sources for the material of interest here.

*

Let us now turn to the content of Albert's *De natura loci* and begin the comparison with Roger Bacon. The first passage to be examined is the first paragraph of the opening chapter, entitled "That it is proper in natural knowledge (*scientia naturalis*) to know the nature of place, and that they err, who do not inquire about this:"[138]

> We will treat the nature of places, which comes from the relation of the place to the heavens (*quae provenit ex habitudine loci ad caelum*); first we will mention those things determined in the *Physics* [IV.1.10]. For there it was proved that place is an active principle of generation, like the father (*locus est generationis principium activum quemadmodum pater*). The cause of which is that every located thing (*omne locatum*) has a relation to its place (*ad locum suum*) as matter to form. And because the *superiora* have a relationship to the *inferiora* as forms to their matter (as we said in *De caelo et mundo*), it is fitting that the *superiora* are always the places of things below (*loca inferiorum*), and, therefore, the prin-

Frühwerken des Albertus Magnus," *Miscellanea Mediaevalia* 25 (1998): 249–286. In the process, Anzulewicz provides the background for Albert's knowledge of *perspectiva* at the time he wrote *De natura loci*, including the influential sources to which he had access.

133 Charles B. Schmitt and Dilwyn Knox, *Pseudo-Aristoteles Latinus: A Guide to Latin Works Falsely Attributed to Aristotle before 1500*, London: Warburg Institute, 1985. See also Steven J. Williams on the pseudo-Aristotelian corpus, "Defining the Corpus Aristotelicum," *Journal of the Warburg and Courtauld Institutes* 58 (1995): 29–51.

134 *Pseudo-Aristoteles Latinus*, Number 14 (20).

135 I quote and discuss this passage just below.

136 For this reason, there should at least be a mention of *De natura loci*, with a reference to Hossfeld's arguments against its attribution, in any future edition of Schmitt and Knox's work.

137 "Ethicus Astronomus" is Aethicus Ister, the assumed author of a curious *Cosmographia* written sometime during the early Middle Ages. My thanks to David Juste for this identification. In addition, an English translation of Albert's entire treatise as the central part of a PhD dissertation in Geography should be noted: Sister Jean Paul Tilmann, O.P., *An Appraisal of the Geographical Works of Albertus Magnus and His Contribution to Geographical Thought*, Michigan Geographical Publication No. 4, Ann Arbor: Dept. of Geography, University of Michigan, 1971.

138 Cap. 1. Quod naturam loci scire oportet in scientia naturali et quod peccant, qui de ipso non quaerunt (1, 6–8).

> ciple of the formation of the *inferiora* flows into them from the *superiora* as from active principles.[139]

It is clear already, I trust, that we are at the very least in the same conceptual ballpark with Bacon's style of astrologizing Aristotelian natural philosophy. Even in these opening lines, we immediately encounter the view (1) that the nature of place has to do with the relationship of the heavens to the earth (*quae provenit ex habitudine loci ad caelum*), and (2) that *locus* is a *principium* of generation, like the father, with the *superiora* informing the *inferiora.*

After providing two more reasons why the natural philosopher should study the nature of place, we come to a second relevant passage (the fourth reason) also in this first chapter:

> Because of which, *locus* is enumerated among the *principia naturae* by the more reputable Aristotelians. They said that, due to the same cause, the nature of place is marvelous (*loci naturam esse mirabilem*), and especially for this reason that, since the prime mover dispenses its nobility to everything that is formed toward being (*motor primus nobilitatem suam largiatur omni ei quod formatur ad esse*), it undoubtedly does not dispense being or form over a longer distance except through that which is nearer to it. Moreover, the container (*continens*) is closer than what is contained (*contentum*), and, thus, being or form (*esse vel forma*) are dispensed to what is located (*locato*) through the containing place (*per locum continentem*).[140]

Bacon also explicitly argued that place is a principle of nature that a natural philosopher must take into account to fully understand generation. Here Albert emphasizes this same feature and provides a glimpse into his understanding of how this works. Nothing in Roger quite fits this in detail, but one can easily imagine Albert's description here also applying more generally to Roger's conceptions. So far, then, we seem to have some basic level of similarity and coherence, but not quite enough detail to make a proper comparison.

With the third passage, still in Chap. 1, we begin to get more detailed information:

> Therefore, from all of these things, it is sufficiently allowed that one ought to know the nature of place. Nor does the treatise in the *Physics* suffice, for this reason, that it only verifies concerning place universally. But we ought to know the diversities of places in particular, and the cause of their diversity, and the accidents of diverse places. For then we would

[139] De natura locorum, quae provenit ex habitudine loci ad caelum, tractaturi primo mentionem de his quae in PHYSICIS determinata sunt; in his enim probatum est, quod locus est generationis principium activum quemadmodum pater. Cuius causa est, quod omne locatum se habet ad locum suum quemadmodum materia ad formam. Et quia superiora ad inferiora se habent sicut formae ad materias suas, sicut in DE CAELO ET MUNDO diximus, oportet, quod superiora semper loca sunt inferiorum, et ideo principium formationis inferiorum ex superioribus influitur eis sicut ex principiis activis (1, 9–20).

[140] Propter quod etiam a PERIPATETICIS probatioribus locus inter principia naturae enumeratur. Dixerunt etiam propter eandem causam loci naturam esse mirabilem et praecipue propter hoc, quia cum motor primus nobilitatem suam largiatur omni ei quod formatur ad esse, absque dubio non largitur longius distanti esse vel formam nisi per id quod sibi est propinquius; est autem propinquius continens quam contentum, et sic per locum continentem largitur locato esse et formam (1, 44–53).

> completely know even those things generated and corrupted in places (*perfecte sciemus etiam ea quae in locis generantur et corrumpuntur*). And just as it does not suffice to determine the nature of animal in general with respect to genus, unless the diversity of animals is also known in generation and food and customs, likewise it does not suffice to treat of place universally, unless the diversity of places is treated, and they come to know the accidents of diverse places, and the cause of accidents.[141]

It is clear that Albert is also very interested in a scientific treatment of place, which requires, as for Roger, both a more universal treatment, and then a treatment *in particulari*. Further, they use much the same language. Still, the similarities are so general that all we can say is that both consider a proper understanding of place central to natural philosophy, primarily because of what we can learn about what is generated and corrupted there.

Albert continues:

> Because of which, those err who treat *scientia naturalis* and do not bring in anything about the diversity of places. They seek solace for their inexperience if they say that it is not fitting to inquire about this, for it does not suffice that this sort of diversity be treated in mathematical [sc. books] (*in MATHEMATICIS*), where which stars rise or fall in this or that manner above which places is treated (*quae stellae super quae loca hoc vel illo modo oriantur vel occidant*), namely, whether obliquely or straight [sc. perpendicular]. And which of them neither rise nor set, but are always apparent, or which never appear, and which rise and set sometimes. Because, although from these the location of the *climata* may be known, nevertheless the nature of places and the diversity of their complexions are not sufficiently known by means of these (*non tamen per eadem sufficienter scitur natura locorum et diversitas complexionis ipsorum*).[142]

Albert's point here is interesting with respect to disciplinary boundaries and competences, especially concerning the study and teaching of mathematical subjects. He says that the study of such things in mathematical works—namely the mathematical astronomical part and what pertains to mathematical geography (concerning the climates)—is not sufficient for a complete understanding of places. Rather, those studying natural knowledge (*scientia naturalis*) must also apply themselves to this problem in order to obtain complete knowledge.[143]

---

[141] Ex omnibus ergo his satis liquet, quod oportet scire naturam loci, nec sufficit tractatus, qui in PHYSICIS habitus est de ipso, eo quod ille non nisi universaliter certificat de loco, et oportet nos scire diversitates locorum in particulari et causam diversitatis ipsorum et accidentia diversorum locorum; tunc enim perfecte sciemus etiam ea quae in locis generantur et corrumpuntur. Et sicut non sufficit determinare naturam animalis in communi secundum genus, nisi sciatur etiam animalium diversitas in generatione et cibo et moribus, ita non sufficit tradere universaliter de loco, nisi tradatur locorum diversitas et innotescant accidentia locorum diversorum et causa accidentium (2, 25–37).

[142] Propter quod tradentes scientiam naturalem et non inducentes aliquid de locorum diversitate peccant et solacium quaerunt suae imperitiae, si dicunt non oportere de ipso quaerere; non enim sufficit, quod in MATHEMATICIS de huiusmodi diversitate traditum est, ubi traditur, quae stellae super quae loca hoc vel illo modo oriantur vel occidant, utrum videlicet oblique vel recte, et quae ipsarum nec oriantur nec occidant, sed semper appareant, vel quae numquam et quae oriantur et occidant aliquando; quia licet ex his sciatur situs climatum, non tamen per eadem sufficienter scitur natura locorum et diversitas complexionis ipsorum (2, 37–49).

[143] In a seminal study, Robert S. Westman insightfully discusses the issue of disciplinary boundaries between mathematics and natural philosophy for the 16th and early 17th centuries; "The

As Albert further articulates what is necessary for fully knowing the nature of place, it becomes increasingly clear just how close he is to Roger here, albeit still in general terms. Albert just added mathematical astronomy and geography to the picture. Unlike Roger, however, Albert has not explicitly argued that this mathematical structure is necessary for the full science of places, but it is surely implied. Regardless, it is quite definitely not sufficient, as he explicitly states. In addition, of course, bells of recognition will have gone off on reading the last phrase in the section quoted above: *non tamen per eadem sufficienter scitur natura locorum et diversitas complexionis ipsorum*, which Albert explicitly relates to the *situs climatum*, the mathematical geographical information derived from the intersection of mathematical astronomy and geography.

Onward with this same passage:

> And therefore, just as in other natures it is fitting for *scientia naturalis* to go down to the species level (*ad specialia*), so also *in locis*. For if we consider only mixed things without complexions (*in mixtis tantum non complexionatis*), we will not completely know their nature, unless we knew all the *differentiae* of mixed things among metals and stones.[144] And it is likewise concerning what has complexions and souls. It is fitting, therefore, that he also speak thus in the science of places. And therefore, we see that the foremost men in philosophy, as Aristotle and Plato were, have treated this sort of thing, although their books on these subjects have not come down to us whole, but only in parts.[145]

We now have a good idea of what Albert thinks is necessary for a science of places, but still only in general terms. He also tells us that both Aristotle and Plato, and perhaps others, wrote treatises on the same subject, which have only partially survived.

Albert mentions Plato and Aristotle once again in the final passage of this chapter in describing the structure of his treatise:

> And we divide this book into three *distinctiones*: in the first we show that all the diversities [sc. of places] proceed from the [sc. celestial] orb into every difference of place;[146] in the second we show the accidents of diverse places; and in the third we shall determine the particular places of rivers and cities and mountains. For in this manner, Aristotle and Plato proceeded in this science.[147]

Astronomer's Role in the Sixteenth Century: A Preliminary Study," *History of Science* 18 (1980): 105–47.

[144] I also refer to Albert's views of metals and stones in relation to talismans in Chap. 7.

[145] Et ideo sicut in aliis naturis oportet scientiam naturalem usque ad specialia deducere, ita et in locis; si enim nos considerationem ponimus in mixtis tantum non complexionatis, non sciemus perfecte naturam ipsorum, nisi sciamus omnes mixtorum differentias in metallis et lapidibus. Et similiter est de consideratione complexionatorum et animatorum. Oportet ergo, quod etiam sic ait in scientia locorum. Et ideo praecipuos viros in philosophia, sicut fuit ARISTOTELES et PLATO, tractasse videmus de huiusmodi, licet libri eorum de his non integri, sed per partem ad nos devenerint (2, 49–60).

[146] [...] in prima ostendentes omnes diversitates ex orbe procedere in omnem loci differentiam [...] (3, 9–11).

[147] Et hunc librum per tres distinctiones dividimus, in prima ostendentes omnes diversitates ex orbe procedere in omnem loci differentiam et in secunda ostendentes accidentia locorum diversorum, et in tertia determinabimus particularia loca fluminum et civitatum et montium; hoc enim modo in hac scientia processerunt ARISTOTELES et PLATO (3, 8–15).

Albert, at any rate, clearly claims both an Aristotelian and a Platonic pedigree for his science of places.

*

Chapter 5, "On the general properties of places, in which there are things mixed from the four elements,"[148] provides detailed material for a proper comparison. Albert here looks at places *in particulari*, as he previously said was necessary for a complete *scientia locorum*:

> Moreover, if anyone wished to understand *particulariter* all the natures and properties of particular places in water and air and on earth, he would know that there is not a point (*punctus*) in them which does not have a special property from the power (*virtus*) of the stars oriented toward the middle habitation of mixed things [sc. the earth] (*ex virtute stellarum mediam habitationem commixtorum respicientium*), for the circle of the horizon is varied in relation to each point of the habitation of animals and plants and stones (*ad quodlibet enim punctum habitationis animalium et plantarum et lapidum variatur circulus horizontis*), and the entire orientation of the heavens to the middle place of habitation [sc. the earth] is varied in relation to the variation of the horizon's circle (*ad variationem circuli horizontis totus respectus caeli ad medium habitationis variatur*). From which cause, their natures, properties, customs, actions and species, which seem to be generated in the same perceptible place, are varied to such an extent that diverse properties and customs are attributed to twins' seeds, both for brute animals and for men, from this different orientation. And this is reasonable because it has been learned that the heavens pour forth formative powers into everything that exists (*caelum diffundere virtutes formativas in omne quod est*). Moreover, it mostly pours them forth by means of rays emitted by the lights of the stars (*diffundit eas per radios emissos a luminibus stellarum*), and therefore it follows that each pattern and angle of rays (*quaelibet figura radiorum et angulus*) causes different powers (*virtutes*) in things below.[149]

There can be no doubt that Albert has just provided exactly the same causal analysis as Roger did, and in much the same language. As we can see, Albert has now introduced a fully developed geometrical-optical model of celestial influences, where the different patterns of celestial rays striking different points on earth cause different powers/virtues in things below. He also uses the very same integrating mathematical astronomical and geographical structure to individuate each place on earth in

[148] De communibus proprietatibus locorum, in quibus sunt commixta ex quattuor elementis.

[149] Si quis autem particulariter velit cognoscere omnes naturas et proprietates particularium locorum in aqua et aere et terra, sciet, quod non est punctus in eis, qui non habeat specialem proprietatem ex virtute stellarum mediam habitationem commixtorum respicientium; ad quodlibet enim punctum habitationis animalium et plantarum et lapidum variatur circulus horizontis, et ad variationem circuli horizontis totus respectus caeli ad medium habitationis variatur. Qua de causa variantur naturae et proprietates et mores et actus et species eorum quae in eodem loco sensibili videnter generari, in tantum, quod et geminis seminibus et animalibus brutis et hominibus ex hoc diverso respectu proprietates et mores diversi attribuantur. Et hoc est rationabile, quia COMPERTUM EST caelum diffundere virtutes formativas in omne quod est. Maxime autem diffundit eas per radios emissos a luminibus stellarum, et ideo consequens est, quod quaelibet figura radiorum et angulus diversas in inferioribus causet virtutes (8, 43–62).

relation to the celestial configuration, namely the horizon, just as Roger did and for the very same reasons.

The passage continues:

> Since, therefore, [1] from a change of the horizon, it is necessary that [a] the entire circle would be changed, and from the change of the circle, [b] the entire configuration of the rays would be changed (*tota mutetur figuratio radiorum*). And, since [2] each point of a habitation constitutes one special center of a horizon, it follows necessarily that each point of the habitation has special powers, by which that which is located in it is informed (*quodlibet punctum habitationis habere virtutes speciales, quibus informatur id quod locatur in ipso*). And this is the cause why no generated thing is found to be completely similar to something else in every detail of its properties. And this is *locus* properly, about which Porphyry says that it is one of the individuating features because, as Boethius tells us, one cannot even imagine or comprehend that the place of any two things could be completely the same. And, as this is the case in whole and completed things (*res*), so it is also in the parts of things because, just as the place of one thing is not the place of another in every respect, so also the place of one part is not the place of another part. Nevertheless, there are corresponding features in places which are near—in common properties—and, therefore, those things which live nearby frequently have similar complexions, and similar things preserve and similar corrupt in general.[150]

As he provides more detail, it becomes increasingly clear how very similar Albert's analysis is here to what we saw in Roger, including using Porphyry as an authority and explicitly responding to anti-astrological arguments about twins.

Albert continues, now discussing matter in relation to place:

> The similarity and dissimilarity of places and of formative powers (*virtutes formativae*) which are from a place can be understood especially in matter by a sign in a certain type of magnet, which at one angle flees iron and at another attracts it. For since one angle does not differ from another by a great amount of place, it is fitting that places particularly close are sometimes of diverse and contrary powers (*virtutes*). For this contrariety is not from the matter because matter is not the cause of *virtus* and form. It is fitting, therefore, that it be from a place informed by the configuration of the rays of the stars (*ex loco informato a figuratione radiorum stellarum*). And this is what Hermes says superlatively in the book, *De virtutibus universalibus*, that a celestial configuration causes the qualitative power of the stars which are poured into the *inferiora* (*constellatio est causans virtutem qualitativam stellarum, quae infunditur inferioribus*), and it is formative of these [sc. the *inferiora*] by means of the qualities of the elements, which are as instruments of the celestial virtues (*formativa ipsorum per qualitates elementorum, quae sunt sicut instrumenta virtutum caelestium*). Moreover, not only is each particular place of a different property than another, but also each place in itself is made of different properties successively [sc. at different times],

[150] Cum igitur ex mutatione horizontis necesse sit mutari totum circulum et ex mutatione circuli tota mutetur figuratio radiorum, cumque quodlibet punctum habitationis unum speciale centrum constituat horizontis, necessario consequitur quodlibet punctum habitationis habere virtutes speciales, quibus informatur id quod locatur in ipso. Et haec causa est, quare nulla rerum generatarum invenitur omnino et per omnia in proprietatibus similis alteri. Et hic locus proprie est, de quo dicit PORPHYRIUS, quod est unum de individuantibus, quia, sicut tradit BOETHIUS, impossibile est etiam imaginari vel intelligi, ut duorum aliquorum sit locus omnino idem. Et sicut hoc est in totis et perfectis rebus, ita est etiam in partibus rerum, quia sicut locus unius non est alterius in toto, ita nec locus unius partis est locus alterius partis. Convenientia tamen est in locis, quae vicina sunt in proprietatibus communibus, et ideo ea quae vicine habitant, similes frequenter habent complexiones et similia conservantia et similia corrumpentia in communi (8, 63–82).

> for this reason, that by means of the rise and fall of the stars above it [sc. the place in question], and the different relationship of these stars to each place, it is necessary for its properties to be varied. And this is the cause why the *res generatae* are transformed by what contains them, sometimes into something better, sometimes into something worse, although this can sometimes be helped or impeded for this reason, that generated things received contrary dispositions deep in themselves, as we see that the bodies of some people do not get ill in places of sickness because of the extraordinarily strong disposition of their bodies, which is contrary to the disposition of the sickness found in the place.[151]

Here we find an allusion to change over time (*successive*), much as in Roger, and a reference to some of the consequences of these causal factors on health, namely, bodily complexions in relation to place, as also in Roger. We have also been introduced to a new authority, along with Plato and Aristotle from Chap. 1, and Porphyry and Boethius just above, namely, Hermes.[152]

Albert now discusses some of the deeper physical processes at work:

> Moreover, such transformations appear immediately in the elements (*in elementis*) even from a moderate variation of the stars, but, in what is composed from elements (*in elementatis*), they are not considered to transform so quickly. And this happens for this reason, that elements (*elementa*) are not far from matter and, therefore, quite easily receive every impression of what is acting on them. But what is composed of elements (*elementata*) have their matter, as it were, bound and determined by their forms, and, therefore, are more resistant to the agents altering them. Nevertheless, it is necessary that these be altered and corrupted, measured by the period of their existence (*ipsa alterentur et corrumpantur finita*

[151] Similitudo et dissimilitudo locorum et virtutum formativarum, quae sunt ex loco, in materia per signum maxime deprehendi potest in quodam magnetis genere, qui in uno angulo fugat ferrum et in alio attrahit ipsum; cum enim unus angulus ab alio non distet per magnum loci spatium, oportuit, quod vicina valde loca aliquando sint diversarum et contrariorum virtutum; haec enim contrarietas non est ex materia, quia materia non est causa virtutis et formae. Oportet igitur, quod sit ex loco informato a figuratione radiorum stellarum. Et hoc est, quod egregie dicit HERMES in LIBRO DE VIRTUTIBUS UNIVERSALIBUS, quod constellatio est causans virtutem qualitativam stellarum, quae infunditur inferioribus et est formativa ipsorum per qualitates elementorum, quae sunt sicut instrumenta virtutum caelestium. Non tantum autem quilibet locus particularis est alterius proprietatis quam alius, sed etiam quilibet per se diversarum successive efficitur proprietatum, eo quod per ortum et occasum stellarum super ipsum et diversam habitudinem earum ad locum quemlibet necesse est variari proprietates ipsius. Et haec causa est, quare mutantur res generatae ex suis continentibus quandoque in melius et quandoque in peius, licet hoc iuvari et impediri possit aliquando per hoc quod generata profunde in se receperint contrarias dispositiones, sicut videmus quorundam corpora non aegrotare in locis infirmis propter fortissimam illorum corporum dispositionem contrariam dispositioni infirmitatis, quae inventa est in loco (8, 82–9, 25).

[152] Hossfeld gives references to Ps.-Apuleius, *Asclepius*, and to Apuleius, *De Mundo*. For Albert's use of Hermes, see Loris Sturlese, "Saints et magiciens: Albert le Grand en face d'Hermes Trismégiste," *Archives de Philosophie* 43 (1980): 615–34; and Paolo Lucentini, "L'Ermetismo magico nel secolo XIII," in his *Platonismo, ermetismo, eresia nel medioevo*, Louvain-la-Neuve: Fédération Internationale des Instituts d'Études Médiévales, 2007, 264–324 (originally published, 2000). See also Alessandro Palazzo, "The Scientific Significance of Fate and Celestial Influences in Some Mature Works by Albert the Great: *De fato, De somno et vigilia, De intellectu et intelligibili, Mineralia*," in Per perscrutationem philosophicam: *Neue Perspektiven der mittelalterlichen Forschung, Loris Sturlese zum 60. Geburtstag gewidmet*, A. Beccarisi, R. Imbach and P. Porro (eds.), Hamburg: Felix Meiner Verlag, 2008, 55–78. At p. 63, he states that scholars have not adequately stressed the relevance of Hermes to Albert's doctrine of divination.

> *periodo sui esse*), which they have from the powers of the celestial circle (*quod habent ex virtutibus circuli caelestis*), as will be explained better at the end of the second book of the *De generatione*.[153]

Albert does indeed discuss "periods" at the end of *De generatione et corruptione*, Book II, as we saw in Chap. 1, as well as in *De fato*, as we will see in a related analysis in Chap. 5.

Albert concludes this passage by refering to something very dear to Roger's program, the practical benefits of this knowledge:

> Therefore, these things noted in general concerning place are similar, which, nevertheless, by accident from the closeness of mountains or their location, and the closeness of seas, are frequently impeded. They sometimes receive an impediment by means of the handiwork (*artificium*) of experienced men who know the locations and powers of the stars and impede their effects because, as Ptolemy says in the *Tetrabiblos*, the effects of the stars can be impeded and expedited through the wisdom of men experienced in the stars (*stellarum effectus et impediri et expediri possunt per sapientiam peritorum virorum in astris*).[154]

This last passage comes close to conjuring up the image of a Renaissance magus so well known from Frances Yates's work, a tradition which seems to stem directly from the *Secretum secretorum.*[155] Albert does not go quite so far here, but he does clearly allude to the practical benefits which can come from deep experience in the science of the stars. Roger does as well, and on many occasions. We saw a passage where the medical benefits of astrology were discussed; it runs throughout his work, although that side has not been emphasized here.[156] This last passage also touches on a major argument used to counter astrological determinism and its concommitant feature, that astrological judgments are certain, namely, that receiving celestial influences is greatly affected both by the receiving matter itself, and by the more

[153] Immutationes autem tales in elementis statim apparent etiam ex modica variatione stellarum, sed in elementatis non adeo cito perpenduntur. Et hoc contingit ideo, quod elementa parum distant a materia et ideo facillime recipiunt impressiones omnes agentium in ea. Sed elementata quasi victas et terminatas habent suas materias per suas formas et ideo magis resistunt alterantibus agentibus in ea; tandem tamen necesse est, quod et ipsa alterentur et corrumpantur finita periodo sui esse, quod habent ex virtutibus circuli caelestis, sicut melius explanabitur in FINE SECUNDI PERI GENESEOS (9, 26–37).

[154] "Haec igitur et similia sunt, quae notantur circa locum in communi, quae tamen per accidens ex vicinitate montium vel situ eorum et ex vicinitate marium frequenter impediuntur et aliquando recipiunt impedimentum per artificium peritorum virorum, qui noverunt situs et virtutes stellarum et impediunt effectus eorum, quia, sicut dicit PTOLEMAEUS in QUADRIPARTITO, stellarum effectus et impediri et expediri possunt per sapientiam peritorum virorum in astris (9, 38–46)." Hossfeld gives a reference to Ptolemy, *Tetrabiblos* I.3 (ed. Venet.), 1493, f.8v ff.

[155] See, for example, her *Giordano Bruno and the Hermetic Tradition*; and *The Occult Philosophy in the Elizabethan Age*, London: Routledge, 1979. For the *Secretum secretorum* and its extraordinary history and influence, see Steven J. Williams, *The Secret of Secrets: The Scholarly Career of a Pseudo-Aristotelian Text in the Latin Middle Ages*, Ann Arbor: The University of Michigan Press, 2003. I discuss the *Secretum secretorum* in relation to Roger Bacon's *Opus maius* in Chap. 8.

[156] He also referred to talismans with respect to their benefits for a republic, which also seems to be what Albert refers to here. I discuss talismans in Albert's and Roger's thought at greater length in Part 3 below.

general circumstances of reception, as we also saw in Chap. 1, and as we will see moreso in what follows.

*

I will next treat *Tractatus* 2,[157] Chap. 1, entitled: "What diversity the elements make on things-composed-of-elements with respect to different places."[158] We pick up the discussion where Albert provides an example that Roger also used, namely, plants:

> Moreover, a fuller sign of this—that a place, with its powers that exist in it, gives much of its property to what is located in it (*locus cum virtutibus, quae sunt in ipso, det multam proprietatem locato*)—we get in the case of plants which have had their roots pulled up. Even with their roots preserved and transplanted to nearby places with much earth adhering to them, they nevertheless languish for a long time before they can grow strong. The cause of which is none other than the dissimilarity of the powers of one place to a different place (*dissimilitudo virtutum unius loci ad locum alium*), even though it is near. And, therefore, it is fitting that the plant becomes assimilated little by little to the powers of that place in which it is transplanted before it would grow fully strong. And this also signifies that no point of the earth (*nullus punctus terrae*) has the same powers that another has, and nothing has all the powers altogether that another has. And the powers that each has are continually transformed in accordance with the variation of the circle (*virtutes, quas habet quidlibet, continue mutantur secundum circuli variationem*), which first makes and generates what is located (*qui primo facit et generat locata*), by means of its powers that flow into each place (*per virtutes, quas influit unicuique loco*).[159]

Athough the details of the example are different, the causal analysis is exactly the same, with the same emphasis on each place's unique qualitative makeup due to the celestial configuration's unique spacial relationship to each particular place on earth. Albert also refers once again to another significant feature mentioned by Roger, namely, that the nature of the influences changes over time, that is, the planets, and the heavens more generally, keep moving in both their diurnal and annual motions.

Albert continues the passage and thus concludes the first chapter of treatise 2:

> Because of which, also, those who are wise concerning the stars (*sapientes in astris*) first [1] diligently consider the [sc. celestial] circle of each place,[160] then they proceed [2] to consider the powers of the places (*virtutes locorum*) as a second [sc. set of] stars (*quasi*

[157] In qua tractatur de natura locatorum causata a locis (23, 14–15).

[158] Quid diversitatis elementa faciunt in elementatis secundum diversa loca (23, 16–7).

[159] Amplius autem, signum eius quod locus cum virtutibus, quae sunt in ipso, det multam proprietatem locato, accipimus in plantis, quae eradicantur, et quamvis salvis radicibus et cum multa terra sibi adhaerente ad vicina loci transplantentur, tamen diu languescunt, antequam possint convalescere. Cuius causa alia non est nisi dissimilitudo virtutum unius loci ad locum alium, quamvis sit eidem vicinus. Et ideo oportet plantam virtutibus illius loci in quem transplantatur, assimilari paulatim, antequam plene convalescat. Et hoc significat etiam nullum punctum terrae cum alio habere easdem virtutes et nihil omnino habere virtutes omnes, quas habet alterum. Et virtutes, quas habet quidlibet, continue mutantur secundum circuli variationem, qui primo facit et generat locata per virtutes, quas influit unicuique loco (24, 53–69).

[160] That is, they determine the planetary positions relative to a given place.

> *stellae secundae*),[161] and from these two joined together, [3] they prognosticate about the nature of the thing generated (*prognosticare de natura rei generatae*).[162] For the stars flow their powers in (*stellae* [...] *influunt virtutes*) by means of something else (*per aliud*), not without an intermediary, and they use a twofold medium for the influx of their powers (*utuntur ad influxum suarum virtutum duplici medio*), of which one is a ray and the second, a containing place (*quorum unum est radius et secundum locus continens*). And by means of these two, they [sc. the wise astrologers] come to the matter of the generated thing, which the flowed-in virtues form and determine (*per haec duo veniunt ad materiam rei generatae, quam formant et terminant virtutes influxae*). Moreover, the variation of the stars' location is the cause of the variation of the figure (*variatio autem situs stellarum est causa variationis figurae*) which the rays describe above the circle of the horizon (*quam describunt radii super circulum horizontis*). And this is the cause of the variation of the powers of places (*haec est causa variationis virtutum locorum*) and of those things which are generated in places (*eorum quae generantur in locis*), and therefore it is necessary that a natural philosopher know these variations.[163]

Albert further reveals here how he conceives the dynamics of generation. In particular, he describes the two means by which the stars influence things on earth, namely rays and places. We have already seen his rich use of places. His use of rays in this context is of the utmost significance. We should emphasize that Albert was a theologian and natural philosopher arguing for the importance of mathematical knowledge—in Roger's sense, namely a mathematical-astronomical and -geographical framework to support a geometrical-optical model of planetary action—as necessary for the most utterly fundamental features of a natural philosopher's understanding. Albert also notes here that this knowledge can be used for prognosticating.

*

We must conclude, then, that the contents of Albert's natural philosophy discussed here could hardly be closer to Roger's as reconstructed earlier in this chapter. This is true both conceptually and terminologically, in the broader patterns of thought and the intricacy of the detailed analyses. Or, we might perhaps more accurately say that the patterns of thought and language in certain central patterns of Roger Bacon's

---

[161] Albert defines '*stellae secundae*' in *De fato* in discussing a text from Ptolemy (66, 50–57): "[…] stellas secundas vocans effectus stellarum, qui apparent in inferioribus elementis, sicut in nubibus vel aliquo huiusmodi […] (66, 52–54)."

[162] This passage is similar to the one completing stage 4 in Chap. 1 from *De generatione et corruptione* (207, 1–6), where Albert offers the natural philosophical foundations for nativities. We will see Albert's analysis of the natural philosophical foundations for revolutions in my Chap. 3.

[163] Propter quod etiam sapientes in astris circulum uniuscuiusque loci diligenter primo considerant et deinde praecipiunt considerare virtutes locorum quasi stellas secundas et ex his duobus coniunctis prognosticare de natura rei generatae; stellae enim per aliud et non immediate influunt virtutes et utuntur ad influxum suarum virtutum duplici medio, quorum unum est radius et secundum locus continens. Et per haec duo veniunt ad materiam rei generatae, quam formant et terminant virtutes influxae. Variatio autem situs stellarum est causa variationis figurae, quam describunt radii super circulum horizontis. Et haec est causa variationis virtutum locorum et eorum quae generantur in locis, et ideo necesse est scire has variationes philosophum naturalem (24, 70–84).

astrologizing Aristotelian natural philosophy bear an extraordinarily close resemblance to Albert's, considering that Albert wrote his influential work over 10 years before Roger wrote his. One might even think that Roger completely took over the substance of Albert's work, and simply refined it by adding Grosseteste's notion of the multiplication of species. If Albert knew Grosseteste's work, he apparently decided not to embrace this characteristic feature of his analysis.

Given this analysis and the comparison with Roger Bacon's views on mathematics, *perspectiva* and astrology, the burden of proof shifts to those who claim that the unnamed master, maligned for his ignorance of *perspectiva*, is Albertus Magnus. That is, unless one believes that Bacon is being perverse here (as is sometimes his wont),[164] asserting, in effect, that ignorance of *perspectiva* can be claimed for simply not using the multiplication of species because this seems to be the only discernable difference between Albertus Magnus's and Roger Bacon's richly astrological analyses of place.

2.8 Conclusion

Finally, although I have not yet discussed Thomas Aquinas in relation to my reconstruction of this geometrical-optical model of celestial influence by Albertus Magnus and Roger Bacon, I would like to conclude this chapter with a brief indication that Thomas also embraced it, and thus that it implicitly lies behind his broader analyses, including those we will explore in Chap. 5. I will briefly refer here to two passages cited by Thomas Litt in his *Les corps celestes* (220–21), in discussing the view that each star, luminary and planet has its own particular nature, and thus its own particular influence or effect, one of the essential bases of astrology's natural philosophical foundations, as we have just seen with Albert and Roger as derived ultimately from al-Kindi.

I will only give two examples here, both from the first book of the *Summa theologiae*. In 67.3, Thomas states: "The rays (*radii*) of different stars have different effects, according to the different natures of their bodies."[165] Litt goes on to clarify that the final clause does not refer to the different natures of the bodies that receive celestial influences, but to the different corporeal natures of the planets themselves (221).

The second example is from 70.1 *ad* 2 and refers to the 6 days of Creation:

> Moreover, if the light (*lux*) made on the first day is understood to be corporeal light (*lux corporalis*), it is fitting to say that the light produced on the first day was made in accordance with the common nature of light (*secundum communem lucis naturam*). But on the

[164] See some examples of this trait of his character in Richard Lemay, "Roger Bacon's Attitude Toward the Latin Translations and Translators of the Twelfth and Thirteenth Centuries," in *Roger Bacon and the Sciences*, 25–47.

[165] Radii diversarum stellarum habent diversos effectus, secundum diversas naturas corporum.

> fourth day, a determinate power (*determinata virtus*) was attributed to the luminaries for determinate effects. Accordingly, we see that the ray of the sun (*radium solis*) has some effects, and the ray of the moon (*radium lunae*) different ones, and so about the others [sc. the five planets, and perhaps also implying the rest of the fixed stars as well].[166]

For Thomas, then, each planetary body has a different nature and thus their rays each have a different effect, precisely as we saw in al-Kindi's *De radiis stellarum* and the texts by Roger and Albert discussed here. As far as I know, Thomas did not articulate this view in greater detail.

[166] Si autem lux primo die facta intelligitur lux corporalis, oportet dicere quod lux primo die fuit producta secundum communem lucis naturam; quarto autem die attributa est luminaribus determinata virtus ad determinatos effectus; secundum quod videmus alios effectus habere radium solis, et alios radium lunae, et sic de aliis.

Chapter 3
The Natural Philosophical Foundations for Astrological Revolutions: Albertus Magnus's Commentary on the Pseudo-Aristotelian *De Causis Proprietatum Elementorum*

Now that I have established the fundamentally similar structures of astrologizing Aristotelian natural philosophy in the works of Albertus Magnus and Roger Bacon, and analyzed how they both provided the natural philosophical foundations for the practice of nativities or birth horoscopes and for elections, I would now like to develop our understanding of Albert's views further by discussing some related sections from the companion text to his commentary on *De natura loci*, namely, his *De causis proprietatum elementorum* (*On the Causes of the Properties of the Elements*) which will help us fill in and sharpen the picture, and thereby complete my reconstruction of the basic structures of Albert's brand of astrologizing Aristotelian natural philosophy.[1]

In the *De causis proprietatum elementorum,* Albert provides (*inter alia*) the natural philosophical foundations for the practice of astrological revolutions—including for the great conjunctions so important to astrological views of history—by using the very same model of astrologizing Aristotelian natural knowledge reconstructed so far. As we will see, the *De causis proprietatum elementorum* shows us further how Albert conceived of the heavens acting on the earth in relation to each of the four elements—water, air, fire and earth (his order)—offering a sort of natural philosophical miscellany that treated, among other things, a broad range of meteorological phenomena.[2] We will explore each of the four in turn, beginning with water.

[1] Texts related to Albert's views on fate and providence, and on magic, especially talismans, will be discussed in Chaps. 5 and 7 respectively. They will further articulate our understanding of his system, including its more metaphysical and epistemological dimensions.

[2] For the study of meteorology in the Renaissance with some discussion of the Middle Ages, see Craig Martin, *Renaissance Meteorology: Pomponazzi to Descartes*, Baltimore: Johns Hopkins University Press, 2011.

H D. Rutkin, *Sapientia Astrologica: Astrology, Magic and Natural Knowledge, ca. 1250–1800*, Archimedes 55, https://doi.org/10.1007/978-3-030-10779-6_3

3.1 Water

In Book I, Tractate 2, Chapter 2 (I.2.2),[3] "On the opinion stating that the sea is shifted from place to place," Albert begins eight chapters (2–9) on questions concerning the motion of water and how it transpires, beginning with whether it is affected by the motion of the stars. In so doing, Albert refers twice to what appears to be Albumasar's *On Great Conjunctions*, as he had also done in the passage from his translation-commentary on *De caelo* discussed at the end of Chap. 1, and once again without explicitly mentioning Albumasar as the author: *On the Great* (or *Greater*) *Accidents of the World* (*De magnis* [or *maioribus*] *accidentibus mundi*, 64, 21–22 and 65, 16–17).[4] I will be very selective in what follows.

In criticizing the work of unnamed natural philosophers (*physiologi*), Albert claims that, although they adduce astrological causes for flooding, their arguments remain unpersuasive because they do not provide geometrical demonstrations, making only simple assertions instead. We should note that Albert here uses "geometrical" in precisely the same way that Roger did in *Opus maius* IV and elsewhere:[5]

> [T]herefore, whether the sea's motion (*permutatio*) is caused by one of these two motions or by both, they ought to have provided demonstrations and observations for this by means of geometric figures of lines drawn out for view, from which they might prove what they say. And indeed this they do not do, but they only profess with a simple assertion that everything which comes to be on earth and in lower [sc. bodies] (omnia *quae fiunt in terra* et in inferioribus), comes to be on account of the motion of the superior bodies (*fiunt propter motum superiorum* corporum) because this is the noble element [sc. the quintessence (*elementum nobile*)], which, alone of all bodies, moves with a perpetual motion of its own, and therefore it is the agent for all the accidents of inferior things (*agens omnia* accidentia *rerum* inferiorum). And they introduce several of its effects (*effectus*), as they say that the

[3] Albertus Magnus, *Opera omnia*, 5.2, *Liber de causis proprietatum elementorum*, Paul Hossfeld (ed.), Monasterii Westfalorum: Aschendorff, 1980, 49–104. Hossfeld's edition contains both Albert's commentary on this pseudo-Aristotelian work, and, at the bottom of each page, Gerard of Cremona's Latin translation of the text from the Arabic, in a critical edition established by Stanley Luis Vodraska, "Pseudo-Aristotle *De causis proprietatum et elementorum:* Critical Study and Edition," PhD thesis, University of London, 1969. My thanks to Irven Resnick for allowing me to use his translation before it was published; it has now appeared as Albert the Great, *On the Causes of the Properties of the Elements* (Liber de causis proprietatum elementorum), translated from the Latin with an introduction and notes by Irven M. Resnick, Milwaukee: Marquette University Press, 2010. His translation is the basis for mine, which I modify where appropriate. I have also used and/or drawn on some of his explanatory footnotes where appropriate.

[4] In *De caelo*, he called it: *De accidentibus magnis universalibus in mundo.* He noted there that it had eight distinctions, and he explicitly attributed it to Ptolemy.

[5] As with the authentic Aristotelian texts, the Cologne edition typographically indicates what is the underlying Aristotelian text and what are Albert's additions. This is reflected in the Latin in parentheses in my translation, and in the text in the footnotes. As with *De generatione et corruptione*, the words in italics indicate that Albert took the words from the existing text of *De causis proprietatum elementorum*, and the non-italicized words indicate what Albert added for clarity and explanation. I give the page reference for Resnick's valuable translation in my slightly modified English version in the text, and to the *Editio Coloniensis* for the Latin in the footnotes.

> flood (*diluvium*) that was upon the earth under Pyrrha and Deucalion[6] would not have existed but for the great conjunction (*coniunctio* magna) of seven planets in the sign of Pisces. There was an inundation (*inundatio*) of waters, and the pestilential wind (*ventus* pestilens) that killed peoples (*gentes*) everywhere (universaliter) in the land of Hadramot[7] was due to the great conjunction of the planets (*coniunctio* planetarum magna) in the sign of Gemini, which is an air sign causing corruptions of the air (quod est signum aëreum causans aëris corruptiones). Moreover, the cause of the plague (*pestilentiae* [...] *causa*) that occurred in the land of Iamen[8] (a region of the Indies) was [sc. a great conjunction] of these same planets in the earth sign of Virgo (*in signo* terrestri Virginis), which causes and produces corruptions of the earth (quae terrae corruptiones causat et producit). Likewise, they talk about the other effects that result from the conjunction of the stars (in *coniunctione* stellarum), which we will mention below. For, a perfect knowledge (perfecta scientia) of these matters belongs to the book, *On the Greater Accidents of the World* (De maioribus accidentibus mundi), which is subalternated to the science of the stars (qui astronomiae subalternatur, 44-45).[9]

Although Albert rejects this account by the *physiologi*, we can still derive insight into the different putative effects the conjunctions have in different elemental triplicities in their respective "mutable" signs. Albert only discusses the mutable water, air and earth signs here, not the fire sign, Sagitarrius. He offers his own analysis of great conjunctions below.

After criticizing others for their views on planetary action, Albert turns to another of their views on great conjunctions:

> Moreover, the stars, when they rise (ascendunt) in different ways (*diversimode*) in the seven climates, are found to be the causes of years of barrenness (*sterilitas*) and fruitfulness (*ubertas*), to such an extent that those wise in the stars (sapientes astrorum) report that,

---

[6] On Deucalion and Pyrrha, see Resnick's translation I.2.9, with note. Deucalion was the Greek Noah. In the sixteenth century, Giuliano Ristori also used this story in the context of great conjunctions in his teaching manuscript on Ptolemy's *Tetrabiblos*, in a course offered at the University of Pisa in the 1540s. See my "Teaching Astrology in the 16th Century: Giuliano Ristori and Filippo Fantoni on Pseudo-Prophets and Other Effects of Great Conjunctions," in *From Masha'allah to Kepler: Theory and Practice in Medieval and Renaissance Astrology*, Charles Burnett and Dorian Gieseler Greenbaum (eds.), Ceredigion, Wales: Sophia Centre Press, 2015, 353–406.

[7] The southern part of the Arabian peninsula; see Vodraska, "Pseudo-Aristotle," 292–93.

[8] For Iamen, an area roughly identical to Yemen, see Resnick's tr. I.1.5 (p. 35 with n. 64).

[9] [S]ive ergo permutatio maris causatur ab altero istorum duorum motuum sive ab utroque, debuissent *attulisse* super hoc *demonstrationes* et observationes per figuras geometricas linearum ad *visum* protractarum, ex quibus probarent, quod dicebant. Et hoc quidem non faciunt, sed simplici confessione *confitentur, quod* omnia *quae fiunt in terra* et in inferioribus, *fiunt propter motum superiorum* corporum, eo quod hoc est *elementum nobile, quod* solum ex se perpetuo cursu *currit ex corporibus, et* ideo *est agens omnia* accidentia *rerum* inferiorum. Et horum quidem effectuum plurimos inducunt, *sicut dicunt, quod diluvium, quod fuit* sub Pyrrha et Deucalione *in terra, non fuit nisi propter coniunctionem* magnam septem planetarum *in signo Piscis*, in quo est inundatio aquarum, et *quod ventus* pestilens, *qui interfecit gentes* universaliter, *in* terra *Hadramot fuit propter coniunctionem* planetarum magnam *in signo Geminorum*, quod est signum aëreum causans aëris corruptiones. *Pestilentiae* autem, *quae fuit in terra Iamen,* quae est regio Indorum, *causa fuit* eorundem planetarum *in signo* terrestri *Virginis*, quae terrae corruptiones causat et producit. *Similiter* autem dicunt de aliis effectibus, qui *fiunt* in *coniunctione* stellarum, de quibus nos INFERIUS faciemus mentionem; nam perfecta scientia horum ad LIBRUM pertinet DE MAIORIBUS ACCIDENTIBUS MUNDI, qui astronomiae subalternatur (63, 64–64, 22).

> when the two stars (*duae stellae*) that are called "heavy" (graves) due to the slowness of their motion, namely Saturn and Jupiter, are conjoined (*coniunguntur*) at one of the points [= degrees] of any [sc. zodiacal] sign in the heavens (in uno punctorum signi alicuius in caelo), then death (*mortalitas*) and laying waste (depopulatio) happen (accidit), such that kingdoms are emptied (*regna evacuantur*). And this should occur when they are conjoined at the change of [sc. one] triplicity to [sc. another] triplicity (cum *permutatione triplicitatis ad triplicitatem*, 46).[10]

Albert here discusses the deleterious effects of what is normally called either a greater or greatest conjunction, namely, when the great conjunction of Saturn and Jupiter (which occurs every 20 years) shifts from its 240 year residency in one elemental triplicity to begin its residency in the next (a "greater" conjunction), or at the overall end of this four-fold 960-year cycle and the beginning of the next with a fire sign (a "greatest" conjunction).

Albert then clearly describes the astronomical component of great and greater conjunctions. This explanatory paragraph is entirely his addition:[11]

> For the signs of the zodiac (*signa zodiaci*)—which are twelve [sc. in number]—have four triplicities, that is four [sc. groups of] three [sc. zodiacal] signs (*quattuor ternarios signorum*) which are of one complexion (*complexio*), as Aries, Leo, and Sagitarrius are fire signs, hot and dry (*signa ignea, calida et sicca*). The earth signs, however, cold and dry (*signa* [...] *terrea, frigida et sicca*), are Taurus, Virgo, and Capricorn. The water [signs], moreover, cold and moist (*aquea* [...] *frigida et umida*), are Cancer, Scorpio, and Pisces, and the air signs, hot and moist (*signa aërea, calida et umida*), are Gemini, Libra and Aquarius, 46-47).[12]

This is all introductory material—basic astrology—as we saw in the discussion of Roger Bacon's texts in the introductory excursus.

With this background on the elemental nature of the four triplicities, Albert turns to the great conjunctions of Jupiter and Saturn and their astronomical permutations:

> And it has been proved that the two stars Saturn and Jupiter are conjoined by their mean motion in one triplicity twelve times before they are conjoined in some other sign. For they are conjoined one time in twenty years, less one-eighth part of a year, and their conjunction occurs over the arc that is in the ninth sign from the first conjunction, with the addition of two and one-half degrees. Moreover, twelve times two and one-half degrees makes thirty degrees, and therefore their thirteenth conjunction is outside the triplicity (*extra triplicitatem*) in the tenth sign. For the ninth sign, reckoned from any sign in the zodiac [i.e. counting inclusively], is of the same triplicity with itself, but the tenth belongs to another. Moreover, twelve times twenty makes 240 years, and this is why for 240 years these stars are always

[10] Inveniuntur autem *stellae* etiam *esse causae annorum sterilitas et ubertatis*, quando ascendunt *diversimode* in *climatibus septem*, in tantum, quod sapientes astrorum tradunt, quod quando *coniunguntur duae stellae*, quae propter tarditatem sui motus graves dicuntur, *scilicet Saturnus et Iuppiter*, in uno punctorum signi alicuius in caelo, quod tunc accidit *mortalitas et* depopulatio, ita quod *regna evacuantur*. Et hoc *fieri* consuevit, quando coniunguntur cum *permutatione triplicitatis ad triplicitatem* (64, 72–81).

[11] I discuss Roger Bacon's views of Great Conjunctions in relation to religion in Chap. 6.

[12] Habent enim signa zodiaci, quae sunt XII, quattuor triplicitates, hoc est quattuor ternarios signorum, quae sunt unius complexionis, sicut signa ignea, calida et sicca sunt Aries, Leo, Sagittarius; signa autem terrea, frigida et sicca sunt Taurus, Virgo et Capricornus; aquea autem sunt frigida et umida Cancer, Scorpio et Piscis, et signa aërea, calida et umida sunt Gemini, Libra et Aquarius (64, 82–89).

> conjoined in one triplicity from one twenty year period to the next. Moreover, when they change triplicity (*permutant triplicitatem*), then the astrologers (*astronomi*) say that a new power is poured into lower things from the heavens (*nova virtus infunditur inferioribus ex caelo*), and ancient places of habitation are destroyed and new ones begin, just as is demonstrated well in the book *On the Great Accidents of the World* (*in libro De magnis accidentibus mundi*, 47).[13]

Albert here provides the basic astronomical structure of great conjunctions, namely, Jupiter and Saturn conjoining once every 20 years in its characteristically triangular pattern. He also informs us that a "new power" (*nova virtus*) enters the lower world from the heavens when the great conjunction changes triplicity, thus becoming a greater or greatest conjunction with its concommitant transformations in the world. Albert discusses great conjunctions further below. He concludes this chapter with the general statement he had also made earlier, namely, that it has now been proved that what exists in the lower world is caused by the motions of the superior [= celestial sc. bodies] (*ea quae sunt in inferioribus, causantur a superiorum motibus*, 47).[14]

*

In another chapter on water (I.2.4), Albert offers a digression in his own voice, in which he discusses eight things the astrologizing natural philosopher should know before trying to understand the tides. The first offers a natural philosophical understanding of how the planets affect things on earth that fills in details of our earlier analysis. It also provides the natural philosophical foundations for the medical view of critical days. Albert's main concern is to explain how planets effect motion in the world below in terms of motion and light, with an emphasis on why the luminaries (the sun and moon) have the greatest influence:

> Moreover, one of these things is that, although all the planets have an effect in common on lower [sc. bodies] (*omnes planetae communiter habeant in inferioribus effectum*), nevertheless, the sun and the moon especially are the planets whose properties and powers lower bodies follow (*proprietates et virtutes sequuntur inferiora*) due to three causes. The first of these is the quantity of their light (*quantitas luminis*), for the other planets are indeed lucent bodies (*lucentia* [...] *corpora*). They move lower [sc. bodies] by both their motion and their light (*et motu et lumine movent inferiora*), but they do not emit rays and noticeable shadows onto inferior bodies (*radios et umbras notabiles non emittunt super inferiora*). The sun and

[13] Et probatum est, quod duae stellae Saturni et Iovis duodecim vicibus coniunguntur per medium motum suum in una triplicitate, antequam in alio aliquo signo coniungantur; coniunguntur enim in viginti annis semel minus octava parte unius anni, et est coniunctio eorum super arcum, qui est in nonum signum a prima coniunctione cum appositione duorum graduum et dimidii; duodecies autem duo gradus et dimidius faciunt XXX gradus, et ideo tertiadecima coniunctio eorum est extra triplicitatem in decimum signum; nonum enim signum, a quolibet signo acceptum in zodiaco, est eiusdem triplicitatis cum ipso, sed decimum est alterius; duodecies autem XX faciunt CCXL annos, et ideo per ducentos XL annos istae stellae de viginti in viginti annos coniunguntur semper in triplicitate una. Quando autem permutant triplicitatem, tunc dicunt ASTRONOMI, quod nova virtus infunditur inferioribus ex caelo et delentur antiquae habitationes et incipiunt novae, sicut bene probatur in LIBRO DE MAGNIS ACCIDENTIBUS MUNDI (64, 89–65, 17).

[14] [E]x his enim sufficienter probatur, quod ea quae sunt in inferioribus, causantur a superiorum motibus (65, 19–21).

> moon, however, are luminaries (*luminaria*) and they move [sc. things below] by both motion (*motus*) and light (*lumen*), and by rays (*radii*). And for this reason, they also make shadows when some opaque body blocks their rays. This is why their influence is very strong on inferior bodies (*impressio eorum fortissima est in inferioribus*, 52).[15]

Here Albert articulates clearly that both the planets and the luminaries act on the world below by means of both motion and light, but then he makes a distinction between the planets and the luminaries concerning rays. Because the luminaries have a greater quantity of light, their rays also effect things below in addition to their light. By contrast, the planets—due to their lesser quantity of light—although they also effect things below by means of both motion and light, somehow that light does not also emit rays at all, or at least rays strong enough to effect things below. Emitting more light, the aptly named luminaries (the sun and moon) thus have a greater influence on the terrestrial world and its inhabitants by means of both light and rays, in addition to motion.

After discussing their nature as agents, Albert gives the second cause for the luminaries' power, namely, their location in relation to the five planets. He begins with the moon:

> Another cause, moreover, is their location (*locatio*) among the planets, since the moon, although it is smaller in size (*quantitas*) than all the planets but one, is nevertheless nearer to lower [sc. bodies] (*vicinior est inferioribus*) and is closer to their natures (*appropinquat magis naturis eorum*).[16] Therefore, it changes (*permutat*) them, and because of this, the critical days (*cretici dies*) are numbered according to the moon.[17] It is called the queen of heaven (*regina caeli*)[18] because it rules the moisture of lower bodies (*regit umiditates corporum inferiorum*). And this is why metals and plants and animals' members—especially the eyes, in whose composition the nature of water (*aquae natura*) abounds—receive the greatest alterations and increases and diminutions in accordance with the moon (*recipiunt maximas alterationes et augmenta et diminutiones secundum lunam*, 52-53).[19]

---

[15] Unum autem horum est, quod licet omnes planetae communiter habeant in inferioribus effectum, tamen sol et luna praecipui planetae sunt, quorum proprietates et virtutes sequuntur inferiora propter tres causas, quarum prima est quantitas luminis ipsorum; alii enim planetae lucentia quidem corpora sunt et motu et lumine movent inferiora, sed radios et umbras notabiles non emittunt super inferiora; luminaria autem, quae sunt sol et luna, et motu et lumine, et radiis movent et ideo etiam umbras faciunt, quando aliquod corporum opacorum obicitur radiis eorum, et ideo impressio eorum fortissima est in inferioribus (67, 39–50).

[16] Both spatially and ontologically, it appears.

[17] For critical days, see also Resnick tr. I.2.4. "'Critical days' are days on which an illness was perceived to undergo a sudden, violent change that either implies recovery or death. They are, then, especially useful in determining a medical prognosis. Galen's *De creticis diebus*, circulating in Latin translation in the second half of the thirteenth century, emphasized the importance of such days for medical judgment. So too did the popular late thirteenth-century *Aggregationes de crisi et creticis diebus* (*Medieval Prognosis and Astrology*, Cornelius O'Boyle, ed., Cambridge: Wellcome Unit, 1991), which relied heavily on Galen's work." See also Giuseppe dell'Anna, *Dies critici: la teoria della ciclicità delle patologie nel XIV secolo*, Galatina (Lecce): Congedo, 1999.

[18] On the moon as the queen of heaven, see also Albert's *De caelo* 2.3.15, *De anima* 18.2.8, and *De vegetabilibus* 7.1.9.

[19] Alia autem causa est locatio eorum inter planetas, quoniam luna, quamvis sit quantitate minor omnibus planetis praeter unum, tamen vicinior est inferioribus et appropinquat magis naturis

The moon thus acts most strongly on watery things due to its similarity in nature and its closeness in spatial proximity to them. We can see at the end that Albert refers explicitly to two of the Aristotelian types of *motus*, namely, changes in quality (*alteratio*) and quantity (*augmentum* and *diminutio*).

Then Albert turns to the sun. To explain how the sun works, Albert employs a microcosm-macrocosm analogy, where the heart in a human being corresponds to the sun's position and function in the *machina mundi*:

> Moreover, the sun is located as the middle planet by glorious God and is just like the heart that administers powers everywhere (*vires undique ministrat*). Although it is farther away than some planets, nevertheless, in terms of size (*quantitas*) and light (*lumen*) it is greater than them all, and therefore it moves and transforms lower bodies most powerfully (*fortissime movet corpora inferiora et transmutat*, 53).[20]

Due primarily to its significantly greater size and quantity of light—and to its location in the center like the human heart—the sun has the greatest celestial influence on the terrestrial world. We will notice that this is the first explicit microcosm-macrocosm analogy encountered so far in this study.

Then Albert offers the third cause, turning from the more generic qualities of light, size and location to their particular natures, namely, their particular properties and virtues:

> The specific properties and powers of the qualities of these luminaries (*proprietates et virtutes qualitatum horum luminarium speciales*) are the third cause. For the moon, because it has an earthy (*terrea*) and watery (*aquea*) property, has the capacity to move all things in which water and earth are dominant by virtue of its connaturality (*ex connaturalitate*). Moreover, the sun, because it is the source of vivifying heat (*fons caloris vivifici*), causes the humors to bubble up (*ebullire*) because moisture is naturally drawn to the heat raising it up and becomes vapor (*vaporet*). For this reason, the ancient Egyptians said that the sun attracts moisture for the nourishment of all the celestial bodies. Therefore, this is one of those things that we should presuppose (53).[21]

Albert has just provided significant information on how the sun and moon act within the economy of nature. We can easily see how this fleshes out and at least at one point complicates the geometrical-optical model discussed in Chap. 2. Rays continue to be central, since the luminaries—which are the most powerful celestial bodies—act by means of motion, light *and* rays. The planets also act by means of

---

eorum et ideo permutat ea, et numerantur propter hoc cretici dies secundum lunam, et vocatur regina coeli, quia regit umiditates corporum inferiorum; et ideo metalla et plantae et animalium membra et praecipue oculi, in quorum compositione abundat aquae natura, recipiunt maximas alterationes et augmenta et diminutiones secundum lunam (67, 50–61).

[20] Sol autem medius planetarum a deo glorioso positus est tamquam cor, quod vires undique ministrat, et licet longinquior sit quibusdam planetis, tamen quantitate et lumine maior est omnibus eis, et ideo fortissime movet corpora inferiorum et transmutat (67, 61–66).

[21] Tertia causa sunt proprietates et virtutes qualitatum horum luminarium speciales; luna enim, eo quod est proprietatis terreae et aqueae, ex connaturalitate habet cuncta movere, in quibus aqua et terra dominantur. Sol autem, eo quod sit fons caloris vivifici, facit ebullire umores, eo quod umidum naturaliter trahatur et vaporet ad calorem se elevantem, propter quod antiqui dixerunt AEGYPTII, quod sol attraheret umidum ad omnium caelestium corporum nutrimentum. Hoc est ergo unum eorum quae nos praesupponere oportet (67, 66–75).

motion and light, but apparently not also by rays, employing a distinction that is not fully clear, and one that we will not encounter elsewhere. Albert will return to the results of this analysis in discussing great conjunctions in the following passage.

*

In I.2.9, Albert offers another digression, this time on the cause of floods, where he returns to discuss great conjunctions in the mutable water sign, Pisces. After discussing the differences between universal and particular floods, namely, those that cover the entire world and those that only affect particular places, Albert offers the outlines of a causal analysis:

> Moreover, the cause of a universal flood (*universale diluvium*) is one in which—at the same time—all the celestial and terrestrial factors that cause an inundation of water (*caelestia omnia et terrestria causantia inundationem aquae*) coincide. A less universal cause is one in which some of the celestial (*quaedam coelestium*) and some of the terrestrial causal factors coincide. And a particular cause (*causa* [...] *particularis*) is one in which either some celestial causes alone coincide, or some terrestrial ones alone (71).[22]

We now have a causal criterion for describing a flood as "universal" full stop, "less universal" or "particular." Albert goes into more detail just below.

After discussing the opinions of others, including the Arabs, Albert offers his own detailed analysis. First he points to the cause, which refers to the views expressed above: "Therefore, abandoning these things that were said both obscurely and imperfectly, we say that the cause of the flood that was on the earth was touched upon above, namely, that it was the conjunction that occurred in the sign of Pisces (*causam fuisse coniunctionem, quae fuit in signo Piscis*)."[23] As it turns out, this seems to have been *a* if not *the* key text inspiring concerns about the next great flood of biblical proportions widely predicted for a later great conjunction in Pisces of 1524. We know that Johannes Stöffler—professor of mathematics at Tübingen and co-author with Jakob Pflaum of the 1499 almanac that first predicted the flood of 1524—was deeply influenced by Albert and knew his works well, including the *De causis proprietatum elementorum.*[24]

Establishing this broader point, Albert turns to a more general description of conjunctions and their effects, thereby providing more information on the natural philosophical foundations for revolutions, and especially for great conjunctions.

---

[22] Causa autem universalis diluvii est, in qua conveniunt simul et caelestia omnia et terrestria causantia inundationem aquae. Causa vero minus universalis est, in qua quaedam caelestium conveniunt et quaedam terrestrium. Causa vero particularis est, in qua conveniunt aut quaedam caelestia sola aut quaedam terrestria sola (76, 55–61).

[23] Haec igitur tamquam obscure et imperfecte dicta relinquentes dicimus SUPRA tactum esse, quod diluvii, quod fuit in terra, causam fuisse coniunctionem, quae fuit in signo Piscis (77, 66–69).

[24] Wilhelm Maurer, *Der Junge Melanchthon zwischen Humanismus und Reformation*, 2 vols., Göttingen: Vandenhoeck and Ruprecht, 1967–69, I: 136: "Dessen Hauptquelle für ihn [Stöffler] ist aber offenbar der Schwäbische Landsmann Albertus Magnus. Ihn, der 'splendor Suevorum' ruft er besonders häufig als Kronzeugen für seine naturwissenschaftlichen Erkenntnisse an; er verrät eine eingehende Kenntnis seine Bücher, die er auß den Handschriften zitiert."

First he runs through the logical possibilities of the number of conjunctions, from all seven planets to as few as two (76, 69–89).[25] Then Albert discusses the two different ways planets can conjoin, namely, where they actually eclipse each other, or where they lie on the same degree of celestial longitude but differ in latitude. He calls the former "true" and the latter "untrue" (77, 90 ff.).

His explanation of how the two work is very interesting:

> Moreover, the conjunction is said to be true (*vera*) because, since the stars (*stellae*) are luminaries (*luminaria*) absorbing[26] the light (*lumen*) that is directed to them, then the lower [sc. planet] absorbs the light of the higher all the way to its center (*inferior imbibit sibi lumen superioris usque ad centrum suum*) and informs it with its own power (*informat ipsum sua virtute*). And we can see something similar to this in the human body, where one member transmits the *spiritus* and humor of another, and yet the receiving member informs them according to its own nature and power (*membrum recipiens informat illa secundum suam naturam et virtutem*). Thus, the brain (*cerebrum*) informs the *spiritus* and humors sent to it from the heart and liver for its own power and animal operation (*ad virtutem et animalem operationem*), and the testicles inform the *spiritus* and humor sent to them towards a generative power formative of the species (*ad virtutem generativam et formativam speciei*).[27] Every planet that receives light from another acts in this way, and we can recognize this in the moon, which receives light from the sun and forms it into its own power (*accipit lumen a sole et format ipsum ad virtutem suam*), so that it is cold and moist, although in the sun it is hot and dry temperately (74).[28]

In addition to another microcosm-macrocosm analogy to clarify his analysis, Albert raises several interesting natural philosophical issues that we have seen before, namely, [1] the differentiation of light in all of the planets, and [2] that what receives a higher influence transforms it by its own nature as the material cause. We will see this fundamental ontological principle again in Chap. 5. We also here for the first time encounter the fundamental physiological entity *spiritus* (along with humors), which will play a much larger role in volume II.[29]

---

[25] This should be compared with chapter seven of the *Speculum astronomiae* on revolutions, where the anonymous author also mentions that there are 120 logical possibilities.

[26] Literally, "drinking in," *imbibentia*.

[27] See Albertus Magnus, *De caelo* 2.3.5 (153, 1–25) and 2.3.6 (154, 9–86).

[28] "Dicitur autem vera coniunctio, quia cum stellae sint luminaria imbibentia sibi lumen, quod ad eas dirigitur, tunc inferior imbibit sibi lumen superioris usque ad centrum suum et informat ipsum sua virtute. Et huius simile videre possumus in corpore humano, ubi unum membrum transmittit alii spiritum et umorem et tamen membrum recipiens informat illa secundum suam naturam et virtutem, quemadmodum cerebrum spiritum et umores sibi a corde et hepate missum informat ad virtutem et operationem animalem et testiculi spiritum et umorem sibi missum informant ad virtutem generativam et formativam speciei. Ita facit quilibet planeta qui recipit lumen ab alio, et hoc cognoscimus in luna, quae accipit lumen a sole et format ipsum ad virtutem suam, ut sit frigidum et umidum, cum in sole sit calidum et siccum temperate (78, 7–23)." See also Albert's *De caelo* 1.1.11 (29, 65–68) and 2.3.3 (157, 55–56).

[29] James J. Bono discusses *spiritus* in Albert, primarily in the *De animalibus*; see "The Languages of Life: Jean Fernel (1497–1558) and *Spiritus* in Pre-Harveian Thought," PhD thesis, Harvard University, 1981, and "Medical Spirits and the Medieval Language of Life," *Traditio* 40 (1984): 91–130. I have learned a great deal from Bono's work. My thanks to Nancy Siraisi for bringing his work to my attention.

Albert then returns to the specific case of great conjunctions of Saturn and Jupiter, and the effects of greater or greatest conjunctions:

> Moreover, it still must be noted, as we said above [in I.2.2], that when the two stars [sc. Saturn and Jupiter]—which are more universal in moving[30] and are stronger because they are superior—are conjoined, they sometimes change the triplicity (*permutant triplicitatem*) in which they were previously, whereas sometimes they are conjoined in the same triplicity in which they were previously conjoined. Moreover, when they change triplicity, then they are, as it were, made new in their moving (*efficiuntur in movendo quasi novae*),[31] and this conjunction signifies great accidents and marvels (*prodigia*), as well as changes in the general state of the elements and of the world (*mutationes generalis status elementorum et mundi*). A natural philosopher (*naturalis*) ought to state the cause of this on his own, which an *astronomus* knows.[32] And this is why the Philosopher says that the science of the stars is another part of natural philosophy (*astronomia est altera pars physicae*), and Ptolemy says that a maker of judgments (*iudex*), a practitioner of elections (*elector*), and an observer of the stars (*observator stellarum*) will err if he is not a natural philosopher (*physicus*, 74-75).[33]

Albert here explains how greater and greatest conjunctions can have such great effects on the world. This happens because their ability to act on the world is renewed and thereby transformed by the shift in triplicity. Albert also argues strongly—and with the authority of both Aristotle and Ptolemy—for the importance of natural knowledge to the astrologer (*astronomus*) in his quest for accuracy in interpretation. Claims for astrology being a part of natural philosophy (*pars physicae*) will recur in volume III with (*inter alios*) Philipp Melanchthon and Francis Bacon using precisely this phraseology.

Then Albert offers his causal analysis for why this happens:

> Moreover, this is surely the cause. For, since we already know from what was said in *De caelo et mundo* that in the heavens there is a cause universally of both figure (*figuratio*) and form (*formatio*),[34] and it is fitting that the heaven (the *caelum*) nearer to the mover be more universal in causing (because that which exists universally in a superior [sc. body] will be determined by inferior [sc. bodies] (*inferiora*), and that which determines first with respect to *figura* and *species* is the circle of the stars (*circulus stellatus*). Moreover, the first [sc. planets] leading toward a complexion (*prima autem ad complexionem ducentia*) are Saturn and Jupiter because one [sc. Saturn] moves the cold and dry, and the other [sc. Jupiter] moves the hot and moist. And when these come together in that triplicity of a sign by which

[30] The phrase "in moving" (*in movendo*) is ambiguous here between meaning "in their motions" and "in their moving other things." Albert means the latter transitive sense here and just below, although this is also ontologically related to their actual motion.

[31] That is, in the qualitative nature of their moving things below.

[32] Resnick notes here: "This would seem the best translation of a sentence that is far from clear. We take 'naturalis' as a substantive, indicating 'natural scientist'. Note that earlier editors emended the text at this point. We have tried to preserve Hossfeld's reading."

[33] Adhuc autem notandum, sicut diximus SUPERIUS, quod duae stellae, quae sunt universaliores in movendo et fortiores, eo quod sunt superiores, aliquando coniunctae permutant triplicitatem, in qua prius fuerunt, aliquando autem coniunguntur in eadem triplicitate in qua prius fuerunt coniunctae. Cum autem permutant triplicitatem, tunc efficiuntur in movendo quasi novae, et significat illa coniunctio magna accidentia et prodigia magna et mutationes generalis status elementorum et mundi. Cuius causam debet dicere naturalis secundum ipsum, quia scit astronomus. Et ideo dicit PHILOSOPHUS, quod astronomia est altera pars physicae, et PTOLEMAEUS dicit, quod iudex, elector et observator astrorum errabit, si non sit physicus (78, 24–37).

[34] *De caelo* 2.2.5 (136, 58–137, 10); 3.15 (178, 28–88); 1.7 (124, 87–125, 7).

> it can be strengthened (*confortari*) and receive influence (*influentiam accipere*), it is fitting that a disposition is poured out into the entire world through these two stars (*per illas duas stellas toti mundo dispositio infundatur*). Moreover, a change of triplicity (*permutatio triplicitatis*) bespeaks a universal change of the primary elemental qualities (*permutatio primarum qualitatum elementalium universaliter*). Therefore, it is fitting that, when these two stars [sc. Saturn and Jupiter] change triplicity, a change of the entire world will be perceived with respect to lower [sc. bodies] (*totius mundi quoad inferiora mutatio sentiatur*, 75).[35]

Jupiter and Saturn have such a universal qualitative influence due primarily to their location in the cosmos, and thus a tremendous cumulative effect on all the lower natures in the world, including the lower planets in the cosmic system and all bodies on earth. In this context, when they conjoin in a different elemental triplicity, it indicates fundamental changes in the world.

With this background, Albert turns to a four-fold causal analysis of the universal flood:

> Therefore, having maintained these things in this manner, I say that the cause of the universal flood is composed of four causes, of which one is a true seven-fold conjunction of the planets (*coniunctio septenaria vera planetarum*). The second is that all or many of them [sc. the planets] were in the lower part of their cycles (*circuli*).[36] The third is that the conjunction was such that it began in the sign of Aquarius near the four stars that are called the Water Pot of Aquarius (*Hydria Aquarii*), and which some call the Pourer of Water (*Effusor aquae*), because it was discovered that they have a particularly wondrous effect in moving waters (*specialem effectum prodigiosum habent in movendo aquas*). And since what crossed over into the sign of Pisces began there, which belongs to the water triplicity and has the greatest power (*virtus*) in that triplicity (more than Cancer or Scorpio),[37] and because there was a change of triplicity in that conjunction, then it was fitting that the triplicity which ruled in the world (*quae regnabat* [...] *in mundo*) before this was airy—and air by its moisture and convertibility is conducive to an abundance of water (*multitudo aquarum*). Moreover, I say that a given conjunction begins in one sign and is completed in another when the centers of the epicycles or the epicycles themselves approach each other earlier in one sign and the planets themselves gradually come together in another (75-76).[38]

---

[35] Est autem profecto haec causa: Cum enim sciamus iam per ea quae in CAELO ET MUNDO dicta sunt, causam esse universaliter et figurationis et formationis esse in caelo et caelum, quod motori propinquius est, esse universalius in causando illud, oportet, quod id quod universaliter est in superiori, determinetur per inferiora, et id quod primo determinat ad figuram et speciem, est circulus stellatus. Prima autem ad complexionem ducentia sunt Saturnus et Iuppiter, eo quod unus movet frigidum siccum et alter calidum et umidum; et cum ista conveniunt cum ea triplicitate signi ex qua confortari habet et influentiam accipere, oportet, quod per illas duas stellas toti mundo dispositio infundatur. Permutatio autem triplicitatis dicit permutationem primarum qualitatum elementalium universaliter. Oportet igitur, quod cum istae duae stellae permutant triplicitatem, totius mundi quoad inferiora mutatio sentiatur (78, 38–55).

[36] We saw in the excursus that the different locations of planets in their cycles affects how strongly they act in the world.

[37] Pisces is the most powerful water sign in this respect presumably because it is the mutable sign, as we saw above.

[38] His ergo sic praehabitis dico, quod causa universalis diluvii componitur ex quattuor causis, quarum una est coniunctio septenaria vera planetarum; secunda est, quod fuerunt omnes vel plures eorum in inferiori parte suorum circulorum. Tertium est, quod talis fuerit coniunctio, quae inceperit in signo Aquarii iuxta quattuor stellas, quae dicuntur Hydria Aquarii et a quibusdam vocantur Effusor aquae, quia de illis compertum est, quod specialem effectum prodigiosum habent in movendo aquas. Et cum ibi inceperit, quod transierit in signum Piscis, quod est triplicitatis aqueae et

Given the relative depth of analysis, the third of the first three causes seems to be the most important to Albert in accounting for the universal flood. He seems particularly keen on analyzing the transition of a greater conjunction of Jupiter and Saturn from the airy to the watery triplicity, relating both triplicities here to the ultimate abundance of water at the flood.

Albert then offers the fourth cause:

> Moreover, the fourth and last cause is that the moon was strengthened in its powers at the hour of conjunction (*luna fuerit confortata viribus suis in hora coniunctionis*), such that [1] it was itself ascending from the circle of the hemisphere,[39] [2] the conjunction itself was directly over the water, and [3] it was at the hour and day of the moon.[40] For then without doubt (*absque dubio*) the moon had in itself whatever light was in every planet (*luna habuit in se, quidquid luminis fuit in omnibus planetis*). And it [sc. the moon] moves with all that light in accordance with the nature of the moon, and therefore the water did not advance gradually at that time, but leapt forth, as it were, toward it [sc. the moon] from the deepest bowels of the earth (76).[41]

In this dramatic scenario, the watery moon was rising over the horizon at the time of the greater conjunction and was thus powerfully placed, with the great conjunction in Pisces itself also occurring above the water, together with significant water dignities. The extremely abundant watery influences are thus strikingly apparent. Albert then turns to the lower terrestrial and meteorological causes, and to an astrological interpretation along these lines of a story from Ovid's *Metamorphoses*, whither we shall not follow him. With this chapter, Albert ends the series of chapters on water. This chapter thus provides the natural philosophical analysis for the famous later great conjunction in Pisces of 1524, with its numerous alarming predictions of a massive flood of biblical proportions, in a vast outpouring of printed almanancs from 1499 to 1524.[42]

3.2 Air

In I.2.10, Albert turns from water to air. I will only select three relevant passages from this chapter here:

---

maximam habet virtutem in illa triplicitate plus quam Cancer vel Scorpius, et quod fuerit permutatio triplicitatis in illa coniunctione, tunc [enim] oportuit, ut triplicitas, quae regnabat ante hoc in mundo, esset aërea, et aër umore et convertibilitate iuvat ad multitudinem aquarum. Dico autem coniunctionem aliquam incipere in uno signo et perfici in alio, quando centra epicyclorum vel ipsi epicycli ad se accedunt prius in uno signo et ipsi planetae conveniunt paulatim in alio (78, 55–76).

[39] That is, the moon was ascending over the horizon.

[40] Thus offering a strong location and the relevant dignities.

[41] Quartum autem et ultimum est, quod luna fuerit confortata viribus suis in hora coniunctionis, sicut quod ipsa sit ascendens a circulo hemisphaerii et quod ipsa coniunctio directe fuerit super aquam et quod fuerit in hora et die lunae; tunc enim absque dubio luna habuit in se, quidquid luminis fuit in omnibus planetis, et movet toto illo lumine secundum naturam lunae, et ideo tunc paulatim processit aqua, sed quasi exilivit ad eam de intimis terrae visceribus (78, 76–85).

[42] I discuss this further in Chap. 11 below, where I will offer richer bibliography.

> Moreover, there seems to be no reason for air not to rise and fall (*accedere et recedere*) based on the property of some star (*ex alicuius stellae proprietate*), since it is itself more movable (*mobilis*) than water. And it is simple (*simplex*), and it is appropriate for a simple [sc. element], which is nearer to the stars (*quod vicinius est stellis*), to be moved more by the stars (*magis moveri a stellis*) than that which is more remote from them (77).[43]

Greater proximity to the stars thus makes an element more capable of being affected by them. Since air is higher than water and simpler in nature, it thus has a greater susceptibility to being moved.

Later, in recounting four views of "the ancients" concerning air, the second is of most interest to us:

> Moreover the second one is true, that [sc. air] indeed brings the powers of the stars to earth (*stellarum vires defert ad terram*), since the power of an extreme only reaches [sc. another] extreme through a medium, and therefore it is fitting that every power of the stars (*omnis stellarum vis*) comes down to earth with fire and air mediating (78).[44]

Albert thus authorizes this ancient interpretation: fire and air always mediate celestial influences on their way down to earth.

Our final passage concerns Albert's response to the fourth view, one that claims that air is the spirit of the gods (*spiritus deorum*, 80, 29 ff):

> Now this has some truth to it, which is, nonetheless, mixed up with myth (*in fabula*). For, it is not true that air reaches heaven (as was held above, and will also be maintained in the second book of this volume [II.1.2]), but the celestial powers are transported through the air in the rays of stars (*virtutes caelestes transportantur per aërem in radiis stellarum*), which travel (*discurrunt*) through the air, and the rays are conjoined with lower [sc. bodies] (*coniunguntur inferioribus radii*). Nevertheless, in these powers (*virtutes*) they are more efficacious than the air itself (79).[45]

Celestial powers are thus explicitly transported through the air in the rays of stars, which rays then conjoin with lower bodies. This seems to contradict what Albert had said earlier at I.2.4 in distinguishing between light and rays vis-à-vis the planets and luminaries. We will see something very similar when we examine cosmic *spiritus* vis-à-vis the rays of the stars in Marsilio Ficino's *De vita* III in volume II.

*

[43] Non autem videretur esse ratio non accedere et recedere aërem ex alicuius stellae proprietate, cum ipse sit magis mobilis quam aqua et sit simplex et conveniens sit simplex, quod vicinius est stellis, magis moveri a stellis quam id quod est remotius ab eis (79, 45–50).

[44] "Secundum autem verum est, quod quidem stellarum vires defert ad terram, quoniam vis extremi non venit in extremum nisi per medium, et ideo oportet, quod mediante igne et aëre omnis stellarum vis ad terram deveniat (79, 72–76)." Just below (79, 86–80, 13), Albert discusses *spiritus* in the air and in the human body, with air as the vehicle of *spiritus*, as it is just below with celestial rays. This material is worth developing in greater depth.

[45] Et hoc aliquid habet veritatis, quae tamen in fabula permixta est; non enim est verum, quod aër pertingat ad caelum, sicut SUPERIUS est habitum et etiam in SECUNDO HUIUS VOLUMINIS habebitur, sed virtutes caelestes transportantur per aërem in radiis stellarum, quae discurrunt per aërem, et coniunguntur inferioribus radii; tamen in his virtutibus efficaciores sunt quam ipse aër (80, 38–45).

In II.2.1, Albert also treats air, but this time its corruption: "Moreover, air is corrupted in two ways: namely, sometimes from something below coming upon and corrupting it, and sometimes again from some superior celestial body corrupting it (*ex superiori corpore aliquo caelesti corrumpente ipsum*, 107)."[46] Later Albert gives his analysis of air's corruption by celestial means, among which planetary conjunctions are included:

> In addition, sometimes air is corrupted from a celestial cause (*ex causa celesti*). For, a conjunction of two stars in particular—which are Jupiter and Mars, with others assisting in the sign of Gemini, which is of an airy triplicity—make pestilent winds and corrupted airs, which suddenly kill a multitude of men and animals, just as there was a wind that killed an army suddenly in Adremoth.[47] For Jupiter, since it is hot and moist in nature, is able to raise winds and vapors and especially in the sign of Gemini, which sign is hot and moist, in the most distant state of air's nature (*in ultimo statu naturae aeris*) [sc. in its mutable state]. Moreover, Mars, because it is intemperately hot and dry, ignites the elevated vapors, and this is why lightning and flashes begin to be multiplied through the air, and pestilential vapors and fires infect the air with a sharp poison. This is also why they often introduce pestilence (110).[48]

Albert here explains why Jupiter conjunct Mars in Gemini, a mutable air sign, generates pestilential air.[49]

Albert then discusses corrupted air's deleterious effects on human health:

> And from other similar causes, an excessive dryness occurs in the air, which sharpens (acuit) the humors generated in human beings. This is why terrible illnesses and deadly pestilences happen to those who live in this air. For corrupted air infects more than does corrupted food or drink, as wise physicians (sapientes medicorum) say, because corrupted air passes unaltered to the lungs and heart and into the entire body through hidden pores (per poros occultos). Food and drink, however, reach the interior vital [sc. organs] (ad interiora vitalia) only once they have been altered through decoction and digestion (110).[50]

We can thus see that Albert here provides the natural philosophical foundations for pestilence in particular, as he also did for critical days just above. We can also see

[46] Corrumpitur autem aër dupliciter, aliquando videlicet ex inferiori invenenante et corrumpente, aliquando autem ex superiori corpore aliquo caelesti corrumpente ipsum (95, 14–17)[.]

[47] I.e., the southern part of the Arabian peninsula.

[48] Corrumpitur etiam aliquando aër ex causa caelesti; coniunctio enim duarum praecipue stellarum, quae sunt Iuppiter et Mars, cum aliis coadiuvantibus in signo Geminorum, quod est triplicitatis aëreae, faciunt ventos pestilentes et aëres corruptos, qui subito necant multitudinem hominum et animalium, sicut ventus fuit, qui in Adremoth interfecit exercitum unum subito; Iuppiter enim, cum calidus et umidus in natura, habet elevare ventos et vapores et praecipue in signo Geminorum, quod signum est calidum et umidum in ultimo statu naturae aëris. Mars autem, cum sit intemperate calidus et siccus, ignit vapores elevatos, et ideo incipiunt per aërem multiplicari fulgura et scintillationes et pestiferi vapores et ignes et veneno peracuto inficiunt aërem et ideo inducunt frequenter pestilentias (96, 53–68).

[49] Corruption of the air is thus a significant factor in analyzing the cause of plagues, as we will see in volume II in relation to Marsilio Ficino's and Girolamo Manfredi's late-Quattrocento plague treatises.

[50] Aliis etiam de causis similibus accidit *aëri superflua siccitas*, quae acuit umores generatos in hominibus, et ideo *habitatoribus* illius aëris *accidunt aegritudines pravae et pestilentiae mortiferae*; plus enim inficit aër corruptus quam corruptus cibus vel potus, sicut dicunt SAPIENTES MEDICORUM, eo quod aër corruptus non alteratus transit ad pulmonem et ad cor et per poros occultos in corpus totum. Cibus autem et potus ad interiora vitalia non pertingunt nisi alterata per decoctiones et digestiones (96, 69–78).

that the geometrical-optical analysis of celestial influences in terms of motion and light—and its rays—continues to be central for Albert.

3.3 Fire

At I.2.12, Albert digresses on fire, its motion, and the flood of fire, which comprise the third property concerning fire. He begins by discussing its motion in relation to the other elements:

> Moreover, let us now discuss the third property we are investigating concerning fire, namely, the motion and flood of fire (*de motu et de diluvio ignis*). It is reported by philosophers that fire is moved by the sun (*ignem a sole moveri*), and that it follows the sun's rays (*sequi ipsum solis radios*). For, just as the moon moves water's moisture (*luna umidum aquae movet*), so too the sun has the ability to move fire (*sol habet movere ignem*), and the remaining five [sc. planets] have the ability to move the air (*quinque residui habent movere aërem*). For this reason, they say that there are many motions (*multi motus*) in air, due to the variety of the five planets' motions, whereas there is only one motion in water and one in fire (83).[51]

Here Albert discusses how the planets move the elements, and which planet is partial to each: the sun is partial to fire, the moon to water, and the remaining five planets to air, which thus has a more complex range of motions. No planet moves the earth, which is, after all, *immobilis* at the center of the cosmos in Aristotle's system. I discuss earth below.

Albert focuses here on fire:

> Moreover, this motion belongs to fire's power (*virtus*) rather than to its substance (*substantia*) because the sun has an ordered course and is not varied either by an epicycle or a difference in latitude, since in its motion it holds to the ecliptic, which is the middle of the zodiac. There cannot be diverse motions in fire due to this alone, namely, that fire's property is more and less poured out (*diffunditur*) upon the earth *with* the rays and *in* the rays of the sun (*cum radiis et in radiis solis*), in accordance with the sun rising (*ascendit*) to the north in the zodiac or descending to the south in the same [sc. zodiac], and in accordance with it striking the earth perpendicularly with a straight ray (*recto radio perpendiculariter incidit*) or with an oblique one (my emphasis, 83).[52]

Here we see the geometrical-optical model of planetary influence in practice in relation to the sun's annual motion.

---

[51] Dicamus autem deinceps de tertio, quod quaerimus de igne, videlicet de motu et de diluvio ignis. Fertur autem a PHILOSOPHIS ignem a sole moveri et sequi ipsum solis radios; sicut enim luna umidum aquae movet, ita sol habet movere ignem et quinque residui habent movere aërem. Propter quod in aëre multos motus esse dicunt propter varietatem motus quinque planetarum, in aqua autem unum et in igne unum (82, 18–25).

[52] Est autem iste motus virtutis ignis potius quam substantiae, quia cum sol ordinatum cursum habeat nec varietur ex epicyclo vel diversa latitudine, eo quod in motu tenet eclipticam, quae medium est zodiaci, non possunt in igne esse motus diversi propter hoc solum, quod cum radiis et in radiis solis ignis proprietas in terras diffunditur plus et minus, secundum quod sol ascendit ad aquilonem in zodiaco vel descendit in meridiem in eodem et secundum quod recto radio perpendiculariter incidit terris vel obliquo (82, 26–35).

Now Albert provides another factor to account for the sun's ability to warm:

> And it warms more in accordance with it being [1] in the lower part of its cycle (*circulus*) than it does in the upper part, where the diameter [sc. of the ray] is longer, and in accordance with it being [2] in the confined space of the sphere of the circle of the signs [= the zodiac] or is distant from it. For in these ways, just as we said in the book *De natura locorum*, the sun warms (*fervet*) more over one region than over another. In sum, there are four modes that the sun alone has: namely, [1] the approach to the zenith overhead (*accessus ad zenith capitum*), and [2] the rightness of the angle or, generally speaking, the size (*quantitas*) of the angle according to which its ray strikes the earth, and [3] the confines of the sphere, as when it is under the tropics where the sun remains around the same place for a long time, and [4] the sun's descent (*depressio*) to the earth from the shortness [sc. of the orbit's] diameter [sc. its perigee] (83-84).[53]

Here we can see how the sun's astronomical motion affects its astrological influence, and how the geometrical-optical model of planetary action can be put to explanatory use to explain the warming of the seasons. This relates to an astrological analysis of—and predictions for—the weather, and is similar to what we see in Kepler's *De fundamentis astrologiae certioribus* (*On the More Certain Foundations of Astrology*), to be discussed in volume III.

Albert then discusses the sun's ability to move fire in relation to Mars (the fiery planet), Jupiter, and some fixed stars:

> Moreover, from the position where the sun is, it sometimes has the assistance (*adiutorium*) of heat from above, due to the hot fixed stars (*per stellas fixas calidas*) under which it is, as when it is in Leo and is aided by the star which is called the Dog star (*Canicula*). And sometimes it has assistance through the planets, especially Mars, which has the greatest heat and dryness (*maxima est caliditas et siccitas*). Moreover, although Mars is hotter than the sun, nevertheless the ability to move fire (*movere ignem*) is not attributed to it, but to the sun. They say that this happens from the magnitude of the sun's body, and because it is nearer to fire.[54] Therefore, it moves it [sc. fire] more powerfully (*fortius*) than Mars does. Moreover, Jupiter also assists (*coadiuvat*), but to burn intemperately, like the burning of Mars, is not in Jupiter's nature. Besides, the sun has light from itself (*habet* [...] *lumen a se*), which Mars does not have, as will be maintained in the second book of this *scientia*. Moreover, every light is hot with respect to its effect (*lumen autem calidum est omne secundum effectum*), and if sometimes it moistens, it does not have this ability insofar as it is light (*lumen*), but insofar as it is the light of this or that body (*inquantum est lumen istius corporis vel illius*) in which it is informed to produce moisture (*in quo ad umorem faciendum informatur*, 84).[55]

---

[53] Et secundum quod est in inferiori parte circuli, magis calefacit quam in superiori, ubi longior est pars diametri, et secundum quod est in loco arto sphaerae circuli signorum vel distat ab illo; his enim modis, sicut diximus in LIBRO DE NATURA LOCORUM, sol magis fervet super unam regionem quam super aliam. Qui sunt modi quattuor in summa, quos sol solus habet, scilicet accessus ad zenith capitum et rectitudo anguli vel universaliter loquendo quantitas anguli, secundum quem radius eius incidit terrae, et artitudo sphaerae, sicut est sub tropicis, in quibus diu manet sol circa locum eundem, et depressio solis ad terram ex brevitate diametri (82, 35–48).

[54] That is, nearer to the element fire in the cosmos than Mars is; i.e. the sun is below Mars in the order of the planets and is thus closer to the four elements of the sublunar cosmos.

[55] Ex loco autem, in quo est sol, habet adiutorium caloris a superiori per stellas fixas calidas aliquando, sub quibus est, sicut quando est in Leone et iuvatur a stella, quae dicitur Canicula, et aliquando habet adiutorium per planetas et maxime per Martem, cuius maxima est caliditas et siccitas. Licet enim Mars calidior sit quam sol, non tamen attribuitur sibi movere ignem, sed soli,

Thus the heat of the sun is augmented by the Dog star and the fire sign Leo. Albert also mentions the critical factor that each celestial body has its own unique nature and thus its own particular quality of light, one of astrology's most fundamental natural philosophical foundations. We also learn something about light's relation to heat, a theme that will also recur in volume II.

Finally, Albert describes the four ways that the sun moves fire by explicating a poetic image of the sun's chariot from Ovid's *Metamorphosis* (II.153–54):

> Therefore, the sun has the ability to move fire due to the causes which have been mentioned, and this was not hidden (*absconditum*) from the ancients because, due to the four causes which the sun has for moving fire by itself (*ad ignem movendum ex se*), they gave four horses to the sun's chariot: Pyroun, Eoun, Aethon and Phlegon. Moreover, they assigned two wheels to the chariot on account of the circle bearing the sun [sc. the deferent] (*circulum deferentem solem*), which is eccentric, and due to its ascent (*ascensum*) and descent (*descensum*) toward the longer and shorter part of the diameter. For, although each of these motions is in the same circle, nevertheless, the sun does not have a single power (*unica* [...] *virtus*) from these two motions, but two, because progressing (*currens*) according to the signs [sc. of the zodiac], it looks upon (*respicit*) the zenith overhead (*zenith capitum*) in different ways (*diversimode*). Both descending and ascending, it is both lifted up (*exaltatur*) to the apogee (*aux*) and pressed down (*depremit*) to its opposition [sc. the perigee] (84-85).[56]

This account further explicates Albert's discussion of the sun's *accessus* and *recessus* in *De generatione et corruptione* II.10. Likewise, Albert here gives the natural philosophical foundations for understanding the dignities "exaltation" and "depression" as grounded in the structures of planetary motion.

Keeping with this same poetic imagery, Albert then turns from the sun's motion to its light and concommitant heat, in particular, the relationship of heat to the sun's rays and their effects:

> Moreover, both the horses and the wheels make heat's burning (*faciunt caloris incendium*) because of which the ability to move fire is justly given to the sun. For in shining (*splendendo*) it heats, and in heating it inflames into the innermost parts (*medullas*), and then it cools somewhat, calling moisture into itself through the vapors extracted from the innermost parts of things. According to the order enumerated above, this is what the sun's horses mean. Moreover, when moving fire, as we said, it is impossible for it [sc. the sun] to move

---

et hoc dicunt contingere ex magnitudine corporis solis et quia magis est propinquus ad ignem, et ideo fortius movet ipsum quam Mars. Iuppiter autem etiam coadiuvat, sed incendere intemperate, sicut est incendium Martis, non est Iovis. Habet praeterea sol lumen a se, quod non habet Mars, sicut in SECUNDO LIBRO HUIUS SCIENTIAE habebitur. Lumen autem calidum est omne secundum effectum, et si aliquando umefacit, hoc non habet, inquantum est lumen, sed inquantum est lumen istius corporis vel illius in quo ad umorem faciendum informatur (82, 48–66).

[56] Sol igitur propter causas, quae DICTAE SUNT, movere habet ignem, et hoc ANTIQUIS non fuit absconditum, quia propter quattuor causas, quas sol habet ad ignem movendum ex se, currui solis dederunt quattuor equos, Pyroun et Eoun et Aethon et Phlegon. Duas autem rotas assignaverunt currui propter circulum deferentem solem, qui est excentricus, et propter ascensum eius et descensum ad longiorem partem diametri et breviorem; licet enim uterque istorum motuum sit in circulo eodem, non tamen sol unicam habet virtutem ex his duobus, sed duas, quia currens secundum signa diversimode respicit zenith capitum et descendens et ascendens ad augem et oppositionem augis exaltatur et deprimitur (82, 66–79)[.]

> it to descend downward, physically, since fire is absolutely light and its characteristic is to be lifted up and positioned above all things. For this reason, if it were to descend physically, this would be altogether contrary to nature (*omnino contra naturam*). But it moves it so that its heat is diffused in the rays (*diffundatur calor eius in radiis*). For this reason, some philosophers have said that the sun's rays are warmer than other rays because they pass through fire, although [sc. their explanation] is inadequate because the moon's rays also pass through fire and yet they are not hot, but cold and moist, as we said above. Let us say, therefore, that just as the sun is hot and dry in its effect (*per effectum*) and not through its informing by heat (*per informationem caloris*), so too the rays have this in their effect (*ita radii habent hoc in effectu*), and, for this reason, it is in their nature to move fire by generating fire in those things in which fire is generated in inferior things. And this method (*modus*) is the one handed down by the philosophers, namely, that fire descends in the sun's rays to the place of generation (*ignis in radiis solis descendit ad locum generationis*). Thus it has been explained how it pertains to the sun to move fire (84).[57]

The sun thus moves fire both by its motion and by its light by means of rays. This is accomplished by the effects the rays have, however, and not because the sun itself is hot.[58]

*

After describing how the sun moves fire—and thus further filling in our picture of how the planets affect changes on earth vis-à-vis the elements—Albert analyzes the flood of fire (*diluvium ignis*).[59] First he describes the basic phenomenon, then he distinguishes universal from particular floods of fire:

> Moreover, there is a flood of fire when the fire that has been activated (*excitatus*) by the sun's light dries out and burns the hot climes, and does not temper but inordinately heats the cold climes. And this occurs sometimes through a universal cause and sometimes through a cause that is partial. And sometimes a flood of fire is universal on the earth and sometimes it is particular, just as we said concerning a flood of water (85).[60]

---

[57] [T]am equi autem quam rotae faciunt caloris incendium, propter quod soli iuste datur movere ignem; splendendo enim calefacit et calefaciendo inflammat usque ad medullas et tunc tepefacit in se vocans umidum per vapores extractos a medullis rerum; hoc enim secundum ordinem SUPERIUS enumeratum sonant equi solis. Movendo autem ignem, ut diximus, impossibile est, quod moveat eum, ut corporaliter descendat deorsum, quoniam ignis simpliciter levis est et illius est super omnia ferri et locari, et ideo si corporaliter descenderet, hoc esset omnino contra naturam; sed movet eum, ut diffundatur calor eius in radiis. Propter quod PHILOSOPHI QUIDAM dixerunt, quod radii solis sunt calidiores aliis, eo quod transeunt per ignem, licet non sit sufficiens dictum eorum, quia etiam radii lunae transeunt per ignem et tamen non sunt calidi, sed frigidi et umidi, sicut diximus SUPERIUS. Dicamus igitur, quod sicut sol calidus est et siccus per effectum et non per informationem caloris, ita radii habent hoc in effectu, et ideo ipsorum est movere ignem generando ignem in his in quibus in inferioribus generatur ignis. Et iste modus est qui traditus est a PHILOSOPHIS, quod ignis in radiis solis descendit ad locum generationis. Sic ergo dictum est, qualiter solis est movere ignem (82, 79–83, 17).

[58] Albert also discusses this issue in *De caelo* II.3.

[59] For a discussion of Albert's role in the debate surrounding the "flood of fire," see Roland Hissette, "Albert le Grand et l'expression *Diluvium ignis*," *Bulletin de philosophie médiévale* 22 (1980): 78–81.

[60] Diluvium autem ignis est, quando excitatus ignis a lumine solis exsiccat et incendit climata calida et non temperat, sed inordinate calefacit climata frigida. Et hoc aliquando accidit universali causa et aliquando causa, quae est secundum partem; et est diluvium ignis aliquando universale in terris et aliquando particulare, sicut etiam DIXIMUS de diluvio aquae (83, 18–24).

As with floods of water, there is the same three-fold range of causes for floods of fire: universal, less universal and particular.

After discussing the views of others, as is his wont (following Aristotle's method), Albert offers his own deeply astrological five-fold causal analysis:

> I say, moreover, that the cause of the burning is an aggregate of five causes, that is [1] from the assembling (*congregatio* = conjunction) of the sun, Mars and Jupiter, and [2] from the place of their assembly (as it may occur, for example, in Cancer, between the Lion's heart [*Cor leonis*] and the Dog),[61] and [3] from the diameter of the sun and Mars (as, for example, both of them and also Jupiter might be in the lower part of their orbs). Now what I said about Cancer I understand with reference not to Cancer's length [sc. in the zodiac circle] (*spatio*) but to its constellation (*imago* [sc. with its actual stars]). Cancer's constellation in the heavens (*imago Cancri* [...] *in caelo*), moreover, is very small (*brevis est valde*). Cor Leonis is not far distant from it on one side, and a certain red star (*quaedam stella rubea*), which is in Gemini and is of the nature and power (*virtus*) of Mars, is not far distant on another side, and the Dog star (*Canicula*) is nearby on a third side a little bit to the south (*ad meridiem*). Moreover, from its very location [sc. in the zodiac], although the constellation is watery (*imago sit aquea* [sc. Cancer is a water sign]), nevertheless, because it is [sc. composed of] small and dark stars, it does not block out (*impedit*) much. And in no other part [sc. of the zodiac] does it accomplish more because the sun fixes its rays there for a long time (*sol ibi diu figit radios*) around the same point on the earth (*idem punctum terrae*), owing to the narrowness of the sphere, which is the greatest cause (*maxima causa*) of the fires. In addition, if at that time the sun were on the lower part of the diameter of its circle, it would be closest to the earth and so too will the other hot planets, and it will bring about (*operabitur*) fire, without Cancer's [sc. watery] nature getting in the way. Beyond that, it is a small constellation (*imago brevis*) to which the constellation Leo (*imago* [...] *Leonis*) is immediately conjoined, and Leo confers its powers (*vires suas*) upon it because the heat in it is at the highest state (*in statu summo*) of the fiery triplicity [sc. because Leo is a fixed sign] (87-88).[62]

From the first three causes, we can see that even though the zodiacal sign Cancer is a water sign, it is also a rather small constellation that does not provide much protection from the heat, especially since the sun—which moves heat par excellence—is in Cancer at the summer solstice and thus is closest to the earth for a very long time due to the slowness of its motion in that location of the zodiac. He also discusses here how the fixed stars work in this context to augment and further inform planetary powers.

Albert then discusses the fourth and fifth causes:

> There is also [4] a fourth [sc. cause], which is that the sun and the hot planets are not impeded (*impediti*) by the cold planets, for example, Saturn, the moon and Venus (and

---

[61] For the *Cor leonis*, see Ptolemy, *Tetrabiblos* 1.9, and Vitali, *Lexicon*, s.v.

[62] Dico autem incendii esse causam congregatam ex quinque, hoc est ex congregatione solis et Martius et Iovis et ex loco congregationis, ut videlicet sit in Cancro inter cor Leonis et Canem, et ex diametro solis et Martis, ut videlicet uterque et etiam Iuppiter sint in inferiori parte orbium suorum; et istud quod dixi de Cancro, intelligo non de spatio Cancri, sed de imagine. Imago autem Cancri in caelo brevis est valde, et non longe distat ab ipsa cor Leonis in parte una et quaedam stella rubea, quae est in Geminis et est de natura et virtute Martis, in parte altera, et Canicula est prope in parte tertia ad meridionem aliquantulum. Ex ipso autem loco, licet imago sit aquea, tamen quia est parvarum stellarum et obscurarum, non multum impedit et promovet in altera parte plus, quod sol ibi diu figit radios circa idem punctum terrae propter spherae artitudinem, quae maxima causa est incendiorum. Praeterea si sol sit tunc in inferiori parte diametri circuli sui, proximus erit terris et alii planetae similiter calidi, et operabitur incendium non impediente natura Cancri. Est insuper imago brevis, cui statim imago coniungitur Leonis, et Leo ingerit ei vires suas, eo quod in ipso est calor in statu summo triplicitatis igneae (84, 14–37).

> especially Saturn). And [5] the fifth [sc. cause] is that this conjunction [sc. of the sun, Mars and Jupiter] could occur with the change of Saturn and Jupiter's triplicity, for then it will produce great events (*magna accidentia*). Moreover, even if perchance all seven [sc. planets] are conjoined (*conjunctae*), still, if the three said [sc. planets: sun, Mars and Jupiter] have their powers (*vires*), they will convert all the other planets to their properties (*omnes alias in suas proprietates convertent*). This will occur if the other cold stars are in the upper part of their apogees, while these are in the lower part because at that time, although Saturn would be higher than Jupiter, nonetheless Jupiter is larger in size (*maioris* [...] *quantitatis*), and the sun is larger in size than any of the others. But this happens very rarely (*rarissime*), and therefore a flood of fire occurs very rarely, and this is what the ancients called the long deviation (*exorbitatio*). In fact, the planets' apogees are moved by one degree beyond the apogee of the moon in 100 years, and this is why the orb moves (*currit*) for a long time before it deviates (*exorbitat*), so that the three apogees of the hot planets will come together (*conveniant*). I think that this is the real cause (*vera* [...] *causa*) of the flood of fire. For that flood has no cause among the lower [sc. elements] as the flood of water did because the lower elements (*inferiora elementa*) cannot move fire, nor is their vapor the matter of fire (*neque vapor eorum ignis materia*), but rather it is opposed to fire (*contrarius igni*). Instead, the lower elements are entirely overcome (*vincuntur*) by the higher elements, and the vapor is destroyed (88).[63]

Although he had earlier said that floods of water and fire can both have a three-fold nature, Albert here informs us that, in fact, the flood of fire has no cause among the lower elements.

After another reference to Ovid, Albert continues, bringing this discussion to a close:

> Moreover, that conflagration (*incendium*) is loosed by two things, of which one is an eclipse of the sun, and the other is the effect of Jupiter and Mars. For an eclipse of the sun does not occur except through the interposition of the moon, and therefore it separates the hot rays (*radii calidi*) from the earth. Then the ember is released (*favor remittitur*), and the waters begin to smoke and to send vapors into the air, which air is tempered by the cold parts from the water and earth rising into it. And this is why they [sc. the poets, meaning Ovid] sing in a mythic fashion that Phaethon's father, who is the Sun, grieved over his son's death and withdrew the light (*lumen*) for one day[64]—not that an eclipse would last for a day, but because its effect often lasts throughout the day from the darkness of the clouds. And no one should be surprised when we say that the eclipse looses the conflagration, although an eclipse lasts for a short time. For it is impossible that all the bodies of that region *not* be changed in quality (*alterentur*) into something cold, no matter how briefly the very bright light (*limpidissimum lumen*) of the sun is separated from some region. This is why the

[63] Est autem quartum, quod sol et planetae calidi non sint impediti a stellis frigidis, sicut a Saturno et luna et Venere et praecipue a Saturno. Et quintum est, quod haec coniunctio sit cum permutatione triplicitatis Saturni et Iovis; tunc enim magna accidentia producet. Licet autem forte omnes septem coniunctae sunt, tamen si tres dictae habuerint vires, omnes alias in suas proprietates convertent, et hoc erit, si aliae stellae frigidae sint in superiori parte augium suarum, istae autem in inferiori, quia tunc licet Saturnus sit altior Iove, Iuppiter tamen maioris est quantitatis et sol maioris quantitatis quam aliquis aliorum. Contingit autem istud rarissime, et ideo rarissime incendii fit diluvium, et hoc est quod vocaverunt ANTIQUI longam exorbitationem; auges enim planetarum moventur in centum annis uno gradu praeter augem lunae, et ideo diu currit orbis, antequam exorbitet, ut conveniant tres auges planetarum calidorum, et hanc veram ignis diluvii causam existimo; nullam enim habet istud diluvium in inferioribus causam, sicut habuit diluvium aquae, quia inferiora elementa non possunt movere ignem neque vapor eorum est ignis materia, set potius contrarius igni; sed vincuntur in toto a superioribus elementis inferiora et destruitur vapor (84, 38–61).

[64] The allusion is to Ovid, *Metamorphoses* II.329–331.

> astrologers (*astronomi*) say that an eclipse's effect lasts for as many years as there are hours of the eclipse. Perhaps its effect will persist for periods of unequal length according to the different locations of the eclipse. But to know that time definitively belongs to the science of the stars (*ad scientiam astrorum*, 89).[65]

Here Albert explains why the effects of eclipses last longer than the eclipse itself and relates this to the flood of fire. As before, Albert confirms his analysis by interpreting an image from Ovid's *Metamorphoses.* He concludes the chapter thus: "These, then, are the things we have to explain philosophically concerning the flood of fire (90)."[66]

3.4 Earth

Finally, in I.2.14, Albert digresses on the properties of earth, explicitly developing here his analysis from that in *De natura loci*:

> Many things have been said about the properties of the earth in the book *De natura locorum* and in general only a few things remain to be said, namely, that the earth (*terra*) is an element (*elementum*) that does not have any local motion by the power of the stars (*quod non habet aliquem motum localem virtute stellarum*), as the other elements have. And this is so for this reason: because it [sc. the earth] is the furthest moveable (*ultimum mobile*), it cannot participate in motion due to its distance from the nobility of the first cause, as was determined in the second book of *De caelo*.[67] But because the earth is the center of the entire universe (*centrum est terra totius mundi*), it is necessary that all the rays of the starry orb be united on it (*in ipsa uniantur omnes radii orbis stellati*), whose center it is. For this reason, in addition, it can be moved according to nature by the very sphere of the fixed stars (*habet moveri ab ipsa sphaera stellarum fixarum*) into the shapes of the fixed stars (*in figuras stellarum fixarum*). And because that orb has many constellations (*imagines*), for this reason the earth can receive many shapes (*terra susceptibilis est multarum figurarum*), which shapes are varied (*figurae variantur*) by the shape of the constellations (*ex figura imaginum*) and by the manner of the rays' diversity (*ex modo diversitatis radiorum*), which are varied by the rising and setting of the stars over the earth (*variantur ex ortu et occasu stellarum super terram*, 94–95).[68]

---

[65] Solvitur autem istud incendium ex duobus, quorum unum est solis eclipsis et alterum est effectus Iovis et Martis; eclipsis enim solis non fit nisi per lunae interpositionem et ideo radios calidos separat a terra, et tunc favor remittitur, et incipiunt fumare aquae et vapores mittere in aërem, qui temperatur partibus frigidis ab aqua et terra in eum ascendentibus. Et hoc est quod cantant fabulose patrem Phaethontis, qui sol est, doluisse de morte filii et subtraxisse lumen per unum diem, non quod die una durat eclipsis, sed quia effectus suus durat in obscuritate nubium per diem ut frequenter. Et nemo miretur, quod dicimus eclipsim solvere incendium, cum eclipsis parum duret; quantumcumque enim parum limpidissimum lumen solis separetur a regione aliqua, impossibile est, quod non alterentur in frigus omnia corpora illius regionis, et ideo ASTRONOMI dicunt, quod effectus eclipsis durat per plures annos, quam sint horae eclipsis, et forte durabit inaequaliter secundum diversa loca eclipsis. Sed scire tempus illud determinate ad scientiam pertinet astrorum (84, 72–85, 6).

[66] Hac igitur sunt, quae dicere habemus philosophice de diluvio ignis (85, 27–8).

[67] See Albert, *De caelo* 2.3.14; 2.4.8 (174, 59–63; 175, 27–30; 195, 25–28).

[68] De proprietatibus vero terrae multa in LIBRO DE NATURA LOCORUM sunt dicta, nec remanent in communi dicenda nisi pauca, quod videlicet terra est elementum, quod non habet aliquam

In discussing the nature of earth here and its propensity to be moved by stellar influences (but not with respect to local motion), Albert here also provides some of the natural philosophical foundations for making talismans, which are closely related to those articulated in his *De mineralibus*. These natural philosophical features are relevant for making talismans in efficient, formal and material respects, regarding both choosing the proper matter for making the talisman, but also for the optimal astrological timing as to when one should be fabricated. I will explore this further in Part 3.

Albert continues by describing the earth's extraordinary fecundity vis-à-vis celestial influences:

> And, therefore, it is the case that the earth produces such multiform variations (*multiformes varietates*), which no other element does, in which the celestial rays do not collect at one point (*in quo radii caelestes ad unum punctum non colliguntur*). For, all the lines coming from the circumference are united (*uniuntur*) at no point but the center, and therefore rays (which are, as it were, lines of celestial light) have almost their entire effect on the earth, which is the perceptible center of the circle.[69] Moreover, there is yet another cause of this, namely, that no element except earth entirely reflects rays (*nullum est elementorum, quod reflectat radios omnino nisi terra*). For water reflects them in some manner, but not entirely, because they pass through it to the bottom of the water. Earth, however, stops them entirely at its surface and reflects them. Therefore, since a ray will only be in its power (*in sua virtute*) where it is reflected, it can only produce the shapes of the species (*figurae specierum*) adequately in the earth on which its rays are reflected and multiplied (*in qua radii eius reflectuntur et multiplicantur*, 95).[70]

Due to its complete reflection, earth has a unique ability to receive and produce the shapes or figures of species from rays, which Albert defines here as lines of celestial light (*lineae caelestis luminis*). Albert even mentions their multiplication, and thus comes ever closer here to Roger's position in *Opus maius* IV.

To complete this section, Albert further contrasts earth with water and fire:

---

motum localem virtute stellarum, quemadmodum habent cetera elementa. Et hoc est ideo: quia ipsa est ultimum mobile, non potest participare movere propter distantiam a nobilitate causae primae, sicut determinatum est in SECUNDO DE CAELO ET MUNDO. Sed quia centrum est terra totius mundi, necesse est, quod in ipsa uniantur omnes radii orbis stellati, cuius ipsa est centrum. Propter quod etiam secundum naturam habet moveri ab ipsa sphaera stellarum fixarum in figuras stellarum fixarum, et quia orbis ille multarum est imaginum, propter hoc terra susceptibilis est multarum figurarum, quae figurae variantur ex figura imaginum et ex modo diversitatis radiorum, qui variantur ex ortu et occasu stellarum super terram (87, 26–43).

[69] [...] in nullo enim puncto omnes lineae a circumferentia venientes uniuntur nisi in centro, et ideo radii, qui sunt quasi lineae caelestis luminis, effectum suum fere totum habent in terra, quae est centrum sensibile orbis.

[70] Et ideo est, quod terra profert tam multiformes varietates, quod non facit aliquod aliud elementum, in quo radii caelestes ad unum punctum non colliguntur; in nullo enim puncto omnes lineae a circumferentia venientes uniuntur nisi in centro, et ideo radii, qui sunt quasi lineae caelestis luminis, effectum suum fere totum habent in terra, quae est centrum sensibile orbis. Huius autem est etiam alia causa, quia videlicet nullum est elementorum, quod reflectat radios omnino nisi terra; aqua enim reflectit aliquo modo, sed non omnino, quia transeunt in profundum aquae. Terra autem omnino terminat eos in sua superficie et reflectit eos. Radius igitur, cum non sit in sua virtute, nisi ubi reflectitur, non potest producere figuras specierum sufficienter nisi in terra, in qua radii eius reflectuntur et multiplicantur (87, 44–60).

> In addition, moreover, the moist (*umidum*) is capable of receiving (*receptibilis*) forms (*formae*), but not of retaining them, whereas fire's dryness does not sit still to receive forms, but evades them with its mobility. It results, therefore, that immoveable dryness can retain shapes (*figurae*), but does not receive them easily. And therefore the cold earth has been inundated (*influxa*) with water so that it will be softened for receiving and retaining forms and shapes (*ad formas et figuras suscipiendas et tenendas*, 95).[71]

This final section on earth has revealed much more information of interest as to how Albert understands the influences of the heavens on the four elements: water, air, fire and earth. In addition to providing the natural philosophical foundations for revolutions and their great conjunctions, and for certain aspects of medicine (including critical days), we will see in Part 3 that Albert here also provides some of the natural philosophical foundations for making talismans.

3.5 Conclusion to Part 1 (Chapters 1–3)

This intensive and detailed investigation in Part 1 permits a proper comparison of central features of Albert's and Roger's thought and, thereby, much more fully articulates the astrologizing Aristotelian natural philosophy they both shared, where astrology—with its geometrical-optical model of planetary influences within an integrated mathematical-astronomical and -geographical framework—is deeply interwoven at the roots of Aristotelian natural philosophy. As we saw, fundamentally astrological concepts are used as important tools of analysis in order to elucidate more fully the thought of Aristotle himself.

The thoroughly mathematicized natural philosophy reconstructed here thus serves to supplement and deepen Grant's picture of medieval natural philosophy and cosmology. This deeply astrologizing Aristotelianism, especially that found in Albertus Magnus and Roger Bacon—developing basic ideas in al-Kindi and Robert Grosseteste—fills a significant gap in Grant's account. At the same time it also serves to revise his views on the relationship of astrology to natural philosophy, even in these most characteristic works, namely, Aristotle commentaries. Likewise, since this astrologizing Aristotelianism continued to be influential throughout the fifteenth and sixteenth and into the seventeenth centuries, this reconstruction also adds an important type of Renaissance Aristotelianism to those reconstructed by Charles B. Schmitt. I have also re-opened and re-framed the question of the unknown master's identity in Roger Bacon's *Opus tertium* within this context. We will now shift gears and explore Roger Bacon's defense of astrology in *Opus maius* IV and that in the *Speculum astronomiae*.

[71] Adhuc autem, umidum receptibile est formarum et non tentivum. Siccum autem ignis non quiescit ad formas suscipiendas, sed evadit sua mobilitate. Relinquitur ergo, quod siccum immobile sit tentivum figurarum, non autem est receptivum de facili. Et ideo frigida est terra influxa aquis, ut mollificetur ad formas et figuras suscipiendas et tenendas (87, 61–67).

Chapter 4
Defending Astrology: Roger Bacon and the *Speculum Astronomiae*

4.1 Introduction

Having explored astrology's relationship to essentially Aristotelian and Ptolemaic natural knowledge in central works by Albertus Magnus and Roger Bacon, and in framing works by Thomas Aquinas and in the *Speculum astronomiae*, I will now broaden the picture and explore two noteworthy defenses of astrology. The first, by Roger Bacon, is also in *Opus maius* IV. It begins to articulate his characteristic and somewhat idiosyncratic understanding of astrology's relationship to theology/religion and to magic. Roger's views on theology/religion will be further explored in Chap. 6, and on magic in Chap. 8.

Although Albert and Roger both articulate a mathematically sophisticated astrologizing Aristotelian natural philosophy and both discuss practical astrology, for Albert, practical astrology takes up a very small percentage of his writings (well under 5%), whereas (as we will see) Roger extensively explains how to interpret and use the heavens for various practical purposes, both personal and societal, including how to be an astrologically-informed political advisor. As we saw in the excursus, Roger showed us in outline how to make an astrological interpretation or judgment. Here in Chap. 4 he defends practical astrology in general, specifically the making of astrological judgments or interpretations, which he explicitly calls '*iudicia astronomiae*'. In Chap. 6 he offers a detailed example of how to interpret astrological information as well as numerous practical benefits that may be derived from this knowledge, especially in the area of religion. Albert, on the other hand, only alludes to practical astrology briefly but significantly, as we saw. The other defense of astrology I will treat here is in the *Speculum astronomiae*, Chaps. 12–15, which is itself oriented in turn towards each of the four types of practical astrology.

In this intermezzo, I will first focus on Roger's defense of astrology in *Opus maius* IV. Once again, we will pay close attention to both his terminology and the conceptual structures underlying his use of these terms. There seems to be only one

H D. Rutkin, *Sapientia Astrologica: Astrology, Magic and Natural Knowledge, ca. 1250–1800*, Archimedes 55, https://doi.org/10.1007/978-3-030-10779-6_4

significant—or even detectable—shift in what is otherwise Roger's clearly articulated and profoundly consistent system, namely, that concerning the etymology of the central term '*mathesis*'. This shift has been used to date related texts as earlier or later in Roger's *oeuvre*.[1] Although the etymology itself shifted 180 degrees, its use in all cases supports precisely the same underlying conceptual structures in both the earlier and the later texts, namely Roger's utterly central distinction between true and false mathematics, which deeply informs his arguments concerning astrology's legitimacy regarding both knowledge and practice. This chapter will also help us make the transition from Part I on the relationship between astrology and natural philosophy/science to both Part II on astrology and theology/religion and Part III on astrology and magic, in that, in his explicit defense of astrological judgments (*iudicia astronomiae*) as mathematics' weakest link, Roger raises many issues pertinent to astrology's relationship to both theology/religion and to magic that will be developed further in Parts II and III.

*

The *Opus maius* was surreptitiously commissioned by Pope Clement IV in 1266.[2] Roger composed it straightaway and sent it to the pope ca. 1267. The section of the *Opus maius* on which we will focus here is in Book IV. It is entitled '*Iudicia astronomiae*' (238–69 of Bridges' edition) and has a three-fold structure. The first part (238–49) introduces Bacon's defense of astrological judgments by distinguishing strongly between true (legitimate) and false (illegitimate) mathematics. Having sufficiently defended astrology, Parts II (249–53) and III (253–69) actively promote astrology by discussing its utility in two distinct spheres: in human affairs, and in religion. In this coherent and extended discussion, Bacon defends and promotes astrology in relation to what he explicitly calls "magic," which he constructs in a characteristically idiosyncratic manner. With respect to its contents, Roger primarily discusses astrology's legitimacy in relation to the theologically sensitive and closely related issues of human free will, the certainty of predictions, and necessity or determinism in nature, themes we will continue to explore here and in what follows. I discuss the first two parts of the section entitled '*Iudicia astronomiae*' here in Chap. 4, and the third in Chap. 6.

[1] George Molland discusses this in his "Roger Bacon's *De laudibus mathematicae*," 69. He gives 1270 as the dividing line. Steven J. Williams is more circumspect. He states that all we can say for certain is that Roger's edition of the *Secretum secretorum* with his significant introduction—including the switched etymology for the first time—was written at some point between 1270 and when Bacon died in 1292; "Roger Bacon and his Edition of the Pseudo-Aristotelian *Secretum secretorum*, 57–63. *Secretum secretorum cum glossis et notulis [...] fratris Rogeri*, Robert Steele (ed), Oxford: Clarendon, 1920.

[2] For the dating, see the informative introduction to A.G. Little, *Part of the Opus tertium of Roger Bacon*.

4.2 Roger Bacon's *Iudiciaria-Operativa* Distinction

To both set the stage and to relate Chap. 4 to the earlier chapters in Part I and to those that follow in Parts II and III, it will be useful first to show in sharper focus how Roger configured astrology within the mathematical disciplines, and then how he further articulates astrology itself. I will begin with his configuration of astrology within the mathematical disciplines in the introduction to his important edition of the *Secretum secretorum*, and especially its bifurcation into [1] an interpretive knowledge-based part (*iudicialis*), and [2] an operative or what we would call magical or technological part (*operativa*) that is concerned with acting in the world.[3]

After introducing the distinction between true and false mathematics within the context of divination and astrology, a topic to which we shall return shortly, Roger discusses astrology's configuration among the "true" mathematical disciplines: "And this *mathematica* contains four sciences (*scientiae*), namely geometry, arithmetic (*arismetrica*), music and *astrologia*, under which *astrologia*, *astronomia iudiciaria* and *operativa* are contained in a common term (*vulgari nomine*)."[4] We can see clearly here that the four *scientiae* contained within true mathematics are the four normal disciplines of the *quadrivium*, namely, geometry, arithmetic, music and the science of the stars.

More interesting for our purposes is that, after normatively configuring the science of the stars within the mathematical disciplines, Roger further distinguishes *astrologia* into two parts. He refers to the first as '*astronomia iudiciaria*', which I will neutrally and loosely translate here as "the judicial or interpretive part of the science of the stars," and the second as '*astronomia operativa*' or "the operative part of the science of the stars."[5] We will see in Chap. 8 that the latter is concerned with two different practices that Roger refers to as the works and words of wisdom (*opera et verba sapientiae*), namely, making works (*opera*)—primarily talismans—or uttering potent and powerful words (*verba*) in order to move or affect things in the world.

We may ultimately wish to relate Roger's distinction here to thirteenth-century knowledge/power dynamics as well as use it to help clarify by historicizing both our normal "natural vs. judicial" astrology distinction, and the terminological and conceptual correlates of Roger's understanding of magic and ours. We will see how this

---

[3] For an excellent and richly informative article on Bacon's version of the *Secretum secretorum*, see Williams, "Roger Bacon and his Edition." There is much of related interest in the introduction to Bacon's edition, but to explore it here would take us too far afield. Essentially, the position there ca. 1270 or later is extremely similar (with minor variations and amplifications) to the earlier *Opus maius* of 1266–67, despite the etymological shift.

[4] Et hec mathematica continet quatuor sciencias, scilicet, Geometriam, Ar*i*smetricam, Musicam, Astrologiam sub qua Astrologia, Astronomia iudiciaria et operativa vulgari nomine continetur (3.1–3).

[5] The terminological shift from *astrologia* as the relevant mathematical discipline to enumerating each part with *astronomia* seems to reflect a terminological *variatio* without a corresponding conceptual shift.

significant and extremely useful distinction plays out in what follows. Although this phrasing appears in Roger's introduction to his later edition of the *Secretum secretorum*, it concisely and precisely captures structures he expounds at much greater length in the three *Opera* composed for Pope Clement.

In addition to this fundamental distinction, we will also pay close attention to several recurring themes: [1] The distinction between true and false mathematics, and the concomitant relationship between astrology and magic, by focusing on magic's two primary manifestations in Roger's work, namely talismans and the power of language (which also relates to prayer). In doing so, we will remain acutely aware of our contemporary terminological and conceptual distinctions between these terms and Roger's rather different ones, pointing out both tensions and overlaps. [2] The relationships between astrology and human free will, necessity in nature, and interpretive certainty. And [3] the relationship between legitimate and illegitimate knowledge and praxis. We will find that what Bacon calls "magic" is always illegitimate and is consistently used as a term of abuse, much as both his and Thomas Aquinas's use of the term '*divinatio*', to be discussed in Chap. 5 below. Thus, the distinction between terminological and conceptual anachronism will continue to be useful here.[6]

Without further ado, let us turn to Roger's defense of astrological interpretations or judgments (*iudicia astronomiae*) in *Opus maius* IV. As indicated by the terminology alone, we can see that our first focus will be on the *iudiciaria* side of the *iudiciaria-operativa* distinction. This will be almost entirely the case throughout Chaps. 4 and 6. In Chap. 8, however, we will shift over to the *operativa* side. We will also see that Roger's usage here can help us to historicize the more modern natural-judicial distinction, which is nowhere apparent in Roger's work examined here and thus should not be imposed thereon, at least not without good reasons.

4.3 Roger Bacon's Defense of Astrology

In *Opus maius* IV (238–69), Roger begins a new section that is centrally concerned, first, with removing all doubts and concerns about mathematics, and ultimately with establishing its utility for theology/religion. Given the centrality of mathematics to his overall natural philosophy, and of philosophy's centrality to theology, Roger devotes significant attention to defending mathematics' weakest because most controversial link, namely, what he calls the judgments of the science of the stars (*iudicia astronomiae*), which the *Speculum astronomiae* calls the *scientia iudiciorum astrorum* and we call astrology.[7] Roger's attention to these issues is fully justified

[6] I discuss this distinction more fully in my "Understanding the History of Astrology Accurately."

[7] This is a standard expression in many sources. In this case, it might also properly be called "judicial" astrology, but only as long as we do not automatically contrast it with what we normally call "natural" astrology, which would immediately set us off on the wrong path. Rather, we shall duly note Roger's usage and try to understand *his* configuration as fully and accurately as possible. Then and only then should we compare Roger's usage to the modern natural-judicial distinction.

within his system of thought, and it is very interesting for our purposes, especially given its centrality to his views concerning philosophical knowledge and its power, which is correspondingly central to his views on theology.

4.4 Framing the Problem in Terms of True vs. False Mathematics

To defend astrological judgments in *Opus maius* IV, Roger begins by introducing an essential, characteristic and utterly fundamental distinction for him between true and false mathematics (his terminology), which leads into a rich discussion of both astrology and magic, and in a way—especially for what *he* explicitly calls "magic"—that we have not yet encountered:

> Having made clear how *mathematica* is necessary for both divine and human wisdom (*sapientia*), it is also necessary for certifying what went before, namely, that certain complaints (*cavillationes in contrarium*) against it [sc. mathematics] be removed, and that certain sayings of the saints (*dicta sanctorum*) be set out, so that every doubt (*omnis dubitatio*) is destroyed concerning the utility of mathematics.[8] That respect in which mathematics is primarily attacked (*percutitur*) is due to astrological judgments (*iudicia astronomiae*).[9]

Roger immediately identifies mathematics' weak link—and thereby the source of its attacks—as what he calls *iudicia astronomiae*.

*

Roger begins by attacking ignorant theologians, who, in his view, created the problem in the first place due to their systematic misunderstandings. Here Roger offers a quasi-historical account of the devolution of knowledge:

> Therefore, since [1] the contention of the many people ignorant of the power of philosophy and its great utility for theology, and [2] the considerations of those who reprove the *mathematici*—both in relationship and absolutely—impede the study of wisdom (*studium sapientiae*) and gravely harm (*laedit gravissime*) it in this part, therefore, I wish at present to verify their intention, and remove the disrespect (*infamia*) from true mathematics (*vera mathematica*).[10]

---

[8] Ptolemy treats astrology's utility rather differently at *Tetrabiblos* I.3.

[9] Manifestato quomodo mathematica necessaria est sapientiae tam divinae quam humanae, adhuc necesse est ad certificationem praecedentium, ut evacuentur quaedam cavillationes in contrarium et exponantur quaedam dicta sanctorum, ut omnis dubitatio tollatur circa mathamaticae utilitatem. Et illud in quo maxime percutitur mathematica est propter iudicia astronomiae (238, 34–239, 3).

[10] Quoniam igitur contentio multorum ignorantium philosophiae potestatem et maximam utilitatem theologiae, tam relate quam absolute reprobantium mathematicorum considerationes, impedit studium sapientiae et laedit gravissime in hac parte; ideo volo in praesenti verificare intentionem eorum, et evacuare infamiam verae mathematicae (239, 3–9).

Because mathematics has power for philosophy and great utility for theology, it must be defended and its bad reputation (*infamia*) removed by a process of reconceptualization. To do so, Roger here introduces the concept of true mathematics (*vera mathematica*), which he will immediately contrast with false mathematics. We will soon see what he means by this fundamental distinction.

After putting forward the concept of true mathematics, Roger immediately and explicitly distinguishes it from false mathematics. He first discusses the statements (*dicta*) of the saints and offers his interpretation of their intentions:

> Therefore, theologians (*theologi*) have found many things uttered by the saints (*a sanctis*) against the *mathematici* (*contra mathematicos*), and some of them [sc. the theologians]—because of their ignorance of true mathematics and false mathematics—do not know how to distinguish true from false, and therefore, as though by the authority of the saints, they blame true mathematics with [sc. the errors of] the false.[11]

Due to their ignorance of the true nature of mathematics, certain theologians have confused true with false mathematics and thus mistakenly attacked the true kind by conflating the two and thus reproaching both. Bacon begins his defense of true mathematics by first interpreting influential statements by religious authorities—the saints—that have previously been (mis)used to support some theologians' ignorant views.

To set the record straight by properly distinguishing true from false mathematics, Bacon turns to an etymological argument on which he places great weight, but which changed diametrically over time:

> For the word (*vocabulum*) for true mathematics (*vera mathematica*) is written with an aspirated 't', and from this, the noun (*nomen*), *mathesis*, with the middle shortened (*correpta*), [sc. is derived], which signifies knowledge (*designat scientiam*). It is said to be derived from many authors [or authorities, *auctores*]. It is certainly from Greek, since '*matheo*' is the same verb (*verbum*) as '*disco*' [= I teach] is, and '*mathetes*' is '*discipulus*' [= student], and '*mathesis*', '*disciplina*' [= teaching, body of knowledge]. Whence *mathematica* is '*disciplinalis scientia et doctrinalis*' as Cassiodorus said above. But the word (*vocabulum*) for false mathematics (*falsa mathematica*) is written without an aspiration and is asserted by the same authorities, and is from *mathesis* with the middle lengthened (*media producta* [vs. *correpta*]), which denotes divination (*divinatio*) and derives [sc. etymologically] either from '*mantos*' (which is more certain) or from '*mantia*', which are both the same as '*divinatio*', as Jerome says in the original of IX Isaiah.[12]

Although the terminology for true and false mathematics is distinctive, it is also deceptively similar, boiling down to the difference between a 't' vs. a 'th' sound, and

[11] Theologi igitur multa invenerunt a sanctis effusa contra mathematicos, et aliqui eorum propter ignorantiam mathematicae verae et mathematicae falsae nesciunt distinguere veram a falsa, et ideo tanquam auctoritate sanctorum culpant veram cum falsa (239, 9–13).

[12] Vocabulum enim verae mathematicae scribi per *t* aspiratum, et ab hoc nomine *mathesis* media correpta, quod scientiam designat, derivari a multis refertur auctoribus, et certum est ex Graeco; quia *matheo* verbum idem est quod disco, et *mathetes* est discipulus, et *mathesis* disciplina. Unde *mathematica* est disciplinalis scientia et doctrinalis, sicut Cassiodorus dixit superius. Sed vocabulum falsae mathematicae sine aspiratione scribi asseritur ab eisdem auctoribus et a *mathesi* media producta, quod divinationem notat, descendere, vel quod certius est, a *mantos* vel a *mantia*, quae sunt idem quod divinatio, sicut Hieronymus dicit in originali ix Isaiae (239, 13–24).

a shortened or lengthened vowel quality with correspondingly different etymologies. True mathematics is thus derived from '*matheo*' and '*mathesis*', and is pronounced with an aspirated 't' and a shortened vowel quality. It means "learning" and "knowledge." False mathematics, on the other hand, derives from '*mantos*' or '*mantia*' and refers to divination. A further distinction here between true and false *knowledge* is also implied. Thus, divination as false knowledge or learning is decidedly negative for Roger, which we will see is also the case with Thomas Aquinas in Chap. 5.[13]

As we saw, Bacon subsequently revised his analysis of the etymology in his edition of the *Secretum secretorum*, which he wrote at some point between 1270 and 1292. Despite the switched terminology with its correlative etymologies, however, the results of his new analysis were exactly the same, in that Roger still conceptually distinguished true from false mathematics in precisely the same manner. The weight of his concern thus falls much more heavily on the conceptual distinction than on the terminology with its respective etymology.

*

To complete this brief initial etymological discussion, Roger makes another major point that he will return to often. In addition to divination (or false knowledge), false mathematics is also closely associated with what he explicitly calls the magical art (*ars magica*). In Roger's usage, what he explicitly calls magic (*ars magica* and cognates) and divination (*divinatio*)—as we will repeatedly see—are *always* negative, just as divination is for Thomas, and necromancy for the Magister Speculi, as we will see in Parts II and III below:

> But whatever the case is about this writing (*scriptura*, spelling) and its [sc. etymological] derivation, false mathematics is, nevertheless, a magical art (*ars magica*). For five species of the magical art are enumerated, namely, *mantice* [= divination], *mathematica* [= mathematics or astrology], *maleficium* [= doing harm], *praestigium* [= illusions][14] and *sortilegium* [= the casting of lots (sometimes translated as "sorcery")].[15] Therefore, false mathematics

[13] Both Williams ("Roger Bacon and his Edition," 61) and Molland ("Roger Bacon's *De laudibus mathematicae*," 69) discuss the distinction. Bridges's note *ad loc.* is very useful. He refers to the *Communia mathematica* and the *Secretum secretorum*, both of which also concern the relationship between true and false mathematics (and thus astrology vis-à-vis divination). Bacon always makes the same distinction regardless of etymology. In the *Communia mathematica* and here in *Opus maius* IV, he uses the same spelling; in the *Secretum secretorum*, however, he distinguishes the two terms orthographically as well. Nevertheless, the etymology itself is not original with Roger. Apparently he took it from Michael Scot's *Liber Introductorius*, as mentioned by Lynn Thorndike in his *Michael Scot* (London: Nelson, 1965) and Caroti's section on Michael Scot in his *L'astrologia in Italia*. The text by Michael Scot is in Glen M Edwards, "The *Liber Introductorius* of Michael Scot," PhD thesis, University of Southern California, 1978, 206, 2–15.

[14] This is closest to the modern use of the term "magic," as in a stage magic show with magic tricks, or the magic shops that sell these sorts of tricks, which are almost always associated with sleight of hand. For many marvelous 19th and 20th century examples (with many photographs), see *Imponderable: The Archives of Tony Oursler*, Ann Wehr (ed), Zurich: JPS/Ringier, 2016. Roger discusses *praestigium* further just below.

[15] Thomas Aquinas has a specific consultation on the use of lots (*De sortibus*) that I will briefly mention in Chap. 5, but which is worthy of further analysis.

> is the second part of the magical art. It usurps to itself a consideration of the celestial bodies (*coelestia*) deformed by characters, charms (*carmina*), conjurations, superstitious sacrifices and various frauds (239, 24–240, 8).[16]

Roger here offers a five-fold typology of the magical arts (*ars magica*), adopting the structure from Hugh of St. Victor's *Didascalicon*.[17] False mathematics is thus included integrally among the magical arts, as is *mantice* or divination. They are all anathema, as Roger proceeds to discuss. Astrology, magic and divination are thus all closely associated here, and we can begin to see how they are configured within Bacon's thought with respect to both his distinctive terminology and its underlying conceptual structures.

*

In order to defend legitimate astrological judgments as a part of true mathematics and thus of true knowledge, Bacon distances them from both magical (and thus false) mathematics as well as from divination, which both usurp a certain relation to the heavens, but a distorted one. He also describes false magical mathematics as employing characters, charms, conjurations and sacrifices joined to the celestial bodies, themes that we will also encounter in the *Speculum astronomiae* in relation to talismans (in Chap. 7 below), which are also considered illegitimate there. Here, however, there is no explicit reference to talismans. They will appear later for Roger in the *operativa* section discussed in Chap. 8.

Bacon then strongly asserts why this false magical mathematics is problematic, and in so doing indicates the main issues for a theologian:

> And it [sc. mathematical magic, or magical and thus illegitimate astrology] posits [1] that everything happens necessarily (*omnia de necessitate contingere*) by the power of celestial configurations (*per virtutem constellationum*), [2] that nothing can go one way or another (*nihil ad utrumlibet*), [3] that nothing happens by chance or fortune (*nihil a casu nec fortuna*), and [4] nothing by counsel (*nihil a consilio*)[.][18]

This false type of mathematics—that is, illegitimate magical astrology—both assumes and implies a deterministic structure of nature, in which everything happens by necessity in such a way that free will, chance and human choice are wholly undermined.

The terminology Roger uses is significant, as we will see. The implications of this view for free will in particular, both divine and human, had long made astrology

---

[16] "Quicquid vero sit de hac scriptura et derivatione, tamen falsa mathematica est ars magica. Nam numerantur quinque species artis magicae, scilicet mantice, mathematica, maleficium, praestigium, sortilegium. Mathematica ergo est secunda pars artis magicae. Haec sibi usurpat considerationem coelestium characteribus, carminibus, coniurationibus, sacrificiis superstitionis, et fraudibus variis deformatam (239, 24–240, 8)." Most of these magical terms will be discussed in due course.

[17] *The* Didascalicon *of Hugh of St. Victor: A Medieval Guide to the Arts*, Jerome Taylor (tr.), New York: Columbia University Press, 1961, 154–55.

[18] Et ponit per virtutem constellationum omnia de necessitate contingere, nihil ad utrumlibet, nihil a casu nec fortuna, nihil a consilio (240, 8–10)[.]

suspect. Roger will discuss this subject at length in what follows and return to it repeatedly as a leitmotif. As we have already seen (and will see repeatedly in what follows), this is one of the most theologically sensitive issues concerning astrology and its legitimacy. We will encounter these very same themes (albeit configured somewhat differently) in Thomas Aquinas's *Summa theologiae* II.II.92–95 with respect to astronomy, astrology and divination, in Thomas's *Summa contra gentiles* III.82–94 in relation to divine providence, as well as in the *Speculum astronomiae*'s defense of astrology, in the second part of this chapter and in Chap. 5 below.

Then Roger turns to the origins of magic:

> [N]evertheless, from the goodness of its [sc. divinely created] essence (*de bonitate essentiae*) and to help make the celestial configuration more efficacious, he [sc. the false magical astrologer] arranged for (*ordinavit*) the invention of characters and the other things mentioned above [sc. *carmina*, sacrifices, etc] proper to individual celestial configurations (*singulis constellationibus*). And these things are explicitly asserted in magical books (*libri magici*). Whence this *scientia* posits that all these things happen necessarily by means of the heavens (*omnia ponit per coelum de necessitate contingere*), and it presumes by this necessity to make judgments infallibly about all future events (*praesumit per hanc necessitatem infallibiliter de omnibus iudicare futuris*).[19]

In Roger's presentation of this magico-astrological world-view, he makes an ontological claim about the deterministic nature of the world, which then has strong implications for the nature of astrological predictions about future events, namely, that the predictions are infallible or certain, especially if the celestial configurations themselves have been augmented with magical additions. Both claims are deeply problematic for a theologian as they undermine human free will, which is fundamental—along with the related ability to make choices—for one's moral development. As we will see, this argument is utterly central to Roger's defense of legitimate astrological judgments by distancing them from their illegitimately usurping counterparts. These themes will recur.

*

After alluding to certain statements by the saints, Roger turns to the authoritative rejection by philosophers of this false magical astrology:

> But this mathematics has been condemned (*damnata*) not only by the saints, but by the philosophers, as Isidore says in his treatise on *astrologia*, asserting that one part of *astronomia* is superstitious (*superstitiosa*), namely, what is magical (*magica*), and is called mendacious mathematics (*mathematica falsidica*). Whence Aristotle and Plato condemned it [sc. this false magical superstitious astrology], as Isidore bears witness, and Pliny[20] in different places in the *Natural History* often abuses (*percutiens*) it due to the errors which this fantasy (*haec fantasia*) inscribed in natural and medical works. Nevertheless, greatly abhorring it, he reveals its origin in Book XXX, and he shows plainly how it ruined (*defoedavit*) the

[19] [D]e bonitate tamen essentiae, et in adiutorium constellationum efficacius ordinavit singulis constellationibus propria figmenta characterum et aliorum praedictorum. Et haec expresse asseruntur in libris magicis. Unde haec scientia ista omnia ponit per coelum de necessitate contingere, et praesumit per hanc necessitatem infallibiliter de omnibus iudicare futuris (240, 10–16).

[20] Bridges's n. 1 here is useful (240).

> entire world. Also [Marcus] Tullius [Cicero] in his book *De divinatione*, descending more to its bad character (*malitia*) in particular, shows that it destroyed divine worship (*cultum divinum*), violated the Republic, infected medicine and natural philosophy, and subverted all good arts.[21]

According to these weighty philosophical authorities—Isidore, Pliny and Cicero, not to mention Plato and Aristotle—false magical mathematics can be profoundly harmful to culture and society as well as to natural knowledge.

*

After these general claims, Roger gets much more specific, in the process introducing further characteristic elements of his position. Here he discusses what the finest philosophical authorities say, and in particular, what they do *not* assert concerning the theologically sensitive issues just mentioned:

> Also Ptolemy, Aristotle, Avicenna, Messahalah, Haly[22] and Albumasar, who before others spoke about these things with a greater authority, do not posit an absolute necessity among lower things through the power of the heavens[23] because free will (*liberum arbitrium*) is not subject to natural things (*res naturales* [sc. including, in this case, the planets, stars and luminaries]),[24] nor do they think that an astrological judgment (*iudicium*) ought to be infallible. Nay rather, they also do not posit any necessity to free will, since they also do not

---

[21] Sed ista mathematica damnata est non solum a sanctis, sed a philosophis, ut dicit Isidorus in tractatu astrologiae, asserens unam partem astronomiae esse superstitiosam, scilicet quae est magica, et dicitur mathematica falsidica. Unde Aristoteles et Plato, testante Isidoro, eam damnaverunt; et Plinius per diversa loca Naturalis Historiae eam saepius percutiens propter errores quos haec fantasia scripsit in naturalibus et medicinalibus, tandem eam nimis abhorrens xxx libro originem illius aperit, et quomodo totum mundum defoedavit evidenter ostendit. Tullius etiam in libro Divinationum magis in particulari ad eius malitiam descendens ostendit quod cultum divinum destruxit, rempublicam violavit, et medicinam infecit et naturalem philosophiam et omnes bonas artes subvertit (240, 16–29).

[22] There are three relevant astrological Haly's, two authentic, the other pseudonymous. The first authentic Haly is the physician and astrologer Haly Rodoan (Abu l-Hasan 'Ali ibn Ridwan) who lived in Cairo from 998-ca. 1061. In addition to being a strict Galenist, he was appointed the chief physician of Egypt by the Fatimid ruler. For our purposes, he is most important for his commentary on Ptolemy's *Tetrabiblos* (Hasse, *Success and Suppression*, 373–74). Pseudo-Haly is credited with a commentary on pseudo-Ptolemy's *Centiloquium*, but we now know that it was written by Abu Ja'far Ahmad ibn Yusuf ibn Ibrahim ibn al-Daya, who lived in Cairo ca. 900 (374–75). The third astrological Haly, finally, is Haly Filius Abenragel (Abu l-Hasan 'Ali ibn abi r-Rigal, early eleventh c.), whose astrological summa, *De iudiciis astrorum*, was translated in the mid-thirteenth century, and thus later than most of the other astrological works translated from Arabic into Latin (371–72). It is unclear to which of the Haly's Roger refers here, but since he later calls Haly a commentator on Ptolemy (242, 15–24), it is most likely to one of the first two discussed here. There is a recent PhD thesis on Haly Rodoan by Jennifer Ann Seymore, "The Life of Ibn Ridwan and his Commentary on Ptolemy's *Tetrabiblos*," PhD thesis, Columbia University, 2001; Seymore's graduate advisor is George Saliba.

[23] [...] necessitatem absolutam in rebus inferioribus non ponunt per virtutem coeli [...] (241, 2–3).

[24] We also see something like this below, including the related notion that the *anima rationalis* has more dignity than the stars. Thomas's analysis in Chap. 5 also treats this fundamental issue.

> ascribe it [sc. necessity] to natural things, as will be made clear. And therefore, philosophers universally condemn the insanities (*insaniae*) of these false mathematicians.[25]

The most authoritative astrologizing philosophers do not claim necessity in nature, nor its concomitant, infallible astrological judgments. Nor, in this view, is free will subordinated to natural things because the soul is not subordinated to the stars, thus sharply distinguishing between the body and the soul registers. In this way, Roger defends legitimate astrological judgments and provides solid safeguards to protect human free will. Thomas Aquinas, Albertus Magnus and the *Speculum astronomiae* all make this very same point, as we will see later in this chapter and in Part II.

*

To further distance legitimate astrology from its illegitimate magical pretender, Bacon removes the false position from the frying pan and casts it fully into the fire, as he moves from the undermining of human free will to now introducing the influence of demons, which are, of course, always illegitimate:[26]

> Not only do they [sc. the philosophers] condemn it insofar as it is related to the principle problem, namely, because of the error which they have concerning the celestial bodies,[27] but because these [sc. false] *mathematici* call on demons (*advocant daemones*) to aid the celestial dispositions [sc. configurations] through conjurations and sacrifices,[28] which is altogether unspeakable (*nefandum*). Nonetheless, they stain their considerations concerning the heavenly bodies (*in coelestibus*) through the use of the vainest circles, figures and characters, and the stupidest charms (*carmina stultissima*) and irrational utterances (*orationes irrationabiles*) in which they place their trust.[29]

Beyond the problems with positing a deterministic universe with its concomitant infallible judgments, these false mathematicians also invoke demons—with the panoply of magical techniques and paraphernalia, including conjurations, sacrifices, magic circles, figures, characters and charms, thus adding figures and magical circles

---

[25] "Ptolemaeus etiam et Aristoteles et Avicenna et Messehalac et Haly et Albumazar, qui prae aliis maiori auctoritate de his locuti sunt, necessitatem absolutam in rebus inferioribus non ponunt per virtutem coeli, quia liberum arbitrium non subiacet rebus naturalibus, nec aestimant iudicium debere fieri infallibile, immo nec aliquam libero arbitrio ponunt necessitatem, cum nec rebus naturalibus eam ascribant, ut patebit. Et ideo philosophi universaliter damnant istorum falsorum mathematicorum insanias (240, 29–241, 8)." Pico also uses the term '*insania*' to describe astrology, but for Pico, all of astrology is to be condemned, as we will see in volume II.

[26] As we will see in volume II, these are precisely the characteristic medieval safeguards that Marsilio Ficino later attempts to subvert, namely, that the planets only operate on the body register and that astrology does not involve demons in any way, shape or form.

[27] This seems to refer to the discussion just above, namely, that the heavenly bodies act necessarily on things below.

[28] To which Roger had also just referred.

[29] Nec solum damnant quantum ad principale, scilicet propter errorem quem habent de coelestibus, sed quia mathematici isti daemones advocant in adiutorium coelestium dispositionum per coniurationes et sacrificia, quod est omnino nefandum, atque nihilominus maculant suas considerationes in coelestibus per circulos et figuras et characteres vanissimos et carmina stultissima, et orationes irrationabiles in quibus confidunt (241, 9–16).

to the list above—in order, they hope, to powerfully augment the celestial circumstances. Roger clearly alludes here to ritual or ceremonial magic, such as one finds in Richard Kieckhefer's marvelous edition of a fifteenth-century necromantic manuscript, replete with nonsensical squiggles and incoherent irrational mutterings.[30]

Roger continues, turning now to *praestigium*, the making of illusions:

> In addition, they adjoin fraudulent works (*fraudes operum*), namely, by means of collusion (*consensum*), shadows, sophistic [sc. delusionary] instruments (*instrumenta sophistica*) and sleight of hand (*per subtilitatem motionis manualis*), in which they know that there is an optical illusion (*illusio*). And they make many things wondrous to fools (*multa stultis miranda*) by these things, in which the power of the heavens does not operate (*in quibus virtus coeli nihil operatur*). And therefore, contradicting themselves, what they attribute to the heavens openly before others, among themselves they know does not have truth.[31]

Besides adding demons, rituals and necessity/determinism to celestial influences, these charlatans also knowingly claim celestial actions when there are none, due to sleight of hand or other types of fraudulent illusion. Both are profoundly problematic because they simulate the appearance of celestial action, but are really deceptions, based on nothing real or true.

Bacon continues describing this sort of misleadingly fraudulent behavior:

> Likewise, although in other contexts they place their trust in certain conjurations, sacrifices and charms, and in various characters and figures, as though they were working together with a celestial configuration (*cooperantibus constellationi*) in accordance with their judgment (*iudicium*), nevertheless, many times they put these things together fraudulently in accordance with the aforementioned species of fraud, and before those who believe in them, they ascribe more power to the celestial configuration than it has.[32] And this does not escape them [sc. the practitioners]. Therefore, because of these stupidities of a sophistic mathematics (*has stultitias mathematicae sophisticae*), the philosophers condemned it [sc. false mathematics], and the saints and catholic people, perceiving these things, reproached it at the same time with the philosophers.[33]

Here Roger rebukes those who deliberately defraud their audiences and/or customers by attributing actions to the heavens that are really just stage tricks or other types

[30] *Forbidden Rites: A Necromancer's Manual of the Fifteenth Century*, University Park, PA: The Pennsylvania State University Press, 1998. This also raises the useful distinction between ritual or ceremonial magic vs. image magic as articulated by Frank Klaassen in his *The Transformations of Magic*. I discuss this distinction further in the introduction to Part III below, and more fully in the overall introduction to volume II.

[31] Praeterea fraudes operum adiungunt, scilicet per consensum, per tenebras, per instrumenta sophistica, per subtilitatem motionis manualis, in quibus sciunt illusionem esse, et multa stultis miranda faciunt per haec in quibus virtus coeli nihil operatur, et ideo sibimetipsis contradicentes, quod coelo attribuunt coram aliis, apud seipsos sciunt non habere veritatem (241, 16–22).

[32] "[...] ascribunt plurima constellationi, respectu quorum ipsa ullam habet potestatem (241, 27–28)." This is also a theme that Pico develops in the *Disputationes*, as we will see in volume II.

[33] Similiter licet alias in quibusdam coniurationibus et sacrificiis et carminibus et characteribus et figuris variis confidunt tanquam cooperantibus constellationi secundum eorum iudicium, tamen pluries fraudulenter ista componunt secundum species fraudis predictas, et coram sibi credentibus ascribunt plurima constellationi, respectu quorum ipsa nullam habet potestatem. Et hoc eos non latet. Propter igitur has stultitias mathematicae sophisticae philosophi eam damnaverunt, et sancti ac viri catholici haec percipientes eam simul cum philosophis reprobaverunt (241, 22–31).

of illusions. They are justly reproved by both philosophers and saints as false and illegitimate mathematician-astrologers. Thus, Roger here does not reject those philosophers (*sapientes*) who actually work with celestial configurations, but, rather, only those who do not in fact do so, but merely appear or pretend to do so.

*

To complete this first stage of his argument, Roger returns briefly to the saints. Here he discusses miracles, presenting the arguments of the saints-fathers against the false *mathematici* while emphasizing the deleterious effects of their actions on the faithful:

> But the most important argument of the saints was to this end, that such [sc. false] *mathematici* impeded from the beginning the entrance of the faith into this world because, not only imbued with this fantasy have they erred in their faith and alleged that their customs were in the celestial bodies, just as, by means of the celestial bodies and other things said about necessity, they became angry or mild, chaste or luxuriant, and thus about other things. But also these (*isti* [sc. false *mathematici*]) ascribed the miraculous works (*opera miraculosa*) proving faith in Christ to astrology (*mathematicae*), saying that Christians are astrologers (*mathematici*) and magicians (*magi*) who seduce the people. For just as by means of this demoniacal skill (*daemoniaca calliditas*) those were able to do many things before an ignorant populace (*coram populo rudi*) by which they occupied them with errors and dominated them, thus they imposed on the apostles, martyrs and other preachers of the faith that they made true miracles (*vera miracula*) not from God's side (*non ex parte Dei*), but by magical art (*per artem magicam*), which they themselves used. And therefore, the saints and the church have profoundly condemned this [sc. false] mathematics and its professors.[34]

False *mathematici* claim that Christian miracles were performed by astrology and magic, and thus they delude the faithful. This is the only mathematics or astrology that the saints and the church condemn, not—Roger loudly implies—true mathematics, by which, of course, he means legitimate astrology and its judgments, at least so far. Later we will also see that works (*opera*) are also involved.[35]

[34] Sed praecipua ratio sanctorum fuit ad hoc, quod tales mathematici impediverunt a principio ingressum fidei in hunc mundum, quia non solum imbuti hac fantasia erraverunt in fide et finxerunt mores suos in coelestibus, tanquam per coelestia et caetera dicta de necessitate fierent iracundi vel mansueti, casti vel luxuriosi, et sic de aliis; sed opera miraculosa fidem Christi probantia ascripserunt isti mathematicae, dicentes, Christianos esse mathematicos et magos populum seducentes. Sicut enim ipsi per hanc daemoniacam calliditatem potuerunt multa coram populo rudi facere per quae detinebant eos ad errores et dominabantur eis, sic imposuerunt apostolis et martyribus et caeteris praedicatoribus fidei quod non ex parte Dei fecerunt vera miracula, sed per artem magicam qua ipsimet usi sunt. Et ideo hanc mathematicam et eius professores sancti et ecclesia penitus damnaverunt (241, 32–242, 9).

[35] I will now skip most of the next section in Bacon's argument (242, 10–249, 3). Although it is very interesting, it must be passed over here due to considerations of space.

4.5 Astrology's Epistemological Limitations

After offering these authoritative attacks on the false magical *mathematici*, whether due to deterministic, fraudulent or demonic views, Bacon now offers his defense of true mathematics and also, in the same process, its positive promotion. Roger does so by offering arguments and authorities supporting astrology more generally *qua* both its value for knowledge and its utility. Then he turns specifically to religious authorities. I will only discuss here his use of the most important astrological authority from Antiquity, Ptolemy himself, who explicitly delimits astrology's epistemological possibilities.

Roger begins by positively describing the true *mathematici*:

> But the true *mathematici*—who in this part we call astronomers and astrologers (*astronomi et astrologi*), because they are thus indifferently called by Ptolemy, Avicenna and many others—are not proven wrong with respect to sacrifices, conjurations, charms or characters, as also the crowd of students (*plebs studentium*) is not unaware, but they are only singled out concerning infallible judgment and the necessary outcome of things (*super iudicio infallibili et rerum necessario eventu*, 242, 10–15).[36]

For true mathematian-astrologers, there are no magical behaviors, whether sacrifices, conjurations, charms or characters. The only concerns—even for the crowd of erring students—are infallible judgments and the necessary outcome of events, i.e. necessity in nature and certainty of prediction with the implied concomitant loss of human free will and choice, precisely the themes Roger had earlier articulated. We will note that Bacon here refers indifferently to these true *mathematici* as both astronomers and astrologers, and that he explicitly follows contemporary usage here as derived, ultimately, from Ptolemy.

Bacon then addresses how we know—and normatively should know—about what these true *mathematici* think, so that we can properly evaluate their arguments concerning necessity in nature and infallible judgments, and thus whether they actually held these positions or not:

> Moreover, we cannot grasp their opinion (*sententia*) except as we bring to light their own testimony (*propria testimonia*) from their books, and thus we either condemn them as erring from their own words (*ex propriis verbis*), or we rescue them as free men, exonerated, from the teeth of the inexperienced crowd (*a dentibus vulgi imperiti*). Therefore, since an error in judgment (*error iudicii*) is especially imposed on them, as if they wish to contend [sc. that their judgments are made] infallibly and with divine certainty (*cum divina certitudine*), as the plebs impose on them, let the more noble philosophers be brought forward,

[36]"Sed veri mathematici, quos in hac parte vocamus astronomos et astrologos quia indifferenter a Ptolomaeo et Avicenna et aliis pluribus sic vocantur, non redarguntur de sacrificiis coniurationibus carminibus characteribus, ut etiam plebs studentium non ignorat, sed solum super iudicio infallibili et rerum necessario eventu notantur (242, 10–15)." Bridges n. 1 is valuable here. In *Communia mathematica* (Sloan MS 2156, f. 8a, c) Roger distinguishes astrology and astronomy and gives them their etymology, making astrology speculative and astronomy practical, and thus the terminological opposites of the modern distinction.

> namely, Aristotle, Avicenna, Ptolemy, Haly his commentator, Messahalah and Albumasar, on whom this burden leans more.[37]

Roger sensibly claims here that judgments about the astrologers' positions should be based on what they actually wrote. He then proceeds to marshall some of the authoritative and programmatic texts of the major philosophico-astrological authorities. Once again, he names the ancients, Aristotle and Ptolemy (with his later commentator Haly), and the more recent authorities who wrote in Arabic: Avicenna, Messahalah and Albumasar. I will only focus here on one argument from Ptolemy's *Tetrabiblos*, where Roger, using Ptolemy, argues for astrology's epistemological limitations which may thus be used to obviate both necessity in nature and infallibility in judgments.[38]

*

The first stage in Bacon's reconstruction of the astrologers' position is to solidly establish that the finest astrologico-philosophical authorities do *not* in fact claim infallible judgments.[39] On the contrary, they are quite explicit about two complementary challenges: the difficulty of the art itself, on the one hand, and the limitations of human intelligence, on the other. Roger does not begin by discussing the supposed implications of necessity in nature here (although he will below), focusing rather on the art side of the ignorant theologians' claims.

Turning from the pseudonymous *Centiloquium* and *De dispositione sphere* (242, 24–243, 34),[40] Roger discusses the authentic *Tetrabiblos* on which I will focus. At the very beginning, Ptolemy states his position clearly and strongly concerning why astrologers err, as Roger now informs us:

> And Ptolemy in the first chapter of the *Tetrabiblos* says: "Moreover, of this excellent art there is such a hidden depth (*occulta profunditas*), and its practice (*exercitatio*) is so lofty and incomprehensible that it is impossible that it (*eam,* sc. *ars*, astrology) can be deeply grasped by the human mind (*ab humano ingenio*) due to its ineffable subtlety (*ex ineffabili subtilitate*), and it seems to exist due to its admirable and quasi divinity (*quasi quadam divinitate*). For it is far removed from human sense perception (*ab humano sensu*), and, transcending bodily nature (*corpoream transcendens naturam*) as if, at a distance, it goes

[37] Non autem possumus eorum sententiam deprehendere nisi ex libris eorum testimonia propria eruamus, ut sic vel errantes damnemus ex propriis verbis, vel excusatos a dentibus vulgi imperiti liberos extrahamus. Quoniam igitur maxime imponitur eis error iudicii, tamquam infallibiliter velint cum divina certitudine contendere, ut plebs eis imponit, adducantur philosophi nobiliores, scilicet Aristoteles, Avicenna, Ptolemaeus, Hali commentator eius et Messehalac, Albumazar, quibus magis incumbit hoc onus (242, 15–24).

[38] In this section, Roger also treats interesting texts from the pseudo-Ptolemaic *Centiloquium* and *De dispositione spherae*, but I must pass over them here.

[39] This harmonizes nicely with Ptolemy's opening statement in the *Tetrabiblos* (I.1) that astronomy is an exact science whereas astrology is conjectural.

[40] Hackett identifies the *De dispositione sphere* as Ptolemy's *Almagestum parvum*, an introduction to the *Almagest* based on Geminus's *Introduction to the Phenomena*; Hackett, "Roger Bacon on *Scientia Experimentalis*," in *Roger Bacon and the Sciences*, 277–316, 285.

> beyond man." And in the third chapter [sc. of Book I] he says that "from the great profundity of this art (*nimia ipsius artis profunditate*) each error happens" (243, 34–244, 9).[41]

According to Ptolemy, astrology as a science is divine and beyond reproach. Due to its profundity and sublimity, however, it poses severe challenges to human understanding and can never be fully understood.

From these passages in Ptolemy's *Tetrabiblos*, I.1–3, Roger then draws a lesson for his defense of astrological judgments:

> Wherefore, according to Ptolemy, to whom at this time now (*nunc temporis*) the understanding of [sc. how to] make an [astrological] judgment (*ratio iudicandi*) is especially ascribed, it is obvious that the astrologer (*astrologus*) cannot provide full certainty for his judgments (*plena certitudo suorum iudiciorum*), especially concerning particulars (*singulis*). But also, not only is it clear from these passages, but from other of his teachings (*sermones*). Although he indicates the possibility of making judgments (*possibilitas iudicandi*) about many things with a reasonable certainty (*cum rationabili certitudine*), nevertheless, he asserts that there is such great difficulty in this art *simpliciter*, that he easily makes plain this deep definition, that an astrologer (*astronomus*) ought not to take pleasure in particular ([sc. interpretations] *in singulis*) with sufficient certainty (*de sufficiente certitudine*).[42]

In accordance with Ptolemy's authoritative rationale for making astrological interpretations, Roger finds support for the conclusion that certainty in judgments is simply unattainable due to the profound depths of the art and/or science, on the one hand, and the limitations of the human intellect, on the other. Nevertheless, within these limitations, relative certainty is greater in proportion to the judgment's generality. Regardless, a full blown and robust certainty is impossible, thus establishing astrology's epistemological limitations and its concommitant fallibility in practice, precisely what he had set out to defend.

---

[41] Et Ptolemaeus in primo capitulo quadripartiti dicit, "huius autem tam excellentis artis occulta profunditas eiusque tam augusta et incomprehensibilis exercitatio, ut ab humano ingenio eam impossibile sit penitus amplecti, ex ineffabili subtilitate, eiusque admiranda quasi quadam divinitate videtur contingere. Longe enim ab humano sensu remota est, et corpoream transcendens naturam quasi ultra hominem eminus transcendetur. Et tertio capitulo dicit, "quod ex nimia ipsius artis profunditate quicunque error incidit."

[42] Quapropter secundum Ptolemaeum, cui nunc temporis maxime ascribitur ratio iudicandi, manifestum est quod astrologus non potest dare plenam certitudinem suorum iudiciorum, praecipue in singulis. Quod etiam non solum ex his liquidum est, sed ex aliis eius sermonibus, quibus licet innuat possibilitatem iudicandi de multis cum rationabili certitudine, tamen tantam difficultatem simpliciter arti inesse pronuntiat, ut facile pateat ipsum penitus definire, quod non debet astronomus de sufficienti certitudine in singulis gloriari (244, 9–18).

4.6 Astrology's Utility in Human Affairs

4.6.1 Individuals

After finishing the first part of his argument defending mathematics and its weakest link, the judicial or interpretive part of astrology—primarily by articulating the strong distinction between true and false mathematics, where he characterized both clearly and defended true mathematics from charges of determinism in nature and certainty in predictions—Roger now turns to defend astrology in relation to human affairs, one of its most controversial areas. He treats the main themes in detail and so shall I, both in themselves, and later in relation to Thomas Aquinas's related analyses in Chap. 5. In particular, Bacon focuses on the key issue of how astrology's role in human affairs does not in any way imply necessity in nature, that is, a deterministic world-view, which would thus undermine both human free will and God's omnipotence.

As we will immediately see, Bacon continues to use the fundamental distinction between true and false mathematics in what follows:

> But in human affairs (*in rebus humanis*) true *mathematici* do not presume to certify (*certificare*),[43] but they consider how a body (*corpus*) is qualitatively changed (*alteratur*) by means of the heavens (*per coelum*). And, when the body has been qualitatively changed, how the soul (*anima*) is excited (*excitatur*) now to private, now to public actions, nevertheless with freedom of the will preserved in everything (*salva tamen in omnibus arbitrii libertate*).[44]

In the excursus and in Chap. 2 we saw Bacon discuss how bodies can be qualitatively changed (*alteratio*), and how that related both to celestial radiation and to astrological dignities. Here Roger goes into greater detail in relating celestial influences on the body to those on the soul. Defending free will in this dynamic context is of central concern for a theologically acceptable legitimate astrology.

First Bacon clarifies the differences—and thereby the relationship—between body and soul with respect to celestial influences:

> For, although the rational soul (*anima rationalis*) is not compelled (*non cogitur*) toward these actions, nevertheless it can be strongly induced and excited (*fortiter induci potest et excitari*) to desire these things freely (*gratis*) towards which the celestial power inclines (*virtus coelestis inclinat*), just as we ourselves see men change (*mutare*) from what they have resolved due to society, counsel, fear and love, and anything of this sort, and to want freely those things that they did not want beforehand, although they are not compelled (*non cogantur*), just like he who casts precious goods into the sea hoping for health ([or salvation?][45] *spe salutis*).[46]

---

[43] That is, to make judgments with certainty.

[44] Sed in rebus humanis veri mathematici non praesumunt certificare, sed considerant quomodo per coelum alteratur corpus, et alterato corpore excitatur anima nunc ad actus privatos nunc publicos, salva tamen in omnibus arbitrii libertate (249, 4–8).

[45] In both religious and purely survival senses.

[46] Quamvis enim anima rationalis non cogitur ad actus suos, tamen fortiter induci potest et excitari ut gratis velit ea ad quae virtus coelestis inclinat, sicut nos videmus homines per societatem consilia timorem et amorem et huiusmodi multum mutare de proposito, et gratis velle ea quae prius non volebant licet non cogantur, sicut ille qui spe salutis proiicit merces in mare carissimas (249, 8–14).

The stars incline but do not compel, as the old saying goes. There is thus a fundamental difference between [1] celestial influences (a) exciting a person towards an action and (b) the person choosing to embrace or defy that sometimes passionate desire, and [2] celestial bodies fully and robustly *determining* and thereby *necessarily* compelling a person's actions, and thus undermining their free will and, thereby, their ability to make choices freely. Roger will articulate the natural philosophical foundations for this view in what follows. We will see in Chap. 5 that Thomas too is concerned with precisely these same issues.

*

Roger begins his argument by discussing the influence of bodies in the world on people. In so doing—and by utilizing his analysis of the multiplication of species—he develops the psychological dimension of his analysis:

> Again, we see that species or powers of things below change our sense perceptions (*species seu virtutes rerum inferiorum immutantes sensus nostros*). Even species of visible and audible things that change (*immutant*) the body weakly (*debiliter*) excite people so strongly (*fortiter*) to desire what they previously did not care about that sometimes they value neither death, bad reputation (*infamia*) nor fear, provided that they fulfill their wishes (*suas compleant voluntates*). Just as there are those who see and hear their enemies approaching them and are borne by every contingent circumstance (*omni casu contingente*) to punish them.[47]

Even where the measurable effect of things in the world on a *body* is weak—as in sense perception—the very same effect on the mind can be very strong indeed, especially when it derives from the senses, as everyone has experienced, both then and now. With the talk of species and powers/virtues, Roger refers here to the general model of action reconstructed in Chap. 2, where everything that exists radiates its own influences-energies by means of species and powers, and ultimately by means of rays.[48] This information would also have been of great interest to the thirteenth-century equivalent of advertising, public relations and propaganda professionals.

Bacon continues along these lines with an illuminating example, namely, the response engendered by an object of desire:[49]

---

[47] Caeterum nos videmus quod species seu virtutes rerum inferiorum immutantes sensus nostros, etiam species visibilium et audibilium quae debiliter immutant corpus, ita fortiter excitant homines ad volendum quae prius non curabant, quod aliquando nec mortem nec infamiam, nec timorem aestimant dummodo suas compleant voluntates, sicut sunt illi qui vident et audiunt suos inimicos eis occurere, et feruntur omni casu contingente ut se vindicent (249, 14–21).

[48] Katherine Tachau articulates some of the deeper structures at play in her "*Et maxime visus, cuius species venit ad stellas et ad quem species stellarum veniunt*: *Perspectiva* and *Astrologia* in Late Medieval Thought," in *Micrologus 5: Natura, scienze e società medevali*, Florence: SISMEL—Galluzzo, 1997, 201–24.

[49] As we will see in volume II, Marsilio Ficino offers a similar analysis in *De vita* III also in an astrological context in analyzing figures and how they work. Ficino also employs there a geometrical-optical model of celestial influences, and a comparison of vision with celestial influences.

> Likewise, when an opportunity for fulfilling their delights (*delicias*) has been seized upon in relation to things seen and heard, pleasure seekers (*voluptuosi*) are moved (*moventur*) against the judgment of their reason (*contra iudicium rationis*) as if they were brute animals, freely choosing (*gratis eligentes*) those things towards which they are excited.[50]

Roger thus establishes how the power of things that exist in the world as transmitted by perception—primarily sight and hearing—can so powerfully affect people and their actions. There is a strongly implied moral dimension here as well.

*

To these influences from terrestrial bodies, Bacon now contrasts those from celestial bodies. He also contrasts their effects on both bodies and minds (and thus on souls). This is utterly central to Bacon's position:

> But far more can the strong powers (*virtutes*) of the heavens and their species and those of the stars impress (*imprimere*) the body and its organs, by which, when they have been powerfully altered (*vehementer alteratis*), a person will be excited strongly (*fortiter*) to actions about which he had not previously cared, while preserving his free will (*salvo suae libertatis arbitrio*). Since the powers of the heavens (*virtutes coelorum*) are stronger (*fortiores*) than those of these lower visible and audible things, and of many other perceptible entities, they can both change (*mutare*) a substance (*substantia*)—not only accidents (*accidentia*)—and corrupt and destroy all lower things (*res omnes inferiores*). Since, in accordance with the movement (*allatio*) of the sun under the oblique circle [sc. the zodiac], generation and corruption happen in things, as Aristotle says.[51]

Here we see Roger sounding very much like Albert in the texts examined in Chap. 1, and using very similar terminology. Roger here relates the movement in place of the heavens to both generation and corruption (change in substances) and to *alteratio* (change in quality), one type of change in accidents. He claims that these celestial influences are also much stronger than those of the terrestrial entities he had just discussed, and thus have a correspondingly greater affect on people: directly on their bodies and indirectly on their minds and desires. Nevertheless, although they are powerful, they are certainly not necessary.

Bacon continues in the same vein, further developing the picture:

> And not only this *allatio* of the sun considered in itself (*absolute*), but with the powers (*virtutes*) of the other planets and stars. Therefore the *alteratio* of our bodies by means of

---

[50] Similiter voluptuosi, opportunitate accepta complendi suas delicias circa res visas et auditas, quasi bruta animalia moventur contra iudicium rationis, gratis eligentes ea ad quae excitantur (249, 21–24).

[51] Sed longe magis possunt virtutes coelorum et species eorum et stellarum fortes imprimere in corpus et organa, quibus vehementer alteratis excitabitur homo fortiter ad actus de quibus non curavit prius, salvo suae libertatis arbitrio; quoniam virtutes coelorum fortiores sunt quam istorum inferiorum visibilium et audibilium et multorum aliorum sensibilium, et possunt mutare substantiam non solum accidentia, et corrumpere et destruere res omnes inferiores; quoniam secundum allationem solis sub obliquo circulo accidunt generatio et corruptio in rebus, ut Aristoteles dicit (249, 25–34).

> celestial powers (*per virtutes coelestes*) is strong (*valida*), and consequently, the mind (*animus* [vs. *anima* above]) is strongly excited (*fortiter excitatur*) towards its actions (*ad actus suos*), although it is not compelled (*non cogatur*)[.][52]

Bacon here expands celestial influences from the sun to include all the planets, as we also saw Albert do in Chap. 1 in the same context of interpreting Aristotle's *De generatione et corruptione* II.10. This characterizes a major shift between Aristotle's thought in itself and that of the later astrologizing Aristotelians, as does the mathematical framework that lies behind their analysis, especially the geometrical-optical model of celestial influences reconstructed in my Chap. 2. Here Bacon claims that the influence of the heavenly bodies on our bodies is strong, and thence on our minds, but without compulsion, a refrain we will encounter time and again. In this way, necessity continues to be obviated in human action. Roger seems to use *anima* and *animus* interchangably here as we will also see Marsilio Ficino do in volume II.

That human beings can act freely is obviously a fundamental tenet of Bacon's analysis. To establish this, he argues that the planets as bodies act directly on our bodies, and through them, indirectly influence our souls, minds and emotions. Crucial to Roger's position is the implicit point that the stars and planets *qua* bodies do not and cannot *directly* influence the soul. This is fundamentally important for obviating the charge of determinism, and thereby the annihilation of human free will, by transforming necessity into probability (even strong probability), which is all one needs to undercut claims for both necessity regarding human nature—and nature at large—and infallibility in astrological judgments. Roger thus delimits direct celestial influences here to the body register as a safeguard of human free will. Thomas Aquinas articulates this view even more fully and clearly in texts to be examined in Chap. 5 below.

*

Bacon then shifts from a natural philosophical analysis with psychological overtones to the astrological side of the equation, namely, the making of astrological judgments or interpretations. The closely connected issues of necessity in nature and the infallibility of astrological judgments remain central to his concerns:

> [I]n accordance with this [sc. natural philosophical analysis] runs an astrologer's judgment (*iudicium astronomi*), and not by means of infallibility and necessity. The astrologer (*astronomus*) is much helped in this because he sees that human beings in their actions much follow their complexions, as the choleric is easily moved to anger, and he cannot bridle all of his initial impulses (*refraenare primos motus omnes*), and thus about the others [sc. sanguine, melancholic and phlegmatic] in accordance with which people are differentiated into their complexions.[53] And therefore the astrologer (*astronomus*), when he sees

[52] "Et non solum haec allatio solis absolute considerata, sed cum virtutibus aliorum planetarum et stellarum; et ideo valida est alteratio corporum nostrorum per virtutes coelestes, et per consequens animus fortiter excitatur ad actus suos licet non cogatur (249, 34–38)[.]" I end the sentence here.

[53] See the classic study by Raymond Klibansky, Erwin Panofsky and Fritz Saxl, *Saturn and Melancholy: Studies in the History of Natural Philosophy, Religion and Art*, London: Nelson, 1964; repr. 1979 (Kraus); and the minimally revised German translation, *Saturn und Melancholie:*

> people follow their complexions that arise from a celestial operation (*oriuntur a celesti operatione*), just as all generation (*tota generatio*) does, there is no wonder that he extends this to the consideration of human actions.[54]

Celestial operations are fundamental more generally to the generation of substances and their complexions. In relation to human beings in particular, this analysis deeply informs our understanding of human action. A key to responsible and legitimate astrological interpretations, then, is for the astrologer to understand the basics of premodern humoral psychology and its implications for the behavior of individuals. There is also a significant medical dimension to this system, as we will see to some extent in Part IV below, and much more fully in Ficino's *De vita* in volume II.

From the more directly astrological dimensions of physiology and psychology, Bacon then shifts to another interpretive factor with a significant astrological dimension, namely, the influence of place on complexions, a theme we will also recall from Chap. 2:

> Again, he himself [sc. the *astronomus*] sees obviously that, in accordance with the diversity of those who live under the different parallels of the heavens, customs (*mores*) are varied, as those who live near the pole, as the Scythians, have different customs than those living toward the south, as the Ethiopians. And they have different customs from those who live in the fourth clima. Nay rather, according to the diversity of each clima, and even of a clima's parts, the customs of the inhabitants are varied, as, namely, the Picards, Gauls, Normans, Flemish and English in the seventh clima. We see manifestly that they differ (*discrepare*) in customs, even though they border each other (*continui ad invicem*) and are close.[55]

Bacon here reiterates and develops structures presented earlier concerning the nature of different places and their influence on the customs and practices of the different peoples who live there.

Roger then integrates the analysis of complexion and place by showing that differences of location deeply influence complexions, here emphasizing the role of celestial influences in both respects. Once again, he highlights the different effects on the body and the soul-mind:

Studien zur Geschichte der Naturphilosophie und Medizin, der Religion und der Kunst, Christa Buschendorf (tr), Frankfurt am Main: Suhrkamp, 1992. This work, and especially the section on Ficino (for which there are no bibliographical references later than 1930), requires a thoroughgoing revision. For a more up-to-date analysis, see E. Ruth Harvey, *The Inward Wits: Psychological Theory in the Middle Ages and the Renaissance*, London: Warburg Institute, 1975.

[54] [E]t secundum hoc currit iudicium astronomi, et non per infallibilitatem et necessitatem. Et astronomus in hoc multum iuvatur, quod videt homines in actibus suis sequi multum suas complexiones quas habent, ut cholericus movetur de facili ad iram, nec potest refraenare primos motus omnes, et sic de aliis, secundum quod homines diversificantur in complexionibus. Et ideo astronomus, cum videt homines sequi suas complexiones, quae oriuntur a coelesti operatione, sicut et tota generatio, non est mirum si se extendat ad considerationem actuum humanorum (249, 38–250, 9).

[55] Caeterum ipse videt manifeste, quod secundum diversitatem habitantium sub diversis coeli parallelis variantur mores, sicut habitantes versus polum, ut Scythae, alios habent mores quam habitantes versus meridiem, sicut Aethiopes; et alios ab his habent illi, qui in quarto climate; immo secundum diversitatem cuiuslibet climatis et etiam partium climatis variantur mores habitantium, ut in climate septimo Picardos, Gallicos, et Normannos, et Flamingos, et Anglicos, videmus manifeste in moribus discrepare, cum tamen sint continui ad invicem et propinqui (250, 10–19).

> Moreover, this [sc. different customs, etc] does not exist for these people on account of the diversity of their *anima rationalis*, but due to the complexions of their bodies—inborn from the nature of the heavens (*innatas a natura coeli*)—under whose different parallels and stars they are situated, and according to the difference of their location with respect to the planets.[56]

As the formal cause, the *anima rationalis* is the same for all people *qua* being members of the human species. Differences arise with individuals and their bodies generated in different places receiving celestial influences differently, with the resultant different root and current complexions impressed on their bodies. As we saw in Chap. 2, celestial influences are thus profoundly conditioned by the place where they are received, since place is a principle of generation.

Roger continues, further differentiating his picture with respect to place:

> And not only is this diversity in accordance with the latitude of the regions from the equinoctial circle to the pole, as has now been touched upon, but in accordance with longitude, although the causes are more hidden (*occultiores*). For we see by experience (*per experimentiam*) that under the same parallel, regions are varied according to their greater distance from the west or east; and not only the regions themselves, but the parts of regions.[57]

Differences in both latitude and longitude matter, thus developing the analysis familiar from Chap. 2. I should also note that, although Roger has not yet specifically discussed these issues here in terms of his geometrical-optical model of celestial influences, we know that this lies behind and beneath his analysis, as we will now see explicitly.

*

In the next step of his analysis, Bacon returns to the deeper mathematical structures of his causal analysis, namely, the geometrical-optical model of planetary action:

> And the primary cause (*causa principalis*) of this thing cannot be discovered on earth nor in people, but it must be sought in the heavens according to all wise people (*omnes sapientes*). Whence, as was said before, to every point on earth falls the cone of one pyramid made virtuous/powerful from the entire heavens (*ad omne punctum terrae incidit conus unius pyramidis virtuosae a toto coelo*). And these cones are diverse in nature, and the pyramids likewise, because they have different bases due to the diversities of horizons, since each point of the earth is the center of its own proper horizon (*quilibet punctus terrae est centrum proprii horizontis*). And therefore, it is fitting that the great diversity of all things arises from this cause, however close they are, as twins in the same womb. And thus about everything, in so far as we see that from two nearby points of the earth arise plants diverse in species. And here the astrologer (*astronomus*) takes the foundations of his judgment

[56] Hoc autem non est ex ipsis hominibus ex parte diversitatis animae rationalis, sed propter complexiones corporum innatas a natura coeli, sub cuius parallelis diversis et stellis situantur, et secundum diversitatem situs eorum respectu planetarum (250, 19–23).

[57] Et non solum est haec diversitas secundum latitudinem regionum ab aequinoctiali circulo versus polum, sicut nunc tactum est, sed secundum longitudinem, licet causae sint occultiores. Per experientiam enim videmus, quod sub eodem parallelo variantur regiones, secundum quod magis ab occidente vel oriente distant, et non solum ipsae regiones sed partes regionum (250, 23–29).

(*fundamenta sui iudicii*), and rightly, because the full diversity of things is thus found by means of the heavens (*diversitas plena rerum per coelum sic invenitur*).[58]

Bacon here takes the deeper structure of his richly mathematicized natural philosophy (as examined before) and relates it explicitly to the exigencies of astrological interpretation. For Roger, these structures provide the natural philosophical foundations for making astrological judgments, precisely as Albert claimed more generally in his commentary on Aristotle's *De caelo*. They also undermine the possibility of necessity in nature and its concomitant infallibility in judgments and thereby safeguard human free will.

4.6.2 From the Individual to the Political

Having completed his causal analysis concerning individual human beings along familiar lines—and thus explicitly offering the natural philosophical foundations for the astrological practice of nativities—Bacon then turns to the next stage of his argument, which embraces a broader range of human activities: "For which reason, an experienced astrologer (*astronomus peritus*) can not only consider many things about the present, future and past in natural matters (*in naturalibus*), but also in human affairs (*in rebus humanis*)[.]"[59] Here Roger takes up what he had set out to address at the beginning of this section (249, 4 ff.), but only after first re-establishing its foundations in nature and relating it to the individual members of human societies. The realms of nature—and of both individual and collective human affairs—are thus intimately related, but also contrasted, ultimately, into the distinct—but still deeply interrelated—realms of nature and politics, and both within the temporal modalities of past, present and future.

To begin articulating astrology's numerous roles in collective human affairs, Bacon discusses the making of astrological judgments related to the practice of general astrology or revolutions, of the sort found in annual prognostications. In doing so, he turns from the realm of individuals to that of collective entities like cities and kingdoms. This also relates directly to the distinction between universal and particular astrological judgments raised earlier:

> [A]nd therefore, he [sc. the experienced astrologer] can make judgments (*potest iudicare*) about realms (*regna*) and cities (*civitates*) by means of celestial bodies (*per coelestia*) and

[58] Et non potest in terra nec in hominibus inveniri causa principalis huius rei, sed in coelo requiritur secundum omnes sapientes; unde, sicut prius dictum est, ad omne punctum terrae incidit conus unius pyramidis virtuosae a toto coelo. Et coni isti sunt diversae in natura, et pyramides similiter, quia diversas habent bases propter diversitates horizontium, quoniam quilibet punctus terrae est centrum proprii horizontis. Et ideo oportet omnium rerum diversitatem magnam ex hac causa oriri, etiam quantumcunque propinqui sunt, ut gemelli in eodem utero; et sic de omnibus, prout videmus quod a duobus punctis terrae proximis oriuntur herbae diversae secundum speciem. Et hic sumit astronomus fundamenta sui iudicii, et merito, quia diversitas plena rerum per coelum sic invenitur (250, 29–251, 4).

[59] Quapropter potest astronomus peritus non solum in naturalibus sed in humanis rebus multa considerare de praesenti et futuro et praeterito[.] (251, 5–7).

> the *secunda coelestium,* which are renewed (*renovantur*) by means of special powers of the heavens (*virtutes speciales coelorum*),[60] as are comets and things of this sort, because a judgment about a collective (*communitas*) is easier (*facilius*) than one about an individual person (*super singulari persona*).[61]

Bacon makes an important claim here about the relative ease of making astrological judgments: general ones about a community are significantly "easier" than those about individuals. In addition to celestial bodies (*coelestia*), what we would call meteorological entities and what Roger calls *secunda coelestium*, including comets, should also now be taken into account.[62]

Next Roger explains why general interpretations are much more likely to be accurate than those on particulars:

> For the judgment on a community is a universal judgment (*iudicium universale*), and an experienced astrologer can do well with universal judgments (*astronomus potest bene in iudicia universalia*). And because an entire surrounding province is inclined towards the customs and practices of any given famous city (*qualibet civitas famosa*), then [1] due to the recourse to it [sc. the famous city] and the communication of the business (*negotia*) of life, and [2] due to the power of cities over nearby places (and likewise is the case of a more powerful kingdom in relation to surrounding kingdoms), and [3] due both to this communication and to violence, the careful astrologer (*prudens astrologus*) can usefully consider many things in this part [sc. of astrology, namely, general astrology or revolutions] with respect to customs, laws, sects, wars and peace, and these sorts of things, which pertain to the republic of cities (*res publica civitatum*), provinces and kingdoms, although he encounters great difficulty in making judgments on the actions of individual persons.[63]

Here we see the progression from a powerful city (or kingdom) towards its surroundings as the foundation for a more universal prognostication. Roger will add the ruler just below. We also see here some of the contents of universal judgments found in annual prognostications, namely, those derived from the astrological practice of revolutions of the years of the world (*revolutiones annorum mundi*) or annual revolutions. Ptolemy had already called these universal judgments in *Tetrabiblos*, Book II.

*

Bacon then turns back from universal judgments and revolutions to the births (and/or conceptions) of individuals, and thus to nativities, that is, precisely,

---

60 Bacon seems to mean God's absolute power here.

61 [E]t ideo saltem super regna et civitates potest iudicare per coelestia et secunda coelestium quae per virtutes speciales coelorum renovantur, ut sunt cometae et huiusmodi, quia facilius iudicium est super communitate quam super singulari persona (251, 7–11).

62 As we saw in Chap. 2, Albert also discussed these phenomena, but he called them *'stellae secundae'* in both *De natura locorum* and *De fato*.

63 Nam iudicium communitatis est iudicium universale, et astronomus potest bene in iudicia universalia. Et quia ad mores et consuetudines cuiuslibet civitatis famosae tota provincia circumiacens inclinatur, tum propter refugium ad eam et communicationem negotiorum vitae, tum propter potentiam civitatum super loca vicina; (et similiter est de regno potentiori respectu regnorum circumiacentium,) tum propter communicationem, tum propter violentiam; prudens astrologus potest multa considerare utiliter in hac parte super moribus et legibus et sectis et guerris et pace et huiusmodi, quae pertinent ad rempublicam civitatum, provinciarum, et regnorum, quamvis difficultatem recipiat maiorem in actibus singularium personarum iudicandis (251, 11–23).

judgments on the actions of individuals. He also explains how this relates to political analysis:

> And [1] if he [sc. the prudent and experienced astrologer] wishes to consider diligently and without error the hours [= times] of the conceptions and the nativity [sc. parturition, birth] of individual persons, in order that the lord of the celestial power (*dominium coelestis virtutis*)[64] be known for those hours; and [2] if he were to diligently consider when the celestial bodies (*coelestia*) come to these dispositions in accordance with the individual parts of any one person's age, then he can make judgments sufficiently about all natural things (*de omnibus naturalibus*), as about sickness and health and things of this sort, and when they ought to die (*occidere*) and how it [sc. a person's life] will be terminated, in accordance with which the authorities [authors, *auctores*] not only of *astronomia*, but also of medicine, as Hippocrates, Galen, Haly, Isaac,[65] and all authors determine. But few physicians (*medici* [...] *pauci*) at this time know *astronomia*, and therefore many do not understand these authors nor *can* they understand them. Therefore they neglect the better part of medicine. But about these things a long and useful discourse can be made, when there is an opportunity.[66]

In the hands of an experienced and prudent astrologer, then, a nativity can provide insight into the natural (vs. volitional) parts of an individual's life throughout its course, concerning (i.a.) sickness, health and death. Pietro d'Abano and other physicians seem to have taken up this or a similar call, as we will see in Part IV below.[67]

Bacon continues with the medical dimension of nativities in the abstract, referring it back here, as one would expect, to complexions. He also expands the focus from the natural to the volitional, and thus to the psychological dimension:

> And since illness and health, and the wills (*voluntates*), desires (*desideria*) and considerations of men are varied according to their complexions—although they are not compelled (*non cogantur*), but strongly induced (*fortiter inducantur*), as is clear [sc. from the previous argument]—then the prudent astrologer can prudently make judgments (*prudenter iudicare*) about the moral actions of an individual person (*de actibus moralibus singularis personae*), while, nevertheless, preserving freedom of the will for everyone (*salva tamen omnibus arbitrii libertate*), and he will be able in many things (*in multis*) to have a certain

[64] The lord of the geniture, as discussed in the excursus.

[65] Isaac refers to Isaac Israeli (Ishaq ibn Sulayman al-Isra'ili, d. probably ca. 932, but certainly before 955/6), the Jewish physician and philosopher, who moved from Egypt to Qayrawan in North Africa around 905/7. Constantine the African translated his medical writings in the late eleventh century (Hasse, *Success and Suppression*, 383–84). It is unclear to which Haly Roger refers here.

[66] Et si velit considerare diligenter et sine errore horas conceptionum et nativitatem singularium personarum, ut sciatur dominium coelestis virtutis ad horas illas, et diligenter consideret quando ad eas dispositiones venient coelestia secundum singulas partes aetatis cuiuslibet, potest de omnibus naturalibus, sicut de infirmitatibus et sanitate et huiusmodi iudicare sufficienter, quandocunque debent occidere et qualiter terminari, secundum quod auctores non solum astronomiae sed medicinae, ut Hippocrates, Galenus, Hali, Isaac, et omnes auctores determinant. Sed medici huius temporis pauci sciunt astronomiam et ideo nec auctores suos multi intelligunt nec possunt intelligere, et ideo negligunt meliorem partem medicinae. Sed de his longus sermo fieri potest et utilis, cum fuerit opportunum (251, 24–37).

[67] This may also end up being a good way to understand "natural" astrology in relation to human beings. Roger, of course, would call both "judgments" (*iudicia*). Nevertheless, there may well be something to be said about making a natural vs. judicial distinction here, namely, between the natural (including medical) and the volitional parts of a person's actions and/or experiences, which is often the way the distinction is construed.

judgment (*iudicium certum*) in relation to the possibility of the matter (*secundum possibilitatem materiae*) which it [sc. the judgment] treats.[68]

Given these influences on people's bodies and thereby on their wills and desires, Bacon indicates the possibility of judgments—even certain ones—on nativities in significant areas of life, both medical and psychological. As ever, preserving free will is of paramount importance; certainty of judgment has also been raised again.

Roger continues, returning to issues of contingency and necessity in relation to prediction:

> For, because this is contingent and not necessary, he cannot say that this or that will happen by necessity. Nevertheless, he can say in many things what will [sc. actually] happen (*potest tamen dicere in multis quod contingent*), and what the truth is about the future (*quod veritas est de futuro*), although it is not necessary (*licet non necessaria*). For it is one thing [sc. for a statement] to be true (*esse verum*), and another thing for it to be verified necessarily (*necessario verificari*). And this judgment is midway between the necessary and the impossible.[69]

For Bacon, then, a statement as a prediction about the future with respect to a person's actions may in fact be true or certain, while at the same time it does not *necessarily* come to pass in actuality. We will see a similar discussion later in this chapter in the *Speculum astronomiae*, and also in Chap. 5 in both Albertus Magnus's *De fato* and Thomas Aquinas's *Summa theologiae.*

Roger then contrasts this particular judgment on an individual's nativity with the easier universal judgments based on revolutions: "And in those things in which one cannot have a judgment of this sort, he will easily have a universal judgment (*iudicium universale*), or a median between the universal and the particular (252, 12–14)."[70] Where particular judgments cannot legitimately be undertaken, universal judgments must suffice. This relationship between universal and particular judgments on both individuals and communities is important for the next more political stage of Roger's analysis. As we will see, this discussion of nativities is moving towards the nativity of one person in particular—and then of the relationship of revolutions to that, both in the short term with annual revolutions, and in the longer term with great conjunctions as a part of historical astrology.

*

[68] Et cum secundum complexiones infirmitates et sanitates varientur voluntates hominum et desideria et considerationes, licet non cogantur sed fortiter inducantur, ut manifestum est, tunc prudens astronomus potest de actibus moralibus singularis personae prudenter iudicare, salva tamen omnibus arbitrii libertate, et poterit in multis habere iudicium certum secundum possibilitatem materiae quam tractat (251, 37–252, 6).

[69] Nam quia ista est contingens et non necessaria, non potest dicere quod de necessitate contingens haec vel illa, potest tamen dicere in multis quod contingent, et quod veritas est de futuro, licet non necessaria. Aliud enim est esse verum, et aliud necessario verificari. Et hoc est iudicium medium inter necessarium et impossibile (252, 6–11).

[70] Et in quibus non potest habere huiusmodi iudicium habebit de facili iudicium universale, aut medium inter universale et particulare[.]

Within this practical astrological framework of nativities and revolutions, Bacon then identifies the nativity with which he is most concerned, namely, that of the prince, by relating its significance for a judgment on the community over which he presides. In so doing, Roger also raises a major issue for himself (and one that we will encounter again), namely, the proper role of the prince's astrologically informed counselor. We will see this role continue in premodern Europe from at least the 13th throughout the seventeenth century,[71] and it still occasionally occurs in the present, even in the West, as we saw with Ronald Reagan and François Mitterand in the 1980s and 1990s.

A prince's nativity directly affects the judgment on a community:

> Nevertheless, by means of a universal judgment and with respect to what is possible concerning a public person, as a prince and his counselor, in a city or a region, [sc. the astrologer] can often have a particular judgment about the doings in a republic (*de factis rei publicae*). Because, as was said, it is easier (*facilius*) to make judgments about a community than about an individual person, and according to the judgment of a prince (*iudicium principis*), cities and kingdoms are ruled; for what pleases a prince has the power of law (*quod enim principi placet, legis habet vigorem*).[72]

In this way, astrologers can gain insight into the activities in a city by examining the leader's nativity (or those of his influential counselors) as well as the relevant revolution of the year and the nature of the place itself. This is the full astrological basis for the political predictions in annual prognostications that will be explored more fully in Part IV.

Bacon continues:

> And therefore, an astrologer has much ability [sc. to make judgments] concerning the business (*negotiis*) of famous cities, provinces and kingdoms, provided that he knows well (*sciat bene*) that lord of the heavens (*dominium coeli* [sc. the lord of the geniture]) at the conception and nativity of the prince, and that he knows well that the complexion varies through this, and how he [sc. the prince] is inclined to manners [customs, *mores*] according to the property of his complexion.[73]

In Roger's view, complexions are utterly central to understanding people, especially princes and their influence on the communities that they rule and on nearby regions.

[71] E.g. see Azzolini, *The Duke and the Stars* on fifteenth century Milan, and for Kepler in the early seventeenth century with Rudolf II, see (eg) Barbara Bauer, "Die Rolle des Hofastrologen und Hofmathematicus als fürstlicher Berater," in *Höfischer Humanismus*, A. Buck (ed.), Wienheim: VCH, 1989, 93–117. Roger's analysis here provides the natural philosophical analysis that grounds such practices. I briefly discuss Roger Bacon in this context in "An Idealized Astrological Courtier at a thirteenth-Century Papal and Royal Court: The Case of Roger Bacon," in *De Frédéric II à Rodolphe II: Astrologie, divination et magie dans les cours (XIIIe-XVIIe siècle)*, Jean-Patrice Boudet, Martine Ostorero and Agostino Paravicini Bagliani (eds), Micrologus Library 85, 2017, 53–69.

[72] [P]er iudicium tamen universale et secundum quod possibile est de persona publica, ut principe et consiliario principis in civitate vel regione potest saepius habere iudicium particulare de factis reipublicae; quia, ut dictum est, facilius est, iudicare de communitate quam de singulari persona, et secundum iudicium principis regulantur civitates et regna; quod enim principi placet, legis habet vigorem (252, 14–20).

[73] Et ideo super negotiis civitatum famosarum et provinciarum et regnorum potest multum astronomus, dummodo sciat bene dominium coeli in conceptione et nativitate principis, et bene sciat quod complexio variatur per hoc, et qualiter ad mores inclinatur iuxta proprietatem complexionis (252, 21–25).

Roger thus indicates how important knowledge of the ruler's nativity could be and the real world value of such knowledge. This also helps to explain the later vogue in the sixteenth and seventeenth centuries for geniture collections, as we will see in volume III.[74]

On this basis, Bacon deepens his analysis by discussing how celestial configurations influence minds by altering and improving, or, alternatively, by negatively affecting complexions. He is still primarily concerned with the relevant prince or king, and the influence of current celestial configurations (i.e. revolutions) on his nativity. Here Roger implicitly relies on the distinction discussed earlier between root and current complexions:

> And then he [sc. the experienced and prudent astrologer] notes when a celestial configuration (*constellatio*) will arrive that stimulates and strengthens the complexion [sc. of the prince] (*quae stimulet complexionem et fortificet*), so that his mind (*animus*) is stimulated and excited to similar acts (*ut stimuletur et excitetur animus ad actus consimiles*). And likewise, when the complexion is transformed (*mutatur*) into its contrary by means of the contrary disposition of the heavens (*quando in contrarium mutatur complexio per contrarium coeli dispositionem*), so that the mind of the prince is excited toward contrary things (*ut ad contraria excitetur animus principis*).[75]

Bacon emphasizes here how significant current celestial configurations can be for stimulating root complexions, and how that affects the mind of the ruler and his subsequent actions. The experienced astrologer can thus understand the prince's nature and potential actions from his nativity (his root complexion) in relation to the current and future states of the heavens. Ultimately, he can work to strengthen or improve the prince's basic nature in relation to current celestial circumstances, as we will see. We may rest assured that Bacon will soon argue that this in no way implies necessity or compromises free will.

Bacon continues developing the picture, in particular, of how the prince's nature affects the entire community:

> For, according to [1] such variations of (a) celestial configurations and (b) complexions (*variationes constellationum et complexionum*), and [2] the will of princes and prelates (*voluntatem principum et praelatorum*), arise innovations of habits, and transformations (*mutationes*) of laws and customs among the people by a higher authority (*auctoritate superiori* [sc. divine providence working through the heavens?]).[76] Then sometimes discords and dissensions arise easily, upon which wars follow, or sometimes, because of the honesty and utility of laws, the concord of citizens and of others arise, and there is peace.[77]

---

[74] For now, see Grafton, *Cardano's Cosmos*, Chap. 4, "The Astrologer," 56–70, especially 64 ff, and his "Geniture Collections: Origins and Use of a Genre," in *Books and the Sciences in History*, M. Frasca-Spada and N. Jardine (eds.), Cambridge: Cambridge University Press, 2000, 49–68.

[75] Et tunc notet quando constellatio veniet quae stimulet complexionem et fortificet ut stimuletur et excitetur animus ad actus consimiles; et similiter quando in contrarium mutatur complexio per contrarium coeli dispositionem, ut ad contraria excitetur animus principis (252, 25–30).

[76] Thomas Aquinas offers an analysis of how this would work in the closely contemporary *Summa contra gentiles* III, as we will see in Chap. 5.

[77] Secundum enim huiusmodi variationes constellationum et complexionum et voluntatem principum et praelatorum oriuntur in populo auctoritate superiori innovationes consuetudinem et mutationes legum et morum. Deinde de facili oriuntur aliquando discordiae et dissensiones ad quas sequuntur bella, vel aliquando propter legum honestatem et utilitatem oritur concordia civium et aliorum, et fit pax (252, 30–37).

Such great political matters as war and peace (and their predictions) thus arise from current celestial configurations in relation to a ruler's root complexion, and his consequent desires and actions. Bacon adds prelates here to his previous analysis of princes. So much depends on the prince's nature. One senses a resonance here with Machiavelli's not unrelated analysis in *Il Principe* that will be composed two and a half centuries later; Machiavelli also offered an astrological analysis of war and peace.[78] One can also easily see why a ruler with a strong horoscope would want to publicize it (like Augustus Caesar, Charles V and Cosimo I de' Medici did), and one with a weak one (like Rudolf II) would downplay it.[79]

Bacon then refines his analysis by relating the celestial situation to that on the ground, and its implications for astrological analysis and thus foreknowledge:

> And therefore, an experienced astrologer (*astronomus peritus*) can easily make judgments (*potest de facili iudicare*) about this sort of common business of cities and regions because not only does he have [sc. information] by means of the pathways proper to them whence he [sc. the astrologer] proceeds [that is, from the heavens], but also by the conditions of the people ruled (*principantur*). And likewise, through the properties of the people who assist princes and prelates [sc. courtiers, counselors, etc] (*per proprietates earum personarum quae principibus et praelatis assistunt*), whose counsels they rely on, because bad princes are led to the goods of a republic by good counselors (*principes mali ducuntur ad bona reipublicae per bonos consiliarios*), and good princes are undermined by bad counselors (*principes boni pervertuntur per malos consiliarios*).[80]

Here Bacon adds the important factors of the city's populace (with its implied political factions, which Machiavelli strikingly called '*umori*' [= humors in the body politic] in *The Prince*) and the prince's counselors to the celestial situation and the prince's own constitution in order for the astrologer to make useful and accurate judgments based on all the relevant factors, astrological and otherwise. The stakes are very high indeed! Roger also emphasizes here the fundamental importance of the prince's political counselors. We will see why just below.

*

Bacon then returns to the fundamental importance of the prince's nature and its astrological discovery, relating it now even more fully to his counselors' natures. In this way, Bacon instantiates his earlier more general statement:

---

[78] I discuss this in detail in volume III as I critically engage with Anthony Parel's useful but problematic interpretation in *The Machiavellian Cosmos*, New Haven: Yale University Press, 1992.

[79] For the three stronger examples, see my "Astrology, Politics and Power in Sixteenth-Century Florence: Giuliano Ristori's Extensive Judgment on Cosimo I's Nativity (1537)," in *Astrologers and their Clients in Medieval and Early Modern Europe*, Wiebke Deimann and David Juste (eds.), Cologne: Böhlau Verlag, 2015, 139–150. For Rudolf's (and Cosimo II's), see my "Celestial Offerings: Astrological Motifs in the Dedicatory Letters of Galileo's *Sidereus Nuncius* and Kepler's *Astronomia Nova*," in *Secrets of Nature*, 133–72.

[80] Et ideo astronomus peritus potest de facili iudicare de huiusmodi negotiis communibus civitatum et regionum, quia non solum per vias proprias eis habeat unde procedat, sed per conditiones personarum, quae principiantur. Et similiter per proprietates earum personarum quae principibus et praelatis assistunt, et quorum consiliis innituntur, quia principes mali ducuntur ad bona reipublicae per bonos consiliarios, et principes boni pervertuntur per malos consiliarios (252, 37–253, 6).

> If, therefore, from [1] the celestial configuration of the nativity and conception (*ex constellatione nativitatis et conceptionis*), the complexion of any prince (*complexio alicuius principis*), or of another person on whom he relies, is discovered to be disposed towards a perversity of morals and discords and wars, and [2] the astrologer (*astronomus*)—at the same time, with this he would see that they [sc. the configurations and complexions] would drag him [sc. the prince] into such a habit (*consuetudo*), and they are excited strongly when a similar configuration occurs (*fortius excitantur quando constellatio consimilis accidit*)—can reasonably make a judgment (*potest rationabiliter iudicare*) about a city's and a kingdom's misfortunes in which they preside, when the disposition of the heavens and of those things that are renewed by the celestial [sc. bodies] as comets and phenomena of this sort, happen accordingly (*quando dispositio coeli* [...] *conformiter accidunt*).[81]

Thus the nativity of the individual prince (with his root complexion) is to be compared with the celestial configuration (with its current qualitative nature) at a particular time in a revolution horoscope. With this information, a skilled—and hopefully also wise and prudent—astrologer can make accurate, but not infallible judgments about the broader city or realm. Given the importance of political counsel for good or ill, it seems likely that Bacon here is trying to fashion a role for himself as the pope's trusted counselor and advisor, or at least as the skilled astrologer who can offer insider access to such information, the value of which is becoming increasingly clear.

Bacon now heads toward his considered conclusion:

> And the best judgments (*optima iudicia*) can be made, in accordance with the opposite condition of princes (*secundum oppositas conditiones principum*) and of those in whom they trust for everything (*eorum quibus per omnia credunt*), when likewise a similar celestial disposition (*coelestis dispositio conformis*) is found. And since such persons in one kingdom are few, and they are public and known to all, whose customs brighten (*relucent*) the entire kingdom, an astrologer can have much certain knowledge about them (*multum potest astronomus certificare de eis*), and usefully make judgments about public deeds through the properties of such persons (*utiliter iudicare de factis publicis per proprietates talium personarum*).[82]

The astrologer and his knowledge are thus held to be of the greatest import and utility for gaining profound insight into both public and private life.

*

Bacon completes this section by relating it back to and thus concluding the larger argument being addressed here, namely, defending mathematics' value in relation to its dubious associations by some with astrological judgments (*iudicia astronomiae*):

---

[81] Si igitur ex constellatione nativitatis et conceptionis complexio alicuius principis, vel alterius cui innititur, inveniatur disposita ad perversitatem morum et discordias et guerras, et astronomus simul cum hoc videat quod trahunt huiusmodi in consuetudinem et fortius excitantur quando constellatio consimilis accidit, potest rationabiliter iudicare de infortuniis civitatis et regni quibus praesunt, quando dispositio coeli et eorum quae per coelestia renovantur, ut cometae et huiusmodi, conformiter accidunt (253, 6–15).

[82] Et optima iudicia fieri possunt, secundum oppositas conditiones principum et eorum quibus per omnia credunt, quando similiter coelestis dispositio conformis invenitur. Et quoniam tales personae in uno regno paucae sunt, et sunt publicae et notae omnibus, quorum mores relucent toti regno, multum potest astronomus certificare de eis, et utiliter iudicare de factis publicis per proprietates talium personarum (253, 15–22).

> These things, therefore, are what I have wished to tell (*recitare*) in order to remove mathematics' bad reputation concerning judgments of this sort (*pro infamia mathematicae tollenda in huiusmodi iudiciis*), from which it is clear to every wise person (*sapiens*) that true mathematics (*vera mathematica*) in this part [sc. judgments] should not be criticized adversely (*vituperanda*), but altogether embraced and loved (*diligenda*) because of the glorious usefulness which can result from the judgments of true mathematics (*propter gloriosas utilitates quae possunt evenire ex iudiciis mathematicae verae*), which is not contradicted by any truth (*quae in nullo veritati contradicit*).[83]

Thus Roger completes this section by utilizing a splendid rhetorical strategy. For what reader, especially a pope, would deny their own wisdom? True mathematics has thus been defended in the central and controversial realm of both individual and collective human affairs, where concerns about the implications of necessity in nature and infallible astrological judgments have both been successfully addressed. In the process, Roger has given us valuable insight into his views on the roles of nativities and revolutions with both medical and political applications, and thus of astrology's profound utility in human affairs.

4.7 The *Speculum astronomiae*'s Defense of Astrology

The *Speculum astronomiae* also offers a defense of astrology in its Chaps. 12–17 that is rather differently structured than Roger Bacon's, as it precisely follows the structure of the four canonical types of practical astrology presented in Chaps. 4–11. As with Roger in the material just explored, the Magister Speculi is primarily concerned in these chapters—especially 12–15—to defend astrology from charges that it undermines human free will by promoting a deterministic world-view. He also explicitly discusses astrology in relation to divine providence.[84] The defense in Chaps. 16 and 17 concerns astrological images or talismans, and will therefore be discussed fully in Chap. 7 below. This material is worth exploring in detail.

Chapter 12 begins with a general introduction to the defense:

> Moreover, since the occasion (*occasio*)[85] has been provided by them, as was said, many of the aforementioned books stand accused, even though some perhaps are innocent. Even though their accusors are our friends (*amici nostri*), we must, nevertheless, honor the truth (*veritatem* [...] *honorare*), as the Philosopher says. I swear, however, that if I say anything that I wish to use in defense of these books (*in defensione eorum*), I do not speak as in a determination (*determinando non dico*), but instead I speak in opposition (*opponendo*),

[83] Haec igitur sunt quae volui recitare pro infamia mathematicae tollenda in huiusmodi iudiciis, ex quibus patet omni sapienti quod non sit vera mathematica in hac parte vituperanda, sed omnino amplectanda et diligenda propter gloriosas utilitates quae possunt evenire ex iudiciis mathematicae verae, quae in nullo veritati contradicit (253, 22–27).

[84] In both respects, then, we will notice many points of contact with the issues just discussed by Roger Bacon, as well as with those discussed by Thomas Aquinas in his analysis of God's providence in Chap. 5, even though Thomas is not explicitly concerned to defend astrology there.

[85] Thomas Aquinas also discusses '*occasiones*' in relation to astrology, as we will see in Chap. 5.

offering exceptions (*excipiendo*), so as to provoke the mind of those reaching a decision to pay careful attention [sc. to the criteria and evidence they are using] for their determination (*ad determinationis animadversionem determinatoris ingenium provocando*, 251).[86]

After these introductory remarks that reveal current and/or recent attacks on astrology—just as Roger's had done with the contemporary ignorant theologians—the anonymous author begins his defense with general astrology or revolutions.

Given that Roger's defense of *iudicia astronomiae* was written in 1266–67 and the *Speculum astronomiae* was most likely composed during the 1260s and certainly by 1271–72—and that Roger's was certainly written in Paris and the *Speculum astronomiae* possibly was, and both were written in theological contexts—this raises the intriguing possibility that they both came to astrology's defense in their different but related ways by responding to the very same (or at least similar) attacks. Even if this question cannot be definitively answered here, it is clear that astrology and its legitimacy was a live issue in Paris at this time.[87]

4.8 Revolutions (or General Astrology)

The Magister Speculi offers his defense of revolutions in Chap. 12 and at the beginning of Chap. 13. We will recall from the introductory excursus that general astrology or revolutions was divided into three parts in Chap. 7: [1] the 120 conjunctions of the planets and their eclipses; [2] the revolutions of the years of the world (*de revolutione annorum mundi*) or annual revolutions; and [3] the changes of times or seasons (*de mutatione temporum*), primarily concerning the weather. He treats the third in some detail in Chap. 12, and the first two much more briefly to begin Chap. 13.

[86] "Quoniam autem occasione eorum, ut dictum est, multi libri praenominati, et fortassis innoxii accusantur, licet accusatores eorum amici nostri sint, veritatem tamen oportet, sicut inquit Philosophus, nihilominus honorare, protestor tamen quod si aliquid dicam quo velim uti in defensione eorum, quoniam determinando non dico, sed potius opponendo vel excipiendo et ad determinationis animadversionem determinatoris ingenium provocando (XII, 2–8)." I use the translation in Zambelli's edition as the basis for mine. Zambelli publishes excerpts from some of the sources for these positions, primarily one extended quotation from Albumasar's *Introductorium maius*; Speculum astronomiae, 297–304.

[87] Jeremiah Hackett treats related issues very differently (but with some overlap) in his "Aristotle, *Astrologia*, and Controversy at the University of Paris (1266–1274)."

4.8.1 Revolutions (1): The Weather

The Magister Speculi begins his defense of astrology with revolutions, and particularly that part concerned with predicting the weather:

> Therefore, I would like to return first to that part of revolutions which deals with the change of seasons (*de temporum mutatione*), whose necessity (*necessitas*) is clear from what was said above, that is, from the obedience of the motion of lower things to the motion of higher ones (*ex oboedientia motus inferiorum ad motum superiorum*).[88] The necessity of this has nothing by which it may be impeded, since it is not subject to free will (*neque libero arbitrio sit subiecta*), but only to the will of its Creator (*sed soli voluntati sui conditoris*), who foresaw it thus from the beginning (*ab initio providit*). And it can be averted by Him alone, since in Him alone is there the plenitude of power (*plenitudo potestatis*). Nevertheless, He does not wish to avert it (for His will is not changeable [*non est* (…) *consilium mutabile*], as is that of children or servants), but He wants this to endure until the end imposed on it by Him alone (as Ptolemy and Albumasar proclaim)—and it is known by Him alone, namely, when motion (*motus*) will stop, as it began, by His command (*ex ipsius praecepto*, 251)[.][89]

The weather may thus directly follow (obey!) natural necessity as created and foreseen by God. It cannot be impeded in any way, since it is not subject to human free will. Rather, it (as everything else) was provided for by God from the beginning, and only He can change it, although He is powerfully disinclined to do so. We should note the use of the term '*providit*', the active verbal form alluding to God's providential creation and governance of the world. The rhetoric here strikes me as much more theological than natural philosophical in tone, which would also resonate with the emphasis in the Magister Speculi's self-description as '*quidam vir zelator fidei et philosophiae*' (208, 10) on initially introducing himself to the reader.

Having mentioned God's creation and ultimate termination of the world, the Magister Speculi turns to criticize Aristotle, who had erred on this central cosmological issue:

> In this alone we find that useful Aristotle (*ille utilis Aristoteles*) has erred. Nevertheless, he is to be thanked for a million other [sc. ideas]. And we already know that there is no cause in the [sc. zodiacal] circle (*causa in circulo*) which is not wisely disposed by the will of God (*quae non sit sapienter disposita nutu Dei*). Therefore, since a wise person (*sapiens*) should not regret what the Wisest One has provided (*quod providit sapientissimus*), it is not in this person's hands to avert or change it. But if this part of the science of the judgments of the stars (*scientia iudiciorum astrorum*) concerning the change of the seasons should stand

[88] This is very similar to Thomas's discussion in *Summa contra gentiles* III.82, as we will see. The Magister Speculi also talked about the obedience of a *motus inferior* to the celestial motions in *Speculum astronomiae*, Chap. 3, as we saw in Chap. 1.

[89] Primum itaque volo reverti ad partem illam revolutionum, quae est de temporum mutatione, cuius necessitas ex praedictis apparet, videlicet ex oboedientia motus inferiorum ad motum superiorum, nec habet unde impediatur eius necessitas, cum neque libero arbitrio sit subiecta, sed soli voluntati sui conditoris, qui ab initio providit sic, et ab ipso solo averti potest, ut apud quem solum plenitudo potestatis habetur, cum tamen nolit avertere, non est enim eius consilium mutabile sicut unius ex pueris aut ancillis, sed vult illud durare usque ad terminum ab ipso ei impositum, sicut clamant Ptolemaeus et Albumasar, et ab ipso solo notum, quando scilicet ex praecepto suo stabit motus, sicut et coepit ex ipsius praecepto (12.9–19)[.]

> (*stare*), should not that part concerning its principles (*principia*) remain as well? Especially since, if that were destroyed, all the other parts [sc. of astrology] would be destroyed as well (251, 253).[90]

After defending revolutions concerning the weather by enthusiastically emphasizing the wisest God's creation of the world which He oversaw with his divine providence, the Magister Speculi expands his purview: If we are going to preserve revolutions concerning the weather, then we also need to preserve their principles as well, namely, astrology's basic doctrines, without which revolutions and the other practices would no longer continue to exist.

*

In an apparent nonsequitur, this defense of astrology's *principia* immediately inspires the Magister Speculi to defend Albumasar, who had written extensively about these principles.[91] This seems to imply that astrology's attackers specifically attacked Albumasar. He begins by isolating some of Albumasar's apparently problematic views:

> But also what can be found in it that is contrary to sound truth (*contrarium veritati sanae*)? What appears to be more worthy of reprehension in the writings of Albumasar, who discusses this [sc. astrology's *principia*], is: [1] that he says in treatise I, difference 5 in the chapter on the third sect, namely, that the planets are animated by a rational soul (*quod planetae sunt animati anima rationali*). But what he says, he seems to say as if he were quoting (*recitando*), since he says that Aristotle had said it (even though it is not found in any of Aristotle's books that we have—but perhaps it is in the 12th or 13th book of the *Metaphysics* which have not yet been translated and which discuss the intelligences [*de intelligentiis*], as he himself promises). But, that Albumasar does not agree with this opinion about the rationality of the planets (*de rationalitate planetarum*) is clear from what he says in the same chapter: "Even if the planets had rational souls (*animae rationales*), nevertheless, they do not choose (*eligunt*), nor do they need choice (*indigent electione*) due to their distance from impediments." And it is more clear from what he says in the same treatise in the third difference, where near the end he writes: "The entrance of the Sun into this degree [sc. of the zodiac] did not happen in accordance with a choice (*electio*) by the Sun itself—nor is the effect of these things, nor their corruption (*neque effectus ipsarum rerum et earum corruptio*)—but through its arrival to this degree [sc. of the zodiac] by its natural motion (*per motum naturalem*)." [2] And further on, he proves that the motion of a circle exists from the power (*virtus*) of the immobile and eternal first cause (*prima causa*). Whence he also says: "One must bless His name, and exalt Him, not the Sun." Therefore, what is the merit of his book? If you study it carefully, you will find many good things, but no bad ones that I know of (253).[92]

---

[90] [I]n quo solo ille utilis Aristoteles invenitur errasse (nihilominus regratiandus in mille milium aliorum). Et iam scimus, quod non est causa in circulo quae non sit sapienter disposita nutu Dei; cum ergo sapiens non poenitteat quod providit sapientissimus, non est eius avertere seu mutare. Quod si pars ista scientiae iudiciorum astrorum, quae scilicet est de mutatione temporum, debet stare, numquid et stare partem illam quae est de principiis oportebit? Cum praesertim ipsa destructa, omnes aliae destruantur (12.19–27)[.]

[91] See Lemay, *Abu Ma'shar and Latin Aristotelianism*, and his critical edition of Albumasar's *Liber introductorii maioris ad scientiam judiciorum astrorum*.

[92] Sed et in ipsa quid invenitur contrarium veritati sanae? Quod apud Albumasar, qui de ea tractat, plenissime reprehensione dignius invenitur, est illud quod dicit in tractatu primo sui libri, differentia

Even if planets have rational souls, they still do not make choices that affect or change their motions, which means that their motions are not due to choice, but are simply their natural motions.[93] Further, according to Albumasar, these natural motions also lead us beyond worshipping the planets as gods (and thus of idolatry) to praise the first cause, namely, God Himself.

The Magister Speculi continues to defend Albumasar and thereby astrology's most fundamental principles as articulated by him:

> [3] You will also find in treatise I, difference 2, in the chapter on the suitability of the times when the motion will cease, where he says: "The planets are not subject to corruption (*non corrumpuntur*), nor do they receive increase (*augmentum*) or decrease (*diminutio*) [sc. in size], nor effect nor detriment until that time which God has wished (*quod Deus voluerit*)."[94]
> [4] And in the same treatise, in difference 5, in the chapter on the first sect [sc. he says]: "From the natural and enduring motions of the planets, a natural and enduring effect is realized (*efficitur*), which will exist until the time that God has wished (*quod Deus voluerit*)." Ptolemy gives a similar testimony in book 1, chapter 1 of the *Almagest*, where it is written: "Moreover, we have laboured so that, in our love for the knowledge of eternal things that remain until the end which their Creator (*conditor*) has imposed on them, in the following chapters of this book we should add, etc (253)."[95]

Thus, for both Albumasar and Ptolemy, God alone determines how long this system will endure.

After defending against potentially negative interpretations, the Magister Speculi now praises the positive in Albumasar:

> You will also find in Albumasar something far more elegant (*longe elegantius*), namely, a testament of faith and eternal life, which is not acquired save by faith. This appears in the sixth treatise, differentia 26, where he assigns the cause (*causa*) as to why the ninth house

quinta, capitulo de secta tertia, scilicet quod planetae sunt animati anima rationali; sed quod dicit, dicere recitando videtur, cum dicat Aristotelem hoc dixisse, licet non inveniatur in universis libris Aristotelis quos habemus, et forte illud est in duodecimo aut in decimo tertio Metaphysicae, qui nondum sunt translati et loquuntur de intelligentiis, sicut ipse promittit. Quod autem sententiae de rationalitate planetarum non consentiat Albumasar, apparet ex hoc quod in eodem capitulo dicit: "Planetis etsi sint animae rationales, non eligunt tamen, nec indigent electione propter longitudinem eorum ab impedimentis." Et apertius ex hoc quod dicit in eodem tractatu, differentia tertia, ubi habetur versus finem: "Et non secundum electionem eiusdem solis fuit introitus eius in hanc partem, neque effectus ipsarum rerum et earum corruptio, sed per adventum eius per motum naturalem in ipsam partem." Et infra probat quod motus circuli est a virtute primae causae immobilis et aeternae. Unde et dicit: "Benedicendum est nomini eius, et ipsum exaltare, non Solem." Quid ergo meruit liber suus? Quem si revolveris, invenies multa bona, mala autem nulla, quod sciam (12.27–48).

[93] I discuss the animation of the heavens later in volume I, and much moreso in volume II.

[94] God thus has the relevant will and choice, not the planets.

[95] Illic invenies tractatu primo, differentia secunda, capitulo de aptatione temporum, quoniam stabit motus, ubi dicit: "Planetae non corrumpuntur, neque recipiunt augmentum, neque diminutionem, neque effectum, neque detrimentum usque ad tempus quod Deus voluerit." Et in eodem differentia quinta, capitulo de prima secta, "Efficitur ex motibus planetarum naturalibus atque durabilibus effectus naturalis et durabilis, qui fit usque ad tempus quod Deus voluerit." Quod similiter testatur Ptolemaeus in *Almagesti* capitulo primo dictionis primae, ubi habetur: "Nos autem laboravimus, ut in amore scientiae sempiternorum manentium usque ad terminum, quem eorum conditor eis imposuit, in sequentibus huius nostri libri addamus, etc. (12.48–59)."

> is the house of faith (*domus fidei*), and he says: "The ninth house is also called the house of pilgrimage and the movement of faith and of good works, on account of its reversion to Jupiter, etc (253, 255)."[96]

As we will also see in Roger Bacon's discussion of astrology and theology/religion in Chap. 6, the ninth house rules both faith and religion.

To develop this idea further, he contrasts the domains in which the two great benefics—Jupiter and Venus—act:

> And further on: "Again, Jupiter and Venus are the fortunes (*fortunae*). But there are two kinds (*species*) of fortune, of which one is the fortune of this world (*fortuna huius mundi*) and the other is the fortune of the future age (*fortuna futuri saeculi*). The fortune of the future age is more worthy than the fortune of this world; and this is sought by faith. And because Jupiter is a greater fortune (*plus fortuna*) than Venus, therefore that signification about faith, by means of which the fortune of the future age is sought, was made for him [sc. Jupiter], which is more worthy. Signification of the fortunes of this world, such as those from games (*ex ludis*), joy (*gaudio*) and happiness (*laetitia*), was made for Venus." What, then, does his [sc. Albumasar's] book deserve in these matters (255)?[97]

Jupiter, which rules the ninth house, thereby rules both the religious life and the greater fortune of the afterlife. Venus, on the other hand, rules earthly fortunes and pleasures. Albumasar here heartily supports this view.

Now the Magister Speculi turns to one of Christianity's greatest mysteries, the Virgin birth:

> But also, what is its value if it was written in it that the birth of Jesus Christ from the Virgin (*nativitatem Iesu Christi de Virgine*), as well as the utterance of the Name announced by the angel, was figured in the heavens from the beginning (*ab initio figuratam esse in caelo*)? For, in treatise six, differentia 1, in the section on the risings of the images (*imagines*) that ascend with Virgo, this is found: "And in its first face (= decan [sc. of Virgo]), there arises a girl whom he calls Celchuis Darostal. She is a beautiful, honorable and pure virgin with long hair and a beautiful face, holding two spikes [sc. of wheat] in her hand. She sits on a covered bench, and nourishes a boy, giving broth to him in a place which is called Abrie. And a certain people (*quaedam gens*) call this child Jesus, which is translated as 'Eice' in Arabic. And rising there with her is a man sitting on the same seat. And there rises with her the star of the virgin […] etc." Also we know that under the ascendent of this same part of heaven, that is of Virgo, the Lord Jesus Christ was born, and also that the equation of the motion of the eighth sphere at the time of His birth was 8° 37' and 2" according to a most certain calculation (*secundum calculationem certissimam*), and that it was then subtracted from the positions of the planets found by means of canons [sc. of astronomical tables].[98]

---

[96] Invenies quoque apud Albumasar longe elegantius scilicet testimonium fidei et vitae aeternae, quae non acquiritur nisi per fidem, tractatu sexto, differentia vicesimasexta, ubi assignat causam, quare nona domus est domus fidei, et dicit: "Domus quoque nona vocata est domus peregrinationis et motionis fidei quoque atque operum bonorum, propter reversionem eius ad Iovem etc. (12.60–66)."

[97] Et infra: "Rursum Iuppiter et Venus sunt fortunae. Fortunae autem duae sunt species, quarum una est fortuna huius mundi et altera fortuna futuri saeculi, et fortuna futuri saeculi dignior est fortuna huius mundi, et hoc quaeritur per fidem; et quia Iuppiter est plus fortuna quam Venus, ideo facta est ei significatio super fidem per quam quaeritur fortuna futuri saeculi, quae est dignior, et facta est Veneri significatio super fortunas huius mundi, ex ludis et gaudio et laetitia. Quid ergo in his meruit liber suus? (12.66–74)"

[98] Sed et quid meruit, si scriptum est in eo ab initio figuratam esse in caelo nativitatem Iesu Christi de Virgine, etiam cum expressione nominis ab angelo nuntiati? In tractatu namque sexto, differen-

Here the Magister Speculi describes Albumasar's discussion of Christ's birth in his volume on Great Conjunctions.[99]

Then he describes how this works, making sure to clarify that religious matters are not subjected to the heavens. In doing so, he again invokes God's providence:

> Not because the most desired of those born (*natorum desideratissimus*) who had created these stars (*qui creaverat ipsas stellas*) was subject to the motion of the stars or to their judgment (*non quia subiaceret stellarum motui aut earum iudicio*), but because when He spread out the heavens, just like vellum (*extenderet caelum sicut pellem*), to form the book of the universe (*formans librum universitatis*), He refused to make the work incomplete. He did not wish there to be missing from its letters, written according to His providence (*secundum providentiam suam*) in the book of eternity, even what was furthest removed from nature—that He would be born from a Virgin—in order that by this means He might be recognized as a natural and true human being, who was not born in the natural manner. Not because the figure of heaven (*caeli figura*) was the cause of His birth, but rather because it was a sign (*significatio*). Or rather, as is truer than truth, He Himself was the cause by which the manner of His wondrous birth was signified by the heavens (*ipse erat causa quare modus admirandae suae nativitatis significaretur per caelum*, 255, 257).[100]

There is much that is worthy of note in this rich passage defending Albumasar by proclaiming how Jesus's birth was written in the book of the heavens by God's providence, without at the same time being subject to the stars, namely, as a sign, not a cause. Albumasar's virtues for promoting the Christian faith have been fully emphasized here. The theological inflection of his defense is pronounced.

Now the Magister Speculi completes his defense of Albumasar:

> You could discover this and many other notable things if you went through the book carefully. But if, perhaps, something suspect (*aliquod suspectum*) happens to be found in addition to this, which I do not remember seeing, it should be corrected (*corrigatur*) rather than be indiscriminately condemned (*indiscrete damnetur*) because many useful things happen to be bound with one erroneous statement. Again, what has the book by Abdilaziz deserved, whom he calls Alcabitius, which was similarly included amongst the iniquitous books? If

---

tia prirna, in capitulo de ascensionibus imaginum quae ascendunt cum Virgine, invenitur: "Et ascendit in prima facie illius (scilicet Virginis) puella quam vocat Celchuis Darostal; et est virgo pulchra atque honesta et munda prolixi capilli, et pulchra facie, habens in manu sua duas spicas, et ipsa sedet super sedem stratam, et nutrit puerum, dans ei ad comedendum ius in loco qui vocatur Abrie. Et vocat ipsum puerum quaedam gens Iesum, cuius interpretatio est arabice Eice. Et ascendit cum ea vir sedens super ipsam sedem. Et ascendit cum ea stella virginis etc." Etiam scimus quod sub ascendente eiusdem partis caeli, scilicet Virginis, natus fuit Dominus Iesus Christus, cum hoc quod aequatio motus octavae sphaerae in tempore nativitatis eiusdem fuit octo graduum et triginta septem minutorum et duorum secundorum secundum calculationem certissimam, et quod ipsa tunc erat minuenda de locis planetarum inventis per canones (12.74–91)[.]

[99] We will see Roger Bacon treat this material somewhat differently and in richer detail in my Chap. 6. Pico attacks this analysis in Book V, Chap. 14 of the *Disputations*. Ornella Pompeo Faracovi treats all of these discussions in her *Gli oroscopi di Cristo*, Venice: Marsilio, 1999.

[100] [N]on quia subiaceret stellarum motui aut earum iudicio natorum desideratissimus, qui creaverat ipsas stellas, sed quia cum extenderet caelum sicut pellem, formans librum universitatis, et dedignaretur opus facere incompletum, noluit litteris eius deesse, ex eis quae secundum providentiam suam in libro aeternitatis sunt scripta, etiam illud elongatissimum a natura quod de Virgine nasceretur, ut profecto per hoc innueretur homo naturalis et verus, qui non naturaliter nascebatur, non quod caeli figura esset causa quare nasceretur, sed potius significatio, immo ad vero verius, ipse erat causa quare modus admirandae suae nativitatis significaretur per caelum (12.92–102).

there are names in an unknown language in his text, their meanings (*interpretationes*) are immediately added to the text itself; but if perhaps the meanings of some of these words should be missing, there is a man prepared to supply them (257).[101]

To complete his defense of Albumasar, the Magister Speculi states that even if one part is faulty, one need not scrap everything else therein that is worthy and thereby throw out the baby with the bathwater. After a much more extensive defense of Albumasar, he then briefly defends Alcabitius, the author of an extremely popular introductory textbook, and thus one also concerned with *principia*.

*

In Chap. 13, the Magister Speculi returns to defending revolutions. This time he focuses briefly on the other two types: [1] the revolutions of the years of the world or annual revolutions, the basis for annual astrological prognostications, and [2] the range of conjunctions and eclipses:

> One can say the same things about the parts of revolutions that concern the revolution of the years of the world, and the conjunctions and eclipses of the planets as about that which concerns the change of seasons. For if an earthquake, or flood or fires—as far as rich men and all the common people are concerned—war or peace, famine or death, and further, the appearance of some great prophet or heretic, or the rise of some horrible universal or particular schism is signified by the figure of the revolution of the year, or an eclipse, or a conjunction (which signifies a [sc. religious] sect), in accordance with which God, the most high, has provided [sc. by means of his providence] (*secundum quod providit Deus altissimus*), what has that to do with free will (*quid ad arbitrium liberum*)? Is it in a person's power (*in potestate hominis*) to change (*immutare*) such things? It seems that these parts ought to be preserved (*stare*) as well, and that they do not demand reprehension, unless something else stands in their way (*aliud obstet*) that I have not yet heard proposed against them (257).[102]

The Magister Speculi here defends annual revolutions along with eclipses and conjunctions, great and small, as part of God's providential plan for the world. What does any of this have to do with free will, he repeatedly asks? People are still free to make their own choices within these larger natural, political and societal contexts,

[101] Haec et multa alia notabilia poteris reperire, si diligenter revolveris librum illum. Quod si forte cum his aliquod suspectum, quod non memini me vidisse, invenire contingat, corrigatur, potius quam, multa utilia cum uno relegando, indiscrete damnetur. Quid iterum meruit liber Abdilaziz, quem vocat Alkabitium, qui similiter cum iniquis deputatus est? Si sunt in textu eius nomina ignotae linguae, statim subduntur in littera interpretationes eorum; quod si forte aliquorum interpretationes defuerint, paratus est vir earum copiam exhibere (12.102–110).

[102] De partibus autem revolutionum, quae sunt de revolutione annorum mundi et de coniunctionibus et eclipsibus planetarum, sicut de illa quae est de mutatione temporum, potest dici. Si enim ex figura revolutionis anni, aut eclipsis, aut coniunctionis, quae significat sectam, significatur terraemotus sive diluvium aut scintillae aut super divites et universitatem vulgi guerra vel pax, fames sive mortalitas, caeterum apparitio alicuius magni prophetae sive haeretici, aut ortus horrendi schismatis universalis vel particularis secundum quod providit deus altissimus, quid ad arbitrium liberum? Numquid est in potestate hominis talia immutare? Apparet quod et istae partes stare meruerint, neque reprehensione indigeant, nisi aliud obstet quod nondum audivi fuisse propositum contra eas (13.2–13).

as well as regarding the weather. None of this impinges upon free will, the Magister Speculi's overriding concern here. The theologico-moral emphases of the defense are, once again, front and center.

4.9 Nativities

For the rest of Chap. 13 (14–60) the Magister Speculi discusses nativities, which he indicates are the most problematic astrological practice in relation to free will. He begins by discussing some of the major themes related to both bodies and souls, beginning with the simpler situation of the former:

> I turn now to nativities, which seems to be the part [sc. of astrology] that pricks free will (*pungere liberum arbitrium*) more sharply (*acrius*) than the others, so much so that it even seems that they might destroy each other (*invicem se destruant*). Nor do they seem compatible in any respect, especially that part which pertains to the character of the soul (*ad mores animi*, 257, 259).[103]

In this dramatic configuration, nativities and free will are constructed as locked in mutually mortal combat. They are diametrically opposed and thus fundamentally incompatible. Can the Magister Speculi save nativities in these circumstances as he has just done with general astrology or revolutions, and the principles on which they are based?

After this stark framing, he proceeds to explore the issue in greater depth. He does so first in relation to the body—realistically the easier case to resolve—beginning with the length of life and its prediction, always a controversial issue:

> For, knowing the length of the native's life by the degree of the hylech and the planetary alchochoden already mentioned, the judgment cannot be made about how long he [sc. the native] ought to live by necessity (*de necessitate*), but that beyond which his life cannot be extended naturally (*ex natura*); for a person's days can be shortened (*abbreviari*), but not increased (*augeri*, 259).[104]

Predicting the length of life is a crucial issue, but the Magister Speculi makes clear that in making such an astrological interpretation, the astrologer is not claiming that a person will necessarily die at a certain time, only that the length of their natural life will not exceed that indicated by the hylech and alchochoden. This analysis is similar to that in Albertus Magnus's discussion of periods in his translation-commentary on Aristotle's *De generatione et corruptione* II.10 as we saw in Chap. 1, and even moreso in his *De fato*, as we will see in Chap. 5.[105]

---

[103] Ad nativitates me transfero, quae pars caeteris acrius videtur pungere liberum arbitrium, ut etiam appareat quod invicem se destruant, neque ullatenus se posse compati videantur, praecipue quantum ad partem illam quae pertinet ad mores animi (13.14–17).

[104] De scienda namque quantitate vitae nati per gradum hylech et planetam alchochoden iam dictum est, quod non iudicatur quantum oporteat vivere de necessitate, sed ultra quod vita eius non protenditur ex natura: abbreviari enim possunt dies hominis, non augeri (13.17–21).

[105] For a valuable discussion of the hyleg and alcocoden, see Hasse, *Success and Suppression*, 262–65.

Then the Magister Speculi turns to health and sickness, i.e. issues related to medicine. Here he is concerned with the value of foreknowledge in such matters:

> Also, on knowing about occurrences (*eventibus*) of the sickness and health of a native's body, one might ask what it profits a person to foreknow (*praesciat*) a future evil (*malum* [...] *futurum*), since he cannot prevent it. Because if he could, then the profession of the stars (*magisterium astrorum*) is false and whoever looks into it is deceived. But I say that every operation of a cause acting on some thing (*omnis operatio causae agentis supra rem aliquam*) exists according to the proportion of matter receiving that operation (*secundum proportionem materiae recipientis ipsam operationem*), as for example, one and the same fire effects (*operatur*) drying in mud and melting in wax (259).[106]

This analysis continues to be very similar to Albert's in *De generatione et corruptione*, this time, however, concerning the variable reception of celestial influences in matter with respect to the occurrences of a person's health and sickness, thereby undermining necessity.

Then he turns to a medical example to address the issue just raised about the value of such foreknowledge:

> Whence, if a person knew in advance (*praescierit*) from the teaching of the stars (*ex astrorum magisterio*) that he would experience in a future summer a superfluity of heat and dryness from the operation of the heavens (*ex operatione caeli*), he could change his complexion (*potest mutare complexionem suam*) a long time beforehand by maintaining his diet until it declines to the side of cold and moisture, so that, when the operation of the heavens is approaching (*operatio caeli adveniens*), which, if it had found him in a moderate temperament, would have drawn him to the side of illness from the excess heat and cold. Now, since it finds him on the opposite side, it brings him back to a moderate health (*ad medium sanitatis*) instead. In this way, therefore, a foreknown impediment (*impedimentum praescitum*) could be removed wholly or in part. And yet, the operation of the heavens is not frustrated but perfected. For the operation of the heavens was not removed, but the quality of the operation (*non enim caeli operatio, sed operationis qualitas est remota*), in accordance with Ptolemy's intention expressed in the fifth sentence (*verbum*) of the *Centiloquium*, where he says: "An astrologer can avert a great deal from the operation of the stars (*avertere de operatione stellarum*) when he is informed about the nature acting on him (*sciens naturae agentis in eum*), and has prepared himself before its descent so that the one who will receive the influence will be able to bear it well (259)."[107]

[106] De sciendis autem eventibus corporis nati ex infirmitate et sanitate, quaeritur quid prodest homini si malum sibi futurum praesciat, cum illud praepedire non possit; quod si potest, est ergo falsum magisterium astrorum et fallitur aspiciens in eodem. Ego autem dico, quod omnis operatio causae agentis supra rem aliquam est secundum proportionem materiae recipientis ipsam operationem, ut unus idemque ignis operatur in luto arefactionem atque liquefactionem in cera (13.21–28).

[107] Unde si homo ex astrorum magisterio praescierit quod in aestate futura ex operatione caeli passurus sit ex superfluitate caloris et siccitatis, in multo tempore ante per exhibitionem diaetae potest mutare complexionem suam, donec declinet ad latus frigiditatis et humiditatis, ut operatio caeli adveniens, quae si ipsum in media consistentia invenisset, ad latus aegritudinis ex calore superfluo et siccitate traxisset, dum ipsum in opposito latere invenit, potius reducit ad medium sanitatis. Hac ergo via potuit removeri in toto aut in parte impedimentum praescitum; nec tamen frustrata fuit caeli operatio, sed perfecta; non enim caeli operatio, sed operationis qualitas est remota, iuxta quam intentionem loquitur Ptolemaeus in verbo quinto, ubi dicit: "Potest astrologus plurimum avertere de operatione stellarum, cum fuerit sciens naturae agentis in eum, et praeparaverit ante suam descensionem recepturum sustinere valentem (13.28–43)."

Here he describes how foreknowledge can be used to inform future actions related to health. Thus, knowing what the future will bring in relation to ones bodily complexion can help prepare one to best cope with what eventually arrives. Like Roger Bacon, the Magister Speculi is also fond of quoting the pseudo-Ptolemy of the *Centiloquium*, as we will also see in Chap. 5 with Thomas Aquinas.

The Magister Speculi then turns to wealth and its acquisition, his last example in the material-corporeal realm:

> Also, regarding knowing about the acquisition of wealth, the question is: what advantage is there in rejoicing in it before it comes, since under that hope he perhaps would do something rashly (*temere*). And I say again that a future good (*bonum futurum*), if it is foreknown (*praesciatur*), can be increased (*augeri*) and its outcomes improved, according to the eighth *verbum* of the same work [sc. the *Centiloquium*] of Ptolemy, which is: "The wise soul can help the celestial operation (*anima sapiens potest adiuvare caelestem operationem*), just as the sower (*seminans*) can help the strength [sc. of plants] through cultivation and weeding (259, 261)."[108]

In each of these cases, foreknowledge of what will happen in the corporeal realm can be a boon for planning for the future by informing a person how to help the heavens in their work. Ptolemy's *Centiloquium* is the principal authority here.

*

After these examples concerning the body, the Magister Speculi now turns to the much more complex issue of the soul, as it was also for Thomas Aquinas as we will see:

> But how should one respond [sc. to those questions] about the character of the soul (*de moribus animi*), except by saying that the native is not judged to be chaste or impure, wrathful or patient, and so on save according to his aptitude or its lack? Whence, nonetheless, he chooses (*eliget*) this or that [sc. conduct]. But it belongs to the operation of the heavens (*ex opere caeli*) whether he is inclined more readily towards choosing that for which he has an aptitude. But if this science (*scientia*) is condemned for this reason—because it seems to destroy free will (*liberum arbitrium destruere*) in this way—then, certainly, the profession of medicine will not stand (*non stabit magisterium medicinae*) for the same reason, since, surely, it is judged from its profession who, due to inferior causes (*secundum causas inferiores*), is fit or unfit for something? But, if the profession of medicine were destroyed, it would detract greatly from the public good. As long as this profession is preserved, the accusors can allege nothing against the part on nativities (261).[109]

---

[108] De scienda quoque acquisitione substantiae quaestio est ad quid prosit ei congaudere antequam veniat, quia sub illa spe fortassis aliquid temere attentabit. Et ego iterum dico, quoniam bonum futurum potest, si presciatur, augeri et effici melioris profectus iuxta verbum eiusdem Ptolemaei octavum, quod est: "Anima sapiens potest adiuvare caelestem operationem, quemadmodum seminans virtutem per cultum et purgationem (13.43–49)."

[109] Sed quid de moribus animi respondendum, nisi quia non iudicatur natus castus, aut incestus, aut iracundus, aut patiens et talia, nisi secundum aptitudinem et ineptitudinem? Unde nihilominus eliget hoc aut illud, sed ex opere caeli est quod, ad eligendum id ad quod aptus est, citius inclinetur. Quod si propter hoc condemnetur ista scientia, eo quod liberum arbitrium destruere videatur hoc modo, certe eadem ratione non stabit magisterium medicinae; numquid enim ex eius magisterio iudicatur quis secundum causas inferiores aptus ad huiusmodi vel ineptus? Quod si magisterium

The Magister Speculi here focuses on the inclinations caused by the celestial bodies. These superior causes incline choices concerning moral behavior, but do not compel them. Likewise, medicine explores who is fit or unfit for anything by means of lower causes. Therefore, if we reject astrology for these reasons, we must then also destroy medicine as well, despite its well established value to society!

4.10 Interrogations

Having now defended revolutions and nativities to his satisfaction, the Magister Speculi then turns in Chap. 14 to the other most controversial astrological practice to defend, namely, interrogations:[110]

> I pass on to interrogations. Those made concerning present affairs, do not appear subject to doubt, as [1] when one asks about someone absent, such as whether he is alive or dead. Or [2] about whether rumours are true or false. Or [3] from what kind of person a received letter has come, whether from a king or some other person. And [4] about a woman whom we know has given birth, asking what kind of child was born, that is, whether masculine or feminine. And [5] whether the work of a man who professes alchemy is true or not.[111] For there is nothing surprising if such things—whose truth has already been determined in the other direction [i.e. the past] among the nature of things—are signified by the heavens (*significentur per caelum*, 261).[112]

Questions concerning the present and the past do not seem to present any serious problems. Here he offers a sample range of possible questions.

Interrogations concerning the future, however, raise many challenging issues. The Magister Speculi begins by narrowing the relevant framework to things that are neither necessary nor impossible, namely, the realm of the possible:

> But those [sc. questions] concerning the future rightly admit a doubt (*dubitatio*), since we do not need to make interrogations about matters that are necessary or impossible. Nevertheless, some of those [sc. interrogations] concerning future possibilities (*futuris possibilibus*) have greater doubt than others, such as, those which concern matters that are deeply subject to free will (*de rebus quae penitus libero arbitrio sunt subiectae*). For some

medicinae destruatur, multum erit utilitati reipublicae derogatum; eo vero stante, non videntur habere quid contra partem nativitatum allegent (13.49–60).

[110] Zambelli discusses interrogations in ch. 10 of her preliminary study, "Divine Providence and the Meaning of 'Interrogationes'," 95–101. She focuses on interrogations at first (95–97), and then moves on to some interesting natural philosophical reflections on two of Albert's authentic texts: the *De causis proprietatum elementorum* and *De caelo*.

[111] I believe that this is the only mention of alchemy in the material explored in volume I. For more on medieval alchemy, see William R. Newman, *The* Summa perfectionis *of Pseudo-Geber*.

[112] Ad interrogationes transeo, et illae quidem quae fiunt de praesentibus, non videntur habere dubitationem, ut quando quaeritur de absente, utrum sit vivus vel mortuus; aut de rumoribus, utrum sint veri vel falsi; et de epistola recepta, a cuiusmodi persona exierit, utrum scilicet a rege aut ab alio; et de muliere quam scimus peperisse, cuiusmodi prolem peperit, masculum scilicet an foeminam; et de homine, qui profitetur alchimiam, utrum sit veritas operis apud eum. Talia enim, quorum veritas determinata est in partem alteram apud rerum naturam, nihil est mirum si significentur per caelum (14.2–11).

> things are possible and in the future, which, nonetheless, no one's will can impede: such as an interrogation about [1] the high or low price of grain in the coming year (although this can be known more certainly from the annual revolution); or a question about [2] whether someone might acquire wealth from his profession or from business; or [3] whether a certain man might acquire this or that kingdom, and other things of this sort (261, 263).[113]

The issue of free will has once again been raised, as has the relationship between interrogations and annual revolutions. He has also raised the fundamental issue of inquiring about future contingent matters, both those kinds deeply subject to free will and those that are not.[114]

Now the Magister Speculi explains how the process of interrogations works; in particular, how a person's nativity inclines him or her to ask the question in the first place:

> For such things happen (*accidunt*) to a person from the signification of his nativity (*ex significatione suae nativitatis*) because, when he makes an interrogation about himself (*interrogaverit de seipso*), he was moved by the heavens with a root intention (*motus per caelum cum intentione radicali*), for he has already come from his nativity [= natal horoscope, natal celestial configuration] to the good or bad that his nativity signified (*significavit*). For the concern of a person at the time (*hora*) of the interrogation will be in accordance with the situation of the [sc. zodiacal] circle [sc. of his nativity] (*secundum habitationem circuli*); and the circle at that time is in accordance with his intention. Wherefore, it is fitting that the figure of the heavens at the time of the interrogation (*figura caeli horae interrogationis*) is proportional (*proportionalis*) to the figure of the heavens at his nativity (*figura caeli suae nativitatis*), otherwise there is no radical intention; and this can also be judged (*perpendi*) from the figure of the interrogation. Whence, if his nativity were known, an interrogation concerning similar things would not be necessary, since the nativity is a stronger root (*radix fortior*), according to Haly: "Nativities are natural things (*res naturales*), and interrogations are things similar to natural ones (*res similes naturalibus*)." In the cases of those people whose nativities are unknown, therefore, the interrogation is necessarily regarded as the root. And for the same reason for which the science of nativities (*nativitatum scientia*) should be saved, also interrogations made about this kind of subject should be preserved, by the will of God.[115]

---

[113] Sed illae, quae de futuro sunt, merito dubitationem admittunt; neque enim super rebus necessariis aut impossibilibus interrogationibus indigemus; illarum tamen, quae fiunt super futuris possibilibus, maiorem habent dubitationem quaedam quam aliae, ut illae quae sunt de rebus quae penitus libero arbitrio sunt subiectae. Nam quaedam res sunt possibiles et futurae, quas nihilominus non potest cuiusquam arbitrium impedire; ut est interrogatio de gravitate vel levitate annonae, utrum futura sit in eodem anno, licet hoc certius possit ex anni revolutione cognosci; ut est interrogatio utrum quis acquiret multam substantiam ex magisterio suo, aut ex negotiatione aut utrum quidam vir adipiscetur hoc regnum vel illud et caetera huiusmodi (14.11–22).

[114] For medieval discussions of future contingents and their historical background, see Christopher D. Schabel, *Theology at Paris, 1316–1345: Peter Aureol and the Problem of Divine Foreknowledge and Future Contingents* Aldershot: Ashgate, 2000. My thanks to Craig Martin for this reference.

[115] Talia namque accidunt homini ex significatione suae nativitatis; quia cum ipse interrogaverit de seipso, et fuerit motus per caelum cum intentione radicali, iam pervenit ex nativitate sua ad bonum seu ad malum quod significavit eius nativitas. Sollicitudo enim hominis in hora interrogationis erit secundum habitudinem circuli, et est circulus in eadem hora secundum intentionem ipsius. Quare oportet figuram caeli horae interrogationis esse proportionalem figurae caeli suae nativitatis, alioquin non est intentio radicalis et hoc etiam potest ex figura interrogationis perpendi. Unde etsi esset nota eius nativitas, non esset nobis necessaria interrogatio super similibus, cum nativitas sit radix fortior, iuxta illud Haly: "Nativitates sunt res naturales, et interrogationes sunt res similes naturalibus." In illis ergo quorum nativitates ignotae sunt, necessario ponitur interrogatio pro

Here the Magister Speculi explains that, concerning many issues on which interrogations are made, the answer is more easily derived from the nativity itself, if it be known, since it is what he calls a stronger root, a turn of phrase of which Roger Bacon is also fond. If it is not known, however, the interrogation itself can provide insight because the time at which the question was posed will be proportional to the nativity. Since nativities should be preserved, therefore, so should interrogations.[116] We hereby learn more about the deeper structural relationship between nativities and interrogations, as we had just done with interrogations and annual revolutions. He also emphasizes the importance of a radical intention—intense focus—when asking the question.

The Magister Speculi then turns to the trickiest issue, namely, interrogations about contingent matters, and especially those subject to free will. He begins by distinguishing two kinds, namely, questions of fact, and questions of advice:

> There are two kinds of interrogations about possible future things (*de rebus futuris possibilibus*) which are subjected to free will (*arbitrio subiacent libero*). For there are [1] interrogations of fact (*interrogationes facti*), such as, what will happen concerning something. And there are [2] interrogations of advice (*interrogationes consilii*), such as, would it be more convenient if this or that happened. And [*ad* 2] those which concern advice do not destroy the freedom of the will (*non destruunt* [...] *arbitrii libertatem*), but rather they correct and direct (*rectificant et dirigunt*) it, such as [a] an interrogation about business, will it be useful to me or not; or [b] about two things, which of them would it be better to buy; and [c] about a route I intend to take, whether it would be better to proceed or delay. To destroy such things would be more against free will than for it (*plus esset contra liberum arbitrium quam pro eo*) because, that it is fitting to take advice and negotiate, is one of the most persuasive means by which it is shown that everything does not exist by necessity (*non omnia esse ex necessitate*), but that some things happen by chance and could go either way (*sed quaedam a casu atque ad utrumlibet*, 263).[117]

The Magister Speculi begins with interrogations of advice and argues not only that they do not undermine free will, but rather, that they strengthen and direct it. Such interrogations—and the need for taking counsel and doing business—also argue persuasively both that there is no necessity, on the one hand, and that both chance exists and things that can go either way (*ad utrumlibet*). We will see in Chap. 5 that Thomas Aquinas was also concerned with precisely these issues.

*

radice, et eadem ratione, qua salvatur nativitatum scientia, salvantur et interrogationes super huiusmodi nutu Dei (14.22–37).

[116] This analysis of the nativity prompting a question is similar to that in Carl Gustav Jung's foreward to the Wilhelm-Baynes edition of the *I Ching* with respect to the question being asked of the oracle; *The I Ching: Or, Book of Changes*, third ed., Princeton, N.J.: Princeton University Press, 1967.

[117] Interrogationum vero de rebus futuris possibilibus, quae arbitrio subiacent libero, duo sunt modi. Sunt enim interrogationes facti ut: quid fiet de aliquo? Et sunt interrogationes consilii ut: quid melius fieri conveniat, hoc an illud? Et illae quae sunt consilii, non destruunt, immo potius rectificant et dirigunt arbitrii libertatem, ut est interrogatio de negotiatione, utrum sit mihi utilis vel non; et de duabus rebus, quam illarum emere sit melius; et de via quam intendo arripere, utrum ire sit mihi melius an morari. Talia destruere plus esset contra liberum arbitrium quam pro eo, quia oportere consiliare et negotiari est unum ex mediis urgentioribus per quae ostenditur non omnia esse ex necessitate, sed quaedam a casu atque ad utrumlibet (14.37–48).

Then he turns to the tricker questions of fact, which he treats as one continuous argument:

> Moreover, it is extremely difficult to determine about interrogations of fact as to how they might remain in existence along with freedom of the will, such as in an interrogation about money being sought from someone, that is, whether he will give it or not. For, even if it will have been signified in a thousand ways that he will not give it, nevertheless, he *can* give it. Likewise, if [sc. it were indicated] that he will give it, he will always be able not to give it. Otherwise, the choice (*electio*) will not remain for him. Not because an astrologer is not found to have affirmed that he will give it, but because it was signified by the figure of the interrogation that he will give it. There still remains the question about whether what has been signified will be realized (*utrum erit*). For, if this will not happen, then the profession (*magisterium*) [sc. of interrogations] is false. But, if it will, therefore, it cannot be the case that it does not exist, or perhaps that does not follow (my emphasis, 263, 265).[118]

With interrogations of fact, we seem, once again, to have a direct clash between an astrological practice and human free will, the protection of which continues to be the Magister Speculi's primary concern in defending astrology. This issue turns in part on the distinction between what is signified by the figure (= horoscope) of the interrogation, and what will actually turn out to be the case in reality. Whatever the horoscope signifies, the person still has free choice concerning his or her actions. The anonymous author could, of course, have rejected interrogations as a practice as later astrologers often did, as we will see in volume III.[119]

The Magister Speculi then tries to save interrogations by discussing contingent matters and their logical analysis:

> For, about something contingent, that it will be (and about which it is true to say that it will be) before it exists, it is always possible for it both to exist and not to exist. But when it [sc. actually] exists, then it cannot not exist (as about that which is white now and about which it was previously true to say that it will be white). It does not follow (*non sequitur*), therefore, that before it exists, it cannot not be, but [sc. only] that when it exists, it cannot not exist. For every contingent thing (*omne* [...] *contingens*), regardless of whether it comes to exist (*sit natum*) often (*in pluribus*) or infrequently ([sc. in few circumstances] *in paucioribus*), or could go either way (*ad utrumlibet*),[120] always, before it exists, can always both exist and not exist (as was said), although some of them not equally. But when it exists, it reverts to the nature of the necessary (*revertitur ad naturam necessarii*), not because it was necessary previously, but because it exists necessarily when it exists. For it is not the same

---

[118] Determinare autem de interrogationibus facti, qualiter maneant cum arbitrii libertate, dificillimum est, ut est interrogatio de substantia ab aliquo petenda: utrum scilicet eam dabit vel non dabit? Nam si millesies significatum fuerit quod non dabit, nihilominus poterit eam dare. Similiter si quod dabit, semper poterit et non dare; alioquin non remaneret electio apud eum, nisi quia non invenitur astrologus affirmasse quod dabit, sed quod per figuram interrogationis significatum est eum daturum, et de eo quod significatum est, adhuc restat quaestio utrum erit. Nam si non erit, magisterium est falsum; si vero erit, ergo non potest non esse, vel forte istud non sequitur (14.48–58)[.]

[119] Hasse is very interesting on interrogations and some of their later history; *Success and Suppression*, 256–58.

[120] That is, whether something exists in nature for the most part (*in pluribus*), by chance or infrequently (*in paucioribus*) or could go either way (*ad utrumlibet*). All three are also distiguished explicitly in Thomas Aquinas's discussion in *Summa theologiae* II.II.95 in discussing astrology in relation to legitimate conjectural predictions about the future, as we will see in Chap. 5.

> thing to exist necessarily when something [sc. already] exists and to exist *simpliciter* "by necessity." Therefore, before it exists, it can not exist, and nevertheless it will exist [sc. if it comes into existence] because it is not necessary that that potentiality be reduced to actuality (265).[121]

The Magister Speculi here discusses the nature of contingency vs. necessity: for every contingent thing, before it actually exists, it is possible for it to both exist and not exist. This is the very nature of contingency, and it is relevant in all three physical modalities: [1] things that come to be for the most part (*in pluribus*), [2] in few circumstance (*in paucioribus*), and [3] what could go either of two ways (*ad utrumlibet*). Thomas Aquinas is also exercised by precisely these issues of contingency vs. necessity, as we will see in some detail in Chap. 5.

He now turns to the all-important temporal dimension:

> The case is similar regarding that about which it is signified that it will not exist at a determinate time, and about which it is true to say that it will not exist then. Nonetheless, it can always exist beforehand, and up until it finally reverts to the nature of the impossible (*revertitur ad naturam impossibilis* [the opposite of the situation just described]). This is Albumasar's opinion, from which, nevertheless, the famous Aristotle seems to depart to some extent, since he [sc. Aristotle] does not concede that one may say something true before it exists. Moreover, I do not regret having said this, but the situation is not the same in the case of those negatives (*in his negativis*) which are signified without the determination of time, as is that about which it is true to say that it will never exist because it does not revert back to the nature of the impossible. Instead, it could always exist until the motion ceases because, already from this, it will not be able not to exist (265).[122]

Here the Magister Speculi discusses some of the complexities of the logic of possibility and necessity in relation to contingency with respect to time. A large part of the complexity arises from trying to logically analyze both the potential and actual ontological situations. Once again, the truth of predictions in themselves in relation to the reality of actual outcomes is of central concern, as we will see is also the case in Chap. 5 in Albert's *De fato*.

---

[121] [N]am de contingenti quod erit, et de quo verum est dicere quoniam erit, antequam sit semper possibile est de esse et non esse; sed quando est, iam non potest non esse, sicut de eo quod est album nunc, et de quo prius erat verum dicere quoniam erit hoc album; non sequitur ergo antequam esset, non potuit non esse, sed quod non potest non esse quando est. Omne enim contingens sive sit natum in pluribus, sive in paucioribus, sive ad utrumlibet, semper antequam sit, potest et esse et non esse, ut dictum est, licet non aequaliter quaedam eorum; sed quando est iam revertitur ad naturam necessarii, non quod prius fuerit necessarium, sed quod necessario est quando est. Non enim idem est esse necessario quando est, et simpliciter esse ex necessitate. Antequam ergo sit potest non esse, et tamen erit, quia non est necesse illam potentiam ad actum reduci (14.58–71).

[122] Similiter de eo de quo significatum est quoniam non erit in tempore determinato, et de quo verum est dicere, quoniam non erit tunc, nihilominus semper ante hoc potest esse, et tandem revertitur ad naturam impossibilis. Et haec est sententia Albumasaris, a qua tamen famosus Aristoteles in aliquo declinare videtur, cum non concedat quod prius sit verum dicere. Me autem nihilominus sic dixisse non piget, sed in his negativis quae absque temporis determinatione significata sunt non similiter, ut est illud de quo verum est dicere, quoniam numquam erit, quia non revertitur ad naturam impossibilis, quin semper possit esse usquequo cesset motus, quia ex hoc iam non poterit non esse (14.71–81).

The Magister Speculi then compares this argument with related issues concerning divine providence itself, which he expresses as a *dubitatio*:

> And perhaps someone considering this matter more closely will have the same doubt or one similar in kind (*similis genere*) to that doubt which concerns divine providence (*de divina providentia*). For, in those things that God does by means of the heavens (*in his quae operatur Dominus per caelum*), the signification of heaven is nothing other than divine providence.[123]

Nihil aliud est caeli significatio quam divina providentia! The Magister Speculi thus explicitly and boldly identifies the signification of the heavens precisely as divine providence, which instantiates and develops his argument in Chap. 3, as discussed in my Chap. 1.

Then the Magister Speculi makes a very interesting argument about the causality of human choice vis-à-vis that in the heavens, utilizing here the distinction between causality and signification:[124]

> But, in those things of which *we* are the first cause (*nos sumus principium*), nothing prevents that there is also not a cause in the heavens, but a signification (*nihil prohibet etiam caelo non causam, sed significationem inesse*). For of the two parts of a contradiction, from which a person can choose (*eligere*) one or the other [sc. possibility (*alterutrum*)], God knew from eternity which he or she would choose. Whence, in the book of the universe (*in libro universitatis*), which is the vellum of heaven (*caeli pellis* [as he had said before]), He was able to figure (*figurare*), if He wished, what He knew. But if He did this, then there is the same determination about the compatibility (*compossibilitas*) of free will with [1] divine providence and [2] the signification of an interrogation. Therefore, if it cannot be denied that divine providence co-exists with free will (*si ergo divinam providentiam stare cum libero arbitrio annullari non possit*), neither will it be denied that the profession (*magisterium*) of interrogations should stand (*stet*) with it as well. But I think the question of how it would not be denied concerning divine providence should be left to a more elevated occasion. Nevertheless, I do not want to say that whichever of those things that are not hidden by divine providence might also be known in the heaven (*cognita apud caelum*); for the heavens are greatly inferior [sc. to divine providence]. Whence, as was said,[125] when the signifiers (*significatores*) are equal in good fortune and bad, the counsel of the profession of the stars is to abandon [sc. the interrogation], since the Lord (*Dominus*) wished to keep it hidden from us (265, 267).[126]

---

[123] Et fortassis attingentius intuenti, eadem aut saltem similis genere est ista dubitatio ei dubitationi, quae est de divina providentia; nam in his quae operatur Dominus per caelum, nihil aliud est caeli significatio quam divina providentia (14.82–85).

[124] Thomas also uses this distinction in *Summa theologiae* II.II.95, but to a different effect, as we will see in Chap. 5.

[125] In Chap. 9, and specifically at 9.16–18.

[126] In his vero quorum nos sumus principium, nihil prohibet etiam caelo non causam, sed significationem inesse: duarum enim partium contradictionis, quarum alterutram potest homo eligere, sciebat Deus ab aeterno quam illarum eligeret. Unde in libro universitatis, quod est caeli pel lis, sicut praedictum est, potuit figurare, si voluit, quod sciebat; quod si fecit, tunc eadem est determinatio de compossibilitate liberi arbitrii cum divina providentia et cum interrogationis significatione. Si ergo divinam providentiam stare cum libero arbitrio annullari non possit, neque annullabitur quin stet magisterium interrogationum cum eo. Qualiter autem non annulletur de divina providentia, relinquendum arbitror negotio altiori; verumtamen non volo dicere, quod quaecumque non latent divinam providentiam, sint etiam cognita apud caelum, longe enim est caelum inferius. Unde sicut

The upshot of this fascinating argument is that, if divine providence is compatible with human free will, then so is the astrological practice of interrogations. Thomas Aquinas addresses closely related issues in *Summa contra gentiles* III.82–94 to be discussed in Chap. 5, and in fact provides a solid analysis of both the ontological and metaphysical dimensions of this position. If Thomas were more bold or more strongly (or explicitly) astrologizing, he too might have spoken exactly like the Magister Speculi does here.

4.11 Elections

Chapters 15 to 17 treat elections. I will discuss Chap. 15 here, and Chaps. 16 and 17 concerning talismans in Chap. 7. After this incisive discussion about the relationship between interrogations and divine providence, the Magister Speculi addresses the final astrological practice—and one somewhat less contentious, at least concerning free will—namely, elections, the choosing of astrologically propitious times. The introduction is very interesting:

> The question of elections is certainly less difficult, for freedom of the will (*libertas arbitrii*) is not coerced by the choice of a favourable hour, but instead, it is a precipitation of the will, not its freedom, to disregard the choice of the hour (*electionem horae contemnere est arbitrii praecipitatio, non libertas*) for the beginnings of great matters.[127] Also someone is found to have said about such men that, as many of them as might happen (*contigerit*) to be saved by such things [sc. good timing], God will save them not as men, but as beasts. Furthermore, all the philosophers are in agreement (*concordati*) on this point, namely, that, when we know the time (*hora*) of a woman's impregnation, we may know what might happen regarding the fetus until the time when it is quickened (*inspiretur*) or until it is delivered from the uterus, and, perhaps, until its death (267).[128]

In fact, using astrological elections should be understood as the realization of free will, not its undermining, which is, ironically, reserved for those who reject elections out of hand.

Then he offers some examples where elections can be helpful, beginning with the timing of impregnation-conception, a notoriously difficult moment to pinpoint, even today:[129]

---

dictum est, cum fuerint significatores aequales in fortuna et malo, consilium magisterii astrorum est supersedere, quia Dominus voluit celare a nobis (14.85–101).

[127] He had just made a similar point with interrogations of advice, as we just saw.

[128] De electionibus vero est quaestio minus difficilis, non enim libertas arbitrii ex electione horae laudabilis coercetur, quin potius in magnarum rerum inceptionibus electionem horae contemnere est arbitrii praecipitatio, non libertas. Etiam invenitur quidam dixisse de huiusmodi hominibus, quod quotquot ex eis salvari contigerit, eos Deus non ut homines, sed ut iumenta salvabit. Caeterum in hoc concordati sunt omnes philosophi, quod cum sciverimus horam impraegnationis alicuius mulieris, sciamus per eam quid fiat de foetu donec inspiretur et quid usquequo egrediatur ab utero, et quid forte usque ad obitum (15.2–10).

[129] Ptolemy discusses this issue in depth at *Tetrabiblos* III.1.

> For astrologers have not judged [sc. these things] by means of nativities, only because the time of conception can rarely be verified. Whence Ptolemy says that the time of the nativity exists according to the judgment. Why, therefore, when the wife of a king or prince or magnate exists in the optimum conditions, do we not choose for her husband the hour for getting a boy from her, if the Creator of all generation approves (*annuerit*), so that good things might come (*eveniant*) to him when he is born, which the astrologer (*astrologus*) could predict (*praenuntiat*) will come to pass from the series of books of nativities (267, 269)?[130]

Since we do not usually know the time of conception in regular sexual practices, it makes much more sense, so he argues, for astrologers themselves to choose the time of conception, especially in the case of rulers. Here the Magister Speculi advocates a species of eugenics based on astrological timing.

Astrological timing can thus be valuable for family planning. Next he turns to medicine, including using astrological timing in choosing when to take medicaments:

> Again, why do we not choose the time for offering medicine (*pharmacum*), if we know that the ascendent and signifiers in the ruminating signs (and especially in Capricorn) provoke vomiting? And if we know in addition that they ought to be struck by an aspect of both malefics (*infortunae*), that is, Saturn and Mars, because Saturn fixes medicine and Mars draws it to the blood. And also [sc. if] we know that when the Moon exists with [= is conjunct with] Jupiter, the difficulty of the operation of purging is reduced. Again, in the profession (*magisterium*) of surgery, why shall I not avoid making an incision in a limb when the Moon is in a sign that has a signification for that limb? For then the limb is very rheumatic and pain provokes rheum (269).[131]

Thus elections in medicine can also be very valuable, including for purgations and surgery. We will see many more examples of elections concerning medicine—and talismans—in treating Ficino's *De vita* in volume II.

Then he discusses his own experience with medical elections:

> And I have the courage to say that I myself have seen, as it were, an infinite number of inconveniences happen (*accidisse*) as a result of this. I have seen a man experienced in the science of the stars and in medicine (*peritus astrorum et medicinae*), who due to the threat of angina (*squinantia*) bled himself from his arm while the Moon was in Gemini, which has a signification for the arms. Without any apparent illness, except for a moderate inflamation of the arm, he died seven days later. I also knew a certain person who was suffering from an ulcer near the head of his gut, and was cut into by some miserable surgeon (*a quodam*

---

[130] Neque enim iudicaverunt astrologi per nativitates, nisi quia hora impraegnationis vix potest certificari. Unde inquit Ptolemaeus horam nativitatis esse secundum iudicium. Quare ergo uxore regis seu principis aut magnatis existente in optimis conditionibus, non eligemus viro eius horam suscipiendi ex ea liberum, si creator universae generationis annuerit, ut scilicet eveniant nato bona, quae ex serie librorum nativitatum astrologus futura praenuntiat (15.10–17)?

[131] Quare iterum non eligemus horam pharmacum exhibendi, si sciverimus quod ascendens et significatores in signis ruminantibus, et praecipue in Capricorno, provocant vomitum? et cum hoc si sciverimus, quod oportet eos proiici ab aspectu utriusque infortunae, Saturni videlicet et Martis, et quod Saturnus constrigit medicinam, Mars vero educit usque ad sanguinem, et cum hoc etiam sciverimus quod Luna existente cum Iove operationis purgatorii minuitur angustia. Rursum in magisterio chirurgiae, quare non cavebo facere incisionem in membro, Luna existente in signo habente significationem super illud membrum. Tunc est enim membrum valde rheumaticum, et dolor rheuma provocat (15.17–28).

> *misero chirurgico*) who was ignorant (*ignarus*) of both professions (namely, of medicine and the stars) while the Moon was in Scorpio (which signifies those parts), and without the cutting of a vein or some other reasonable cause, he was found dead in the arms of the men holding him within that same hour. His death was ascribed to the operation of the heavens (*operatio caeli*), since it did not seem to have occurred (*accidisse*) from any cause that kills suddenly, such as obstructions in the ventricles of the brain or a lesion or failure of the air passages (*spiritualia* [i.e. the respiratory system], 269).[132]

Astrological timing thus has a valuable place within medicine also. In the last case, it was ignorance of both medicine and astrology that brought this poor man to his unfortunate end. Medical and astrological malpractice, not the working of the heavens, was the true fatal cause. This evidence also seems to indicate our author's significant medical experience.

The Magister Speculi then offers his considered conclusion:

> But if something appearing to be frivolous (*aliquid apparens frivolum*) is found within elections, such as that new clothes should be put on when the Moon is in Leo, etc., it must be noted that Ptolemy, since he was a man of great authority (*vir tantae auctoritatis*), did not say this except to signify greater matters (*maiora*), and by this he meant that fixed [sc. zodiacal] signs (*signa fixa*) are useful for matters about which we want stability, just like the houses (*domus*) called cardines (*anguli*). But the mobile [= cardinal] signs (*signa* [...] *mobilia*), just like the houses cadent from the cardines, are useful for things that change quickly (*ad res cito mutabiles*) whose departure is expected and not the other way around (269).[133]

With these reflections on the differences between fixed and mobile signs and their corresponding fixed and mobile houses or places,[134] the Magister Speculi completes his deeper analysis of the legitimacy of the four canonical types of astrological practice: revolutions or general astrology, nativities, interrogations and elections. Having completed the first more astrological part of elections, he then turns to the more controversial—what we would call "magical"—part of elections concerning *imagines*-talismans, which I will discuss in Chap. 7 below.

---

[132] Et audeo dicere me vidisse ex hoc quasi infinita inconvenientia accidisse. Vidi hominem peritum astrorum et medicinae, qui pro periculo squinantiae minuerat sibi de brachio, Luna existente in Geminis qui habent significationem super brachia, et absque ulla manifesta aegritudine, excepta modica brachii inflatione mortuus est in die septimo. Scivi quoque quendam patientem fistulam iuxta caput longaonis fuisse incisum, Luna in Scorpione, qui significat super partes illas, a quodam misero chirurgico, qui erat ignarus utriusque magisterii, medicinae scilicet et astrorum, et absque venae incisione, aut alia causa rationabili, inter manus eum tenentium inventus est mortuus ipsa hora, fuitque caeli operationi adscriptum, cum non videretur ab aliqua causa interficiente subito accidisse, ut sunt oppilationes ventriculorum cerebri, aut laesio spiritualium seu defectus (15.28–41).

[133] "Quod si inveniatur inter electiones aliquid apparens frivolum, ut est indumenta nova induere Luna in Leone, etc., attendi debet quod Ptolemaeus, cum esset vir tantae auctoritatis, non dixit hoc nisi ut significaret maiora, et per hoc innuit quod signa fixa utilia sunt ad res quarum stabilitatem volumus, sicut et domus quae dicuntur anguli; signa vero mobilia, sicut et domus cadentes ab angulis, ad res cito mutabiles, quarum recessio expectatur et non e converso (15.41–48)." I discuss a related question in relation to architecture in my "Astrological Timing and Architectural Sites: The Deborah Loeb Brice *Loggiato* in One of its Historical Contexts," *Renaissance Studies in Honor of Joseph Connors*, Machtelt Israels and Louis Waldman (eds.), Milan: Officina Libraria, 2013, 133–37.

[134] For the different types of houses: cardines/angles, cadent and succedent, see Burnett, "Astrology," 373, and Hand, "Use of Military Astrology," 350–57.

4.12 Conclusion

These texts by Roger Bacon and the Magister Speculi were both most likely composed in the 1260s and most likely in Paris to support and defend astrology in response to her detractors' attacks, although unfortunately we do not possess the texts of those attacks. Despite treating similar issues, they were both structured quite differently: Rebuffing the ignorant recent (contemporary?) theologians' attacks, Roger defended what we may now call judicial—i.e. interpretive, knowledge-based—astrology as mathematics' weakest link (in a configuration of astrology with the mathematical disciplines), whereas the *Speculum astronomiae* defended the four types of practical astrology from detractors who attacked it on primarily theological grounds, including in relation to issues that Roger addressed as well, namely, free will, determinism, and related topics, as well as magic in Chaps. 16 and 17, as we will see in Chap. 7. In Chap. 5 we will see that Thomas Aquinas explicitly defended astrology as a predictive art that is not to be conflated with the theologically problematic practices that he calls "divinatory"; protecting free will and avoiding demons are the main issues he addressed. I do not know of any particular text where the authentic Albertus Magnus defends astrology per se from particular attacks.

Roger also used different terminology in his defense, but there remain, nevertheless, many points of contact: although true vs. false mathematics is configured differently, the themes he treats—infallibility of judgment with respect to human free will and determinism in nature—are precisely those treated in the *Speculum astronomiae*, and (as we will see) in Thomas's *Summa theologiae*. Roger's use of the term "magic," finally, is very different than anything we have seen so far, but is in fact structurally similar to Thomas's use of "divination," as we will see in Chap. 5. Regardless of his attacks against what he explicitly calls "magic," however, Roger strongly advocates the use of talismans and powerful words (the *opera et verba sapientiae*, in his terms), as we will see in Chap. 8.

*

Having focused so far on astrology's natural philosophical foundations and how these structures of legitimate natural knowledge were used to defend practical astrology from theologically motivated attacks, we will now turn to exploring astrology's also at times controversial border regions with theology/religion and magic in Parts II and III, with a particular emphasis on three fundamental themes: fate, divination and divine providence on the theologico-religious side, and talismans (*imagines astronomicae*) on the magical. I will examine these themes first in the works of Albertus Magnus, Thomas Aquinas and the *Speculum astronomiae* (Chaps. 5 and 7), and then in central works of Roger Bacon (Chaps. 6 and 8). Exploring astrology's normative structures at its often hotly contested boundaries with what we call theology/religion and magic can help us to articulate their distinctive borders. We will continue to pay close attention to medieval conceptual structures and their distinctive and characteristic terminology.

Part II
Conceptual Structures (2): Astrology and Theology/Religion

Introduction: Astrology and Theology/Religion

What was astrology's relationship to theology in particular, and to religion more generally in the Middle Ages? As with astrology's relationship to natural philosophy/science, many modern and postmodern preconceptions are currently at play. As with most related issues, I think it is methodologically imperative to begin by asking the relevant basic questions about their relationship rather than simply assuming some sort of understanding and then imposing it on the historical material. To begin to ask this question, I will focus on three central themes in particular—fate, divination and divine providence—in the work of two extremely influential Dominican philosopher-theologians: Albertus Magnus and Thomas Aquinas. On this basis, I will build up a picture of their understanding of astrology's relationship to some fundamental theologically-inflected issues in Chap. 5. In the process, I will emphasise how these analyses relate to the astrologizing Aristotelian natural philosophy that both authors embrace. Chapter 6 will be devoted to the Franciscan Roger Bacon's rather different and far more practical-astrological approach to religion and theology, albeit with some related and recurring themes.

As ever, complexity will reign. In this and the following chapter we will explore patterns of astrology's relationship to theology and religion within the medieval Roman Catholic and Aristotelian-Ptolemaic thirteenth-century mental universe. In volume II, some of these themes—with respect to both their continuities and transformations—will be further explored in relation to the expanded philosophical and theological contexts initiated by the new translations of Plato and the Neoplatonists into Latin in the 1480s and beyond, and their integration into a distinctively Renaissance-style Platonizing Christian synthesis. Finally, in volume III, I will also explore developments related to Protestant expressions of theology during the Reformation and beyond, both Lutheran and Calvinist. These patterns of thought and analysis will help sharpen our understanding of how deeply astrology was woven into both the natural philosophico-scientific and the theologico-religious roots of premodern European culture: medieval, Renaissance and early modern.

*

Part I reconstructed fundamental features of astrology's conceptual structure *qua* natural knowledge in the thirteenth century, primarily in the works of Albertus Magnus and Roger Bacon. The Intermezzo focused on two related but differently structured defenses of astrology. Part II now takes over this material and relates it primarily to the realm of theology/religion, in part by discussing the controversial issue of astrology's relationship to divination on the border with both theology/religion and magic. Boundary criteria for distinguishing between licit and illicit knowledge and practice will be of central concern, as will terminology. We will see that clarifying the relationship between astrology and divination can help us to sharpen our understanding of these significant but controversial and often misunderstood domains, as well as to clarify our own usage. Exploring the relationship between divine providence and fate will also prove to be illuminating. Furthermore, the texts to be discussed here are extremely useful for exploring the next set of fundamental conceptual structures, as well as for introducing more central themes to be traced out and/or developed over the remainder of this investigation (ca. 1250–1800) in addressing both continuities and transformations.

Many of the major themes raised in Roger Bacon's defense of legitimate astrology as true mathematics will be explored further here, including astrology's relationship to human free will and determinism in nature vis-à-vis the certainty of predictions. We will also treat an important theme raised in the *Speculum astronomiae's* defense of astrology, namely, astrology's relationship to God's providential creation and ongoing governance of the world. As ever, the dynamics both within and between the different positions will reveal themselves to be extremely complex, and thus well worth exploring in rich detail. We will now turn from astrology's well-established natural philosophical foundations to establish its overarching and well-integrated theological foundations, as grounded primarily in God's providential ordering, creation and governance of the world vis-à-vis both His ordained and absolute power.

*

In my reading, I have found that in most of the rich and penetrating scholarship on medieval philosophy and theology—although there are numerous insightful analyses on a nexus of related issues, including divine providence, God's foreknowledge, future contingents, fate and free will—there is usually little to no explicit discussion of the astrological context in which at least some of these issues arise.[1] Some, how-

[1] I am thinking in particular of valuable works by Chris Schabel (especially his extremely interesting book on future contingents, *Theology at Paris*), and relevant chapters in the *Cambridge History of Later Medieval Philosophy* (1982) and the *Cambridge History of Medieval Philosophy* (2009). A counterexample for Schabel is his and Fritz S. Pederson's "Miraculous, Natural, or Jewish Conspiracy?: Pierre Ceffons' Question on the Black Death, with Astrological Predictions by Gersonides and Jean de Murs/ Firman de Beauval," *Recherches de Théologie et Philosophie Médiévales* 81 (2014): 137–79. My Part II is but a sounding in extremely vast, deep, complex, and at times turbulent intellectual-historical waters that I have only recently begun to explore in depth.

ever, do treat astrology explicitly,[2] including important synthetic studies, such as Tullio Gregory's "Astrologia e teologia nella cultura medievale."[3] I suggest that historians of astrology join forces with historians of medieval theology and philosophy in order to more fully explore these issues in their proper depth. I hope that my Part II can help contribute to this discussion.

[2] Here I am thinking of work by Jeremiah Hackett on Roger Bacon, Alessandro Palazzo and Henryk Anzulewicz on Albertus Magnus, and Thomas Litt, Pasquale Porro and Valérie Cordonier on Thomas Aquinas. This is also the case for Alain de Libera's chapter on astrology, "Le philosophe et les astres," in his *Penser au Moyen Age*, Paris: Éditions du Seuil, 1991, 246–98.

[3] In his Mundana Sapientia: *Forme di conoscenza nella cultura medievale*, Rome: Edizioni di Storia e Letteratura, 1992, 291–328. His essay appears for the first time in this volume. He discusses several other interesting and related themes there as well, including natural prophecy and the role of angels.

Chapter 5
Astrology and Theology in Albertus Magnus and Thomas Aquinas: Fate, Divination and Providence

5.1 Introduction

What is astrology's undoubtedly complex relationship to theology? Three valuable themes for approaching this question are fate, divination and divine providence. I will explore these and related topics in the works of Albertus Magnus and Thomas Aquinas in this chapter. I will first address the issue of fate in Albertus Magnus.[1]

5.2 Astrology and Theology (1): Fate

5.2.1 Albertus Magnus, De fato

To begin to delineate astrology's relationship to theology more precisely, I will analyze Albertus Magnus's *De fato*, which he disputed in 1256 while at the papal curia for the first time (in Anagni, Italy), and thus soon after composing his major translation-commentaries on the basic texts of Aristotle's natural philosophy, ca. 1250–54.[2] In his disputation on fate, Albert further refined his astrologizing Aristotelian natural philosophy, thereby taking us to a deeper level of

[1] Boudet treats these issues valuably in the extensive chapter 5 of his *Entre science et nigromance*, "La recherche d'une norme théologique et juridique aux XIIe et XIIIe siècles: ajustements, contradictions, condamnations," 205–78.

[2] The text is in Albertus Magnus, *Opera omnia* 17.1, *De fato*, Paul Simon (ed.), Monasterii Westfalorum: Aschendorff, 1975, with a rich introduction. We do not know much in detail about the circumstances of the disputation, but it seems likely that the pope (Alexander IV, r. 1254–61) and the members of the curia were in attendance. I have published some of this material in my "Astrology and Magic." Albert also discussed fate in his somewhat earlier translation-commentary on Aristotle's *Physics* (II.2.19-20), which I will discuss in the final section of this chapter.

H D. Rutkin, *Sapientia Astrologica: Astrology, Magic and Natural Knowledge, ca. 1250–1800*, Archimedes 55, https://doi.org/10.1007/978-3-030-10779-6_5

understanding. Fate is here treated within the context of one of the most central dividing issues concerning astrology's legitimacy or otherwise, namely, necessity in nature vis-à-vis human free will (among other topics), and thus within an explicitly moral and theological context, as well as in an institutional venue—the papal curia—where high-level theological issues were often and authoritatively debated.[3] Albert uses the basic structures of both practical astrology and of astrologizing Aristotelian natural philosophy to explicate his understanding of fate, and will thus permit us a deeper insight into the uses of astrology in his work, and more generally into how astrology fit within his broader map of knowledge.[4]

5.2.2 The Metaphysics of Fate (the Formal Dimension)

Albert begins his disputation by raising five questions about fate, which become the titles of its five articles: [1] Does it exist? [2] What is it? [3] Does it impose necessity on things? [4] Is it knowable? [5] Into which genus of cause does it fall?[5] After floating several possible definitions of fate in article 1, Albert narrows the field to three in article 2, focusing in particular on the third:[6]

> [F]ate is said to be the form of the order of the existence and life of things below (*dicitur fatum forma ordinis esse et vitae inferiorum*), caused in them from the period of the celestial circle (*causata in ipsis ex periodo caelestis circuli*), which surrounds their births with its radiations (*qui suis radiationibus ambit nativitates eorum*). And in this manner Hermes speaks about fate, calling the stars gods and the immovable disposition of the existence and life of things below the sacrament of the gods.[7]

[3] Zambelli treats issues of free will and determinism in Albert's authentic works; Speculum astronomiae *and its Enigma*, 64–71.

[4] In the introduction to his edition of *De fato* (xxxiii-xxxiv), Paul Simon argues strongly for *De fato's* authenticity, indicating the range of undoubtedly authentic texts on which it drew and with which it strongly resonates. It was also often attributed to Thomas Aquinas, including by Marsilio Ficino, as we will see in volume II. I assume that the text we have reflects at least in part an actual disputation held at the papal curia and before Pope Alexander. Albert was there in September 1256 in order to defend his Order before the pope against its detractors, in particular, William of Saint-Amour, which he did successfully. The disputation was most likely held soon after, but we do not know this for certain.

[5] Quaeritur de fato, an sit, quid sit, utrum necessitatem imponat rebus et an scibile sit et in quo genere causae incidat (65, 2–4).

[6] Albert explicitly says that he will not inquire about the first two: [1] that fate is equivalent to death, as in Roman literature (68, 5–13), and [2] that "fate is the disposition of divine providence about the future development of the existence and life of things below" (68, 14–30). I will discuss this second definition of fate from Boethius's *Consolation of Philosophy* in the last section of this chapter.

[7] "Tertio modo dicitur fatum forma ordinis esse et vitae inferiorum, causata in ipsis ex periodo caelestis circuli, qui suis radiationibus ambit nativitates eorum; et hoc modo Hermes loquitur de fato, deos vocans stellas et sacramentum deorum immobilem dispositionem esse et vitae inferiorum (68, 31–36)." Simon does not give a source for the statement by Hermes.

Here Albert immediately configures fate astrologically with precisely the same conceptual structures and terminology as in his discussion of periods at *De generatione et corruptione* II.10, and with the same geometrical-optical model of planetary action articulated in *De natura loci* and elsewhere, as I analyzed in detail in Chaps. 1, 2 and 3.

Assuming this framework, Albert characterizes the type of form fate is, and in the process offers insight into deeper causal structures, this time at the metaphysical level, treating both form and matter, and their deep relationship.[8] Generation is still utterly central:

> This form, moreover, is not the form which gives existence (*non forma dans esse*),[9] but, rather, it is the form of a universal order of existence and life (*forma cuiusdam universalis ordinis esse et vitae*), simple *in essentia*, multifold *in virtute*. It has the simplicity of essence from the simplicity of the circulation of the common circle [sc. the zodiac], but it has its multiplicity of power (*virtus*) from the multitude of what is contained in the circle. For it [sc. the multiplicity of its power] flows from many stars (*fluit enim a multis stellis*), locations (*sitibus*), spaces (*spatiis*), constellations (*imaginibus*), radiations (*radiationibus*), conjunctions (*coniunctionibus*), eclipses (*praeventionibus*) and multiple angles described by the intersecting rays of celestial bodies (*ex intersecationibus radiorum caelestium corporum*), and by the emission of rays over the center (*productione radiorum super centrum*), in which alone, as Ptolemy says, all the powers of those [sc. celestial bodies] in the celestial circle are gathered together and united (*omnes virtutes eorum quae sunt in caelesti circulo congregantur et adunantur*).[10]

[8] I have found Alessandro Palazzo's essays very helpful, especially on the metaphysical dimension of Albert's analysis. The three essays I have learned most from are: "The Scientific Significance of Fate and Celestial Influences in some Mature Works by Albert the Great: *De fato, De somno et vigilia, De intellectu et intelligibilia, Mineralia*," in Per perscrutationem philosophicam: *Neue Perspektiven der mittelalterlichen Forschung. Loris Sturlese zum 60. Geburtstag gewidmet*, A. Beccarisi, R. Imbach and P. Porro (eds.), Hamburg: Felix Meiner Verlag, 2008, 55–78; "Albert the Great's Doctrine of Fate," in *Mantik, Schicksal und Freiheit im Mittelalter*, Loris Sturlese (ed.), Cologne: Böhlau Verlag, 2011, 65–95; and, on a slightly different but still relevant topic, "Albert the Great's Doctrine of Fascination in the Context of his Philosophical System," in *Via Alberti: Texte—Quellen—Interpretationen*, L. Honnefelder, H. Möhle, S. Bullido del Barrio (eds.), Münster: Aschendorff Verlag, 2009, 135–215. Among other things, Palazzo discusses Albert's use of Hermes, as in the passage just above. Palazzo also argues strongly and persuasively that Albert's view of fate is utterly central to understanding his overall theologico-philosophical project; "Scientific Significance," 56: "Connecting the superior world of the celestial spheres with the sublunar world, the so-called sphere of the active and passive qualities, fate turns out to be the real pivot around which the other doctrines rotate." He argues that Albert's treatment here in *De fato* is concerned primarily with explaining the scientifico-astrological structures of his position, whereas his treatment of fate in *Physica* II.2.19-20 is more concerned with differentiating fate from providence (57), and is thus more metaphysically oriented. I have also found Henryk Anzulewicz's "*Fatum*: Das Phänomen des Schicksals und die Freiheit des Menschen nach Albertus Magnus" of great interest; in *Nach der Verurteilung von 1277: Philosophie und Theologie an der Universität von Paris im letzten Viertel des 13. Jahrhunderts. Studien und Texte*, J.A. Aertsen, K. Emery, Jr. and A. Speer (eds.), Berlin: De Gruyter, 2001, 507–34, and his "Der Einfluß." For relevant historical background, see Joerg O. Fichte, "Providentia—Fatum—Fortuna," *Das Mittelalter* I (1996): 5–20.

[9] I.e., the soul.

[10] Est autem haec forma non forma dans esse, sed potius forma cuiusdam universalis ordinis esse

In this system, unity is derived from the celestial circle and its simple regular diurnal motion, and diversity from the multiplicity of powers contained therein, including all the planets, their various locations, rays and aspects. This is precisely the complex celestial situation represented in a horoscope as fixed for a specific time at a particular place. In *De fato*, then, this astrologically articulated form with its unique celestial configuration profoundly affects the earth and its inhabitants by means of rays, especially here at their birth/generation, and thus deeply informs by imprinting everything that exists. In this way we derive a clearer understanding of the formal causal role that celestial influences play, in addition to their earlier established efficient causality.

*

To articulate his view more fully, Albert then relates the form in itself to its instantiation in an actually existing thing with its material nature, thus differentiating two fundamental ontological moments:

> Moreover, such a form is in the middle between the necessary and the possible. For whatever is in the motion of the celestial circle is necessary, but whatever is in the matter of generable and corruptible things is possible and changeable. But this form [1] caused by the celestial circle and [2] inhering in generable and corruptible things is in the middle between each.[11]

Albert here distinguishes three levels in his system: [1] What is in the motion of the celestial circle (*quidquid est in motu caelestis circuli*), namely, the planets and luminaries. These exist necessarily. [2] What is in the matter of generable and corruptible things (*quidquid est in materia generabilium et corruptibilium*), that is, of everything on earth, including people and *res naturales*. These are possible and changeable (*possibile et mutabile*), that is, they do not exist necessarily as the celestial bodies do. They may or may not exist and they are subject to change. Between these two extremes are the forms that Albert is describing here, which participate in both, that is, they are [1] caused by the celestial circle (which necessarily exists) and [2] inhere in actual generable and corruptible things, thus providing an essential link between the heavens and the earth, on the one hand, and the necessary and the possible, on the other.[12] This provides an utterly central pivot in Albert's system.

---

et vitae, simplex in essentia, multiplex in virtute; et simplicitatem essentiae habet a simplicitate circulationis circuli communis, multiplicitatem autem virtutis habet a multitudine eorum quae continentur in circulo. Fluit enim a multis stellis et sitibus et spatiis et imaginibus et radiationibus et coniunctionibus et praeventionibus et multiplicibus angulis, qui describuntur ex intersecationibus radiorum caelestium corporum et productione radiorum super centrum, in quo solo, sicut dicit Ptolemaeus, omnes virtutes eorum quae sunt in caelesti circulo, congregantur et adunantur (68, 36–49).

[11] Haec enim talis forma media est inter necessarium et possibile; necessarium enim est, quiquid est in motu caelestis circuli, possibile autem et mutabile, quiquid est in materia generabilium et corruptibilium. Forma autem ista causata ex caelesti circulo et inhaerens generabilibus et corruptibilibus, media est inter utrumque (68, 50–56).

[12] This is the very same ontological middle ground explored in the logically and ontologically challenging passages in *Speculum astronomiae*, chapter 14, that were just explored in my Chap. 4. Those concerned interrogations of fact regarding contingent entities that also inhabit this same

*

Next Albert reveals an underlying metaphysical principle relevant to the scale of causality and its concommitant hierarchy of being, which also further relates the two levels of existence mediated by these forms:

> For everything that proceeds from a noble cause [sc. the heavenly bodies] into something ignoble that has been caused [sc. with its terrestrial matter], although in some respect it retains the property of the cause, nevertheless, it is not its existence (*esse suum*), except insofar as the potential of the subject permits, in which it exists. For everything received—as Boethius says, and Aristotle in the 6th book of the *Ethics*—exists in what receives it in accordance with the ability of the receiver (*secundum potestatem recipientis*) and not of the cause (*causa*) from which it exists.[13]

That causes can only be fully realized to the extent that the receiving substance permits is central to Albert's ontology. This also relates directly to his discussion in *De generatione et corruptione* II.10, where Albert explained why some periods of life are longer or shorter for members of the same species. This concerns both the variable strength of the celestial influences *and* the variability of how they are received in each individual generated thing, thus intimately integrating this element of the formal cause with both the efficient and material causes for an increasingly rich causal analysis.

Albert then uses this principle to develop an analogy for fate and its status within the hierarchy of being:

> We can see this in what are called "divine processions" by Dionysius,[14] just as "life," "reason" and "wisdom," etc., which, inasmuch as they proceed further [sc. ontologically] from God on the scale of being (*gradus entium*), are made more temporal, mutable, and mixed with material potential and privation, even though in God they are extremely simple, eternal, immutable, and immaterial. And it is likewise concerning the form of the order of existence and life (*forma ordinis esse et vitae* [the definition of fate above]), which in the celestial circle is necessary and immutable, that is, unalterable, but in generated things, because of the mutability of their existence, it[15] is received contingently and mutably.[16]

---

logical and ontological "no man's land" between the necessary and the possible. The clarity of Albert's analysis here should help to clarify the *Speculum astronomiae's* analysis there. Just below, Albert treats some of the same issues regarding the tensions surrounding the logical "truth" of an astrological interpretation about the future, regardless of whether what was predicted actually comes to pass.

[13] Omne enim quod procedit a causa nobili in causatum ignobile, licet in aliquo teneat proprietatem causae, tamen esse suum non est, nisi quantum permittit possibilitas subiecti, in quo est; omne enim quod recipitur, ut dicut BOETHIUS et ARISTOTELES in VI ETHICORUM, est in eo in quo recipitur, secundum potestatem recipientis et non secundum potestatem causae (68, 56–69, 3).

[14] [Ps.-]Dionysius the Areopagite. The influential pseudonymous texts thought to be written by a contemporary of St. Paul were actually composed by a contemporary of Proclus five centuries later. See (e.g.) *The Cambridge History of Late Antique and Early Medieval Philosophy*, A. H. Armstrong (ed.), Cambridge: Cambridge University Press, 1970. Albert produced commentaries for all of his major works. Thomas did so for his *On Divine Names*.

[15] Sc. the *forma ordinis esse et vitae*, namely, fate understood as the particular celestial configuration imprinting the thing generated at the moment of birth.

[16] Hoc possumus videre in his quae a DIONYSIO dicuntur processiones divinae, sicut est vita et ratio et sapientia et huiusmodi, quae secundum quod procedunt longius a deo secundum gradus

The form of the order of existence and life—derived from the celestial circle by means of motion and light, and analyzed in terms of the geometrical-optical model of celestial influences—is circumscribed in its actual effects by the limitations imposed by both the receptivity and the mutability of the particular generated thing in which the form inheres.

Albert then uses Boethius to develop his analysis further:

> Whence, in the 4th book of the *Consolation of Philosophy* (IV.6), Boethius makes a figure (*figurat*) of many circles, in the center of which is a pivot (*cardo*) that is the cause of fate and of fated things (*fatalia*). In the first circle next to the center is the immutability of fate, insofar as it is referred to its cause, and in the distant circle, in which the generable and corruptible things of the same fate are contained, is contingency and mutability through the existence of generated and corrupted things. Thus, the center is the disposition of the existence and life in the mind of the prime mover (*in mente motoris primi*); the circle next to the center is the same form of the disposition, insofar as it exists in the celestial period; and the circle distant from the center designates the same form of the disposition, insofar as it adheres mutably to generated and corrupted things.[17]

Albert uses this famous image from Boethius's *Consolation* to illustrate his own more properly astrological analysis, and also to insert God as prime mover into the equation.[18] As we will see in the final section of this chapter, the first two movements described in this image reflect the basic structure of the relationship between providence and fate, and thus of fate metaphysically inflected. The second and third articulate the structure of fate scientifically inflected, namely, as embodied in the ever unfolding nature of the celestial bodies and their motions in the zodiac.[19]

*

entium, efficiuntur magis temporalia et mutabilia et potentiae materiali et privationi permixta, cum tamen in deo sint simplicissima et aeterna et immutabilia et immaterialia. Et similiter est de forma ordinis esse et vitae, quae in caelesti circulo est necessaria et immutabilis, hoc est inalterabilis, in rebus vero generatis propter mutabilitatem esse ipsarum est recepta contingenter et mutabiliter (69, 3–14).

[17] Unde Boethius in IV *De consolatione philosophiae* figurat multos circulos, in quorum centro cardo et causa est fati et fatalium. Et in circulo primo propinquo centro est immutabilitas fati, secundum quod refertur ad causam, et in circulo distante, in quo continentur generabilia et corruptibilia eiusdem fati, est contingentia et mutabilitas per esse generatorum et corruptorum. Et sic centrum est dispositio esse et vitae in mente motoris primi; circulus iuxta centrum propinquus est eadem forma dispositionis, prout est in periodo caelesti, circulus autem a centro distans designat eandem formam dispositionis, prout mutabiliter adhaeret rebus generatis et corruptis (69, 14–27)

[18] For more on Boethius on fate (including a discussion of the passage that Albert discusses here [IV.6.15, pp. 108–9]), see Robert W. Sharples's editions and translations with a valuable introduction to and commentaries on Cicero, *On Fate* (*De fato*) and Boethius, *The Consolation of Philosophy* (*Philosophiae consolationis*), Westminster: Aris and Phillips, 1991. There will be more discussion of Boethius where relevant throughout this chapter.

[19] Thomas Aquinas will develop an influential analysis of divine providence in relation to astrology on this ontological basis in his *Summa contra gentiles* III.82-94, as we will see in detail below.

Albert develops his analysis further with reference to Ptolemy. First he recapitulates the general point:

> Moreover, this form, since it is an image of the period (*imago periodi*), has, in advance, potentially and virtually, the entire existence (*esse*) and activity (*operatio*) of the duration of what is generated and corrupted. And thus, although it exists [sc. in itself] necessarily, nevertheless it is mutable and contingent [sc. with respect to its actual expression in a generable and corruptible thing].[20]

Albert's formulation here of the form as an "image" of the "period of life" is worth emphasizing, especially since it has in itself *in potentia* information that represents the entire course of each particular individual generated thing's existence. This analysis thus picks up and develops Albert's recent discussion of the "period of life" in his translation-commentary on Aristotle's *De generatione et corruptione*.

One can easily see how valuable access to such insider information would be in a range of circumstances, medical, political and otherwise. This also relates directly to the possibility of prognostication, as we also saw in Chap. 1. And it clearly articulates the formal dimension of all astrological practices, especially nativities. This "form of the order of existence and life" from the heavens thus informs by imprinting the specific form of any entity with its particular structures and range of latitudes—i.e. its root complexion—by means of stellar rays at its birth-generation. As we will see, this also relates to the causal analysis of artificial or man-made objects, including talismans, in both the thirteenth and fifteenth centuries.[21]

Albert then turns to Ptolemy for an explanation:

> Ptolemy in the *Tetrabiblos* assigns the cause of this excellently, saying that the powers of the stars (*virtutes stellarum*) come to be in what is below *per aliud* and *per accidens*. Through something else (*per aliud*) because through the sphere of actives and passives [i.e., the four elements], through whose active and passive qualities they [sc. the *virtutes stellarum*] inhere in things below. By accident (*per accidens*), moreover, because although this form flows out (*effluat*) from a necessary and immutable cause, it happens[22] that it has existence in contingent and mutable things. Therefore, it has mutability from these two, namely, [1] from the qualities of the elements, through which it is brought down to what is generated, and [2] from the existence of what is generated, in which it exists as in a subject.[23] Therefore, this is fate.[24]

---

[20] Forma autem haec, cum sit imago periodi, potentialiter et virtualiter praehabet totum esse et operationem durationis generatorum et corruptorum; et sic, licet sit ex necessario, tamen est mutabilis et contingens (69, 27–31).

[21] This material may thus be used to supplement Quinlan-McGrath's analysis in *Influences*.

[22] *Accidit* picks up *per accidens*.

[23] That is, as in an Aristotelian substance, namely, an actually existing thing.

[24] Cuius causam optime assignat PTOLEMAEUS in QUADRIPARTITO, dicens, quod virtutes stellarum per aliud et accidens fiunt in inferioribus, per aliud quidem, quia per sphaeram activorum et passivorum, per quorum qualitates activas et passivas inhaerent inferioribus; per accidens autem, quia cum haec forma effluat a causa necessaria et immutabili, accidit ei habere esse in rebus contingentibus et mutabilibus. Ex duobus ergo habet mutabilitatem, scilicet ex qualitatibus elementorum, per quas defertur ad generata, et ex esse generatorum, in quo est sicut in subiecto (69, 32–43).

Albert has here very deliberately analyzed fate, which he defines as the form of the order of existence and life, in an explicitly astrologizing manner, and using Hermes, Aristotle, Ptolemy and Boethius as his primary authorities. Although fate as the form of something generated's order and existence is necessary in itself (because that of which it is an image exists necessarily, namely, the celestial bodies), when it has been instantiated in an actually existing thing, it is modified in the expression of its potentialities by the limitations of its actual matter and existence. Although celestial influences and the potential for the practice of nativities are woven deeply into this analysis, necessity and its concomitant determinism are fully undermined ontologically at the deepest levels, a conclusion Albert states explicitly in article 3, and which he uses to good effect in what follows.

5.2.3 The Epistemological Dimension

This penultimate stage here reconstructing Albert's astrologizing Aristotelian analysis of fate within an overtly theological context[25] addresses the epistemological nature of astrological analysis and some of the implications for prognosticating, a topic I will also address in relation to Thomas Aquinas's *Summa theologiae* in the following section. In *De fato*, article 4, Albert turns from ontological to related epistemological concerns, asking "Is fate knowable?" (*An fatum sit scibile*?). He begins the *solutio* with Ptolemy's distinction between the two parts of the science of the stars (73, 36–56). Concerned with planetary motions, what we call "astronomy" admits of certain knowledge, whereas "astrology" is concerned with celestial influences and their effects, and operates, by contrast, with conjecture.[26]

Albert then discusses the logical structure of conjecture:[27]

> The making of conjectures (*coniecturatio*), moreover, since it is from mutable signs (*ex signis mutabilibus*), generates a condition of less certainty than either knowledge (*scientia*) or opinion (*opinio*) does. For since signs of this sort are common and mutable (*communia et mutabilia*), the *via syllogistica* cannot be had from them for this reason, that neither in every (*nec in omnibus*) nor in most (*nec in pluribus*) cases do they include what has been signified (*significatum*). But insofar as it comes from itself, certain judgments are mutable from many causes (*iudicia quaedam multis de causis mutabilia*), as is clear from what was said before. And therefore, an astrologer often says something true (*saepe astronomus dicit*

[25] The final stage will be in the final section of this chapter.

[26] Solutio: Dicendum, quod duae partes sunt astronomiae, sicut dicit PTOLEMAEUS: una est de sitibus superiorum et quantitatibus eorum et passionibus propriis; et ad hanc per demonstrationem pervenitur. Alia est de effectibus astrorum in inferioribus, qui in rebus mutabilibus mutabiliter recipiuntur; et ideo ad hanc non pervenitur nisi per coniecturam, et oportet astronomum in ista parte secundum aliquid physicum esse, et ex signis physicis coniecturari (73, 36–44).

[27] For a history of conjecture that discusses astrology and divination (albeit not as fully as they deserve), see James Franklin, *The Science of Conjecture: Evidence and Probability before Pascal*, Baltimore: Johns Hopkins University Press, 2001. See also my related discussion in volume III on Placido Titi. In *De divinatione*, Cicero also discusses different types of divination as interpretive arts in terms of conjecture. This important topic is worthy of fuller study.

> *verum*) and, nevertheless, what he says does not come to pass (*non evenit*) because his statement was extremely true with respect to the disposition of the celestial bodies (*quoad dispositionem caelestium verissimum*), but this disposition was excluded by the mutability of things below (*haec dispositio a mutabilitate inferiorum exclusa est*).[28]

Qua logic, astrology does not admit of demonstrative syllogistic knowledge for essentially the same reasons—*qua* ontology—that it does not imply necessity in nature. Thus, although the astrologer can make a true prediction based entirely on the celestial dynamics, matter's opposing dispositions regularly exclude the predicted outcome from actually coming to pass. For Albert, then, the epistemology of astrology's conjectural predictions is grounded solidly on its astrologizing Aristotelian natural philosophical foundations.

*

Responding to an objection that fate cannot be known due to the vast number of interpretive factors, Albert develops his analysis further:[29]

> It seems that fate is not knowable (*quod non*). For, since it is an effect of the celestial circle (*effectus caelestis circuli*) and is a certain likeness of it (*similitudo quaedam ipsius*),[30] just as the form of the order of something is similar to the cause itself of the same order (*sicut forma ordinis alicuius similis est ipsi causae ordinis eiusdem*)—and in the celestial circle, as far as we are concerned, there are an infinite number of factors to consider, as stars (*stellae*) in number, type (*species*), and powers (*virtutes*), and their locations in the inclined circle (*in circulo declivi* [sc. the zodiac]) and beyond it, and distances and conjunctions and the quantity of the angle, under which a ray strikes (*sub quo indicit radius*), and the part of fortune, and the degrees of light and shadow in wells and towers, and an infinite number of this sort, as far as we are concerned—it will also seem that its effects cannot be known by us (*effectus a nobis sciri non possit*).[31]

---

[28] "Coniecturatio autem, cum sit ex signis mutabilibus, generat habitum minoris certitudinis, quam sit scientia vel opinio. Cum enim huiusmodi signa sint communia et mutabilia, non potest haberi ex ipsis via syllogistica, eo quod nec in omnibus nec in pluribus includunt significatum, sed quantum est de se, sunt iudicia quaedam multis de causis mutabilia, sicut patet per ANTEDICTA. Et ideo saepe astronomus dicit verum et tamen non evenit, quod dicit, quia dictum suum fuit quoad dispositionem caelestium verissimum, sed haec dispositio a mutabilitate inferiorum exclusa est (73, 45–56)." As we saw in Chap. 4, the *Speculum astronomiae* also treats this issue in ch. 14 in relation to future contingents.

[29] As we will see in Part IV below, Pietro d'Abano also treats this issue in an epistemological context.

[30] Albert used '*imago*' earlier in *De fato* and now '*similitudo*' here. We will recall that these are two of the same terms Roger Bacon used as synonyms for '*species*' in his discussion of their multiplication, as we saw in Chap. 2.

[31] Et videtur, quod non; cum enim sit effectus caelestis circuli et similitudo quaedam ipsius, sicut forma ordinis alicuius similis est ipsi causae ordinis eiusdem—et in caelesti circulo quoad nos infinita consideranda sunt, sicut stellae in numero et specie et virtutibus et situs earum in circulo declivi et extra ipsum et distantiae et coniunctiones et quantitas anguli, sub quo incidit radius, et pars fortunae et gradus lucidi et umbrosi in puteis et in turribus existentes et huiusmodi infinita quoad nos—, videbitur etiam, quod effectus a nobis sciri non possit (72, 56–66).

Alluding to the multiplicity of factors discussed above, the objection precludes the knowledge (or "knowability") of fate due to our inability to interpret the excessive number of celestial causal factors, of which he here indicates only a few.

As we would expect by now, Albert disagrees with this assessment:

> [W]e should say that indeed there are many factors to consider and that, as far as we [sc. human beings] are concerned, they are infinite, but only a very few are to be considered [sc. in an interpretation], which the others obey, and from these a prognosticatory conjecture (*pronosticabilis* [...] *coniecturatio*) can be had. For this reason, Ptolemy says that the practitioner of elections (*elector*) ought not to make a judgment unless it is probable and general, that is, by means of superior general causes (*per causas superiores communes*), which the proper [= proximate] causes of things (*propriae rerum causae*) frequently exclude.[32]

There are indeed many causal factors, but only a very few are required to make a prognosticatory conjecture. Albert explicitly mentions elections here. He also discusses universal celestial causes in relation to the proximate causes of actually existing things, a distinction Giovanni Pico della Mirandola exploited repeatedly and to good effect (albeit inconclusively) in his attempt to undermine astrology's natural philosophical foundations, and thus the possibility of prognostication, as we will see in detail in volume II.[33]

*

We have now seen Albert once again graft explicitly astrologizing features onto central Aristotelian and now Boethian conceptual structures that causally relate the heavens to life on earth in general, and to the generation of individuals in particular. In their turn, these ontological and metaphysical structures provide the basis for the knowledge-oriented practices of interpretation, and thus of understanding and prediction. Albert integrally linked this astrologizing Aristotelian natural philosophical analysis to the particular structure of the horoscope, and so to an individual's astrologically conditioned (but by no means inexorably determined) fate, for which we now have a better sense of its formal-metaphysical dimension and its relation to some significant epistemological issues. Finally, we should note that Albert did not at all shy away from analyzing fate with deeply astrological conceptual structures, even in a disputation enacted at the papal curia and most likely before the pope himself.[34] We will explore further aspects of Albert's position when I analyze his

---

[32] Ad primum dicendum, quod quidem multa et quoad nos infinita consideranda essent, sed considerantur paucissima, quibus oboediunt alia, et ex illis pronosticabilis habetur coniecturatio. Propter hoc dicit PTOLEMAEUS, quod elector non nisi probabiliter et communiter iudicare debet, hoc est per causas superiores communes, quas propriae rerum causae frequentissime excludunt (73, 57–64).

[33] This theme runs throughout Pico's *Disputationes adversus astrologiam divinatricem*. I have studied it particularly in book III, chapters 1–6, in chapter five of my PhD thesis, to which I may refer the interested reader. I also discuss it in volume II. This is precisely the *vera causa*, whose discovery Ernst Cassirer attributes to Pico. See his *Individual and Cosmos in Renaissance Philosophy*, 115–20.

[34] For the high-level scholarly and scientific interests of the papal curia at this time—including investigations into astrology, *perspectiva*, alchemy and medicine—see Paravicini Bagliani's rich

views on the more metaphysically-oriented relationship between providence and fate in the final section of this chapter.

Although Albert further articulates his system in greater depth and detail in these and other authentic works, this reconstruction must suffice for now, as we turn to compare the epistemological, ontological and metaphysical issues raised here with those raised by Thomas Aquinas in a text also deeply connected to theological concerns, and one that provided a major authority for the next several centuries—even into the eighteenth century—as we will see in volume III. We should also recall that *De fato* was regularly attributed to Thomas Aquinas in the fifteenth century and beyond, as we will see with Marsilio Ficino in volume II.[35] We will now move from fate to divination, another topic with a significant theological resonance. The next few sections of this chapter will be devoted to Albert's most famous student, Thomas Aquinas.

5.3 Astrology and Theology (2): Divination

5.3.1 Thomas Aquinas, Summa Theologiae, *II.II.92-95*

What, then, is astrology's relationship to divination? Astrology is often configured with divination, even today,[36] but these and related conceptual and terminological structures should be problematized and more precisely historicized.[37] A useful and

---

volume of studies *Medicina e scienze della natura alle corte dei papi nel Duecento*, and Lindberg, "Lines of Influence."

[35] In Simon's prolegomena to his critical edition, he shows that, of the 14 known manuscripts, seven are attributed to Albert and seven to Thomas, but none of the ancient catalogues of Albert's works list it (p. XXXIII). We are now fully persuaded that it is an authentic work by Albert by both internal and external criteria as Simon argues in detail in the prolegomena to his edition (pp. XXXIII-IV).

[36] See (e.g.) Patrick Curry and Roy G. Willis, *Astrology, Science and Culture: Pulling Down the Moon*, Oxford: Berg, 2004; *Seeing with Different Eyes: Essays on Astrology and Divination*, Patrick Curry and Angela Voss (eds.), Newcastle: Cambridge Scholars Publishing, 2007; *Divination: Perspectives for a New Millenium*, Patrick Curry (ed.), Burlington: Ashgate, 2010, and Geoffrey Cornelius, "Interpreting Interpretations: The Aphorism in the Practice of the Renaissance Astrologers," in *From Masha'Allah to Kepler*, 101–21. The foundational modern work in this tradition is Geoffrey Cornelius, *The Moment of Astrology: Origins in Divination* (revised and expanded second edition, Bournemouth, England: The Wessex Astrologer, 2003; originally published 1994), a copy of which the author kindly gave me after a marvelous conference in Erlangen (2016).

[37] There is, of course, an extensive bibliography on divination. In additon to Boudet's relevant chapters in *Entre science et nigromance*, and Dieter Harmening, *Superstitio: Überlieferungs- und theoriegeschichtliche Untersuchung zur kirchlich-theologischen Aberglaubensliteratur des Mittelalters*, Berlin: E. Schmidt, 1979, I have found these useful: William E. Klingshirn, "Isidore of Seville's Taxonomy of Magicians and Diviners," *Traditio* 58 (2003): 59–90, his "Divination and the Disciplines of Knowledge according to Augustine," in *Augustine and the Disciplines: From*

influential text for doing so is Thomas Aquinas's illuminating and authoritative discussion in an explicitly theological context in *Summa theologiae* II^{a} II^{ae} 92-95. Written between January 1271 and Easter 1272, this text came to be a major authority for the next several centuries.[38]

In clarifying the theological context of superstition and idolatry,[39] Thomas analyzed legitimate and illegitimate modes of knowing and predicting the future, only the latter of which are properly called "divination" (*divinatio*). In the process, he clearly explains that what he calls '*astrologia*', which we may neutrally translate as "the science of the stars," in its two main respects—what we call "astronomy" and "astrology"—are perfectly legitimate modes of knowledge, and thus of prediction. He distinguishes them sharply from divination proper, which he goes on to clarify in itself, and in language clearly reflected in the 1557, 1559, 1564 and later *Indexes of Prohibited Books*, and in Sixtus V's and Urban VIII's closely related anti-astrological (or better, anti-divinatory) Bulls, *Coeli et Terrae Creator* (1586) and *Inscrutabilis* (1631), which are all discussed in detail in volume III. This discussion provides essential background thereto.[40]

*

In Question 92, article 1 of the Secunda Secundae, Thomas discusses and defines "superstition" (*superstitio*), which he concludes thus: "[S]uperstition is a vice opposed to religion (*religio*) by excess, not because it offers more divine worship (*cultus divinus*) than true religion, but rather because it offers divine worship either

---

Cassiciacum to Confessions, K. Pollmann and M. Vessey (eds.), Oxford: Oxford University Press, 2005, 113–40, and Valerie I.J. Flint, *The Rise of Magic in Early Medieval Europe*, Princeton: Princeton University Press, 1991, although I find the definition of magic in her introduction deeply problematic conceptually. In future research, I hope to more fully explore Thomas's analysis here vis-à-vis Augustine's authoritative discussions of astrology within a similar context of divination, fate, superstition and divine providence.

[38] For the dating, see Torrell, *Saint Thomas Aquinas*, 145–48 and 333. I discuss the following question (II.II.96) on *imagines astronomicae* in Chap. 7. Boudet treats Thomas's views on astrology and magic at *Entre science et nigromance*, 228–34. He briefly but insightfully discusses Thomas's views in *Summa contra gentiles* III.82-92 and 103-10, *De iudiciis astrorum, De operationibus occultis naturae* and *Summa theologiae* II.II.92-96. He says that, although Thomas was inspired by his teacher Albert, Thomas himself offered "un synthèse neuve" (228).

[39] On superstition in particular, in addition to Harmening, *Superstitio*, see Edward Peters, *The Magician, The Witch and the Law*, Philadelphia: University of Pennsylvania Press, 1978, and his "The Medieval Church and State on Superstition, Magic and Witchcraft: From Augustine to the Sixteenth Century," in *Witchcraft and Magic in Europe: The Middle Ages*, Bengt Ankarloo and Stuart Clark (eds.), Philadelphia: University of Pennsylvania Press, 2002, 173–245. See now also Euan Cameron, *Enchanted Europe: Superstition, Reason, and Religion, 1250–1750*, Oxford: Oxford University Press, 2010, and Michael D. Bailey, *Fearful Spirits, Reasoned Follies: The Boundaries of Superstition in Late Medieval Europe*, Ithaca: Cornell University Press, 2013, which should be used with caution concerning the issues treated here.

[40] In his "Thomas von Aquin und die Mantik," Loris Sturlese persists in calling astrology "astrologisches Mantik." He does not discuss *Summa theologiae* II.II.92-95, although he cross-references (cfr's) it in nn. 14 and 18, but without any discussion; see *Mantik, Schicksal und Freiheit*, 97–107.

to something not deserving it, or in an improper manner (5)."[41] In article 2, Thomas discusses the different species of superstition:

> Therefore, the species of superstition can be differentiated, first, with respect to the object (*ex parte objecti*). Divine worship can be offered to whom it should be offered, namely the true God, but in an improper manner; this is the first species of superstition. Or it can be offered to whom it should not be offered, namely to any creature (*cuicumque creaturae*). This is another genus of superstition, which is divided into many species corresponding to the different ends (*fines*) of divine worship: [1] [Idolatry] First, divine worship is oriented to showing reverence to God. Accordingly, the first species of this genus is idolatry (*idololatria*), which offers divine worship undeservedly to a creature [i.e. to something created, *not* to the Creator Himself]. [2] [Divination] Secondly, divine worship is oriented to that which human beings (*homo*) learn from God whom they worship. To this belongs divinatory superstition (*superstitio divinitiva*), which consults demons through certain pacts—tacit or explicit—entered into with them. [3] [Practices] Thirdly, divine worship is oriented towards a certain arrangement of human actions according to God's prescriptions, whom we worship. To this belongs the superstition of certain practices (*observationes*, 7–9).[42]

The ends towards which—the final causes—thus differentiate the species of this genus.

Of these three types of superstition, the second subdivision—*superstitio divinitiva*—is our principal concern, which Thomas clarifies further just below: "[S]ome [2] divinations and [3] practices belong to superstition to the extent that they depend on the actions of demons (*operationes daemonum*), and thus require compacts made with them (9)."[43] For Thomas, then, the pact with demons is a decisive criterion for characterizing superstition and, thus, a practice's legitimacy. Demons will be discussed more fully below.

*

[41] "Sic igitur superstitio est vitium religioni oppositum secundum excessum, non quia plus exhibeat in cultum divinum quam vera religio: sed quia exhibet cultum divinum vel cui non debet, vel eo modo quo non debet (298b7-12)." I use the readily available bilingual Blackfriar edition for the English translation, which provides the basis for my often significantly modified translation; Summa theologia: *Latin text and English Translation*, 61 vols., London: Blackfriars, 1964–80, Vol. 40 (1968), 1–69. I cite the Latin from the Leonine edition of Thomas's *Opera omnia*, Vol. 9 (1897), 298–322; I number the lines with 'a' and 'b' for the two columns respectively, beginning with '1' at the top of each column.

[42] Diversificatur ergo superstitionis species, primo quidem, ex parte objecti. Potest enim divinus cultus exhiberi vel cui exhibendus est, scilicet Deo vero, *modo tamen indebito*: et haec est prima superstitionis species. Vel ei cui non debet exhiberi, scilicet cuicumque creaturae. Et hoc est aliud superstitionis genus, quod in multas species dividitur, secundum diversos fines divini cultus. Ordinatur enim, primo, divinus cultus ad reverentiam Deo exhibendam. Et secundum hoc, prima species huius generis est *idololatria*, quae divinam reverentiam indebite exhibet creaturae. Secundo, ordinatur ad hoc quod homo instruatur a Deo, quem colit. Et ad hoc pertinet superstitio *divinitiva*, quae daemones consulit per aliqua pacta cum eis inita, tacita vel expressa. Tertio, ordinatur divinus cultus ad quandam directionem humanorum actuum secundum instituta Dei, qui colitur. Et ad hoc pertinet superstitio quarundam *observationum* (299a31-b11).

[43] Ad secundum dicendum quod divinationes et observationes aliquae pertinent ad superstitionem inquantum dependent ex aliquibus operationibus daemonum. Et sic pertinent ad quaedam pacta cum ipsis inita (299b28-32).

Question 95 on *superstitio divinitiva* treats divination extensively, including that through dreams, demons, augury, the stars, and the casting of lots.[44] Articles 1 and 5 are of great interest here, in that Thomas explicitly and in detail shows how astrology is a legitimate form of knowledge and why. Article 1 addresses whether divination is a sin (*utrum divinatio sit peccatum*?). Thomas's response is illuminating:

> I respond by saying that by the term '*divinatio*' is understood a certain foretelling of future things (*praenuntiatio futurorum*). Moreover, future things can be foreknown (*praenosci possunt*) in two ways: in one way, in their causes (*in suis causis*); in the other, in themselves (*in seipsis*). The causes of future things are threefold: [1] Some necessarily and always (*ex necessitate et semper*) produce their effects. Future effects of this sort can be foreknown and foretold with certainty (*per certitudinem praenosci possunt et praenuntiari*) from the consideration of their causes, as when *astrologi* predict future eclipses. [2] But some causes produce their effects not necessarily and always (*non ex necessitate et semper*), but in most cases (*in pluribus*), and rarely fail. From such causes their future effects can be foreknown, not with certainty (*non* […] *per certitudinem*) but by some conjecture (*per quandam conjecturam*), as when *astrologi*, by considering the stars, are able to foreknow and foretell some things (*quaedam*) about rains or droughts, and physicians (*medici*) about health or death (37, 39).[45]

Thomas here discusses what can causally (and thus legitimately) be foreknown and foretold. He distinguishes two modes. In the first, the causes always and necessarily produce their effects. These can be foreknown and thus foretold with certainty. In the second, the causes only produce their effects for the most part, that is, in the normal course of nature. These too can be foreknown and foretold, but only conjecturally, that is, with a certain probability. Regardless of the degree of certainty, then, both are perfectly legitimate modes of predicting the future because the foreknowing and foretelling are done by means of *causal* knowledge.

Thomas's examples are informative. In the first, '*astrologi*' predict eclipses with certainty from causes that always and necessarily produce the same effects. *Astrologi* here should be translated "astronomers," since they are concerned with planetary motions and their predictions. In the second case, *astrologi* (once again) make predictions about rains and droughts, namely weather phenomena, but since the relevant causes do not always and necessarily produce the same effects, the predictions can only be probable. Here *astrologi* refer to "astrologers" and their revolutions or general astrology that predict celestial influences on the world at large and in particular regions—including in annual prognostications—and especially to the subgroup of revolutions devoted to the weather: *mutatio temporum*.

---

[44] For the range of divinatory practices, see Harmening, *Superstitio*, 178–216.

[45] Respondeo dicendum quod in nomine divinationis intelligitur quaedam praenuntiatio futurorum. Futura autem dupliciter praenosci possunt: uno quidem modo, in suis causis; alio modo, in seipsis. Causae autem futurorum tripliciter se habent. Quaedam enim producunt ex necessitate et semper suos effectus. Et huiusmodi effectus futuri per certitudinem praenosci possunt et praenuntiari ex consideratione suarum causarum: sicut astrologi praenuntiant eclipses futuras. Quaedam vero causae producunt suos effectus non ex necessitate et semper, sed ut in pluribus, raro tamen deficiunt. Et per huiusmodi causas possunt praenosci futuri effectus, non quidem per certitudinem, sed per quandam conjecturam: sicut astrologi per considerationem stellarum quaedam praenoscere et praenuntiare possunt de pluviis et siccitatibus, et medici de sanitate vel morte (311a27-b9).

As we will recall, in *Tetrabiblos* I.1, Ptolemy made the very same conceptual distinction within the science of the stars (*astronomia*) between what we call "astronomy" and "astrology" with respect to certainty vs. conjecture, where astronomy is an exact science and astrology, conjectural. In *Tetrabiblos* I.2-3, Ptolemy also discussed why, although not an exact science, astrology is still useful, comparing astrology and medicine in this respect, which are both identified explicitly as conjectural sciences, as they also are here. Although Thomas uses the same term '*astrologi*' to refer to both practitioners, he refers here to two different sets of practices that he distinguishes conceptually but not terminologically.

Thomas then draws a telling contrast:

> [3] But there are some causes which, if they are considered *in themselves* (*secundum se*), can go either way (*ad utrumlibet*). This is the case particularly (a) with rational powers (*potentiae rationales*) which are poised before opposites, according to Aristotle.[46] And such effects, or also (b) if any effects come about by chance in few cases (*in paucioribus casu accidunt*),[47] they cannot be foreknown by considering causes because their causes do not have a determinate inclination (*inclinatio determinata*) toward such effects. Consequently, these effects cannot be foreknown unless they are considered in themselves (*in seipsis*, 39).[48]

The causal knowledge in [1] and [2] is thus sharply contrasted with [3a] and [3b], the two types of knowledge *in itself*, namely [a] rational choice, which has the ability to choose different outcomes (and thus relates to free will), and [b] irregular and infrequent events that come about by chance. Human beings cannot legitimately foreknow or thus foretell these because they do not have consistent and determinative, and thus knowable causes. These latter cases, therefore, can only be known in themselves, the second type of knowledge. We have already seen this significant phrase '*ad utrumlibet*' in related contexts in both Roger Bacon and the *Speculum astronomiae* in Chap. 4.

Thomas goes on to clarify an epistemological dimension of these issues by sharply distinguishing between God's cognitive abilities and ours:

> Moreover, human beings (*homines*) can consider effects of this sort in themselves only when they are present, as when a man sees Socrates running or walking. But to consider such in themselves before they come to be is proper to God (*Dei proprium*), who alone in his eternity sees what will be as present [...]: whence Isaiah says: "Announce what will come in the future and we will know that you are gods." Therefore, if someone presumes to foreknow or foretell future things of this sort in any manner whatsoever, except with God's revelation (*nisi Deo revelante*),[49] he manifestly usurps to himself

[46] *Metaphysics* IX.5. The references are all identified in the margins of the Leonine edition and in the footnotes of the Blackfriars edition. I only note the most significant ones here.

[47] *In paucioribus*, as opposed to *in pluribus* from [2] just above.

[48] Quaedam vero causae sunt quae, si secundum se considerentur, se habent ad utrumlibet: quod praecipue videtur de potentiis rationalibus, quae se habent ad opposita, secundum Philosophum. Et tales effectus, vel etiam si qui effectus ut in paucioribus casu accidunt ex naturalibus causis, per considerationem causarum praenosci non possunt: quia eorum causae non habent inclinationem determinatam ad huiusmodi effectus. Et ideo effectus huiusmodi praenosci non possunt nisi in seipsis considerentur (311b10-20).

[49] Girolamo Savonarola made much of this caveat in his treatise against the astrologers. See Claudio Gigante's informative introduction to his edition of Savonarola's *Contro gli astrologi*, Rome: Salerno Editrice, 2000, 7–23.

> what is God's. And from this, some are called diviners (*divini*). Whence Isidore says in the *Etymologies*: "Men are called diviners as though they are full of God. They pretend to be filled with divinity (*divinitate plenos*) and they forecast the future for men by shrewd fraud (39)."[50]

Here Thomas relates this illegitimate foreknowledge for human beings to the earlier discussion of *superstitio*, namely, the usurpation of what is properly God's.

Thomas concludes by discussing the proper use of the term '*divinatio*' and evaluating its sinfulness:

> Therefore, it is *not* called "divination" (*divinatio*) if someone foretells those things which come about [1] by necessity or [2] for the most part, which can be foreknown by human reason (*ratio humana* [that is, by causal analysis]). Nor is it divination if someone knows other contingent future matters (*futura alia contingentia*)[51] when God reveals them (*Deo revelante*). For then he does not divine (*divinat*), that is, do what is divine (*quod divinum est facit*), but rather he receives (*suscipit*) what is divine. It is only called divining (*divinare*), then, when one usurps to himself in an inappropriate manner (*indebito modo*) the foretelling of future events. This is considered a sin (*peccatum*), whence *divinatio* is always a sin. And because of this, Jerome (*Super Michaeam*) says that *divinatio* always has a negative sense (my emphasis, 39).[52]

Thomas's use of *divinatio* and its negative evaluation is crystal clear: Foretelling the future from necessary or regular causes or by divine revelation is *not* divination. Only the inappropriate foretelling of events is divination and a sin. Although he does not identify them as such here, the two types of future contingent matters that can only be known in themselves—namely, those arising from rational choice and by chance—become central to the description of what is later called "judicial astrology," as we will see more fully in volume III. In this first strong statement of Thomas's position (T1), then, they are both associated with divination and categorically rejected.

---

[50] Homines autem in seipsis huiusmodi effectus considerare possunt solum dum sunt praesentes, sicut cum homo videt Socratem currere vel ambulare. Sed considerare huiusmodi in seipsis antequam fiant, est Dei proprium, qui solus in sua aeternitate videt ea quae futura sunt quasi praesentia, ut in Primo habitum est: unde dicitur Isaiae XLI, *Annuntiate quae futura sunt in futurum, et sciemus quoniam dii estis vos*. Si quis ergo huiusmodi futura praenoscere aut praenuntiare quocumque modo praesumpserit, nisi Deo revelante, manifeste usurpat sibi quod Dei est. Et ex hoc aliqui *divini* dicuntur: unde dicit Isidorus, in Libro Etymol (VIII, 9): *Divini dicti quasi Deo pleni: divinitate enim se plenos simulant, et astutia quadam fraudulentiae hominibus futura conjectant* (311b20-36).

[51] What can be known about future contingent events is, of course, a major issue in both scholastic logic and theology; see e.g. Schabel, *Theology at Paris*, with much further bibliography, including relevant historical background.

[52] Divinatio ergo non dicitur si quis praenuntiet ea quae ex necessario eveniunt vel ut in pluribus, quae humana ratione praenosci possunt. Neque etiam si quis futura alia contingentia, Deo revelante, cognoscat: tunc enim non ipse *divinat*, idest quod divinum est facit, sed magis quod divinum est suscipit. Tunc autem solum dicitur divinare quando sibi indebito modo usurpat praenuntiationem futurorum eventuum. Hoc autem constat esse peccatum. Unde divinatio semper est peccatum. Et propter hoc Hieronymus dicit, *super Michaeam,* quod *divinatio semper in malam partem accipitur* (312a1-13).

*

In responding to the second objection *quod non*, Thomas reiterates this distinction and returns to the demonic dimension:

> *Ad* 2: There are certain arts (*artes quaedam*) for foreknowing future events that happen necessarily or frequently; this does not belong to *divinatio*. But for knowing other future events, there are no other true arts or disciplines (*verae artes seu disciplinae*), but they are false and vain (*fallaces et vanae*), introduced by the deception of demons; as Augustine says in Book XXI of *De civitate Dei* (41).[53]

Here Thomas closely follows Augustine.[54]

Thomas clarifies precisely why just below (article 2), asking whether *divinatio* is a species of superstition:

> Accordingly, it belongs to superstition not only when a sacrifice is offered to demons in idolatry (*non solum cum sacrificium daemonibus offertur per idololatriam*), but also when someone takes on the assistance of demons for doing or knowing something (*cum aliquis assumit auxilium daemonum ad aliquid faciendum vel cognoscendum*). Now all divination comes about from the activity of demons (*operatio daemonum*), either because demons are expressly invoked to manifest [sc. what will happen] in the future (*vel quia expresse daemones invocantur ad futura manifestanda*), or because demons impose themselves on these futile searchings into the future in order to entangle the minds of human beings with vain conceits. [...] Moreover, the inquiry after future matters is vain when someone attempts to foreknow the future about what cannot be foreknown. Clearly, then, *divinatio* is a species of superstition (41, 43).[55]

For Thomas, divination—the attempt to know the future about what cannot legitimately be foreknown—always involves demons (whether implicitly or explicitly), and is thus a species of superstition and a sin.

Responding to article 3, Thomas reiterates the central point that "all *divinatio* uses the advice and assistance of demons for the foreknowledge of a future event (43)."[56] Here are his conclusions:

---

[53] Ad secundum dicendum quod artes quaedam sunt ad praecognoscendum futuros eventus qui ex necessitate vel frequenter proveniunt, quod ad divinationem non pertinet. Sed ad alios futuros eventus cognoscendos non sunt aliquae verae artes seu disciplinae, sed fallaces et vanae, ex deceptione daemonum introductae; ut dicit Augustinus, in XXI de Civ. Dei (312b3-10).

[54] On Augustine's views, see Klingshirn, "Divination and the Disciplines of Knowledge."

[55] Et ideo ad superstitionem pertinet non solum cum sacrificium daemonibus offertur per idololatriam, sed etiam cum aliquis assumit auxilium daemonum ad aliquid faciendum vel cognoscendum. Omnis autem divinatio ex operatione daemonum provenit: vel quia expresse daemones invocantur ad futura manifestanda; vel quia daemones se ingerunt vanis inquisitionibus futurorum, ut mentes hominum implicent vanitate [...]. Vana autem inquisitio futurorum est quando aliquis futurum praenoscere tentat unde praenosci non potest. Unde manifestum est quod divinatio species superstitionis est (313a14-b3).

[56] Omnis divinatio utitur ad praecognitionem futuri eventus aliquo daemonum consilio et auxilio (315a23-25).

> *Ad* 3: Therefore, it is clear that the genus of divination (*genus divinationis*) is threefold. The first is by open invocation of demons (*per manifestam daemonum invocationem*); this belongs to necromancers (*nigromantici*). The second is only by observing the disposition and movement of something else; this belongs to augurs (*augures*). The third is when we do something in order to make something hidden (*aliquid occultum*) manifest to us: this belongs to lots (*sortes*).[57] Within each of these categories are many subtypes (47)[.][58]

Thus causal knowledge of the future is legitimate and is not to be called '*divinatio'*, which both has truck with demons and is a sin, as Thomas here states emphatically and in no uncertain terms.

*

In Question 95, article 5, Thomas directly and extensively adresses our issue of central concern, namely, whether divination through the stars is illicit (*utrum divinatio quae fit per astra sit illicita*). He begins with a positive quote from Augustine:

> But against this [sc. that astrology is illicit], Augustine says in Book IV of the *Confessions*: 'I never hesitated to consult those astrologers (*planetarios*) whom they call *mathematici* because they required no sacrifice or prayers to any spirit for divination, which true Christian piety rejects and condemns' (51).[59]

Even for Augustine, a normally notorious anti-astrological authority, at least the astrologers required no sacrifices or prayers, and thus did not invoke demons.[60]

Thomas then begins his reply proper, which I will treat in full, beginning with demons:

> I respond, saying that, as already remarked, a demonic act (*operatio daemonum*) rushes upon divination that derives from false or vain opinion, so that our souls become mazed in vanity and falsehood. Moreover, one uses vain or false opinion if, by considering the stars, he wishes to foreknow future events which cannot be foreknown by their means. Therefore, we should consider what about the future can be foreknown from observing celestial bodies (*quid per caelestium corporum inspectionem de futuris possit praenosci*). And, concerning

[57] The Blackfriars translate this as "sorcery," which seems extreme, but is also etymologically related. Thomas has a contemporary self-standing treatise (a consultation) on lots, *De sortibus*, which is worth comparing to the arguments here, but would take us too far afield.

[58] Sic igitur patet triplex esse divinationis genus. Quorum primum est per manifestam daemonum invocationem: quod pertinet ad *nigromanticos*. Secundum autem est per solam considerationem dispositionis vel motus alterius rei: quod pertinet ad *augures*. Tertium est dum facimus aliquid ut nobis manifestetur aliquid occultum: quod pertinet ad *sortes*. Sub quolibet autem horum multa continentur, ut patet ex dictis (316a19-b4).

[59] Sed contra est quod Augustinus dicit, in IV Confess: *Illos planetarios quos mathematicos vocant, consulere non desistebam; quod quasi nullum esset eis sacrificium, et nullae preces ad aliquem spiritum ob divinationem dirigerentur. Quod tamen Christiana et vera pietas expellit et damnat* (328a4-10).

[60] Filippo Fantoni (ca. 1530–91) also used Augustine as a pro-astrological authority in the teaching manuscript of his sixteenth-century course on Ptolemy's *Tetrabiblos* at the University of Pisa. On this manuscript, see my "The Use and Abuse of Ptolemy's *Tetrabiblos* in Renaissance and Early Modern Europe: Two Case Studies (Giovanni Pico della Mirandola and Filippo Fantoni)," in *Ptolemy in Perspective: Use and Criticism of his Work from Antiquity to the Nineteenth Century*, Alexander Jones (ed.), Dordrecht: Springer, 2010, 135–149, 143, and in my "Teaching Astrology in the 16th Century."

> what comes about necessarily, it is obvious that they can be foreknown through consideration of the stars, as *astrologi* foretell future eclipses (51, 53).[61]

In order to clarify his position on what can legitimately be known about the future by considering celestial bodies, Thomas first reiterates the point made earlier in 95.1 concerning what comes about from necessary causes, namely, the prediction of eclipses. He contrasts this pointedly with what can be derived from false and vain opinion with demonic assistance.

Thomas now recounts several different positions on the matter, beginning with the view that the stars signify but do not cause:

> But concerning the foreknowledge of future events from a consideration of the stars (*praecognitio futurorum eventuum ex consideratione stellarum*), there have been different opinions: Some have said that the stars signify (*significant*) rather than do/make (*faciant*)[62] what is foretold from their consideration. But this is unreasonable, for every corporeal sign (*omne* [...] *corporale signum*) is either [1] an effect of that of which it is a sign, as smoke signifies fire, from which it is caused, or [2] it proceeds from the same cause, and thus, when it signifies the cause, by consequence it also signifies the effect, as a rainbow when it signifies fair weather (*serenitas*), in that its cause is also the cause of fair weather. But, it cannot be said that the dispositions of the celestial bodies and their motions are the effects of future events (*non autem potest dici quod dispositiones caelestium corporum et motus sint effectus futurorum eventuum*). Nor again can they be referred [sc. causally (*reduci*)] to some common higher cause that is corporeal. They can, however, be referred to one common cause, which is divine providence (*providentia divina*). But [1] the motions and locations (*situs*) of the celestial bodies are arranged (*disponuntur*) by divine providence by one causal structure (*ratio*), and [2] the outcomes of future contingents (*eventus contingentium futurorum*) by another, since the former [1] are arranged according to the structure of necessity (*secundum rationem necessitatis*), so that they always turn out in the same manner. Whereas the latter [2] are arranged according to the structure of contingency (*secundum rationem contingentiae*) so that they turn out differently (*variabiliter contingant*, 53).[63]

---

[61] Respondeo dicendum quod, sicut dictum est, divinationi quae ex opinione falsa vel vana procedit, ingerit se operatio demonis, ut hominum animos implicet vanitati aut falsitati. Vana autem aut falsa opinione utitur si quis ex consideratione stellarum futura velit praecognoscere quae per ea praecognosci non possunt. Est igitur considerandum quid per caelestium corporum inspectionem de futuris possit praenosci. Et de his quidem quae ex necessitate eveniunt, manifestum est quod per considerationem stellarum possunt praenosci: sicut astrologi praenuntiant eclipses futuras (320a11-23).

[62] The verb '*facere*' here directly reflects the Aristotelian notion of efficient causality, to which Thomas implicitly refers in this entire discussion of causal knowledge derived from the stars.

[63] Circa praecognitionem vero futurorum eventuum ex consideratione stellarum, diversi diversa dixerunt. Fuerunt enim qui dicerent quod stellae significant potius quam faciant ea quae ex earum consideratione praenuntiantur. Sed hoc irrationabiliter dicitur. Omne enim corporale signum vel est effectus eius cuius est signum, sicut fumus significat ignem, a quo causatur: vel procedit ab eadem causa, et sic, dum significat causam, per consequens significat effectum, sicut iris quandoque significat serenitatem, inquantum causa eius est causa serenitatis. Non autem potest dici quod dispositiones caelestium corporum et motus sint effectus futurorum eventuum. Nec iterum possunt reduci in aliquam superiorem causam communem quae sit corporalis. Possunt autem reduci in unam causam communem quae est providentia divina: sed alia ratione disponuntur a divina providentia motus et situs caelestium corporum, et alia ratione eventus contingentium futurorum; quia illa disponuntur secundum rationem necessitatis, ut semper eodem modo proveniant; haec autem secundum rationem contingentiae, ut variabiliter contingant (320a23-47).

Thomas's causal analysis thus removes the possibility that the stars signify without also being causes. He also indicates that, although both celestial bodies and contingent events are ultimately arranged by the same common incorporeal superior cause, namely, divine providence, nevertheless, they are arranged with two significantly different causal structures. The celestial bodies are arranged by divine providence with a necessary and regular causal structure—which thus allows for legitimate foreknowledge and prediction—in contrast to the outcome of future contingent events, which have a strikingly different, non-necessary contingent structure that results in variable outcomes.[64] I discuss Thomas's views on divine providence further below.

Thomas then draws a broader conclusion relating legitimate foreknowledge to the knowledge of causes. In particular, one must bear in mind that not everything is caused by celestial bodies:[65]

> Whence, it cannot be the case that foreknowledge of the future (*praecognitio futurorum*) is drawn from an inspection of the stars (*sidera*) except as effects are foreknown from causes. Moreover, two kinds of effects are withdrawn from the causality of celestial bodies (*duplices autem effectus subtrahuntur causalitati caelestium corporum*). The first are all the effects that happen by accident (*contingentes per accidens*), both in human affairs and in the world of nature (*sive in rebus humanis sive in rebus naturalibus*), because, as proved in [Aristotle's] *Metaphysics*, a chance event does not have a cause (*ens per accidens non habet causam*) and especially a natural one, such as the power (*virtus*) of celestial bodies. Because what comes to be by accident is not strictly speaking a single entity, as when a boulder falls and a landslide ensues, or as when a man digging a grave finds a treasure. For these occurences and others of this sort are not one but many *simpliciter*. Whereas, the operation of nature always finishes at something single, just as it starts from a single principle, namely the form of a natural thing (*forma rei naturalis*, 53).[66]

In the realm of contingent events, accidents (chance events) in both human affairs and in nature do not have single causes with determinative inclinations toward particular effects (as Thomas explained in 95.1); therefore, they cannot be causally foreknown by human beings.

---

[64] For a penetrating analysis of Thomas's views on necessity and contingency, see Pasquale Porro, "*Lex necessitatis vel contingentiae*: Necessità, contingenza e provvidenza nell'universo di Tommaso d'Aquino," *Revue des Sciences Philosophiques et Théologiques* 96 (2012): 401–450. My thanks to Alessandro Palazzo for this reference. Albertus Magnus's analysis in *De fato* is closely related to Thomas's here, as we just saw.

[65] Ptolemy also makes this point in *Tetrabiblos* I.1-3.

[66] Unde non potest esse quod ex inspectione siderum accipiatur praecognitio futurorum nisi sicut ex causis praecognoscuntur effectus. Duplices autem effectus subtrahuntur causalitati caelestium corporum. Primo quidem, omnes effectus per accidens contingentes, sive in rebus humanis sive in rebus naturalibus. Quia, ut probatur in VI *Metaphys.*, ens per accidens non habet causam: et praecipue naturalem, cuiusmodi est virtus caelestium corporum. Quia quod per accidens fit neque est ens proprie neque unum: sicut quod, lapide cadente, fiat terraemotus, vel quod, homine fodiente sepulcrum, inveniatur thesaurus; haec enim, et huiusmodi, non sunt unum, sed simpliciter multa. Operatio autem naturae semper terminatur ad aliquid unum: sicut et procedit ab uno principio, quod est forma rei naturalis (320a48-b9).

The other effects not caused by celestial bodies involve free will actions by human beings. Here Thomas makes an argument from the nature of the human psyche according to Aristotle's *De anima*:

> Secondly, acts of free choice (*actus liberi arbitrii*), which is [sc. a part of] the faculty of will (*voluntas*) and reason (*ratio*), are removed from the causality of the celestial bodies (*subtrahuntur causalitati caelestium corporum*). For intellect (*intellectus*) or reason (*ratio*) is not a body, nor the act of a bodily organ (*non est corpus nec actus organi corporei*). Consequently, neither is the will, which is in [sc. the faculty of] reason, as Aristotle shows in *De anima* Book III.[67] Moreover, no body can make an impression on an incorporeal thing (*nullum autem corpus potest imprimere in rem incorpoream*).[68] Whence it is impossible for the heavenly bodies to make an impression directly on intellect or will (*unde impossibile est quod corpora caelestia directe imprimant in intellectum et voluntatem*). To allow that would deny the difference between intellect and sense (53, 55).[69]

The heavenly bodies *qua* bodies, then, cannot *directly* impress (= influence) the intellect and will, since they are part of the *ratio*, namely the rational soul (*anima rationalis*), and are thus incorporeal. In this way, Thomas sharply distinguishes here between the body and the soul register within human beings. This is one of the main strategies he used to establish a widely influential medieval safeguard to protect human free will. We will see in volume II how Marsilio Ficino subverted this and other well-established medieval safeguards in his *De vita libri tres* (1489).

Thomas continues, refining his analysis by showing how celestial bodies act *indirectly* on the human mind:

> Therefore, the celestial bodies cannot *per se* be the cause of free will acts (*causa operum liberi arbitrii*). Nevertheless, they can incline a person dispositively (*dispositive inclinare*) to this [sc. rather than that action], inasmuch as they make an impression on the human body (*inquantum imprimunt in corpus humanum*), and consequently on the sense powers (*in vires sensitivas*), which are the actions of bodily organs (*actus corporalium organorum*), which incline to human acts (*quae inclinant ad humanos actus*). Yet, as Aristotle makes clear in *De anima* III and *Ethics* I, since [sc. people's] sense powers obey reason (*oboediunt ratione*), this imposes no necessity on free will (*nulla necessitas ex hoc libero arbitrio imponitur*), for by [sc. his or her] reason, a human being (*homo*) can act counter to the inclination of the celestial bodies (*contra inclinationem caelestium corporum* [...] *operari*, 55).[70]

---

[67] III.4, 429b4.

[68] This is a central feature of Thomas's ontology.

[69] Secundo autem, subtrahuntur causalitati caelestium corporum actus liberi arbitrii, quod est *facultas voluntatis et rationis*. Intellectus enim, sive ratio, non est corpus nec actus organi corporei; et per consequens nec voluntas, quae est in ratione: ut patet per Philosophum, in III *De anima*. Nullum autem corpus potest imprimere in rem incorpoream. Unde impossibile est quod corpora caelestia directe imprimant in intellectum et voluntatem: hoc enim esset ponere intellectum non differe a sensu[.] (320b10-20).

[70] Unde corpora caelestia non possunt esse per se causa operum liberi arbitrii. Possunt tamen ad hoc dispositive inclinare, inquantum imprimunt in corpus humanum, et per consequens in vires sensitivas, quae sunt actus corporalium organorum, quae inclinant ad humanos actus. Quia tamen vires sensitivae obediunt rationi, ut patet per Philosophum, in III *de Anima* et in I *Ethic.*, nulla necessitas ex hoc libero arbitrio imponitur, sed contra inclinationem caelestium corporum homo potest per rationem operari (320b24-34).

Thomas here shows that there is no celestial causal basis for making predictions about chance events or those dependent on human free will, the two areas isolated in 95.1, because these two realms have been removed from celestial causality. There he argued strongly that predictions in these areas do not admit of astrological causal analysis, and are thus divinatory, demonic and illegitimate (T1). With the last point here about the stars *indirect* influence on the mind by means of its influence on the body, however, the solidity of Thomas's position begins to modulate, allowing at least some (albeit indirect) celestial influences on a person's mind, and thus on their subsequent actions.

Thomas's response to the second introductory argument *quod non* develops this analysis further in addressing how astrologers make true predictions:

> *Ad* 2: That *astrologi* frequently make true predictions (*vera praenuntiant*) from consideration of the stars happens in two ways: First, since most people follow their bodily passions (*plures hominum passiones corporales sequuntur*), therefore their acts are disposed for the most part according to the inclination of the celestial bodies (*actus eorum disponuntur, ut in pluribus, secundum inclinationem caelestium corporum*) [as we just saw]. But there are few, namely the wise (*sapientes*) alone, whose inclinations of this sort are moderated by reason. Therefore, astrologers (*astrologi*) make many true forecasts, and especially in communal events that depend on a multitude [sc. of people] (*praecipue in communibus eventibus, qui dependent ex multitudine*, 55).[71]

Celestial *bodies* thus act *directly* on human *bodies*, their organs and faculties, and thereby *indirectly* on the rational soul, its intellect and will. Necessity is thus removed from the equation: the stars incline but do not compel, as the old saying goes. Nevertheless, astrologers often make true predictions because most people follow their bodily passions, which are almost always unmitigated by reason. Here Thomas directly offers support for broader based social and political predictions derived from general astrology or revolutions, but also (and perhaps inadvertantly) from nativities, as he also did in his extremely interesting contemporary work, *De operationibus occultis naturae*, as we saw in Chap. 1.[72]

Thomas concludes:

> Therefore, if anyone uses a consideration of the stars to foreknow future chance or fortuitous events, or also to know with certainty (*per certitudinem*) the future works (*opera*) of men, this proceeds from a false and vain opinion. Thus, the act of a demon (*operatio daemonum*) is mixed in. Whence this *divinatio* will be superstitious and illicit. But if someone were to use a consideration of the stars to foreknow future things caused by celestial bodies, for instance, droughts and rainfall and other things of this sort, it will be neither illicit nor superstitous divination.[73]

---

[71] [Ad secundum dicendum quod] hoc quod astrologi ex consideratione astrorum frequenter vera praenuntiant, contingit dupliciter. Uno quidem modo, quia plures hominum passiones corporales sequuntur, et ideo actus eorum disponuntur, ut in pluribus, secundum inclinationem caelestium corporum: pauci autem sunt, idest soli sapientes, qui ratione huiusmodi inclinationes moderentur. Et ideo astrologi in multis vera praenuntiant: et praecipue in communibus eventibus, qui dependent ex multitudine (320b47-321a2).

[72] Thomas's argument here is similar to Roger's on the role of astrology in human affairs explored in Chap. 4.

[73] Si quis ergo consideratione astrorum utatur ad praecognoscendos futuros casuales vel fortuitos eventus, aut etiam ad cognoscendum per certitudinem futura opera hominum, procedet hoc ex falsa

Here we have Thomas's more nuanced second statement (T2) of his position concerning the knowledge and prediction of future contingent and chance events and those dependent on free will.

In both statements of his position, Thomas is clear and consistent for the most part: With causal analysis, whether exact or conjectural, human beings can foreknow and foretell the future by means of the stars. The two statements are not as consistent, however, concerning the more controversial prediction of future contingent events, whether accidental and irregular, or those deriving from free will. In the first stronger statement (95.1 [T1]), Thomas relegates predictions on all of these matters to divination, illegitimacy and sin, saying that there is no causal basis whatsoever for either their foreknowledge or prediction. In the second modified statement of his position (95.5 *ad* 2 [T2]), however, Thomas seems to maintain this strong position only concerning accidental events. With actions dependent on free will choices, on the other hand, Thomas opens the door to legitimate foreknowledge and prediction, but only of a conjectural nature. '*Per certitudinem*' is the key phrase here, a certainty caveat, which in T2 seems clearly to apply only to predictions about acts dependent on free will, due to the stars' direct action on the body, and thereby indirect action on the mind.

*

Both of Thomas's positions are clearly reflected in the sixteenth century material to be discussed in volume III. We will see that Rule IX of the *Index* (1564) follows on and further develops Thomas's second more tolerant position (T2), and that Sixtus V's Bull *Coeli et terrae creator* (1586) follows the more restrictive first statement (T1). Both sixteenth-century texts thus find implicit support in Thomas's much earlier but increasingly authoritative writings, although neither explicitly cites him to support their respective positions. We should also note that the terms natural and judicial astrology did not occur in Thomas's texts discussed here. We do, however, find them explicitly mentioned in Rule IX of the *Index*, which I will also discuss in volume III. In sum, for Thomas, astronomy and astrology are perfectly legitimate disciplines for knowing about and predicting the future if practiced within their proper limitations. Needless to say, negotiating these boundaries could prove to be tricky.

Finally, Thomas's safeguard for free will is based on human nature: celestial influences act directly on the body and thereby only *indirectly* on the will, which need not follow its bodily inclination. For Thomas, celestial *bodies* can only act directly on terrestrial *bodies*. Celestial influences per se can thus only occur in the bodily register. The soul register is only indirectly affected, if at all.[74] By contrast,

et vana opinione. Et sic operatio daemonis se immiscet. Unde erit divinatio superstitiosa et illicita. Si vero aliquis utatur consideratione astrorum ad praecognoscendum futura quae ex caelestibus causantur corporibus, puta siccitates et pluvias et alia huiusmodi, non erit illicita divinatio nec superstitiosa (320b35-45).

[74] We will also see further developments of this analysis in Thomas's somewhat earlier *Summa contra gentiles* to be explored just below.

in the *De fato* and elsewhere, Albert safeguarded human free will from the other direction by ontologically undermining the possibility of necessity in nature due to the variable receptivity of celestial influences in matter (as we will also see Thomas do below). The upshot is the same, however. These two great Dominican authorities both offered solid medieval safeguards to both protect human free will and to undermine determinism in nature (human and otherwise), both essential criteria for astrology's legitimacy.[75] Before exploring Thomas's views of divine providence in relation to astrology in greater detail—which also treats both of these legitimating themes directly—I will first discuss a brief consultation on the judgments of the stars that extends some of Thomas's analysis treated here.[76]

5.3.2 *Thomas Aquinas,* De iudiciis astrorum

In this contemporary short treatise also written during Thomas's second sojourn in Paris (ca. 1268–72)[77]—a consultation addressing a particular question in response to a petition (like the *De operationibus occultis naturae* and *De sortibus*)—Thomas discussed the legitimacy of making astrological judgments or interpretations. In it, he offered a more expanded range of astrological examples than he did in *Summa theologiae* II.II.92 and 95. Thomas begins by directly addressing his interlocutor's question: "Because you have asked, I will write to you about whether it is licit to use the judgments of the stars (*iudicia astrorum*), wishing to satisfy your petition about which I have taken care to write what has been passed down by the sacred doctors."[78] Thomas then begins to offer his response: "In the first place, therefore, it is fitting to know that the power of the celestial bodies (*virtus celestium corporum*) extends itself to changing lower bodies (*ad immutanda inferiora corpora*)."[79] Here we can

[75] And as we saw especially in Chap. 4, both Roger Bacon and the Magister Speculi were also exercised by this same central issue.

[76] There is also much of interest for our purposes in chapter four of the consultation *De sortibus* written also during Thomas's second sojourn in Paris (1268–72). It treats the legitimacy of predicting the future by means of lots (*sortes*) that Thomas had also treated in his contemporary *Summa theologiae* II.II.92–95, and especially at 95.8. Thomas also discusses God's providence there. This text should be treated in depth in a comprehensive study of these issues in Thomas, but will only be mentioned here. For '*sortes*' more generally and their use in a slightly later period, see Marie-Cécile Van Hasselt, "Les livres des sortes en Italie de 1482 à 1551: L'imaginaire astrologique, les systemes de causalité et le marge de liberté, accordée à l'individu," PhD thesis, 2 vols., Université de la Sorbonne Nouvelle (Paris III), 1997, and *Studi per le "sorti": Gioco, immagini, poesia oracolare a Venezia nel Cinquecento*, Treviso: Edizione Fondazione Benetton, 2007.

[77] See Torrell, *Saint Thomas Aquinas*, 215 and 356.

[78] "Quia petisti ut tibi scriberem an liceret iudiciis astrorum uti, tue petitioni satisfacere volens, super ea que a sacris doctoribus traduntur scribere curavi (1–4)." The critical edition of this text is in vol. 43 (1976) of the Leonine Edition, 201.

[79] In primis ergo oportet te scire quod virtus celestium corporum ad immutanda inferiora corpora se extendit (5–7).

see Thomas's customary emphasis on both celestial and terrestrial *bodies* as he provides the barest essentials of astrology's natural philosophical foundations.

After this initial response, Thomas again discusses a passage from Augustine, a great authority normally used against astrology:

> For Augustine says in the fifth book of the *De civitate Dei*: "It is not absurd to say that certain stellar breezes [= influences] (*afflatus quosdam sydereos*) continuously reach to the individual differences of bodies (*ad solas corporum differentias*)." And therefore, if someone uses the judgments of the stars (*iudicia astrorum*) for foreknowing *bodily* effects (*ad prenoscendum corporales effectus*) like [1] stormy or serene air, [2] the health or sickness of a body, or [3] the abundance or barrenness of the harvest, and other things of this sort that depend on bodily and natural causes (*ex corporalibus et naturalibus causis*), it does not seem that there is any sin (*peccatum*, my emphasis).[80]

These are all examples of what can legitimately be foreknown by means of annual revolutions, the kind of astrological practice used to make annual prognostications. The examples Thomas offers pertain to the weather, medicine and agriculture. Once again, Thomas emphasizes the corporeal and causal dimension of his analysis.

To expand his purview, Thomas then offers examples of some legitimate uses of astrological elections, a type of astrological practice not mentioned in the *Summa theologiae*, presumably because it does not involve predicting the future, the theme treated there:

> For all human beings use some observation (*observatio*) of the celestial bodies concerning effects of this sort as [1] farmers (*agricole*) sow and reap at a specific time (*certo tempore*) that is observed in accordance with the sun's motion; [2] seamen (*naute*) avoid sailing (*navigationes*) at full or new moon; and [3] physicians (*medici*) observe critical days (*creticos dies observant*) in relation to illnesses, which are determined according to the course of the sun and moon.[81]

Thomas here mentions three types of elections involving the observation of celestial bodies associated with three professions: farmers (*agricole*), seamen (*naute*) and physicians (*medici*). These are the very same professions mentioned 300 years later in Rule IX of the 1564 and later editions of the *Index of Prohibited Books* for permitting legitimate uses of astrology, and thus for indicating the types of astrological books that can legitimately be sold and owned. Precisely as here, the term '*observatio*' was used in Rule IX, but there they were influentially referred to as '*naturales observationes*', as we will see in volume III.

---

[80] Dicit enim Augustinus V De civitate Dei, "Non usquequaque absurde dici potest ad solas corporum differentias afflatus quosdam sydereos pervenire." Et ideo, si aliquis iudiciis astrorum utatur ad prenoscendum corporales effectus, puta tempestatem et serenitatem aeris, sanitatem vel infirmitatem corporis, vel ubertatem et sterilitatem fructuum, et cetera huiusmodi que ex corporalibus et naturalibus causis dependent, nullum videtur esse peccatum (7–16).

[81] Nam omnes homines circa huiusmodi effectus aliqua observatione utuntur celestium corporum: sicut agricole seminant et metunt certo tempore quod observatur secundum motum solis; naute navigationes vitant in plenilunio vel in lune defectu; medici circa egritudines creticos dies observant, qui determinantur secundum cursum solis et lune (17–24).

Then Thomas draws his conclusion: "Whence, it is not unbefitting (*inconveniens*) concerning some other more obscure observations of the stars (*aliquas alias occultiores observationes stellarum*) to use a judgment of the stars (*astrorum iudicium*) with respect to bodily effects (*corporales effectus*)."[82] Thomas here indicates that there are two main legitimate uses of astrological judgments and thus of their relevant astrological practices, namely, general astrology or revolutions and elections, for which he provides an informative but still rather limited range of examples that expand upon those mentioned in *Summa theologiae* II.II.92 and 95. Those were, we will recall, eclipses for the astronomers and weather for the astrologers. Here he also seems to allude to the legitimacy of other more obscure and unspecified astrological practices as long as they too are concerned solely with bodily effects.[83]

Having established these patterns of legitimacy, Thomas now confronts a theme of special concern to him (as it was also to Albertus Magnus, Roger Bacon and the Magister Speculi), namely human free will:

> Moreover, one needs to grasp this completely: that man's will (*voluntas hominis*) is not subjected to the necessity of the stars (*non est subiecta necessitati astrorum*). Otherwise, free will would perish (*alioquin periret liberum arbitrium*), which, when it has been removed, good works for praise and bad for blame are not accounted. And therefore it should be held as most certain (*certissime*) by every Christian that those things that depend on a person's will (*ea que ex voluntate hominis dependent*), such as all human works (*omnia humana opera*), are not subordinated to the stars by necessity (*non ex necessitate astris subduntur*). And therefore, X Jeremiah says: "Do not fear what the gentiles (*gentes*) fear from the signs of heaven."[84]

Thomas uses a quotation from Jeremiah to authoritatively support his argument that human free will is not dependent *by necessity* on the heavens. This theme will recur in greater detail in discussing Thomas's views of divine providence in *Summa contra gentiles*, Book III.

The third and final section of Thomas's short treatise then turns to the devil and his wily ways:

> But, in order to draw everyone into error, the devil (*dyabolus*) insinuates himself into the works of those who attend to the judgments of the stars (*immiscet se operibus eorum qui iudiciis astrorum intendunt*). Therefore, Augustine says in the second book of his *Super Genesim ad litteram*: "It should be said that when true things are said by astrologers (*quando ab astrologis vera dicuntur*), it is said by a certain very hidden impulse (*instinctu quodam occultissimo*) which unknowing human minds (*nescientes humane mentes*) receive. Since it exists for deceiving people, it is the work (*operatio*) of filthy and seductive spirts,

[82] Unde non est inconveniens, secundum aliquas alias occultiores observationes stellarum, circa corporales effectus uti astrorum iudicio (24–27).

[83] One would presume that he means nativities and/or interrogations, the other two main types of astrological practice.

[84] Hoc autem omnino tenere oportet, quod voluntas hominis non est subiecta necessitati astrorum; alioquin periret liberum arbitrium, quo sublato non deputarentur homini neque bona opera ad meritum, neque mala ad culpam. Et ideo certissime tenendum est cuilibet christiano, quod ea que ex voluntate hominis dependent, qualia sunt omnia humana opera, non ex necessitate astris subduntur; et ideo dicitur Ier. X: "A signis celi nolite metuere que gentes timent" (28–37).

> by which one is permitted [sc. by God] to know certain true things about temporal matters (*quibus quedam vera de temporalibus rebus nosse permittitur*)."[85]

We will notice that this demonically-inflected negative assessment of why some astrological statements are true is very different than both Albert's in the *De fato* and Thomas's in the contemporary *Summa theologiae* II.II.95.

Thomas now offers another quotation from Augustine:

> And therefore Augustine says in Book II of *De doctrina christiana* that observations of the stars (*observationes astrorum*) of this sort should be referred to certain pacts held with demons (*quaedam pacta cum demonibus habita*). Moreover, having a pact or association (*societas*) with demons is to be altogether avoided by a Christian, in accordance with that saying of the Apostle [Paul] in I *Corinthians* x: "I do not want you to become associates of demons (*socii demoniorum*). And therefore, it should be held as certain (*pro certo*) that it is a grave sin (*grave peccatum*) to use a judgment of the stars (*iudicio astrorum uti*) concerning those things that depend on human will (*a voluntate hominis dependent*).[86]

Thus, after exemplifying what types of astrological judgments may legitimately be used and in what realms, Thomas vehemently rejects that they may be used for those matters that depend on human free will. This aligns his analysis here with the more restrictive interpretation (TI) in *Summa theologiae* II.II.95. We will now turn from Thomas's analysis of astrology and divination to explore astrology's relationship to divine providence, and thus, as I will argue, his provisions for—and thereby construction of—astrology's theological foundations.

5.4 Astrology's Theological Foundations: The Celestial Bodies in Thomas Aquinas's Analysis of Divine Providence

5.4.1 [1] Summa contra gentiles *III.82-94*

Foundational features of Thomas Aquinas's analysis of divine providence are developed in his slightly earlier *Summa Contra Gentiles* III.82-94—written most likely between 1263 and 1265[87]—in which he analyzed God's providence in rela-

[85] Sed dyabolus, ut omnes pertrahat in errorem, immiscet se operibus eorum qui iudiciis astrorum intendunt; et ideo Augustinus dicit in II Super Genesim ad litteram: "Fatendum, quando ab astrologis vera dicuntur, instinctu quodam occultissimo dici quem nescientes humane mentes patiuntur; quod cum ad decipiendos homines fit, spirituum immundorum et seductorum operatio est, quibus quedam vera de temporalibus rebus nosse permittitur" (38–47).

[86] Et ideo Augustinus dicit in II De doctrina christiana quod huiusmodi observationes astrorum referende sunt ad quedam pacta cum demonibus habita. Est autem omnino christiano vitandum pactum vel societatum cum demonibus habere, secundum illud Apostoli I Cor. x: "Nolo vos fieri socios demoniorum." Et ideo pro certo tenendum est grave peccatum esse, circa ea que a voluntate hominis dependent iudicio astrorum uti (47–56).

[87] See Torrell, *Saint Thomas Aquinas*, 101–104 and 332–33. He dates *SCG* III.85 in particular to 1263–64 (102). Given the closeness in content of III.85 to the other chapters treated here, I am

tion to the planets, human free will and necessity in nature, to which I will now turn.[88] I will then discuss related passages from the even earlier *De veritate*, which was written between 1256 and 1259.[89]

One of the essential insights that seems to motivate Thomas's view of divine providence is that, in the beginning, God created the heavens and the earth *ex nihilo*. Then He used the heavens—and especially the celestial bodies—as the corporeal instruments to manage/govern His creation.[90] Much of the subsequent system follows from this initial insight as buttressed by both Roman Catholic theology and Aristotelian natural philosophy, issuing in an integrated and holistic world-view, as we will see, although one not without its tensions. The celestial bodies play an utterly central role among the fundamental structures of Thomas's deeply-influential philosophically-informed theological system explored here. This is a rich and complex topic, so I will perforce be highly selective, focusing solely on essential structures and dynamics.[91]

Given that the celestial bodies are the central corporeal instruments by which God governs everything on earth within the context of His providential care for the world, it becomes imperative for Thomas to both preserve human free will, on the

---

inclined to date them all together to this time. Brian J. Shanley, O.P. states that the *Summa contra gentiles* "constitutes Thomas's longest sustained treatment of divine providence"; "Thomas Aquinas on Demonstrating God's Providence," in *The Science of Being as Being: Metaphysical Investigations*, Gregory T. Doolan (ed.), Washington, D.C.: Catholic University of America Press, 2012, 221–42.

[88] Boudet states strongly that Thomas expressed his personal and overall point of view about astrology and magic for the first time in Book III of the *Summa contra gentiles*; *Entre science et nigromance*, 228. Valérie Cordonier emphasises the originality of Thomas's program concerning divine providence, and that he approached it exceptionally in the *Summa contra gentiles*; "Sauver le Dieu du Philosophe: Albert le Grand, Thomas d'Aquin, Guillaume de Moerbeke et l'invention du *Liber de bona fortuna* comme alternative autorisée à l'interprétation averroïste de la théorie aristotélicienne de la providence divine," in *Christian Readings of Aristotle from the Middle Ages to the Renaissance*, Luca Bianchi (ed.), Turnhout: Brepols, 2011, 65–114, 97. Stefano Caroti's comments in his all-too-brief section on Thomas are valuable here too; *L'astrologia in Italia*, 177–82.

[89] Torrell, *Saint Thomas Aquinas*, 62 and 334.

[90] We saw something similar in *Speculum astronomiae*, chapter 3, and in its defense of astrology. Thomas will articulate his views here in much greater detail.

[91] In addition to Litt, *Les corps célestes*, and Cordonier, "Sauver le Dieu," I have found these works of scholarship of most use in what follows: Porro, "*Lex necessitatis vel contingentiae*"; Valérie Cordonier, "La doctrine Aristotélicienne de la Providence selon Thomas d'Aquin," in *Fate, Providence and Moral Responsibility in Ancient, Medieval and Early Modern Thought. Studies in Honour of Carlos Steel*, P. D'Hoine and G. van Riel (eds.), Leuven: Peeters, 2014, 495–515, and Nicholas Kahm, "Divine Providence in Aquinas's Commentaries on Aristotle's *Physics* and *Metaphysics*, and its Relevance to the Question of Evolution and Creation," *American Catholic Philosophical Quarterly* 87 (2013): 637–56. More generally on providence, see Hester Gelber, "Providence," in *The Cambridge History of Medieval Philosophy*, Pasnau and Van Dyke (eds.), Cambridge: Cambridge University Press, 2010, 761–72, who brings out the importance of both Augustine and especially Boethius to later medieval analyses of providence. Cordonier argues that Thomas himself devised an original Aristotelianizing version of providence, since Aristotle did not himself have a complete one. She does not, however, discuss the influence of Boethius on Thomas's position.

one hand, and to protect against the concomitant claim that this view of providence entails necessity in nature, on the other. Thus, these essential themes will continue to recur here. In this section, we will explore the astrologizing mechanics and dynamics of Thomas Aquinas's view of divine providence within the divinely ordered and regulated economy of nature.

*

The extensive Book III of the *Summa contra gentiles* is devoted entirely to explicating Thomas's systematic and influential view of divine providence. I will first focus in detail on the central article, III.82, in which Thomas established the fundamental structure announced in its title: "That lower bodies are ruled by God by means of celestial bodies" (*quod inferiora corpora reguntur a Deo per corpora caelestia*). In this circumstance, as he asserts in III.80, a distinct subset of the spectrum of separated intellectual substances—one of the orders of angels, the Virtues (*Virtutes*)—directly controls planetary motions: "Whence it seems that to this order of angels pertains the motion of celestial bodies, from which, as from certain universal causes, particular effects in nature follow (*ex quibusdam universalibus causis, consequuntur particulares effectus in natura*); therefore, they are called the '*Virtutes caelorum*'."[92] Here is the precise link between universal causes and particular effects in nature, an extremely important causal nexus in Thomas's system. In their turn, these Virtues are ruled indirectly by God (as the first cause) by means of the four previous sets of intellectual substances that Thomas explicitly identifies as angels. In fact, the Virtues are located exactly in the middle of the ninefold hierarchy of angels as delineated by ps.-Dionysius the Areopagite, whom Thomas explicitly follows here.

There are three orders of three angels each. The Virtues are in the middle of the second group, and thus precisely in the middle of all nine, perhaps reflecting the planets' pivotal middle linking roles in the providentially designed economy of nature, as Thomas establishes here.[93] For the rest of this section, I will ignore the angelic dimension and focus on the corporeal, and in particular, on Thomas's view as to how God rules all bodies on earth—the *inferiora corpora*, living and otherwise—and their motions or changes by means of the celestial bodies and their

[92] "Unde videtur quod ad hunc ordinem [sc. of angels = separated intellectual substances] pertineat motus caelestium corporum, ex quibus, sicut ex quibusdam universalibus causis, consequuntur particulares effectus in natura: et ideo *Virtutes caelorum* nominantur (234a1-5)[.]" I use the online translation by Vernon J. Bourke, which I have modified where necessary, and the Latin text in the Leonine edition. For the English translation, see *On the Truth of the Catholic Faith (Summa Contra Gentiles),* Anton C. Pegis, James F. Anderson, Vernon J. Bourke, and Charles J. O'Neil, transs., 5 vols., New York: Doubleday, 1955–57; reprinted as *Summa contra gentiles,* Notre Dame, Ind.: University of Notre Dame Press, 1975. III.82 and 83 are in volume 3:1; III.84-105 are in 3:2 (both originally published 1956). E-text with some revision at dhspriory.org/thomas/ContraGentiles.htm

[93] This is very similar to the *scientia iudiciorum astrorum*'s role as the link between metaphysics (theology) and natural philosophy in *Speculum astronomiae*, chapter 3. The planets there play this middle role as well, and thus the science of the stars becomes the best human science for understanding both the divine nature and nature in the world, as we saw in Chap. 1.

motions. Whether he intended to or not, Thomas thereby established astrology's theological foundations, but without explicitly defending or supporting astrology itself, as he will later do in *Summa theologiae* II.II.92-95 and the *De iudiciis astrorum*, as we just saw, although he does use the pseudo-Ptolemaic *Centiloquium* as one of his main authorities here.

Once Thomas establishes this foundation in III.82, he then defends this providential structure against charges of physico-ontological determinism, as well as its moral and psychological counterpart, namely, the undermining of human free will. He does this in III.83-88, as we will see. This section effectively ends with III.93, a brief statement on fate, and III.94 on providence and necessity. There is much more to this rich topic, but this brief treatment will have to suffice.

5.4.1.1 III.82: How God Uses the Celestial Bodies to Rule Lower Bodies

To establish the most central and fundamental structure of his system, in which the divine and the natural are bridged as are the heavens and the earth, Thomas argues in seven slightly different ways for essentialy the same compound conclusion (whether expressed actively or passively), namely, that (i) the celestial bodies and (ii) their motions rule, dispose and move all (i) lower bodies and (ii) their motions here on earth.[94] As it turns out, this conclusion seems so glaringly obvious to him and thus so profoundly irrefutable that he offers no opposing or even alternative positions. I will treat III.82 in full due to its central importance in Thomas's system.

Thomas begins III.82 by proposing a hierarchy of substances in two different registers, namely, intellectual and corporeal substances:

> [1] Moreover, just as there is a higher and a lower among intellectual substances, so also there is a higher and a lower among corporeal substances. But intellectual substances are ruled by the higher ones (*reguntur a superioribus*), since the disposition of divine providence (*dispositio divinae providentiae*) descends proportionately to the lowest (*ad infima*), as we have said already [III.78].[95] Therefore, by the same reasoning, lower bodies are disposed through higher ones (*inferiora corpora per superiora disponuntur*).[96]

---

[94] In fact, the arguments often come off more as assertions in the guise of arguments than as truly debatable propositions. The sense of motion here is precisely the four-fold Aristotelian kind that I discussed in Chap. 1.

[95] This is apparently the cornerstone of Thomas's influential contribution to developments in Aristotelian analyses of divine providence, since for Aristotle himself, divine providence only descended to the level of the celestial bodies. See Cordonier, "Saveur le Dieu," 88 ff., and Robert Sharples, "Alexander of Aphrodisias on Divine Providence: Two Problems," *The Classical Quarterly* 32 (1982): 198–211.

[96] Sicut autem in substantiis intellectualibus est superius et inferius, ita etiam in substantiis corporalibus. Substantiae autem intellectuales reguntur a superioribus, ut dispositio divinae providentiae proportionaliter descendat usque ad infima, sicut iam dictum est. Ergo, pari ratione, inferiora corpora per superiora disponuntur (243a1-8).

Intellectual substances have higher and lower dimensions in the same way that bodies do. In both cases, the higher ones in each register rule and dispose the lower. This principle will recur.

Now Thomas refines this basic parallel structure:

> [2] Again, the higher a body is in place (*quanto aliquod corpus est superius loco*), the more formal it is found to be (*tanto invenitur esse formalius*). Because of this, it easily becomes the place of the lower [sc. body] (*locus inferioris*), since it is the nature of form to contain (*formae est continere*), just as it also is of place (*sicut et loci*). In fact, water is more formal than earth, air than water, and fire than air. But celestial bodies are superior in place to all [sc. bodies] (*corpora caelestia sunt omnibus loco superiora*). Therefore, they are more formal than all the others, and, therefore, more active. Therefore, they act on the lower bodies (*agunt ergo in inferiora corpora*), and thus lower [sc. bodies] are disposed by them (*sic per ea inferiora disponuntur*).[97]

Thomas now further articulates this structure by relating formality to place.[98] The celestial bodies as the bodies highest in place thereby also become both the most formal, and also the most active corporeal entities. The more passive lower terrestrial bodies are thus disposed by the higher celestial bodies acting upon them. We will continue to see the language of disposition used throughout this analysis, as we also will later with both Thomas's and Albert's analyses of fate in relation to providence.

Thomas then turns to corporeal perfection in relation to elemental contrariety:

> [3] Besides, what is perfected in its nature and without contrariety is of a more universal power (*virtus*) than what is not perfected in its nature except with contrariety. For contrariety exists from the different things that determine and contract a genus, whence, in the reception of the intellect, because it is universal, the species of contraries are not contrary, since they exist at the same time. But celestial bodies are perfected in their natures without any contrariety (*corpora autem caelestia sunt in suis naturis absque omni contrarietate perfecta*), for they are neither light nor heavy, neither hot nor cold. But lower bodies are not perfected in their natures except with some contrariety. Their motions also demonstrate this, for nothing is contrary to the circular motion of the celestial bodies, whence there can be no violent [sc. motion] in them. But the motions of lower bodies are contraries, namely, downward motion as opposed to upward motion. Therefore, celestial bodies are of a more universal power than lower bodies are. But universal powers are the movers of particular [sc. powers] (*universales autem virtutes sunt motivae particularium*), as is clear from what was said. Therefore, celestial bodies move and dispose lower bodies (*corpora igitur caelestia movent et disponunt corpora inferiora*).[99]

---

[97] Amplius. Quanto aliquod corpus est superius loco, tanto invenitur esse formalius; et propter hoc etiam rationabiliter est locus inferioris, nam formae est continere, sicut et loci; aqua enim est formalior terra, aer aqua, ignis aere. Sed corpora caelestia sunt omnibus loco superiora. Ipsa igitur sunt magis formalia omnibus aliis. Ergo magis activa. Agunt ergo in inferiora corpora. Et sic per ea inferiora disponuntur (243a9-17).

[98] Albertus Magnus makes a similar point about place functioning as form in *De natura loci* (1, 9–20), as discussed in Chap. 2.

[99] Item. Quod est in sua natura perfectum absque contrarietate, est universalioris virtutis quam illud quod in sua natura non perficitur nisi cum contrarietate: contrarietas enim est ex differentiis determinantibus et contrahentibus genus; unde in acceptione intellectus, quia est universalis, species contrariorum non sunt contrariae, cum sint simul. Corpora autem caelestia sunt in suis naturis absque omni contrarietate perfecta: non enim sunt levia neque gravia, neque calida neque frigida.

Because the celestial bodies are more perfect and universal than lower bodies, celestial bodies move and dispose the lower bodies. This is, of course, basically the same conclusion repeated with minor variations from different but related perspectives.

Motion has now been mentioned here, and in two senses: [1] local motion with respect to both celestial and terrestrial bodies, and [2] that the celestial bodies actively "move" and "dispose" the more passive and contrary lower bodies. This foreshadows the second mode of celestial governance, namely, that the celestial bodies *and their motions* rule, dispose and move terrestrial bodies *and their motions*. This view in III.82 will ultimately be cashed out in terms of celestial bodies and their one-dimensional (= local) type of motion vis-à-vis the terrestrial bodies and their fourfold motions within the context of Aristotle's distinctive view of kinesis/*motus* (motion), as I discuss below. We will see the issue of contrariety arise again below.

Thomas then associates the celestial bodies more closely with intellectual substances:

> [4] Moreover, it was shown above [III.78] that all other things are ruled by intellectual substances (*per substantias intellectuales alia omnia reguntur*). But celestial bodies are more similar to intellectual substances than other bodies are, insofar as they are incorruptible. They are also nearer (*propinquiora*) to them [sc. in place], insofar as they are moved immediately by them [i.e. with nothing in between] (*inquantum ab eis immediate moventur*), as we showed above [III.80]. Therefore, lower bodies are ruled by them (*per ipsa igitur reguntur inferiora corpora*).[100]

Not only are celestial bodies incorruptible and thus more like intellectual substances than other bodies, they are also moved directly by the intellectual substances and thus rule lower bodies.

For the last of the first set of five arguments that are more or less minor variations on a theme, Thomas addresses the central theme of motion and immobility—that is, the ontological inability to be moved—understanding motion in the normal fourfold Aristotelian sense articulated in my Chap. 1:

> [5] Furthermore, it is fitting that the first principle of motion (*primum principium motus*) be something immovable. Therefore, what are nearer to immobility should be the movers of other things. But celestial bodies are nearer to the immobility of the first principle than are lower bodies, for they are not moved except in one kind (*species*) of motion, namely, local motion (*motus localis*). But other bodies are moved in all the [sc. four] species of motion.

Corpora vero inferiora non perficiuntur in suis naturis nisi cum aliqua contrarietate. Et hoc etiam motus eorum demonstrant: nam motui circulari corporum caelestium non est aliquid contrarium, unde nec in eis violentia esse potest; motui autem inferiorum corporum contrarii sunt, scilicet motus deorsum motui sursum. Corpora ergo caelestia sunt universalioris virtutis quam corpora inferiora. Universales autem virtutes sunt motivae particularium, sicut ex dictis patet. Corpora igitur caelestia movent et disponunt corpora inferiora (243a18-39).

[100] Adhuc. Ostensum est supra quod per substantias intellectuales alia omnia reguntur. Corpora autem caelestia sunt similiora substantiis intellectualibus quam alia corpora, inquantum sunt incorruptibilia. Sunt etiam eis propinquiora, inquantum ab eis immediate moventur, ut supra ostensum est. Per ipsa igitur reguntur inferiora corpora (243a40-b2).

> Therefore, celestial bodies are the movers and rulers of lower bodies (*corpora igitur caelestia sunt motiva et regitiva inferiorum corporum*).[101]

Thomas here argues that the celestial bodies are closer in nature to the immovable first principle (*principium*) of motion because they only exhibit one type of motion, namely, locomotion or motion in place, unlike bodies in sublunar nature that exhibit all four kinds: generation and corruption (changes in the Aristotelian category "substance"), augmentation or growth (changes in "quantity") and alteration (changes in "quality") in addition to locomotion. The first five paragraphs thus all move together towards the same conclusion, but with slightly different approaches: that higher bodies (i.e. the celestial bodies, which he does not further differentiate) move, rule and dispose all lower bodies (*inferiora corpora*), that is, all bodies in nature. This is the first major conclusion. The complementary fuller conclusion is that celestial bodies *and their motions* are the movers, rulers and disposers of all lower bodies *and their motions*.

*

After this fivefold introduction, Thomas then offers an explicit '*primum in quolibet genere*' argument, which we will also find used by Albertus Magnus in Chap. 7, and by both Marsilio Ficino and Giovanni Pico della Mirandola in volume II.[102] Thomas's central concern remains *motus*, and thus all changes, transformations and/or motions on earth. I will treat this three-part argument in detail:

> [6] Again, the first [sc. entity] in any genus (*primum in quolibet genere*) is the cause of what come after (*causa eorum quae sunt post*). But among all the other motions, the first is the motion of the heavens (*motus caeli*). [a] First of all because local motion is first among all motions (*quia motus localis est primus inter omnes motus*). This is so [b] in regard to time, for it alone can be perpetual, as is proved in *Physics* VIII [7: 260b 29]. It is also so [c] in regard to nature (*naturaliter*), for without it there cannot be any of the other kinds of motion. For something is not increased (*augmentatur*) unless there be a preceding alteration (*alteratio*) through which what was formerly unlike is changed (*convertatur*) and becomes similar. Nor can alteration exist unless there be a preceding local change (*loci mutatio*), since for alteration to be achieved the agent of alteration must now become closer to the thing altered

[101] Praeterea. Oportet primum principium motus esse aliquid immobile. Quae ergo magis accedunt ad immobilitatem, debent esse aliorum motiva. Corpora autem caelestia magis accedunt ad immobilitatem primi principii quam inferiora: quia non moventur nisi una specie motus, scilicet motu locali; alia vero corpora moventur omnibus speciebus motus. Corpora igitur caelestia sunt motiva et regitiva inferiorum corporum (243b3-11).

[102] Paul O. Kristeller discusses this theme in *The Philosophy of Marsilio Ficino* (New York: Columbia University Press, 1943, repr. Gloucester, MA: Peter Smith, 1964, 146–70), but he does not seem to fully grasp how important this principle is to Thomas. In fact, all Kristeller says is: "This term and its explicit formulation seem rather to originate in scholastic philosophy. In any case, the phrase *perfectum in aliquo genere*, which occurs in Thomas Aquinas and in other Scholastics, is very close to Ficino's concept. The examples and the context, however, show that the Scholastics merely wished to give a general emphasis to certain objects, in particular God, and to attribute to them certain qualities of their genus. However, so far as I am aware there had been no comprehensive interpretation or application of the concept until Ficino's day (152)." We will see more examples below.

> (*alterato*) than it was before. It is also prior [d] in perfection because local motion does not change (*variat*) a thing in regard to something inherent in it, but only according to something extrinsic. For this reason, it belongs to an already perfected thing.[103]

Celestial motion is thus first in motion, time, nature and perfection, and local motion is prior to the other three types of motion. Although he does not mention generation and corruption here, he does do so later.

Thomas then identifies celestial circular motion as the first of the local motions:

> Secondly, because even among local motions circular motion is prior. And again, in regard to time, because it alone can be perpetual, as is proved in the *Physics* [VIII, 8: 261b 27]. And in regard to nature because it is more simple and unified, since it is not divided into beginning, middle and end, but the whole motion is like a middle. And even in perfection because it is brought back (*reflectitur*) to its origin (*principium*).[104]

Thomas here reiterates that celestial motion is also prior with respect to motion, time, nature and perfection with specific regard to its circularity.

He then turns to the regularity of celestial motions:

> Thirdly, because only the motion of the heavens is found to always be regular and uniform. For, among the natural motions of heavy and light things there is an increase (*additio*) of velocity at the end, but among violent motions, there is an increase of slowness. Therefore, it is fitting that the motion of the heavens is the cause of all other motions (*motus caeli sit causa omnium aliorum motuum*).[105]

Here again we see Thomas differentiate regular (= uniform circular) celestial motions from changing terrestrial motions. He thus reaches his intended conclusion, namely, that the motion of the heavens as the first in the genus of motions must be the cause of all other motions or changes in the world. This is the second major conclusion of this section. The first is that celestial bodies rule, dispose and move all lower bodies. Thus, we now know in the integrated version of these conclusions that celestial bodies *and their motions* rule, dispose and move all lower bodies *and their motions*. These arguments provide the essential structure for astrology's integrated theological and natural philosophical foundations.

---

[103] Amplius. *Primum in quolibet genere est causa eorum quae sunt post.* Inter omnes autem alios motus, primus est motus caeli. Primo quidem, quia motus localis est primus inter omnes motus. Et tempore: quia solus potest esse perpetuus, ut probatur in VIII *Phys*. Et naturaliter: quia sine eo non potest esse aliquis aliorum; non enim augmentatur aliquid nisi praeexistente alteratione, per quam quod prius erat dissimile, convertatur et fiat simile; neque alteratio potest esse nisi praeexistente loci mutatione, quia ad hoc quod fiat alteratio, oportet quod alterans magis sit propinquum alterato nunc quam prius. Est etiam perfectione prior: quia motus localis non variat rem secundum aliquid ei inhaerens, sed solum secundum aliquid extrinsecum; et propter hoc est rei iam perfectae (243b12-28).

[104] Secundo, quia etiam inter motus locales est motus circularis prior. Et tempore: quia solus ipse potest esse perpetuus, ut probatur in VIII *Phys*. Et naturaliter: quia est magis simplex et unus, cum non distinguatur in principium, medium et finem, sed totus sit quasi medium. Et etiam perfectione: quia reflectitur ad principium (243b28-35).

[105] Tertio, quia solus motus caeli invenitur semper regularis et uniformis: in motibus enim naturalibus gravium et levium fit additio velocitatis in fine, in violentis autem additio tarditatis. Oportet ergo quod motus caeli sit causa omnium aliorum motuum (243b35-40).

*

Thomas now reaches the final stage of this argument, which focuses on the necessity of immobility as the source of motion. Here he focuses on the ontological immobility of the celestial bodies. They may move in place, but otherwise they do not change at all in any respect:

> [7] Besides, just as the immobile *simpliciter* is to motion *simpliciter*, so is the immobile with respect to a given motion related to such a motion. But that which is immobile *simpliciter* is the *principium* of all motion, as we proved above [I.13]. Therefore, what is immobile with respect to *alteratio* is the *principium* of all changes in quality. But the celestial bodies, alone among bodily things (*corporalia*), are unalterable (*inalterabilia*). Their condition (*dispositio*) demonstrates this, which is discovered to always be the same (*semper eadem*). Therefore, the celestial body is the cause of every change in quality in those things that are changed in quality (*est ergo corpus caeleste causa omnis alterationis in his quae alterantur*). But, in these lower bodies, *alteratio* is the source of every motion (*alteratio autem in his inferioribus est principium omnis motus*), for through alteration, increase (*augmentum*, change in quantity) and generation are achieved (*per alterationem pervenitur ad augmentum et generationem*), but the agent of generation (*generans*) is a self-mover (*motor per se*) in the local motion of heavy and light things. Therefore, it is fitting that the heavens (*caelum*) are the cause of every motion in these lower bodies (*caelum sit causa omnis motus in istis inferioribus corporibus*).[106]

This last sentence is almost a complete statement of Thomas's compound conclusion. For the last part of this argument, Thomas focuses on the fact that the celestial bodies do not undergo '*alteratio*' or change in quality in addition to moving locally. Therefore, celestial bodies are the cause of all '*alteratio*' in the economy of nature.

And because change in quality (*alteratio*) is the source of all other changes and/or motions among lower bodies (including generation and growth), the celestial bodies and their motions are the cause of every motion in the world, in formulations similar to what we saw in Roger Bacon's works, as well as in Albertus Magnus and Aristotle concerning local motion being prior to generation. Thomas then completes III.82 by reiterating the fundamental point of this exercise: "Thus, it is clear that lower bodies [sc. and their motions or changes] are ruled by God through the celestial bodies [sc. and their motions]" (*corpora inferiora a Deo per corpora caelestia reguntur*).[107]

Thomas thus offers seven strong, relatively self-evident and often virtually tautological argument-assertions to this end, which are basically all variations on a theme that he is adamant in promoting as he states here once again at the very end. Thomas

[106] Adhuc. Sicut se habet immobile simpliciter ad motum simpliciter, ita se habet immobile secundum hunc motum ad motum talem. Id autem quod est immobile simpliciter, est principium omnis motus, ut supra probatum est. Quod ergo est immobile secundum alterationem, est principium omnis alterationis. Corpora autem caelestia sola inter corporalia sunt inalterabilia: quod demonstrat dispositio eorum, quae semper eadem invenitur. Est ergo corpus caeleste causa omnis alterationis in his quae alterantur. Alteratio autem in his inferioribus est principium omnis motus: nam per alterationem pervenitur ad augmentum et generationem; generans autem est motor per se in motu locali gravium et levium. Oportet ergo quod caelum sit causa omnis motus in istis inferioribus corporibus (243b41-244b5).

[107] Sic ergo patet quod corpora inferiora a Deo per corpora caelestia reguntur (244b6-7).

here very deliberately establishes the celestial bodies and their motions as utterly central to his understanding of the providentially designed economy of nature, thus providing an utterly central structure of astrology's integrated theological and scientific foundations. The role of the celestial bodies and their motions could not be more fundamental, located, as they are, at the heart of the nexus between the divine and nature. The most basic structure of Thomas's construction of astrology's theological foundations has now been established.[108]

5.4.1.2 III.84 and 85: On Protecting Human Free Will

Once Thomas has established to his satisfaction this central and utterly fundamental structure of his system, namely, that God rules all sublunar/lower bodies and their motions by means of the celestial bodies and their motions, he then turns to protect human free will as a central concern for the legitimate establishment of his view. I will only treat the heart of his arguments here, focusing primarily on what parts of a human being the celestial bodies influence. Once again, Thomas here embraces the fundamental distinction between the body and soul registers.

In the first sections of III.84—entitled, "That celestial bodies do not make impressions on our intellects"[109]—Thomas is concerned to argue that the celestial bodies do not *directly* affect the human *intellectus*, primarily because the celestial bodies are, after all, *bodies* (however exalted). The intellect, on the other hand, is neither a bodily organ nor the expression of one, and bodies (whether celestial or terrestrial) can only affect other bodies. In fact, the intellect is an intellectual substance, although not one wholly separated from bodies like angels are.

We will now look at the heart of his argument in detail:

> Nevertheless, one should know that, although celestial bodies cannot directly be the causes of our understanding (*licet corpora caelestia directe intelligentiae nostrae causae esse non possint*), yet they may do something indirectly in regard to it (*aliquid tamen ad hoc operantur indirecte*). For, although the intellect is not a corporeal power (*virtus corporea*), nevertheless, the operation of intellect cannot be completed in us without the operation of corporeal powers, which are the imagination, the power of memory, and the cogitative power (*imaginatio et vis memorativa et cogitativa*), as is evident from preceding explanations [II.68]. And as a result, if the operations of these powers are impeded by some indisposition of the body, the operation of the intellect is impeded, as is evident in frenetic [= manic] and lethargic [= depressive] people, and in others similarly affected. And that is also why the goodness of the human bodily disposition makes one able to understand well (*bonitas dispositionis corporis humani facit aptum ad bene intelligendum*), insofar as the aforesaid powers are in a stronger condition because of this. Whence it is said in Book II of *On the Soul* [*De anima* II.9] that we observe that "men with soft flesh are well endowed mentally."[110]

[108] Thomas supports this analysis in the very brief III.83 by providing a few theological authorities for this view, including the pseudo-Dionysius, Boethius and Augustine, which I will pass over here.

[109] Quod corpora coelestia non imprimant in intellectus nostros.

[110] Sciendum est tamen quod, licet corpora caelestia directe intelligentiae nostrae causae esse non possint, aliquid tamen ad hoc operantur indirecte. Licet enim intellectus non sit virtus corporea, tamen in nobis operatio intellectus compleri non potest sine operatione virtutum corporearum,

Here Thomas employs one of his normal gambits, which he articulates more fully here. Although celestial influences can have no *direct* impact on the intellect, they can, nevertheless, affect it indirectly. He isolates imagination, memory and cogitation as the operations of corporeal powers.

Then Thomas turns to authoritative support for his position:

> Moreover, the disposition of the human body is subjected to the celestial motions (*dispositio autem corporis humani subiacet caelestibus motibus*). For Augustine says, in the *City of God* V, that "it is not utterly absurd to say that certain stellar breezes [= influences] (*afflatus quosdam* [...] *sidereos*) are able to produce differences in bodies only." And the Damascene says, in Book II [of *De fide orthodoxa*], that "different planets establish in us diverse complexions, habits and dispositions (*alii et alii planetae diversas complexiones et habitus et dispositiones in nobis constituunt*)." Therefore, the celestial bodies work indirectly on the goodness of intellect (*indirecte corpora caelestia ad bonitatem intelligentiae operantur*). Thus, just as physicians can make judgments about the goodness of an intellect from the complexion of its body (*sicut medici possunt iudicare de bonitate intellectus ex corporis complexione*), as from a proximate disposition (*sicut ex dispositione proxima*), so also can an astrologer (*astrologus*) judge from the celestial motions as from the remote cause of such a disposition (*ita astrologus ex motibus caelestibus sicut ex causa remota talis dispositionis* [sc. *iudicare*]). In this way, then, what Ptolemy says in the *Centiloquium* can be verified [*verbum* 38]: "Since Mercury was in one of Saturn's houses (*in aliqua domorum Saturni* [sc. in Capricorn or Aquarius]) at someone's birth (*in nativitate alicuius*) and is itself in a strong condition, it gives inwardly to things the goodness of intelligence (*dat bonitatem intelligentiae medullitus in rebus*)."[111]

Thomas cites Augustine, John of Damascus and the pseudo-Ptolemy of the *Centiloquium*. In fact, the quotation from Ptolemy is explicitly astrological, referring to nativities, and Thomas uses it to support the validity of astrological judgments or interpretations. Indeed, he indicates explicitly that his natural philosophical analysis and the analogy with medical judgments itself verifies Ptolemy's astrological assertion![112] Here he notes that medicine is concerned with proximate causes

---

quae sunt imaginatio et vis memorativa et cogitativa, ut ex superioribus patet. Et inde est quod, impeditis harum virtutum operationibus propter aliquam corporis indispositionem, impeditur operatio intellectus: sicut patet in phreneticis et lethargicis, et aliis huiusmodi. Et propter hoc etiam bonitas dispositionis corporis humani facit aptum ad bene intelligendum, inquantum ex hoc praedictae vires fortiores existunt: unde dicitur in II *De anima* quod *molles carne bene aptos mente videmus* (249b34-50).

[111] Dispositio autem corporis humani subiacet caelestibus motibus. Dicit enim Augustinus, in V *De civitate Dei*, quod *non usquequaque absurde dici potest ad solas corporum differentias afflatus quosdam valere sidereos*. Et Damascenus dicit in secundo libro, quod alii et alii planetae *diversas complexiones et habitus et dispositiones in nobis constituunt*. Et ideo indirecte corpora caelestia ad bonitatem intelligentiae operantur. Et sic, sicut medici possunt iudicare de bonitate intellectus ex corporis complexione sicut ex dispositione proxima, ita astrologus ex motibus caelestibus sicut ex causa remota talis dispositionis. Et per hunc modum potest verificari quod Ptolomaeus in *Centilogio* dicit: *Cum fuerit Mercurius in nativitate alicuius in aliqua domorum Saturni, et ipse fortis in esse suo, dat bonitatem intelligentiae medullitus in rebus* (249b50-250b7).

[112] In Francesco Sylvestri of Ferrara's extensive early sixteenth-century analysis of Thomas's text here—especially the interpretation of Ptolemy's text—he explicitly discusses Giovanni Pico della Mirandola's analysis of astrology's natural philosophical foundations in *Disputationes* III.3 at 252a6-254b42. Sylvestri defends Thomas's astrologizing analysis against Pico's anti-astrological

and astrology with remote ones. Thomas had also cited the same passage by Augustine in his *De iudiciis astrorum*, as we saw above.

*

In III.85, Thomas is concerned to argue—as he states in the title—"That celestial bodies are not the causes of our acts of will and our choices" (*quod corpora caelestia non sunt causae voluntatum et electionum nostrarum*). Drawing on the position established in III.84 that the will is in the intellective part of our soul, celestial bodies thus do not *directly* cause our acts of will or our choices. Rather, they do so indirectly by means of the '*occasio*' or opportunities that arise, in relation to which we may make choices. Both III.84 and 85 may thus be compared with the later *Summa theologiae* II.II.92-95, where Thomas also used arguments employing the distinction between the direct and indirect effects of celestial influences in relation to both the mind and body. Once again, Thomas here cites ps.-Ptolemy at the end as an authoritative support.

In the heart of III.85 with respect to our interests, Thomas explains how the '*occasio*' arises:

> Nevertheless, one should know that, although celestial bodies are not *directly* the cause of our choices (*licet corpora caelestia non sint directe causa electionum nostrarum*), as if they directly make impressions on our wills (*quasi directe in voluntates nostras imprimentia*), nevertheless, some occasion (*aliqua occasio*) for our choices may be indirectly offered by them because they do make an impression on bodies (*habent impressionem super corpora*), and in a twofold sense. [1] In one way, the impressions of the celestial bodies on external bodies (*in exteriora corpora*) are for us the occasion of a certain choice. For instance, when the air is disposed to severe cold by the celestial bodies, we choose to get warmed near a fire or to perform other such acts which suit the weather. [2] In a second way, they make an impression on *our* bodies (*imprimunt in corpora nostra*). When a change (*immutatio*) occurs in them, certain movements of the passions (*aliqui motus passionum*) arise (*insurgunt*) in us. Or we are made prone by their impression to certain passions, as choleric people are prone to anger. Or again, some bodily disposition (*aliqua dispositio corporalis*) that is an occasion for an act of choice (*occasio alicuius electionis*) may be caused in us by their impression, as when, being sick, we choose to take medicine. At times, too, a human act (*actus humanus*) is caused by the celestial bodies, in the sense that some people lose their minds (*amentes*) as a result of a bodily indisposition and are deprived of the use of reason. Strictly speaking, there is no act of choice for such people, but they are moved by a natural instinct, as are brutes (my emphasis).[113]

---

criticisms, as I will discuss in volume III. I discuss these features of Pico's analysis in volume II. Sylvestri's commentary is published along with the *Summa contra gentiles* in the Leonine edition.

[113] Sciendum tamen est quod, licet corpora caelestia non sint directe causa electionum nostrarum quasi directe in voluntates nostras imprimentia, indirecte tamen ex eis aliqua occasio nostris electionibus praestatur, secundum quod habent impressionem super corpora. Et hoc dupliciter. Uno quidem modo, secundum quod impressiones corporum caelestium in exteriora corpora est nobis occasio alicuius electionis: sicut, cum per corpora caelestia disponitur aer ad frigus intensum; eligimus calefieri ad ignem, vel aliqua huiusmodi facere quae congruunt tempori. — Alio modo, secundum quod imprimunt in corpora nostra: ad quorum immutationem insurgunt in nobis aliqui motus passionum; vel per eorum impressionem efficimur habiles ad aliquas passiones, sicut cho-

The two ways that *occasiones* for choice arise are thus by the celestial bodies acting, first, on other bodies in the world, and, secondly, on our own bodies. Thomas continues to refine his analysis in such a way that further nuances and fine-tunes our understanding of astrology's natural philosophical and theological foundations.

To complete this article, Thomas avers here to both experience and authority:

> Moreover, it is plain and known to experience (*experimento cognitum*) that such occasions, whether they be external or internal, are not a necessary cause of choice (*non sunt causa necessaria electionis*), since a person is able by means of reason to resist or obey them (*homo per rationem possit eis resistere vel obedire*). But there are many who follow their natural impulses (*plures sunt qui impetus naturales sequuntur*), while but few, the wise only (*soli sapientes*), do not take these occasions of acting badly and of following their natural impulses. This is why Ptolemy says in his *Centiloquium* that "the wise soul assists the work of the stars" (*anima sapiens adiuvat opus stellarum*); and that "the astrologer (*astrologus*) cannot make judgments based on the stars (*iudicia secundum stellas*), unless he knows well the power of the soul (*vis animae*) and the natural temperament (*complexio naturalis*)"; and that "the astrologer (*astrologus*) should not speak about particulars (*specialiter*) on a matter, but in general (*universaliter*)." Because, namely, the impression from the stars alots its effect on most people (*scilicet impressio stellarum in pluribus sortitur effectum*) who do not resist the inclination that comes from their body (*qui non resistunt inclinationi quae est ex corpore*). Moreover, it is not always effective in this or that [sc. particular] case (*in hoc vel in illo*), since a person may perchance resist their natural inclination through reason (*qui forte per rationem naturali inclinationi resistit*).[114]

As ever, Thomas argues that people have free will *not* to follow their bodily impulses as inclined or disposed by the celestial bodies, even though most people end up doing so, as he had also argued in *Summa theologiae* II.II.95, thus further establishing the medieval safeguards for protecting human free will. Although Thomas is not directly supporting or defending astrology here, it is striking to note that he once again buttresses his arguments with the authority of Ptolemy's explicitly astrological *Centiloquium*, and in the process explicitly describes how astrologers should legitimately make their judgments.

lerici sunt proni ad iram; vel etiam secundum quod ex eorum impressione causatur in nobis aliqua dispositio corporalis quae est occasio alicuius electionis, sicut cum, nobis infirmantibus, eligimus accipere medicinam. — Interdum etiam ex corporibus caelestibus actus humanus causatur inquantum ex indispositione corporis aliqui amentes efficiuntur, usu rationis privati. In quibus proprie electio non est, sed moventur aliquo naturali instinctu, sicut et bruta (256b12-37).

[114] Manifestum autem est, et experimento cognitum, quod tales occasiones, sive sint exteriores sive sint interiores, non sunt causa necessaria electionis: cum homo per rationem possit eis resistere vel obedire. Sed plures sunt qui impetus naturales sequuntur, pauciores autem, scilicet soli sapientes, qui occasiones male agendi et naturales impetus non sequuntur. Et propter hoc dicit Ptolomaeus in *Centilogio* quod *anima sapiens adiuvat opus stellarum* [*verbum* 8]; et quod *non poterit astrologus dare iudicia secundum stellas nisi vim animae et complexionem naturalem bene cognoverit* [*verbum* 7]; et quod *astrologus non debet dicere rem specialiter, sed universaliter* [*verbum* 1]: quia scilicet impressio stellarum in pluribus sortitur effectum, qui non resistunt inclinationi quae est ex corpore; non autem semper in hoc vel in illo, qui forte per rationem naturali inclinationi resistit (256b38-55).

5.4.1.3 III.86: Celestial Bodies Do Not Require Necessity in Nature

After defending human free will from the direct effects of celestial influences in III.84 and 85, Thomas now turns to the other side of the equation in III.86, namely, that bodily effects here below also do not follow necessarily on the influence of celestial bodies, mainly due to the complexity and thus imperfection of their material reception. In Thomas's view, then, celestial bodies do not impose necessity either on human will and choice or on bodily effects, and thus on either the soul or body registers. This chapter also concludes with a set of quotations from Ptolemy, who seems to be one of Thomas's favorite authorities in this context, but this time he cites the *Tetrabiblos* as well as the *Centiloquium*. It seems obvious that Thomas must be aware here of at least some of the astrological dimensions and ramifications of his analysis.

Thomas begins III.86 thus:

> Moreover, not only is it impossible for the celestial bodies to impose necessity on human choice (*non solum autem corpora caelestia humanae electioni necessitatem inferre non possunt*), but not even do corporeal effects in things here below *necessarily* result from them (*sed nec etiam corporales effectus in istis inferioribus ex necessitate ab eis procedunt*). For the impressions of universal causes are received in their effects according to the mode of the recipients (*impressiones enim causarum universalium recipiuntur in effectibus secundum recipientium modum*). Now, these lower things are in a state of flux (*haec autem inferiora sunt fluxibilia*) and do not always maintain the same condition (*non semper eodem modo se habentia*): [1] because of matter which exists *in potentia* to many forms (*propter materiam, quae est in potentia ad plures formas*), and [2] because of the contrariety of forms and powers [sc. which can be received by material/corporeal entities] (*propter contrarietatem formarum et virtutum*). Therefore, the impressions of celestial bodies are not received in these lower things in the manner of necessity (*non igitur impressiones corporum caelestium recipiuntur in istis inferioribus per modum necessitatis*, my emphasis).[115]

Offering a clear and detailed analysis, Thomas argues here that the reception of celestial influences is complex and variable, which thus ontologically obviates even the possibility of material necessity. Celestial influences are received differently in different entities due to their different and ever changing material natures, and are thus not received in the manner of necessity, but rather (he implies) in the manner of contingency.[116] As with Albert in *De fato*—and towards the same end—Thomas argues that the mode of reception of celestial influences depends fundamentally on the recipient, a basic feature of *his* ontology as well.

Thomas later discusses the material dimension in greater depth, and thus develops his analysis in III.82 with respect to lower bodies:

---

[115] Non solum autem corpora caelestia humanae electioni necessitatem inferre non possunt, sed nec etiam corporales effectus in istis inferioribus ex necessitate ab eis procedunt. Impressiones enim causarum universalium recipiuntur in effectibus secundum recipientium modum. Haec autem inferiora sunt fluxibilia et non semper eodem modo se habentia: propter materiam, quae est in potentia ad plures formas; et propter contrarietatem formarum et virtutum. Non igitur impressiones corporum caelestium recipiuntur in istis inferioribus per modum necessitatis (261a1-14).

[116] Porro, "*Lex necessitatis vel contingentiae*," is very helpful here.

> Moreover, the celestial bodies are agents naturally (*corpora caelestia sunt agentia naturaliter*), which require matter on which to act (*quae requirunt materiam in quam agant*). Therefore, what matter requires is not removed from the action of celestial bodies (*non igitur ex actione corporum caelestium tollitur id quod materia requirit*). But the matter on which the celestial bodies act are the lower bodies (*corpora inferiora*) which, being corruptible in their nature (*cum sint corruptibilia secundum suam naturam*), are just as able to fail in their existence as they are in their operations (*sicut deficere possunt ab esse, ita ab operari*). Thus, their nature has this characteristic: they do not produce their effects by necessity (*non ex necessitate producant effectus*). Therefore, the effects of the celestial bodies do not come about by necessity, even in lower bodies (*non igitur ex necessitate proveniunt effectus caelestium corporum etiam in corporibus inferioribus*).[117]

Because matter is essentially corruptible, it receives celestial influences differently, and not always successfully. Here Thomas continues to sound like his master Albertus Magnus in *De fato*, and for the same reasons, as well as in the discussion of the variable periods of life in his translation-commentary on Aristotle's *De generatione et corruptione* II.10, as discussed in Chap. 1. The material dimension thus continues to systematically obviate necessity and thereby also provide solid medieval safeguards therefor. This provides the natural limitations for Thomas's main theological claim that God rules all lower bodies and their motions by the celestial bodies and theirs.[118]

Thomas completes this article with two more authoritative quotations from Ptolemy:

> Ptolemy, too, in his *Tetrabiblos* [I.2], says: "Again, we should not think that *superiora* proceed inevitably (*quod superiora procedant inevitabiliter*), like things that happen under divine control (*ut ea quae divina dispositione contingunt*), and which can in no way be avoided (*nullatenus sunt vitanda*); nor as things which come about truly and of necessity (*necnon quae veraciter et ex necessitate proveniunt*)." He also says in the *Centiloquium*: "These astrological judgments (*iudicia*) that I give you are midway between the necessary and the possible."[119]

The upshot here as elsewhere is that celestial influences simply do not entail or even imply necessity in nature.[120]

---

[117] Amplius. Corpora caelestia sunt agentia naturaliter, quae requirunt materiam in quam agant. Non igitur ex actione corporum caelestium tollitur id quod materia requirit. Materia autem in quam agunt corpora caelestia, sunt corpora inferiora: quae, cum sint corruptibilia secundum suam naturam, sicut deficere possunt ab esse, ita ab operari; et sic eorum natura hoc habet ut non ex necessitate producant effectus. Non igitur ex necessitate proveniunt effectus caelestium corporum etiam in corporibus inferioribus (261a44-b8).

[118] We will see this same problem with necessity arise below in III.94 because some people fear that God's providence also implies determinism.

[119] Ptolemaeus etiam, in *Quadripartito* [I.2], dicit: *Rursus, nec aestimare debemus quod superiora procedant inevitabiliter, ut ea quae divina dispositione contingunt et quae nullatenus sunt vitanda, necnon quae veraciter et ex necessitate proveniunt.* — In *Centilogio* etiam dicit: *Haec iudicia quae tibi trado, sunt media inter necessarium et possibile* (262b37-44).

[120] At the end of his commentary on this chapter (266b12-33), Sylvestri provides an analysis that has exactly the same structure as Albert's argument referring to the same statement by Ptolemy at *De fato* (68, 50–56, as discussed above), but without referring explicitly to that discussion.

We have just learned more about the variable receptability of celestial influences in material-corporeal entities, namely, Aristotelian hylomorphic substances, that is, actually existing things in the world. Thomas's explanation here is much more straightforward and direct than Albert's more metaphysically-inflected analysis in *De fato*. Successfully combatting and thereby undermining necessity is essential for the success—including the theological legitimacy—of Thomas's endeavor here, and his analysis seems quite persuasive.

Thomas has in effect offered an elegant three-part argument here, based on astrology's natural philosophical foundations, for the most fundamental structure in his view of divine providence, namely, that God rules, disposes and moves every body here on earth in relation to their fourfold motion/changes by means of the celestial bodies and their local motions in the heavens. This is the central structure of Thomas's construction of astrology's theological and scientific foundations in his understanding of God's providential governance of the world and of human beings.

The structure of the argument is straightforward. First (in III.82), Thomas establishes the most fundamental structure, namely, that God uses the celestial bodies and their motions to rule, dispose and move all of the lower bodies and their motions on the earth and in its atmosphere. Once he establishes this to his satisfaction, he then shows in the next step (III.84 and 85) that this does not undermine human free will, on the one hand. Finally (in III.86), on the other hand, Thomas shows that the material reception of celestial influences is itself highly complex, and is utterly dependent on the particulars of reception in each and every case. The outcome of this analysis is that the second and third parts of the argument effectively and systematically undermine charges of necessity/determinism with respect to both human beings and their free will and with respect to nature, both of which are required for his views to be legitimate. Thomas has thus successfully used natural philosophical building blocks to erect solid and persuasive theological structures that provide essential ontological and epistemological foundations—with their correlative legitimations—for astrological theory and practice, whether Thomas explicitly intended them to or not. They also provided solid medieval safeguards both for protecting human free will and for obviating necessity in nature.

5.4.1.4 III.87-88: Are Celestial Bodies Animated?

Thomas then shifts gears and returns to the celestial bodies, addressing whether they are ensouled or not, and, if so, how that would affect the fundamental positions established in III.82-86. In this case, he is particularly concerned to defend against the positions articulated by Avicenna and Albumasar that celestial souls can be a direct cause of our choices. Thomas begins III.87 thus:

> [W]e should note that Avicenna [*Metaph.* X] wants the motions of the celestial bodies to also be the causes of our acts of choice (*motus caelestium corporum sint etiam nostrarum electionum causae*), not simply as an occasion (*per occasionem*), as was said above, but in themselves (*per se*). For he claims that the celestial bodies are animated (*corpora caelestia esse animata*). Whence, it is fitting that, since celestial motion (*motus caelestis*) is from a

> soul (*ab anima*) and is the motion of a body (*motus corporis*), therefore, just as, insofar as it is the motion of a body (*inquantum est motus corporis*), it has the power of causing change in bodies (*virtus transmutandi corpora*), so insofar as it is from a soul (*inquantum est ab anima*), it has the power to make an impression [sc. directly] on our souls (*virtus imprimendi in animas nostras*). Thus, the celestial motion is the cause of our acts of will and choice (*sic motus caelestis sit causa nostrarum voluntatum et electionum*). To this point also Albumasar's position in the first book of his *Introduction* seems to return.[121]

According to Avicenna and Albumasar, celestial motions directly cause our acts of will and our choices because the celestial bodies are animated, and thus their souls directly affect our souls. The *Speculum astronomiae* also discussed this aspect of Albumasar's writings, as we saw in Chap. 4.

As we should expect by now, Thomas does not embrace this position:

> But this position is irrational. For, every effect that proceeds through an instrument from an efficient [sc. cause] ought to be proportionate to the instrument, as to an agent, for we do not use just any instrument for any effect. Whence, a result cannot be accomplished by means of an instrument (*illud non potest fieri per aliquod instrumentum*) if the action of the instrument in no way extends to that [sc. result] (*ad quod nullo modo se extendit actio instrumenti*). But the action of a body in no way extends to the change of intellect and will, as we showed, unless perchance by accident, insofar as a body is changed by these, as was said [III.84-85]. Therefore, it is impossible that the soul of a celestial body (*anima coelestis corporis*), if it be animated (*si sit animatum*), would make an impression [sc. directly] on the intellect and will with the motion of a celestial body mediating (*in intellectum et voluntatem imprimat mediante motu caelestis corporis*).[122]

Thomas criticizes Avicenna here for saying that celestial souls can act directly on a person's intellect and will by means of the celestial body. This is impossible because the motion of the celestial body—as Thomas showed—cannot act directly on our will and intellect, but only indirectly through our bodies. Thus, if a celestial soul were to act on our soul, it could only do so with a celestial body and our body mediating. Thomas does not even entertain the possibility of direct soul-to-soul interaction.

We can see this in the human realm as well in the next passage:

[121] Est tamen attendendum quod Avicenna vult quod motus caelestium corporum sint etiam nostrarum electionum causae, non quidem per occasionem tantum, sicut supra dictum est, sed per se. Ponit enim corpora caelestia esse animata. Unde oportet, cum motus caelestis sit ab anima et sit motus corporis, quod sicut, inquantum est motus corporis, habet virtutem transmutandi corpora, ita, inquantum est ab anima, habeat virtutem imprimendi in animas nostras, et sic motus caelestis sit causa nostrarum voluntatum et electionum. Ad quod etiam redire videtur positio Albumasar, in Primo sui *Introductorii* (266a1-14).

[122] Haec autem positio irrationabilis est. Omnem enim effectum qui est per instrumentum aliquod ab efficiente procedens, oportet esse proportionatum instrumento, sicut et agenti: non enim quolibet instrumento utimur ad quemlibet effectum. Unde illud non potest fieri per aliquod instrumentum ad quod nullo modo se extendit actio instrumenti. Actio autem corporis nullo modo se extendit ad immutationem intellectus et voluntatis, ut ostensum est: nisi forte per accidens, inquantum ex his immutatur corpus, sicut praedictum est. Impossibile est ergo quod anima caelestis corporis, si sit animatum, in intellectum et voluntatem imprimat mediante motu caelestis corporis (266a14-b3).

> Again, a particular agent cause (*causa agens particularis*), when acting, bears a likeness to the universal agent cause (*causa agens universalis*) and is its exemplum. But, if a human soul were to impress another human soul through a corporeal operation (*anima humana in aliam animam humanam aliquid per operationem corporalem imprimeret*), as when it reveals its understanding (*intelligentia*) by the signification of a voice (*per significationem vocis*), the bodily action which is from one soul does not reach the other soul without a body mediating (*actio corporalis quae est ab una anima non pervenit ad aliam nisi mediante corpore*). For the uttered voice (*vox prolata*) changes the auditory organ, and then, having been perceived by a sense [sc. organ] (*a sensu*), it brings its signification to the intellect (*pervenit eius significatum usque ad intellectum*). Therefore, if the celestial soul makes an impression on our souls through a bodily motion (*anima caelestis aliquid imprimat in animas nostras per motum corporeum*), that action will not reach our soul except through a change in our body (*actio illa non perveniet ad animam nostram nisi per immutationem corporis nostri*). Now, this is not a cause of our acts of choice, but only an occasion, as is clear from the foregoing. Therefore, celestial motion will not be the cause of our act of choice, except as an occasion alone.[123]

For Thomas, souls act on other souls (whether celestial or human) *only* by means of their respective bodies, which then transmit their influences (whatever they may be) directly only to the *body* of the recipient, and thereby indirectly to their soul (if they have one). This is not a cause of the choosing, Thomas reiterates, but merely an occasion for making a choice. Here he emphasizes the arguments he had made in III.85.

Thomas continues by referring to the hierarchical structure of nature:

> Besides, since the mover (*movens*) and the thing moved (*motum*) should exist at the same time (as is proved in *Physics* VII), it is fitting that the motion should extend in a certain order (*quodam ordine*), from the first mover (*a primo movente*) to the last thing that is moved (*ad ultimum quod movetur*); that is, such that the mover moves what is far away from it by means of what is near to it. But our body (*corpus nostrum*) is closer than our soul is to the celestial body which is asserted to be moved by the soul joined to it, for our soul has no relation (*ordo*) to a celestial body except with a body mediating (*nisi mediante corpore*). This is evident from the fact that separated intelligences (*intellectus separati*) have no relation (*nullus ordo*) to a celestial body, unless, perhaps, that of a mover to something moved. Therefore, the change in a celestial body proceeding from its soul (*immutatio igitur corporis caelestis ab anima eius procedens*) does not reach our soul except through the mediation of our body (*non pertingit ad animam nostram nisi mediante corpore*). But our soul is not moved when our body is moved, except accidentally. Nor does choice result from a change in our body, except by way of an occasion, as we said. Therefore, celestial motion cannot be the [sc. direct] cause of our act of choice because it is from a soul (*motus igitur caelestis non potest esse causa electionis nostrae per hoc quod est ab anima*).[124]

---

[123] Amplius. Causa agens particularis similitudinem in agendo gerit causae agentis universalis, et est exemplum eius. Si autem anima humana in aliam animam humanam aliquid per operationem corporalem imprimeret, sicut cum per significationem vocis suam intelligentiam pandit, actio corporalis quae est ab una anima non pervenit ad aliam nisi mediante corpore: vox enim prolata immutat organum auditus, et, sic a sensu percepta, pervenit eius significatum usque ad intellectum. Si igitur anima caelestis aliquid imprimat in animas nostras per motum corporeum, actio illa non perveniet ad animam nostram nisi per immutationem corporis nostri. Quae quidem non est causa electionum nostrarum, sed occasio tantum, sicut ex praemissis patet. Non igitur erit motus caelestis causa nostrae electionis nisi per occasionem tantum (266b4-21).

[124] Item. Cum movens et motum oporteat esse simul, ut probatur in VII *Phys.*, oportet quod a primo movente perveniat motus usque ad ultimum quod movetur, quodam ordine: ut scilicet movens per

As we can see, Thomas cannot even conceive of a celestial soul acting directly on a human soul. For him, such contact can *only* take place with intermediating bodies, whether celestial, human or both. As we will see in volume II, however, Marsilio Ficino does not share these limits to his deeply Platonizing imagination. This is, therefore, another medieval safeguard for human free will that Ficino attempts to subvert in the *De vita*.

*

To complete this diptych of articles on celestial souls, Thomas asserts at the beginning of III.88 that "we must not think that the souls of the heavens (*animae caelorum*), if there be any—or any other created separated intellectual substances—can directly insert an act of will into us (*possint directe voluntatem nobis immittere*) or be the [sc. direct] cause of our act of choice (*electio*)."[125] Only God Himself can do this, as Thomas states in the title to III.88: "That created separated substances cannot directly be the cause of our choices and acts of will, but only God [sc. can],"[126] as he will further clarify in III.92. We will not explore this argument here beyond noting that it closely coheres with the sentiment of III.87.

5.4.1.5 III.92: What Makes a Person Fortunate?

Thomas offers more valuable information in III.92, where he discusses how we can understand "how someone is said to be fortunate, and how a person (*homo*) is assisted by higher causes," as he states in the title:[127]

> Moreover, man can obtain from higher causes (*ex superioribus causis*) other help in regard to the outcomes (*exitus*) of his actions. For, since a person has both the ability to choose (*eligere habeat*) and to carry out what he chooses (*prosequi quae eligit*), he may at times be assisted (*adiuvatur*) or impeded by higher causes in regard to both. Regarding choice, as we

id quod est sibi proximum, moveat illud quod est ab eo distans. Corpori autem caelesti, quod moveri ponitur ab anima sibi coniuncta, propinquius est corpus nostrum quam anima, quae non habet ordinem ad corpus caeleste nisi mediante corpore: — quod ex hoc patet, quia intellectus separati nullum ordinem habent ad corpus caeleste, nisi forte moventis ad motum. Immutatio igitur corporis caelestis ab anima eius procedens non pertingit ad animam nostram nisi mediante corpore. Ad motum autem corporis non movetur anima nisi per accidens, nec immutationem corporis sequitur electio nisi per occasionem, ut dictum est. Motus igitur caelestis non potest esse causa electionis nostrae per hoc quod est ab anima (266b22-267a15).

[125] Non est autem aestimandum quod animae caelorum, si quae sint, vel quaecumque aliae intellectuales substantiae separatae creatae, possint directe voluntatem nobis immittere, aut electionis nostrae causa esse (269a1-6).

[126] Quod substantiae separatae creatae non possunt esse causa directe electionum et voluntatum nostrarum, sed solus Deus.

[127] "Quomodo dicitur aliquis bene fortunatus, et quomodo adiuvetur homo ex superioribus causis." For recent insightful scholarship on this chapter and on divine providence in Thomas more generally, see Cordonier, "Sauver le Dieu," and "La doctrine Aristotélicienne de la providence divine selon Thomas Aquinas."

> said, [sc. a person is helped] insofar as he [or she] is either disposed by the celestial bodies to choose something (*disponitur ad aliquid eligendum per caelestia corpora*), or he is enlightened (*illustratur*) by the guardianship of angels, or even he is inclined by divine action. But regarding the carrying out (*executio*), [sc. this can happen] only insofar as a person may obtain from some higher cause the strength and efficacy (*robur et efficacia*) for accomplishing what he chooses (*ad implendum quod elegit*). Now, this can come not only from God and the angels, but also from the celestial bodies (*a corporibus caelestibus*), to the extent that such efficacy (*efficacia* = efficient causality) is located in a body (*inquantum talis efficacia in corpore sita est*).[128]

The higher causes here are [1] God Himself, [2] the angels (= separated intellectual substances) and [3] the celestial bodies that can work on a person's will, reason and body respectively to provide assistance concerning both the choosing itself, and realizing the results of ones choices.[129]

Thomas then turns to celestial influences on inanimate bodies, including occult properties:

> For it is obvious that inanimate bodies (*inanimata corpora*) also obtain certain powers and abilities to act from the celestial bodies (*quasdam vires et efficacias a caelestibus corporibus consequuntur*), even beyond those that follow [sc. ontologically] on the active and passive qualities of the elements, which, doubtless, are also subject to the celestial bodies (*caelestibus corporibus esse subiectas*). Thus, the fact that a magnet attracts iron is due to the power of a celestial body (*quod magnes attrahat ferrum, habet ex virtute caelestis corporis*), and also certain stones and herbs have other hidden powers (*occultae virtutes*). Whence, nothing prevents a person, too, from getting, as a result of the influence of a celestial body (*ex impressione caelestis corporis*), a certain efficacy [= heightened ability] in performing some bodily actions (*aliquam efficaciam in aliquibus corporalibus faciendis*), which another person does not have (*quas alius non habet*): for example, a physician (*medicus*) in healing, a farmer in planting, and a soldier in fighting.[130]

Drawing on an analogy with occult influences on inanimate bodies, Thomas extends this analysis to account for why individual human beings have particular skills or

---

[128] Consequitur autem homo ex superioribus causis et aliud auxilium, quantum ad exitus suarum actionum. Cum enim homo et eligere habeat, et prosequi quae eligit, in utroque a causis superioribus adiuvatur interdum, vel etiam impeditur. Secundum electionem quidem, ut dictum est, inquantum homo vel disponitur ad aliquid eligendum per caelestia corpora; vel quasi illustratur per angelorum custodiam; vel etiam inclinatur per operationem divinam. — Secundum executionem vero, inquantum homo consequitur ex aliqua superiori causa robur et efficaciam ad implendum quod elegit. Quae quidem non solum a Deo et ab angelis esse potest, sed etiam a corporibus caelestibus, inquantum talis efficacia in corpore sita est (280b44-281a4).

[129] Cordonier emphasises that this threefold causality corresponding to three different aspects of a person is a significant innovation by Thomas, and is closely related to an influential text that was discovered, composed and/or compiled at precisely this time, namely, the *De bona fortuna*; "Sauver le Dieu," 88 ff.

[130] Manifestum est enim quod etiam inanimata corpora quasdam vires et efficacias a caelestibus corporibus consequuntur, etiam praeter eas quae ad qualitates activas et passivas elementorum consequuntur, quas etiam non est dubium caelestibus corporibus esse subiectas: sicut quod magnes attrahat ferrum, habet ex virtute caelestis corporis, et lapides quidam et herbae alias occultas virtutes. Unde nihil prohibet quod etiam aliquis homo habeat ex impressione caelestis corporis aliquam efficaciam in aliquibus corporalibus faciendis, quas alius non habet: puta medicus in sanando, et agricola in plantando, et miles in pugnando (281a4-17).

talents. He seems to derive them from the occult—i.e. non elemental—realm of celestial influences. This seems to supplement his analysis in *De operationibus occultis naturae* discussed in Chap. 1 above, where Thomas attributed different human qualities to the range in latitude permitted by their specific form as indicated by the unique celestial configuration at their birth, and thus by extension in their astrological nativity. Here he extends it even further to also include people's natural abilities and vocational aptitudes, once again implicitly offering the natural philosophical foundations for nativities. This could also be read as an instantiation and further development of the astrological quotation from the pseudo-Ptolemaic *Centiloquium* in III.84 (250b3-7) discussed above.

5.4.1.6 III.93: *De fato*

In III.93, Thomas briefly discusses fate, asking "whether it exists and what it is" (*de fato, an sit et quid sit*). Thomas's treatment is very different than Albert's deeply astrological view presented at the papal curia a few years earlier, as we have seen. After mentioning three definitions that he rejects, Thomas seems to embrace the fourth here, which he treats in much greater depth than the others:

> But some (*quidam*) have wished to [sc. causally] refer (*reducere*) everything (*omnia*) to the disposition of divine providence, whatever seems to happen by chance (*a casu contingere*) among lower things (*in his inferioribus*). Whence they said that everything is acted upon by fate (*omnia fato agi*), calling "fate" an order (*ordinatio*) which exists in things from divine providence (*quae est in rebus ex divina providentia*). Whence Boethius says that "fate is the disposition inhering in movable things, by means of which providence binds each thing to its orders." In this description of fate, *dispositio* is put down for *ordinatio*; — "moreover, inhering in things" is put down in order that fate be distinguished from providence: for this *ordinatio*—in accordance with what exists in the divine mind (*in mente divina*), but not yet impressed in/on things—is providence: but, in accordance with what has already been unfolded in things (*explicata in rebus*), it is called fate; — moreover, he says "in *mobilibus*" in order to show that the order of providence (*ordo providentiae*) does not remove contingency and mobility from things, as some have proposed.[131]

In this view, Thomas discusses fate in relation to divine providence, primarily following the structure of Boethius's deeply influential analysis in the *Consolation of Philosophy*. In it, Thomas (following and quoting Boethius) defines fate directly in relation to, and indeed, in terms of divine providence. In the process, he further explains how he conceives of providence by explicitly contrasting it with fate.

---

[131] Quidam vero in dispositionem divinae providentiae omnia reducere voluerunt, quaecumque in his inferioribus a casu contingere videntur. Unde omnia fato agi dixerunt, ordinationem quae est in rebus ex divina providentia *fatum* nominantes. Unde Boetius dicut quod *fatum est inhaerens rebus mobilibus dispositio, per quam providentia suis quaeque nectit ordinibus.* In qua fati descriptione, *dispositio* pro *ordinatione* ponitur: — *in rebus autem inhaerens* ponitur ut distinguatur fatum a providentia: nam ipsa ordinatio secundum quod in mente divina est, nondum rebus impressa, *providentia* est: secundum vero quod iam est explicata in rebus, *fatum* nominatur; — *mobilibus* autem dicit ut ostendat quod ordo providentiae rebus contingentiam et mobilitatem non aufert, ut quidam posuerunt (126a22-b13).

Regardless, in the next and final passage, Thomas concludes the brief discussion here by rejecting the term entirely so as not to mislead anyone:

> Therefore, according to this understanding (*acceptio*), to deny fate is to deny divine providence (*negare fatum est providentiam divinam negare*). But, since we should not even have terms (*nomina*) in common with unbelievers, lest occasion for error could be taken from their association, the term "fate" should not be used by the faithful lest we appear to agree with those who have held a wrong opinion about fate, by subjecting all things to the necessity of the stars (*omnia necessitati siderum subiicientes*). Consequently, Augustine says, in Book V of the *City of God*: "If any man calls the will or power of God by the name "fate," let him hold his view, but correct his way of speaking (*sententiam teneat, linguam corrigat*)." And also Gregory, in accordance with the same understanding of it, says: "Far be it from the minds of the faithful to say that there is any fate."[132]

For Thomas, then, regardless of how it is parsed, we should reject the term "fate" altogether, lest it confuse the minds of the faithful who might think that it implies a deterministic universe.[133] In the final section of this chapter, I will explore Thomas's further views on fate in the later *Summa theologiae* I.116, where he more fully discusses the relationship between fate and providence. I will also compare it there with another of Albertus Magnus's treatments of fate in his translation-commentary on Aristotle's *Physics*. Albert there is also concerned to compare and contrast fate with providence.

*

Finally (for us, for now), in III.94, Thomas poses a potentially fatal difficulty for the analysis of divine providence that he has been developing, namely, that if the foregoing is true, then [1] either divine providence is not certain, or [2] everything in the world happens by necessity, neither of which is an acceptable conclusion.[134] In this chapter, Thomas argues against the view that his analysis of divine providence supports a view of necessity in nature and thus undermines human free will. From this perspective, providence itself raises many of the same problematic issues that astrology does (to which he had just referred in the discussion on fate), thus revealing some of the interesting tensions within Thomas's system. This vast and fundamental issue is worthy of further investigation.

---

[132] Secundum hanc ergo acceptionem, negare fatum est providentiam divinam negare. Sed quia cum infidelibus nec nomina debemus habere communia, ne ex consortio nominum possit sumi erroris occasio; nomine *fati* non est a fidelibus utendum, ne videamur illis assentire qui male de fato senserunt, omnia necessitati siderum subiicientes. Unde Augustinus dicit, in V De civitate Dei: *Si quis voluntatem vel potestatem Dei fati nomine appellat, sententiam teneat, linguam corrigat.* Et Gregorius, secundum eundem intellectum, dicit: *Absit a fidelium mentibus ut fatum aliquid esse dicant* (286b14-25).

[133] Sylvestri's commentary is very interesting here, including his response to Pico's arguments in Book IV of the *Disputations*. I will discuss it further in my volume III.

[134] Difficultas autem quaedam ex praemissis suboritur. Si enim omnia quae hic inferius aguntur, etiam contingentia, providentiae divinae subduntur, oportet, ut videtur, vel providentiam non esse certam; vel omnia ex necessitate contingere (287a1-6).

5.4.2 [2] De veritate

With Thomas's *De veritate* written between 1256 and 1259,[135] I will be even more selective, treating only his official *Responsio* to the first of the two relevant articles. Question V is on Providence. Articles 9 and 10 treat the role of the celestial bodies, as we can readily see by their titles: [9] "Does divine providence dispose bodies here below by means of the celestial bodies?" [10] "Are human actions governed by divine providence with the celestial bodies mediating?"[136] It is obvious that Thomas will treat similar themes here to those in the later *Summa contra gentiles*. I will discuss the arguments *contra* and the *Responsio* to article 9.

I will first simply set out the five arguments *contra*, which in fact represent Thomas's view:

> [1] Augustine says: "Denser (*crassiora*) and weaker (*infirmiora*) bodies are ruled in a certain order (*quodam ordine*) by bodies more subtle and powerful." But, the celestial bodies (*corpora caelestia*) are more subtle and powerful (*subtiliora et potentiora*) than lower bodies (*inferiora*). Therefore, lower bodies are ruled by them (*haec ergo inferiora reguntur per illa*).[137]

> [2] Furthermore, Dionysius says in the fourth chapter of *On Divine Names* that the solar ray (*radius solaris*) "confers the generation of visible bodies (*generationem visibilium corporum confert*), moves them towards life (*ad vitam ipsa movet*), nourishes them, and makes them grow." Moreover, these are the more noble effects in lower bodies (*nobiliores effectus in his inferioribus*). Consequently, all other bodily effects (*omnes alii effectus corporales*) are produced by divine providence (*producuntur a divina providentia*) with the celestial bodies mediating (*mediantibus corporibus caelestibus*).[138]

> [3] Additionally, according to the Philosopher in Book II of the *Metaphysics*, that which is first in any genus (*primum in aliquo genere*) is the cause of those things that exist afterwards in that genus (*causa eorum quae sunt post in illo genere*). But the celestial bodies are the first in the genus of bodies (*corpora caelestia sunt prima in genere corporum*), and their motions are first among the other bodily motions (*motus eorum sunt primi inter alios motus corporales*). They are, therefore, the cause of bodily things that are acted upon here (*causa corporalium quae hic aguntur*). Hence, the conclusion is the same as before.[139]

---

[135] Torrell, *Saint Thomas Aquinas*, 62 and 334.

[136] [9] "Nono quaeritur utrum per corpora caelestia disponat divina providentia inferiora corpora (V.9.1-3)." [10] "Decimo quaeritur utrum humani actus gubernentur a divina providentia mediantibus corporibus caelestibus (V.10.1-3)." The text is from vol. XXII (1/2) of the Leonine edition, 1970.

[137] Sed contra est quod Augustinus dicit III De trinitate, "Corpora crassiora et infirmiora per subtiliora et potentiora quodam ordine reguntur"; sed corpora caelestia sunt subtiliora et potentiora quam inferiora; haec ergo inferiora reguntur per illa (V.9.169-74).

[138] Praeterea, Dionysius dicit IV cap. De divinis nominibus quod radius solaris "generationem visibilium corporum confert et ad vitam ipsa movet et nutrit et auget"; hi autem sunt nobiliores effectus in his inferioribus; ergo et omnes alii effectus corporales producuntur a divina providentia mediantibus corporibus caelestibus (175–82).

[139] "Praeterea, secundum Philosophum in II Metaphysicae, illud quod est primum in aliquo genere est causa eorum quae sunt post in illo genere; sed corpora caelestia sunt prima in genere corporum

> [4] In addition, the Philosopher says in Book II of *De generatione* that the motion of the sun on the inclined circle (*allatio solis in circulo declivi*) is the cause of generation and corruption in things here below (*causa generationis et corruptionis in inferioribus*). Consequently, generations and corruptions are measured by this motion (*generationes et corruptiones mensurantur per motum praedictum*). Aristotle also says in the book *De animalibus* that all the varieties that exist in things conceived (*omnes diversitates quae sunt in conceptibus*) are due to the celestial bodies (*sunt ex corporibus caelestibus*). Therefore, these lower things are disposed (*haec inferiora disponuntur*) with them [sc. the celestial bodies] mediating (*eis mediantibus*).[140]

> [5] Additionally, Rabbi Moses [Maimonides] says that the heavens are in the world (*caelum est in mundo*) as the heart is in an animal. But all the other members are governed by the soul with the heart mediating (*omnia alia membra gubernantur ab anima mediante corde*). Therefore, all the other bodies (*omnia corpora alia*) are governed by God (*gubernantur a Deo*) with the heavens mediating (*mediante caelo*).[141]

Although Thomas offers some interestingly different examples here, we can easily see that the sentiment is precisely the same as—and perhaps even more strongly stated than—what we found in the slightly later *Summa contra gentiles* III.82-94, especially in III.82. We should also note the '*primum in aliquo genere*' argument, the allusion to *De generatione et corruptione* II.10, and the microcosm-macrocosm analogy, one of the very few we have encountered so far, and the first by Thomas.

*

I will now treat the essential argument of Thomas's *Responsio* to the arguments *pro* and *contra*:

> Whence, it is fitting to propose that in [sc. bodies] here below (*in his inferioribus*), there are some principles of generation and corruption, and of the other motions that follow [sc. ontologically] upon these, that do not always remain the same. Nevertheless, it is fitting that these always remain as the first principles of generation (*prima generationis principia*) so that continual generation can exist. Therefore, it is fitting that they be unchangeable with respect to substance (*invariabilia secundum substantiam*), and moved only in place (*moveri* […] *solum secundum locum*), so that, by means of their approach or withdrawal (*per accessum et recessum*), they make contrary and diverse motions in lower bodies (*contrarios et diversos motus in his inferioribus efficiant*). The celestial bodies (*corpora coelestia*) are of this nature. Therefore, it is fitting to [sc. causally] refer (*reducere*) all bodily effects to them

---

et motus eorum sunt primi inter alios motus corporales; ergo sunt causa corporalium quae hic aguntur, et sic idem quod prius (183–89)." Thomas quoted this passage in the slightly later *Summa contra gentiles* III.82, but without the reference there to Aristotle's *Metaphysics*.

[140] Praeterea, Philosophus dicit in II De generatione quod allatio solis in circulo declivi est causa generationis et corruptionis in his inferioribus, unde et generationes et corruptiones mensurantur per motum praedictum; in libro etiam De animalibus dicit quod omnes diversitates quae sunt in conceptibus sunt ex corporibus caelestibus; ergo eis mediantibus haec inferiora disponuntur (190–98).

[141] Praeterea, Rabbi Moyses dicit quod caelum est in mundo sicut cor in animali; sed omnia alia membra gubernantur ab anima mediante corde; ergo omnia corpora alia gubernantur a Deo mediante caelo (199–203).

> as to their causes (*ideo omnes effectus corporales oportet reducere in ea [sc. corpora coelestia] sicut in causas*).[142]

As we saw in Chap. 1, the planets in their annual motion in place around the zodiac provide the foundations for our understanding of all the other motions on earth, including and especially generation and corruption. In this passage, Thomas sounds very much like his teacher Albert, but also like Roger Bacon with his strong emphasis on the *coelestia* being the causes of things below.

Thomas then defends against two common errors that result from this understanding:

> But in this [sc. causal] *reductio*, there has been a twofold error. [1] For some have [sc. causally] reduced the lower bodies to the celestial bodies (*quidam enim haec inferiora in corpora caelestia reduxerunt*) as to their first causes *simpliciter* (*sicut in causas simpliciter primas*) for this reason, that they thought that no incorporeal substances exist (*nullas substantias incorporeas arbitrabantur*). Consequently, they said that what is prior among bodies (*priora in corporibus*) is first among entities (*prima inter entia*). But this is clearly false! For, everything moved should be reduced causally to a first immovable principle (*in principium immobile*), since nothing is moved by itself, and one cannot keep going back infinitely. Now, even though a celestial body is not changed (*non varietur*) with respect to generation or corruption, or by any motion that would change what belongs to its substance, it is nevertheless moved in place. Consequently, the causal *reductio* should be made to some prior principle (*in aliquod prius principium*), so that things undergoing qualitative change (*ea quae alterantur*) are referred causally in a certain order to that which causes this change without being changed itself (*reducantur in alterans non alteratum*), although it is moved in place; and then further back (*ulterius*) to that which is not changed/moved in any way.[143]

Thomas asserts here that it is an error to stop one's causal analysis at the level of the celestial bodies. Rather, one should trace the causes further back into the divine realm, as he set out in detail in *Summa contra gentiles* III.78-81, and more schematically in *De operationibus occultis naturae* and elsewhere, namely, to the realm of intellectual substances, and ultimately to the first cause, God Himself.

Thomas then responds to the second error, that somehow the planets themselves are the essential cause of what is moved on earth. Here again he discusses Avicenna:

---

[142] Unde oportet ponere principia generationis et corruptionis et aliorum motuum consequentium in his inferioribus esse aliqua quae non semper eodem modo se habeant; oportet tamen ea semper manere quasi prima generationis principia ut generatio continua esse possit: et ideo oportet ea esse invariabilia secundum substantiam, moveri autem solum secundum locum ut, per accessum et recessum contrarios et diversos motus in his inferioribus efficiant; et huiusmodi sunt corpora caelestia; et ideo omnes effectus corporales oportet reducere in ea sicut in causas (237–49).

[143] Sed in hac reductione duplex error fuit. Quidam enim haec inferiora in corpora caelestia reduxerunt sicut in causas simpliciter primas eo quod nullas substantias incorporeas arbitrabantur, unde priora in corporibus dixerunt esse prima inter entia. Sed hoc manifeste apparet esse falsum: omne quod enim movetur oportet in principium immobile reduci cum nihil a se ipso moveatur et non sit abire in infinitum; corpus autem caeleste, quamvis non varietur secundum generationem et corruptionem aut secundum aliquem motum qui variet aliquid quod insit substantiae eius, movetur tamen secundum locum; unde oportet in aliquod prius principium reductionem fieri ut sic ea quae alterantur quodam ordine reducantur in alterans non alteratum, motum tamen secundum locum, et ulterius in id quod nullo modo movetur (250–68).

> [2] But others have proposed that the celestial bodies are the causes of these lower bodies not only with respect to motion/change (*quantum ad motum*), but also with respect to their first establishment [= creation (*prima* […] *institutio*)]. Avicenna says in his *Metaphysics* that, from what is common to all celestial bodies, namely, their circular motion by nature, is caused in these lower bodies what is common to them, namely, first matter (*materia prima*); and, from those things which differentiate one celestial body from another, the difference in the forms of lower bodies is caused. Thus, the celestial bodies would be the intermediaries (*media*) between God and these lower bodies (*ista inferiora*) in some manner in the path of creation (*in via creationis*). But this position is alien to the faith, which posits that all nature was produced directly (*immediate*) by God (*a Deo condita*) in its first establishment. But that one creature [= created entity] should be moved by another, presupposing that natural powers are attributed to each creature as a result of God's work (*ex divino opere*) [sc. is not alien to the faith]. Therefore, we propose that the celestial bodies are the causes of lower [sc. bodies] (*causae inferiorum*) only by the path of motion (*per viam motus*). Thus, [sc. these celestial bodies] are intermediaries (*media*) in the work of governance (*in opere gubernationis*), but not in the work of creation (*in opere creationis*).[144]

According to Thomas, not the planets but God Himself has directly created this glorious system *ex nihilo*, both the heavens and the earth. The planets then—as physical celestial bodies—play an essential and central role as God's physical formal and efficient instruments in the work of *governing* life on earth,[145] but certainly not in *creating* it, which would be contrary to the faith and therefore something to be assiduously avoided. Nevertheless, this role of the planets in God's providential governance of the world is utterly central, as we also saw in the *Summa contra gentiles*.

*

Thus, although Thomas does not overtly defend astrology in *Summa contra gentiles* III.82-94 or in *De veritate* V as he will later do in both *Summa theologiae* II.II.92–95 and the *De iudiciis astrorum*, he does something even more profound here. In fact, Thomas seems to be establishing nothing less than the fundamental structures of astrology's integrated theological and scientific foundations by weaving central patterns of astrology's deeply Aristotelian natural philosophical foundations into the heart of his extremely influential theology in articulating God's providential governance of the world and of man.

---

[144] Quidam vero posuerunt corpora caelestia esse causas istorum inferiorum non solum quantum ad motum sed etiam quantum ad primam eorum institutionem, sicut Avicenna dicit in sua Metaphysica quod ex eo quod est commune omnibus corporibus caelestibus, scilicet natura motus circularis, causatur in his inferioribus id quod est eis commune, scilicet materia prima, et ex his in quibus corpora caelestia differunt ad invicem causatur diversitas formarum in his inferioribus; ut sic caelestia corpora sint media inter Deum et ista inferiora etiam in via creationis quodam modo. Sed hoc est alienum a fide, quae ponit omnem naturam immediate esse a Deo conditam secundum sui primam institutionem, unam autem creaturam moveri ab altera praesuppositis virtutibus naturalibus utrique creaturae ex divino opere attributis: et ideo ponimus corpora caelestia esse causas inferiorum per viam motus tantum, et sic esse media in opere gubernationis non autem in opere creationis (269–90).

[145] We also saw this view, albeit less fully articulated, in *Speculum astronomiae*, Chap. 3.

5.5 Divine Providence and Fate in Albertus Magnus and Thomas Aquinas

5.5.1 *Albertus Magnus,* Physica *II.2.19*

I would like to complete this investigation/sounding of astrology's complex and interesting relationship to theology by returning to Albertus Magnus and exploring his analysis of fate and providence—and their relationship—in his translation-commentary on Aristotle's *Physica* (II.2.19-20).[146] As Albert tells us at the beginning of II.2.19, he is concerned with two main questions in this digression in his own voice: [1] to define what fate is, and [2] to explore whether fate imposes necessity on things.[147] Although this analysis is more metaphysico-theological than that in the *De fato* due to Albert's explication here of divine providence in relation to fate, nevertheless, it still ties in nicely with the material treated in the first section of this chapter, and it supplies more background information for understanding the material reconstructed in Part I. As ever, I will be highly selective here, focusing on two passages in particular.

Before discussing Albert's analysis in the *Physics*, however, I would first like to mention his dismissal of the second definition of fate in the slightly later *De fato*:

> In the second manner, fate is called the disposition of divine providence about the future progress of the existence and life of things below (*dispositio providentiae divinae de futuro progressu esse et vitae inferiorum*). Since it is eternal, this disposition endures, which puts nothing in things (*nihil ponit in rebus*). But when it is unfolded in its effect (*explicatur in effectu*), then that effect is distributed through things (*per res*), and realized (*expletur*) in fitting times and places. This is just like the fore-ordering and predetermination in someone's mind, for conducting their business by means of some messenger (*nuntius*), places nothing *in* the messenger. Nevertheless, it is accomplished *by* the messenger, since it directs the messenger and joins its business to it. In this manner, Boethius speaks about fate in the *Consolation of Philosophy.* And once again, we will not inquire here about fate in this manner. The form of this fore-ordering existing in the divine mind is simple, divine, eternal, immaterial and unchangable. Nevertheless, when it is unfolded through temporal things, it becomes temporal, material, multiplied, movable and contingent (my emphases).[148]

---

[146] One section of the chapter "Albert's Physics" by Steven Baldner in *A Companion to Albert the Great* (173–219) is entitled "Chance, Fate and Providence" (182–84).

[147] Antequam autem nos solvamus rationes primo INDUCTAS contra casum et fortunam, oportet nos hic inducere duo capitula de fato, in quorum primo tangemus, quid sit fatum. In secundo autem inquiremus, utrum ponat rebus necessitatem (126, 25–29).

[148] Secundo modo dicitur fatum dispositio providentiae divinae de futuro progressu esse et vitae inferiorum. Quae dispositio cum sit aeterna, constat, quod nihil ponit in rebus, sed cum explicatur in effectu, tunc effectus ille per res digeritur et expletur temporibus et locis opportunis, sicut etiam praeordinatio et praedeterminatio alicuius in mente de gerendis negotiis suis per aliquem nuntium nihil ponit in nuntio, sed tamen per nuntium expletur, quando nuntium dirigit et negotia sibi iniungit; et hoc modo Boethius in consolatione philosophiae loquitur de fato. Et hoc modo iterum non quaeritur hic de fato. Forma tamen huius praeordinationis in mente divina existens simplex est et divina et aeterna et immaterialis et incommutabilis, et tamen cum per res temporales explicatur, temporalis fit et materialis et multiplicata, mobilis et contingens (68, 14–30).

As we will see, the second definition in *De fato* is very similar to the analysis of fate in *Physica* II.2.19, to which we will now turn.

*

Chapter 19 is entitled: "This is a digression declaring what fate is (*quid sit fatum*), and that this exists should not be denied."[149] Albert first offers some earlier definitions, including those by Hermes, Firmicus Maternus, Ptolemy and Seneca (126, 25–46). Then he turns to Boethius: "In the fourth [book] of the *Consolation of Philosophy*, our Boethius says that 'fate is the disposition inherent in movable [= changeable] things, through which providence binds each thing to its orders'."[150]

After this initial definition from Boethius's extremely influential *Consolation of Philosophy* IV.6, Albert presents an extensive passage from this same chapter, which I will treat in full:[151]

> [10] For providence exceeds (*excedit*) fate in such a way that providence embraces (*complectitur*) all things (*cuncta*), however diverse and infinite; but fate arranges individual things (*singula*) in their movement (*in motu*) distributed as to places, forms and times, so that this whole temporal unfolding, when united in the view of the divine mind, is providence, but the same uniting (*adunatio*) of things, when arranged and unfolded (*explicata*) in time, is called fate. [11] Although these are different, one [sc. ontologically] depends on the other (107)."[152]

Fate thus depends ontologically on providence, and they seem to be two metaphysically distinct yet utterly complementary sides of the very same coin, as it were, functioning in fact as a fundamental pivot in the emanation/realization of God's providential conception in the world. This is also the precise and critical inflection point in the system where eternity transmutes into time and unity into multiplicity.

Boethius continues with an analogy:

> [12] For just as the craftsman (*artifex*) grasps in advance in his mind the form (*forma*) of the thing to be made first and sets the realisation of the work into motion, carrying out in

---

[149] Et est digressio declarans, quid sit fatum et quod ipsum esse non est negandum (126, 22–24).

[150] Noster autem BOETHIUS in QUARTO DE CONSOLATIONE PHILOSOPHIAE dicit, quod 'fatum est inhaerens rebus mobilibus dispositio, per quam providentia suis quaeque nectit ordinibus' (126, 47–50).

[151] I use Robert Sharples's translation of the passage from Boethius as the basis for mine in the text with translation cited above. I have slightly modified it where his text of Boethius differs from that of Albert's in the Cologne edition. I have also indicated the most significant differences in the Latin texts, which are very close. Sharples adopts the sentence numbers from the critical edition in the *Corpus Christanorum, Series Latina* XCIV, pars I (1957); I give the page number of Sharples's translation with the English translation; the Latin text is from the *Editio Coloniensis.*

[152] "[10] Providentia enim excedit fatum, eo quod 'providentia cuncta pariter, quamvis diversa infinita, complectitur, fatum vero singula dirigit [digerit, Sh.] in motu, locis, formis et temporibus distributa, ut haec temporalis ordinis explicatio in divinae mentis adunata prospectu providentia sit, eadem vero adunatio digesta atque explicata temporibus fatum vocetur, [11] quae licet diversa sint, alterum tamen pendet ex altero [ordo namque fatalis ex providentiae simplicitate procedit, add. Sh] (126, 50–57)." This extra sentence is in Sharples's edition: "for the sequence of fate (*ordo* […] *fatalis*) proceeds from the singleness of providence."

> temporal sequence (*per temporales ordines*) what he had foreseen (*prospexerat*) simply and in present time, just so God by his providence arranges (*disponit*) in a single and fixed way what is to be done, but by fate he manages (*administrat*), in their plurality and in time, these very same things which he arranges.[153]

Using an analogy from how a craftsman/artisan works in transforming a conception in his mind—that is, the formal cause—into an actually existing thing in the world (and here the Platonic resonances with God as demiurge in the *Timaeus* become pronounced), Boethius further clarifies the relationship between providence and fate. By fate, God manages what he had previously—and providentially—arranged in His mind.

Boethius continues by discussing a range of possible mediating agents by which God may unfold his providential conceptions into the world:

> [13] Therefore, whether [sc. fate] is put into practice (*exercetur*) by certain spirits (*quibusdam* [...] *spiritibus*) which are servants of divine providence, or whether the sequence of fate (*fatalis series*) is woven together (*contexitur*) through the service of the soul[154] or the whole of nature, or by the motions of the heavenly bodies (*caelestibus siderum motibus*) or by the power of angels (*angelica virtute*) or by the varied offices of *daemones*—by some of these things or by all of them—what at any rate is certain is that the unmovable (*immobilis*) and simple (*simplex*) pattern (*forma*) of things to be brought about is providence, while fate is the movable connection (*mobilis nexus*) and sequence in time (*ordo temporalis*) of the things whose bringing-about has been arranged (*disposuit*) by the divine simplicity (107)."[155]

Providence is thus fundamentally concerned with the formal design and planning in God's mind, whereas fate has the responsibility to execute these plans and govern their results as explicated/unfolded in time and place. Also, Boethius here offers a range of possible agents/instruments for how God's providence works its way into the world by means of fate. He mentions spirits, soul, nature, the celestial bodies, angels and demons (*daemones*),[156] and he wonders whether some or all of them are thus employed.

Although Boethius mentions nothing explicitly astrological here, the fact that he mentions the celestial bodies as one of the potential intermediate or secondary

[153] [12] Sicut enim artifex faciendae rei formam mente praecipiens movet ad operis effectum, et quod simpliciter praesentialiterque prospexerat, per temporales ordines ducit, ita deus providentia quidem singulariter stabiliterque facienda disponit, fato vero haec ipsa quae disponit, multipliciter ac temporaliter administrat (126, 57–63).

[154] Sharples identifies the soul here as the *anima mundi*.

[155] [13] "Sive igitur famulantibus quibusdam divinae providentiae spiritibus [fatum, add. Sh.] exercetur seu anima seu tota inserviente natura seu caelestibus siderum motibus seu angelica virtute seu daemonum varia solertia seu aliquibus horum seu omnibus fatalis series contexitur, illud certe manifestum est immobilem simplicemque gerendarum formam rerum esse providentiam, fatum vero eorum quae divina simplicitas gerenda disposuit, mobilem nexum atque ordinem temporalem' (126, 63–72)." Sharples adds 'fatum' as does Thomas in his treatment *De fato* in *Summa theologiae* I.116.2 (554b3) to be briefly discussed just below.

[156] These seem to be much more Neoplatonically-inflected '*daemones*' than proper Catholic demons. We will see in volume II that Ficino's use of '*daemones*' is also normally much more Neoplatonic than Catholic, but not always.

causes permits this possibility, and allows the metaphysically-inflected analysis of fate in relation to providence here to tie in directly to—by integrating structurally with—the more scientifically-inflected analysis of fate in Albert's *De fato*.[157] Thomas seems to have much more fully articulated Boethius's suggestion here in *Summa contra gentiles* III.82-94—embracing angels, the celestial bodies and nature as the relevant intermediary agents—perhaps based at least in part on the more fully articulated causal structures of Albert's astrologizing Aristotelian natural philosophy.[158]

*

After this extensive quotation from Boethius, Albert then explains *in propria persona* how fate works:

> From all these things, what the truth is about this matter concerning fate is gathered. For, since God is the first of the causes, and all the things of which He (*ipse*) is the cause have providence, what he causes through his intellect and by his knowledge (*scientia*) is the cause of things. He causes these things in such a way that, with universal causes mediating, He produces particular causes, and, with particular [sc. causes] mediating, he produces each thing (*quaeque*) that comes to be by nature and from the design of [His] will (*proposito voluntatis*, 126, 73–80).[159]

Here Albert explicates Boethius's position by sketching out the broader causal structures of his system, much as we saw in the broader framing texts I used to set the scene at the beginning of Chap. 1 as well as in the first section of this chapter on his *De fato* when Albert explicated Boethius's image of the several circles.

This structure unfolds in three distinct but closely related stages: First, God, the Cause of causes, produces universal causes. Then, in the second stage, with universal causes mediating, He produces particular or proximate causes. Finally, with particular causes mediating, He produces each particular thing (*quaeque* = each *res naturalis* = each actually existing thing in the world). Thus, there are four levels in Albert's system articulated here: [1] God, the Cause of causes, with his providential conception of everything united in His mind (the Neoplatonists' 'nous' [Gk]); [2] the first of three levels of the *series fatalis*—universal causes—which are explicated in *De fato* as the planets within the celestial circle; [3] particular or proximate causes (= specific causes/specific forms); [4] each actually existing thing/entity in the world, much as we found, albeit less fully articulated, in Thomas's *De operationibus occultis naturae*, as discussed in my Chap. 1.

Albert now explains further what he means:

---

[157] Palazzo's analysis is very helpful here; "Fate and Celestial Influences."

[158] We will see in volume II that Ficino embraced soul and daemons as well, especially the *anima mundi*.

[159] Ex his omnibus colligitur de fato, quid sit secundum rei veritatem. Cum enim prima causarum sit deus et omnia quorum causa est ipse, providentiam habeant, eo quod ipse causat per intellectum et sua scientia est causa rerum, ipse causat ea sic, quod mediantibus causis universalibus producit causas particulares et mediantibus particularibus producit quaeque, quae fiunt a natura et a proposito voluntatis (126, 73–80).

> Therefore, that which proceeds first from Him is necessary, such as the [sc. celestial] orbs and their motions [= the universal causes]. But that which proceeds under it—from Him and from the motion of the orbs—has something contrary, and therefore can be impeded, and therefore lacks necessity, but comes to be (*evenit*) frequently [= for the most part, i.e. in the normal course of nature]; and this exists in the mixture of elements and the seminal [= specific] causes of things. Moreover, the things of nature (*naturae res*) and all its works (*opera omnia*) come to be from these causes because the effects and the motions of lower things depend on the superiors (*pendet ex superioribus*) as from their causes (*sicut ex causis*), and the superiors depend on providence. Therefore, the disposition of providence (*dispositio providentiae*) infused into this entire web of causes (*toti isti contexioni causarum*) is called fate (126, 80–127, 5).[160]

Here Albert explicitly astrologizes this system by identifying the celestial bodies and their motions as the universal causes, and by reiterating that the motions of lower things depend on that of the superiors—i.e. the celestial bodies— as from their causes, and that these depend ultimately—both formally and ontologically—on divine providence. Thus, the disposition of providence within this entire causal structure—the *series fatalis*—is called fate. In *De fato*, Albert mainly looked as high as the celestial bodies, but here in the *Physics* he takes it all the way up to God Himself as the Cause of causes in this metaphysically-inflected definition, much as we find Him described in *Speculum astronomiae*, chapter 3. We can also easily see the similarity in structure here with that in Thomas's analysis of divine providence in the later *Summa contra gentiles* III.82 ff. and in *De veritate*. We should also note that Albert's causal analysis here explicitly obviates necessity in nature just as he also does in *De fato*.

*

The final passage I will explore here is Albert's treatment of the Peripatetics' position after discussing the Epicureans' and the Stoics', both of which he rejects. This further articulates the position just established:

> But the Peripatetics, considering this subtly, say [1] that the entire disposition of the world (*mundi tota dispositio*) depends on the motion of the heavens, and [2] that disposition by which the world is moved and ruled (*movetur et regitur*) is infused in the heavens from the first cause. Moreover, from the heavens [sc. it, the world] is infused with forms by means of motion, which [sc. forms] are the movers in the matter of generable things. Therefore, what exists in the first cause simply and as one and indifferently from the essense of the first cause, this—infused in the celestial [sc. bodies] and the motors of the heavens—is multiplied, diversified/differentiated and arranged/ordered. Nevertheless, it does not lose (*ammitit*) [sc. its] necessity, for this reason that neither the celestial body (*caeleste corpus*) as a whole or in its parts, nor even in its motion, has something contrary, by which an effect would follow less, but this has been infused in the heavens into the matter of generable and corruptible things. In addition to this, and in accordance with its existence (*suum esse*), it is

[160] Id ergo quod primo procedit ab ipso, est necessarium sicut orbes et motus eorum. Quod autem sub illo procedit ab ipso et ab orbium motu, contrarium habet et ideo impediri potest et ideo caret necessitate, sed evenit ut frequenter; et hoc est in elementorum mixtione et causis seminalibus rerum. Naturae autem res et opera omnia fiunt ex his causis, quia effectus et motus inferiorum pendet ex superioribus sicut ex causis, et superiora pendent ex providentia, et ideo dispositio providentiae infusa toti isti contexioni causarum dicitur fatum (126, 80–127, 5).

> varied and distributed in many things due to the many variations of its subject, and it acquires something contrary, and therefore can be impeded, according to the opinion of all the Peripatetics.[161]

Then Albert returns to the Stoics, whither we shall not follow him. In this analysis attributed explicitly but generally to the Peripatetics, Albert describes this causal structure clearly. Here he takes us from the One (= God, the first cause) to the many by means of the fundamental formal and efficient causal nexus operating between God and nature with the celestial bodies and their motions mediating. Once again, we can see how Thomas articulated the utter fundaments of these very same structures in an overtly theological idiom in his *Summa contra gentiles* III.82-94, and how this causal analysis provided a powerful ontological safeguard against implying or requiring determinism in nature. We shall now turn to a later statement by Thomas.

5.5.2 *Thomas Aquinas,* Summa theologiae, *I.116*

Thomas also treated fate in a chapter entitled *De fato* at *Summa Theologiae* 1.116 (composed between 1265 and '68),[162] which was immediately preceeded by four articles in I.115 (3–6) treating the causal role (if any) of the celestial bodies on lower bodies. Here are their titles: [3] Whether celestial bodies are the cause of those things which come to be here among lower bodies? [4] Whether celestial bodies are the cause of human actions? [5] Whether celestial bodies can make impressions on demons themselves? And, [6] whether celestial bodies impose necessity on those things which are subordinate to their action?[163] Articles 3, 4 and 6 are much like those in *Summa contra gentiles* III.82-86. Article 5 concerns demons and is much different than those treated above.

*

[161] Sed PERIPATETICI subtiliter hoc considerantes dicunt, quod mundi tota dispositio pendet ex motu caeli et dispositio illa qua mundus movetur et regitur, ex prima causa infunditur caelo. Ex caelo autem per motum infunditur formis, quae sunt in materia generabilium moventes, et ideo quod in prima causa est simpliciter et ut unum et indifferens ab essentia causae primae, hoc infusum caelestibus et motoribus caelorum multiplicatur et diversificatur et ordinatur; et tamen non amittit necessitatem, eo quod nec caeleste corpus in toto vel in partibus nec etiam in motu suo habet contrarium, quo minus consequitur effectum, sed infusum in caelo in materiam generabilium et corruptibilium. Adhuc amplius, secundum suum esse variatur et distribuitur in plura propter plurimas sui subiecti variationes et acquirit contrarium et ideo valet impediri secundum sententiam omnium Peripateticorum (127, 45–62).

[162] Torrell, *Saint Thomas Aquinas*, 146 and 333.

[163] [3] Utrum corpora caelestia sint causa eorum quae hic in inferioribus fiunt. [4] Utrum corpora caelestia sint causa humanorum actuum. [5] Utrum corpora caelestia possint imprimere in ipsos daemones. [6] Utrum corpora caelestia imponant necessitatem his quae eorum actioni subduntur.

Thomas treats four topics in *Summa theologiae* I.116: [1] Whether fate exists? [2] In what does it exist? [3] Whether it is unmovable? And, [4] whether everything is subordinate to fate?[164] I will be very brief here and only treat the responses to articles [2] and [3]. After making one argument *contra* and thus offering his own position, Thomas responds to the second question thus:

> But *contra* is what Boethius says in the fourth [book] of the *Consolation*: that "fate is the disposition inhering in movable things." I respond, saying that [...] divine providence carries out (*exequitur*) its effects by means of intermediate causes. Therefore, this ordering/arrangement (*ordinatio*) of effects can be understood in two ways. In one way, according to what exists in God Himself: and thus this arrangement of effects is called "providence." But according to the aforementioned arrangement considered in intermediate causes arranged by God for producing some effects, thus it has the structure of fate (*ratio fati*).

Then, after quoting at some length the statement in Boethius about the various different ways in which these intermediate causes could be conceived of as operating—including by spirits, '*daemones*' and the celestial bodies, as we saw just above in a quotation of the same passage by Albert—Thomas completes the heart of his response thus: "Therefore, it is clear that fate exists in these created causes (*fatum est in ipsis causis creatis*), insofar as they are arranged/ordered by God (*inquantum sunt ordinatae a Deo*) for producing effects (*ad effectus producendos*)."[165]

*

Finally, I will offer Thomas's response to the question raised in article 3: "Whether fate is unmovable?" Thomas responds thus, beginning with his position *contra*:

> But *contra* is what Boethius said, that fate is an unmovable disposition. I respond by saying that the disposition of secondary causes, which we call fate, can be considered in two ways: in one way, in accordance with those secondary causes, which are disposed or arranged thus; in another way, through the relationship to the first principle by Whom they are arranged, namely, God.[166]

In both of these responses clarifying the relationship of fate to providence, although Thomas did not mention the celestial bodies explicitly here (except once within a

[164] Deinde considerandum est de fato. Et circa hoc quaeruntur quatuor. Primo: an fatum sit. Secundo: in quo sit. Tertio: utrum sit immobile. Quarto: utrum omnia subsint fato (552a1-b3).

[165] Sed contra est quod Boetius dicit, in IV *De consol.*, quod *fatum est dispositio rebus mobilibus inhaerens*. Respondeo dicendum quod, sicut ex praedictis patet, divina providentia per causas medias suos effectus exequitur. Potest ergo ipsa ordinatio effectuum dupliciter considerari. Uno modo, secundum quod est in ipso Deo: et sic ipsa ordinatio effectuum vocatur providentia. — Secundum vero quod praedicta ordinatio consideratur in mediis causis a Deo ordinatis ad aliquos effectus producendos, sic habet rationem fati. [Boethius quotation with brief comment.] Sic ergo est manifestum quod fatum est in ipsis causis creatis, inquantum sunt ordinatae a Deo ad effectus producendos (554a20-b11)

[166] Sed contra est quod Boetius dicit, quod fatum est immobilis dispositio. Respondeo dicendum quod dispositio secundarum causarum, quam fatum dicimus, potest dupliciter considerari: uno modo, secundum ipsas causas secundas, quae sic disponuntur seu ordinantur; alio modo, per relationem ad primum principium a quo ordinantur.

group), it seems clear that the main physical secondary = intermediate causes are the very celestial bodies and their motions by which God governs, rules, disposes and moves the sublunar world with its terrestrial bodies and their motions, including human beings, as we saw explicitly in *Summa contra gentiles* III.82-94 and in *De veritate* V.9. We can also see that Boethius here is a major authority on these issues for both Thomas Aquinas and Albertus Magnus.

*

As it turns out, Boethius influentially articulated the broader conceptual nexus that includes providence and fate in the *Consolatio philosophiae* (IV.6.1-6) in his eponymous character's discussion with a personified Philosophy:[167]

> [1] "Yes," I said; "but since it is your task to lay open the causes of hidden things (*latentium rerum causas*) and to unfold reasons that are hidden in obscurity (*velatasque caligine explicare rationes*), I ask you to explain to me what conclusions you draw concerning this, since the strangeness of this disturbs me most of all." [2] Then, smiling briefly, she said: "You summon me to a matter which is the greatest of all (*ad rem* [...] *maximam*) to enquire into, for which even completeness would scarcely be enough. [3] For the subject is such that, when one doubt has been removed, countless others grow up like the heads of the Hydra; nor is there any limit to them, except to check them with the most lively fire of the mind. [4] In this connection there is customarily enquiry about [i] the singleness of providence (*de providentiae simplicitate*), [ii] the sequence of fate (*de fati serie*), [iii] sudden chances (*de repentinis casibus*), [iv] divine knowledge and [v] predestination (*de cognitione et praedestinatione divina*), and [vi] free will (*de arbitrii libertate*). How weighty these are you yourself judge. [5] But since for you to know these things too is a part of your therapy (*medicina tua*), we will try to consider them to some extent even though hemmed in by narrow limits of time. [6] If the delights of musical song charm you, you must put off this pleasure for a time, while I weave arguments bound together in sequence (*dum nexas sibi ordine contexo rationes*).[168]

Providence, fate, chance, divine knowledge, predestination and free will are all explicitly mentioned here as interconnected themes.[169] As we have now seen in

[167] In addition to Sharples's valuable edition, translation and commentary on this text (from which I take this translation), see also "Fate, Prescience and Free Will," his contribution to the *Cambridge Companion to Boethius*, John Marenbon (ed.), Cambridge: Cambridge University Press, 2009, 207–27, in which there is a section entitled, "Providence and Fate," 214–15.

[168] "[1] Ita est, inquam; sed cum tui muneris sit latentium rerum causas evolvere velatasque caligine explicare rationes, quaeso, uti, quae hinc decernas, quoniam hoc me miraculum maxime perturbat, edisseras. — [2] Tum illa paulisper arridens: Ad rem me, inquit, omnium quaesitu maximam vocas, cui vix exhausti quicquam satis sit. [3] Talis namque materia est, ut una dubitatione succisa innumerabiles aliae velut hydrae capita succrescant; nec ullus fuerit modus, nisi quis eas vivacissimo mentis igne coherceat. [4] In hac enim de providentiae simplicitate, de fati serie, de repentinis casibus, de cognitione ac praedestinatione divina, de arbitrii libertate quaeri solet, quae quanti oneris sint ipse perpendis. [5] Sed quoniam haec quoque te nosse quaedam medicinae tuae portio est, quamquam angusto limite temporis saepti tamen aliquid deliberare conabimur. [6] Quodsi te musici carminis oblectamenta delectant, hanc oportet paulisper differas voluptatem, dum nexas sibi ordine contexo rationes (IV.6.1-17)." Both the text and translation are Sharples's. I have made minor modifications in the latter.

[169] The only relevant theme that Boethius/Philosophy does not explicitly mention here is the knowledge and/or prediction of future contingents, which Boethius treated influentially in the

detail in this and the previous chapters, astrology in both practical and natural philosophical respects is interwoven with all of these themes in this rich conceptual nexus. Although Boethius mentions the celestial bodies, he does not explicitly discuss astrology. Albert and Thomas, on the other hand, both explicitly employ astrological structures and the celestial bodies within their penetrating analyses in these richly fertile and sometimes contentious border regions where natural philosophy and theology interact. So too does Roger Bacon, as we will see in the next chapter.

*

Finally, according to Valérie Cordonier, Thomas Aquinas himself influentially constructed an innovative Aristotelian account of divine providence where there had been none as such in Aristotle. In it, the celestial bodies play a fundamental role in God's governance of the world below by mediating from the divine through secondary causes—both universal and proximate—down to individual things here on earth.[170] From the material reconstructed in Part I, it seems that Thomas received the astrologizing interpretation of the universal secondary causes as precisely the celestial bodies from his esteemed teacher, Albertus Magnus. Thomas thus constructed an astrologizing Aristotelian interpretation of divine providence that turned out to be very influential indeed!

5.6 Conclusion

In contrasting astronomy and astrology with divination in *Summa theologiae* II.II.92-95, we saw what Thomas thought of astrology per se. Although his discussion there was not very detailed, astrology's legitimacy as a predictive practice due to its natural causality is perfectly clear from his analysis. Thomas offered more details in the *De iudiciis astrorum*, but he was still cautious there. In the earlier *Summa contra gentiles* III.82-94, however, we can see him construct both of astrology's theological and scientific foundations by locating the celestial bodies at the very center of God's providential governance of the world. We also saw him use Ptolemy's authentic *Tetrabiblos* and especially the pseudonymous *Centiloquium* as his major authorities. I wonder if Thomas personally held a more expansive view about practical astrology, but did not feel comfortable expressing it. Was he perhaps "in the closet" about astrology, as we might describe it today?

---

two versions of his commentary on Aristotle's *De interpretatione* 9. It is most likely implied in section [iv] on divine knowledge. For more on this central topic, see *Ammonius on Aristotle* On interpretation, David Blank (tr), with *Boethius on Aristotle* On interpretation 9, Norman Kretzmann (tr), with essays by Richard Sorabji, Norman Kretzmann and Mario Mignucci, London: Duckworth, 1998.

[170] See Cordonier, "Saveur le Dieu," 88 ff., and Sharples, "Alexander of Aphrodisias on Divine Providence," with a discussion of Aristotle's own relatively limited analysis of providence.

*

Given the description of God's providence in the *Speculum astronomiae*—where the heavens and their particular celestial configurations are identified precisely as God's providence[171]—together with the material just reconstructed with Thomas arguing strongly that God rules all of the motions in the world below by means of the celestial bodies (in *SCG* III.82-94) and Albert's and Thomas's most recent claims concerning the relationship between Providence and Fate, all of this richly evocative material ultimately begs at least this one fundamental question: How then can people read/interpret God's providential intentions for His governance of the world and of human beings, however imperfectly?[172] Although Roger Bacon does not use the language of providence as fully in the material discussed in Chap. 6 below, his description of how astrology may be used to support and defend theology/religion in *Opus maius* IV may present itself as a way to fulfill that desideratum. So too would the interpretation of annual prognostications, especially in relation to Great Conjunctions.

*

I have now explored astrology's natural philosophical foundations in Aristotle's *De generatione et corruptione* (and elsewhere) as developed by Albertus Magnus and Roger Bacon in Part I. Now here in this chapter, I have reconstructed astrology's theological foundations primarily in Thomas Aquinas's view of divine providence—and Albert's view of fate—in both of which, as in generation, the celestial bodies play an utterly central and fundamental mediating causal role. After thus focusing on fate, divination and providence in Albertus Magnus and Thomas Aquinas, we will now explore astrology's rather different configuration with theology/religion in Roger Bacon.

---

[171] "For, in those things that God does by means of the heavens, the signification of heaven is nothing other than divine providence." [*N*]*am in his quae operatur Dominus per caelum, nihil aliud est caeli significatio quam divina providentia* (14.84–85).

[172] This is also related to natural theology as a natural correlate and complement to divine revelation in scripture; see Smoller, *History, Prophecy and the Stars*, chapter 7, Sybils: John of Legnano's *De adventu Christi* and the Natural Theology of the Middle Ages," *Science in Context* 20 (2007): 423–50. Unfortunately, Smoller does not treat this very interesting material in relation to their views on divine providence, which would have enriched her analysis, and tied it more directly to the material analysed here.

Chapter 6
Astrology and Theology/Religion in Roger Bacon

6.1 Introduction

6.1.1 Astrology's Utility for Religion

Having just explored three facets of astrology's relationship to theology/religion via the rich conceptual structures articulating the central themes of fate, divination and providence in influential works by Albertus Magnus and Thomas Aquinas, we will now examine Roger Bacon's quite different expression of astrology's relation to theology/religion. As we will see, Roger's analysis in *Opus maius* IV is much more engaged with astrology's practical dimensions.[1]

Having defended astrological judgments (and thus mathematics) from its ignorant attackers and having shown its utility in human affairs (as we saw in Chap. 4), where he defended it from charges of implying determinism in nature and thus undermining human free will, Roger turns next in the same section of *Opus maius* IV—*Iudicia astronomiae*—to an extensive discussion of astrology's utility for theology/religion. In offering such a case study, Roger employs astrological dignities and great conjunctions (among other practical astrological features) to establish astrological arguments for bolstering the faith. As we can now see more clearly, Roger is deliberately constructing a complex and cumulative argument with a solidly astrological core for Pope Clement's benefit.

Roger switches his focus here from discussing human affairs—both personal and political—to concerns with and for the Church by asserting that, whereas he had just been discussing astrology in itself, now he will discuss it "in relationship" (*relative*), namely, in relation to theology/religion:

[1] Roger also discusses mathematics' (and thus in part astrology's) value for the study of theology in *Opus maius* IV, 180–82. See a discussion of these passages in Gregory, "Astrologia e teologia," 291–92.

H D. Rutkin, *Sapientia Astrologica: Astrology, Magic and Natural Knowledge, ca. 1250–1800*, Archimedes 55, https://doi.org/10.1007/978-3-030-10779-6_6

> After the power of mathematics with respect to the sciences of philosophy and the things of this world (*res istius mundi*), and of theology, and thus with respect to all wisdom (*tota sapientia*), has been shown insofar as this wisdom (*ipsa sapientia*) is considered absolutely in itself (*secundum se*), now I wish to show that the same is the case concerning this wisdom in relation to the church of God (*ecclesia Dei*) and the republic of the faithful, and the conversion of the infidel and the repression of those who cannot be converted.[2]

To develop these themes, Bacon discusses (among other things) the astrological bases of the different religions.

To begin constructing this phase of his argument, Roger starts with the basics of practical astrology, about which he provides fundamental information, but towards a wholly interested end, namely, towards answering the question: how can astrology strengthen the faith? We will pay close attention to how he builds his argument. We should also note that Roger provides all of the technical astrological information necessary to make his case, thus assuming no specialized knowledge on his readers' part. All they need to follow him is a willingness to listen and to think about what he is saying. Of course, he is mainly concerned with persuading his primary audience, Pope Clement himself:

> And because in many things (which cannot now be enumerated) there is something necessary for the Church, I now want to propose three cases (*casus*) which are, as it were, infinite miracles of ineffable utility. The first consists in certifying the faith (*in certificatione fidei*) which the Church holds. For we can take great solace in our faith after the philosophers, who have been led by the motion of reason alone (*solo motu rationis*), agree with us, and both confirm our religion (*secta*) or the profession of the Christian faith, and harmonize with us in stabilizing this religion. Not because we seek reason before faith, but after faith, so that, certified with a two-fold confirmation, we will praise God concerning our salvation, which we hold without any doubt. And by this path of mathematics, not only are we certified concerning our profession of faith, but we are also forearmed against the religion (*secta*) of Antichrist, about which—together with the religion (*secta*) of Christ—there is a consideration in mathematics.[3]

Reason in the particular form of a philosophical astrology can thus strengthen faith, which will then be supported on the complementary pillars of both reason and revelation. We should note that Roger uses the term *secta* neutrally to refer to a religion

[2] Postquam potestas mathematicae respectu scientiarum philosophiae et rerum istius mundi et theologiae, et sic respectu totius sapientiae ostensa est, prout ipsa sapientia secundum se consideratur absolute, nunc volo ostendere illud idem secundum quod refertur haec sapientia ad ecclesiam Dei et rempublicam fidelium et conversionem infidelium et repressionum eorum qui converti non possunt (253, 28–34).

[3] Et quia in multis quae nunc numerari non possunt necessaria est ecclesiae, nunc volo tres casus proponere, qui sunt quasi infiniti miraculi et utilitatis ineffabilis. Primus consistit in certificatione fidei quam tenet ecclesia. Magnum enim solatium fidei nostrae possumus habere, postquam philosophi qui ducti sunt solo motu rationis nobis consentiunt, et sectam seu professionem fidei Christianae confirmant et nobiscum concordant in stabilitate huius sectae; non quia quaeramus rationem ante fidem, sed post fidem, ut duplici confirmatione certificati laudemus Deum de nostra salute quam indubitanter tenemus. Et per hanc viam mathematicae non solum certificamur de professione nostra, sed praemunimur contra sectam Antichristi, de qua simul cum secta Christi fit consideratio in mathematica (253, 34–254, 9).

regardless of whether it is evaluated as positive or negative.[4] This is, of course, strikingly different from our contemporary usage where "sect" almost always has a negative connotation. In what follows, Roger offers a sustained argument for astrology's utility to religion with three main sections. I will present them in two case studies that show three different but related ways that Roger used astrology to support theology/religion. We will see another case of a very different tenor in Chap. 8.

6.2 First Case Study: History of Religion

6.2.1 Stage 1: The Astrological Structure of Religion

The first case study concerns the history of religion. In the first stage of his four-part argument, Roger offers an abstract taxonomical discussion of the six main world religions vis-à-vis their planetary natures in terms, primarily, of Jupiter's relationship with the other six premodern planets. He offers an ahistorical structural argument about the nature and number of what he considers to be the six main historical religions, even though the particular religions actually came into existence at specific times in history. Then Roger historicizes this analysis and puts it into practice in stage 2 by utilizing a properly historical structure—namely, great conjunctions—to demonstrate astrology's (and thus true mathematics') value for theology/religion. This is a complex cumulative argument.

Roger begins his first case study certifying the faith by offering an astrological analysis of the history of religion that links it directly to the heavens:

> And the noblest investigation of this sort comes about through the revolution (*revolutio*) of all the principle religions from the beginning of the world. There cannot be more, and they are the religions of [1] the Hebrews, [2] the Chaldaeans, [3] the Egyptians, [4] the Agarenae or Saracens [= Muslims], who came to be from Hagar and Ishmael, [5] the religion of Christ, and [6] the religion of Antichrist.[5] It is not surprising that philosophers have spoken about these things, since they came to exist after the patriarchs and prophets, and were instructed by their sons and their books, as was shown before. Therefore, as much as I can explain this more plainly and fully at present, I will recite the opinions of the *mathematici* in which the authorities (*auctores*) agree.[6]

---

[4] In what follows, I always translate '*secta*' as religion. Where Roger uses another term, I will note it.

[5] In n. 1 (254), Bridges notes that a different list is found in the fourth part of Bacon's *Moralis Philosophia*: Pagans, Tartars, Idolators, Saracens, Jews and Christians.

[6] Et huiusmodi nobilissima perscrutatio fit per revolutionem omnium sectarum principalium a principio mundi, nec possunt esse plures, et sunt sectae Hebraeorum, et Chaldaeorum, et Aegyptiorum, et Agarenarum seu Saracenorum, qui fuerunt de Agar et Ismaele, secta Christi, ac secta Antichristi. Nec mirum si locuti sunt philosophi de his, quoniam fuerunt post patriarchas et prophetas et instructi per filios et libros eorum, ut prius ostensum est. Quanto igitur planius et plenius possum ad praesens, recitabo sententias mathematicorum in quibus auctores concordant (254, 9–20).

There are thus six primary religions since the world was created, about the history of which Roger now offers the opinions of the mathematicians (i.e. astrologers), especially where they all agree. This structure here of six major world religions sets the stage for the later discussion of great conjunctions and the actual history of the different religions, especially concerning the timing of their respective beginnings and endings, that is, their coming-into-being (= generation) and passing-away (= corruption).

*

Roger begins explaining the astrologers' view of religion by first discussing the nature of the planets. Although he has not yet told us where he is going, we will come to see that this is the first step of the first stage in building up his astrological analysis of religious history. We will learn a great deal about how Roger uses astrology in the process. In fact, he will develop here some of the material already discussed in the excursus to my overall introduction:

> Therefore, they [sc. the authorities] say that Jupiter and Venus are the benevolent and fortunate planets, and that Saturn and Mars are malevolent and unfortunate. They say that Mercury holds to a middle manner (*medio modo*), since with good [sc. planets] it is good, and with bad planets it is bad, because it is of a changeable nature (*convertibilis natura*). But of the benevolent and fortunate planets, they say that Jupiter is better, and the greater fortune belongs to it and the lesser to Venus.[7]

That Jupiter and Venus are benefic planets, Saturn and Mars malefic, and Mercury neutral is basic astrological doctrine, stemming from Ptolemy's *Tetrabiblos* I.5, although their beneficence and maleficence there are cashed out in terms of the four primary qualities.

Roger now relates the differences in planetary beneficence and maleficence to religious concerns:

> And therefore, since there are two lives (*duae vitae*), present and future, and the future is more important than the present, as the eternal is more valuable than the temporal, they say that Venus signifies the fortunes of this life concerning games (*ludi*), joy (*gaudia*) and pleasure (*laetitia*) and things of this sort, and that Jupiter respects the goods of the other life (*ad bona alterius vitae*) which are greater. And it [sc. Jupiter] signifies concerning wisdom (*sapientia*) and intellect (*intellectus*), the interpretation of dreams (*solutio somniorum*) and divine worship (*divinum cultum*), faith and the teachings of law, religion (*religio*), veneration and the fear of God, and the application of morals, and many such things, as the *astronomi* tell.[8]

---

[7] Dicunt igitur Iovem et Venerem esse planetas benivolos et fortunatos, Saturnum et Martem malivolos et infortunatos. Mercurium dicunt medio modo se habere, quia cum bonis est bonus, cum malis malus, quia convertibilis naturae est. De benevolis vero fortunatis dicunt Iovem meliorem esse, et maiorem fortunam ei deberi, minoremque Veneri (254, 20–25).

[8] Et ideo cum duae vitae sunt, praesens et futura, et plus valet futura quam praesens, sicut aeternum quam temporale, dicunt Venerem significare super fortunas huius vitae, quantum ad ludos et gaudia atque laetitiam et huiusmodi, et Iupiter respectum habet ad bona alterius vitae, quae maiora sunt. Et significat super sapientiam et intellectum et solutionem somniorum et divinum cultum fidem et legis doctrinam, religionem et venerationem et Dei timorem et aptationem morum et multa talia ut astronomi narrant (254, 26–34).

Thus Venus is beneficial for this earthly life and its pleasures, and Jupiter for the next. Roger clearly emphasizes that Jupiter signifies religion, among several other more exalted things, and this provides the first step in Roger's analysis. The bottom line is that Jupiter rules and thus also signifies religion.[9]

*

For the next step in his analysis, Roger adduces the "terrestrial" houses or places:

> In addition, they [sc. the astrological authorities] divide (*distinguunt*) the entire heavens (*totum coelum*) into twelve parts (*partes*), which are called houses (*domus*), and which are divided by means of the meridian circle and the horizon, and four other [sc. great] circles intersecting them in their sections;[10] such that the first house begins from the horizon and is established under it,[11] then the second and third house follow, up to the '*angulum terrae*', which is the point in the heavens under the earth opposite to the point of the meridian above the earth.[12]

This lower point is also called the "lower mid-heaven" (*imum medium coelum*) to correspond with the mid-heaven (*medium coelum*). It is one of the four primary angles or '*cardines*' corresponding to the four fixed points where the horizon and meridian cut the zodiac relative to a given place on earth. I should note that these are the very same astrological structures that Albert described in Chap. 1, where he linked the periods of life to the basic structures of the horoscope.

*

After describing the locations of the remaining "terrestrial" houses in outline, but without any discussion of how they would be used for interpretation (255, 7–10), Roger integrates the first two steps by discussing the planetary rulership of each house in step three. Here we can begin to see how the previous discussion of planetary natures in step one will be used in the broader argument about astrology's value for religion:

> Therefore, they [sc. the astrological authorities] give the first house to Saturn, the second to Jupiter, and thus further according to the order of the planets, such that the eighth is given

---

[9] The Magister Speculi makes a similar argument in *Speculum astronomiae* 12 in relation to defending Albumasar. In fact, both he and Roger here use the same exact terms for Venus's pleasures: *ludi, gaudia* and *laetitia*. Whereas the sentiment is similar in the rest, the language is not quite as precise.

[10] Each of the four quarters of the circle cut by the meridian and horizon is a section.

[11] Literally below the line of the horizon.

[12] Praeterea distinguunt totum coelum in xii partes, quae vocantur domus, quae distinguuntur per meridianum circulum et horizontem, et alios quatuor circulos intersecantes se in eorum sectionibus, ita ut prima domus incipiat ab horizonte et sub eo sit constituta, deinde succedunt ei secunda domus et tertia usque ad angulum terrae, qui est punctus in coelo sub terra oppositus puncto meridei super terram (254, 35–255, 7),

> again to Saturn, and the ninth to Jupiter.[13] And they all agreed that the ninth house is the house of religion and faith. And Ptolemy, Albumasar, Altavicus [= Alcabitius],[14] Messahalah, and all the others assigned their properties to these houses because the first is the house of life (*domus vitae*); the second, the house of wealth (*substantia*), and thus concerning the others in accordance with their properties and natural dispositions.[15]

Roger informs us here that each house has its own planetary ruler, whose properties and natural disposition inform the nature of the house, emphasizing that the ninth house concerns religion and faith. This follows given Jupiter's concerns with the next life, as discussed in step one. We should notice that Roger does not provide us with all the details about each house, as he would if he were writing a textbook account. Rather, as we will repeatedly see, he primarily offers information directly relevant for supporting his broader argument about how astrology can strengthen the faith. He also names his primary astrological authorities: Ptolemy, Albumasar, Messahalah and Alcabitius.

Roger continues by developing his account of the ninth house and its religious significance:

> Whence the ninth, as they say, is the house of pilgrimages (*peregrinationes*) and of the pathways of faith, deity and religion, and the house of the worship of God, of wisdom, books, letters and bequests, stories, rumors and dreams. And therefore, as they say, it is rightly attributed to Jupiter, who has a relationship to the goods of the other life, since faith, religion, the worship of God, the consideration of wisdom, and the multitude of books and letters are owed to these goods as is clear from divine law [sc. as revealed in Holy Scripture], and the wealth (*copia*) of bequests, as of prophets, apostles, and preachers, and the proper reports (*rumores*) of tales about the noble circumstances of that life, and the frequent revelations about that life of those who have them in dreams, ecstasy and raptures.[16]

Roger completes the first stage of his argument by confirming and describing the religious character of both Jupiter and the ninth house, which Jupiter rules.

[13] This is a different structure than the planetary rulership of the signs of the zodiac, the "celestial" houses to be discussed below.

[14] At XII.106, the *Speculum astronomiae* calls Alcabitius Abdilaziz.

[15] Primam igitur donum dant Saturno, secundam Iovi, et sic ulterius secundum ordinem planetarum, ita quod octava iterum datur Saturno, et nona Iovi. Consideraveruntque omnes concorditer quod domus nona est domus religionis et fidei. Et Ptolemaeus et Albumazar et Altavicus et Messehalac et alii omnes assignaverunt istis domibus proprietates suas, quia prima est domus vitae, secunda domus substantiae, et sic de caeteris, iuxta suas proprietates et dispositiones naturales (255, 10–18).

[16] Unde nona domus, ut dicunt, est peregrinationum atque itinerum fidei et deitatis et religionis, ac domus culturae Dei sapientiae librorum epistolarum et legatorum narrationum ac rumorum et somniorum. Et ideo merito, ut dicunt, attribuitur Iovi, qui habet respectum ad bona alterius vitae, quia illis bonis debentur fides et religio et cultura Dei et consideratio sapientalis, et librorum et epistolarum multitudo, ut patet ex lege divina; et legatorum, ut prophetarum et apostolorum et praedicatorum copia narrantium rumores idoneos de nobilibus conditionibus illius vitae, et revelationes frequentes habentium in somnis et extasi et raptibus de hac vita (255, 19–30).

6.2.2 Stage 2: Planetary Conjunctions

To begin the second stage of his argument, Roger discusses the relationships between the planets themselves, and especially their conjoinings or conjunctions, which are one of the most fundamental astrological aspects: "Therefore, they say that planets are conjoined and embrace themselves in turn, and that this is when they are in the same sign, and especially when in the same degree, and within the 16th minute of that degree."[17] We should also recall that conjunctions are considered to be a part of the astrological practice of general astrology or revolutions, as we saw in the *Speculum astronomiae*. We will see why more clearly as we continue with Roger's analysis.

Bacon now relates this general discussion of planetary conjunctions to the history of religion by addressing the conjunctions of Jupiter, which rules religion, with each of the six other planets and luminaries. We should notice that it is the philosophers and not the astrologers who make this argument:

> Therefore, the philosophers want Jupiter from its conjunction with the other planets to signify concerning the sects of religions and of faith (*secta religionum et fidei*). And since there are six planets with which it [sc. Jupiter] can be embraced and conjoined, therefore, they [sc. the philosophers] assert that there ought to be six principle religions in the world.[18]

Here Roger is setting up preliminary structures before properly entering the world of historical astrology with respect to the history of religion.

Bacon now turns to the particular case of each religion, beginning with Jupiter's conjunction with Saturn.[19] We should note that this is an abstract structure, relating generalized conjunctions of Jupiter with each planet in turn, and then applying the resultant nature of the conjunction to a particular historical religion. It does not seem to relate to any actual historical conjunction of the planets in question. In fact, the structure seems to be based primarily on the order of the planets in the heavens, which order is then transposed onto the temporal dimension, with the outermost planet, Saturn, becoming the most ancient. We will see in our second case study below that Roger also uses actual conjunctions in a more strictly temporal approach to analyzing the past and present, and for predicting future events.

First Roger discusses Jupiter's conjunction with Saturn:

---

[17] "Dicunt igitur planetas coniungi et complecti sibi invicem, et hoc est quando fuerunt in eodem signo et praecipue quando in eodem gradu et in xvi minuto illius gradus et infra (255, 31–33)." This extreme precision is not usually required. Bacon here discusses an idealized possibility, all the while knowing full well that each planet has its own normal "sphere of influence," so that conjunctions (and the other aspects) are said to occur even when the planets at issue are not precisely conjoined, or aspected more generally.

[18] Volunt ergo philosophi Iovem ex sua coniunctione cum aliis planetis significare super sectam religionum et fidei. Et quia sunt sex planetae quibus complecti et coniungi potest, ideo asserunt sex fore debere in mundo sectas principales (255, 34–37).

[19] Here we have a traditional "great conjunction" proper, although Roger is not here referring to a particular historical instance. None of the others are, however. Bacon treats great conjunctions specifically in the next case study.

> Whence experienced authors have predicted and others say that if Jupiter is embraced by Saturn, it signifies divine books; and of the religions, it signifies Judaism because it is more ancient than the others and prior, just as Saturn is the father of the planets (*pater planetarum*), and more removed and prior in the going outward of the planets, and in the order of their existence (*ordine in esse*).[20] And they all admit this, and this embraces none other, just as all the planets embrace Saturn and he himself none, due to the slowness of his motion. Because when a planet is before it to the east, Saturn never overtakes any other planet, but only insofar as the other planet grows so powerful that it sometimes catches Saturn and conjoins it. Indeed, all religions (*omnes sectae*) conform themselves to the religion of the Jews because this was the first [sc. religion] and is the root of the others (*radix aliarum*), from which all religions have had some type of witness and establishment (*aliquod genus testimonii et constitutionis*). Whence philosophy received from it [sc. Judaism] many testimonies and modes of establishing a religion, as was explained before.[21]

Jupiter conjoining Saturn thus engendered Judaism, the oldest, and therefore the tap-root of all the other religions.

Roger now discusses the other religions in turn, beginning with that deriving from Jupiter's conjunction with Mars, the next planet in order:

> But if Jupiter embraces Mars, they say that it signifies the Chaldaean law (*lex Chaldaica*), which teaches the worship of fire (*quae docet adorare ignem*), of whose nature Mars is in its natural power and effect (*in naturali potentia et effectu*). If Jupiter embraces the sun, the Egyptian law is signified, which assumes that the militia of the heavens, whose leader is the sun, is to be worshipped.[22]

Roger thus makes short work of Jupiter's conjunctions with Mars and the sun, and their meaning for religion, namely, that they signify the Chaldaean and Egyptian religions respectively.

Roger focuses much more attention on the results of Jupiter's conjunction with Venus, since it signifies Islam:

> If Jupiter embraces Venus, it is said to signify the law of the Saracens, which is entirely pleasure seeking [sc. sensual] and sexual (*tota voluptuosa et venerea*). Although Mohammed limited it in his writings, nevertheless, it has been held for a long time in normal practice (*in usu vitae*) by its worshippers. Whence in the book which is ascribed to Ovid, *On the Transformation of his Life* (*De vitae suae mutatione*),[23] when he speaks about the Venusian

[20] Bacon seems to mean that Saturn was the first planet created.

[21] Unde periti auctores praedicti et alii dicunt, si complectatur Iupiter Saturno, significat de sectis Iudaicam, quia est antiquior aliis et prior, sicut Saturnus pater planetarum et remotior et prior in exitu planetarum et ordine in esse. Et ipsam omnes confitentur, et ipsa nullam aliam, sicut Saturno omnes planetae complectuntur et ipse nulli, propter tarditatem sui motus. Quia quando planeta est ante eum ad orientem, nunquam ipse Saturnus consequitur aliquem, sed in tantum invalescit alius planeta, quod consequitur aliquando Saturnum et conjungitur ei. Omnes quidem sectae appodiant se ad sectam Iudaeorum, quia haec fuit prima et est radix aliarum, a qua omnes aliquod genus testimonii et constitutionis sectae habuerunt: unde philosophia accepit ab illa multa testimonia et multos modos constituendi sectam, sicut prius patuit (255, 38–256, 14).

[22] Si vero Iupiter complectatur Marti, tunc dicunt ipsum significare super legem Chaldaicam, quae docet adorare ignem, cuius naturae Mars est in naturali potentia et effectu. Si Soli, significatur lex Aegyptia, quae ponit coli militiam coeli, cuius princeps est Sol (256, 14–18).

[23] This is a paraphrase of the title to Ovid's authentic poem, the *Metamorphoses*. Bridges n. 1 (256) is useful.

> religion (*de secta venerea*), which he said is the law of his own time (*sui temporis lex*), he says in his verse: "In which, anything pleasing is considered lawful,// Although a written law regarding it is not yet found." Mohammed wrote this in a book called Alcoran [sc. The Koran] 600 years and more later. For Ovid lived before and during Christ's times, and the religion of Mohammed began 600 years and more after the incarnation of Christ, as is clear from the difference of the years of Christ's [sc. epoch] and the Arabs', which is 621 years (dcxxi) and 195 (cxcv) [sc. days].[24] But the years of the Arabs are computed from Mohammed, as al-Farghani says and others likewise.[25]

Roger here uses pseudo-Ovid's *De vetula* to support the view that Jupiter conjoining Venus resulted in the origins of Islam. He thinks that this work is authentically by Ovid, which informs some of his historical arguments.[26]

Roger then discusses Jupiter's conjunction with Mercury, which is of the utmost significance for his purposes, as we will see:

> But if Jupiter embraces Mercury, then there is a Mercurial law, for Mercury has a relationship, as they say, to the deity, and to the oracles of prophets (*oracula prophetarum*), and to credulity and speech, and especially when Jupiter conjoins it, since then it signifies the number of the psalm (*numerum psallendi*) and the number of the sacred books. And they say that the Mercurial law is more difficult to believe than others, and that it has many challenges (*difficultates*) with respect to human understanding (*supra humanum intellectum*).[27]

Without yet identifying the religion, Roger mentions some of its Mercurial characteristics.

Bacon then gives a reason for this difficulty concerning belief, which is based primarily on Mercury's physical nature as a planet; in particular, the peculiar nature of Mercury's orbit, and thus the complex model of its motion. Roger had just given

---

[24] Sorge and Seller have 671 years and 195 days; *Filosofia, scienza, teologia*, 113.

[25] "Si Veneri, significare dicitur super legem Saracenorum, quae est tota voluptuosa et venerea, quam licet in scriptis Mahometus redegit, ipsa tamen per longa tempora in usu vitae habebatur a suis cultoribus; unde in libro qui ascribitur Ovidio de vitae suae mutatione cum loqueretur de secta venerea, quam hominibus sui temporis legem dixit esse, dicit in metro suo. 'In qua, si libeat, quod cunque licere putatur,/ Scripta licet super hoc nondum lex inveniatur;' Quam postea per sexcentos annos et amplius scripsit Mahometus in libro qui dicitur Alcoran. Ovidius enim ante Christum et in temporibus Christi fuit, et secta Mahometi incepit per sexcentos annos et amplius post incarnationem Christi, sicut patet ex differentia annorum Christi et Arabum, quae est dcxxi annorum, et cxcv dies. Sed anni Arabum computantur a Mahometo, ut dicit Alfraganus et alii similiter (256, 18–34)." Alfraganus (Ahmad ibn Muhammad ibn Katir al-Fargani, fl. 830–60) was the author of a compendium of Ptolemaic astronomy, which was the most widely read work of Arabic astronomy in the Latin West (Hasse, *Success and Suppression*, 331–33).

[26] For the *De vetula*, see Paul Klopsch, *Pseudo-Ovidius* De vetula: *Untersuchungen und Text*, Leiden: Brill, 1967, and Dorothy Mae Robathan, *The Pseudo-Ovidian* De vetula, Amsterdam: Hakkert, 1969.

[27] Si vero complectatur Mercurio, tunc est lex Mercurialis. Mercurius enim habet respectum, ut dicunt, ad Deitatem et oracula prophetarum et credulitatem et orationem, et maxime quando coniungitur ei Iupiter; quoniam tunc significat numerum psallendi et numerum librorum divinorum. Et dicunt, quod lex Mercurialis est difficilior ad credendum quam aliae, et habet multas difficultates supra humanum intellectum (256, 35–257, 6).

a somewhat similar analysis for Saturn and Judaism by comparing its place in the heavens as the Ur-planet with Judaism's nature as the Ur-religion:[28]

> This is fitting because of Mercury's difficult motions, whose orbit (*circuitus*) has an epicycle, eccentric and equant, in which its motions in longitude are considered, and the *inflexus* and *reflexus* of the motions in latitude, through the declination of the eccentric from the orb of the signs to the north and south. And [sc. the declination] of the epicycle from the eccentric in the part of the north and south, and they [sc. these motions] are more wondrous (*mirabilior*) and difficult than all the motions of the other planets, as is clear from the sayings (*dicta*) of Ptolemy [sc. in the *Almagest*], and more clearly from the opinions of al-Battani, Thebit, Archaselis (al-Ghazzali?) and probably from the sayings of al-Farghani.[29]

Thus Roger supports his analysis for the difficulty in understanding the as-yet-unidentified Mercurial religion based on an anology with the difficulty of its model of planetary motion.

Roger now identifies the Mercurial religion in question:

> And because of this, it signifies, as they say, a law which has difficult articles and hidden truths (*occultae veritates*), of which sort is the *Lex Christiana.* But [sc. this is] because Mercury is the signifier of writing and writers (*significator scripturae et scriptorum*), and of profound knowledge in profound books (*profunditas scientiae in libris profundis*), and eloquence or the sweetness of expression and language (*dulcedo locutionis et linguae*), and of rhetoric and its swiftness, and the explanation of opinions. And it signifies what is defended by such authentic writings and such profound knowledge and such powerful eloquence, which will always stand in its strength until the final law of the moon disturbs it in its time. And they say that this law is of a prophet born of a virgin, in accordance with which all the ancient Indians, Chaldaeans and Babylonians taught that in the first face (*facies* = decan) of Virgo arises (*ascendit*) a most beautiful maiden (*virgo mundissima*), who will nurture a boy in the land of the Hebrews, whose name is Jesus Christus, as Albumasar says in the *Great Introduction to the Science of the Stars* (*Maior introductorius astronomiae*).[30]

---

[28] For a contemporary representation of the complexities of Mercury's orbit, see Campanus of Novara's *Theorica planetarum*; *Campanus of Novara and Medieval Planetary Theory*: Theorica planetarum, Francis S. Benjamin, Jr. and G.J. Toomer (eds. and trs.), Madison: University of Wisconsin Press, 1971, 212–97.

[29] Et hoc convenit propter motus Mercurii difficiles, cuius circuitus est in epicyclo et eccentrico et aequante, in quibus considerantur sui motus longitudinis et inflexus et reflexus in motibus latitudinis, per declinationem eccentrici ab orbe signorum ad septentrionem et meridiem, et epicycli ab eccentrico in partem septentrionis et meridiei, et sunt mirabiliores et difficiliores omnibus motibus planetarum, sicut patet ex dictis Ptolemaei, planius ex sententiis Albategni, Thebit, et Archaselis, et probabiliter ex dictis Alfragani (257, 7–15).

[30] "Et propter hoc significat, ut dicunt, super legem quae habe[n]t difficiles articulos et occultas veritates, cuiusmodi est lex Christiana. Sed quia Mercurius est significator scripturae et scriptorum, et profunditatis scientiae in libris profundis, atque fecunditatis, sive dulcedinis locutionis et linguae, et rhetoricae et velocitatis eius et explanationis sententiarum, significat quod tam authenticis scripturis et tot profundis scientiis et tanta potestate eloquentiae defendetur, quod stabit semper in robore suo, donec ultima lex Lunae perturbet eam ad tempus. Et dicunt, quod haec lex est prophetae nascituri de virgine, secundum quod omnes antiqui Indi, Chaldaei, Babylonii, docuerunt quod in prima facie Virginis ascendit virgo mundissima nutritura puerum in terra Hebraeorum, cui nomen Jesus Christus, ut dicit Albumazar in majori introductorio astronomiae (257, 15–30)." This passage is also cited in *Speculum astronomiae* XII in defending Albumasar by showing that his work both confirms and increases faith and religion, as we saw in Chap. 4.

With the identification of the Mercurial law as Christianity, we are now approaching the heart of Bacon's analysis. He will explain what he means by the "face" of a sign just below as he explores and describes the planetary dignities, an essential feature of astrological interpretation as we saw in the excursus, and one that will help further inform his analysis. Roger also mentions here the lunar religion as the end of the Mercurial law, but he does not yet go into detail.

6.2.3 Stage 3: Dignities for Interpreting Albumasar

After a brief mention of moral philosophy (257, 30–34), Roger begins the third stage of his argument by focusing on planetary dignities—especially those of Mercury—in order to explicate the text just cited from Albumasar. As we go deeper into Roger's analysis, we can see that various elements of practical astrology are utterly central, integrating well into and more deeply informing his overall argument. They are by no means a gratuitous or superficial overlay:

> And the birth (*ortus*) of a prophet from a virgin coheres well with the Mercurial law, since Mercury has the greatest power (*maxima potestas*) in Virgo according to the judgment of all astrologers (*astronomi*). For it [sc. Mercury] was created (*creatus fuit*) in Virgo, and of dignities (*dignitates*), powers (*potestates*), witnesses (*testimonia*), virtues (*virtutes*) or strengths (*fortitudines*),[31] which belong to planets on account of signs, Mercury has five in Virgo, such as, namely: house (*domus* [= zodiacal sign]), exaltation (*exaltatio*), triplicity (*triplicitas*), term (*terminus*) and face (*facies*).[32]

In order to explicate the curious prediction that the prophet will arise from a virgin, Roger turns to an analysis based on astrological dignities—that is, strengths (increases in power) that a planet receives relative to the zodiacal sign and its relevant degrees that a planet happens to be in at a given time. He claims that Mercury has five such dignities in Virgo. He will now proceed to spell this out.

To support this analysis, Bacon provides relevant astrological background by discussing each dignity in turn, beginning with '*domus*'. Here Bacon discusses the still equivocal and often confusing meaning of the term "house": "And 'house' (*domus*) is now and earlier said equivocally, since these (*istae*) houses are called 'essential and natural' and the earlier (*priores*) are called 'accidental and locational' (*situales*)."[33] There are thus two distinct astrological entities that are often called "houses," which regularly leads to confusion both then and now. The ones discussed earlier—the so-called "terrestrial" houses—are accidental and locational. Those he will discuss now—the so-called "celestial" houses or zodiacal signs—are essential and natural.

---

[31] These are all virtually synonyms here, like species, image, likeness, etc. in Chap. 2.

[32] Et ortus prophetae de virgine multum convenit legi Mercuriali, quia Mercurius habet maximam potestatem in Virgine, secundum iudicium astronomorum omnium. Creatus enim fuit in Virgine, et dignitates, seu potestates seu testimonia seu virtutes seu fortitudines quinque quae debentur planetis ratione signorum habet Mercurius in Virgine, ut sunt scilicet domus, exaltatio, triplicitas, terminus, facies (257, 34–258, 2).

[33] Et domus nunc et prius dicitur aequivoce; quoniam istae domus vocantur essentiales et naturales, priores vocantur accidentales et situales (258, 3–259, 2)[.]

Roger now discusses the "celestial houses," that is, the signs of the zodiac:

> Since these houses (*hae domus*) are the twelve signs (*signa*) [sc. of the zodiac], whose division is natural, since the sections of the zodiac and the heavens remain in their places from the celestial circle, that is, from the firmament, which division of the signs six [sc. great] circles make, intersecting at the poles of the zodiac, and they divide the entire heavens and the earth (*caelum et mundum*) into twelve equal parts [sc. of 30 degrees each]. [1] The parts that can be considered only in the zodiac are then the signs properly, as Aries and the others, or [2] these circles can be extended in the imagination (*imaginarie*) to the poles, where they intersect and then divide the entire heavens (*caelum*) into twelve equal parts, being narrow at the extremes above the poles and wide in the middle in the manner of a ship's hull (*ad modum fundi naviculae*), such that the width (*latitudo*) contains the ends of the degrees of the zodiac, [1] which, with a common term (*communis nomen*), we call "signs," whence it is properly called a sign, as Aries, or Taurus, etc. [2] And otherwise it is taken for the entire part of the heavens contained between the two extending circles, for example, through the boundaries of Aries,[34] which circles meet (*concurrunt*) at the poles of the world. And that sign is said to be the sign of Aries because its latitude consists in the extension of Aries, and thus the stars which are beyond the body of Aries are said to be in the sign of Aries, although they are near the poles of the world.[35]

For Roger, the celestial houses thus normatively include the signs of the zodiac, but also their extensions around the celestial sphere. In this way, the entire heavenly sphere is sliced into twelve signs of equal size and shape, and fixed in the heavens.

Roger then contrasts these celestial houses or zodiacal signs with the terrestrial houses mentioned above in the first stage of his argument:

> But the other houses are called accidental, since their division is accidental and the sections do not remain in the same place in the heavens, since they do not follow the motion of the heaven, and therefore their [sc. the sections'] places (*loca*) are changed (*mutantur*) in the [sc. zodiacal] circle or heavens every hour. And these sections are drawn, as was said, through the meridian circle and the horizon with the other four circles.[36]

Terrestrial houses are accidental and locational because they are not fixed in the heavens, but change over time with the diurnal rotation of the heavens in a geocentric cosmos. Although the fourfold horizon-meridian framework is fixed for the location of the place in question, the heavens and the signs of the zodiac revolve

[34] That is, at the very beginning and end of a sign: e.g. 0° Aries-0° Taurus

[35] [Q]uoniam hae domus sunt xii signa, quorum divisio naturalis est, quia sectiones zodiaci et coeli manent in suis locis de circulo coelesti, hoc est de firmamento, quam divisionem signorum faciunt sex circuli sese intersecantes in polis zodiaci, et dividunt totum coelum et mundum in xii partes aequales. Quae partes possunt considerari in zodiaco solum, et tunc sunt proprie signa, ut Aries et alia, aut possunt illi circuli imaginarie extendi ad polos, in quibus se intersecant et tunc dividunt totum coelum in xii partes aequales habentes angustiam in extremitatibus circa polos et latitudinem in medio ad modum fundi naviculae, ita quod illa latitudo continet extremitates partium zodiaci, quas communi nomine vocamus signa, unde signum dicitur proprie, ut Aries, vel Taurus, etc. Et sumitur aliter pro tota coeli parte contenta inter duos circulos transeuntes, verbi gratia per fines Arietis, qui circuli concurrunt in polis mundi; et istud signum dicitur esse signum Arietis quia eius latitudo consistit in extensione Arietis, et sic stellae quae sunt extra corpus Arietis dicuntur esse in signo Arietis, quamvis sint iuxta polos mundi (259, 2–21).

[36] Sed aliae domus dicuntur accidentales, quia divisio earum est accidentalis, et non manent sectiones in eodem loco coeli, quia non sequuntur motum coeli, et ideo mutantur earum loca in circulo seu coelo in omni hora. Et sumuntur hae sectiones, ut dictum est, per circulum meridianum, et horizontem cum aliis quatuor (259, 21–26).

through this fixed grid during the course of every day at the rate of one sign of 30° every 2 hours, thus 15° per hour.

With this understanding of celestial houses as zodiacal signs, Roger turns to their planetary rulership as he had done before with the terrestrial houses. This part of his explanation is directly relevant for understanding '*domus*' as the first planetary dignity. Roger continues with another distinction: "Moreover, the house called natural so far is twofold: a certain one (*quaedam*) is primary (*principalis*), the other (*quaedam*) is not primary, whence it is called accidental with respect to the primary one."[37] First Roger offered the distinction between essential (= celestial) and accidental (= terrestrial) houses. Now he uses the same distinction to differentiate two different types of celestial houses. Here is his explanation:

> Moreover, the primary house of a planet is that in which it was created, as Leo is the house of the sun, Cancer of the moon, Virgo of Mercury, Libra of Venus, Aries of Mars according to some; according to others, Scorpio, Sagittarius of Jupiter and Capricorn of Saturn.[38]

Primary planetary rulerships of signs are thus fixed as the sign in which the planet was created, and Bacon provides a list here of the relevant signs in which each planet was created. The authorities agree for all of the planets except Mars.

What, then, of the other natural but also accidental houses?:

> The less primary houses [sc. of the planets] are as Aquarius is given to Saturn, Pisces to Jupiter, Scorpio to Mars, according to one opinion; according to another, Aries, Taurus to Venus and Gemini to Mercury, such that each of the five planets has two houses, but the sun and moon only have individual [sc. houses]. Thus the antiquity of wise men (*antiquitas sapientum*) decreed.[39]

Ptolemy offers this exact same structure of house rulership, and discusses its elegance at *Tetrabiblos*, I.17.[40] Ptolemy also discusses the range of planetary dignities overall at *Tetrabiblos* I.17–24, although with a slightly different order. Roger has now shown us the structure of the first planetary dignity, namely, its celestial house or zodiacal sign. He will show us below why this matters in an astrological interpretation.

*

Having described the dignity '*domus*', Roger turns to the second dignity, namely, exaltations, which have a complementary *descensio* or *depressio*:

[37] Domus autem quae dicitur naturalis adhuc est duplex; quaedam est principalis, quaedam non principalis, unde dicitur accidentalis respectu principalis (259, 26–29).

[38] Principalis autem domus planetae est in qua creatus fuit, ut Leo est domus Solis, Cancer Lunae, Virgo Mercurii, Libra Veneris, Aries Martis secundum quosdam, secundum alios Scorpius, Sagittarius Iovis, Capricornus Saturni (259, 29–260, 1).

[39] Domus autem minus principales sunt, ut Aquarius Saturno datur, Pisces Iovi, Scorpius Marti secundum unam opinionem, secundum aliam Aries, Taurus Veneri, Gemini Mercurio; ita quod quilibet quinque planetarum habet duas domos sed Sol et Luna non nisi singulas. Ita decrevit antiquitas sapientium (260, 1–6).

[40] Roger's description makes the relationships less clear, whereas in Ptolemy it is much clearer, and he describes a lovely symmetry in the arrangement.

> And these are the '*exaltationes*': The sun is exalted in Aries, the moon in Taurus, Saturn in Libra, Jupiter in Cancer, Mars in Capricorn, Venus in Pisces, and Mercury in Virgo. And as the sun is exalted in Aries, so its '*descensio*' (depression) is in Libra, and thus concerning the rest. And as the '*depressio*' of Mercury is in Pisces, therefore the exaltation of Mercury is in Virgo, just as its house (*domus*) [sc. is Virgo], and this exaltation is at 15 degrees of Virgo.[41]

As Roger indicates, the *descensio* or *depressio* is the complement of (= the opposite sign to, that is, 180° away from) the exaltation, so that the moon is dejected or depressed in Scorpio, Saturn in Aries, etc. As with '*domus*', Bacon here provides all the relevant information for all of the planets and signs, although he focuses, once again, primarily on Mercury.[42] For all the other planets, he only mentions the sign in which they are exalted, but for Mercury, he also gives the degree. We now know that Mercury's '*domus*' and '*exaltatio*' are both in Virgo.

*

The third dignity is "triplicity," which is also highly relevant for the doctrine of great conjunctions, as we will see below:

> It is called a planet's triplicity when it [sc. the planet] is in the sign in which it was created, or in a sign of the same nature with the sign in which it was created. Whence one should know that there are four triplicities of signs: One is hot and dry [= fire], which contains three hot and dry signs, of which sort are Aries, Leo and Sagittarius. Whence, when the sun [sc. whose celestial "house" (= zodiacal sign in which it was created) is Leo] is in some one of those three, it is said to be in its triplicity. And there is a different second triplicity: of Taurus, Virgo and Capricorn, and this is cold and dry [= earth]; and Mercury, when it is in some one of these, is in its triplicity.[43]

Since Mercury was created in Virgo, Virgo is its celestial house or '*domus*'. Since Virgo is a cold and dry (= earth) sign, then, whenever Mercury is in either Virgo, Taurus or Capricorn (the other earth signs), it is in its triplicity. If Mercury is in Virgo, however, it gets additional points (as we will see) for being in its house, exaltation *and* triplicity.

---

[41] "Exaltationes vero sunt hae. Sol exaltatur in Ariete, Luna in Tauro, Saturnus in Libra, Iupiter in Cancro, Mars in Capricorno, Venus in Piscibus, Mercurius in Virgine. Et sicut Sol exaltatur, sic eius descensio est in Libra, et sic de reliquis; et similiter depressio Mercurii est in Piscibus, et ideo exaltatio Mercurii est in Virgine, sicut eius domus, et est haec exaltatio in xv gradu Virginis (260, 7–13)." Ptolemy describes exaltation-depression at *Tetrabiblos* I.19.

[42] Ptolemy attempts to explain why planets are thus exalted or depressed, but Roger does not bother to do so here. He simply accepts the astrological structures in order to use them for his purposes.

[43] Triplicitas planetae dicitur, cum sit in signo in quo creatus est, vel in aliquo eiusdem naturae cum signo in quo creatus est. Unde sciendum est quod quattuor sunt triplicitates signorum. Una est calida et sicca quae continet tria signa calida et sicca cuiusmodi sunt Aries, Leo, Sagittarius. Unde cum est Sol in aliquo istorum trium dicitur esse in sua triplicitate. Et alia est triplicitas secunda, ex Tauro, Virgine, et Capricorno, et haec est frigida et sicca; et Mercurius, quando est in aliquo istorum, est in triplicitate sua (260, 13–22).

Bacon continues with another distinction within triplicities, that of their diurnal and nocturnal planetary rulerships, which is also found in Ptolemy's *Tetrabiblos* I.18. Roger uses the earth signs to illustrate how this works:

> Although, since the rulers [= lords, *domini*] of this [sc. earth] triplicity during the day are first Venus and then the moon, and at night first the moon and afterwards Venus, their partner both at night and during the day is Mars. Nevertheless, Mercury participates with them properly in Virgo, as the astrologers (*astronomi*) say, and therefore it has its triplicity in Virgo, just as its exaltation and *domus*. The third triplicity is of Gemini, Libra and Aquarius, which is hot and moist [= air], and the fourth is from Cancer, Scorpio and Pisces, which is cold and moist [= water].[44]

We can see, once again, that Roger focuses here primarily on Mercury and its dignities. So, although he also informs us of the nature of these dignities in themselves, he is doing so principally as a continuing and integral part of his overall argument.

*

The fourth dignity is the "terms." Roger discusses two systems:[45]

> Moreover, the more famous terms (*termini*) are those of the Egyptians: Jupiter has the first 6 degrees of Aries, Venus the 6 following, Mercury 8, Mars 5, Saturn 5 [= 30 degrees]; Venus then has the 8 first degrees of Taurus, Mercury the 6 following. And thus these terms are varied with a wonderful diversity (*mira diversitas*), as is clear in the table of terms [not printed], such that Mercury has the first seven degrees of Virgo for its term, not only according to the Egyptians but also according to Ptolemy, and this is what we are seeking now.[46]

Roger here describes all of the terms for Aries and some for Taurus, but then he leaps to his terms of greatest concern, namely—as we would now expect—of Mercury's terms in the first seven degrees of Virgo.

*

The last dignity Roger discusses is "faces" or "decans":

> Moreover, the faces (*facies*) of the signs are received through the division of each sign into three equal parts; and each consists of ten degrees, which are called "faces" (*facies*), and in another manner, "decans" (*decani*). Of these faces, the beginning is from the first degree of

---

[44] Quia licet domini istius triplicitatis in die sint primo Venus, deinde Luna, et in nocte primo Luna, postea Venus, et eorum particeps in nocte et die sit Mars, tamen Mercurius participatur eis in Virgine proprie, ut dicunt astronomi, et ideo triplicitatem habet in Virgine sicut exaltationem et domum. Tertia triplicitas est ex Geminis, Libra, Aquario, quae est calida et humida. Et quarta est ex Cancro, Scorpione, et Pisce, quae est frigida et humida (260, 22–30).

[45] Stephan Heilen treats the history of the "terms" extensively in his "Ptolemy's Doctrine of the Terms and its Reception," in *Ptolemy in Perspective: Use and Criticism of his Work from Antiquity to the Nineteenth Century*, Alexander Jones (ed.), Dordrecht: Springer, 2010, 45–93. He treats Bacon briefly at 71–72.

[46] Famosiores autem termini sunt Aegyptiorum. Iupiter habet sex primos gradus Arietis, Venus sex sequentes, Mercurius octo, Mars quinque, Saturnus quinque, Venus adhuc octo primos Tauri, Mercurius sex sequentes. Et sic mira diversitate variantur isti termini, ut patet in tabula terminorum, ita quod Mercurius habeat septem primos gradus Virginis pro termino, non solum secundum Aegyptios, sed secundum Ptolemaeum, et hoc est quod nunc quaerimus (260, 31–38).

> Aries and terminates at the tenth degree of the same [sc. sign], and is called [sc. the face] of Mars. The second is up to the 20th degree and is called the face of the sun because the sun follows it in the order of the circles [i.e. in the order of the planets, moving downwards towards the earth]. The third is at the end of Aries and is called the face of Venus, and thus of the others in order, as is clear in the table of faces [also not printed], such that Mercury has the last ten degrees of Virgo for its face.[47]

Thus, with terms and faces, Roger continues with his earlier emphasis on Mercury and Virgo. Once again, he defines what the dignity is, but not how it is actually used in an interpretation; this will follow shortly.

*

In a slight departure, Roger then explains how these dignities (called *potestates* here) work by providing analogies from human society. This type of analysis is striking in its constrast to all of the earlier discussions so far, where analogy is virtually never invoked, thus providing strong evidence against Brian Vickers's definition that analogy fundamentally characterizes an "occult" science, among which he included astrology.[48] There also seem to be residual traces here of the planets' earlier careers as gods.[49] Once again, Bacon focuses on Mercury:

> And thus it is clear that Mercury has all these powers (*potestates*) in Virgo, and these are called powers by analogy (*per similitudinem*). Whence a planet in its house is compared to a king in his royal house (*in domo sua regia*) and in his dominion (*in dominatione sua*). And when [sc. a planet] is in its exaltation, it is like a man (*vir*) in his kingdom and glory (*in regno suo et gloria*). And when it is in its triplicity, it is like a man in his honor and among his helpers and ministers (*inter auxiliatores et ministros*). And when it is in its term, it is just like a man among his parents, family and clan. And when it is in its face, it is just like a man in his governance (*in magisterio*).[50]

A planet's location in various dignities or powers in the different signs is thus understood on analogy with a king's and other peoples' varying powers in different

[47] Facies autem signorum accipiuntur per divisionem cuiuslibet signi in tres partes aequales; et unaquaeque constat ex decem gradibus, quae vocantur facies, et alio modo decani; quarum facierum initium est a primo gradu Arietis, et terminatur in decimo gradu eiusdem, et dicitur Martis. Secunda usque a vicesimum, et dicitur facies Solis, quia Sol succedit ei in ordine circulorum. Tertia est in finem Arietis et dicitur facies Veneris, et sic de caeteris secundum ordinem, ut patet in tabula facierum; ita quod Mercurius habet decem gradus Virginis ultimos pro facie (261, 1–10).

[48] See the editor's introduction to *Occult and Scientific Mentalities in the Renaissance*, Brian Vickers (ed.), Cambridge: Cambridge University Press, 1984. For a strong critique of Vicker's analysis, see now William R. Newman, "Brian Vickers on Alchemy and the Occult: A Response," *Perspectives on Science* 17 (2009): 482–506. Although he focuses on alchemy, the criticism is directly relevant to astrology as well.

[49] For this part of the story, see Jean Seznec, *The Survival of the Pagan Gods: The Mythological Tradition and its Place in Renaissance Humanism and Art*, Barbara F. Sessions (tr), New York: Pantheon, 1953.

[50] Et sic patet, quod Mercurius habet omnes istas potestates in Virgine. Et vocantur istae potestates per similitudinem. Unde planeta in domo sua comparatur regi in domo sua regia et in dominatione sua; et cum fuerit in exaltatione sua, est sicut vir in regno suo et gloria; et cum fuerit in triplicitate sua, est sicut vir in honore suo et inter auxiliatores atque ministros; et cum fuerit in termino suo, est sicut vir inter parentes suos et cognatos et gentem suam; et cum est in facie sua, est sicut vir in magisterio suo (261, 10–19).

contexts of public and private life. Roger here draws these analogies (with slight variations) from Alcabitius's *Liber introductorius*.[51]

*

Having established the five different planetary dignities, Roger now completes his treatment thereof by offering a critical feature for interpretations or judgments, namely, he now tells us how much relative value each planetary dignity is worth: "*Domus* is said to have 5 strengths (*fortitudines*); *exaltatio* 4; triplicity 3; terms 2; and face 1. Whence, *domus* has in itself the strength of five faces, and *exaltatio* has the strength (*fortitudines*) of four faces, and thus further."[52] This value system in terms of points is utterly central for interpreting a horoscope, in particular, for calculating relative planetary strengths in order to determine which planet "rules" the nativity or any other sort of horoscope, that is, the so-called "lord of the geniture."[53] I call this "dignity accounting." Roger concludes: "From these [sc. dignities], therefore, it is clear that Mercury's essential and principle *potestates* are all in Virgo."[54] This is, of course, the conclusion towards which the entire discussion about dignities had been leading.

*

Bacon then turns from the five essential dignities to the accidental dignities, among which he only mentions a planet's '*gaudium*':

> And one should add that each planet has in addition a certain accidental power (*potestas*) in a sign to which it belongs (*in signo sibi debito*), which is called '*gaudium*'. Whence Saturn, when it enters Aquarius, is said to rejoice (*gaudere*), as Jupiter in Sagittarius, Mars in Scorpio, Venus in Taurus, and Mercury in Virgo. And therefore, Mercury never rules (*dominatur*) alone, as in Virgo. Nor does any planet have so many [sc. dignities] in its [sc. Virgo's] domain, because of which, Mercury is proper to Virgo. And therefore, for this reason (*ex hac causa*), they say that the *Lex Mercurialis* ought to be the religion of a prophet born of a virgin (*sectam prophetae nascituri de virgine*). And therefore, this Mercurial sect is posited by them to be the *Lex Christiana*.[55]

[51] Here is the text from Diff. I.24: "Quidam etiam de hac re talem dederunt comparationem, dicentes quia planeta cum fuerit in domo sua est similis viro in domo atque in dominatione sua, et cum fuerit in exaltatione sua est similis viro in regno suo atque gloria, et cum fuerit in termino suo est sicut vir inter parentes suos et cognatos *atque gentes*, et dum fuerit in triplicitate sua erit sicut vir in honore suo et inter auxiliatores suos atque ministros, et cum fuerit in facie sua erit sicut vir in magisterio suo. Hee sunt universe potestates planetarum essentiales in signis." I use the text from Burnett (et al), *Al-Qabisi*, 240.

[52] Et domus dicitur habere quinque fortitudines, exaltatio quatuor, triplicitas tres, terminus duas, facies unam. Unde domus habet in se fortitudines quinque facierum, et exaltatio habet fortitudinem quattuor facierum, et sic alterius (261, 19–23).

[53] We see such tables of dignities clearly in Galileo's horoscopes at the turn of the seventeenth century.

[54] Ex his igitur patet quod hae potestates Mercurii essentiales et principales sunt omnes in Virgine (261, 24–25).

[55] Atque addendum est, quod unusquisque planeta habet adhuc potestatem accidentalem quandam in signo sibi debito, quae vocatur gaudium. Unde Saturnus, cum intrat Aquarium, gaudere dicitur,

With this last dignity (*gaudium*) and Mercury's privileged position in Virgo now solidly established, Roger has amassed more astrologically derived evidence in support of his overall argument. Since Mercury rules Virgo and has most of its dignities there, the prophet of the Mercurial Law (Christianity) should therefore be born of a virgin.

6.2.4 *Stage 4: The Lunar Religion*

Returning from this detailed discussion of planetary dignities, Roger turns—in the fourth and final stage of his argument—to discuss the final historical religion, and one currently weighing heavily on his mind, namely, that derived from Jupiter's conjunction with the moon. Here he takes up where he left off at the end of stage 2:

> But if it [sc. Jupiter] embraces [= conjoins] the moon, the masters of astrology say (*dicunt domini astronomiae*) that there will arise the law of the moon and the last [sc. in the series] because the circle of the moon is the last, and this will be the law of corruption and filth (*foeda*), which will violate all the other laws and suspend them—even the Mercurial law—in time.[56]

Jupiter's conjunction with the moon thus signifies the lunar law and the ominous undoing of the Christian as of all other religious laws in the ultimate religious corruption.

Before identifying it, however, Roger associates the lunar law with necromancy (*nigromantia*) and magic, thus linking this part of his analysis to the earlier discussion in his defense of astrological judgments in Chap. 4. Here we can see another dimension of Roger's distinction between true and false mathematics, even though he had not yet mentioned "necromancy" per se:

> For the moon, as they say, signifies "necromancy" (*nigromantia*) and lying (*mendacium*), and therefore the law of the moon will be necromantic, magical and mendacious (*nigromantica, magica et mendosa*). And due to the corruption of the lunar motion and lunar shapes (*figurationes lunares*), it signifies the corruption of that law which will be corrupted in itself and corrupt others. Nevertheless, it will not remain long, as they say, because the moon changes quickly (*velociter mutatur*) in its appearance (*figuratio*)—and in its light and motion—due to the brevity of its circle.[57]

---

ut Iupiter in Sagitarrio, Mars in Scorpione, Venus in Tauro, et Mercurius in Virgine. Et ideo nusquam dominatur Mercurius tantum, sicut in Virgine. Nec aliquis planeta habet tot in ea dominia, propter quod appropriatur Virgini Mercurius. Et ideo ex hac causa dicunt legem Mercurialem debere esse sectam prophetae nascituri de virgine: et ideo haec secta Mercurialis ponitur ab eis esse lex Christiana (261, 25–36).

[56] Si vero complectatur Lunae, dicunt domini astronomiae, quod erit lex Lunae et ultima, quia circulus Lunae est ultimus, et haec erit lex corruptionis et foeda quae violabit omnes alias leges et suspendet eas, etiam Mercurialem ad tempus (261, 36–262, 2).

[57] Luna enim, ut dicunt, significat super nigromantiam et mendacium, et ideo lex Lunae erit nigromantica et magica et mendosa. Et propter corruptionem Lunaris motus et figurationem Lunarium significat super corruptionem istius legis, quae in se erit corrupta et alias corrumpens. Non tamen multum durabit, ut dicunt, quia Luna velociter mutatur a figuratione et luce sua et motu propter brevitatem sui circuli (262, 2–9).

Once again, Bacon uses the particular structure of the planetary body's astronomical orbit and physical nature to clarify its astrological nature, as we also saw with Mercury. In this case, however, it indicates corruption. He also pointed to the moon's motion and light, the two main expressions of astrology's natural philosophical foundations. The good news is that the lunar law, however deleterious, will not last long.

Bacon now identifies the lunar law as the religion of Antichrist:

> And this, as they say, is established by someone great and powerful who will prevail over others. And astrologers of the faith (*astronomi fideles*), both modern and ancient, believe that this is the law of Antichrist, since he will come (*adveniet*) at the very end of the world (*ultimo in fine mundi*), and he will usher in the law of corruption, and he will deceive the world through the magical art (*per artem magicam*) and his lies.[58]

These unnamed astrologers of the faith thus identify the *Lex Lunaris* as the law of Antichrist that will arrive at the end of the world, bewitching the world through the art of magic and various other deceptions. Roger discusses the timing of Antichrist's advent in the next case study.

*

Having now completed his description of this abstract astrologico-religious structure, Roger concludes stage 4—and thus the first major part of his argument (and our first case study)—by turning to Albumasar's book on great conjunctions:

> Therefore, astrologers (*astronomi*) discuss religions, and especially Albumasar in his *Book of Conjunctions*, and especially in the first and second book, in such a way that the six principle religions with which human beings are occupied in this world are discovered. And by this, one holds that the religion of Christ is one of the principle religions. And if we compare it to the others, it is obvious that, through the noble conditions of its lawgiver and of the religion, no other religion is worthy, but they are [sc. merely] people's inventions (*hominum figmenta*).[59]

For Roger, Albumasar provides—among other things—evidence for Christianity's special status among the six major world religions.

Roger then goes through the six religions each in turn, providing a sort of astrologico-philosophical analysis-*slash*-critique of their natures, all towards the end of vindicating Christianity's exceptional status:

> And this is immediately clear concerning the final law, since there is no truth (*veritas*) there. And in the Venusian law, which is of the Saracens, the delight in sin (*delectatio peccati*) abounds in accordance with its aforementioned tenor. But philosophy excludes sin (*peccatum*) from law. Likewise, there is no law of the Egyptians nor of the Chaldaeans because

---

[58] Et hoc, ut dicunt, statuetur ab aliquo magno et potente qui praevalebit aliis, et aestimant astronomi fideles tam moderni quam antiqui quod haec est lex Antichristi, quia ille ultimo in fine mundi adveniet, et inducet legem corruptionis, et infatuabit mundum per artem magicam et mendacia sua (262, 9–14).

[59] Sic igitur astronomi discutiunt sectas et praecipue Albumazar in libro coniunctionum, et maxime primo et secundo libro, ut inveniantur sex sectae principales in quibus homines occupantur in hoc mundo. Et per hoc habetur quod secta Christi sit una de principalibus. Et si comparemus eam ad alias, manifestum est per nobiles conditiones legislatoris et ipsius sectae quod nulla alia digna est, sed sunt hominum figmenta (262, 15–22).

> they teach the worship of a creature (*docent colere creaturam* [= idolatry]), and philosophy denies this. For worship is owed to God alone (*soli Deo cultus debetur*), as is set out in moral [sc. philosophy].[60]

Here we see a reference to idolatry, as part of a broader analysis of the unphilosophical nature of four of the six religions. We should also note that Bacon never seems to equate the Chaldaeans with astrologers, even in Antiquity.

Now Roger treats the Jews:

> But the religion of the Jews is less distant from truth (*minus elongatur a veritate*). But its [sc. law] giver was not the son of a virgin, as in Christian law. And it does not have confirmation through so many authentic writings, nor does it have such noble articles according to what was set out before according to the philosophers. Therefore, it is fitting that the law of Christ holds the first place (*principatum*).[61]

As Roger would have it, Christianity is the best because it is the most philosophical world religion, with Judaism a not-too-distant runner-up. Roger thus completes the first major part of his argument claiming astrology's utility for theology/religion, and our first case study of his use of astrology towards a theologico-religious end.

6.3 Second Case Study in Two Major Stages

6.3.1 Historical Astrology (1): Great Conjunctions

The second case study on great conjunctions per se will present the second part of Roger's larger argument here. To begin its more explicitly historical stage, Roger moves from more general astrologically-derived information to more specific knowledge concerning the actual history of religion by now discussing the timing of the beginnings and endings of each particular religion he had just discussed. In this context, Bacon articulates the major structure of historical astrology, namely, great conjunctions:

> But in moral philosophy, this will be clear from its own [sc. tenets/principles]. For, not only do they thus investigate [sc. religious] laws in general (*in universali*),[62] but they determine the times of their beginnings and the end of some of them. They investigate this by means of the conjunctions of the planets and the revolutions of their motions (*per coniunctiones planetarum et per revolutiones motuum eorum*). Therefore, Albumasar in the *Book of Conjunctions* and other astrologers (*astronomi*) determine that there are three conjunctions of Saturn and Jupiter, namely, great (*magna*), greater (*maior*) and greatest (*maxima*).[63]

---

[60] Et hoc de ultima lege statim patet, quia non est ibi veritas. Et in lege Venerea, quae est Saracenorum, delectatio peccati abundat secundum tenorem eius praedictum. Sed philosophia excludit peccatum a lege. Similiter lex Aegyptiorum nulla est nec Chaldaeorum, quia docent colere creaturam, et hoc negat philosophia. Nam soli Deo cultus debetur, sicut exponetur in moralibus (262, 22–28).

[61] Sed lator non fuit filius virginis, sicut in lege Christiana; atque non habet confirmationem per tot scripturas authenticas, nec habet tam nobiles articulos secundum quod prius expositum est secundum philosophos. Et ideo oportet quod lex Christi obtineat principatum (262, 28–263, 4).

[62] As Roger had just done in the previous section.

[63] Sed in morali philosophia hoc ex propriis erit planum. Nam non solum sic in universali investigant leges, sed determinant tempora inceptionum earum et finem aliquarum. Et hoc investigant per

The timing of the coming-into-being and passing-away of a religion as well as of the dynamics of religious change more generally can thus be understood in terms of the three types of great conjunctions of Saturn and Jupiter, namely, great, greater and greatest.[64] We will see how in what follows. Roger here closely and explicitly associates great conjunctions and revolutions, although they offer two rather different techniques for historical astrology.[65]

Roger begins by describing clearly and in detail the most basic type, namely, great conjunctions *simpliciter*:

> A great [sc. conjunction] is that which conjoins every twenty years, in whatever sign it is. For Jupiter completes its course in twelve years, and Saturn in almost thirty years. And therefore, it comes to pass that after twenty years they are joined (*iunguntur*) in the ninth sign from that in which they had been joined before.[66] And after another twenty years they are joined in the fifth sign from the first; and after the third twenty years they are joined again in that first sign. And this is a great conjunction (*coniunctio magna*), which comes to pass in that triplicity twelve times, or sometimes thirteen. For the first, fifth and ninth sign make a [sc. an elemental] triplicity. And this conjunction is said to signify (*significare*) many things about the elevation of kings and powerful people, and about the weight of the harvest (*gravitas annonae*) and the rise of prophets (*ortus prophetarum*).[67]

Due to the length of Jupiter's and Saturn's astronomical cycles through the zodiac—12 and 30 years respectively—they conjoin every 20 years in the strikingly regular triangulating patterns represented so clearly 350 years later in Johannes Kepler's and Giambattista Riccioli's diagrams. These 20-year conjunctions—one type of revolution—signify significant religious, political and economic transformations. As we saw in Chap. 3, Albertus Magnus provided his analysis of the natural philosophical foundations for great conjunctions in the *De causis proprietatum elementorum*. Roger does not discuss them at the natural philosophical/scientific level here, but only at the theoretical (in our terms).

coniunctiones planetarum, et per revolutiones motuum eorum. Albumazar igitur in libro coniunctionum et caeteri astronomi determinant tres esse Saturni et Iovis coniunctiones, magnam scilicet, maiorem, et maximam (263, 4–11).

[64] Laura Smoller is very helpful on all of this; *History, Prophecy and the Stars*, 21–22 and 52–57, and also on astrologizing patterns in history writing, 77–80. Girolamo Cardano uses a different terminology: small, middle and great. In addition to this tidbit, there is much more of interest in Hasse's extensive discussion of great conjunctions; *Success and Suppression*, 272–89.

[65] Pico attacks both of these cycles in Book V of the *Disputations*.

[66] Counting inclusively, with the first as one of the nine.

[67] Magna est, qua coniungitur in omnibus viginti annis in quocunque signo hoc sit. Iupiter enim perficit suum cursum in duodecim annis, et Saturnus quasi in triginta annis, et ideo fit ut post viginti annos iunguntur in nono signo ab eo, in quo prius iuncti fuerunt; et post alios viginti in quinto a primo; et post tertios viginti iterum in illo primo. Et haec est coniunctio magna, quae fit in hac triplicitate duodecies, vel aliquando terdecies. Primum enim signum, quintum, et nonum faciunt triplicitatem. Et haec coniunctio dicitur significare pluries super sublimationem regum et potentum, et super gravitatem annonae, et super ortus prophetarum (263, 11–22).

DE TRIGONO IGNEO. 25

masculina. Aries igitur, & cum eo triangulantia, Leo & Sagittarius sunt ignea. In primo quadrante principium Aries, in secundo medium Leo, in tertio finis Sagittarius. Quartus quadrans hyemalis caret signo igneo.

Sed cur tria succedentia, Taurus, Virgo, Capricornus, terrea potius quàm aerea, cùm aer immediatè succedat igni? Causam refert Albumasar hanc: Calori quidem actionem esse vivificandi: frigori verò itidem actionem, sed mortificandi, & corrumpendi: Omnem verò activam qualitatem passivâ esse nobiliorem: itaq; Tauro, qui proximè sequitur Arietem, activam qualitatem alteram obtigisse, frigus nempe, & inter elementa terram, frigoris sedem. Itaq; quæ proportio est vitæ ad mortem, diei ad noctem, maris ad fœminam; hæc quoque videbatur caloris ad frigus. Nam & fœmina in generandi actu aliquid agit (juvandus enim est Albumasar miserè laborans) nec merè passivè se habet.

Dixit Albumasar, cur igneo signo censeat succedi oportere à frigido: Dicat etiàm, cur id siccum potius, quàm humidum?

Primùm inquit, si Taurus constitutus fuisset humidus, Cancrum oportuisset esse siccum. At melius erat Taurum esse siccum, ut bina conjugata signa in una passiva & materiali qualitate convenirent; Aries & Taurus in sicco, Gemini & Cancer in humido; Leo & Virgo in sicco; & sic consequenter.

D Videtur

Image 6.1 Kepler, *De stella nova*, Prague: Paulus Sessius, 1606, p. 25 (Image courtesy of History of Science Collections, University of Oklahoma Libraries)

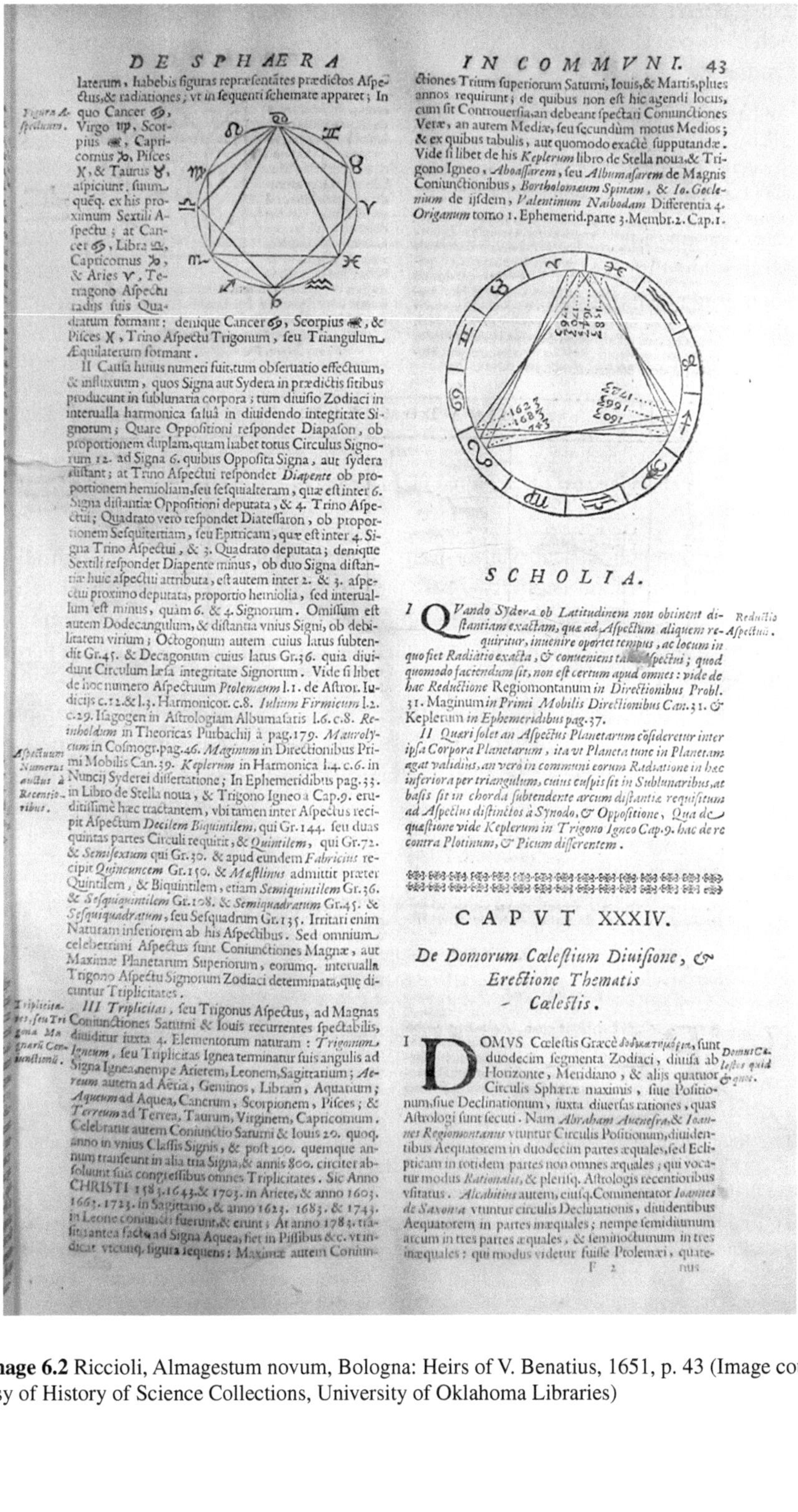

DE SPHAERA

laterum, habebis figuras repræsentātes prædictos Aspectus, & radiationes; vt in sequenti schemate apparet; In quo Cancer ♋, Virgo ♍, Scorpius ♏, Capricornus ♑, Pisces ♓, & Taurus ♉, aspiciunt suum quęq. ex his proximum Sextili Aspectu; at Cancer ♋, Libra ♎, Capricornus ♑, & Aries ♈, Tetragono Aspectu radijs suis Quadratum formant: denique Cancer ♋, Scorpius ♏, & Pisces ♓, Trino Aspectu Trigonum, seu Triangulum Æquilaterum formant.

Figura Aspectuum.

II Caussa huius numeri fuit, tum obseruatio effectuum, & influxuum, quos Signa aut Sydera in prædictis sitibus producunt in sublunaria corpora; tum diuisio Zodiaci in interualla harmonica saluâ in diuidendo integritate Signorum; Quare Oppositioni respondet Diapason, ob proportionem duplam, quam habet totus Circulus Signorum 12. ad Signa 6. quibus Opposita Signa, aut sydera distant; at Trino Aspectui respondet *Diapente* ob proportionem hemioliam, seu sesquialteram, quæ est inter 6. Signa distantiæ Oppositioni deputata, & 4. Trino Aspectui; Quadrato verò respondet Diatessaron, ob proportionem Sesquitertiam, seu Epitricam, quæ est inter 4. Signa Trino Aspectui, & 3. Quadrato deputata; denique Sextili respondet Diapente minus, ob duo Signa distantiæ huic aspectui attributa, est autem inter 2. & 3. aspectui proximo deputata, proportio hemiolia, sed interuallum est minus, quàm 6. & 4. Signorum. Omissum est autem Dodecangulum, & distantia vnius Signi, ob debilitatem virium; Octogonum autem cuius latus subtendit Gr.45. & Decagonum cuius latus Gr.36. quia diuidunt Circulum læsa integritate Signorum. Vide si libet de hoc numero Aspectuum *Ptolemæum* l.1. de Astror. Iudicijs c.12.& l.3. Harmonicor. c.8. *Iulium Firmicum* l.2. c.29. Isagogen in Astrologiam Albumasaris l.6. c.8. *Reinholdum* in Theoricas Purbachij à pag.179. *Maurolycum* in Cosmogr. pag.46. *Maginum* in Directionibus Primi Mobilis Can.59. *Keplerum* in Harmonica l.4. c.6. in Nuncij Syderei dissertatione; In Ephemeridibus pag.33. in Libro de Stella noua, & Trigono Igneo à Cap.9. eruditissimè hæc tractantem, vbi tamen inter Aspectus recipit Aspectum *Decilem Biquintilem*, qui Gr.144. seu duas quintas partes Circuli requirit, & *Quintilem*, qui Gr.72. & *Semisextum* qui Gr.30. & apud eundem *Fabricius* recipit *Quincuncem* Gr.150. & *Mæstlinus* admittit præter Quintilem, & Biquintilem, etiam *Semiquintilem* Gr.36. & *Sesquiquintilem* Gr.108. & *Semiquadratum* Gr.45. & *Sesquiquadratum*, seu Sesquadrum Gr.135. Irritari enim Naturam inferiorem ab his Aspectibus. Sed omnium celeberrimi Aspectus sunt Coniunctiones Magnæ, aut Maximæ Planetarum Superiorum, eorumq. interualla Trigono Aspectu Signorum Zodiaci determinata, quę dicuntur Triplicitates.

Aspectuum Numerus auctus à Recentioribus.

III *Triplicitas*, seu Trigonus Aspectus, ad Magnas Coniunctiones Saturni & Iouis recurrentes spectabilis, diuiditur iuxta 4. Elementorum naturam: *Trigonum Igneum*, seu Triplicitas Ignea terminatur suis angulis ad Signa Ignea, nempe Arietem, Leonem, Sagittarium; *Aereum* autem ad Aëria, Geminos, Libram, Aquarium; *Aqueum* ad Aquea, Cancrum, Scorpionem, Pisces; & *Terreum* ad Terrea, Taurum, Virginem, Capricornum. Celebratur autem Coniunctio Saturni & Iouis 20. quoq. anno in vnius Classis Signis, & post 200. quemque annum transeunt in alia tria Signa, & annis 800. circiter absoluunt suis congressibus omnes Triplicitates. Sic Anno CHRISTI 1583.1643.& 1703. in Ariete, & anno 1603. 1663. 1723. in Sagittario, & anno 1623. 1683. & 1743. in Leone coniuncti fuerunt, & erunt; At anno 1783. transitu antea facto ad Signa Aquea, fiet in Piscibus &c. vt indicat vtcumq. figura sequens: Maximæ autem Coniun-

Triplicitates, seu Trigona Magnarū Coniunctionū.

IN COMMVNI. 43

ctiones Trium superiorum Saturni, Iouis, & Martis, plures annos requirunt; de quibus non est hic agendi locus, cum sit Controuersia, an debeant spectari Coniunctiones Veræ, an autem Mediæ, seu secundùm motus Medios; & ex quibus tabulis, aut quomodo exactè supputandæ. Vide si libet de his *Keplerum* libro de Stella noua, & Trigono Igneo, *Aboassarem*, seu *Albumasarem* de Magnis Coniunctionibus, *Bartholomæum Spinam*, & *Io. Goclenium* de ijsdem, *Valentinum Naibodam* Differentia 4. *Origanum* tomo 1. Ephemerid. parte 3. Membr.2. Cap.1.

SCHOLIA.

I Quando Sydera ob Latitudinem non obtinent distantiam exactam, quæ ad Aspectum aliquem requiritur, inuenire oportet tempus, ac locum in quo fiet Radiatio exacta, & conueniens tali Aspectui; quod quomodo faciendum sit, non est certum apud omnes: vide de hac Reductione Regiomontanum *in Directionibus Probl.* 31. Maginum *in Primi Mobilis Directionibus Can.* 31. & Keplerum *in Ephemeridibus pag.* 37.

Reductio Aspectuū.

II *Quæri solet an Aspectus Planetarum cōsideretur inter ipsa Corpora Planetarum, ita vt Planeta tunc in Planetam agat validiùs, an verò in communi eorum Radiatione in hæc inferiora per triangulum, cuius cuspis sit in Sublunaribus, at basis sit in chorda subtendente arcum distantiæ requisitum ad Aspectus distinctos à Synodo, & Oppositione, Qua de quæstione vide Keplerum in Trigono Igneo Cap.9. hac de re contra Plotinum, & Picum disserentem.*

CAPVT XXXIV.

De Domorum Cœlestium Diuisione, & Erectione Thematis Cœlestis.

I DOMVS Cœlestis Græcè δωδεκατημόρια, sunt duodecim segmenta Zodiaci, diuisa ab Horizonte, Meridiano, & alijs quatuor Circulis Sphæræ maximis, siue Positionum, siue Declinationum, iuxta diuersas rationes, quas Astrologi sunt secuti. Nam *Abraham Auenesra*, & *Ioannes Regiomontanus* vtuntur Circulis Positionum, diuidentibus Aequatorem in duodecim partes æquales, sed Eclipticam in totidem partes non omnes æquales; qui vocatur modus *Rationalis*, & plerisq. Astrologis recentioribus vsitatus. *Alcabitius* autem, eiusq. Commentator *Ioannes de Saxonia* vtuntur circulis Declinationis, diuidentibus Aequatorem in partes inæquales; nempe semidiurnum arcum in tres partes æquales, & seminocturnum in tres inæquales: qui modus videtur fuisse Ptolemæi, quate-

Domus Cœlestes quid & quot.

F 2 nus

Image 6.2 Riccioli, Almagestum novum, Bologna: Heirs of V. Benatius, 1651, p. 43 (Image courtesy of History of Science Collections, University of Oklahoma Libraries)

Next Roger discusses *greater* conjunctions, namely, when the conjunctions switch from one elemental triplicity to the next after 12 or 13 triangular trips around the zodiac:

> And after they have been conjoined so many times [namely, twelve or thirteen] in that triplicity, and they change to another [sc. triplicity] (*ad aliam mutentur*), then it is called a greater conjunction (*coniunctio maior*). And this comes to pass every 240 years or thereabouts, and signifies religion (*secta*) and its transformation (*mutatio*) in certain regions.[68]

Greater conjunctions, then, are primarily concerned with religious change. We will recall the four different elemental triplicities of the signs from Roger's recent discussion of planetary dignities. Great conjunction theory is thus an astrological technique in which the elemental triplicities of zodiacal signs are highly significant.

What about *greatest* conjunctions?:

> And when the conjunction has changed from that triplicity into another, as from the end of Cancer to the beginning of Aries, then it is called "greatest" (*maxima*) by the revolution of Saturn in 32 turns, and it takes place every 960 years, and signifies changes (*mutationes*) of empires and kingdoms, fiery impressions in the air, floods and earthquakes, and the weight of the harvest.[69]

A greatest conjunction occurs after four 240-year triangular treks through all the signs of each elemental triplicity. Profound transformations, political and natural, are then signified. The greatest conjunction occurs with the return of Saturn and Jupiter from water signs back into fire signs, the first triplicity, and Aries in particular, the first sign.

Having provided these basic structures, Roger now offers one telling example of great conjunctions and their profound historical effects, thus linking the previous discussion of the Mercurial Law (Christianity) directly to the basic structures of historical astrology:[70]

> And one greater or almost a greatest was in the 24th year of [sc. the reign of] Augustus Caesar, which the wise astrologers (*sapientes astronomi*) said signified the future Mercurial law. And in the book which speaks about the transformation of Ovid's life, which is entitled *De vetula*, because of which a *mutatio* had taken place, Ovidius Naso is said to have spoken about this conjunction. From its disposition, he is said to have burst forth in admiration of the Mercurial religion to be brought forth into the world by means of a prophet who will be born of a virgin without "commingling" with a male (*absque maris commixtione*). He predicted (*praedixit*) that it would come into being six years after the conjunction, such that he (*ipsum* [the prophet (m.), not the sect (f.)]) would be born in the 30th year of Octavian

---

[68] Et postquam totiens in ista triplicitate coniuncti fuerint, ut ad aliam mutentur, tunc vocatur coniunctio maior. Et hoc fit in omnibus ducentis quadraginta annis vel circiter, et significat super sectam et mutationem eius in quibusdam regionibus (263, 22–26).

[69] Et quando mutata fuerit coniunctio ab ista triplicitate in aliam, ut a fine Cancri ad initium Arietis, tunc dicitur maxima, per revolutionem Saturni triginta duabus vicibus, et fit omnibus nongentis sexaginta annis, et significat super mutationes imperiorum et regnorum, et super impressiones ignitas in aere, et super diluvium, et super terrae motum et gravitatem annonae (263, 26–32).

[70] I have found helpful here the translations into Italian by Ornella Pompeo Faracovi in her *Gli oroscopo di Christo*, 151–55, and that in Sorge and Seller, *Filosofia, scienza, teologia*, 116–18 with their notes.

> Augustus [sc. of his reign, not his life]. For twelve years passed from the death of Julius [Caesar] to the battle of Actium (*Actium bellum*), in which Octavian Augustus fully obtained '*imperium*' [sc. official power].[71] For beforehand, he had labored more to acquire the kingdom (*regnum*) that he possessed.[72]

Roger here used the pseudo-Ovidian *De vetula* to argue that wise astrologers in Antiquity predicted on the basis of a greater or greatest conjunction in 7/6 BCE that Christ's birth would occur 6 years hence, thus marking a profound *mutatio* of religions and kingdoms, the hallmarks of greater and greatest conjunctions.

*

Roger now discusses the dating of Christ's birth in relation to Augustus Caesar's reign, comparing two possible dates:

> Then Christ was born in the 30th year of Octavian. But if these twelve years of his reign were calculated,[73] then Christ was born in the 42nd year of Augustus, according to what others calculate. But it comes to the same thing. For Augustus had undertaken five great civil wars within those twelve years, as the histories tell (*ut narrant historiae*), and especially Orosius in the book *De ormesta mundi*.[74] The last was at Actium, in which he defeated Anthony and Cleopatra, and quieted the empire in peace [sc. the famous *Pax Augusta*]. Therefore, Ovid, talking about a greater conjunction and almost a greatest, speaks in this manner in his verse: "Indeed, one such [sc. conjunction] (*una talis*) took place recently at a fatal time (*tempus fatale*) in the 24th year of Caesar Augustus from the beginning (*novitas*) of his rule, which signified that after the sixth year a prophet ought to be born of a virgin apart from *coitus* with a man, who would be held to be a model (*typus*), such that the power of Mercury would be multiplied more, whose concourse [= conjunction] will be the first embrace of the future religion, for Mercury never rules signs as it does in the sign of Virgo."[75]

---

[71] From the Ides of March 44 BCE to 2 September 31 BCE.

[72] Et una maior vel fere maxima fuit xxiv anno Augusti Caesaris, quam dixerunt sapientes astronomi significare super legem Mercurialem futuram. Et in libro, qui dicitur de mutatione vitae Ovidii, qui inscribitur de Vetula, propter quam mutatio fuerat facta, refertur Ovidius Naso locutus fuisse de hac coniunctione, et ex eius dispositione prorupisse in admirationem sectae Mercurialis producendae in mundum per prophetam nasciturum de virgine absque maris commixtione, quam futuram esse praedixit post illam coniunctionem per annos sex, ita quod secundum ipsum nasceretur xxx anno Octaviani Augusti. Nam xii anni fluxerunt a morte Iulii usque ad Actium bellum, in quo plene obtinuit imperium Octavianus Augustus. Nam ante magis laboravit ut acquireret regnum, quam possedit (263, 32–264, 8).

[73] That is, those just mentioned between Julius Caesar's death and the battle of Actium, namely, in the period before Octavian posessed full *imperium*.

[74] Bridges's n. 1 (264) is informative here.

[75] Et tunc Christus fuit natus xxx anno Octaviani. Si vero computentur illi xii anni de regno eius, tunc Christus fuit natus xlii anno Augusti, secundum quod alii computant. Sed in idem redit. Gessit enim Augustus quinque bella civilia magna infra illos xii annos, ut narrant historiae, et maxime Orosius in libro de Ormesta Mundi. Ultimum vero fuit ad Actium, in quo Antonium et Cleopatram devicit et quievit imperium in pace. Loquens igitur Ovidius de coniunctione maiore et fere maxima dicit in metro suo hoc modo, 'Una quidem talis fatali tempore nuper/ Caesaris Augusti fuit anno bis duodeno/ A regni novitate sui, quae significavit/ Post annum sextum nasci debere prophetam,/ Absque maris coitu de virgine, cuius habetur/ Typus, uti plus Mercurii vis multiplicatur./ Cuius erit concors complexio prima futurae/ Sectae nam nusquam de signis sic dominatur/ Mercurius sicut in signo Virginis' (264, 8–26).

De vetula thus emphasizes Mercury by claiming that the prophet will be born of a virgin. Of course, the pseudonymous author had the two great benefits of hindsight in making his "prophecy" as well as of having access to Albumasar's writings, neither of which would have been available to the authentic Publius Ovidius Naso, who lived from 43 BCE to 17/18 CE. Nor indeed would the theory of great conjunctions itself even have existed, since it only arose in Sasanian Persia (224–651 CE).[76]

Roger continues by further articulating the astrological circumstances at Christ's birth. We should not be surprised to find Bacon continuing to emphasize Mercury's roles by discussing the dignities at that time, thus instantiating and developing his earlier analysis:

> And the first face [= decan] of Virgo was ascending in the East when that conjunction (*coniunctio*) took place. And that conjunction was near the head [sc. beginning] of Aries (*caput Arietis*). For if we revolve the motions of Saturn and Jupiter to that time, we will find that they were conjoined through their mean motions (*per medios cursos suos*) at six years, five days and three hours before the birth of Christ. And the mean motion (*medius cursus*) of each [sc. Saturn and Jupiter] was at 10° 56' 52" of Aries.[77] But the mean motion of the eighth sphere was 10° 5' 51" 27"' from the signs of the small circle (*circulus parvus*), and it should be subtracted (*minuenda*) from the places of all the planets; whence they remained at 2° 14' 42" of Aries. Therefore, since the difference between the two conjunctions by their mean motions adds 8 signs, 2° 25' 17", it follows that the preceding conjunction was at 29° 51' 25" of Cancer, and thus the triplicity had changed from a water to a fire sign. But if this conjunction had been closer to the beginning of Aries, it would have been a greatest, and then there would have been 305 complete Greek years and 9 months, and almost 18 days, which can be proved by annual tables (*per tabulas annorum*).[78]

[76] For the origins of great conjunctions, see Pingree, *From Astral Omens to Astrology*. For the actual astrology of Ovid's time, see Tamsyn Barton, *Ancient Astrology*, London: Routledge, 1994, and her *Power and Knowledge*. See also Katherina Volk, *Manilius and His Intellectual Background*, Oxford: Oxford University Press, 2009; *Forgotten Stars: Rediscovering Manilius's* Astronomica, K. Volk (ed.), Oxford: Oxford University Press, 2011, and Steven J. Green, *Disclosure and Discretion in Roman Astrology: Manilius and his Augustan Contemporaries*, Oxford: Oxford University Press, 2014, which all draw upon the formidable labors of a host of formidable scholars, including Joseph Scaliger and A.E. Housman. For Scaliger's contribution, see Anthony Grafton, *Joseph Scaliger: A Study in the History of Classical Scholarship,* vol. I: *Textual Criticism and Exegesis*, Oxford: Clarendon, 1983, ch. VII: "Scaliger's Manilius: From Philology to Cultural History," 180–226.

[77] Faracovi's numbers here and the next number are quite different than those in Bridges's text, although she cites the 1964 Frankfurt a.M. reprint of his works. She has 10° 14′ 42″ for the first and 10° 5′ 21″ 26‴ for the second. The other numbers are the same as in Bridges's text. Sorge and Seller's numbers, on the other hand, agree with both Bridges and with my numbers.

[78] Et prima facies Virginis ascendebat in oriente, quando coniunctio illa facta fuit. Et fuit coniunctio illa prope caput Arietis. Si enim revolvamus motus Saturni et Iovis ad tempus illud, inveniemus eos fuisse coniunctos per medios cursos suos ante nativitatem Christi per sex annos, quinque dies, et tres horas; et erat medius cursus utriusque in Ariete decem gradus, lvi minuta, lii secunda. Medius vero motus octavae sphaerae erat ex signis circuli parvi decem gradus, quinque minuta, li secunda, xxvii tertia, et erat minuenda a locis omnium planetarum; unde remanserunt de Ariete ii gradus, xiv minuta, xlii secunda. Cum ergo differentia inter duas coniunctiones per cursus medios addat viii signa, ii gradus, xxv minuta, xvii secunda, sequitur quod praecedens coniunctio fuerat in Cancro xxix gradibus, li minutis, xxv secundis, et ita mutata fuit triplicitas a signo aquatico ad igneum. Si vero haec coniunctio fuisset propinquior capiti Arietis, fuisset maxima, et tunc erant anni

According to Roger's somewhat complex calculations, this conjunction of Saturn and Jupiter announcing Christ's birth—and thus the coming into being of Christianity—is most likely a greatest conjunction.[79]

6.3.2 *Historical Astrology (2): Ten Saturn Cycles*

After discussing the theory of great conjunctions to gain insight into the patterns of history, and in particular into the birth/generation of the most important religion and its founder, Roger offers another technique of historical astrology, namely, that involving ten revolutions of Saturn on its own, in order to discern a religion's duration. Once again, Albumasar is Roger's guide:

> But they consider the same thing through the revolutions of the planetary motions. For Albumasar in the 8th difference of the 2nd book on conjunctions says that the duration of a religion and kingdom and their transformation happen (*accidunt*) especially in accordance with the quantity of ten revolutions of Saturn. This is especially the case if the change (*mutatio*) to *signa mobilia*, which are Cancer, Libra, Capricorn and Aries, is fitted to Saturn, provided that Jupiter is cadent from it. But if Jupiter were with it [sc. thus resulting in a great conjunction], or aspecting it, it would much diminish the badness [sc. of Saturn's influence] due to its [sc. Jupiter's] goodness.[80]

According to Albumasar, changes in religions and kingdoms also happen after ten revolutions of Saturn, that is, approximately every 300 years, especially if they occur in the cardinal (*mobilia*) signs of the zodiac. Roger also discusses how this influence is conditioned by Jupiter's relation to Saturn.

Roger then offers a series of historical examples to illustrate these patterns:

> [1] For when ten revolutions of Saturn were completed in the days of Darius, Alexander the Great appeared, and the destruction of the kingdom of the Persians [ca. 330-20 BCE]. [2] And after around ten other complete revolutions [sc. of Saturn], Jesus appeared, the son of Mary, about whom there are speeches (*orationes*) on the transformation (*permutatio*) of religion (ca. 0). [3] And when ten other [sc. revolutions of Saturn] were completed, Mani (*Meni*) appeared, and he came with the law that is between the pagans and the Christians (*Nazarenos*, ca. 300 CE). [4] And after ten others, Mohammed (*Mahometus*) came, and perhaps, that is, before the completion of ten revolutions, as in the ninth revolution, and

---

Graecorum perfecti trecenti quinque et novem menses, et fere xviii dies, quod potest probari per tabulas annorum (264, 27–265, 12).

[79] On Bacon's astronomical and chronological knowledge with much valuable analysis, see now C. Philipp E. Nothaft, *Dating the Passion: The Life of Jesus and the Emergence of Scientific Chronology (200–1600)*, Leiden: Brill, 2012. Chapter six is devoted to Roger Bacon and his successors.

[80] Per revolutiones vero motuum planetarum considerant illud idem. Nam Albumazar octava differentia libri secundi de coniunctionibus dicit, quod mora sectae et regni et permutatio accidunt praecipue secundum quantitatem decem revolutionum Saturniarum, praecipue si Saturno conveniat mutatio ad signa mobilia, quae sunt Cancer, Libra, Capricornus, Aries, dummodo Iupiter fuerit cadens ab eo. Sed si Iupiter cum eo, aut aspiciat eum, minuet multum de malo propter eius bonitatem (265, 13–21).

> perhaps, after, in the 11th (ca. 600 CE). And that is in accordance with its quantity, which the aforementioned [sc. great] conjunctions complete, which are stronger than these revolutions.[81]

After this loose but suggestive historical analysis, Roger informs us that great conjunctions are more powerful agents of change than these sets of ten revolutions of Saturn.

*

After these two temporal locators of change, in which he discussed the beginnings of religions (with great conjunctions) and their durations (with ten Saturn cycles), in the next step (a brief one), Bacon offers another significant interpretive factor—namely, the planetary rulership of places—in order to spatially locate in what region of the earth such transformations will occur:

> And likewise, a religion is changed faster or slower in accordance with the properties of the planets ruling in diverse kingdoms, as Saturn rules (*dominatur*) India, Jupiter Babylon, Mars Thrace, the Sun the Romans (*Romani*) and their empire, Mercury Egypt and the moon Asia.[82]

Thus both the effects of these celestial influences and our knowledge thereof can be localized in place as well as in time. This interpretive factor also relates directly to our earlier discussion of the nature of place. More detail would have been useful for our understanding of how to fully use this information.

*

In his next step analyzing the timing of religions, Bacon addresses specific interpretations for how to time the fall of Islam, the religion he would most like to see end. Once again, Albumasar is Roger's authority, and specifically II.8, which is obviously of great importance to him. We should recall that the *Speculum astronomiae* also made a point of defending Albumasar's writings in particular in its Chap. 12, and especially its value for validating the Christian religion.

In fact, timing Islam's fall is the other primary end towards which Roger's rich and complex analysis of astrology's utility for religion has been aiming, along with the validation of Christianity:

---

[81] Quando enim fuerunt completae decem revolutiones Saturni in diebus Darii, fuit apparitio Alexandri magni, et destructio regni Persarum. Et circiter post decem alias revolutiones completas apparuit Iesus filius Mariae, super quam sunt orationes cum permutatione sectae. Et quando completae sunt decem aliae apparuit Meni, et venit cum lege quae est inter Paganos et Nazarenos. Et post decem alias venit Mahometus, et fortasse illud est ante complementum decem revolutionum, ut in revolutione nona, et forsan post, ut in undecima. Et illud est secundum quantitatem eius, quod exigunt coniunctiones praemissae, quae sunt fortiores istis revolutionibus (265, 21–32).

[82] "Et similiter mutatur citius vel tardius secta secundum proprietates planetarum dominantium regnis diversis, ut Saturnus Indiae dominatur, Iupiter Babyloniae, Mars Thraciae, Sol Romanis et imperio eorum, Mercurius Aegypto, Luna Asiae (265, 32–36)." Roger (or Bridges) seem to have left Venus out of this scheme.

> And they [sc. the astrologers] speak about the destruction of Mohammed's law beautifully and with certainty. For, according to what Albumasar says in the 8th chapter of the 2nd book, the law of Mohammed cannot endure beyond 693 years. But it will only be strong to endure and will [sc. actually] endure unless, due to some coinciding cause, the time will be shortened in accordance with what was touched on earlier, which abbreviation can be made more or less from different causes.[83] And now is the 665th year of the Arabs from the time of Mohammed,[84] and therefore, it will be destroyed quickly by the grace of God, which ought to be a great solace to Christians. Because of this, God should be praised, who gave philosophers the Light of Wisdom (*lumen sapientiae*) through which the Law of Truth (*lex veritatis*) is confirmed and strengthened, and through which we perceive that the enemies of faith should be destroyed.[85]

Here Roger shows how historical astrology can profoundly aid the Christian faith by providing analyses that predict the end of Islam and thus provide comfort to the Christian faithful.

*

To complete his argument, Bacon shifts the interpretive register by comparing the results of his astrological analysis to scriptural prophecies in order to confirm both, thus using reason and revelation to mutual benefit. The light of wisdom that God gave philosophers to confirm the truth of scripture seems to be none other than astrology! Roger here adduces the *Book of Revelations*:

> And with this opinion agrees the 13th chapter of the *Apocalypse*. For it says that the number of the beast (*numerus bestiae*) is 663, which number is less than that predicted [sc. by astrology] by 30 years.[86] But in many places scripture does not say anything about the complete number. For this is the manner of scripture (*mos scripturae*), as Bede says. And here perhaps God wished that it not be expressed completely (*totaliter*), but hidden a little (*aliquantulum occultaretur*), as other things written in the *Apocalypse* are. Whence, before the last time determined for this religion in accordance with its principle cause, in so far as Albumasar determines, perhaps it will happen that the Saracens will be destroyed either by the Tartars or the Christians. And already the greater part of the Saracens has been destroyed by the Tartars, both the head of the kingdom, which was Baldac [= Baghdad], and the caliph who was as their pope. These things already happened twelve years ago.[87]

---

[83] This discussion is similar to that on the changing of a person or thing's natural period in Chap. 1, but timing can also change by the exercise of God's absolute power.

[84] Bridges (n. 1 [266]) shows how these numbers result in 1267, the year in which Bacon is writing.

[85] Et de destructione legis Mahometi pulchre et certitudinaliter loquuntur. Nam secundum quod Albumazar dicit viii capitulo secundi libri, non potest lex Mahometi durare ultra sexcentos nonaginta tres annos. Sed tantum valet durare et durabit nisi propter aliquam causam coincidentem abbrevietur tempus secundum quod prius tactum est, quod abbreviatio potest fieri maior et minor ex causis diversis. Et nunc est annus Arabum sexcentesimus sexagesimus quintus a tempore Mahometi, et ideo cito destruetur per gratiam Dei, quod debet esse magnum solatium Christianis. Propter quod laudandus est Deus, qui philosophis dedit lumen sapientiae, per quod lex veritatis confirmatur et roboratur, et per quod percipimus inimicos fidei destrui debere (266, 1–13).

[86] That is, the length of Islam predicted by Albumasar in the previous passage, namely, 693 years.

[87] "Et huic sententiae concordat Apocalypsis xiii capitulo. Nam dicit quod numerus bestiae est 663, qui numerus est minor praedicto per xxx annos. Sed scriptura in multis locis subticet aliquid de numero completo, nam hic est mos scripturae, ut dicit Beda. Et hic forsan voluit Deus, quod non

Albumasar seems to be Roger's guide for interpreting biblical prophecy as well! From the other direction, however, Roger also brings biblical revelation to bear on the interpretation of historical astrology, especially for the timing of Islam's demise. Thus historical astrology and biblical prophecy may be used to mutually inform each others' interpretations.

6.4 Conclusion of *Iudicia astronomiae*

To conclude this rich section—and now with the background of our more fully developed understanding of his system—Roger returns to some of the central themes raised earlier and treated in my Chap. 4 and earlier in Chap. 6, beginning with human free will. He also broaches a new theme: planetary motions as signs vs. causes. In this final stage of his presentation, Bacon takes a more explicitly theological turn:

> And although they speak about religions (*de sectis*), and religions depend on freedom of thought (*libertas rationis*), nevertheless, they do not impose any necessity on free will (*non imponunt aliquam necessitatem libero arbitrio*), saying that planets are signs (*signa*) that intimate to us those things that God has disposed from eternity (*quae Deus disposuit ab aeterno*) to happen either through nature or human volition (*voluntas humana*), or reason proper (*ratio propria*), according to the pleasure of His will (*secundum beneplacitum suae voluntatis*). Thus it is said in the book *On the Courses of the Planets* (*De cursibus planetarum*).[88]

In this view, planetary patterns are understood as signs indicating God's providential ordering and governance of the world and of man, which may happen either by nature, human volition or reason proper in accordance with God's will. The religions thus engendered are not imposed by necessity, but incline people in their direction in accordance with God's will.

To explain how this works, Bacon returns to the natural philosophical themes raised earlier:

> And because of these things, they say that the will is not compelled (*voluntas non cogitur*), but, nevertheless, the body is altered by the powers of the heavens (*corpus alteratur per virtutes coelorum*), and then the soul-united-to-the-body (*anima corpori unita*) is excited

exprimeretur totaliter, sed aliquantulum occultaretur, sicut caetera quae in Apocalypsi scribuntur. Unde ante tempus ultimum quod isti sectae determinatur, secundum eius causam principalem, prout determinat Albumazar, forsan continget quod Saraceni destruentur aut per Tartaros aut Christianos. Et iam maior pars Saracenorum destructa est per Tartaros, et caput regni quod fuit Baldac, et Caliph qui fuit sicut papa eorum. Iam haec facta sunt xii annis elapsis (266, 13–26)." John D. North, "Roger Bacon and the Saracens," is very helpful on this; *Filosofia e scienza classica, arabo-latina medievale e l'eta moderna*, Graziella Federici Vescovini (ed), Turnhout: Brepols, 1999, 129–60.

[88] Et quamvis loquantur de sectis, et sectae dependent ex libertate rationis, tamen non imponunt aliquam necessitatem libero arbitrio, dicentes planetas esse signa innuentia nobis ea quae Deus disposuit ab aeterno fieri sive per naturam, sive per voluntatem humanam, sive per rationem propriam secundum beneplacitum suae voluntatis. Ita dicitur in libro de cursibus planetarum (266, 27–267, 1).

> strongly and induced effectively (*excitatur fortiter et inducitur efficaciter*)—although in nothing is it compelled—so that it wishes freely (*gratis*) to follow the inclinations of the body (*ut velit gratis sequi inclinationes corporis*) toward private or public actions, and to good as to bad actions, so that thus opinions and religions and transformations of practices are induced among people by someone famous and powerful, in accordance with what was foreseen (*praevisum*) and foreknown (*praecognitum*) by God, in such a way that planets are not only signs, but also do something in the exciting (*aliquid faciant in excitando*).[89]

Thus the celestial bodies may function as signs, but they are also efficient causes. They incline but do not compel, acting directly on the body and thereby indirectly affecting the soul connected to it, thus providing a systematic and solid safeguard for free will. Roger also indicates how God's foreknowledge and omniscience also in no way undermine peoples' free will. Roger's analysis here is thus wholly in line with the basic structures of Thomas's more detailed analysis explored in Chap. 5.

Roger continues, developing his argument by discussing Christ's nature and returning to the pseudo-Ovidian *De vetula*. Here he uses astrology to explain certain features of the Christian mysteries, in particular, the Virgin birth and God-made man:

> And when they proposed that the Lord Jesus Christ is both God and man, as Ethicus the Astrologer (*Ethicus Astronomus*) says manifestly in the *Cosmographia*, and Alchimus likewise, and also in that book which is entitled Ovid's *De vetula*, which concludes that God is incarnated in Christ. Attributing to him what is denied to a mere person (*soli homini denegatur*), they wish that the celestial disposition (*dispositio coelestis*) could be in the sign of the conception of a virgin, and of the nativity of that man, in so far as he is a man (*homo*), just as a star presides as a sign in his nativity (*sicut stella praestitit signum in eius nativitate*). This is in accordance with what is said in the book *On the Courses of the Planets*, namely, that all the planets and the other stars serve a God-made man (*Deo homini facto*) in one way and a "pure" human being in another, and in one way for the Creator and in another for creatures. Therefore, God wished to arrange (*ordinare*) his affairs (*suas res*) in such a way that certain future things, which he had foreseen or predestined (*quaedam quae futura praeviderit vel praedestinaverit*), could be shown to rational ([sc. creatures = human beings] *rationabilibus*) by means of the planets (*per planetas*), for this reason, namely, that the human mind (*mens humana*), recognizing the wonders of God (*Dei mirifica recognoscens*), would grow inflamed in the love of its Creator (*in amorem sui conditoris succensa excresceret*).[90]

---

[89] Et praeterea dicunt quod voluntas non cogitur, sed tamen corpus alteratur per virtutes coelorum, et tunc anima corpori unita excitatur fortiter et inducitur efficaciter, licet in nullo cogatur, ut velit gratis sequi inclinationes corporis ad actus privatos vel publicos, et ad bonos sicut ad malos, ut sic opinones et sectae et mutationes consuetudinem inducantur per aliquem famosum in populo et potentem, secundum quod praevisum fuit et praecognitum a Deo; ita quod planetae sic non solum sint signa, sed aliquid faciant in excitando (267, 1–9).

[90] Et cum posuerunt Dominum Iesum Christum esse Deum et hominem, ut Ethicus astronomus manifeste dicit in Cosmographia, et Alchimus similiter, necnon et in illo libro qui inscribitur, *Ovidius de vetula*, Deum incarnari in Christo colligitur, attribuentes ei quod soli homini denegatur, volunt quod dispositio coelestis potuit esse in signum conceptionis Virginis, et nativitatis illius Hominis, in quantum homo, sicut stella praestitit signum in eius nativitate, secundum quod dicitur in libro cursuum planetarum, quod planetae omnes et caeterae stellae aliter Deo homini facto, aliter puro homini, aliter creatori, aliter creaturis famulantur. Voluit ergo Deus res suas sic ordinare, ut quaedam quae futura praeviderit vel praedestinaverit rationabilibus per planetas ostenderentur, ideo scilicet ut mens humana Dei mirifica recognoscens in amorem sui conditoris succensa excresceret (267, 10–24).

As we have already learned, celestial influences work differently on different members of the same species due to their differences of matter and the places where the influences are received.

By contrast, Roger asserts here that the divine and the human also receive celestial influences differently—or are served by them differently—as if they were members of different species. Here Roger identifies two such species: A God-made divine man, namely Christ, in relation to pure natural human beings, which Roger then generalizes, referring to the Creator, on the one hand, and His creatures, on the other. Roger here also relates astrology to God's foreknowledge with respect to his providence, and thus to another aspect of the problem of the knowledge of future developments—including contingents—although he does not mention contingents here specifically. Sometimes this happens with the planets functioning as efficient causes and sometimes as signs, but always under God's ordered and absolute power. All of this is towards the end of communicating these matters to human beings and thus inspiring us to love God more, much as we have already seen in the *Speculum astronomiae.*

*

Roger then further develops this theological strain:

> And therefore they wish that the celestial ([sc. bodies] *coelestia*) can indicate and signify this God-made work (*hoc opus deificum*) of conception and nativity, insofar as a creature attests to its creator coming into flesh, just as against nature (*contra naturam*) the sun was obscured at the passion. Because of which, the philosophers, seeing these things, said that either God experienced [suffered?] something in nature (*aliquid Deus naturae patiebatur*), or the entire system of the world (*machina mundi*) would have disintegrated. And doubtless they want it to be impossible that God could be subjected to a creature [sc. something created], and not because that divine work (*istud opus divinum*), insofar as it came to be by an infinite power (*a virtute infinita*) and beyond nature (*supra naturam*), was in some manner subjected to the celestial disposition, but only because that was the case in a sign.[91]

Although astrology is primarily concerned with the planets as efficient causes, there are occasions for which God uses the planets as signs alone as an expression of his absolute power, in order, among other things, so that divine works (and thus God's power itself) are not seen to be subjected to the stars and thus to something created. As in *Speculum astronomiae*, Chap. 3, we can also see here that Roger thinks that God reveals information about his providential creation and ordering of the world to rational creatures, namely, human beings, by means of the heavens, and that he does this, among other reasons, to inspire people to love Him more.

---

[91] Et ideo volunt quod innuere et significare possunt coelestia hoc opus deificum conceptionis et nativitatis, quatenus creatura attestetur suo creatori in carne venienti. Sicut contra naturam sol obscuratus fuit in passione; propter quod philosophi haec videntes dixerunt, quod aut aliquid Deus naturae patiebatur, aut tota mundi machina dissolveretur. Et proculdubio volunt quod impossibile sit Deum subiici creaturae, nec quod istud opus divinum in quantum fuit a virtute infinita et supra naturam, aliquo modo fuerit subiectum dispositioni coelesti: sed solum quod illud fuerat in signum (267, 24–34).

Roger now brings this rich section to a close by returning to his broader argument of astrology's support for religious faith, once again drawing on earlier themes:

> Nevertheless, in so far as the most beautiful virgin (*virgo mundissima*) was the true and natural mother of our lord Jesus Christ, and she was made (*operata*) from a natural power in preparing the matter (*ex virtute naturali in praeparando materiam*), and in nurturing it after conception (*in fovendo post conceptionem*), and in matters of this sort, they proposed that the power of the heavens had cooperated with the natural power of the Virgin (*virtutem coeli cooperatam fuisse virtuti naturali Virginis gloriosae*), and that it [sc. the heavens] had excited her in so far as it acted naturally (*excitasse eam in quantum naturaliter operabatur*), because a man and the sun generate a man (*homo generat hominem et sol*).[92]

Thus, with respect to his human side, celestial influences worked together with the natural—but specially prepared—material substrate to form and influence Mary in her role as Jesus's mother. In this way, her natural parts worked together with the sun (and the other planets) in Jesus's generation and growth, in accordance with the same Aristotelian dictum encountered in Chap. 1.

Roger continues articulating the deeper structures of Mary's giving birth:

> For if there was something natural in that conception by the preparation of the matter and the kindling (*fomentum*) in the womb, and of this sort of thing, insofar as she was his natural and true mother, they do not judge it unfitting to assume that the celestial disposition is more than a sign, when things are considered as purely natural. But whatever they say in this part is to be referred back to the rule of faith, so that it is not discordant with Catholic truth. And although all these things do not fully suffice to show the secrets of this religion (*secreta istius sectae*), nevertheless, whether this religion exists and what sort (*qualis*) it is in general (*in universali*) are attested beautifully (*pulchre attestantur*), so that sufficiently admiring the wisdom (*sapientia*) given to them, we easily excuse their ignorance because they have fallen off from a full certification of the Christian rite (*ritus Christiani*), since they were not instructed in it.[93] And we ought to praise them [sc. the Arabic astrologers] because they agree with us and confirm our profession of faith.[94] But in the third part of moral philosophy there will be a fuller discourse about this confirmation.[95]

---

[92] In quantum tamen Virgo mundissima fuit vera mater et naturalis Domini Iesu Christi, et operata fuerit ex virtute naturali in praeparando materiam et in fovendo post conceptionem et in huiusmodi posuerunt virtutem coeli cooperatam fuisse virtuti naturali Virginis gloriosae, et excitasse eam in quantum naturaliter operabatur, quia homo generat hominem et sol (267, 34–268, 2).

[93] Roger seems to refer here to Albumasar and perhaps other Arabic astrologers, as well as to the pseudo-Ovid of the *De vetula*.

[94] We saw this view also in Chap. 4 in discussing *Speculum astronomiae* 12 in relation to Albumasar's understanding of the ninth house.

[95] Si enim aliquid naturale fuerit in illa conceptione per materiae praeparationem et fomentum in utero et huiusmodi, in quantum fuerate naturalis et vera mater, non aestimant inconveniens ponere coelestem dispositionis esse plus quam signum, consideratis pure naturalibus. Sed quicquid dicunt in hac parte, hoc ad regulam fidei reducendum est, ut a catholica veritate non discordet. Et licet omnia et ad plenum non sufficient ostendere secreta istius sectae, tamen an sit haec secta, et qualis sit in universali, pulchre attestantur, ut satis admirantes sapientiam eis datam facile excusemus eorum ignorantiam, quia defecerunt a plena certificatione ritus Christiani, cum in eo non fuerant instructi. Et laudare debemus, quod nobiscum concordant et confirmant nostram professionem. Sed in tertia parte moralis philosophiae de hac confirmatione abundantior fiet sermo (268, 2–16).

Astrology thus helps to clarify the mysteries of the Christian faith, however imperfectly, at least in some more general respects. Nevertheless, at a delicate point of tension between astrology and theology/religion, Roger steps back from pushing the astrological analysis any further, falling back instead on the rule of faith as the ultimate bulwark. Regardless, astrology beautifully reveals and clarifies the general nature of the Christian religion, including the mystery of the virgin birth and Mary's nature as Jesus's mother.

*

Roger then returns to historical astrology and the reign of Antichrist:

> And since after the law of Mohammed we do not believe that any religion will come except the Law of Antichrist—and the astrologers (*astronomi*) likewise agree on this, that there will be someone powerful who will establish a filthy and magical law (*lex foeda et magica*) after Mohammed, which law will destroy all others—it would be very useful to the Church of God to consider the timing of this law, whether it will come quickly after the destruction of Mohammed's law or much later.[96]

Can we determine the timing of Antichrist's advent, and, in particular, how long it will arrive after the destruction of Islam? Developing an earlier argument, this is a fundamentally important question for Roger that illustrates precisely where astrology can potentially be extremely useful to religion. Cardinal Pierre d'Ailly was duly exercised 150 years later by these very same questions, and precisely by reading these Baconian texts.[97]

Finally, Roger turns to contemporary evidence for Islam's imminent demise:

> And Ethicus the philosopher in his *Cosmographia* says explicitly that the people (*gens*) who were closed within the Caspian gates will rush out into the world and run up to Antichrist and call him the god of gods (*deus deorum*). And doubtless the Tartars were within these gates and they left, for now the gates have been broken, as we are certain. For the Friars Minor [sc. Franciscans], who the currently ruling Lord King of France [Louis IX] sent, went across with the Tartars through the middle of the gates far beyond and between the mountains, where they were enclosed. And not only to all nations of the East is it known that the Tartars left these places, but it is also known to those who know well the disposition of the world, and have known the habitable parts, and the diversities of regions by means of the science of the stars (*astronomia*), and through other authorities, such as Pliny, Martianus [Capella] and others who describe the regions of the world, and through histories. I do not here wish to place my mouth in the heavens (*ponere os meum in coelum*), but I know that if the Church wishes to return to the sacred text and sacred prophecies, and the prophecies of the Sybils, Merlin, Aquila, Sextus, Joachim [of Fiore] and many others, in addition to the histories and books of the philosophers, and would order that the pathways of the science

[96] Et quoniam post legem Mahometi non credimus quod aliqua secta veniet nisi lex Antichristi, et astronomi similiter concordant in hoc, quod erit aliquis potens qui legem foedam et magicam constituet post Mahometum, quae lex suspendet omnes alias, multum esset utile ecclesiae Dei considerare de tempore istius legis, an cito veniet post destructionem legis Mahometi, an multum longe (268, 17–23).

[97] See Smoller, *History, Prophecy and the Stars, passim.*

> of the stars (*vias astronomiae*) be considered, it [sc. the Church] would find sufficient suspicion or greater certitude about the time of Antichrist![98]

Thus Roger ends this extensive section on astrology's benefits for theology/religion, concluding that astrology may be used with great profit to clarify prophecies of many types—from the Sybils to Merlin—but here he is primarily concerned with significant questions of timing.[99]

6.5 Conclusion of Chap. 6 and Part 2

In order to strengthen, confirm and certify the faith, Roger Bacon used several major elements of astrological practice—including dignities and great conjunctions as well as nativities and revolutions—to build complex, cumulative and deeply astrological arguments towards this end. As mentioned, Pierre d'Ailly was deeply influenced by Bacon's arguments at the beginning of the fifteenth century in this same religio-apocalyptic framework. On the other hand, Pico found both Bacon's and d'Ailly's analysis deeply problematic, primarily because it seemed to subordinate religion to astrology, even though Roger explicitly denied such a conclusion due to God's providential creation, ordering and governance of the world in relation to his absolute power, including His ability to use the heavens as signs as well as causes.

We should also note that Roger's discussion so far concerning astrology's utility in human affairs (Chap. 4) and in theology/religion (here in Chap. 6) has revolved entirely around the judicial or interpretive part of astrological practice, that is, the "knowledge" dimension. What we call "magic," namely, the use of talismans and powerful words, has not yet occurred in the texts examined here. The only time what Roger calls "magic" has arisen so far was in the context of describing false mathematics and Antichrist's advent, and involved conjurations, charms, incoherent mumblings and demons, but no talismans. The magic in this discussion was entirely

---

[98] Et Ethicus philosophus in sua Cosmographia dicit expresse, quod gens quae fuit clausa infra portas Caspias irruet in mundum et obviabit Antichristo et eum vocabit Deum Deorum. Et proculdubio Tartari fuerunt infra portas illas et exiverunt. Iam enim fractae sunt portae, sicut certi sumus. Nam fratres minores, quos dominus rex Franciae Ludovicus qui nunc regnat misit, transiverunt cum Tartaris per medium portarum ultra longe inter montes, ubi fuerunt inclusi. Et notum est non solum omnibus nationibus orientis quod Tartari exiverunt a locis eis, sed et eis qui bene sciunt mundi dispositionem, et noverunt partes habitabiles et regionum diversitates per astronomiam, et per auctores alios ut Plinium et Martianum et caeteros qui mundi regiones describunt, et per historias. Nolo hic ponere os meum in coelum, sed scio quod si ecclesia vellet revolvere textum sacrum et prophetias sacras, atque prophetias Sibyllae, et Merlini et Aquilae, et Sestonis, Ioachim et multorum aliorum, insuper historias et libros philosophorum, atque iuberet considerari vias astronomiae, inveniretur sufficiens suspicio vel magis certitudo de tempore Antichristi (268, 23–269, 6).

[99] Regarding astrology and prophecy, Jonathan Green treats one of the most famous examples in Johannes Lichtenberger's late fifteenth century *Prognosticatio* (1488); *Printing and Prophecy: Prognostication and Media Change 1450–1550*, Ann Arbor: The University of Michigan Press, 2012. He also discusses the Sybils.

negative, being either fraudulent or demonic. By contrast, in part 3 I will discuss Roger's views on the *operative* part of astrology, including astrological images or talismans, thus instantiating Roger's distinction between the two parts of *astronomia—iudiciaria* and *operativa*, including the works and words of wisdom (*opera et verba sapientiae*)—that we found in his edition of the *Secretum secretorum.*

A further contrast in astrology's relationship to theology/religion that we have discovered between Roger Bacon and Thomas Aquinas is that Roger is much more overtly astrological in his argumentation, using practical astrological techniques and structures in detail to support and promote his religious views. For Thomas, on the other hand, although he used fundamental structures of astrologizing Aristotelian natural philosophy to articulate his view of divine providence—and he used Ptolemy's two astrological texts as his primary authorities—he did not in fact make explicitly astrological arguments in his *Summa contra gentiles*. He did, however, explicitly note that astrology could be practiced legitimately in the *De iudiciis astrorum*, and that astrology could legitimately be used to predict the future in *Summa theologiae* II.II.92–95, although both he and Roger strongly rejected what they both explicitly call "divination." The relationship of their views on what we call magic is also complex, as we will now see. In addition, we saw that Albertus Magnus's analysis of fate was also deeply astrological, and further refined the formal dimension of his natural philosophical foundations for nativities as reconstructed in Chap. 1.

*

We have now investigated some deep structures in astrology's relationship to theology by exploring Albertus Magnus's and Thomas Aquinas's analyses of fate, divination and providence in Chap. 5. There we found a more philosophical approach, and one that ultimately provided astrology's theological foundations in Thomas's views of divine providence. By contrast, we also examined a much more practically-oriented astrological analysis in Chap. 6 with Roger Bacon. Despite these differences, we have encountered many similar themes treated with similar terminology that collectively reveal a fundamentally shared discourse with shared language and concerns, and with related conceptual and disciplinary structures. We will now segue from knowledge concerns to those of power/action, or, as Roger Bacon would have it, from *astronomia iudiciaria* to *astronomia operativa*, as we move from Part 2 to Part 3.

Part III
Conceptual Structures (3): Astrology and Magic

Introduction

Astrology and Magic: Introduction to Magic in the Middle Ages

After discussing some central features of astrology's relationship to natural knowledge/science in Part I and to theology/religion in Part II and thereby describing their contours more precisely, in Part III I will now discuss some interesting features of astrology's relationship to what we call magic, and especially to the richly significant issue of astrological images or talismans.[1] Unlike with natural philosophy and theology, however, astrology does not seem to have had "magical foundations," whatever that phrase might mean. Rather, astrology and its natural philosophical foundations seem in their turn to have provided magic—and especially *imagines astronomicae* or talismans—with *their* natural philosophical foundations as well as whatever legimacy they may have acquired, at least with some philosopher-theologians, as we will see.

*

To begin, we should first make a basic distinction in the broader field that we can reasonably call magic in the Middle Ages (or medieval magic).[2] Following the

[1] I will use both terms more or less equivalently in what follows. For an encyclopedic treatment of talismans in the Middle Ages, see Weill-Parot, *Les "images astrologique," passim.* In *Influences*, Mary Quinlan-McGrath has recently expanded and enriched our understanding of this phrase to also include works of art and architecture. She focuses on the Renaissance. Although it is unclear to what extent her arguments are relevant for thirteenth-century artistic practices, it is certainly a question worth pursuing, although I will not do so here.

[2] Wouter J. Hanegraaff suggests that we jettison the term "magic" entirely, but I think it still serves a useful purpose. He discusses this in Chap. 3 of his *Esotericism and the Academy*. Chap. 3 is entitled, "The Error of History: Imagining the Occult." He addresses the term magic in a section entitled, "Tainted Terminologies 2: Magic," 164–77. The first tainted term is "superstition," 156–64. The third of three is "occult" (177–91). I will critically discuss his arguments, especially on the use of the term "magic," in the overall introduction to volume II.

Speculum astronomiae in the thirteenth century and Frank Klaassen in the 21st, we can divide the field of surviving texts of "illicit learned magic" (in Klaassen's terminology) into [1] *imagines astronomicae* or talismans ("astrological image magic," in Klaassen's phrasing), a small percentage of which some considered to be legitimate because they dealt (or claimed to deal) only in all-natural celestial influences, and [2] ceremonial or ritual magic (or necromancy), all of which were deemed illegitimate because these practices were powered by and often explicitly invoked and addressed demons.[3] In the numerous medieval manuscripts that Klaassen has examined, the image magic texts are almost always found to be associated with texts concerned with natural knowledge, whereas the ritual or ceremonial texts almost never are. Thus, image magic inflects more strongly towards natural philosophy/science, and ritual magic more towards religion.[4]

*

In the *Speculum astronomiae*, astrological images (both legitimate and illegitimate) are subsumed under the rubric of astrological elections that are concerned, we will recall, with the choosing of astrologically propitious times. There they are subdivided into legitimate (natural) and illegitimate (demonic) types of texts and practices. For Roger Bacon, on the other hand, the primary distinction is between true (natural) and false (magical, and thus either deterministic or demonic) mathematics in relation to his distinction between *astronomia iudiciaria* and *operativa*, with the latter concerned with what we would call magic, but which Roger emphatically did not. These include the *opera et verba sapientiae*, with talismans representing the *opera*. These are some initial distinctions that will be refined throughout the following two chapters. My primary concern will be with image magic or talismans, thereby sharpening our understanding of the relationship between astrology and magic. I will also treat powerful words in Roger Bacon as derived, primarily, from al-Kindi's *De radiis stellarum*.

We will begin Chap. 7 by exploring the influential *Speculum astronomiae*, whose treatment of talismans I will then compare with relevant material in Albertus Magnus's authentic works and then with Thomas Aquinas. Once again, the patterns—both their harmonies and tensions—will provide the structural foundations for later views on the subject, as we will see in volumes II and III. Natural causal

---

[3] Within ritual magic, Klaassen includes "demon conjuring, angel magic and theurgic arts, such as the *Ars Notoria*, that emphasized the mechanisms of religious rites, dreams and visions (3)." "[T]hese texts involve the explicit binding, invoking, and employing of demons, the ritual engagement of angels or the Holy Spirit, the use of extensive Christian rituals, a heavy emphasis on the affective state of the operator, and a desire to engage directly with the divine or numinous (57)." Klaassen treats these in his Part II; *The Transformations of Magic*.

[4] *Transformations of Magic*, 1–6. Klaassen discusses image magic in the Middle Ages in Part I. See also Kieckhefer, *Magic in the Middle Ages*, which, although valuable as a synthesis, strikes me as conceptually muddy; and Vescovini, *Medioevo Magico*. Sophie Page also treats image magic in her book, *Magic in the Cloister: Pious Motives, Illicit Interests, and Occult Approaches to the Medieval Universe*, University Park, PA: The Pennsylvania State University Press, 2013, 73–92.

analysis and demonic intervention vis-à-vis legitimacy will continue to be central themes encountered here. In Part III, we are moving from astrology considered primarily as knowledge-oriented (*iudiciaria*) to its operative or magical side, to use Roger Bacon's valuable distinction (although he would vehemently reject the second part of my terminology). We will explore Roger's conceptual patterns and related terminology more fully in Chap. 8. I focus on this subset of magic precisely because it is normally configured as a controversial part of astrological elections, as we find it in the deeply influential *Speculum astronomiae*.

Chapter 7
Imagines astronomicae (talismans) in the *Speculum astronomiae*, Albertus Magnus and Thomas Aquinas

7.1 The *Speculum astronomiae* and the Problem of Magical Images

As we saw in Chaps. 1 and 4, the *Speculum astronomiae* was composed primarily to identify and describe legitimate astrological texts and practices and to defend them from theological criticisms and concerns, in part by articulating clear boundaries and criteria for distinguishing legitimate from illegitimate practices, and to indicate likewise which books are legitimate and which are not. To get there, the deliberately anonymous author first articulated the four canonical types of practical astrology—general astrology or revolutions, nativities, interrogations and elections—in the first part of the text, to which he added extensive bibliographies (Chaps. 1–11). In the second part (12–17), he discussed controversial areas as *quaestiones* to be investigated and explored, in particular, the more problematic areas where astrology touched on theologically and morally sensitive issues, such as determinism in nature, human free will, and demons (as discussed in Chap. 4). He offered his opinions, but not as authoritative determinations. This explicitly and very effectively anonymous work was very influential indeed, as we will see.

In setting out astrology's basic structures, the *Magister Speculi* distinguished the two great wisdoms both called *astronomia*, namely, what we call "astronomy" and "astrology," which study, respectively, the motions of the planets and their influence on the world (Chaps. 1, 2 and 3), as we also saw with the authentic Albertus Magnus, Thomas Aquinas and Roger Bacon. Most of the *Speculum astronomiae*'s first part (Chaps. 4–11) characterized the four types of astrological practice. The most controversial practice for the *Magister Speculi* was elections, because *imagines astro-*

H D. Rutkin, *Sapientia Astrologica: Astrology, Magic and Natural Knowledge, ca. 1250–1800*, Archimedes 55, https://doi.org/10.1007/978-3-030-10779-6_7

nomicae—magical objects, often called "talismans"[1]—were subsumed under this rubric, which he discussed at some length. As it turns out, the Magister Speculi called talismans the highest part of *astronomia* (*sublimitas astronomiae est imaginum scientia*) in Chap. 4 (IV.8). This quotation comes from Thabit ibn Qurra's *De imaginibus* as cited and quoted in Zambelli's useful appendix of sources.[2]

I will focus on astrological images (*imagines astronomicae*) or talismans in this and the following chapter. The discussion here provided authoritative support over several centuries for astrologically informed magical practices, which the author here considers perfectly legitimate with respect to theological concerns, at least for one particular all-natural type of talisman. This authoritative support also arose in part because the *Speculum astronomiae* was often later attributed to Albertus Magnus.

For the anonymous author, certain astrological images are kosher, if you will, because they act naturally and, in particular, *not* through demonic intervention (themes to be developed below). He sharply contrasts these natural and thus legitimate *imagines astronomicae* with what he explicitly calls necromantic talismans (*imagines necromanticae*), which he considers to be either abominable or detestable, and whose manufacture and use he emphatically rejects.[3] After treating talismans in the *Speculum astronomiae*, I will compare the analysis there with discussions of magic in general and talismans in particular in Albert's undoubtedly authentic writings, primarily *De mineralibus*. These I will contrast with Thomas Aquinas's equally authoritative but strikingly different views. I will explore Roger Bacon's views on relevant issues in Chap. 8.[4]

---

[1] Don Skemer usefully discusses the respective etymologies of amulets and talismans in his extremely informative introductory, "Note on Terminology"; *Binding Words: Textual Amulets in the Middle Ages*, University Park, PA: The Pennsylvania State University Press, 2006, 6–19, esp. 6–9 (with further bibliography). Brian Copenhaver too offers a useful distinction between amulets and talismans; "Scholastic Philosophy and Renaissance Magic," 530.

[2] Speculum astronomiae *and its Enigma*, 282. Klaassen treats the *Speculum astronomiae* extensively in *The Transformations of Magic*, as does Weill-Parot, *Les "images astrologiques*." Thebit ben Corat (Thabit ibn Qurra) was a famous astronomer, mathematician, physician and magician, as well as translator from the Greek, who was born in Harran in 826 and died in Baghdad in 901. Only a small part of his large oeuvre was translated into Latin (Hasse, *Success and Suppression*, 405–6).

[3] According to Burnett, although "necromancy" etymologically and historically means divination upon bringing a dead person back to life, it is to be contrasted with "nigromancy," which became the normal term to translate 'sihr', the Arabic term for magic; "Talismans: Magic as Science? Necromancy among the Seven Liberal Arts," in his *Magic and Divination in the Middle Ages: Texts and Techniques in the Islamic and Christian Worlds*, Aldershot: Ashgate, 1996, 1–13, 3. Drawing on Jean-Patrice Boudet's discussion in "La genèse médiévale de la chasse aux sorcières: Jalons en vue d'une relecture" (in *Le mal et le diable: Leurs figures à la fin du Moyen Age*, N. Nabert (ed.), Paris: Beauchesne, 1996, 35–52, but especially 38), Weill-Parot (32, n. 22) defines nigromancy as "all magical practices grounded on the invocation of demons." Boudet also discusses this distinction in *Entre science et nigromance*, 92–94. I simply follow the usage of the primary texts discussed here, which all seem to refer to the same body of practices, none of which are "necromancy" in the strict sense. There is an increasing body of excellent scholarship in this murky domain, including works by Thorndike, Pingree, Burnett, Kieckhefer, Vescovini, Weill-Parot, Boudet, Page and Klaassen, some of which will be cited below.

[4] I discuss Albert's views also in my chapter, "Astrology and Magic," in *A Companion to Albert the Great*, 451–505, on which this chapter is partially based.

Caveat lector: We are about to enter the equivalent of a conceptual swampland—fetid, dank, and dark—and with very little *terra firma* on which to build solid conceptual structures. The flashlights, and the mosquito netting to protect us from the conceptual malaria that most investigators fall prey to—namely, a belief in "the occult sciences" as a useful analytic category—will be a close attention to terminology. Comparatively speaking, astrology is the firmest of conceptual *terrae*. Thus, we will use a well defined understanding of astrology and its conceptual domain to shore up our understanding of where astrology borders on (and in part overlaps with) magic and necromancy, and their familiar denizens, namely, demons.[5]

*

Elections are discussed in Chaps. 10 and 11 of the *Speculum astronomiae*; relevant issues are also treated in Chaps. 16 and 17. Chapter 10 is very brief (238–9), describing the practice of elections as the choosing of propitious times to begin any sort of venture, which are then related to both nativities and interrogations, as we saw in the Excursus. Chapter 11, on the other hand, is very long (240–50)[6] and addresses the issue of *imagines* with respect to astrology, which are translated indifferently as "astrological images" or "talismans" in what follows.[7] As it turns out, the *Magister Speculi* seems to have coined the phrase '*imago astronomica*' with this text.[8]

Chapter 11 begins by stating that the science of talismans is subordinated to the part of practical astrology on elections, and that talismans can be made in three ways, which he discusses in turn.[9] The first kind is abominable (*abominabilis*), and requires the use of incense (*suffumigationes*) and prayer (*invocatio*), including the names of demons, prayers to Venus, etc. These sorts of improper rites and practices he calls the worst sort of idolatry, that is, exhibiting to a creature (i.e. something created, whether planet or demon) the honor owed to the Creator, to whom alone all

[5] For an extraordinarily interesting study with much related material and reflection, see Stuart Clark, *Thinking with Demons: The Idea of Witchcraft in Early Modern Europe,* New York: Oxford University Press, 1997. My main criticism of his work, which I discuss more fully in volume II, is his complete reliance on D.P. Walker's classic but now significantly outdated *Spiritual and Demonic Magic from Ficino to Campanella* (1958) for his understanding of "Renaissance magic." It is equally striking that Westman also relies primarily on Walker and Frances Yates for his understanding of Marsilio Ficino in his *Copernican Question*, despite the fact that there has been much valuable scholarship over the last 50 plus years.

[6] With five pages of Latin and five of translation in this edition with facing text.

[7] '*Imagines*' in this context can almost always be translated as "talismans," as Burnett does consistently in "Talismans." Likewise, Weill-Parot uses them synomymously: "'images' ou talismans (33)," as will I. Quinlan-McGrath has valuably complicated this; *Influences*. She always simply and neutrally translates them as "astronomical images," as does the translation in Zambelli, Speculum astronomiae *and its Enigma*.

[8] For an extensive and persuasive argument for this claim, see Weill-Parot, *Les "images astrologiques*," 32–38.

[9] For a close reading of Chap. 11, see Weill-Parot, *Les* "images astrologiques," 34–38.

such honor is due.[10] This resonates closely with Thomas Aquinas's penetrating and influential analysis of idolatry in relation to superstition, divination and astrology in *Summa Theologiae* IIª IIªᵉ 92–95, as we saw in Chap. 5.

Astrological images of the second type are slightly less problematic, but they are still detestable (*detestabilis*), and should therefore be rejected. These concern the writing of characters (*inscriptio characterum*), and they are suspect because the content hidden in an unknown language could surreptitiously harm the Catholic faith.[11] Both the first and second types use the language of exorcism, and the Magister explicitly distances himself from both types of practices: "These are two modes of necromantic talismans (*imagines necromanticae*), which (as I have said) presume to usurp to themselves the noble name of *astronomia* (11, 34–35)."[12]

Before offering bibliography here, the author describes the intense distress he felt on reading these works, which excuse he used to justify his inadequate recall of their bibliographical information. Nevertheless, he still lists three full pages of titles, authors and incipits (11, 34–102).[13] The anonymous author of the *Speculum astronomiae* thus erects these three safeguards, namely, that legitimate talismans not use incense, prayer/invocation, or the inscription of characters.[14] We should also note that the Magister Speculi's distinction between abominable and detestable talismans conceptually corresponds to—but has a different terminology than—Thomas's parallel distinction between explicit and implicit pacts with demons, as we will see below.

I will now more closely examine the third, legitimate type of talismans:

> The third type is that of astrological images (*imagines astronomicae* [sc. as opposed to *imagines necromanticae*]), which eliminates this filth, does not have suffumigations or invocations, and does not allow exorcisms or the inscription of characters, but obtains its power (*virtus*) solely (*solummodo*) from the celestial figure (*a celesti figura*).[15]

For the Magister Speculi, legitimate astrological images have no demonic taint, overt or tacit. Rather, they explicitly derive their power naturally—and

---

[10] A nobis longe sit iste modus: absit enim ut exhibeamus creaturae honorem debitum creatori (11, 19–20).

[11] In *Spiritual and Demonic Magic from Ficino to Campanella,* Walker makes the point that such characters could only be understood by an intelligent being, hence they are to be altogether avoided, in order to rule out the possibility that the intelligences addressed were demons (48, 80). Weill-Parot discusses this in detail as "addressative magic," a neologism that he himself coined, in "Astral Magic and Intellectual Changes (Twelfth-Fifteenth Centuries): 'Astrological Images' and the Concept of 'Addressative' Magic," in *The Metamorphosis of Magic from Late Antiquity to the Early Modern Period*, Jan Bremmer and Jan R. Veenstra (eds), Louvain: Peeters, 2002, 167–87.

[12] Isti sunt duo modi imaginum necromanticarum, quae nobile nomen astronomiae (sicut dixi) sibi usurpare praesument[.]

[13] Weill-Parot describes the texts mentioned by the *Magister Speculi*, and how such abominable and detestable talismans were made and towards what ends (*Les "images astrologiques,"* 41–60). There is also much talk of exorcism in these texts.

[14] We will see in volume II that Marsilio Ficino flouts all three of these safeguards in *De vita* III.

[15] Tertius enim modus est imaginum astronomicarum, qui eliminat istas spurcitias, suffumigationes et invocationes non habet, neque exorcizantes aut characterum inscriptiones admittit, sed virtutem nanciscitur solummodo a figura celesti (11, 103–106) [.]

solely[16]—from the celestial realm.[17] Examples follow. The first aims to eliminate an unwanted species from a place by a magical form of pest control: "For example, if there were a talisman (*imago*) for eliminating some species from some place"—the example he gives is for the eradication of scorpions—"about which we might be asked (11, 106–8)."[18] This is legitimate because the talisman's power is natural, deriving from the heavens, and not from demons. The emphasis on the celestial figure should be duly noted.

The Magister Speculi then explains how the talisman should be made, that is, under what astrological circumstances. For example, under the same ascendant (rising sign) as that which rules the species (in this case, Scorpio), under the ascendant of a relevant interrogation if those signs indicate removal,[19] or under other baleful aspects. After it has been made—and other undisclosed conditions observed—the talisman should be buried in the middle of the place from which the species in question is to be banished, and its stomach filled with dirt from the place's four quarters (11, 108–121).

The second example concerns love and money, issues of perennial concern:

> But if the talisman (*imago*) is to be made in order to attain love and profit, let it be made according to the opposite way to what I have said [i.e. from the instructions above, that is, not to repel something but to attract it], with the addition that its shape is to be engraved under an elected hour (*forma eius sculpenda est sub hora electa*); and it will have an effect from celestial power (*a virtute caelesti*) by God's command (*iussu Dei*) because the images found in this sensible world [sc. made] from the four elements (*imagines quae inveniuntur in hoc mundo sensibili ex quatuor elementis*) obey the celestial images (*oboediunt caelestibus imaginibus*, 11, 121–26).[20]

For bibliography on these legitimate talismans, only one book is listed, Thabit ibn Qurra's *De imaginibus*.[21] I should also note that we are told here that things on

---

[16] Weill-Parot (*Les "images astrologiques,"* 37) rightly emphasizes the importance of the adverb '*solummodo*' (solely) here. Astrological images are *all*-natural and thus legitimate.

[17] This precise distinction also occurs in the *Ut testatur Ergaphalau*, the text of which is edited as an appendix in Burnett, "Adelard, Ergaphalau and the Science of the Stars," also published in his *Magic and Divination*, 2 (133–45, originally published, 1987): "Dividitur autem imaginaria [*scientia*, sc. the science of images] in puram et exorcismalem. Pura est que sine incantationibus et exorcismis imagines fundi, inspecto statu tantummodo supernorum [*superiorum* is perhaps a superior variant reading] docet. Exorcismalis est que exorcismis et incantationibus imaginibus cooperandis spiritus [= demon] includere docet (144)."

[18] [U]t si fuerit imago destructionis alicuius speciei ab aliquo loco, de qua scilicet fuerimus requisiti [...].

[19] This seems to refer to an astrological interrogation made to discover the success of such an extermination. If the horoscope constructed for the time of the interrogation were propitious, the talisman could then be made under the influence of *its* rising sign.

[20] Si vero fuerit imago cuius opere quaeritur dilectio et profectus, fiat e contrario horum quae dixi, addito quod forma eius sculpenda est sub hora electa et habebit effectum iussu Dei a virtute caelesti, eo quod imagines quae inveniuntur in hoc mundo sensibili ex quatuor elementis, oboediunt caelestibus imaginibus [...] (11, 121–26)[.]

[21] For an inadequate Latin text of this work, see Francis M. Carmody, *The Astronomical Works of Thabit b. Qurra*, Berkeley: University of California Press, 1960. Weill-Parot (pp. 63–77) discusses

earth—images found in this sensible world—obey celestial images, but the *Speculum* offers no natural philosophical account of the sort we find in the authentic Albert's *De mineralibus*, to be discussed below. The concern with astrological timing in their making justifies talismans being included within the astrological practice of elections.

Imagines astronomicae, then, are legitimate and operate in some underdescribed manner by means of celestial powers derived all and only from celestial figures. In terms of Roger Bacon's distinction between *astronomia iudiciaria* and *operativa*, we have now moved from the realm of knowledge *per se* to that used for acting in and on the world, that is, from a more gnostic or prognostic (knowledge-based) use of astrology to a more operational or technological use, which we may also call magical.[22] We can see from Burnett's discussion that talismans were sometimes considered the most exalted part of astrology precisely because they were the most practical, that is, the most able to affect changes in the world.[23]

*

The Magister Speculi addresses these controversial issues further in chapters 16 and 17, the last two chapters in the text. Chapter 16 is short and worth quoting in full, since it affirms the basic structures as it offers more detail:

> I do not defend that section concerning *imagines astronomicae* on account of the nearness they have to necromantic images, beyond what is said above in the chapter devoted to them [ch. 11], namely, that they take their power from a celestial figure (*a figura caelesti*) according to the ninth *verbum* of Ptolemy [sc. in the *Centiloquium*] which is touched upon there (that is, that "Images which are, etc.").[24] And [sc. I would not defend them] unless [sc. it were the case (as it is)] that nothing prohibits one from defending them in accordance with what can be denied or defended. So, let it be that when a talisman should be cast (*fundatur imago*) with the conditions mentioned previously for expelling scorpions from some place, if God should wish it, it does not appear [1] to be an exorcism or an invocation if it is said

this text in detail and the Magister Speculi's use of it. Burnett ("Talismans," 9) discusses a closely related text, Adelard of Bath's translation of the *Liber praestigiorum Thebidis secundum Ptolomeum et Hermetem* (133–34). Weill-Parot (86–88) discusses their differences. Boudet also discusses this treatise in his "Un traité de magie astrale Arabo-Latin: Le *Liber de imaginibus* du Pseudo-Ptolémée," in *Natura, scienze e società medievali: Studi in onore di Agostino Paravicini Bagliani*, Claudio Leonardi and Francesco Santi (eds.), Florence: SISMEL—Galluzzo, 2008, 17–35.

[22] A related distinction was made in the mid-12th century in an anonymous introduction to the science of the stars edited by Charles Burnett in "A New Source for Dominicus Gundissalinus's Account of the Science of the Stars," *Annals of Science* 47 (1990): 361–74: "Contemplativa [sc. astronomia] est que habet in se solo cursu planetarum [...]. Activa est que habet se in opere sigillorum et imaginum."

[23] "Talismans," 8–13, especially 13. Thabit also expresses this sentiment in version 1: "Sublimitas autem et altitudo astronomie est imaginum scientia," a phrase quoted word for word in Chap. 4 of the *Speculum astronomiae*, as we saw; Weill-Parot, Les "images astrologiques," 65 (nn. 125–26). Kieckhefer awkwardly handles this distinction in discussing talismans in *Magic in the Middle Ages* (132), where he follows Frances Yates's decision to call talismans a form of "astral magic" as opposed to "astrological magic," because for both Yates and Kieckhefer, astrology only has a knowledge dimension, not an operative one.

[24] Weill-Parot (*Les "images astrologiques"*) discusses this text at 80–83.

> during its casting: "This is a talisman for the destruction of scorpions from that place as long as the image is preserved in it." Nor, again, does it seem [2] to be an inscription of characters if the word "Destruction" (*destructio*) is engraved on its back [sc. any more than] if the word "Love" (*amor*) is written on the heart and on the back on talismans for love (*in imaginibus ad amorem*). Nor if on its forehead the word "Scorpion" (*Scorpius*) were inscribed (that is, the name of the species to be banished) and the name of the ascendant or the name of its [sc. planetary] Lord (which is Mars) or the name of the moon were written on its breast. Again, what cult [= worship, religious practice] is shown by this, if in the middle of the place from which you want some species banished, the talisman were buried with its head facing down and its feet turned upwards? Not that I recommend them (*non commendo eas*),[25] but there is no reason for these to carry the iniquity of the other [sc. type of talismans, namely, necromantic].[26]

Useful astrological images of this sort are perfectly legitimate, the author argues, because the sorts of utterances and writing (i.e. powerful words) he permits do not cross the line into *invocatio* (prayer) or *inscriptio characterum*, a problematic practice with mysterious symbols, including letters in foreign languages.[27] The examples given are of perfectly straightforward Latin terms and phrases. Although these practices for pest control and love are legitimate, the author takes an explicitly neutral position: he describes them, to be sure, but neither commends nor condemns them. We should also recall from the excursus that it was normal practice to call the standard glyphs for the planets, signs and aspects "characters" as well, as we saw in Regiomontanus's *Almanach*. These, it seems, would be legitimate to use according to the *Speculum astronomiae*.

Chapter 17 discusses a range of necromantic practices, and also distinguishes legitimate from illegitimate forms of divination. Geomancy,[28] and chiromancy (as a

---

[25] This sounds similar to Ficino's apotropaic phrase '*narro non probo*' ("I describe, I do not approve") that he repeatedly states concerning talismans in *De vita* III.

[26] Partem vero quae est de imaginibus astronomicis propter vicinitatem quam habent ad necromanticas, non defendo aliter quam secundum quod superius in earum capitulo dictum est, eas nancisci virtutem a figura caelesti iuxta verbum Ptolemaei nonum, quod ibi tactum est, scilicet quod imagines quae sunt etc., et nisi quia nihil prohibet eas defendere secundum quod possunt negari vel defendi. Esto itaque exempli gratia, quod cum praedictis conditionibus fundatur imago ad scorpiones fugandos ab aliquo loco, si Deus voluerit, non videtur esse exorcismus aut invocatio, si dicatur in fusione illius "haec est imago destructionis scorpionum a loco illo quamdiu fuerit in eo imago servata." Non videtur iterum inscriptio esse characterum, si in dorso sculpatur hoc nomen: "Destructio"; sicut et in imaginibus ad amorem hoc nomen "Amor" scriberetur in ventre et in ante scilicet, neque si in fronte eius scribatur hoc nomen "Scorpius," quod est nomen speciei fugandae, et in pectore nomen ascendentis et nomen eius domini qui est Mars et nomen Lunae. Quis iterum cultus exhibetur ei si in medio loci, a quo ipsam speciem fugare volueris, fuerit imago sepulta capite deorsum et sursum pedibus elevatis? Non commendo eas, sed neque videtur quod absque ratione debeant aliarum iniquitatem portare (16.1–21).

[27] For later examples in the same tradition (with illustrations), see Richard Kieckhefer, *Forbidden Rites*, *passim.*

[28] Geomancy is a divinatory technique loosely related to astrology, but one that does not require knowledge of actual planetary configurations. Albert's and Thomas's Dominican confrere, William of Moerbeke, the great translator, composed a geomantic treatise, his only preserved original work. See Lorenzo Minio-Paluello's informative article in the *Dictionary of Scientific Biography*, 18 vols., New York: Scribner, 1970–90, 9: 434–40, and especially 435. For a penetrating study of

part of physiognomy),[29] seem to be acceptable, in contrast to aeromancy, pyromancy and hydromancy, that is, divination by means of air, fire and water, which have an idolatrous demeanor, apparently because there is no natural (= causal) way that these elements could legitimately signify anything concerning the future without the assistance of demons.

On the basis of such texts, we derive a more accurate sense of the late thirteenth-century map of knowledge, with an authoritative and influential delineation of which practices were considered legitimate and which not. Necromantic texts—including problematic talismans as well as the enumerated problematic divinatory practices—are idolatrous and/or demonic (whether implicitly or explicitly), and thus illegitimate for theological reasons. The four canonical types of practical astrology (including *imagines astronomicae*), on the other hand, are legitimate because they rely only on natural celestial powers and thus have no taint of idolatry.

For the Magister Speculi, "necromancy" is always a disparaging term (whether abominable or detestable) in the same way that "divination" is for Thomas Aquinas, and both "magic" and "divination" are for Roger Bacon, and for the same reasons, as I discussed in Chaps. 4 and 5, and as I will discuss further below and in the following chapter. Nevertheless, we will soon see that the Magister Speculi's clear terminological distinction does not hold for Albert's authentic *De mineralibus*, thus offering strong evidence against Albert's authorship thereof. We have also seen that, although the Magister Speculi claims that all-natural talismans are perfectly legitimate, he never offers a detailed natural philosophical analysis to show how this works, unlike the undoubtedly authentic texts by Albertus Magnus that we will now explore, and later texts by Marsilio Ficino to be analyzed in volume II.

7.2 Magic in Albertus Magnus's Authentic Writings

So much for the *Speculum astronomiae's* influential and authoritative treatment of astrological and necromantic images,[30] and several divinatory practices, most of which fall within modern historiographic notions of "magic" and are thus discussed

geomancy in the Middle Ages, see Thérèse Charmasson, *Recherches sur une technique divinatoire: La géomancie dans l'Occident médiéval*, Geneva: Droz, 1980. See now also *Geomancy and Other Forms of Divination*, Alessandro Palazzo and Irene Zavaterro (eds), (Micrologus Library, 87), Florence: SISMEL—Galluzzo, 2017.

[29] On the origins of chiromancy in the West, see Burnett, "The Earliest Chiromancy in the West," (X, originally published 1987), 189–195, and "Chiromancy: Supplement. The Principal Latin Texts on Chiromancy Extant in the Middle Ages," (X, 1–29), both in *Magic and Divination*. For much on physiognomy within the map of knowledge (where it forms part of the link between astrology and medicine), and as taught at the universities in the 13th and 14th centuries, including in the work of Albertus Magnus himself, see Jole Agrimi's groundbreaking studies collected in her *Ingeniosa scientia nature: Studi sulla fisiognomica medievale*, Florence: SISMEL—Galluzzo, 2002.

[30] Weill-Parot reconstructs the history of astrological images in great detail from its thirteenth-century inception (with some prehistory) throughout the fifteenth century, ending with Hieronymus

here, even though we will have noticed a striking lack of the term "magic" itself in the *Speculum astronomiae*. I now turn to Albertus Magnus's authentic views on magic, which, according to Thorndike, he did not discuss much at all.[31] He discusses talismans in *De fato*, *De natural loci* and, most extensively, in *De mineralibus*, on which I will focus.

In *De fato*, Albert explicitly refers to "magical" images-talismans (*imagines magicae*) while drawing parallels between art and nature, a theme he also treats in detail in *De mineralibus*.[32] In article 4 (*ad* 7), Albert discusses talismans:

> For, just as the radiation of the period (*radiatio periodi*) impresses the disposition of the order of existence and its duration on natural things (*dispositionem ordinis esse et durationis imprimit rebus naturalibus*), so it impresses it [sc. the *dispositio*] on man-made [sc. objects] (*artificiata*). Because of which, the figures of magical talismans (*figurae imaginum magicarum*) are taught to be made in relation to an aspect of the stars.[33]

Man-made magical talismans (*imagines magicae*)—which seem very similar indeed to the *Speculum astronomiae*'s *'imagines astronomicae'*, albeit with a slightly different terminology—are thus informed by the heavens with both formal and efficient causality just as natural things are.

Albert thus indicates that he is using precisely the same natural philosophical foundations for making talismans that we saw for astrological practice in Part I, and to articulate his deeply astrologizing understanding of fate as discussed in Chap. 5. This should be contrasted sharply with the *Speculum astronomiae,* which offers no

---

Torella, *Opus praeclarum de imaginibus astrologicis* (1496). Weill-Parot's critical edition of Torella's *Opus* will be discussed in the overall conclusion to volume II; Jérôme Torrella (Hieronymus Torrella), *Opus praeclarum de imaginibus astrologicis*, Nicolas Weill-Parot (ed.), Florence: SISMEL–Galluzzo, 2008.

[31] *HMES*, II: 555: "He nowhere in his commentaries on Aristotle or other works of natural science really stops and discusses magic at any length." Weill-Parot examines five texts where the authentic Albert discussed talismans. In addition to *De caelo et mundo, De mineralibus* and *De fato* (discussed here), he also discussed talismans in two theological works: the early commentary on Peter Lombard's *Sentences* (1246–49) and the late *Summa theologiae seu de mirabili scientia Dei* (after 1270). Weill-Parot discusses Albert's views at *Les "Images astrologiques,"* 260–78; he addresses and rejects Albert's possible authorship of the *Speculum astronomiae* at 278–80. He also compares Albert's position with Thomas's at 281–302. Boudet discusses the authentic Albert and the *Speculum astronomiae* together; *Entre science et nigromance*, 220–27. Henryk Anzulewicz also valuably discusses magic throughout Albert's writings in "Magie im Verständnis Alberts des Großen," in *Mots médiévaux offerts à Ruedi Imbach*, I. Atucha et al. (eds.), Porto: Fédération Internationale des Instituts d'Études Médiévales, 2011, 419–31, as does Sturlese, "Saints et magiciens." See also David J. Collins, "Albertus, *Magnus* or *Magus?* Magic, Natural Philosophy, and Religious Reform in the Late Middle Ages," *Renaissance Quarterly* 63 (2010): 1–44 for later attempts by Dominican's to come to terms with Albert's reputation as a magician.

[32] Weill-Parot discusses the fundamental theme of art vs. nature extensively; *Les "images astrologiques,"* 303–39. So does William R. Newman from the perspective of alchemy, in (e.g.) *Promethean Ambitions: Alchemy and the Quest to Perfect Nature,* Chicago: University of Chicago Press, 2004.

[33] [S]icut enim radiatio periodi dispositionem ordinis esse et durationis imprimit rebus naturalibus, ita imprimit artificiatis. Propter quod figurae imaginum magicarum ad aspectum stellarum fieri praecipiuntur (76, 39–43).

natural philosophical discussion at all beyond the assertion that legitimacy requires natural celestial influences alone, and that terrestrial images composed of the four elements somehow obey celestial figures, as we saw in discussing Chap. 11. For both the authentic Albert and the Magister Speculi, the figure is central.

Further, in a conceptually related passage of *De natura loci*, Albert refers to the practical benefits of astrological knowledge, as we saw before:

> Therefore, these things noted in general concerning place are similar, which, nevertheless, by accident from the closeness of mountains or their location, and the closeness of seas, are frequently impeded. They also sometimes receive an impediment by means of the handiwork of experienced men (*per artificium peritorum virorum*), who know the locations and powers of the stars (*qui noverunt situs et virtutes stellarum*) and impede their effects (*impediunt effectus eorum*), because, as Ptolemy says in the *Tetrabiblos*, the effects of the stars can be both impeded and expedited (*et impediri et expediri*) through the wisdom of men experienced in the stars (*per sapientiam peritorum virorum in astris*).[34]

Albert seems to allude here to talismans made in accordance with astrological timing, precisely as we found in the *De fato* and *Speculum astronomiae*, and in very similar language.

Thus, from the texts examined here, the authentic Albert and the author of the *Speculum astronomiae* seem to hold substantially similar views, although the authentic Albert never uses the terminology '*imago astronomica*', the phrase coined in the *Speculum astronomiae*. Albert does, however, call them '*imagines magicae*' in *De fato*. We have still only seen the term '*imago*' (whether astrological, magical or necromantic) used for these astrologically informed and celestially empowered magical objects with a range of practical uses. Furthermore, as noted, the authentic Albert also clearly indicates here his embrace of the astrologizing Aristotelian natural philosophical foundations for talismans, and his willingness to explicitly call these objects "magical." We will see this even moreso in the *De mineralibus*.

7.3 *De mineralibus*

According to Weill-Parot, *De mineralibus* (also completed by 1254) is the authentic Albert's most important treatise for understanding his views on talismans, and should thus be compared closely with the views expressed in the *Speculum astronomiae* in order to further assess its attribution to Albert.[35] I will be very selective here.[36] At the beginning of book II, tractate 3 on images and seals (sigils) in stone

[34] Haec igitur et similia sunt, quae notantur circa locum in communi, quae tamen per accidens ex vicinitate montium vel situ eorum et ex vicinitate marium frequenter impediuntur et aliquando recipiunt impedimentum per artificium peritorum virorum, qui noverunt situs et virtutes stellarum et impediunt effectus eorum, quia, sicut dicit PTOLEMAEUS in QUADRIPARTITO, stellarum effectus et impediri et expediri possunt per sapientiam peritorum virorum in astris (9, 38–46).

[35] Weill-Parot, *Les "images astrologiques,"* 268.

[36] For further discussion, see Wyckoff's translation with much explanatory material in Albertus Magnus, *Book of Minerals*, Dorothy Wyckoff (tr), Oxford: Clarendon, 1967, hereafter sometimes

(*De imaginibus autem lapidum et sigillis*), in defending why he is writing about such things—namely, because of the goodness (*bonitas*) of the doctrine (!) and the insistence of his associates—Albert locates these practices on the map of knowledge. He states explicitly that this part of the study of minerals is a part of necromancy (*pars necromantiae*), and concerns that species of necromancy which is subalternated to (= conceptually dependent on) the science of the stars (*quae astronomiae subalternatur*) and treats images/talismans and seals.[37]

Talismans and seals are thus located within the part of necromancy associated with the science of the stars, just as we saw in the *Speculum astronomiae*, where talismans were placed within the context of astrological elections. There, however, the Magister Speculi also distinguished sharply between two illegitimate types of talismans, both of which he explicitly called necromantic, and the one legitimate type, for which he coined the term '*imago astronomica*'. Thus, in *De mineralibus*, the undoubtedly authentic Albert used the term necromancy more broadly and without a negative connotation to include what he calls talismans and seals made of stone. There is, thus, a strikingly different terminology between that in Albert's authentic *De mineralibus* and the *Speculum astronomiae*, which does not bode well for arguments making Albert its author.[38]

To further articulate his position, Albert states immediately below that this subject cannot be fully known unless at the same time the science of the stars (*astronomia*), magic (*magica*) and the necromantic sciences (*necromantiae scientiae*) are also known, thus locating talismans and seals as somehow overlapping within these bodies of knowledge (II.3.1; W, 127).[39] Likewise, after an intensive chapter on the meanings (*significationes*) of talismans in stone (*De mineralibus* II.3.5, to be discussed just below), Albert reprises the point: "These matters cannot be proved from the principles of natural philosophy (*ex principiis physicis*): For this one should know *astronomia*, *magica* and the necromantic sciences (*necromanticae*

---

referred to as 'W'. Also John M. Riddle and James Mulholland, "Albert on Stones and Minerals," in *Albertus Magnus and the Sciences*, 203–34; Robert Halleux, "Albert le Grand et l'alchimie," *Revue des Sciences Philosophiques et Théologiques* 66 (1982): 57–80; Weill-Parot, *Les "Images astrologiques,"* 268–69, and now Quinlan-McGrath, *Influences*, 127–32. Apparently the attribution to Albert of the *De mineralibus* has never been questioned. In this section, I rely on Wyckoff's translation, but modify it where appropriate. The textual references are to *De mineralibus* in Albertus Magnus, *Opera omnia*, E. Borgnet (ed.), 38 vols., Paris: Vives, 1890–99, vol. 5. The Cologne critical edition is eagerly anticipated.

[37] *De mineralibus* II.3.1; W. 127: "[L]icet enim pars ista sit pars necromantiae secundum illam speciem necromantiae quae astronomiae subalternatur, et quae de imaginibus et sigillis vocatur: tamen propter bonitatem doctrinae, et quia illud cupiunt a nobis scire nostri socii, aliquid de hoc hic dicemus, omnino imperfecta et falsa reputantes quidquid de his a multis scriptum invenitur (48a2-11)."

[38] Quinlan-McGrath identifies Albert as the author of the *Speculum astronomiae* in discussing *De mineralibus*; *Influences*, (e.g.) 127, 130.

[39] Antiquorum enim sapientium scripturam de sigillis lapidum pauci sciunt, nec sciri potest nisi simul et astronomia et magica et necromantiae scientiae sciantur (48a11-15).

scientiae, W. 145)."[40] Thus for the authentic Albert, the science of the stars, magic and necromancy are all somehow interrelated (at least concerning talismans), and, together with natural knowledge, can offer a complete understanding of images and seals in stone. I should note that none of this has a negative connotation.

*

Albert sheds light on what he means by *magica* at *De mineralibus* II.3.3, where he discusses why gems were taught to be engraved (*insculpi*) in the first place, and what sort of aid these seals can offer. The cause is known, he says, from the science of the Magi, among whom he names Magor the Greek, Germa (Iorma) the Babylonian and Hermes the Egyptian in the first group, and Ptolemy, Geber Hispalensis and Tebith (Thabit) in the second.[41] I will treat the material in this chapter in some depth.

After naming these early authorities, Albert states the "first tenet (*principium*) in this science: that everything which comes to be by nature or art is moved at first by celestial powers (*a virtutibus coelestibus*, 134–35)."[42] We find ourselves immediately in familiar territory, but here it is stated as a first principle of *ipsa scientia*, that is, the *scientia Magorum*. Celestial virtues are, once again, utterly central. Since what-comes-to-be naturally is not controversial in this respect (*et hic de natura non est dubium*), he extends the analysis to engage what is made by man (*arte*):[43]

> In art also it is established for this reason, that something now and not before (*modo et non ante*) stimulates a person's heart to make (or do) [sc. something] (*incitat cor hominis ad faciendum*): And this cannot be the case unless a celestial power (*virtus coelestis*) does so, as the aforementioned wise people (*sapientes*) say (135).[44]

Celestial powers thus inspire people to make or do something at a particular time. Although Albert does not mention it here, this also provides the natural philosophical foundations for interrogations, an astrological practice we have not yet encountered in Albert's authentic works, but only thus far in the *Speculum astronomiae*.[45]

Albert supports this claim and shows how it works by further analyzing how people act:

---

[40] [E]t non possunt haec ex principiis physicis probari: sed oportet ad hoc scire astronomiam et magicam et necromanticas scientias, de quibus in aliis considerandum est (55a47-b2).

[41] Wyckoff discusses the identities of these figures (272–78); Weill-Parot (*Les "images astrologiques*," 41–42) brings our knowledge up to date.

[42] "Est autem principium in ipsa scientia omnia quaecumque fiunt a natura vel arte, moveri a virtutibus coelestibus primo: et hic de natura non est dubium (51a42-46)." This has a strong resonance with the last quotation from *De fato*.

[43] We will recall the talk of man-made objects just above (*artificium, artificiata*), and we will see below that this is precisely the sticking point in Thomas's disagreements with Albert.

[44] In arte etiam constat, eo quod aliquid modo et non ante incitat cor hominis ad faciendum: et hoc esse non potest nisi virtus coelestis, ut dicunt sapientes praenominati (51a37-b1).

[45] As we saw in my Chap. 4, the anonymous author there makes a related point in chapter 14 concerning the relationship between interrogations and nativities; in particular, how the structures of the nativity (whether known or not) stimulates a person to ask a question.

> For there is in man a twofold principle of works (*duplex principium operum*), namely nature (*natura*) and will (*voluntas*). Nature is ruled by the stars (*natura* [...] *regitur sideribus*), but the will is free (*voluntas* [...] *libera est*). Unless it resists, however, the will is drawn along by nature and becomes less flexible. And since nature is moved by the motions of the stars (*natura moveatur motibus siderum*), the will then begins to be inclined to the motions and figures of the stars (*incipit voluntas tunc ad motus siderum et figuras inclinari*, 135).[46]

Although the will is ultimately free, in most cases people follow the inclinations of nature and thus of the stars. At the same time, this argument provides both the natural philosophical foundations for several astrological practices and the safeguards to properly protect free will with respect to human action. We will recall, of course, that we just saw Albert's student, Thomas Aquinas, also utilize this very same analysis (i.a.) in the *Summa theologiae* and *Summa contra gentiles*, and Roger Bacon in *Opus maius* IV.

How, then, do celestial powers/virtues enter a work of art? Since the heavens ultimately incline people to do this or that by first providing them with their particular abilities:

> [T]herefore, if the force and effluence of the stars flows some influence causing art into the artisan (*vis et afflatus siderum influit quamdam causalitatem artis in artefice*), it is certain, unless it is impeded,[47] that it will flow something of its power into all works of art (*influet omnibus operibus artis aliquid suae virtutis*, 135).[48]

Celestial virtues thus enter a work of art by means of the artisan and the timing of his inspiration. These virtues can then be further focused and particularized toward specific ends by means of conscious astrological timing (elections), as we will see. Focus can also be attained by the maker's psychological attitude, as we will see in the cases of al-Kindi's *De radiis stellarum* and in Roger Bacon, as I discuss in Chap. 8.

Having established this first principle, Albert turns to the second: "That the figures of the heavens are the first figures, and [sc. that these exist] before the figures of everything generated by nature and [sc. made by] art (135)."[49] On this basis, Albert establishes the metaphysical principle that "what is first in the genus and order of generators without a doubt flows its causal ability in a fitting manner into everything that follows (135)."[50] Thus, following the principle of *primum in aliquo*

[46] Est in homine duplex principium operum, natura scilicet et voluntas: et natura quidem regitur sideribus: voluntas quidem libera est: sed nisi renitatur, trahitur a natura et induratur: et cum natura moveatur motibus siderum, incipit voluntas tunc ad motus siderum et figuras inclinari (51b1-8)

[47] We will recall talk of "impeding" these virtues in the text recently mentioned from *De natura loci.*

[48] Si igitur vis et afflatus siderum influit quamdam causalitatem artis in artefice, pro certo nisi impediatur, influet omnibus operibus artis aliquid suae virtutis (51b20-24).

[49] "[...] figuras coelorum primas esse figuras, et ante omnium generatorum natura et arte figuras (51b27-29)." We should recall that, according to *Speculum astronomiae* 11.121–26, figures on earth obey celestial figures. Marsilio Ficino also uses "figures" in this context in *De vita* III, as we will see in volume II.

[50] "Quod autem primum est genere et ordine generantium, absque dubio causalitatem suam per modum cuique congruum, omnibus influit sequentibus (51b29-33)." We can thus see that Albert too (like Thomas) also explicitly uses this principle.

genere, the celestial figure will have its causality in everything generated by nature. Moreover, because art is resolved into a principle of nature (*ars resolvitur in principium naturae*), this also follows for what is made by art.[51]

Albert then draws his conclusions:

> Therefore we must (*ex necessitate*) conclude that if a figure is impressed upon matter (*imprimatur figura in materia*), either by nature or by art (*per naturam vel artem*), with due regard to the celestial figure (*observare ad coelestem figuram*), some force of that figure flows into the work of nature or art (*coelestis figurae aliquis vis influitur operi naturae et artis*). And this is the reason why Ptolemy[52] recommends that all actions, comings and goings, and even the putting on and taking off of clothing, be performed with due regard to the images of the heavens (*observare ad imagines coeli*). And therefore, too, in the science of geomancy it is recommended that the figures made up of points be referred to those [sc. of celestial images]; for otherwise they are of no use.[53] And therefore, having also considered this craft (*industria*) of making gems and metallic talismans (*imagines metallicae*) in relation to the *imagines* of the stars, the first teachers and professors of natural knowledge (*physici*) recommended that the carving (*sculpi*) be done at duly observed times (*temporibus observatis*), when the heavenly force is thought to influence the talisman most strongly (*vis coelestis fortissima ad imaginem eamdem esse probatur*), as for example when many heavenly powers combine in it (*coelestibus multis virtutibus admixta*). And they worked wonders (*mira* [...] *operabuntur*) by means of such talismans (136).[54]

Albert's language is very strong (*absque dubio, ex necessitate*) and leaves no room for doubt as to the main thrust of his analysis. Talismans made at astrologically chosen times can be very powerful indeed!

*

After this general analysis, Albert lists the five most important astrologico-astronomical features to be observed for strengthening talismans, including the

---

[51] Nos enim non intendimus hic de figuris mathematice sumptis, sed de figuris prout inducunt diversitatem generantium et generatorum in ordine et specibus et natura formae et materiae suae: habebit igitur figura coelestis causalitatem in omni figura generata a natura, eo quod ars resolvitur in principium naturae: quia principium artis prout diximus, natura est secundum quod exivit a suo coelesti principio, cuius principium est intellectus practicus, sicut idem intellectus est principium artis, sicut diximus saepius in *Coelo et mundo physicis* (51b33-46; W, 135–36).

[52] This is the Ptolemy of the *Centiloquium*; see Wyckoff's note at 136 (n. 4).

[53] Charmasson discusses how geomancy was considered to be derived from (and thus dependent on) astrology; *Recherches sur une technique divinatoire*, 9–14.

[54] "Ex his autem de necessitate concluditur, quod si observare ad coelestem figuram imprimatur figura in materia per naturam vel artem, quod coelestis figurae aliquis vis influitur operi naturae et artis: et inde est, quod observare ad imagines coeli praecipiuntur fieri opera et exitus et introitus et incisio vestium et vestitura a Ptolemaeo sapiente. Hinc est quod in scientia geomantiae figurae, punctorum ad imagines tales reduci praecipiuntur: quia aliter non sunt utiles. Hac ergo industria considerata primi praeceptores et professores physici gemmas et imagines metallicas ad imagines astrorum observatis temporibus quando vis coelestis fortissima ad imaginem eamdem esse probatur, ut puta coelestibus multis virtutibus admixta, sculpi praecipiebant, et mira per tales imagines operabantur (51b47-52a17)." Roger Bacon also uses the language of wondrous works, as we will see in Chap. 8.

signs of the zodiac, the actual constellations, and the location of planetary positions in zodiacal signs that strengthen those signs, namely, the relevant astrological dignities. Here are the first three:

> But the heavenly images (*imagines coeli*) are helped (*adiuvantur*) by many things. Nevertheless, there are five things that are especially to be regarded (*observanda*). [1] The first is the image of the starless sphere (*imago circuli non stellati*) because the circle of the signs without stars (*circulus signorum non stellatus*) is the first having motion for the figure and for life (*primus habens motum figurae et vitae*). [2] Secondly, help comes from the actual constellations (*ex imaginibus stellatis*), which must be properly observed. [3] And third, from the position of the planets in certain signs of the zodiac which strengthen those signs (*ex situ planetarum in signis confortantibus signa*, 136).[55]

Albert focuses on the fourth and fifth:

> [4] From the amount of elevation and elongation, according to the latitude and longitude measured from the equinoctial and the ascendant (136).[56]

> [5] From the relation of all these to the latitude of the clime [sc. where the observation is made (136–37)].[57]

The relationship of the place of making to the celestial situation thus reappears as directly relevant to the strength of talismans.

Albert's conclusions are utterly central to our concerns:

> And the last must be carefully observed, since from this and the fourth the entire quality of the angles is varied (*variatur tota qualitas angulorum*), which the rays describe on the figure (*figura*) of a [= any] thing generated [sc. by nature] or made by art (*quos describunt radii super figuram rei generatae vel factae per artem*). And it is in accordance with this quantity of angles (*secundum quantitatem illam angulorum*) that celestial powers are poured into things (*infunduntur rebus virtutes caelestes*). Few people make these observations, and fewer still know how to make them (136–37).[58]

Albert has now explicitly provided a geometrical-optical analysis of how to strengthen the power of talismans that is exactly the same as what we found in *De natura loci* and elsewhere. From this treatment, we can clearly see how Albert's astrologizing natural philosophy deeply informed his analysis of the making of what he explicitly calls magical talismans (*imagines magicae*), and thereby provided their natural philosophical/scientific foundations. We will see below that this

---

[55] Imagines autem coeli ex multis quidem adiuvantur. Sunt tamen praecipue quinque observanda, scilicet imago circuli non stellati: quia circulus signorum non stellatus est primus habens motum figurae et vitae. Secundo iuvatur ex imaginibus stellatis, quae etiam observare oportet debite. Tertio autem est ex situ planetarum in signis confortantibus signa (52a18-26).

[56] Quarto autem ex quantitate elevationis et elongationis secundum longitudinem et latitudinem a linea aequinoctiali et ascendente (52a27-30).

[57] Et quinto ex respectu omnium horum ad latitudinem climatis (52a30-31).

[58] [H]oc enim multum est observandum, quia ex hoc et quarto variatur tota qualitas angulorum quos describunt radii super figuram rei generatae vel factae per artem, et secundum quantitatem illam angulorum infunduntur rebus virtutes coelestes: hoc enim pauci observant, et pauciores observare sciunt (52a31-39)[.]

is precisely the analysis that Thomas rejected in his later works, particularly in relation to man-made objects (*artificiata*).

One last passage will confirm this interpretation:

> But we are not unaware that, just as the natural powers (*virtutes naturales*) endure for a certain time and no longer (*perdurant in quodam tempore, et non ultra*), so it is also with the power of talismans (*de virtutibus imaginum*). For a certain power (*aliquis virtus*) is poured down from the heavens (*influitur* [...] *de caelo*) only at a certain time of the period (*in quodam tempore periodi*), as we have said at the end of *On generation* [sc. *and corruption*]. And afterwards, the empty useless talisman (*imago*) remains cold and dead (*frigida et mortua*, 137).[59]

Here Albert also explicitly uses the language of "periods" from *De generatione et corruptione* II.10 to inform his understanding of astrological talismans, precisely as he had done in explicating his deeply astrological understanding of fate in *De fato*. The geometrical-optical model of celestial influences is thus central to Albert's analysis in all three of these texts.

*

Albert concludes book II (3.4–5) on a practical note by providing detailed descriptions of how to make images that represent all twelve signs of the zodiac and all seven of the planets, but we will not follow him there, beyond giving one example, that for Saturn:

> The engraving (*inscriptio*) of Saturn is that of an old man holding a curved sickle in his hand. He is not cheerful and smiling, but dark, with a scanty beard. This, by reason of its cold and dryness, is said to confer a steadily increasing power, especially if it is on a stone of the same power (*virtus*, i.e., cold and dry).[60] And you may know that it confers this more quickly on the ignoble than the noble, since Saturn, according to the art of the stars (*ars astrorum*), does not love nobles (143).[61]

Wyckoff notes (following Haskins) that whoever composed the lapidary Albert used here seems to have had an illustrated manuscript of the constellations before him.[62]

---

[59] Sed non lateat nos, quod sicut virtutes naturales perdurant in quodam tempore, et non ultra, ita est etiam de virtutibus imaginum: non enim influitur aliqua virtus de coelo nisi in quodam tempore periodi, sicut diximus in fine *Peri geneseos*: et postea cassa et inutilis remanet imago frigida et mortua (52a44-b3)[.]"

[60] This is a reference to the material dimension that we see emphasized in Thomas Aquinas, and later and even moreso in Ficino.

[61] "Saturni autem inscriptio, hoc est, viri senis falcem decurvam in manu habentis, et non hilaris neque ridentis, fusci paucos pilos in barba habentis, propter frigiditatem et siccitatem dicitur conferre potestatem crescentem et stabilem, praecipue si sit in lapide eiusdem virtutis: et scias quod haec citius confert ignobili quam nobili, quia Saturnos non amat nobiles secundum artem astrorum (54b5-14)." We will see a very similar analysis of talismans in discussing Marsilio Ficino's *De vita* in volume II, where he attends to both the material and celestial dimensions in designing the most powerful talismans. Ficino's analysis, however, goes well beyond Albert's, and in a strikingly Neoplatonic manner. Nevertheless, the basic structures are very much the same.

[62] Wyckoff, *On Minerals*, 140: "This chapter (sc. II.3.5) incorporates an astrological lapidary of engraved gems which is found also in Arnold of Saxony and Thomas of Cantimpré and elsewhere."

She also notes that Albert in the *De mineralibus* offers exactly the same authorities as does the Magister Speculi, except for Geber Hispalensis (275). Finally, every benefit Albert claims in the *De mineralibus* is all-natural, even for ligatures and suspensions, namely, stones bound to any part of the body or hung around the neck, which are the subject of Book II's final chapter (3.6): "in these, medicine and aid are conferred solely by natural powers (146)."[63]

Albert's treatment of what we would call magical images or talismans in *De mineralibus* thus closely coheres with—and is in fact grounded on—his astrologizing Aristotelian natural philosophy with its geometrical-optical model of celestial influences. The authentic Albert thus discusses in some detail their natural philosophical foundations, unlike the auther of the *Speculum astronomiae*, one of their distinctive differences. Furthermore, *De mineralibus* offers some of the practical benefits to be derived from studying natural philosophy in general, and the science of the stars in particular. Nevertheless, and despite minor but significant differences in terminology, the legitimate images Albert discussed in *De mineralibus*, which he refers to there as necromantic, all fall within the legitimate type of images demarcated in the *Speculum astronomiae*, where they are called astrological (but decidedly not necromantic) images. The fundamental distinction turns on whether the talisman is powered by natural and thus legitimate celestial means, or by illegitimate demonic ones, and in this significant respect both the *Speculum astronomiae* and the authentic Albertus Magnus are completely aligned.

7.4 Did Albertus Magnus Write the *Speculum Astronomiae*?

How, then, shall we assess the case for Albert's authorship of the *Speculum astronomiae*?[64] As we have just seen, Albert's theory and practice concerning talismans (with their natural-philosophical foundations) are strikingly similar to—and deeply harmonious with—the views articulated in the deliberately anonymous *Speculum astronomiae*, thus making a strong argument for Albert as its author. This is true even though the Magister Speculi is not nearly as interested in discussing natural philosophical structures as the authentic Albert is. We can also strongly contrast both Albert and the *Speculum astronomiae* with Thomas Aquinas's radical rejection of talismans, as we will see in the next section.

She discusses this further in appendix C ("Astrology and Magic"), Sect. 4, on astrological seals, especially at 276–77. She refers the reader to Charles H. Haskins, *Studies in the History of Medieval Science,* Cambridge, MA: Harvard University Press, 1924, 285–88 and 336–45, and to the photographs in the Houghton Library of Harvard University. Weill-Parot discusses this iconographic tradition at *Les "images astrologiques,"* 105–109.

[63] [I]n illis non nisi naturaliter ex virtutibus conferunt medicinam et iuvamen (55b15-17).

[64] Weill-Parot discusses the arguments for Albert as the author of the *Speculum astronomiae* at *Les "images astrologiques*," 278–80, and reviews the scholarship on this issue at 27–32.

Nevertheless, and despite significant continuities with talismans and other more general points of comparison, the evidence offered in this and earlier chapters presents some major challenges to Albert's authorship, including the striking discrepancy in terminology concerning talismans and necromancy. As we just saw, Albert in *De mineralibus* described legitimate astrological talismans as being a part of necromancy, whereas the *Magister Speculi* radically rejects that term, reserving it only for illegitimate demonic practices.

Furthermore, in the *De caelo*, the two astrological books that Albert refers to betray a remarkable ignorance of practical astrology.[65] The fact that he attributes the *De magnis coniunctionibus* (one of the most influential astrological texts ever written in Latin) to Ptolemy, and that he refers to an obscure text, also supposedly by Ptolemy, sharply contrasts with the author of the *Speculum astronomiae*, who has an unparalleled knowledge of the literature on all branches of astrology, who is obsessed with accurate and detailed references, who knows perfectly well that Albumasar is the author of the *De magnis coniunctionibus*,[66] and who says nothing about the other book supposedly by Ptolemy (although the *Magister Speculi* was certainly aware of many obscure and rare texts).

Likewise, Albert calls the seventh house the house of death, and uses an uncommon technical terminology, for example, *elector, scientia electorum*, and *geneatici*. That a single author wrote both texts is hard to believe. But, of course, *De caelo* (between 1250 and 1254) was probably written several years before the *Speculum astronomiae* (most likely in the 1260s and certainly by 1271–72), thus giving Albert up to 15 years to catch up on his astrological homework and to modify his terminology concerning talismans and necromancy.

*

Regardless, from the mid-fourteenth century, Albert was increasingly held to be the author of the *Speculum astronomiae*, as we can see continuously through at least the middle of the seventeenth century. Unfortunately, the influence of both Albert's authentic astrologizing Aristotelian natural philosophy and the *Speculum astronomiae* have yet to be fully studied.[67] The examples offered here—mainly from the University of Padua, where Albert studied in the early 1220s and the site of his

---

[65] This paragraph is deeply indebted to David Juste's insightful comments.

[66] Zambelli, Speculum astronomiae, 228; the incipit "Scientia significationum […]" is correct.

[67] Zambelli discusses the influence of the *Speculum astronomiae* in various respects, as does Paravicini Bagliani in *Le* Speculum astronomiae; Weill-Parot specifically discusses the influence of *imago astronomica* and related topics. These three all trace influences through about 1500, as traditional medievalists (at least in this respect) are wont to do. Graziella Federici-Vescovini provides some evidence for the influence of Albert's authentic works for the fourteenth century in *Astrologia e scienza*, and in the article cited just below. Although Edward Mahoney mainly treats the influence of Albert's natural philosophy (including *De natura loci*) at Padua in the fifteenth century, there is characteristically a much broader range of useful bibliography in his "Albert the Great and the Studio Patavino."

joining the Friars Preacher—merely point to a much larger story that deserves further attention.

In his early fourteenth century *Conciliator*, the influential professor of astrology, philosophy and medicine, Pietro d'Abano, precisely replicated the *Speculum astronomiae's* structure of practical astrology, but without citing the title nor naming its author, as we will see in Part IV below. In the late fourteenth century, Biagio Pelicani da Parma, professor of mathematics, astrology and philosophy at the universities of Padua, Bologna and Pavia,[68] in addition to citing his authentic (and deeply astrologizing) *De natura loci, De causis proprietatum elementorum* and *De caelo et mundo*, calls Albert the author of the *Speculum astronomiae*.[69]

Further, in his mid-fifteenth-century inaugural oration for a course on al-Farghani's *De scientia stellarum* at the University of Padua (1464), Regiomontanus also explicitly referred to Albert, his fellow countryman, as the author of the *Speculum astronomiae*: '*Albertus Magnus in speculo Astronomiae*', as we will also see in Part IV. At the end of the fifteenth century in his influential *De vita libri tres* (1489), Marsilio Ficino refers to Albert as author of the *Speculum astronomiae*, as does Jerome Torrella in his *Opus praeclarum de imaginibus astrologicis* (1496–1500), although he calls it there the *Speculum scientiarum*.[70] Even as late as 1648 we can see the *Speculum astronomiae* still being explicitly attributed to Albert, as in Andrea Argoli's *Ephemerides* for 1648–1700.[71] I will discuss Argoli further in volume III.

7.5 Thomas Aquinas on Images-Talismans

What, then, is the relationship between Albert's views of talismans and those of his most famous and influential student, Thomas Aquinas? Thomas's earlier views as expressed in the *Summa contra gentiles* III.105 (written most likely between 1263

[68] For his career, see Federici-Vescovini, *Astrologia e scienza*, 24–39.

[69] Vescovini, "L'influenza di Alberto Magno come metafisico, scienziato e astrologo sul pensiero di Angelo da Fossombrone e Biagio Pelicani," in her *"Arti" e filosofia nel secolo XIV: Studi sulla tradizione Aristotelica e i "moderni,"* Florence: Vallecchi, 1983, 77–100; see esp. 97, note 69: "*dicens Albertus in Speculo astronomiae* [...]."

[70] Ficino: '*Albertus quoque Magnus inquit in Speculo*' (III.12.121); Torrella: '*Albertum Magnum in Speculo scientiarum*' (16). For Ficino's text, see *Three Books on Life*, Carol V. Kaske and John R. Clark (ed. and trans.), Binghamton, N.Y.: Medieval and Renaissance Texts and Studies, 1989; for Torrella's, see *Opus praeclarum de imaginibus astrologicis*. I discuss Ficino's treatment of talismans in volume II, and Torrella's in its conclusion.

[71] *Exactissimae coelestium motuum ephemerides* [...], Padua: P. Frambotti: "Aeterna Dei Optimi Maximi providentia Inferiorem hanc machinam tradidit secundis causis gubernandam et inferiora ad usum hominis venientia a Deo mediantibus Angelis et corporibus caelestibus dispensantur quorum fere omnia ostendit nobis (ut bene Albert. Mag. in Speculo) in libro universitatis qui est pellis Caeli (259)." We can also see that Argoli here discussed divine providence in an explicitly astrological context in the mid-seventeenth century, with an analysis that was most likely taken directly from Thomas's *Summa contra gentiles* III.82-94.

and 1265) were much closer to Albert's.[72] His final views, however, as expressed in *Summa theologiae* II.II.96.2 (written between January 1271 and Easter 1272) and *De operationibus occultis naturae* (written at some point during his second sojourn in Paris, 1268–72) turned strongly away from what seems to have been his more tolerant early position, and thus inspired Weill-Parot to characterize Thomas's position on talismans overall as "un rejet global" (227). This also informed Weill-Parot's intention to reconstruct Thomas's normative (= negative) position synthetically (228–47). Weill-Parot then treats *Summa contra gentiles* III.105 and what he calls Thomas's "ambiguity"—what I call his "loophole"—as well as later attempts at reconciling Thomas's texts by significant Thomist interpreters, including Tommaso de Vio (Cardinal Caietanus, 249–57). We will look briefly at these later attempts in the conclusion below, and more fully in volume III.

Although he considered both astronomy and astrology to be perfectly legitimate modes of knowledge (as we saw in Chap. 5), Thomas's mature views on talismans utterly rejected the position that *imagines astronomicae* could ever be legitimate and natural, claiming instead that seemingly natural talismans always concealed a tacit pact with demons, as we will see.[73] Thomas extensively discussed talismans in his mature *Summa Theologiae* II.II.96 (*ad* 2), which follows immediately on the texts treated in my Chap. 5 from II.II.92-95 on astrology and divination. It also explicitly uses the phrase '*imagines astronomicae*', and responds precisely to the argument in *Speculum astronomiae*, Chap. 11.[74] I will begin with Thomas's final words on the subject. According to Weill-Parot (223–27), Thomas treated talismans in five works. I will discuss three of them here: *Summa theologiae* II.II.96.2; *De operationibus occultis naturae*, and *Summa contra gentiles* III.105. The other two are: *Quaestiones disputatae De potentia* VI.10 and the 12th *Quodlibet*. The earliest discussion is in the *Summa contra gentiles* (1263–65), which I will treat last.

---

[72] Weill-Parot says that here Thomas is much closer to the Albert of the *De mineralibus* than his later positions are; *Les "images astrologiques,"* 257.

[73] See Weill-Parot, *Les "images astrologiques,"* 38–39 and 223–27. On 72–77 and especially 77, he also attributes some level of subterfuge to the Magister Speculi's presentation.

[74] Since (1) Thomas here seems to respond directly to the *Speculum astronomiae*, and (2) this section of the *Summa theologiae* is firmly datable to 1271–72, 1271–2 thus provides a *terminus ante quem* for the composition of the *Speculum astronomiae*. Also to the same effect, the 12th *Quodlibet* (dated 1270–72) is the only other text by Thomas to explicitly use the phrase '*imago astronomica'*. Weill-Parot insightfully discusses the dating of both texts at *Les "Images astrologiques,"* 38–40, and he too relies on Torrell, *Saint Thomas Aquinas*. In *De operationibus occultis naturae*, on the other hand, Thomas only uses the term '*imago*' and not the phrase '*imago astronomica*', thus perhaps indicating that *De operationibus occultis naturae* was composed before *Summa theologiae* II.II.96.

7.6 [1] *Summa theologiae* II.II.96.2

Thomas begins II.II.96.2 by stating his question in the title: "Whether practices designed for a change of bodies (*observationes ordinatae ad corporum immutationem*), for example, towards health or something of this sort, are licit."[75] First Thomas offers three arguments for legitimacy (*pro*):

> [1] It seems that practices designed for the change of bodies, for example, towards health or something of this sort, are licit. For it is licit to use the natural powers of bodies (*naturales virtutes corporum*) for inducing their proper effects (*ad proprios effectus inducendos*). But natural things (*res naturales*) have certain hidden powers (*quasdam virtutes occultae*), for which a reason (*ratio*) cannot be assigned by men, just as that a magnet (*adamas*) draws iron (*ferrum*), and many other things that Augustine enumerates (*De civ Dei* XXI). Therefore, it seems that to use things of this sort for changing bodies is not illicit.[76]

Here Thomas refers to occult properties as a normal part of natural things, which are thus fully licit. Here he follows his own analysis articulated clearly and concisely in the contemporary *De operationibus occultis naturae* to be treated just below.

Thomas then turns from natural to man-made bodies as he more fully articulates his position in offering the second argument for. Here he mentions talismans explicitly:

> [2] In addition, just as natural bodies (*corpora naturalia*) are subordinated to celestial bodies, so also are artificial bodies (*corpora artificialia*). But natural bodies are allotted certain hidden powers (*quasdam virtutes occultas*) that [sc. ontologically] follow on a species from the impression of celestial bodies (*ex impressione caelestium corporum*). Therefore, also artificial bodies (*corpora artificialia*)—for example, talismans (*imagines*)—are allotted some hidden power from celestial bodies (*sortiuntur aliquam virtutem occultam a corporibus caelestibus*) for causing certain effects (*ad aliquos effectus causandos*). Therefore, to use them and others of this sort is not illicit.[77]

In the second argument for the licit use of magic (in our terminology), Thomas explicitly mentions talismans (*imagines*), using an analogy from natural to man-made things, drawing, once again, on the role of hidden/occult powers derived from celestial bodies. Here Thomas seems to be referring to something like Albert's

---

[75] Utrum observationes ordinatae ad corporum immutationem, puta ad sanitatem vel ad aliquid huiusmodi, sint licitae.

[76] Ad secundum sic proceditur. Videtur quod observationes ordinatae ad corporum immutationem, puta ad sanitatem vel ad aliquid huiusmodi, sint licitae. Licitum enim est uti naturalibus virtutibus corporum ad proprios effectus inducendos. Res autem naturales habent quasdam virtutes occultas, quarum ratio ab homine assignari non potest: sicut quod adamas trahit ferrum, et multa alia quae Augustinus enumerat, XXI *de Civ. Dei*. Ergo videtur quod uti huiusmodi rebus ad corpora immutanda non sit illicitum (331a1-12).

[77] Praeterea, sicut corpora naturalia subduntur corporibus caelestibus, ita etiam corpora artificialia. Sed corpora naturalia sortiuntur quasdam virtutes occultas, speciem consequentes, ex impressione caelestium corporum. Ergo etiam corpora artificialia, puta imagines, sortiuntur aliquam virtutem occultam a corporibus caelestibus ad aliquos effectus causandos. Ergo uti eis, et aliis huiusmodi, non est illicitum (331a13-21).

argument from the *De mineralibus* just discussed, although Albert himself did not use the concept of *virtutes occultae*.

For the third argument *pro*, Thomas turns to demons: [3] "Further, demons (*daemones*) can also transform (*transmutare*) bodies in many ways, as Augustine says (*De trinitate* III). But their power (*virtus*) is from God. Therefore, it is licit to use their power for making some changes (*immutationes*) of this sort."[78] Since demonic power derives ultimately from God, their use is also legitimate.

After these three pro-magical arguments, Thomas presents only one argument *contra*:

> But *contra* is what Augustine says in Book II of *De doctrina Christiana*, that the great exertions of the magical arts (*molimina magicarum artium*), and ligatures, and those remedies that the medicine of physicians condemns, whether in incantations (*praecantationes*), or in certain notations (*notae*), which they call characters (*characteres*), or in certain suspended or signifying things (*in quibuscumque rebus suspendendis atque insignandis*)—these pertain to superstition.[79]

For Augustine, one of the greatest theological authorities, these magical practices are fully illicit and fall under the problematic rubric of superstition, much as divination did in the closely related texts from the immediately prior *Summa theologiae* II.II.92-95 concerning knowledge, which were explored in Chap. 5. Augustine here mentions (i.a.) ligatures, incantations and characters, and he associates them with the magical arts as well as with medicine.

Now Thomas offers his response:

> I respond that one should say that, in those things that are made (*fiunt*) for inducing certain bodily effects (*ad aliquos effectus corporales inducendos*), one should consider whether they seem to be able to cause such effects *naturally* (*naturaliter*). For thus it will not be illicit, for one may bring natural causes to their proper effects (*licet enim causas naturales adhibere ad proprios effectus*, my emphasis).[80]

In approaching his response, Thomas characteristically links causal naturalness with legitimacy, as he had just done in II.II.92-95. This will be the focus of what follows.

He then turns to non-natural actions:

> But if they do not seem to be able to cause such effects naturally (*naturaliter non videantur posse tales effectus causare*), it follows that they are not brought to bear for causing these

[78] Praeterea, daemones etiam multipliciter possunt corpora transmutare: ut dicit Augustinus, III *de Trin*. Sed eorum virtus a Deo est. Ergo licet uti eorum virtute ad aliquas huiusmodi immutationes faciendas (331a22-26).

[79] Sed contra est quod Augustinus dicit, in II *de Doct. Christ.*, quod ad superstitionem pertinent *molimina magicarum artium, et ligaturae, et remedia quae medicorum quoque medicina condemnat, sive in praecantationibus, sive in quibusdam notis, quas characteres vocant, sive in quibuscumque rebus suspendendis atque insignandis* (331b1-7).

[80] Respondeo dicendum quod in his quae fiunt ad aliquos effectus corporales inducendos, considerandum est utrum naturaliter videantur posse tales effectus causare. Sic enim non erit illicitum: licet enim causas naturales adhibere ad proprios effectus (331b8-13).

> effects as *causes*, but only as *signs*. And this pertains to pacts of significations entered into with demons (*ad pacta significationum cum daemonibus inita*, my emphasis again).[81]

Here Thomas distinguishes causes from signs and thereby introduces demons. Natural causation is fine. Signs, on the contrary, are deeply problematic, as we will see more fully in what follows.

Thomas confirms this view with another quotation from Augustine:

> Whence Augustine says in Book XXI of the *De civitate Dei*: "Demons are attracted (*illiciuntur*) through creatures which not themselves but God established with different delights (*delectabiles*) on behalf of their diversity, not as animals to food, but as spirits to signs, which cohere with the pleasure (*delectatio*) of each through the different types of stones, herbs, woods, animals, charms (*carmina*) and rites (*ritus*)."[82]

With this quotation, Thomas informs us through Augustine how demons might be appealed to on various levels, including by a range of natural objects in addition to charms and rites. With the last two, we enter more fully into the deeply problematic realm of ritual or ceremonial magic.

*

With this background, Thomas now discusses the three pro-magical arguments. The first concerns natural objects:

> To the first, therefore, one should say that if natural things (*res naturales*) are brought to bear for producing some effects towards which they are thought to have a natural power (*naturalis* [...] *virtus*), it is neither superstitious nor illicit. But if either some characters (*characteres*) or names (*nomina*) or whatever other practices (*observationes*) are adjoined, which it is clear do not have their efficacy naturally (*naturaliter efficaciam non habere*), it will be superstitious and illicit.[83]

Thomas here confirms his fundamental distinction between natural and demonic action, namely, that "all-natural" practices are fine, but when non-natural types of practices—whether with characters, names or certain other unnamed practices—are adjoined, these are both superstitious and illicit. We should also notice that Thomas does not make any exceptions for straightforward and prosaic characters and names as the Magister Speculi had done in *Speculum astronomiae*, Chap. 11.

---

[81] Si autem naturaliter non videantur posse tales effectus causare, consequens est quod non adhibeantur ad hos effectus causandos tanquam causae, sed solum quasi signa. Et sic pertinent *ad pacta significationum cum daemonibus inita* (331b13-18).

[82] "Unde Augustinus dicit, XXI *de Civ. Dei*: *Illiciuntur daemones per creaturas, quas non ipsi, sed Deus condidit, delectabilibus pro sua diversitate diversis, non ut animalia cibis, sed ut spiritus signis, quae cuiusque delectationi congruunt, per varia genera lapidum, herbarum, lignorum, animalium, carminum, rituum* (331b18-24)." Ficino too is fond of the term '*illicium*' and its cognates in *De vita* III.

[83] Ad primum ergo dicendum quod si simpliciter adhibeantur res naturales ad aliquos effectus producendos ad quos putantur naturalem habere virtutem, non est superstitiosum neque illicitum. Si vero adiungantur vel characteres aliqui, vel aliqua nomina, vel aliae quaecumque variae observationes, quae manifestum est naturaliter efficaciam non habere, erit superstitiosum et illicitum (331b25-332a6).

Thomas now responds to the second argument on man-made objects (including talismans) and goes into greater detail:

> With respect to the second [sc. argument], we should say [1] that the natural powers of natural bodies follow on their substantial forms (*virtutes naturales corporum naturalium consequuntur eorum formas substantiales*), which are allotted from an impression of the celestial bodies (*quas sortiuntur ex impressione caelestium corporum*). Therefore, [2] from the impression of these same [sc. celestial bodies], certain active powers (*quaedam virtutes activae*) are allotted. But the forms of artificial [i.e. man-made] bodies (*corpora artificialia*) come from the idea of the artisan (*ex conceptione artificis*). Since they [sc. artificial bodies] are nothing other than composition (*compositio*), order (*ordo*) and shape (*figura*), as is said in [Aristotle's] *Physica* I, they cannot have a natural power for acting (*non possunt habere naturalem virtutem ad agendum*). From this [sc. it follows] that from an impression of celestial bodies, no power is allotted (*nullam virtutem sortiuntur*), insofar as they are man-made, except only with respect to their natural matter (*solum secundum materiam naturalem*, 77).[84]

Thomas stakes his claim strongly here: No power is allotted to man-made objects from the celestial bodies, beyond what exists naturally in their material substrate. Thomas here continues to employ the natural philosophical analysis more fully articulated in *De operationibus occultis naturae.*

Thomas supports his argument by pitting Augustine's great authority against Porphyry's:

> Therefore, what seemed to be the case for Porphyry is false, as Augustine says (*De civitate Dei*): "with herbs and stones (*lapidibus*) and living things (*animantibus*), and certain specific sounds (*sonis*) and voices (*vocibus*), and representations (*figurationibus*) and images (*figmentis*)—and with the motions of the stars observed in the turning of the heavens (*observatis in caeli conversione motibus siderum*)—that suitable powers of the stars (*potestates idoneas siderum*) can be fabricated on earth by men for achieving various effects (*fabricari in terra ab hominibus potestates idoneas siderum variis effectibus exequendis*)": as if the effects of magical arts come forth from the power of the celestial bodies (*quasi effectus magicarum artium ex virtute caelestium corporum provenirent*). But as Augustine adds there: "this entire matter pertains to demons (*ad daemones*), who play games (*ludificatores*) with the souls under them (77)."[85]

Not nature through the heavenly bodies but demons are responsible for whatever powers magical arts may have. With this argument, Thomas clearly shows his

[84] Ad secundum dicendum quod virtutes naturales corporum naturalium consequuntur eorum formas substantiales, quas sortiuntur ex impressione caelestium corporum: et ideo ex eorundem impressione sortiuntur quasdam virtutes activas. Sed corporum artificialium formae procedunt ex conceptione artificis: et cum nihil aliud sint quam compositio, ordo et figura, ut dicitur in I *Physic.*, non possunt habere naturalem virtutem ad agendum. Et inde est quod ex impressione caelestium corporum nullam virtutem sortiuntur inquantum sunt artificialia, sed solum secundum materiam naturalem (332a7-19).

[85] Falsum est ergo quod Porphyrio videbatur, ut Augustinus dicit, X *De civ. Dei*, *herbis et lapidibus et animantibus, et sonis certis quibusdam et vocibus, et figurationibus atque figmentis quibusdam etiam observatis in caeli conversione motibus siderum, fabricari in terra ab hominibus potestates idoneas siderum variis effectibus exequendis*: quasi effectus magicarum artium ex virtute caelestium corporum provenirent. Sed sicut Augustinus ibidem subdit, *totum hoc ad daemones pertinet, ludificatores animarum sibi subditarum* (332a19-b5).

awareness of the central issues at stake from both Albert's authentic works and the *Speculum astronomiae*, as he takes a strong general stance *contra*, supported by Augustine's powerful authority.

Thomas then turns directly to astrological images or talismans:

> Whence also, what they call *imagines astronomicae* have their effect from the operation of demons (*ex operatione daemonum habent effectum*). The sign of this is that it is necessary for certain characters to be inscribed (*inscribi*) on them, which do not act naturally on anything (*naturaliter ad nihil operantur*); for a figure is not a principle of natural action (*non enim est figura actionis naturalis principium*). But astrological talismans differ from nigromantic [sc. talismans] in this (*in hoc distant astronomicae imagines a nigromanticis*), that in nigromantic talismans, express invocations and certain deceptions are made, whence they pertain to express pacts entered into with demons; but in other talismans there are certain tacit pacts by means of certain signs of figures or characters (*quaedam tacita pacta per quaedam figurarum seu characterum signa*, 77).[86]

Although astrological talismans are *claimed* to be all-natural and thus legitimate, Thomas vehemently disagrees, asserting strongly that they indicate instead a tacit pact with demons, thus undermining the *Speculum astronomiae*'s third legitimate type of images by folding it into the second detestable class. He also thereby rejects Albert's analysis in the *De mineralibus* concerning man-made entities.

The crucial evidence for Thomas is that these talismans use characters and figures that have no natural effect. Rather, they communicate as signs to demons and should thus be considered addressative, in Weill-Parot's useful terminology. The crucial statement is that a figure is not a principle of natural action. I should also note that Thomas explicitly used the phrase '*imago astronomica*' here, and thus, as Weill-Parot has argued (38–40), responds directly to the *Speculum astronomiae* in which it was initially coined. Indeed, Thomas contrasts '*imagines astronomicae*' explicitly with "necromantic talismans," precisely as in the *Speculum astronomiae*, albeit towards significantly different ends, and despite Thomas using the terminological variant "nigromantic."

Finally, Thomas turns to the third argument and thus also to demons:

> To the third it should be said that it pertains to the domain of divine majesty (*divina maiestas*) to which demons are subordinate, so that God uses them for whatever he wishes. But for a person, no power over demons has been arranged to make it possible to use them licitly for whatever he wishes. But for him, a war has been declared against demons. Whence, in no way is it fitting for a person to use the help of demons by means of either tacit or expressed pacts.[87]

---

[86] Unde etiam imagines quas *astronomicas* vocant, ex operatione daemonum habent effectum. Cuius signum est quod necesse est eis inscribi quosdam *characteres*, qui naturaliter ad nihil operantur: non enim est figura actionis naturalis principium. Sed in hoc distant astronomicae imagines a nigromanticis, quod in nigromanticis fiunt expressae invocationes et praestigia quaedam, unde pertinent ad expressa pacta cum daemonibus inita: sed in aliis imaginibus sunt quaedam tacita pacta per quaedam figurarum seu characterum signa (332b6-17).

[87] Ad tertium dicendum quod ad dominium pertinet divinae maiestatis, cui daemones subsunt, ut eis utatur Deus ad quodcumque voluerit. Sed homini non est potestas super daemones commissa, ut eis licite uti possit ad quodcumque voluerit: sed est ei contra daemones bellum indictum. Unde nullo modo licet homini daemonum auxilio uti per pacta tacita vel expressa (332b17-25).

Thomas thus ends II.II.96.2 by strongly and universally rejecting the possibility of demonic aid for human beings, yet another medieval safeguard that Marsilio Ficino flouts in his strikingly radical *De vita libri tres*, as I will argue in volume II.

With this analysis, Thomas shows full awareness of the issues at stake, since, in the previous chapter of the *Summa Theologiae* (II.II.95), as we saw, he discussed in detail astrology's relation to divinatory practices, arguing forcefully for astrology's legitimacy *qua* knowledge precisely because it is all-natural. Regardless, for Thomas, astrological talismans of all types cross that very same threshold of legitimacy, which both Albert and the Magister Speculi (despite some striking differences in terminology) had extended to one particular type of all-natural talisman. In so doing, Thomas intended to wholly remove astrology's operative or magical dimension, rendering it fully and only a knowledge-oriented practice. Albert, Thomas and the Magister Speculi thus all completely agree that astronomy and astrology are legitimate practices offering conjectural knowledge of the past, present and future. In this respect, the only controversial issue concerns astrological images or talismans, one non-demonic, natural part of which Albert and the Magister Speculi consider legitimate, and Thomas rejects entirely. This is his fully mature normative position.

7.7 [2] *De operationibus occultis naturae*

The other mature text on magic by Thomas that we will explore is the last section of his *De operationibus occultis naturae*, which was written around the same time as *Summa theologiae* II.II.96. At the beginning of my Chap. 1, I discussed some earlier framing passages from this informative text to articulate some of the planets' fundamental integrating and regularizing roles vis-à-vis form in Thomas's understanding of the economy of nature. At the end of this same consultation, Thomas used the same conceptual structures to discuss what we call magic, and talismans in particular. As we just saw, he used the very same conceptual analysis in *Summa theologiae* II.II.96.2.

The passages to be treated here occur just after the texts I treated in Chap. 1, namely, those in which Thomas articulated astrology's natural philosophical justification—especially for nativities—even though that was not explicitly his intention there. More generally, Thomas had just analyzed the relationship of forms and their powers to the hierarchy of being, focusing on natural bodies from the elements to human beings. These natural entities, and especially mixed bodies—including stones, metals and minerals, namely the material bases for talismans—have powers and actions that go beyond those of the four elements. These additional powers derive from their specific forms as mediated by the celestial bodies, ultimately deriving from the separated substances or intelligences that generated the forms in the first place. Thomas calls these *virtutes occultae* as he also did in *Summa theologiae* II.II.96.2.

In this broader context, Thomas then turns to artificial bodies:

> Moreover, from this it is clear further that because artificial forms [sc. the forms of man-made things] are accidents that do not follow [sc. ontologically] (*consequntur*) on a species,[88] it is not possible that some man-made thing (*aliquod artificiatum*) can be allotted (*sortiatur*) some power (*virtus*) and action (*operatio*) of this sort from a celestial body (*a celesti corpore*) in its composition[89] for producing—from an internal power (*ex virtute indita*)—some natural effects that go beyond the powers of the elements (*transcendentes elementorum virtutes*). For powers (*virtutes*) of this sort—if there be any in man-made objects (*artificiatis* [sc. including talismans]) from the celestial bodies—do not follow upon any form (*nullam formam consequerentur*), since the form of artificial things (*forma artificialium*) is nothing other than order (*ordo*), composition (*compositio*) and figure (*figura*) [precisely as in *Summa Theologiae* II.II.96.2], from which such powers and actions (*actiones = operationes*) cannot come forth (*prodire = provenire*). Whence it is clear that, if some artificial things (*artificiata*) accomplish (*perficiant*) any actions of this sort—for example, that serpents die in relation to some sculpture (*ad aliquam sculpturam*), or animals are immobilized or harmed—this does not proceed from some internal (*indita* [as above]) and permanent power [sc. from a natural specific form], but only from the power of an external agent (*ex virtute agentis extrinseci*) which uses such [sc. sculptures] as instruments for its effect (*sicut instrumentis ad suum effectum*).[90]

According to Thomas, artificial objects do not receive specific forms and their correlative powers from the celestial bodies. Rather, their forms are accidents—derived from the order, composition and figure imposed by an artisan—that explicitly do not follow on a natural species. Whatever additional power they might have, then, derives not from their naturally imbued specific form, but rather from an external agent, employing a distinction that Thomas had made at the very beginning of the *De operationibus occultis naturae*. For Thomas there are only two ways that an object can act by the power of a superior agent: with a specific form imprinted on it at generation, and without such an imprinted form (23–70). Artificial objects including talismans represent the latter. He will explain what the relevant external superior agent is just below.

Now Thomas discusses these issues in relation to the power of the celestial bodies, and here he explicitly discusses man-made magical objects, including talismans. He focuses on figures in particular:

[88] I.e. they are not natural specific forms.

[89] Roger Bacon also uses this term in discussing the making of talismans, as I discuss in Chap. 8 below.

[90] Ex hoc autem apparet ulterius quod, quia forme artificiales sunt accidentia que non consequntur speciem, non est possibile quod aliquod artificiatum aliquam huiusmodi virtutem et operationem a celesti corpore in sua compositione sortiatur ad producendum ex virtute indita aliquos effectus naturales transcendentes elementorum virtutes; huiusmodi enim virtutes, si que essent artificiatis ex celestibus corporibus, nullam formam consequerentur, cum forma artificialium nichil aliud sit quam ordo, compositio et figura, ex quibus prodire non possunt tales virtutes et actiones. Unde manifestum est quod si quas huiusmodi actiones aliqua artificiata perficiant, puta quod ad aliquam sculpturam moriantur serpentes aut immobilitentur animalia vel ledantur, non procedit hoc ex aliqua virtute indita et permanenti, sed solum ex virtute agentis extrinseci quod utitur talibus sicut instrumentis ad suum effectum (259–77).

> And it cannot be said that actions of this sort come forth (*proveniant*) from the power of celestial bodies (*ex virtute celestium corporum*)[91] because celestial bodies act naturally on these lower things (*quia celestia corpora naturaliter agunt in ista inferiora*). And from this [sc. it follows] that some body (*aliquod corpus*) figured one way or another (*sic vel aliter figuratur*) has no greater or less fitness for receiving the impression of a natural agent (*ad recipiendum impressionem naturalis agentis*). Whence it is not possible that *imagines* or sculptures (*sculpture*), which are made for producing some singular effects (*ad aliquos effectus singulares producendos*), have [sc. any] efficacy [= efficient causality] from the celestial bodies (*efficaciam habeant ex celestibus corporibus*), even though they seem to be made under specific celestial configurations (*sub certis constellationibus fieri*), but [sc. they derive this efficacy] only by some spirits (*ab aliquibus spiritibus*) which do such things (*tales operantur*) by means of *imagines* and sculptures.[92]

Thus Thomas rejects a natural analysis of talismans, and here he seems to be responding directly to both the *Speculum astronomiae* and to Albert's authentic *De mineralibus*, as well as to his own earlier statement to the contrary in *Summa contra gentiles* III.105, as we will see.

Although these objects—which he explicitly identifies as talismans and sculptures—may be made under specific celestial configurations (and thus using astrological elections), this does not imply that they are actually empowered in any way by these celestial influences. Rather, because they have no naturally given specific form, they cannot receive and retransmit celestial influences by means of their form. Therefore, they can only have such effects by means of spirits, by which he means demons, who use the talismans and sculptures as instruments. Thomas thereby deliberately and explicitly seems to shut down his own earlier loophole here, as we will soon see.

*

Now Thomas turns to the central issue of formal vs. material causes as he shifts gears, transitioning from *opera* to *verba*:

> Moreover, just as [A] *imagines* are made [1] from natural matter (*ex materia naturali*), but [2] their form (*forma*) is alloted by art (*ex arte*), so also [B] human words (*verba humana*) have [1] a natural matter (*materia* [...] *naturalis*), namely the sounds (*soni*) produced from a human mouth (*ab hominis ore*), but they have signification as if [sc. it were] [2] the form from the intellect that expresses its conceptions by sounds of this sort (*quasi formam habent ab intellectu suas conceptiones per huiusmodi sonos exprimente*). Whence, by the same argument (*pari ratione*), human words (*humana verba*) also do not have efficacy for any change of a natural body (*nec* [...] *habent efficaciam ad aliquam immutationem corporis*

[91] With respect to identifying the external vs. internal natural agents.

[92] Nec potest dici quod huiusmodi actiones proveniant ex virtute celestium corporum, quia celestia corpora naturaliter agunt in ista infereriora; et ex hoc quod aliquod corpus sic vel aliter figuratur, nullam ydoneitatem vel maiorem vel minorem habet ad recipiendum impressionem naturalis agentis; unde non est possibile quod ymagines vel sculpture que fiunt ad aliquos effectus singulares producendos efficaciam habeant ex celestibus corporibus, quamvis sub certis constellationibus fieri videantur, sed solum ab aliquibus spiritibus qui per ymagines et sculpturas tales operantur (278–90).

> *naturalis*) from the power of some natural cause (*ex virtute alicuius cause naturalis*), but only from some spiritual substance (*sed solum ex aliqua spirituali substantia*).[93]

As we will see, Thomas's analysis of words (*verba*) is similar to his analysis of significative *voces* in *Summa contra gentiles* III.105, although he uses a slightly different terminology here: *soni* and *verba* here vs. *voces* there. He also discusses the material substrate here. Although Thomas offers a somewhat different argument here than in *Summa contra gentiles* III.105, he draws exactly the same conclusion, namely, that the effects of magical sounds are not natural, but derive from spirits, i.e. demons. We will explore the power of words more fully in Chap. 8 in relation to Roger Bacon's views, and also provide the fundamental theoretical and historical background from al-Kindi's *De radiis stellarum*, to both of which Thomas seems to be responding here.

Thomas concludes his argument and *De operationibus occultis naturae* overall by discussing both the *opera* and *verba* together:

> Therefore, these actions which come to be by means of words (*verba*) of this sort, or by means of any *imagines* or sculptures, or anything else of this sort, are not natural, since they do not proceed from an internal (intrinsic) power, but are empirical and pertain to superstition. But actions which we said above follow [sc. ontologically] on (*consequi*) the forms of bodies are natural, since they proceed from internal principles (*ex principiis intrinsecis*). And these things said about occult operations and actions (*de operationibus et actionibus occultis*) are sufficient at present.[94]

Here Thomas returns to his central position that an art or practice can only be legitimate if it derives from natural structures and/or powers.

Thus, astrology *qua* knowledge is perfectly natural (and can be causally analyzed), and is thus legitimate. What we call "magic," on the other hand—namely, artificial man-made entities—whether talismans (*opera*) or powerful words (*verba*), can only act either by means of their material substrate—which is perfectly fine—or by spirits, by which Thomas means demons. Therefore, these objects may be legitimate and used effectually by means of their matter alone. By their artificially-imposed figures, however, they can only function as signs to and thereby be made effective by spirits. They are thus deemed illegitimate and superstitious, as he had

---

[93] Sicut autem ymagines ex materia naturali fiunt sed formam sortiuntur ex arte, ita etiam verba humana materiam quidem habent naturalem, scilicet sonos ab hominis ore prolatos, sed significationem quasi formam habent ab intellectu suas conceptiones per huiusmodi sonos exprimente. Unde pari ratione nec verba humana habent efficaciam ad aliquam immutationem corporis naturalis ex virtute alicuius cause naturalis, sed solum ex aliqua spirituali substantia (291–300).

[94] "Hee igitur actiones que per huiusmodi verba fiunt, vel per quascumque ymagines vel sculpturas vel quecumque alia huiusmodi, non sunt naturales, utpote non procedentes a virtute intrinseca, sed sunt emperice et ad superstitionem pertinentes. Actiones vero quas supra diximus consequi corporum formas sunt naturales, utpote ex principiis intrinsecis procedentes. Et hec de operationibus et actionibus occultis ad presens dicta sufficiant (300–10)." Thomas also draws the same conclusion earlier in this same text, where he explicitly calls the *imagines* discussed there "nigromantic": "Apparent etiam nigromanticarum ymaginum quidam effectus qui procedunt non ex aliquibus formis quas susceperint predicte ymagines, sed a demonum actione qui in predictis ymaginibus operantur (58–62)[.]"

also argued in relation to divination in *Summa theologiae* II.II.92-95. Both of these later texts, therefore, offer Thomas's normative—and decidedly negative—position about both talismans and powerful words. In his earlier *Summa contra gentiles* III.105, on the other hand, Thomas seems to have created an opening for legitimate all-natural talismans, in a text to which we will now turn.

7.8 [3] *Summa contra gentiles* III.105

The final text by Thomas to be explored here is the earlier *Summa contra gentiles* III.105 that was written between 1263 and 1265, and follows soon after the chapters I treated in Chap. 5 (III.82–94) on Thomas's analysis of divine providence.[95] In it, Thomas poses a problem related to those treated in both *Summa theologiae* II.II.96.2 and *De operationibus occultis naturae*: "Whence do the operations of magicians (*magorum operationes*) have their efficacy?"[96] He begins as follows by discussing a magician's words-sounds:

> Moreover, whence magical arts (*artes magicae*) have their efficacy remains to be investigated. This can easily be examined if we attend to their *modus operationis*. For, in their operations, they use certain words that are significative (*vocibus quibusdam significativis*) for producing determinate effects (*ad determinatos effectus producendos*). But *vox* [= sound, voice], insofar as it is significative (*significativa*), only has its power (*virtus*) from some intellect, either [1] from the intellect of the speaker, or [2] from the intellect of him/that to whom/which it is uttered.[97]

Thomas begins his answer by referring to magical utterances, namely, sounds that signify. He clarifies that these can only have power from an intellect in two ways, either from the intellect of the speaker or that of the hearer.

In *Summa theologiae* II.II.96.2, on the other hand, Thomas was concerned only with the effects of bodies on bodies, whereas in the *De operationibus occultis naturae* he also discussed *verba.* Here he first treats sounds and their power, which come only from and to intellects. Thus Thomas addresses magical practices related to the soul and body registers respectively. Both are forms of magical practices to us, but they are approached very differently. Signification and addressivity, however, are relevant to both. As we will see further in Chap. 8, these two modes correspond to

---

[95] Weill-Parot discusses this text and especially the loophole and its later attempts at interpretation; *Les "images astrologiques,"* 248–59. Brian Copenhaver discusses the tensions in Thomas's thought; "Scholastic Philosophy and Renaissance Magic," 531–34. He does not discuss the dating or order of the texts and thereby possible changes in Thomas's views. He discusses *Summa Theologiae* II.II.96.2, *De fato* and *Summa contra gentiles* III.105 all too briefly at 533.

[96] Unde magorum operationes efficaciam habeant.

[97] Investigandum autem reliquitur unde artes magicae efficaciam habeant. Quod quidem facile perpendi potest si modus operationis earum attendatur. In suis enim operationibus utuntur vocibus quibusdam significativis ad determinatos effectus producendos. Vox autem, inquantum est significativa, non habet virtutem nisi ex aliquo intellectu: vel ex intellectu proferentis; vel ex intellectu eius ad quem profertur (330a1–10).

Roger Bacon's bifurcation of *astronomia operativa* into the works (*opera*) and words (*verba*) of the wise. We will also explore the roots of both in al-Kindi's *De radiis stellarum*.

Thomas responds to each in turn:

> Indeed, from the intellect of the speaker, as if there were an intellect (*aliquis intellectus*) of such power (*tantae virtutis*) that could cause things by its conception (*sua conceptione res possit causare*), which conception it presents in producing effects in some manner (*producendis effectibus quodammodo*) by the function of a voice (*vocis officio*).[98]

Here Thomas describes how a magical speech-act in our terms could be performed by a speaker by the power of his own conception and its utterance.

Then he turns to the power derived from the hearer: "Moreover, from the intellect of him/that towards whom/which the speech (*sermo*) is directed, just as when, through the signification of a sound received in the intellect, hearing it, he is led (*inducitur*) to do something (*ad aliquid faciendum*)."[99] In describing the second circumstance, Thomas seems to allude to a more mundane rhetorical power of speech, namely, that intended to persuade and inspire actions of various sorts.[100]

Thomas then rejects the first possibility:

> But it cannot be said that these significative sounds (*voces illae significativae*) spoken by magicians (*magi*) have efficacy from the intellect of the speaker. For, since power follows on essence (*virtus essentiam consequatur*),[101] the diversity of the power shows forth the diversity of the essential principles. But the intellect of human beings of this disposition are commonly found whose understanding (*cognitio*) is caused from things (*ex rebus causatur*), rather than that by his conception he can cause things (*sua conceptione res causare possit*). Therefore, if there are some people who can transform (*transmutare*) things by their own power (*propria virtute*) with words expressing a conception of their intellect (*verbis conceptionem sui intellectus exprimentibus*), they will be of a different species, and would be called human beings (*homines*) equivocally.[102]

Here Thomas states clearly that in the normal course of events for human beings, things cause understanding and not the other way around: conceptions do not cause things! Thus, for Thomas, if a person has the power to cause things by their

---

[98] Ex proferentis quidem intellectu, sicut si aliquis intellectus sit tantae virtutis quod sua conceptione res possit causare, quam quidem conceptionem vocis officio producendis effectibus quodammodo praesentat (330a10-14).

[99] Ex intellectu autem eius ad quem sermo dirigitur, sicut cum per significationem vocis in intellectu receptam, audiens inducitur ad aliquid faciendum (330a14-17).

[100] Although rhetoric seems like a normal practice to us, in the ancient world it was virtually considered a branch of magic. See (e.g.) Jacqueline de Romilly, *Magic and Rhetoric in Ancient Greece*, Cambridge, MA: Harvard University Press, 1975.

[101] This refers to power following on something's specific or substantial form, as analyzed in both *De operationibus occultis naturae* and *Summa theologiae* II.II.96.2, as just discussed.

[102] Non autem potest dici quod voces illae significativae a magis prolatae efficaciam habeant ex intellectu proferentis. Cum enim virtus essentiam consequatur, virtutis diversitas essentialium principiorum diversitatem ostendit. Intellectus autem communiter hominum huius dispositionis invenitur quod eius cognitio ex rebus causatur, magis quam sua conceptione res causare possit. Si igitur sint aliqui homines qui verbis conceptionem sui intellectus exprimentibus res possint transmutare propria virtute, erunt alterius speciei, et dicentur aequivoce homines (330a17-29).

conceptions, they are not in essence a human being, that is, their specific form has somehow been modified and expanded beyond the range of human capabilities, since power follows essence, in this case, their substantial (= specific) form.

Thomas continues:

> Further, the power of making/doing (*virtus faciendi*) is not acquired through teaching (*per disciplinam*), but only the understanding for making (or doing) something (*cognitio aliquid faciendi*). But through teaching, some people acquire what makes/does (*efficiant*) magical operations (*operationes magicae*) of this sort. Therefore, there is no power (*virtus*) in them for producing effects of this sort, but understanding (*cognitio*) only.[103]

The distinction Thomas is trying to make here seems to be that only the *understanding* (*cognitio*) of explicitly magical procedures may be imparted by teaching, not the actual power itself to perform them. So, Thomas concludes, this power for making or doing—performing magical operations, whatever they are precisely—does not reside in the speaker's intellect, but only, rather, the very human capacity for understanding.

*

Thomas then presents two objections. The first uses astrology as a part of the analysis:

> But [sc. what] if someone were to say that people of this sort [sc. with magical powers] are alloted—by their nativity (*sua nativitate*, sc. their natal astrological configuration) from the power of the stars (*ex virtute stellarum*)—the aforesaid power (*virtus*) before others, such that, however much other people are instructed, whoever does not have this from their nativity cannot be effective (*efficaces*) in works (*opera*) of this sort?[104]

The anonymous objector thus inquires whether such magical powers derive from the magicians' natal astrological makeup, and thus from the stars.

This should be compared with Thomas's discussion of how the particular configuration of the planets at a person's birth can affect the particulars of their being, thus accounting for the differences between the individual members of a given species. In *De operationibus occultis naturae*, these differences derive from the latitude provided by the specific form of human beings for those categories that can change in a person or be different from another person, beyond which they cease to be human, as with Thomas's aforementioned magicians here. This should also be compared with the analysis in *Summa contra gentiles* III.92, where celestial influences help some people become better physicians, farmers and soldiers.

Thomas now analyzes this objection:

---

[103] Amplius. Virtus faciendi non acquiritur per disciplinam, sed solum cognitio aliquid faciendi. Per disciplinam autem aliqui acquirunt quod huiusmodi operationes magicas efficiant. Non igitur est in eis ad huiusmodi effectus producendos virtus aliqua, sed cognitio sola (330a30-35).

[104] Si quis autem dicat quod huiusmodi homines sua nativitate, ex virtute stellarum, sortiuntur prae ceteris virtutem praedictam, ita quod, quantumcumque alii instruantur, qui hoc ex nativitate non habent, efficaces in huiusmodi operibus esse non possunt (330a36-41):—

> First, one should say that celestial bodies (*corpora caelestia*) cannot make an impression (*imprimere* [sc. have an influence]) on the intellect, as was shown above [III.84]. Therefore, no one's intellect can be allotted this power from the power of the stars (*non igitur ex virtute stellarum sortiri potest intellectus alicuius hanc virtutem*), namely, that the representation [sc. utterance, etc.] of his conception by means of his voice is effective of anything (*repraesentatio suae conceptionis per vocem sit alicuius effectiva*).[105]

Celestial *bodies* only act directly on human *bodies*, as we frequently saw in Chap. 5 in earlier passages from this same text. Therefore, celestial bodies cannot bestow this power on anyone's intellect, at least not directly.

Thomas now offers a second objection, this time concerning the imagination:

> Moreover, [sc. what] if it is said that the imagination—on which the celestial bodies *can* make impressions, since its action (*operatio*) exists by means of a bodily organ (*per organum corporale*)—also does something (*aliquid* [...] *operatur*) in the utterance of signifying sounds (my emphasis)?[106]

The operation of the imagination exists by means of bodily organs, unlike the intellect. Can it then receive a special magical power derived from the celestial bodies?

Thomas rejects this possibility also:

> With respect to all the effects that come to be by such arts (*artes*), this cannot be the case. For it was shown that all the effects of this sort cannot be produced (*produci*) by the power of the stars (*virtute stellarum*). Therefore, someone cannot be alloted this power (*virtus*) from the power of the stars to produce these same effects.[107]

As he argued above, signifying words can only derive from and effect intellects. This is their mode of operation. True, the imagination's action exists by means of a bodily organ, but if celestial powers do not produce this effect in the first place, their influence on the imagination can have no influence in this respect. Thomas's argument here seems directed towards a central structure concerning the role of the imagination (and its *spiritus*) in the utterance of powerful signifying words in al-Kindi's *De radiis stellarum* that we will explore briefly in Chap. 8.

*

Having rejected every argument placing the power in the speaker, Thomas now turns to the hearer, or as Weill-Parot calls him, the addressee of this sort of addressative magic undertaken by means of powerful words:

---

[105] — primo quidem dicendum est quod corpora caelestia super intellectum imprimere non possunt, ut supra [cap. 84] ostensum est. Non igitur ex virtute stellarum sortiri potest intellectus alicuius hanc virtutem quod repraesentatio suae conceptionis per vocem sit alicuius effectiva (330a41-46).

[106] Si autem dicatur quod etiam imaginatio aliquid in prolatione vocum significativarum operatur, super quam possunt corpora caelestia imprimere, cum eius operatio sit per organum corporale (330b1-4):

[107] — hoc non potest esse quantum ad omnes effectus qui per huiusmodi artes fiunt. Ostensum est enim quod non possunt omnes huiusmodi effectus virtute stellarum produci. Ergo neque ex virtute stellarum aliquis sortiri potest hanc virtutem ut eosdem effectus producat (330b4-10).

> It remains, therefore, that effects of this sort are accomplished by some intellect towards whom/which the speech (*sermo* [signifying speech]) of the speaker directed sounds of this sort. And the indication (*signum*) of this is: for signifying sounds (*significativae voces*) of this sort which the magicians (*magi*) use are invocations, supplications, adjurations, or even commands (*imperia*), as if they were speaking with someone/something else (*alterum* [sc. something other than themselves]).[108]

With this argument and his listing of several types of speech acts—performative utterances—Thomas completes the first part of his account by showing that the power for realizing these effects must depend on the intellect of the recipient of these kinds of speech. He has yet to identify these addressees.

*

Now Thomas turns from words to works, that is, from *voces* to *opera* and *their* modes of signifying. Signification—whether by sound or writing—is at issue here, as we will also recall from *Summa theologiae* II.II.96.2 and *De operationibus occultis naturae*. Thomas also treats characters and figures here:

> Again, in the practices (*observationes*) of this art, they use certain specific characters and figures. But a figure is not a principle of any action or passion; otherwise, mathematical bodies (*mathematica corpora*) would be active and passive. Therefore, matter cannot be disposed (*disponi*) by means of specific figures (*figuras determinatas*) to receive some natural effect (*ad aliquem effectum naturalem suscipiendum*). Therefore, magicians (*magi*) do not use any figures as if they were dispositions. It remains, therefore, that they use them only as if they were signs. For there is no third [sc. option] to offer.[109]

Thomas now fully switches from words/sounds (*voces*) to works (*opera*), and he seems to refer here to talismans, or at least to material-corporeal objects that have both characters and shapes/figures. Here he makes a strong general statement about figures that is repeated several years later in *Summa theologiae* II.II.96.2, as we saw. Here he states that "a figure is not a principle of any action or passion." There he stated it thus: "a figure is not a principle of natural action." The upshot for both is exactly the same: Because matter cannot be disposed by means of figures to receive natural effects, figures must be used as signs, a conclusion he also drew there.

Now Thomas draws a conclusion that directly concerns addressative practices: "But we do not use signs except towards other intelligences (*intelligentes*). Therefore, magical arts (*magicae artes*) have their efficacy by the other intelligence towards whom/which the magician's speech is directed."[110] Here Thomas seems to

---

[108] Relinquitur igitur quod effectus huiusmodi compleantur per aliquem intellectum ad quem sermo proferentis huiusmodi voces dirigitur. Huius autem signum est: nam huiusmodi significativae voces quibus magi utuntur, *invocationes* sunt, *supplicationes, adiurationes*, aut etiam *imperia*, quasi ad alterum colloquentis (330b11-17).

[109] Item. In observationibus huius artis utuntur quibusdam characteribus et figuris determinatis. Figura autem nullius actionis principium est neque passionis: alias, mathematica corpora essent activa et passiva. Non ergo potest per figuras determinatas disponi materia ad aliquem effectum naturalem suscipiendum. Non ergo utuntur magi figuris aliquibus quasi dispositionibus. Relinquitur ergo quod utantur eis solum quasi signis: non enim est aliquid tertium dare (330b18-27).

[110] Signis autem non utimur nisi ad alios intelligentes. Habent igitur magicae artes efficaciam ab alio intelligente, ad quem sermo magi dirigitur (330b27-30).

functionally equate the signs inscribed by means of characters and figures with speech. Both are addressed to intelligences.

*

Thomas makes his next move concerning figures as he offers someone else's opinion:

> But if someone were to say that [1] some figures are closer to some of the celestial bodies (*figurae aliquae appropriantur aliquibus caelestium corporum*), and that [2] lower bodies (*corpora inferiora*) are determined by means of some figures (*determinantur per aliquas figuras*) to receive the impressions of some celestial bodies (*ad aliquorum caelestium corporum impressiones suscipiendas*), it seems that this is not said rationally. For no *patiens* is arranged (*ordinatur*) to receive the impression of an *agens*, except through that which exists '*in potentia*'. Therefore, those things [sc. the *agentes*] alone determine it [sc. the *patiens*] to receive a specific (*specialis*) impression by means of what exists somehow '*in potentia*' [sc. in the *patiens*].[111]

This opinion thus rejects the natural account of talismans as articulated by Albert in the *De mineralibus* and implied in the *Speculum astronomiae*—that figures in themselves can affect lower bodies—unless the bodies are naturally predisposed in the first place to receive celestial influences, that is, by means of a naturally-endowed specific form. Thomas here is thus perforce discussing artificial man-made objects that have no such innate specific form.

Thomas responds to this possibility thus:

> But matter is not disposed by figures in such a way that it would exist '*in potentia*' in relation to a form because a figure abstracts in accordance with its nature (*ratio*) from all perceptible material and form, since it is something mathematical (*quoddam mathematicum*). Therefore, no body is determined by figures or characters (*non ergo per figuras vel characteres determinatur aliquod corpus*) to receive an influence of a celestial body (*ad suscipiendam aliquam influentiam caelestis corporis*).[112]

Thus, the '*patiens*' cannot receive celestial influences due to such a disposition by a figure because figures are fundamentally mathematical and thereby abstract from all perceptible matter and form.[113] The issue concerns figures and what they can affect in matter. The upshot of this not altogether clear argument will be resolved in the loophole, which is concerned precisely with this issue.

---

[111] Si quis autem dicat quod figurae aliquae appropriantur aliquibus caelestium corporum; et ita corpora inferiora determinantur per aliquas figuras ad aliquorum caelestium corporum impressiones suscipiendas: videtur non rationabiliter dici. Non enim ordinatur aliquod patiens ad suscipiendum impressiones agentis nisi per hoc quod est in potentia. Illa ergo tantum determinant ipsum ad specialem impressionem suscipiendum, per quae in potentia fit quodammodo (330b31-41).

[112] Per figuras autem non disponitur materia ut sit in potentia ad aliquam formam: quia figura abstrahit, secundum suam rationem, ab omni materia et forma sensibili, cum sit quoddam mathematicum. Non ergo per figuras vel characteres determinatur aliquod corpus ad suscipiendam aliquam influentiam caelestis corporis (330b41-331a2).

[113] Perhaps the most interesting and innovative feature of Quinlan-McGrath's argument in *Influences* is her analysis of the mathematical dimension of figures in her chapter 7.

Thomas continues:

> In addition, some figures are closer to celestial bodies as their effects (*figurae aliquae appropriantur corporibus caelestibus ut effectus ipsorum*). For the shapes of lower bodies are caused by celestial bodies (*figurae inferiorum corporum causantur a corporibus caelestibus*). But, the aforementioned arts do not use characters or figures as the effects of celestial bodies; rather, they are the effects of the person operating through art (*effectus hominis operantis per artem*). Therefore, the closeness (*appropriatio*) of the figures to some celestial bodies seem to do nothing with respect to what was proposed.[114]

Bodies in general are affected and shaped by celestial bodies, but this has nothing to do with characters or figures, which are instead the effects of the magicians and not, therefore, of the celestial bodies themselves.

Thomas continues, developing this point concerning matter:

> Again, natural matter (*materia naturalis*) is not disposed in any way toward form by means of figures, as was shown. Therefore, bodies (*corpora*) in which figures of this sort have been impressed (*impressae*) have the same ability (*sunt eiusdem habilitatis*) to receive celestial influences as do other bodies of the same species.[115]

Artificially inscribed figures thus have no effect whatsoever on an object's ability to receive celestial influences, thereby further undermining talismanic theory. All bodies of the same species receive celestial influences in exactly the same way regardless of anything inscribed on them.

Magical objects do not work by means of nature, however, but via intellect:

> But, that something acts on one of the things that are equally disposed—because of something close to it found there, and not in something else—is not [sc. the result/effect] of something operating through the necessity of nature (*necessitas naturae*), but by a choice (*electio*). Therefore, it is clear that arts of this sort using figures for producing some effects (*artes figuris utentes ad effectus aliquos producendos*) do not have their efficacy from some agent through nature (*non habent efficaciam ab aliquo agente per naturam*), but from some intellectual substance acting by means of its intellect (*ab aliqua intellectuali substantia per intellectum agente*). The name which they impose on such figures—calling them "characters"—shows this too. For a character is a sign, in which it can be understood that they do not use figures of this sort, except as signs exhibited to some intellectual nature (*ut signis exhibitis alicui intellectuali naturae*).[116]

---

[114] Praeterea. Figurae aliquae appropriantur corporibus caelestibus ut effectus ipsorum: nam figurae inferiorum corporum causantur a corporibus caelestibus. Praedictae autem artes non utuntur characteribus aut figuris quasi effectibus caelestium corporum, sed sunt effectus hominis operantis per artem. Appropriatio igitur figurarum ad aliqua caelestia corpora nihil ad propositum facere videtur (331a3-11).

[115] Item. Per figuras non disponitur aliqualiter materia naturalis ad formam, ut ostensum est. Corpora igitur in quibus sunt impressae huiusmodi figurae, sunt eiusdem habilitatis ad recipiendum influentiam caelestium cum aliis corporibus eiusdem speciei (331a12-17).

[116] Quod autem aliquid agat in unum eorum quae sunt aequaliter disposita, propter aliquid sibi appropriatum ibi inventum, et non in aliud, non est operantis per necessitatem naturae, sed per electionem. Patet ergo quod huiusmodi artes figuris utentes ad effectus aliquos producendos, non habent efficaciam ab aliquo agente per naturam, sed ab aliqua intellectuali substantia per intellectum agente. Hoc etiam demonstrat et ipsum nomen quod talibus figuris imponunt, *characteres* eos dicentes. *Character* enim signum est. In quo datur intelligi quod figuris huiusmodi non utuntur nisi ut signis exhibitis alicui intellectuali naturae (331a17-b8).

The term "character" itself further indicates that these are in the nature of signs to intellects. Thus magical acts act by means of intelligences, not nature, as Thomas had argued earlier.

7.9 Thomas's Loophole

Now Thomas comes to a critical passage in his argument that seems to undermine his earlier arguments and render them moot. And here he refers specifically to talismans (*imagines*):[117]

> But because figures exist in artificial [sc. man-made] things (*figurae in artificialibus sunt*) as if they were specific forms (*quasi formae specificae*), someone can say (*potest aliquis dicere*) that nothing prohibits the construction of a figure that gives a species to a talisman (*constitutionem figurae, quae dat speciem imagini*), and that some power (*aliqua virtus*) from a celestial influence (*ex influentia caelestia*) could follow upon it [sc. ontologically] (*consequatur*), not because it is a figure (*non secundum quod figura est*), but because it causes the species of the artificial thing (*secundum quod causat speciem artificiati*), which acquires the power from the stars (*virtutem ex stellis*).[118]

But someone might say that the figures function in man-made objects precisely as specific forms that give their species to an object. Therefore, they could then "naturally" receive the relevant celestial influences.[119]

Given its nature, it is striking that Thomas does not even bother to respond to this potentially devestating counter-argument. Indeed, by not doing so, Thomas seems to leave a gaping loophole here to justify the production of natural talismans of the type that Albertus Magnus and the *Speculum astronomiae* defended as legitimate. As we will see, Thomas's influential commentator Cardinal Caietanus (Tommaso de Vio) interpreted him in precisely this way. Did Thomas try to atone for this surprising laxity in the later works already discussed, namely, *Summa theologiae* II.II.96.2 and *De operationibus occultis naturae*, which nowhere mention this argument and which present a staunchly anti-talismanic position? I am inclined to answer in the affirmative.

Thomas continues: "But about the letters and other characters with which something is inscribed on a talisman (*de literis quibus inscribitur aliquid in imagine, et aliis characteribus*), nothing else can be said than that they are signs. Whence

---

[117]At *Les "images astrologiques,"* 248, Weill-Parot calls this passage "curieux et extremement surprenant," especially in light of Thomas's otherwise staunchly consistent anti-talismanic position that he had just systematically reconstructed (228–47).

[118]Quia vero figurae in artificialibus sunt quasi formae specificae, potest aliquis dicere quod nihil prohibet quin constitutionem figurae, quae dat speciem imagini, consequatur aliqua virtus ex influentia caelesti, non secundum quod figura est, sed secundum quod causat speciem artificiati, quod adipiscitur virtutem ex stellis (331b9-15).

[119]Brian Copenhaver treats this passage usefully in his "Scholastic Philosophy and Renaissance Magic," 533. Weill-Parot treats it in much richer detail, including the interesting ways that later interpreters came to understand it; *Les "images astrologiques,"* 249–57.

they do not have an *ordo* [sc. a structuro-ontological relationship] except to an intellect."[120] Of letters and characters, however, there is no doubt that these are signs to communicate with other intelligences. With figures that function as the specific forms of man-made objects, on the other hand, Thomas seems to leave an opening here, perhaps as a conciliatory gesture to his much esteemed teacher, Albertus Magnus.

Thomas completes his analysis with a more straightforward example to make his point: "This is also shown by sacrifices, prostrations and other things of this sort that they use, which cannot be anything except signs of reverence exhibited to an intellectual nature (*signa reverentiae exhibitae alicui intellectuali naturae*) [sc. and are thus idolatrous]."[121] Thomas alludes here to ritual magical practices at the end, to which there is also no doubt of their superstitious and illegitimate nature. Thus he moves from talismans per se to ritual magic, thereby emphasizing his analysis of the nature of addressative practices.

In the end, there seems to be a striking tension within Thomas's thought, perhaps indicating an earlier lenience in his total rejection of talismans, potentially permitting all-natural ones with figures, but in no way allowing any with signs or words—even straightforward Latin terms, as in the *Speculum astronomiae*—designed to communicate with unidentified intelligences, which we may confidently assume are demons. Whatever ambiguity in his position is indicated by the earlier *Summa contra gentiles*, however, Thomas seems to have decidedly shut and locked this gate in his relevant later works.

7.10 Conclusion

With the analyses discussed here, Thomas shows full awareness of the issues at stake, since, in the previous chapter of the *Summa Theologiae* (II.II.95), as we saw, he discussed in detail astrology's relation to divinatory practices, arguing forcefully for astrology's legitimacy *qua* knowledge precisely because it is all-natural. Nevertheless, for the later Thomas, astrological talismans of all types cross that very same threshold of legitimacy, which both Albert and the Magister Speculi (despite some striking differences in terminology) had extended to one particular all-natural type of talisman. In so doing, at the end of his career, Thomas intended to wholly remove astrology's operative or magical dimension, rendering it fully and only a knowledge-oriented practice, even though he seems to have offered a significant loophole specifically for talismans in the earlier *Summa contra gentiles*. Perhaps it was the publication of the *Speculum astronomiae* itself—to which he unmistakably

[120] Sed de literis quibus inscribitur aliquid in imagine, et aliis characteribus, nihil aliud potest dici quam quod signa sunt. Unde non habent ordinem nisi ad aliquem intellectum (331b15-19).

[121] — Quod etiam ostenditur per sacrificia, prostrationes, et alia huiusmodi quibus utuntur, quae non possunt esse nisi signa reverentiae exhibitae alicui intellectuali naturae (331b19-22).

refers in *Summa Theologiae* II.II.96.2 with the phrase '*imagines astronomicae*'—that pushed Thomas to close this loophole.

Albert, Thomas and the Magister Speculi thus all completely agree that astronomy and astrology are legitimate practices offering conjectural knowledge of the past, present and future. In this respect, the only controversial issue concerned astrological images or talismans, one non-demonic, natural part of which Albert and the Magister Speculi considered legitimate. The later Thomas, however, rejected it entirely. In Chap. 8, I will compare these positions with Roger Bacon's.

*

Not surprisingly, the tensions between these diametrically opposed positions on talismans by two of the most prominent Dominican authorities continued to have traction with their followers. This bifurcation was in fact used by Jerome Torrella to structure his *Opus praeclarum de imaginibus astronomicis* of 1496, as we will see more fully in the conclusion to volume II.[122] The remaining knowledge-oriented (*iudiciaria*) parts of astrology, however, were perfectly legitimate for both Thomas and Albert—a point worth emphasizing—as long as certainty in prediction was not claimed, free will not undermined, nor necessity in nature entailed.

Furthermore, although Thomas's mature position as represented here seems adamantly set against any sort of natural and thus legitimate astrological talismans, he was not always interpreted in this way, at least in the 16th and 17th centuries.[123] In fact, in the officially endorsed Tridentine edition of Thomas's *Summa theologiae* with Cardinal Caietanus's (Tommaso de Vio's [1469–1520]) commentary (composed between 1507 and 1520), Thomas was interpreted to have a pro-natural view of at least some kinds of talismans.[124]

Among other things, Caietanus approached Thomas's loophole in *Summa contra gentiles* III.105 by arguing that the discussion in the later *Summa theologiae* II.II.96.2 was not a retraction of the earlier statement. Rather, it was a refinement.[125] First of all, he argues, Thomas did not state the loophole *in propria persona*, so there is no question of a later retraction of his own views. Likewise, the loophole should not be taken as a blanket statement concerning all figures in relation to all man-made objects. Rather, it should be read only in relation to particular figures that have

[122] Weill-Parot, *Les "Images astrologiques,"* 223.

[123] A full study of the later history would also be of great interest. Weill-Parot discusses some of this story in *Les "images astrologiques,"* 249–57, including Marsilio Ficino, Pedro Garsias and Cardinal Caietanus.

[124] Weill-Parot (249–57) discusses Caietanus's interpretation of Thomas. He also discusses the quasi-official commentary on the *Summa contra gentiles* by Francesco Silvestri Ferrariensis, who was appointed the commentator precisely by Caietanus.

[125] Ad hominem vero, quia aut Auctor hic retractat quod concessit in III *Contra Gent.*, cap. cv: aut non. Si non, sequitur quod imagines astronomicae sint tolerabiles si sine characteribus fiant: et quod conclusio istarum rationum non ut sonat hoc in loco sit accipienda, sed moderanda, et oportet glossam invenire non destruentem hunc textum (332b26-31).

this function with particular *artificialia*—namely, talismans with particular types of figures that have a close relationship with particular celestial influences.[126]

Finally, in his extremely valuable 1668 *Lexicon Mathematicum Astronomicum Geometricum*, Gerolamo Vitali (1623–98) also discussed the tensions in Thomas's thought in his richly informative article '*Imagines astronomicae*', where Vitali also interprets Thomas in this way.[127] After briefly discussing Ficino's *De vita*, Vitali turns to the theologians:

> All the theologians as a group (*communiter*) reprove these [sc. *imagines astronomicae*] as superstitious except one Caietanus, who both in his commentary on Thomas's II^{a} II^{ae}—question 96, article 2—and in the *Summa*, doggedly (*mordicus*) defends them. He discusses them as licit and natural whenever it does not admit specific letters, characters, numbers or points, which are obviously signs for a tacit and implicit invocation of demons, as Thomas well understood. Furthermore, purely artificial images (*imagines pure artificiales*), which are formed with celestial signs, as of Leo, Virgo, etc., and are fabricated under a specific celestial configuration (*fabricatas in certa constellatione*) he makes wholly immune from all superstition. And in confirmation of this statement, he offers the same Angelic Doctor's *Summa contra gentiles* III.105, where he seems to approve talismans (*imagines*) of this sort, when they are made without characters and unknown marks (*notis ignotis*).[128]

According to Vitali, Caietanus interprets Thomas—in his official sixteenth-century commentary—as accepting all-natural talismans as licit.

As part of his article, Vitali also quotes verbatim from Albert's positive discussion in *De fato* of the *radiatio periodi* (discussed above) as *Opusculum* 28 (art. 4) by

---

[126] I will explore Caietanus's fascinating and influential argument more fully in volume III. It is printed on pp. 332–33 of the Leonine edition. In the meantime, Weill-Parot gives a full interpretation at *Les "images astrologiques,"* 250–55. Weill-Parot does not discuss Vitali, so I treat him in greater detail here.

[127] Paris: Billaine, 1668; repr. Giuseppe Bezza (ed.), La Spezia: Agora, 2003, 237–40.

[128] "Theologi omnes communiter eas improbant, uti superstitiosas, excepto uno Caietano, qui tum in Commentar. ad secundum secundae D. Thomae quest. 96. art. 2. tum in Summa mordicus eas defendit, aut licitas et naturales afferit quandocumque non admittat certas literas, characteres, numeros, aut puncta, quae plane signa sunt ad tacitam et implicitam daemonis invocationem, ut bene habet D. Thomas. Caeterum imagines pure artificiales, quae conformentur cum caelestibus signis, ut Leonis, Virginis, etc., et fabricatas in certa constellatione, immunes facit ab omni prorsus superstitione. Atque in confirmationem sui dicti affert ipsum Angelicum Doctorum *3. contra Gentes cap. 105.* ubi videtur approbare huiusmodi imagines, quando sine characteribus, et notis incognitis fiant. Quod putat Caietanus sanctum Doctorum non retractasse in dicto articulo secundo, sermonem de dictis imaginibus inducendo, quas dicit continere pactum tacitum cum daemone, eo quia *necesse est,* ait, *eis inscribi quosdam characteres, qui naturaliter ad nil operantur.* Ergo, infert Caietanus, quae hos characteres non habent, sed puras imagines caelestes praeseferunt, eas non excludit Angelicus Doctor, alias debuisset aliam rationem adducere, et alterius loci ubi de iis sermonem habuit meminisse, et retractasse (237–38, my transcription, as for all the Vitali texts cited here)." Vitali also has this to say: "Tandem concludam cum ii quae habet idem Caietanus in summa. *Imaginibus,* inquit, *pure Astronomicis uti ad effectus subiectos caelestibus corporibus nullum peccatum esse videtur, quia, ut res pure naturales habentur, nec debemus secreta naturae, praesertim caelestes influxus nostri ingenii viribus metiri; ut priscis astronomis non deferamus. Dixi autem, pure Astronomicas, ut nulla superstitiosa observatio sit adiuncta, aut suffumigationum, aut characterum; quoniam hac constat non pertinere ad caelestes influxus, sed ad secretos intellectus, quorum societas nobis est inderdicta.* Hucusque Caietanus (239–40)."

Thomas himself.[129] Strangely, Albert's name does not appear at all in Vitali's four-page discussion.[130] As we will see in volume II, Ficino in the *De vita* also thought that *De fato* was by Thomas, and that the *Speculum astronomiae* was actually by Albert. *Fata sua habent libelli*! We will now turn to Roger Bacon's differently configured understanding of astrology vis-à-vis what we would call magic. As we will see, there are significant continuities between Bacon's thought and the works discussed here in this chapter as well as pointed differences.

[129] Quapropter optime dixit D. Thomas *opusc. 28. art. 4.* Anceps adhuc de naturali istarum imaginum virtute deque efficientia caelestium corporum in isthaec artificiata, quod *sicut radiatio periodi dispositionem ordinis, et durationis imprimit rebus naturalibus; ita etiam imprimit artificiatis: propter quod figura imaginum magicarum ad aspectum stellarum imprimuntur, et fieri permittuntur* (239).

[130] I discuss Caietanus's and Vitali's positions in greater detail in volume III.

Chapter 8
Opera et verba sapientiae: Astrology and Magic in Roger Bacon

8.1 Introduction

In order to explore the particular configuration of Roger Bacon's views on what we call magic and thus complete Part III of volume I, I will now treat the second half of what Bridges printed as the last section of *Opus Maius*, Book IV, the section entitled '*Astrologia*' (390–403). As we will recall, I discussed the first part of this section in the Excursus and in Chap. 2. The part of this section to be discussed here treats the uses of astrology that go beyond mere knowledge—what we would call "magic," but which Roger decidedly does not, using instead the phrase '*opera et verba sapientiae*' ("the works and words of wisdom"). It thus falls within the operative (*operativa*) part of the science of the stars, not the interpretive or judicial (*iudiciaria*) part that was explored in Chaps. 4 and 6. For Roger, as we saw, what he calls "magic" is entirely negative, as what they call "divination" is also for both him and Thomas, and "necromancy" is for the Magister Speculi and "nigromancy" for Thomas. This chapter should be read in relation to the contemporary discussion of talismans and the power of words in Albert's authentic works, the *Speculum astronomiae*, and Thomas Aquinas's discussions in *Summa contra gentiles* III.105, *Summa Theologiae* II.II.96.2 and the *De operationibus occultis naturae* as just discussed in Chap. 7.

In this and related texts, Roger treats the operative, technological, or what I have been calling the magical side of *astronomia*, namely the making of astrological images or talismans and other related activities, including another type of what we and Thomas would call magical practice deriving from the power of words (*de virtute verborum*), which sharply differentiates Roger's magic from that in Albertus Magnus and the *Speculum astronomiae*. It also aligns Roger much more closely with al-Kindi's *De radiis stellarum*, a text we know that he explicitly mentioned, as we will see below. Although we would call these practices magical, Roger himself would vehemently resist this usage, as we will continue to see. This text in particular will help us to differentiate *our* use of the term "magic" from Roger's strongly

H D. Rutkin, *Sapientia Astrologica: Astrology, Magic and Natural Knowledge, ca. 1250–1800*, Archimedes 55, https://doi.org/10.1007/978-3-030-10779-6_8

and consistently held practice. This discussion will also help us to further clarify the useful distinction between terminological and conceptual anachronism.

We will repeatedly see how closely related Roger's texts discussed in Chaps. 4, 6, and in this chapter are in both subject matter and in purpose, all either promoting or defending his consistent and strongly held positions. Although he treats what he considers to be important issues in an extremely repetitive manner, Roger adds new information with each iteration, thus ultimately allowing a fuller picture to emerge. Changes and developments over time, at least in the passages under review here, are difficult to discern beyond the etymological shift discussed in Chap. 4 above.[1]

8.2 *Astronomia Operativa*

Roger introduces this extensive operative section by asserting that it is not sufficient for a republic merely to know what knowledge and insight astrology has to offer. Rather, astrology should also be put into action for promoting what is beneficial and preventing or undermining what is harmful, precisely the criteria that we had encountered earlier, namely, the profound practical benefits to be derived from this hard-earned knowledge:

> But the fifth thing announced above is the most important (*principale*) among these. For all were named because of this one because it is not sufficient for a republic that everything be understood (*cognoscantur*), but it is fitting that useful things that are understood (*utilia cognita*) be promoted (*promoveantur*), and harmful things removed (*nociva evacuentur*).[2]

We have now arrived at the goal toward which this broader section in Roger's *Opus minus* (and also *Opus maius* IV) had been heading: Mathematico-astrological knowledge should become operational, and thus be put to good use.

In discussing mathematics (the *quadrivium*), Roger claims that, although the other mathematical disciplines make great contributions, the science of the stars

[1]Although the '*Astrologia*' section of the *Opus minus* will be the only text I treat in this chapter, a full treatment of Roger's views on what we call magic would also discuss four other texts: Roger's concise and informative introduction to his edition of the *Secretum secretorum*, his *De secretis operibus artis et naturae, et de nullitate magiae*, his *Opus tertium*, chapter 26, and *De laudibus mathematicae*, which is known from a fascinating but not widely used study by George Molland. With these texts and against the background of those discussed in Chaps. 4 and 6, we will get a much better sense of Roger Bacon's characteristic and strongly held views on astrology in itself, on the one hand, and in relation to natural knowledge, theology/religion and magic, on the other. Boudet discusses Roger at *Entre science et nigromance*, 234–39. Weill-Parot does at *Les "images astrologiques*," 316–38.

[2]"Quintum vero annunciatum superius est principale inter haec, nam nominata omnia sunt propter hoc; quia non sufficit rei publicae ut cognoscantur omnia, sed oportet quod utilia cognita promoveantur, et nociva evacuentur (390, 3–6)." Sorge and Seller also translate this entire '*Astrologia*' section into Italian; *Filosofia, scienza, teologia* (139–153). I draw on it where appropriate to help clarify my translation.

(*astronomia*) rules everything, within which astrological elections make the primary contribution:

> But here, although geometry, arithmetic and music give the greatest and most aid, nevertheless, the science of the stars rules everything (*astronomia regulat omnia*), for this reason, that every excellent work (*omne opus magnificum*) ought to be made (*fieri*) at chosen times (*in temporibus electis* [i.e. times chosen by astrological elections]). And therefore, the works (*operata*) of the other sciences require fitting times (*tempora idonea*), which, by considering these things, the astrologers understand how to choose (*astronomi eligi cognoscuntur*). Therefore, it surpasses (*praeest*) every body of knowledge (*omnibus scientiis*) in this part [sc. the *quadrivium*], even though there are an infinite number of specific remedies (*specialia remedia*), in which the other sciences do not have power. And because the *potentia* [power, potential] of this science [sc. *astronomia*] is so universal, therefore Aristotle, the wisest philosopher, teaches Alexander in the *Book of Secrets* (*in libro secretorum* [sc. the *Secretum secretorum*]) that he should neither eat nor drink nor do anything else without the counsel of an astrologer (*sine consilio astronomi*) because times are chosen for everything; for every thing has a time (*omnia* [...] *tempus habent*), as Solomon says, who is wiser than Aristotle.[3]

Knowledge of *astronomia* is thus of the greatest practical importance, primarily by virtue of its insight into the timing of actions and events. Here elections join Roger's discussion of nativities and revolutions (including great conjunctions) in the sections of *Opus maius* IV discussed in my Chaps. 4 and 6. Roger thus employs three of the four canonical types of astrological practice. I do not know of him using the fourth, interrogations. Furthermore, two very big guns—one secular and one sacred: Aristotle and Solomon—are marshaled here as Roger's authorities.

Having alluded to astrology's power and the *Secretum secretorum* (towards which he will turn in earnest momentarily), Roger returns to his earlier concern with determinism and necessity, especially as they relate to both God's absolute power and human free will:

> And Aristotle objects, saying to him [sc. Alexander] that God foresees (*praevidit*) everything from eternity, therefore the astrologer (*astronomus*) cannot impede or change these things.[4] And he responds that those things which God foresees are immutable. Nevertheless, the effect of those things which God placed in the power of human beings (*in potestate hominis*) from his eternal foresight (*ex sua praevisione aeterna*) is that a human being can change something as he wishes (*potest homo mutare ut vult*) because it is [sc. in the nature of] contingent things to go one way or another (*in contingentibus ad utrumlibet*).[5] And in

[3] Sed hic licet geometria, et arithmetica, et musica dent maxima et quamplurima iuvamenta, tamen astronomia regulat omnia, propter hoc quod omne opus magnificum debet fieri in temporibus electis. Et ideo operata aliarum scientiarum requirunt tempora idonea, quae consderatione astronomi eligi cognoscuntur, et ideo praeest omnibus scientiis in hac parte, quamvis etiam specialia remedia habeat infinita, in quibus aliae scientiae non habent potestatem. Et quia tam universalis est potentia istius scientiae, ideo Aristoteles sapientissimus philosophus docet Alexandrum in libro Secretorum, quod nec comedat, nec bibat, nec aliquid faciat sine consilio astronomi, quia tempora electa sunt ad omnia; omnia enim tempus habent, sicut dicit Salomon sapientior Aristotele (390, 6–19).

[4] Deus praevidit omnia ab aeterno, ergo non potest astronomus ea impedire vel mutare (390, 19–21).

[5] This is the same terminology that we saw also in Thomas's *Summa theologiae* and in the *Speculum astronomiae*.

> human affairs (*in rebus humanis* [as we saw above in chapter 4 on *Opus maius* IV, 249–53]), God did not establish necessity (*necessitas*)—although he will have known from eternity (*sciverit ab aeterno*) in which part something contingent ought to end up—and [sc. there exists] freedom of the human will (*humani libertas arbitrii*).[6]

God's providential creation, ordering and governance of the world together with man's free will provide the overarching normative framework in which astrology should be understood, and which preserves its legitimacy. Both structures are required within Bacon's astrologizing Aristotelian system, on the one hand, and neither requires necessity in the world or in human affairs, on the other. This is, of course, wholly consistent with what we have seen so far in Thomas, Albert and the *Speculum astronomiae.*

According to the "Aristotle" of the *Secretum secretorum*, astrology *qua* knowledge can be extremely valuable—even life saving—as a resource for informing people and thus their actions. One of the main areas is medical:

> And therefore a person can attend to all their useful things (*utilitates*) and remove their impediments, if someone were experienced in this body of knowledge (*peritus in hac scientia*). And he gives an example: for if, as he says, there will necessarily be a superfluity of cold at a future time, a foreseeing astrologers (*astronomus praevidens*) can apply remedies, so that he can endure that cold without hardship, which others, unprepared, could not endure. For an astrologer (*astronomus*) can prepare hot places, hot food, hot clothing and many medicines (*medicamina*), such that the cold will not harm him in any way, although others who will not have foreseen these things would die of the cold. And therefore it is within a person's power experienced in this knowledge (*in potestate hominis periti in hac scientia*) to evade harmful things and pursue useful ones.[7]

Astrological foreknowledge, then, does not undermine free will. Rather, it helps one to make more fully informed decisions that may well contribute beneficially in the medical as in other human realms by informing peoples' actions. This way, if one knows about a problem in the offing, preparations can be made in a timely manner. We saw a similar argument in *Speculum astronomiae*, chapter 13, concerning nativities in Chap. 4.

Roger now offers another medical example:

> And Isaac [Israeli] teaches this excellently in his book *On fevers*, in the chapter on pestilence. And not only about illnesses, but also universally concerning everything, saying that

[6] Et Aristoteles sibi obiicit dicens, Deus praevidit omnia ab aeterno, ergo non potest astronomus ea impedire vel mutare. Et ipse respondet, quod ea quae Deus praevidit sunt immutabilia. Sed tamen effectus eorum quae Deus posuit in potestate hominis ex sua provisione aeterna, potest homo mutare ut vult, quia in contingentibus ad utramlibet. Et in rebus humanis Deus non posuit necessitatem, licet sciverit ab aeterno in quam partem debet terminari contingens, et humani libertas arbitrii (390, 19–27).

[7] Et ideo potest homo omnes utilitates suas procurare, et impedimenta removere, si fuerit peritus in hac scientia. Et ponit exemplum. Si enim, ut dicit, necessario erit superfluitas frigoris in tempore futuro, astronomus praevidens potest apponere remedia, ut sine gravamine patiatur frigus illud, quod alii incaute non potuerunt sustinere: potest enim astronomus praeparare loca calida, nutrimenta calida, vestimenta calida et medicamina multa, ita ut frigus ei in nullo nocebit, quamvis alii qui haec non praeviderint prae frigore moriantur. Et ideo in potestate hominis periti in hac scientia est ut evadat nociva et consequatur utilia (390, 27–38).

> the rational soul is more worthy than the stars (*anima rationalis dignior est stellis*). For their action [sc. the stars'] harms it [sc. the *anima rationalis*] minimally, unless it is held with ignorance.[8] Provided that he will have known their [sc. the stars'] action, he will be able to take care in general. And he adds, moreover, that what we see done by the greatest astrologers (*ab optimis* [...] *mathematicis*) bears witness, who, although they will most certainly (*certissime*) know the day of death among those who will set forth on the sea, they can sufficiently look after them; likewise also those who will die in war. And he adds, "A human being (*homo*) differs from the animals, for it is the nature of reason and discretion to inquire and desire good things, and to reject bad things and resist them, because he [sc. a human being] is not subject to natural action (*actio naturalis*), unless it is held with ignorance."[9]

Astrological foreknowledge can thus be profoundly beneficial in medicine as well as in every other part of life. Roger also takes pains to emphasize here that the rational soul is not subordinated to natural action, in this case, the action of the stars which are, after all, celestial *bodies*, thus implicitly distinguishing between the body and soul registers. He will develop this fundamental distinction further below.

After another medical example, the next authority to whom Roger turns is the pseudo-Ptolemy of the *Centiloquium*:

> And therefore, Ptolemy says in the fifth *verbum* of the *Centiloquium*: "the best astrologer (*astrologus optimus*) can greatly prohibit what will happen in accordance with the stars"; and in the eighth *verbum*: "the wise soul (*anima sapiens*) will help the work of the stars (*adiuvabit opus stellarum*) as a sower (*seminator*) helps natural powers (*fortitudines naturales*), whence he can also repel harmful things, and promote useful ones."[10]

According to the *Centiloquium*, astrological knowledge can be used to promote positive things and repel harmful ones. In this way, human beings can work harmoniously with the stars. Thus, Roger seems to be planting the seeds of an ideology for what *we* would call magic (and he decidedly does not), where human beings experienced in the science of the stars can help further the work of the stars. The pseudo-Aristotelian *Secretum secretorum* and the pseudo-Ptolemaic *Centiloquium* are central authorities for Roger here.

Roger then turns to another medical example from Isaac and draws the same conclusions:

> And Isaac offers the example of future pestilence, and poses a question to himself in accordance with the ignorance of physicians (*ignorantia medicorum*), as Aristotle did. Therefore,

---

[8] That is, unless one is ignorant of these stellar actions.

[9] Et hoc egregie docet Isaac in libro Febrium, capitulo de pestilentia; nec solum de infirmitatibus, sed etiam universaliter in omnibus, dicens quod anima rationalis dignior est stellis, nam actio earum minime ei nocet nisi cum ignorantia teneatur. Dum enim ipse sciverit earum actionem omnino cavere poterit; et infert, Testificatur autem id quod videmus ab optimis factum mathematicis, qui cum certissime noverint dies mortis in eis qui mare ingressuri sunt, satis possunt eos tueri; similiter et eos qui in bello morituri sunt; et addit, Homo differt ab animalibus, nam rationis et discretionis est super inquirere et desiderare bona et respuere mala, et eis repugnare, quia non est subditus actioni naturali, nisi cum ignorantia tentus sit (391, 1–13).

[10] Et ideo dicit Ptolemaeus in quinto verbo Centilogii, astrologus optimus multo magis prohibere poterit quod secundum stellas venturum est; et in octavo dicit, anima sapiens adiuvabit opus stellarum, quemadmodum seminator fortitudines naturales, unde et nociva poterit repellere, et utilia promovere (391, 13–18).

> he asks whether putrefaction of this sort will come necessarily due to the heavens because then a physician (*medicus*) cannot prohibit it, and therefore it seems useless to know this in advance (*praecognosci*). But he resolves [sc. this question], saying that corruption of the air does not harm bodies prepared for it.[11] Therefore, when a physician will see the signification of corrupt air at a future time of year (*in unoquoque tempore anni venturo*), he will be able to improve (*mundificare*) nature, and remove the bad disposition of the body, as he knows how to fight against the coming corruption. For, if he will see by the motion of the stars that a too great heat—dryness and burning—will come to pass in the air, he will hasten to purge the choler of those in whom there is a choleric complexion (*complexio colerica*), giving cold and moist things that fight against the coming heat. And Haly gives a similar opinion in his exposition of the fifth proposition of the *Centiloquium*.[12]

Foreknowledge gained through experience in the science of the stars can thus allow the skilled physician to prepare appropriate remedies to counteract what would otherwise be straightforwardly negative influences on someone's unprepared body in relation to their complexion.

Now Bacon offers his conclusion:

> From these things it is clear that the astrologer (*astronomus*) can give remedies against harmful things and promote useful ones, not only with sicknesses, but also in other things generally (*universaliter*), by reason of the election of a time fitted to every work (*cum ratione electionis temporis apti ad omnia opera*), and by reason of useful works, with the removal of harmful ones. And this is the case in particular (*specialiter*)[13] in the care of human bodies, than which there is nothing better concerning the particular good of citizens (*ad bonum civium particulare*).[14]

Knowledge of astrological timing is thus crucial for a range of benefits, including in medicine, but also throughout the domain of human affairs. In this way, experienced astrologers can work with and thus improve nature in order to promote what is good and eliminate what is harmful. None of this implies necessity. Roger thus begins to construct an ideology of practical action for working on and improving the world

---

[11] As we will recall from Chap. 3, Albertus Magnus discussed the corruption of the air and its natural philosophical foundations in his translation-commentary on the pseudo-Aristotelian *De causis proprietatum elementorum*. This is also relevant for the natural philosophical analyses in plague treatises to be discussed in volume II.

[12] Et Isaac ponit exemplum de pestilentia futura, et facit sibi quaestionem iuxta ignorantiam medicorum, sicut Aristoteles fecit. Quaerit ergo utrum huiusmodi putrefactio veniet necessario propter coelum, quia tunc medicus non potest prohibere, et ideo inutile videtur hoc praecognosci. Sed solvit, dicens quod corruptio aeris non nocet nisi corporibus praeparatis ad eam; et ideo cum medicus videret significationem aeris corrupti in unoquoque tempore anni venturo, poterit mundificare naturam, et malam corporis dispositionem auferre, ut sciat venienti corruptioni repugnare. Nam si viderit per motum stellarum fieri in aere calorem et siccitatem nimiam et ustionem, festinet purgare choleram eorum in quibus inest complexio cholerica, det frigida et humida calori venturo repugnantia. Et Haly dat consimilem sententiam in expositione quintae propositionis Centilogii (391, 18–31).

[13] In contrast to '*universaliter*' just above.

[14] Ex his patet quod astronomus potest dare remedia contra nociva et promovere utilia, non solum in infirmitatibus, sed in aliis universaliter, cum ratione electionis temporis apti ad omnia opera, cum ratione operum utilium, cum remotione nocivorum; et hoc specialiter in curis corporum humanorum, quo nihil melius est quantum ad bonum civium particulare (391, 31–38).

through various types of technologies, medical and magical, by utilizing the power of astrological timing in the world and on man.

8.3 *Opera*: Talismans

After discussing the good that can come to individuals from the astrological choosing of propitious times, including in medical matters, Roger turns to the common good and raises the stakes. Here he introduces the most secret works of wisdom (*secretissima opera sapientiae*), which include the '*imagines*' made by Moses, who was experienced in the science of the stars (*peritus astronomiae*), as we will see. Roger is very keen here to inform the pope that he himself possesses this extraordinarily valuable secret knowledge. We should always remember who Roger's primary intended audience was, which will perforce influence our interpretation of his intentions in writing these texts:

> Nevertheless, with respect to the common good (*bonus communis*) of a city and of kingdoms by the power of astrology (*per virtutem astronomiae*), greater works (*maiora* [...] *opera*) can be made than anyone can say and more than anyone wishes to explain in writing; for there are many of these most secret works of wisdom (*multa de secretissimis operibus sapientiae*). But because of the magnitude of these matters and the ignorance of the multitude of students who do not care about works of wisdom (*opera sapientiae*), and due to the frauds and malice of the many who abused these [sc. the secret works of wisdom], they are always hidden (*occultantur*) from the crowd and their leaders. Nevertheless, Moses and Aaron, Solomon, Aristotle, Ptolemy and other wise men (*sapientes*) have done (*operati sunt*) so much regarding these things that is wondrous (*admiranda*).[15]

These most secret works of wisdom come with a terrific pedigree, secular and sacred. Since their open knowledge would also render their great power open to broad abuse, secrecy is very much in order.[16] We can immediately see the emphasis on *opera* made through the power of astrology, thus confirming that Bacon is here discussing *astronomia operativa*.

Despite this potential for abuse, Bacon turns to address his primary audience directly, namely, Pope Clement, from whom, he says, he will not keep this power secret:

> Whence, since this is one of the greatest [sc. secrets], and powerful in some way in everything,[17] it is not right that it be hidden (*occultetur*) from Your Glory. And because you

[15] Sed tamen respectu boni communis civitatis et regnorum, maiora possunt fieri opera per virtutem astronomiae quam aliquis potest dicere, et quam aliquis vult per scripturam explanare; nam haec sunt multa de secretissimis operibus sapientiae. Sed propter rerum magnitudinem et ignorantiam multitudinis studentium, qui de operibus sapientiae non curant, et propter fraudes multorum et malitias qui abusi sunt his, occultantur semper a vulgo et a capitibus eius; de quibus tamen Moyses et Aaron, Solomon, Aristoteles, Ptolemaeus et caeteri sapientes operati sunt nimis admirandae (392, 1–10).

[16] We will learn more about the dangers below.

[17] [...] quasi potens quodammodo in omnia [...]

> have commanded that I write about the wisdom of philosophy (*de sapientia philosophiae*), I will tell Your Clemency[18] the opinions of the wise (*sententias sapientium*), especially since it [sc. this knowledge] is altogether necessary for the Church of God against the fury of Antichrist.[19]

Here Bacon makes his claim for the importance of this knowledge in service to the Church against Antichrist, and thus reveals another feature of astrology's utility for religion.[20]

We will now get a sense of the power at stake in the possession of astrological knowledge towards an operative end by means of five historical (or at least quasi-historical) examples. The first two concern major religious figures (Moses and Solomon), and the last three, major figures in philosophy ("Aristotle" and "Ptolemy"). I must emphasize here that Bacon is hardly a disinterested reporter. Although no one would doubt his sincerity or zeal, at least in this case, neither is his self-interest to be doubted. Bacon is keen to inform the pope in no uncertain terms both [1] that this knowledge is extremely powerful, and [2] that he himself possesses it and is thereby offering the pope his services. Bacon thus promotes the value of a learned astrological counselor with respect to both knowledge and power.

8.4 Moses

Roger now explains in greater detail what he means by this power by providing a significant biblical example of an *opus sapientiae* fabricated and employed by a major Old Testament figure, namely Moses, the first *sapiens* mentioned just above in Roger's list of authorities. *We* would call this a magical act, but, as we saw, Roger only uses the term "magic" and its cognates with a strongly negative connotation. If he calls it a "work of wisdom" (*opus sapientiae*), as he does, then it is fine to discuss it. We should also note that Roger actually uses the term '*imagines*' in this context:

> Moses in his youth had been the leader of the Egyptian army (*dux exercitus Aegyptiorum*) against the Ethiopians, and for the sake of peace, had received Aethiopissa, the queen of the Ethiopians—on behalf of whom, and against him, Aaron and Miriam were censured in the 12th book of *Numbers*—and he did not wish to let her return into Egypt due to the overabundance of love (*propter amoris superfluitatem*) [sc. she felt for him]. Since he [sc. Moses] was an experienced astrologer (*peritus astronomus*), he made two *imagines* in rings (*in*

[18] *Vestrae Clementiae*, with a play on Clement's name.

[19] Unde cum hoc sit unum de maximis, et quasi potens quodammodo in omnia, non decet ut Vestrae Gloriae occultetur. Et quia praecepistis ut scriberem de sapientia philosophiae, recitabo Vestrae Clementiae sententias sapientium, praecipue cum ecclesiae Dei sit omnino necessarium contra furiam Antichristi (392, 10–15).

[20] Mark T. Abate has much of interest to say about Bacon's view of Antichrist; "Roger Bacon and the Rage of Antichrist: The Apocalypse of a Thirteenth Century Natural Philosopher," PhD thesis, Boston University, 2000.

> *annulis*), one of forgetting which he gave to the woman,[21] and another of memory, which he kept for himself, and thus he left her in freedom with his army and without war. Josephus tells us these things in the first book of his *Antiquities*. He is a master among historians, and many confirm it. A wondrous thing (*mira res*) this was that changed the mind of a woman (*quae animum mulieris immutavit*).[22]

In this story explicitly attributed to the Jewish historian who wrote in Greek, Flavius Josephus (37 CE-ca. 100), Moses as an experienced astrologer made two rings with *imagines* in order to diffuse a woman's passions and allow him to depart Egypt in peace. This use of an *imago* seems to place us directly within the domain of talismans, and thus of what we would call magical objects. Roger explicitly calls Moses a *peritus astronomus*, so we are also in the operative part of *astronomia*, in which *opera sapientiae* are made. In fact, however, only the marriage itself was mentioned by Josephus. The business about the rings comes entirely from Peter Comestor's *Historia Scholastica*, as Weill-Parot clearly shows.[23]

8.5 Solomon

For his second biblical example, Roger discusses Solomon in another story from Josephus, this time from Book VIII of the *Antiquities*:

> But Solomon arranged (*ordinavit*) many things of this sort, which were done beyond the accustomed course of nature (*praeter solitum cursum naturae*), as Josephus says in the eighth book. Since he [sc. Solomon] was wiser (*sapientior*) than everyone who preceded and followed him, as Sacred Scripture and Josephus teach, he was unable to neglect these works (*haec opera*), and therefore, he left many enigmatic writings (*scripta multa in aenigmate*), which afterwords were turned by magicians (*per magicos*) towards bad uses (*in malos usus conversa*) and were badly understood ([translated?] *male interpretata*), into which many enormities were mixed by fraudulent practitioners. But wise men (*sapientes*) know how to choose the grain from the husk, and to separate the theriac from the serpent.[24]

---

[21] [...] unam [sc. imaginem] oblivionis quam dedit mulieri [...].

[22] Cum vero Moyses in iuventute sua fuisset dux exercitus Aegyptiorum contra Aethiopes, et ob bonum pacis acciperet Aethiopissam, pro qua contra eum duodecimo Numerorum obiurgati sunt Aaron et Maria, quae fuit regina Aethiopum, nolens dimittere eam ut rediret in Aegyptum propter amoris superfluitatem, cum esset peritus astronomus, fecit duas imagines in annulis, unam oblivionis quam dedit mulieri, et alteram memoriae quam sibi detinuit, et sic ab ea libere cum exercitu et sine bello recessit. Haec Iosephus primo Antiquitatum libro, et magister in historiis, et multi confirmant. Mira res fuit haec, quae animum mulieris immutavit (392, 15–26).

[23] *Les "images astrologiques,"* 149–52.

[24] "Salomon vero ordinavit de huiusmodi multis, quae praeter solitum cursum naturae facta sunt, ut dicit Iosephus octovo libro, qui cum fuit sapientior omnibus praecedentibus et sequentibus, ut scriptura sacra et Iosephus docent, non potuit haec opera negligere, et ideo scripta multa in aenigmate reliquit, quae postea per magicos in malos usus conversa sunt atque male interpretata, in quibus multa enormia sunt per fraudulentos admixta. Sed sapientes sciunt eligere grana de paleis, et theriacam de serpenti separare (392, 26–34)." Josephus discussed Solomon's magical powers including incantations at *Jewish Antiquities* VIII.2.5.

Due to his great wisdom, Solomon was simply unable to neglect the *opera sapientiae*. Unfortunately, Roger does not provide any examples here, mentioning only that Solomon left many enigmatic writings on the subject and that his writings with all their power have been badly misunderstood and used for ill by magicians, who have thus returned to Roger's account. Presumably Roger is referring here to the widely dispersed writings of Solomonic magic, including the famous *Claviculae Salomonis*.[25] Fortunately, wise men know how to locate and extract the pearls of wisdom from the dungheap of magical nonsense, among whose company we may assume that Roger includes himself, as he argues strongly for a legitimate purified and powerful use of the Salomonic magical tradition!

Thus, embracing Josephus's account, Roger discusses Solomon and Moses, two potent and authoritative examples from scripture and sacred history, who made and used *opera sapientiae* for good. He also makes it very clear that this powerful knowledge must not fall into the wrong hands, namely, into the hands of magicians who will pervert these *opera* and use them for ill. Roger will provide more color as he ramps up his fearmongering rhetoric below, in the process turning the deleterious use of magical objects into the thirteenth-century equivalent of weapons of mass destruction.

8.6 "Aristotle" (1)

From two of the greatest figures of the Old Testament, Roger now returns to the pseudo-Aristotle of the *Secretum secretorum* as he shifts from sacred to secular history. Here Roger recounts a story of power in warfare, relating Aristotle, the greatest ancient philosopher, to one of the most powerful rulers in antiquity, namely, Alexander the Great, a relationship that Bacon hopes to capitalize on here by replicating it in the present with Pope Clement as Alexander and Bacon himself as Aristotle. Roger also hoped to do this again later with his learned scholarly edition of the *Secretum secretorum*, after his high hopes for papal patronage were dashed soon after on Clement's death on 29 November 1268. Roger unfortunately does not tell us what the *opera* were:

> But Aristotle, the greatest of philosophers, teaches how to make these works (*haec opera*) in the *Book of Secrets*, so that each magnificent thing (*quaeque magnifica*) can be usefully promoted and all harmful things destroyed. For with these works (*opera*), Aristotle made Alexander the Great conquer the world with less than 40,000 armed soldiers. When he [sc. Alexander] was lying on his death bed (*in lecto mortali*) and was already on death's door,[26] he snatched the city and its region from his enemies. The first deed is known from the acts of Alexander (*in gestis Alexandri*); the second in Valerius Maximus, Book V. But it is agreed that he could not have done this with bodily power (*potentia corporali*), but by the great power of

[25] For an insightful analysis of Solomonic magic in itself, and in contrast to Hermetic magic, see Boudet, *Entre science et nigromance*, 137–55, and especially, 145 ff.

[26] Literally, *in* the doors (*in ianuis*) of death.

> wisdom (*magna sapientiae virtus*), which he left behind in the books of secrets [sc. the *Secret of Secrets*] to be discovered only by the wisest people (*solum sapientissimis indagandam*).[27]

The upshot of this story is that great military victories are also obtainable by the great power of wisdom as recorded in the *Secretum secretorum.* Not works of bodily strength, but those of wisdom alone are sufficiently powerful. Roger will also provide a frighteningly negative scenario of this power later, also in a military context, with the same Tartars and Saracens he had discussed at the end of the section on astrological judgments in Chap. 6.

8.7 "Aristotle" (2)

Providing another example from the *Secretum secretorum*—and this time one that goes into greater detail—Roger now tells us more about the actual composition of an *opus sapientiae*:

> And when, due to the weight of old age, after Darius had been defeated, he [sc. Aristotle] delivered the world to Alexander, and he [sc. Aristotle] retired to his own land. He said to Alexander that he would write for him whatever he desired, and he would give timely counsel (*consilium* [...] *opportunum*). And when Alexander discovered that the people had terrible customs (*mores pessimos*), he wrote to Aristotle concerning what he should do about them. This prince of philosophy (*princeps philosophiae*) responded: "If you are able to qualitatively change the nature of their air (*alterare aerem ipsorum*), permit them to live; if not, kill them all (*interfice omnes*)." O, how obscure (*occultissima*) a response is this, but full of the power of wisdom (*plena sapientiae potestate*)! For he understood that in accordance with the transformation of the air (*mutatio aeris*), which contains celestial powers (*coelestes virtutes*), the customs of men are transformed (*mutantur mores hominum*); and for this reason, it is the case that the Gauls have different customs, the Romans others, the Spanish others, and thus concerning the different regions.[28]

Here Roger interprets Aristotle's extremely gnomic utterance in terms familiar from his astrologizing Aristotelian natural philosophy, where celestial powers/virtues borne in the air have different effects on different places in the world. Thus, if the

---

[27] Aristoteles vero summus philosophorum docet haec opera fieri in libro Secretorum, ut quaeque magnifica utiliter promoveantur et omnia nociva destruantur. His enim operibus Alexandrum magnum fecit Aristoteles cum minus quadraginta millibus armatorum mundum superare. Is cum in lecto mortali iacebat, et iam esset in ianuis mortis, civitatem suam et regionem ab inimicis eripuit. De primo facto notum est in gestis Alexandri, de secundo in Valerio Maximo, libro quinto. Sed constat eum non posse hoc fecisse potentia corporali, sed magna sapientiae virtute, quam in libris Secretorum reliquit solum sapientissimis indagandam (392, 35–393, 7).

[28] Et cum propter pondus senectutis, devicto Dario, mundum tradidit Alexandro, recessit in terram suam, dixitque Alexandro quod ei scriberet quae vellet, et ipse consilium daret opportunum. Et cum Alexander invenit gentes habentes mores pessimos, scribens Aristoteli quid faceret de eis, ipse princeps philosophiae respondit, Si potes alterare aerem ipsorum, permitte eos vivere; si non, interfice omnes. O quam occultissima responsio est, sed plena sapientiae potestate! Nam intellexit quod secundum mutationem aeris, qui continet coelestes virtutes, mutantur mores hominum; eo quod alios mores habent Gallici, alios Romani, alios Hispani, et sic de singulis regionibus (393, 7–19).

air can be changed—he does not say how—the deeply problematic but underdescribed customs of the inhabitants may be transformed. Otherwise they should all be killed.

Bacon continues with his interpretation:

> Therefore, he [sc. Aristotle] wished that Alexander would change the quality of the air (*mutaret qualitatem aeris*) of these peoples [= nations, *gentes*] into something good (*in bonum*), so that, in accordance with this transformation (*secundum mutationem illam*), their customs would be transformed (*mutarentur mores*), and they would be excited towards honorable customs (*excitarentur ad honestatem morum*) without contradicting their free will (*sine contradictione liberi arbitrii*); just as each nation is excited to its proper customs by means of its proper air, which has the powers of the stars that are over mens' heads (*per aerem proprium habentem virtutes stellarum quae sunt super capita hominum*), and in accordance with which signs and planets rule individual regions (*secundum quod signa et planetae dominantur singulis regionibus*).[29]

As Bacon continues, his analysis becomes more familiar from our discussion in Chap. 2.

Then Roger mentions a different treatise he himself has written, referring most likely to *Opus maius*, Book IV. Here he discusses how people can be led into different sorts of behaviors. As ever, he is concerned to protect free will and thus astrology's legitimacy:

> For I show in a treatise which I am sending that someone can be led (*induci*) towards good and bad, both public and private, by means of the powers of the heavens without compulsion (*per virtutes coelorum sine coactione*); as we see, for example, that human beings transform their wishes (*mutant suas voluntates*) due to their lords, friends and associates (*per dominos, amicos, socios*), and the presence of new things (*rerum praesentiam novarum*), and in innumerable ways (*infinitis modis*) without compulsion (*sine coactione*). I explain this sufficiently in the chapter on the power of making judgments on human affairs by means of the heavens (*de potestate iudicandi de rebus humanis per coelum*), in which chapter there is an account (*ratio*) of everything that touches on this material.[30]

Thus Bacon refers here to the analysis in *Opus maius* IV (249–53) discussed above in my Chap. 4, which he here refers to as a different work. This is among the most persuasive evidence we have that the *Astrologia* section in Bridges is not actually from the *Opus maius*.

Roger continues, now discussing how to effect such a change in the air. To do so, he once again employs an '*opus sapientiae*', and in so doing refers to the previous discussion of Moses's talismanic ring:

---

[29] Voluit ergo quod Alexander in bonum mutaret qualitatem aeris illarum gentium, ut secundum mutationem illam mutarentur mores, et excitarentur ad honestatem morum sine contradictione liberi arbitrii; sicut quaelibet natio excitatur ad proprios mores per aerem proprium habentem virtutes stellarum quae sunt super capita hominum, et secundum quod signa et planetae dominantur singulis regionibus (393, 19–25).

[30] Ostendo enim in tractatu quam mitto, quod induci potest aliquis ad bonum et malum, tam publicum quam privatum, per virtutes coelorum sine coactione, sicut exemplariter videmus quod homines mutant suas voluntates per dominos, amicos et socios, et rerum praesentiam novarum, et infinitis modis sine coactione. Quod satis explano in capitulo de potestate iudicandi de rebus humanis per coelum, in quo capitulo est ratio omnium quae tanguntur de hac materia (393, 25–394, 1).

> Therefore, the philosopher wished that he [sc. Alexander] would make works of wisdom (*opera sapientiae*) by means of their proper celestial configurations (*per debitas constellationes*) in the manner of Moses (*ad modum Moysis* [i.e. with a talismanic ring]), which excited the soul of a woman by means of celestial powers received in matter.[31] For on what account (*qua ratione*) could that woman be transformed (*mutari*) towards chastity and the forgetting of a man by means of *imagines*? Not only she, but anyone can be moved (*moveri*) towards other customs.[32]

In suggesting that Aristotle was promoting talismanic ring-making to Alexander in the manner of Josephus's story about Moses, Roger explicitly identifies the *opera sapientiae* with powerful *imagines* that can thus excite a person's soul by means of the celestial powers/virtues received in matter. Here Roger seems to be discussing the sorts of talismans described in both the *Speculum astronomiae* and in Albert's authentic *De mineralibus*, and later in Marsilio Ficino's *De vita* III (as we will see in volume II). This sort of power would have been extremely appealing to Pope Clement, especially given the current powerful threats against Christendom, for example, from the Tartars and Saracens, which Roger discusses again below in the context of the negative use of these powers in a military context.

8.8 "Ptolemy"

Developing this picture, Roger now provides more information on *imagines* and how to make them, as he turns from "Aristotle," his first authoritative philosopher, to "Ptolemy," his second:

> In the ninth proposition of the *Centiloquium*, Ptolemy teaches that the faces of this age (*vultus huius seculi*) are subjected to the celestial faces (*subiecti vultibus coelestibus*).[33] And Haly says that in this chapter, Ptolemy wishes to lay open the secrets of images (*secreta imaginum*). And their intention is clear in general because if it happens that they sculpt (*sculpere*)[34] these (*eas*, sc. *imagines*) at astrologically chosen times in relation to the face of the heavens (*si contingit temporibus electis eas sculpere ad vultus coelorum*), everything harmful can be repelled and all useful things promoted (*omnia nociva repelli possunt et utilia promoveri*). The greatest philosopher among all the Christians, Thabit [ibn Qurra]—who in many things added to Ptolemy's works and to those of other practitioners of the science of the stars (*astronomi*), both in speculative and in practical matters—increased this

---

[31] [...] qui excitavit animam mulieris per coelestes virtutes receptas in materia (394, 3–4).

[32] Philosophus igitur voluit quod faceret opera sapientiae per debitas constellationes ad modum Moysis, qui excitavit animam mulieris per coelestes virtutes receptas in materia. Nam qua ratione potuit illa mulier mutari ad castitatem et oblivionem veri per imagines, potuit et ad alios mores moveri non solum ipsa, sed quicunque (394, 1–7).

[33] As we saw, faces (*facies* or decans) are one of the astrological dignities. Although Roger does not use the term *vultus* in his description of dignities either in the *Opus maius* or in the *Secretum secretorum*, it seems that he might be doing so here. Nevertheless, he probably uses the term more generally here, drawing on Ptolemy's vocabulary: in the face of (= under the gaze of) the heavens.

[34] '*Sculpere*' is the same term Albert used in *De mineralibus*, and it is also used in the *Speculum astronomiae*, as we saw in Chap. 7.

body of knowledge (*scientia*) in particular, and other extremely wise people (*sapientissimi*) relied on these writings.[35]

As for Albert and the author of the *Speculum astronomiae*, to be effective, *imagines* need to be "sculpted" (engraved) at the proper astrologically chosen time, in a passage that could have been borrowed from Albert's *De mineralibus*. For authorities on *imagines*, Roger discusses the pseudo-Ptolemaic *Centiloquium* and Haly's commentary thereon. He also has much praise for Thabit ibn Qurra, whom he strikingly considers the greatest philosopher among the Christians, no small accolade. As we saw in Chap. 7, Thabit's *De imaginibus* was also the primary—and only—source for legitimate talismans mentioned in the *Speculum astronomiae*, and it was mentioned as an authority by the authentic Albertus Magnus in his *De mineralibus*.

*

After these five highly authoritative examples, Bacon concludes this section by emphasizing why this knowledge is justifiably secret. He also returns to using the term "magic" in his characteristic and consistently negative sense:

> But since these works (*haec opera* [sc. of wisdom, that is, *imagines*]) seem to the crowd of students to be beyond the capacity of the human intellect, since the crowd with their doctors [= learned teachers] do not busy themselves with the works of wisdom (*opera sapientiae*), therefore, hardly anyone has dared to speak about these works (*de his operibus*) in public. For immediately they are called "magicians" (*magici*), even though they are the wisest people (*sapientissimi*) who know these things. And, as the theologians and the masters of decretals (*decretistae*) are not instructed in such things, and at the same time seeing that bad things can be done as well as good, they neglect and abhor these things, and count them among the magical (*inter magica*).[36]

Here Roger explains why this great and powerful knowledge has been kept secret for the most part: whoever discusses it openly will immediately be branded a magician with all the negative connotations that Bacon himself attaches to this term and its cognates.[37]

To complete this section, Bacon employs an early version of the "guns don't kill people, people do" argument to confirm the importance of this material:

[35] Ptolemaeus in nona propositione Centilogii docet, quod vultus huius seculi sunt subiecti vultibus coelestibus. Et dicit Haly, quod in hoc capitulo vult Ptolemaeus imaginum secreta patefacere. Et patet intentio eorum in universali, quia si contingit temporibus electis eas sculpere ad vultus coelorum, omnia nociva repelli possunt et utilia promoveri. Thebit inter omnes Christianos summus philosophus, qui in multis addidit ad opera Ptolemaei et aliorum astronomorum tam in speculativis quam in practicis, specialiter hanc scientiam ampliavit, et alii sapientissimi super his insistebant (394, 7–16).

[36] Sed quia haec opera videntur vulgo studentium esse supra humanum intellectum, quia vulgus cum suis doctoribus non vacat operibus sapientiae, ideo vix est aliquis ausus loqui de his operibus in publico. Statim enim vocantur magici, cum tamen sint sapientissimi qui haec sciunt. Utique theologi et decretistae non instructi in talibus, simulque videntes quod mala possunt fieri sicut et bona, negligunt haec et abhorrent, et computant inter magica (394, 17–24).

[37] We will see resonances with this argument in Pico's *Apologia* in volume II.

> For they see that magicians (*magici*) and abusers of the texts of the wise (*abusores documentorum sapientiae*) use these things [sc. *imagines*], and therefore, they esteem them to be unworthy of Christians. But the truth should not be condemned due to ignorance, nor utility due to malice, although the same things can be done. For then human beings ought to be without knives (*cultellis*) at table, since people eating can kill their neighbors (*interficere socios*) with the same blade with which they cut bread. Then arms should be destroyed by the Church and the secular arm [sc. the state] because many bad things can be done with these, and daily do happen. Nonetheless, laws (*iura*) ought to exist for the use of good men, although due to cavilling and fraud, many jurists abuse the laws. [...] Nevertheless, not because of this should the study of wisdom (*studium sapientiae*) be condemned. For bad people are always turned towards bad things, from which good people produce various useful things.[38]

This powerful knowledge should neither be neglected nor rejected because it is capable of being abused and causing harm. Thus, Roger completes his fullest treatment of talismans as *opera sapientiae*, which has several significant resonances with my discussion in Chap. 7: Celestial virtues and astrological timing are central, but Roger very deliberately does not call them magical.

8.9 On the Power of Words (1): *De virtute verborum*

Roger now turns abruptly from the use of talismans to a theme also dear to al-Kindi, namely, the power of words (*de virtute verborum*).[39] The power of words provides the second major feature of Bacon's *astronomia operativa*, namely, the *verba* com-

---

[38] Vident enim quod magici et abusores documentorum sapientiae his utuntur, et ideo aestimant indigna Christianis. Sed non debet veritas damnari pro ignorantia, nec utilitas pro malitia, quamvis eadem fieri valeant; nam tunc deberent homines carere cultellis in mensa, quoniam comedentes possunt interficere socios eodem ferro quo scindunt cibaria, et tunc arma debent deleri ab ecclesia et brachium seculare, quia multa mala possunt per haec fieri, et quotidie peraguntur. Et nihilominus iura debent esse in usu bonorum hominum, quamvis per cavillationes et fraudes multi iuristae iuribus abutuntur. Et nunc videmus quod plus est in usu studentium quicquid de cavillationibus sit scriptum quam verus usus demonstrationum; non tamen propter hoc studium sapientiae damnandum est. Nam semper mali in malis convertuntur, ex quibus boni utilitates varias producunt (394, 24–395, 7).

[39] The works of scholarship I have found most useful for understanding how Bacon used powerful words and his use of al-Kindi are: Charles Burnett, "The Theory and Practice of Powerful Words in Medieval Magical Texts"; Clare Fanger, "Things Done Wisely by a Wise Enchanter: Negotiating the Power of Words in the Thirteenth Century," in *Esoterica* I (1999): 97–132; Irène Rosier, *La parole comme acte: sur la grammaire et la sémantique au XIIIe siècle,* Paris: Vrin, 1994, especially Chap. 6; Béatrice Delaurenti, *La Pouissance des mots,* <<Virtus verborum>>: *Débats doctrinaux sur le pouvoir des incantations au Moyen Age*, Paris: Editions du Cerf, 2007, with an extensive chapter on Bacon; and more generally, Skemer, *Binding Words*. Katherine Tachau's analysis is also very relevant here; "*Et maxime visus*," although there is much more to say about both *perspectiva* and astrology in both Roger Bacon and late medieval thought. Ficino also takes up this theme in his *De vita,* with exactly the same phrasing as in Bacon and al-Kindi, but following Bacon's structure, not al-Kindi's. Nevertheless, Ficino like al-Kindi goes into much more detail than Bacon does, as we will explore in volume II.

ponent of his *opera et verba sapientiae*. Like talismans (as we will see), the power of words also centrally involves astrological timing, which integrates the treatment of both *imagines* and *virtutes verborum*, and thus further confirms the disciplinary location of *astronomia operativa* within elections. It also brings the soul register and its power to the fore.

Unlike talismans, the power of words derives from the practitioner's psychological or soul dimension, as we shift from the bodily to the soul register:[40]

> And since the work of the rational soul (*opus animae rationalis*) is especially to use words effectively (*verbis uti efficaciter*),[41] that is, words formed with intention (*verbis* [...] *ex intentione formatis*), [sc. therefore] the astrologer (*astronomus*) can form words at [sc. astrologically] chosen times (*formare verba temporibus electis*) which will have an ineffable [= unspeakable, indescribable] power. For where the intention, desire and power of the rational soul (*intentio, desiderium et virtus animae rationalis*)—which is more worthy (*dignior*) than the stars—harmonize (*concurrunt*, concur, co-ordinate) with the power of the heavens (*cum potestate coeli*), it is necessary that either the word (*verbum*) or a different work (*aliud opus* [a talisman?]) of a wondrous power is produced for altering the things of this world (*mirae virtutis in alteratione rerum mundi huius*), so that not only natural things (*naturalia*), but minds (*animi*) would be inclined towards those things that the wise *operator* wishes, while preserving freedom of the will (*salva arbitrii libertate*), because the mind can follow celestial powers freely without compulsion, as has been treated and declared in its place [i.e. *Opus maius* IV, 249-53].[42]

This material is utterly central for understanding Bacon's position. With respect to altering things and minds in the world, there are two main factors: [1] the non-material soul's (*anima rationalis*) intention, desire and power, and [2] the physical power of the heavens. These two factors combined in words (*verba*) uttered or works (*opera*) made at astrologically chosen times can have a great effect in the world on both minds and bodies. As ever, Bacon takes care to protect human free will within this system.

*

The most important source for Bacon's system here is al-Kindi's *De radiis stellarum*, as we can see in a passage from Charles Burnett's extremely rich article, "The Theory and Practice of Powerful Words in Medieval Magical Texts":

---

[40] Unlike what we will find in Ficino, however, the soul register here refers only to a human soul focusing and intensifying itself—as in prayer—in order to help with and thereby amplify the relevant celestial influences. By contrast, Ficino explicitly also animated the planets and the heavens overall by ensouling them, thus fully and explicitly developing a soul register on both human and cosmic levels, as we will see in detail in volume II.

[41] I.e. as efficient causes, namely, to get things moving, precisely as al-Kindi describes it.

[42] Et cum opus animae rationalis praecipuum sit verbis uti efficaciter et ex intentione formatis, potest astronomus formare verba temporibus electis quae ineffabilem habebunt potestatem. Nam ubi intentio desiderium et virtus animae rationalis, quae dignior est stellis, concurrunt cum potestate coeli, necesse est sive verbum sive aliud opus produci mirae virtutis in alteratione rerum mundi huius, ut non solum naturalia, sed animi inclinentur ad ea quae velit sapiens operator, salva arbitrii libertate, quia potest animus sequi coelestes virtutes gratis sine coactione, ut tractatum est suo loco et declaratum (395, 8–17).

When a man wishes to operate on something, he first imagines (*ymaginari*) the form that he wishes to impress on some matter. Then he has to have a great desire (*desiderium*) for that effect (the word *intentio* sometimes substitutes for *desiderium*). Thirdly, confidence (*fides*) is necessary. But even these are not sufficient, since imagination, desire and confidence are all internal faculties and do not have actual existence and spatial dimensions. The final thing to be added is 'action' (*opus*). This is of two kinds: the "speaking of the mouth" (*oris locutio*) and "the operation of the hand" (*manus operatio*). The conclusion of chapter five of the *De radiis* is: "Some form of speech (*sermo*) spoken by the mouth of a man together with the imagination, confidence and desire of the speaker, sometimes produces actual movements of individual objects in the world. This has been proved by frequent experiences (*experimenta*) among men of almost all nations [...]. Hence the prayers to God, to spirits, and to various creatures are made for the sake of achieving good and banishing evil, through the movement of the objects concerned initiated by the utterance of words (*verborum prolatio*, 217)."

It is plain to see how greatly Bacon was influenced by al-Kindi here.

For al-Kindi's theory of magic, rays too are utterly essential, as we can see from Burnett's description:

Sounds and colors too emit rays. Recalling the traditional definition of sound, the author writes that "the collision of bodies" makes a sound which flows out in all directions, but adds "by means of its own kind of rays." Above all, man produces rays. For he is a microcosm, and the species of all mundane things are imprinted on his imagination. Therefore, his "imagining spirit" (*spiritus ymaginarius*) has rays that conform to the rays of the objects in the world. Hence he can achieve the power to move and change things outside him by his rays just as the objects in the world (both upper and lower) move things by *their* rays, especially if his operation is in conformity with the operation of the celestial harmony (his emphasis).[43]

In a revealing aside (n. 9), Burnett states that rays are not normally a part of Aristotelian natural philosophy. Indeed, al-Kindi himself seems to have brought rays into it, and Robert Grosseteste, Roger Bacon and Albertus Magnus seem to have been the main thirteenth-century figures to integrate them fully into mainstream Aristotelian natural knowledge, as I have reconstructed it in Part I. We will see this tradition continue well into the Early Modern period in volumes II and III.

*

We also know from a very interesting article by George Molland that Roger Bacon explicitly knew and mentioned al-Kindi's *De radiis stellarum* in a small work entitled *De laudibus mathematicae* (*On the Praises of Mathematics*), which was written around the time of the three *Opera* for the pope.[44] As Molland argues, it is very similar to parts of *Opus maius* IV in four of its eight sections, but in the other four it adds new material, including the mention of *De radiis stellarum*:

Soon afterward there occurs a significant deviation from the *Opus maius*. Both texts refer to false mathematics' superstitious use of characters, charms, conjurations, and sacrifices, its

[43] Burnett, "Powerful Words," 217. See also Burnett's valuable translation of most of *De radiis*, chapter 6 "On the Power of Words," (*De virtute verborum*), on pp. 218–24.

[44] "Roger Bacon's *De laudibus mathematicae*."

> deploying of various frauds, and its appeal to celestial necessitarianism. The *Opus maius* merely tells us that these things were expressly asserted "in magical books"; the *De laudibus* is more expansive and says that they occur "in a magical book, whose title is infamous, namely *Theory of the Magical Art* (*Theorica artis magicae*), which several have changed into *Book of Rays* (*librum De radiis*), because the book's author puts forward many excellent things about the multiplication of rays, which are philosophical and true, so that he may better draw the reader' minds to the poison of falsity that is his principal intent."[45] The reference is unmistakably to al-Kindi's *De radiis stellicis*, which at an uncertain date had been translated into Latin and was widely diffused. It clearly had a great influence on Bacon's own doctrine of the multiplication of species, in which substances and qualities send out likenesses of themselves in all directions, which are then transmitted through the medium in accordance with the laws of optics, until they impinge and act upon a suitable patient. Hence Bacon's reference to the "excellent things" that the book contained about the multiplication of rays, which for both writers provided theoretical underpinnings for the ontological aspect of astrology (72-73).

The material in this text is an addition to what is found in the *Opus maius* at p. 246 of Bridges's edition as I treated in Chap. 4. Molland ends his valuable article with a strong statement about the chaotic state of Bacon's textual Nachlaß (83).

*

Having established the power of both *imagines* (or talismans) and words—the *opera et verba sapientiae*—Bacon now treats their respective representatives, namely, characters and charms (*carmina*):[46]

> From this root, the use of characters and *carmina* [= charms/songs] among the wise (*apud sapientes*) began. For characters are just like images (*sicut imagines*), and *carmina* are words uttered from an intention of a rational soul (*verba ex intentione animae rationalis prolata*) that receive the power of the heavens in their expression (*virtutem coeli in ipsa pronunciatione recipientia* [sc. *verba*]); whence I mention the wondrous power of letters (*de mira potestate literarum*) in the third part [i.e. *Opus maius* Book III].[47]

Characters are thus like images (i.e. visual representations) and *carmina* are like words (i.e. verbal expressions), both of which receive celestial influences in their making or utterance, and thus require thoughtful astrological timing. Bacon had discussed both characters and *carmina* negatively in relation to magic in *Opus Maius* IV (238–49). By contrast, he discusses them here as positively used by the wise to effect profound transformations (*alterationes*) in the things of the world and

[45] In libro magico, cuius tetulus infamis est, scilicet Theorica Artis Magicae, quem plures mutaverunt in Librum de Radiis, quia auctor illius libri multa praeclara premittit de radiorum multiplicatione, que phisica sunt et vera, ut magis alliciat animos legentium ad venenum falsitatis quod principaliter intendit.

[46] Skemer usefully discusses how the term '*carmina*' was used in a magical context; *Binding Words*, 9–10.

[47] Ex hac radice incepit apud sapientes usus characterum et carminum. Nam characteres sunt sicut imagines, et carmina sunt verba ex intentione animae rationalis prolata, virtutem coeli in ipsa pronunciatione recipientia; unde de mira potestate literarum ego facio mentionem in tertia parte (395, 18–22).

on the minds of human beings. The planets themselves as celestial *bodies* cannot do this directly, but words can because they derive from a rational soul.

Bacon then offers examples of what we would call the magical powers of language: "For by this power, bodies are healed, poisonous animals put to flight,[48] any brute [sc. animal] is called to one's hand, serpents are called out of caverns, and fish from the depths of the waters."[49] In addition to medical ends on human bodies, the power of words can also significantly affect the animal kingdom. With this promotion of the use of characters and charms as representatives of images (*opera*) and words (*verba*), Roger Bacon goes beyond the authentic Albert's and the *Speculum astronomiae*'s strictures on the composition and use of all-natural talismans, except insofar as the characters and words are prosaic and straightforward.

*

Bacon now compares what we would describe as good vs. bad magic, but which he does not do, of course. Rather, he calls them good or bad "works of wisdom," as we saw above. What distinguishes good from bad magic for Roger is not techniques, but intentions, thus implicitly distinguishing between means and ends. This discussion centers on a subject of the greatest interest to him, namely, performing great works to benefit the Church:

> For the matter of the world (*mundi materia*) is altered towards many wonderous experiences (*alteratur* [...] *ad multas passiones mirables*), if they [sc. the *verba*] are made correctly, and therefore can be wondrously powerful (*mirabiliter valere*) against malevolent people and the enemies of the republic, just as also other "starified" works (*caetera opera stellificata*) can do, as is fitting. But the reproached [= bad] magicians (*magici maledicti*) have brought the worst reputation (*summa infamia*) on this part [sc. on the power of words and associated practices], since not only have they abused the characters and *carmina* written by the wise against harmful things, and for the greatest utility, but they have added deceitful charms (*mendosa carmina*) and vain and fraudulent characters by which people are seduced.[50]

These tools and techniques are powerful but value neutral, and may be put to good *or* bad uses, depending fundamentally on the intention of the practitioners. Their use by bad magicians has given such endeavors a terrible reputation, which Bacon here is doing his best to resuscitate, much as he had tried to do in *Opus maius* IV in defending astrological judgments, the weakest part of mathematics.

Continuing with this bad magic, Roger turns to demons:

---

[48] Like the examples of talismanic pest control we saw in the *Speculum astronomiae*.

[49] Per hanc enim potestatem sanantur corpora, fugantur animalia venenosa, advocantur ad manum bruta quaecunque, et serpentes de cavernis et pisces de profundis aquarum (395, 22–25).

[50] Alteratur enim mundi materia ad multas passiones mirabiles, si recte fierent ista, et ideo contra malevolos homines et inimicos reipublicae possunt mirabiliter valere, sicut et caetera opera stellificata, ut oportet. Sed magici maledicti induxerunt summam infamiam in hac parte, quum non solum in malis abusi sunt characteribus et carminibus scriptis a sapientibus contra nociva, et pro utilibus maximis, sed adiunxerunt mendosa carmina et characteres vanos et fraudulentos quibus homines seducuntur (395, 25–34).

> In addition, demons have tempted many people, and both women and demons have taught many superstitious things (*multa superstitiosa*), of which every nation is full. For these *vetulae*[51] everywhere make characters, *carmina* and conjurations, and these magicians (*ipsi magici*) use invocations and conjurations of demons, and they make sacrifices to them. But these are all abuses (*maledicta*) and outside the pathways of the philosophers, nay, rather, against their opinions; and through these things, the power of philosophy is defamed (*defamatur*).[52]

Roger here offers an evocative picture of bad magicians and *vetulae*, old women, in a manner strikingly reminiscent of later witchcraft tracts.[53] The bottom line for Roger is that such shameful practices greatly harm the reputation of philosophy's power. Roger does, however, seem to distinguish between the two negative practices on the basis of gender: *Vetulae*—old women; should we call them "witches"?—make characters, *carmina* and conjurations, whereas magicians invoke and conjure demons and make sacrifices to them, thus distinguishing between what we would call witchcraft and ritual or ceremonial magic.

Roger now draws some conclusions from this *infamia*, and in the process offers a distinction between what *he* calls "magic" and "philosophy":

> And therefore, theologians of the present time and Gratian and many saints have reproached many useful and excellent sciences (*multas utiles et magnificas scientias*) with these magical ones (*magicis*), not attending to the difference between magic (*magica*) and the truth of philosophy (*philosophiae veritas*) due to the five causes [sc. the five species of the *ars magica*] about which we spoke in the first part. This chapter is necessary for every person aspiring to the *magnalia sapientiae*, and wanting to separate true from false [sc. *scientia*]. For Gratian, just as he wrote many laws which have now been abrogated, with a sounder opinion prevailing, thus, when one speaks about the sciences (*de scientiis*), he says many things which ought to be transfered to another part, as I will explain more fully below.[54]

[51] For more on *vetulae*, see Jole Agrimi and Chiara Crisciani, "Savoir médical et anthropologie religieuse: Les représentations et les fonctions de la vetula (XIIIe-XVe siècle)," Annales ESC 5 (1993): 1281–1308. See also the Italian version: Jole Agrimi and Chiara Crisciani, "Immagini e ruoli della vetula tra sapere medico e antropologia religiosa (secoli XIII-XV)," in *Poteri carismatici e informali: chiesa e società medioevali*, ed. A. Vauchez and A. Paravicini Bagliani (Palermo: Sellerio, 1992), 224–61. My thanks to Tamar Herzig for these references.

[52] Insuper daemones temptaverunt multos et tam mulieres quam daemones docuerunt multa superstitiosa, quibus omnis natio plena est. Nam ipsae vetulae ubicunque faciunt characteres et carmina et coniurationes, ac ipsi magici utuntur invocationibus daemonum et coniurationibus eorum, et sacrificia eis faciunt. Sed haec omnia sunt maledicta et extra vias philosophorum, immo contra sententias eorum; et per haec defamatur philosophiae potestas (395, 34–396, 4).

[53] See e.g. Charles Zika, *The Appearance of Witchcraft: Print and Visual Culture in Sixteenth-Century Europe*, Abingdon: Routledge, 2007, and Henricus Institoris and Jacobus Sprenger, Malleus Maleficarum, Christopher S. Mackay (ed. and tr.), Latin text and English translation, 2 vols., Cambridge: Cambridge University Press, 2006.

[54] Itaque theologi nunc temporis et Gratianus et sancti plures multas utiles et magnificas scientias reprobaverunt cum his magicis, non attendentes differentiam inter magicam et philosophiae veritatem propter causas quinque de quibus in prima parte dictum est; quod capitulum necessarium est omni homini ad magnalia sapientiae aspiranti et volenti verum a falso separare. Nam Gratianus, sicut multa scripsit iura quae nunc abrogata sunt, sententia saniore praevalente, sic, cum de scientii locutus est, multa dixit quae debent in partem alteram commutari, ut inferius abundantius explicabo (396, 4–14).

In discussing the sciences, then, true knowledge should thus be separated from false, and outdated and/or misinformed laws overturned. In this way, Bacon wishes to defend the words and works of wisdom from the taint of magic and its malevolent practitioners, both male and female.

*

After clarifying the glaring differences between true and useful sciences and bad and false magic, whether by magicians or *vetulae*, Bacon returns to his central theme. Here he offers a general account of his view, and discusses both the *verba* and *opera* together:

> I return, therefore, to the words and works of the wise (*verba et opera sapientium*), formed (*formata*) by the power of the stars (*stellarum virtus*) and the power of the rational soul (*animae rationalis potestas*), offering an account in general (*in summa*) about these things according to what the wise have taught. For just as a child born and exposed to new air as to a new world (*sicut enim puer natus, et aeri novo tanquam mundo recenti expositus*) receives the impression of celestial powers (*recepit impressionem virtutum coelestium*)—from which it has a root complexion (*complexio radicalis*) that can never be lost because '*quod nova testa capit, inveterata sapit*'[55]—this is the case concerning every thing newly made (*de omni re noviter facta*) that receives the power of the heavens at its initial existence (*quae recipit virtutem coeli in prima eius existentia* [i.e. when it first comes into being]). What it receives in the beginning it never loses until it desists from its natural existence and is corrupted (*donec a suo esse naturali destituatur et corrumpatur*).[56]

Here Bacon puts to good use the natural philosophical structures established in detail in *Opus maius* IV by extending the significance of celestial influences at birth to the making (generating) of everything, both by nature and by man's artifice, in a passage reminiscent of Albertus Magnus's *De mineralibus*. In both Roger's and Albert's views, things made by man receive celestial influences at their making just as natural things do at their coming into being or generation, only to lose these same powers on their corruption. We will also notice strong echoes here of some of Roger's evocative phrasing where he most fully articulated his geometrical-optical analysis of celestial influences with its central use of rays, as I discussed in chapter two.

Roger continues, integrating this natural philosophical analysis with the words and works of the wise, thereby offering us further insight into how he conceives of these fundamental matters:

> And therefore, in these talismans (*imagines*), *carmina* and characters composed under the proper [sc. celestial] configuration (*constellatione debita compositis*), the powers of the

[55] As we will recall, Bacon used this exact same tag in my Chap. 2.

[56] Redeo igitur ad verba et opera sapientium, stellarum virtute et animae rationalis potestate formata, redens rationem in summa de his secundum quod docuerunt sapientes. Sicut enim puer natus, et aeri novo tanquam mundo recenti expositus, recepit impressionem virtutum coelestium, a quibus habet complexionem radicalem quam nunquam potest amittere, quia quod nova testa capit, inveterata sapit; sic est de omni re noviter facta, quae recipit virtutem coeli in prima eius existentia, et illam quam in principio recipit nunquam amittit, donec a suo esse naturali destituatur et corrumpatur (396, 15–24).

> stars are received and retained (*recipiuntur virtutes stellarum et retinentur*), so that by means of these [sc. powers (*virtutes*)] they can act on the things of this world (*possint per eas agere in res huius mundi*), and when the celestial configuration in which they were made recedes (*recedente constellatione, in qua fuerunt res huiusmodi compositae*), the composed things of this sort [sc. the *imagines, carmina* and characters] recede [sc. in power] (*recedunt*).[57]

The astrological dimension and its timing are thus utterly central for Roger's views on the making and use of talismans, *carmina* (chants) and characters for acting in and on the world. As we saw in Chap. 7, the *Speculum astronomiae* would have considered such techniques illegitimate regardless of the underlying intentions, due primarily to the *carmina* and characters, unless strictly circumscribed. And in *De fato* and especially *De mineralibus*, as we will recall, Albert similarly analyzed the power of talismans in terms of the period derived from the heavens, as well as their loss of power over the course of time.

After showing the fundamental role of celestial influences in making certain objects, Roger returns to the soul and its power, emphasizing once again that it is more dignified ontologically than the stars:

> And since the rational soul is more worthy than the stars, therefore just as the stars and everything make their powers and species in things outside [sc. themselves], about which species and powers I write sufficiently in the fourth part [i.e. Book IV of the *Opus maius*], where geometry is treated, therefore the rational soul (*anima rationalis*), which is the substance most active among everything after God and the angels (*quae est substantia maxime activa inter omnia post Deum et angelos*), can make and does continually make its species and power (*facit continue speciem suam et virtutem*) [1] in the body, of which it is the realization (*actus* [= "entelecheia" (Gk)], *in corpus, cuius est actus*),[58] and [2] in things outside it (*in res extra*); and especially when it acts (*operatur*) from a strong desire (*ex forti desiderio*), specific intention (*intentione certa*) and great confidence (*confidentia magna*). About which works (*opera*) Avicenna speaks powerfully (*potenter*) in the sixth book of the *Naturalia*.[59]

Here Roger makes an explicit analogy between how the stars act on things of the world and how the soul acts on both its own body and on those outside it,[60] utilizing

---

[57] Et ideo in istis imaginibus, carminibus, et characteribus, constellatione debita compositis, recipiuntur virtutes stellarum et retinentur, ut possint per eas agere in res huius mundi, et recedente constellatione, in qua fuerunt res huiusmodi compositae, recedunt (396, 24–29).

[58] For Aristotle, of course, soul (*anima*, psyche [Gk]) is the form of the body and its realization or '*entelecheia*' [Gk], as he discusses at length in his profoundly influential *De anima*.

[59] "Et quoniam anima rationalis dignior est stellis, ideo sicut stellae et omnia faciunt virtutes suas et species in rebus extra, de quibus speciebus et virtutibus scribo satis in quarta parte, ubi de geometricis agitur, potest ergo anima rationalis, quae est substantia maxime activa inter omnia post Deum et angelos, facere et facit continue speciem suam et virtutem in corpus, cuius est actus, et in res extra; et maxime quum ex forti desiderio et intentione certa et confidentia magna operatur (396, 29–37)." James Hankins discusses related texts from Avicenna's *De anima* in relation to Ficino in his "Ficino, Avicenna and the Occult Powers of the Rational Soul," in *Tra antica sapienza e filosofia naturale: La magia nell'Europa moderna*, F. Meroi (ed.), 3 vols., Florence: Olschki, 2007, I: 35–52. I discuss his argument in volume II.

[60] This corresponds to DP Walker's distinction between intransitive and transitive magic in his *Spiritual and Demonic Magic*, 75–84.

the geometrical-optical model of planetary and other influences from *Opus maius* IV (to which he refers explicitly), which he also derived from al-Kindi, who also says that the power of words—along with talismans—act precisely by means of rays. But since the soul is more worthy and more active than the stars, its power is even stronger in the world, especially when focused and fortified by a person's intention, desire and confidence, namely, by the operator's emotional and mental state.

Roger concludes this step in his argument with the most important point of his analysis:

> And therefore, the works (*opera*) and words (*verba*) of this sort about which I am speaking, not only receive power from the heavens (*non solum recipiunt virtutem a coelo*), but also from the rational soul (*sed ab anima rationali*) which is nobler; and because of this, they can have a great power for altering the things of this world (*propter hoc possunt habere magnam virtutem alterandi res mundi huius*).[61]

The most powerful words and works thus result from using the power of both [1] the celestial powers/virtues and [2] the rational soul, which, of course, only human beings have. They may thus be used together to alter the things of the world as well as peoples' minds. Here Roger indicates that one should use both features together, but he conspicuously does not develop this view further here. Perhaps it is among the *magnalia sapientiae* that he hopes to keep secret and disclose only verbally and in person to the pope himself. Regardless, for Roger these are the works and words of wisdom that should not be considered or called magic in any way, shape or form, except insofar as they are abused by *magici maledicti*, as we will see further below.

8.10 Two Objections

With these basic structures now in place, Roger responds to two possible objections, in which he explicitly uses the language of rays, which had thus far only been implied here. He also discusses the central issues of place and time. Here Roger offers the fullest analysis I know of for his view of the natural philosophical foundations for talismans, which is articulated, as we would expect, in terms of his astrologizing Aristotelian natural philosophy:

> And if it is said that, just as works of this sort receive the power of the heavens (*huiusmodi opera recipiunt virtutem coeli*), so do all other things (*alia omnia*) [1] in the same region [2] at the time of putting together (*compositio*) the words (*verba*) and works (*opera*) of this sort.[62] And so everything ought to have these powers: men and cattle, horses and trees;[63] for

[61] Et ideo huiusmodi opera et verba de quibus loquitur non solum recipiunt virtutem a coelo, sed ab anima rationali, quae est nobilior, et propter hoc possunt habere magnam virtutem alterandi res mundi huius (396, 38–397, 3).

[62] That is, those just described in the last two sections, including talismans.

[63] These are all examples of natural living things, i.e. none are artificial.

> the rays of a celestial configuration (*radii coelestis constellationis*) touch an infinite [sc. number of things] at the same time (*attingunt infinita in eodem tempore*).[64]

The objection here seems to be: if everything that exists in a given area receives the same natural influences at the same time, what makes the astrological timing of the words and works of wisdom so special? Thus, if everything made or done at every time and place receives the same celestial influences, then the previous discussion about something being made or said at a given time having a special power is nonsense. Therefore, there would be nothing special at all about either talismans or the utterance of powerful language.

Bacon responds thus:

> One should say that this objection is not strong because [sc. all] things (*res*) are not on the same horizon (*non sunt res in eodem horizonte*). For individual points of the earth are the centers of different horizons (*singula puncta terrae sunt centra diversorum horizontium*), towards which the cones of diverse pyramids of celestial powers come (*ad quae coni diversarum pyramidum virtutum coelestium veniunt*) in order [1] to produce plants of different species in the same small patches of land (*particula terrae minima*), and [2] to differentiate twins in the same womb in both complexion and customs (*gemellos in eadem matrice diversificare in complexione et moribus*), and [3] in the use of sciences, and languages, and business, and in everything else.[65]

According to Bacon, the force of the objection is weak because it does not take the central issue of place vis-à-vis timing fully into account.

Even things very close to each other—what could be closer, after all, than twins in the same woman's womb?—do *not* receive *exactly* the same force of celestial influences at the same time, due to the slightly different angular orientation of the celestial pyramid for each entity, as we learned in *Opus maius* IV (as discussed in my Chap. 2). Particular places and their unique horizons must be taken into account because, in this way, each place is unique. This means in practice that it receives a given celestial influence in a unique angular configuration with its unique mix of qualitative strengths, namely, with its unique complexion. Here Bacon fully returns to his earlier analysis to reintroduce the great individuating mathematico-geographical device: the horizon. Thus, all things everywhere do not receive the same celestial virtues with the same force, even at the same time and in very closely located places. Given Bacon's insistence on place being a principle of generation (as did Albert), one can easily see the central importance of this concept in this context.

---

[64] Et si dicatur, quod sicut huiusmodi opera recipiunt virtutem coeli, sic alia omnia quae sunt in eadem regione, et in tempore compositionis huiusmodi verborum et operum; et ita omnia debent habere has virtutes, et homines, et boves, et equi, et arbores, nam radii coelestis constellationis attingunt infinita in eodem tempore (397, 4–9).

[65] [D]icendum quod non valet obiectum, quia non sunt res in eodem horizonte. Nam singula puncta terrae sunt centra diversorum horizontium, ad quae coni diversarum pyramidum virtutum coelestium veniunt, ut possint producere herbas diversarum specierum in eadem particula terrae minima, et gemellos in eadem matrice diversificare in complexione et moribus, et in usu scientiarum, et linguarum, et negotiorum, et caeteris omnibus (397, 9–16)[.]

Having articulated the natural philosophical foundations of his reply by using the geometrical-optical model of celestial influences, Roger deepens his response by instantiating it using the case of talismans, focusing here on astrological timing or elections, and again explicitly using the term '*imago*':

> Then, because other things that were already made earlier before composing the talisman (*res aliae iam factae prius ante compositionem imaginis*), although they exist with it [sc. with that *imago* at the same time and in the same region], nevertheless at their origin (*ab origine sua* [= generation]) they received *their* proper root influence (*receperunt influentiam propriam radicalem*) in accordance with which they were made (*operantur*). Therefore, the power of the heavens (*virtus coeli*) at that time (*ad hanc horam*) about which we are speaking does not have a [sc. the same] natural effect (*effectus naturalis*) on things of this sort made before, as it has on these works (*opera*) and words (*verba*) now newly made (*nunc de novo factis*, my emphasis).[66]

Although things already existing at the astrologically chosen time a talisman is made are also in existence, they do not receive the same celestial influences *in the same way* that the newly made talisman does.

This is because the talisman comes into being at a particular moment (or over a short period of time) and thus receives the current celestial influences as its root complexion, whereas the other things also existing at that time had already been made or came-into-being earlier, and thus had already received their different root complexions at the time of *their* generation. Thus, *they* receive the celestial influences at the time (and place) of the talisman's generation as a current or running complexion—as we will recall from discussing Bacon's natural philosophy in Chap. 2—but this is an effect of a different order, and thus affects those other things differently. From our discussion in Chap. 7, we can see that Thomas in his later normative position would disagree strongly with Roger's analysis here, but only with respect to man-made objects, the *opera* in question.

To complete his response to the first objection, Bacon turns from *opera* (talismans) to the other "magical" pole, namely, the rational soul, in relation to the power of words:

> Then, because they were not made by a rational soul [i.e. they were made by nature], or not at that time; or if [sc. they were made] then [sc. by a rational soul], many other things were made by a [sc. different] person, but [1] not with that intention, nor desire, nor confidence, nor [2] arranged (*ordinantur*) for actions of this sort, and therefore either they will not have the power of qualitative altering (*virtus alterandi*) of this sort, or they will not have such notable operations/effects.[67]

[66] [T]um quia res aliae iam factae prius ante compositionem imaginis, licet sint cum illa, tamen ab origine sua receperunt influentiam propriam radicalem secundum quam operantur, et ideo virtus coeli ad hanc horam, de qua loquimur, non habet effectum naturalem in rebus huiusmodi prius factis, sicut habet in istis operibus et verbis nunc de novo factis (397, 16–22).

[67] [T]um quia ab anima rationali non sunt facta, aut non tunc; aut si tunc, multa fiunt ab homine alia, non tamen ea intentione, nec desiderio, nec confidentia, nec ordinantur ad huiusmodi actiones, et ideo huiusmodi aut non habebunt virtutem alterandi, aut non ita notabiles operationes (397, 22–27).

All of these psychological-efficient—as a significant part of the "artisanal-efficient"—causal factors must also be taken into account to explain why certain words or works have different and more powerful effects than others. This is based in part on the mental/psychological attitude of the different makers or speakers, namely, the focused intention of the "artisanal-efficient" cause. This is, of course, the artisan-magician himself together with his intention, the final cause, namely, the end for which the talisman was made or the words uttered.[68] Thus Bacon responds to the first objection.

*

The second objection concerns how everything acts in the world in relation to its root complexion. This is articulated in now familiar natural philosophical terms:

> At any rate, if it is said that every thing (*omnis res*)—by the power that it receives at its origin from a celestial configuration (*per virtutem quam recipit in sua origine a coelesti constellatione*), although the root complexion remains in it [sc. in every *res*] in accordance with the strength of the configuration (*maneat complexio radicalis iuxta fortitudinem constellationis*)—will alter things outside itself (*alterabit res extra se*) and will change them perceptibly (*immutabit sensibiliter*), especially when it is new (*in novitate sua*). And that [1] due to the advancing of time (*per processum temporis* [sc. in the course of time]), and [2] the continually different virtues of the heavens (*per continuas virtutes coelorum alias*),[69] the first power is remitted and weakened little by little until it gives out (*remittatur et debilitetur virtus prima paulatim donec deficiat*).[70]

The second objection refers to the weakening of everything's (including a talisman's) initial power over time, namely, that the power inherent in root complexions eventually wears out. The upshot of this not fully clear objection seems to be that there would then be no point in making such objects if they eventually lose their charge and thus their ability to affect things in the world. As we saw in Chap. 7, Albertus Magnus makes a similar point about talismans in *De mineralibus* (52a44-b3): that they too will eventually lose their charge.

In fact, Bacon agrees with this statement:

> It should be said that this is true, and that, in accordance with [sc. how] such things arise [sc. come into being] (*oriuntur tales res*), there sometimes come to be great transformations (*magnae mutationes*), although we should not contemplate whence such alterations happen (*tales alterationes contingunt*), just as happen from comets (*accidit de cometis*) and other such things.[71]

[68] I offer the "artisanal-efficient" cause here in contradistinction to the "celestial-efficient" cause. I develop this distinction more fully in volume II in relation to Ficino's *De vita*. Ficino seems to build on the analysis here in *De vita* III.11 and following.

[69] I.e. the incessant radiation of celestial virtues in different celestial configurations.

[70] Si dicatur, quod saltem omnis res per virtutem quam recipit in sua origine a coelesti constellatione, cum in illa maneat complexio radicalis iuxta fortitudinem constellationis, alterabit res extra se et immutabit sensibiliter, et maxime in novitate sua; et quod processum temporis, per continuas virtutes coelorum alias, remittatur et debilitetur virtus prima paulatim donec deficiat (397, 27–33)[.]

[71] [D]icendum est, quod hoc est verum, et secundum quod oriuntur tales res, fiunt aliquando magnas mutationes, licet non consideremus unde tales alterationes contingunt, sicut accidit de cometis et aliquibus aliis (397, 33–398, 2).

Bacon seems to be saying that, even though both natural and artificial things lose their power over time, they can still bring forth major transformations in the world on analogy with the effect of comets. In this way, the *opera et verba sapientiae* can still be very worthwhile. We will see towards what ends just below. This is the fullest and deepest treatment of Bacon's natural philosophical analysis of the *opera et verba sapientiae* that I know of. Roger Bacon has thus fully embraced al-Kindi's deeply influential system—integrating rays, Aristotelian natural philosophy and magic—on the one hand, and entirely erased al-Kindi from his account here, on the other. We should also note that the details of the natural philosophical part of his analysis are very similar to those in Albert's *De mineralibus*.

8.11 On the Power of Words (2): *Fascinatio*

Then Roger addresses the topic of "fascination," which seems similar (at least in part) to our use of hypnosis, albeit for the negative purposes of, among other things, gaining power over a person and/or bewitching them.[72] Roger's analysis here offers much material of interest, with strong medical and astrological dimensions. It will also help us to more fully understand the material just discussed, especially on the *virtus verborum*:

> And again, one can bring out what should be understood about fascination (*fascinatio*), which, although it has a suspect term (*verbum suspectum*), can also now be understood as the *vetulae*, the female casters of lots (*sortilegae*),[73] and magicians (*magici*) consider it.[74] Nevertheless, it is certain that many people are of a bad complexion (*mala complexio*),[75] and of a corrupted and infirm composition (*corruptae compositionis et infirmae*),[76] such that they become infected (*contagiosi*), just as people sick with many contagious illnesses.[77] And an infective and contagious power (*virtus infectiva et contagiosa*) happens (*accidit*) especially to a complexion of this sort by means of a bad celestial configuration in which a child was conceived or born (*per malam constellationem in qua conceptus est puer vel*

[72] For '*fascinatio*' in Albertus Magnus, see Palazzo, "Albert the Great's Doctrine of Fascination."

[73] Or both together (*vetulae sortilegae*), old female casters of lots, sorcerers, witches?

[74] Does he imply by this phrasing that there is also a positive interpretation of '*fascinatio*'?

[75] That is, they have a bad astrological nature as translated into a qualitative bodily state, either root or current.

[76] '*Compositio*' here seems to be a synonym for '*complexio*'. It seems strange that Roger would here use the term '*compositio*' in what seems a very similar way to how he used it just before in relation to the making ('*compositio*') of an *imago/opus*. What does this imply? That people too are '*compositiones*' in some analogous way? It is certainly conceived as such in Aristotle, namely, that humans are a composition of body and soul (matter and form), each of which had just been emphasized, namely body with respect to an *opus* and soul with respect to the *virtus verborum*, which is the act of an *anima rationalis*.

[77] This seems to be similar to "contagion" in our sense.

> *natus*).[78] For an infinite [sc. number of people] of the worst complexion (*pessima complexio*) were born from the healthiest parents of the finest complexion (*a sanissimis parentibus et optimae complexionis nascuntur*)[.][79]

A bad complexion results from being conceived or born under a malign celestial configuration, and thus being impressed with and thereby receiving bad celestial influences at one's conception or parturition. This happens even to children born to parents with excellent complexions, which are thus not hereditary in an astrological sense. Here Roger alludes to the importance of the natal horoscope in understanding the root complexion, which was mentioned just above, and treated more fully in my Chap. 2.

Roger then discusses how these badly complected people affect others:

> And these [sc. people (*isti*)]—just as all things (*omnes res*) that produce their *species* and *virtus*[80]—contaminate nearby things, especially [sc. those things (sc. *res*)] weak in age and complexion (*teneras aetate et complexione*), and [sc. they contaminate them] especially through the eyes (*per oculos*). Because they are porous and fine (*rari*), both corrupt vapors and spirits (*vapores et spiriti corrupti*) go forth and infect things. Whence Virgil [says]: "I do not know which eye fascinates my weak goats."[81] Just as a menstruating woman (*mulier menstruosa*), if she looks at a new and polished mirror, stains (*maculat*) it with a bloody cloud (*nube sanguinea*), as Aristotle says in the 2nd book of the *De somno et vigilia* (and experience teaches), and so infects other things even though it does not appear to be the case: and thus lepers (*leprosi*) infect. And all these things are natural (*naturalia*) [i.e. things having nothing to do with human volition].[82]

Thus, a person with a badly complected body affects and infects other nearby things transitively and naturally by means of noxious vapors and spirits escaping through their eyes. Bacon cites Virgil, Aristotle and experience in support.[83] We should also

---

[78] Thus concerning their root complexion.

[79] Et iterum potest elici quid intelligendum est de fascinatione, quae licet habeat verbum suspectum, et potest modo intelligi ut vetulae sortilegae et magici considerant; tamen certum est quod multi homines sunt malae complexionis et corruptae compositionis et infirmae, ita quod contagiosi sint, sicut homines infirmi multis morbis contagiosis. Et praecipue accidit complexioni huiusmodi virtus infectiva et contagiosa per malam constellationem in qua conceptus est puer vel natus. Nam a sanissimis parentibus et optimae complexionis nascuntur infiniti qui sunt pessimae complexionis (398, 3–12)[.]

[80] As in the earlier treatment of the multiplication of species.

[81] This quotation is from Virgil's *Eclogue* III.103.

[82] [E]t isti, sicut omnes res suam speciem et virtutem producentes, contaminant res praesentes, maxime teneras aetate et complexione, et praecipue per oculos; quia porosi sunt et rari, et exeunt vapores et spiritus corrupti et inficiunt res. Unde Virgilius. 'Nescio quis teneros oculus mihi fascinat agnos.' Sicut mulier menstruosa, si aspiciat speculum novum et politum, maculat ipsum nube sanguinea, ut Aristoteles dicit secundo De somno et vigilia, et experientia docet, et ita inficit alias res, licet non appareat: et sic leprosi inficiunt. Et haec omnia naturalia sunt (398, 12–21).

[83] For this topos in Aristotle of a menstruating woman staining a mirror by looking at it—and its influence—see Berthold Hub, "Aristotle's 'Bloody Mirror' and Natural Science in Medieval and Early Modern Europe," in *The Mirror in Medieval and Early Modern Culture: Specular Reflections*, Nancy M. Frelick (ed.), Turnhout, Belgium: Brepols, 2016, 31–71.

note that Roger refers here to human-biomedical *spiritus*, which I will discuss much more fully in volume II.[84]

As in discussing the *verba* and *opera sapientiae* above, Roger now joins the purely physiological analysis of *fascinatio* just offered with one that also includes a volitional and thus psychological component:

> But if, further, some malign soul (*aliqua anima maligna*) thinks strongly about infecting someone [or something] else (*cogitet fortiter de infectione alterius*), and ardently desires (*ardenter desideret*), intends with certainty (*certitudinaliter intendat*), and vehemently considers (*vehementer consideret*) that he can do harm, there is no doubt but that nature obeys the cogitations of the soul (*natura obediet cogitationibus animae*), as Avicenna teaches in Book VIII of *De animalibus*, and in Book IV of *De anima*. Insofar as the multiplication of species is stronger (*fortior sit multiplicatio speciei*), also the infection will be more violent (*infectio violentior*). Just as if a leper (*leprosus*) intends, desires and has much confidence to harm someone whom he hates, he will harm him far more (*longe plus*) than another for whom he did not contemplate something bad (*malum non cogitat*).[85]

Here Roger analyzes the power of malignant intentions by adding human desire to purely natural processes, both astrological and physiological. He explicitly uses the multiplication of species as well as referring, once again, to Avicenna. Roger here seems to also be describing the proper mental attitude for *maleficium*. As so often, we will see that Roger's argument and analysis here is cumulative.

Bacon then generalizes the point, and in the process more fully articulates the psychological dimension for both good and bad magic in our parlance:

> For every operation (*operatio*) of a human being is stronger and more passionate when he is very anxious (*sollicitus*) and willful (*voluntarius*) toward it, and when he strengthens his intention with a fixed purpose (*fixo proposito firmat intentionem*), and he hopes firmly (*sperat firmiter*) that he can accomplish what he intends. For wise men (*sapientes*) say that just as '*scammonea*' sharpens medicine (*acuit medicinam*) and gives it vigor (*dat ei vigorem*), so intention (*intentio*), desire (*desiderium*) and confidence (*confidentia*) are disposed in human works (*in operibus humanis*); and up to this point, it is not magical (*adhuc non est magicum*).[86]

Thus, intention, desire and confidence sharpen/focus/intensify all human works, whether for good or ill, as scammony focuses and intensifies medicines. For Bacon, however, this is explicitly *not* magic, at least not yet.

---

[84] For medieval views of *spiritus*, see Bono, "The Languages of Life," and "Medical Spirits and the Medieval Language of Life."

[85] Quod si ulterius aliqua anima maligna cogitet fortiter de infectione alterius, atque ardenter desideret et certitudinaliter intendat, atque vehementer consideret se posse nocere, non est dubium quin natura obediet cogitationibus animae, ut docet Avicenna octavo de Animalibus et quarto de Anima, quatenus fortior sit multiplicatio speciei et infectio violentior; sicut si leprosus intendat desideret et confidat multum nocere alicui quem odio habet, longe plus laedit eum quam alium cui malum non cogitat (398, 21–29).

[86] Omnis enim operatio hominis est fortior et impetuosior, quando ad eam est multum sollicitus et voluntarius, et fixo proposito firmat intentionem, et sperat firmiter se posse consequi quod intendit. Nam dicunt sapientes, quod sicut scammonea acuit medicinam et dat ei vigorem, sic intentio et desiderium et confidentia se habent in operibus humanis; et adhuc non est magicum (398, 29–36).

Then Roger turns to Pliny, and thus refines his own analysis. Here Pliny discusses acting by means of words:

> For Pliny teaches in the seventh book of the [*Historia*] *Naturalium* that many people are of such a complexion who infect and harm others from their presence both through vision and through word[s] (*ex sua praesentia,*[87] *et per visum et per verbum*). For, since a word (*verbum*) [1] is generated from natural internal parts [sc. of the body] (*ab interioribus membris naturalibus generatur*) and [2] is formed from thinking and concern, and a person delights in this, and [3] it [sc. the word] is the most proper instrument of the rational soul (*propriissimum est instrumentum animae rationalis*), therefore it has the greatest efficacy among all those things that come to be from a person (*maximam efficaciam habet inter omnia quae fiunt ab homine*), especially when it comes forth (*profertur*) from a well-defined intention (*ex intentione certa*), a great desire (*desiderio magno*) and profound confidence (*vehementi confidentia*).[88]

For Roger, then, intention, desire and confidence as generated from a person's inner psychological depths are significant keys to understanding how powerful human actions work by uttering words, the rational soul's most proper instrument. This discussion further articulates the earlier analysis, concerning, in particular, the *verba sapientiae*. Here Roger explicitly cites Pliny as his authority, whereas he continues to erase al-Kindi—a much more controversial authority—from his text to the pope.

Bacon then uses this analysis of speaking with intention and desire to explain miracles. A complete analysis, however, also requires astrological timing, which Roger seems to impute here to the saints:

> An indication of this is that almost every miracle (*omnia fere miracula*) performed by the saints from the beginning were done by the power of words (*fiebant per virtutem verborum*), whence there is the greatest power in words (*in verbis summa est potestas*), as I have explained. And if a multiplication of species of this sort and the utterance of a word (*prolatio verbi*) came to be under [sc. at the time of] a fitting celestial configuration (*in constellatione debita*), it is necessary that a powerful operation would follow (*operatio valida consequatur*); and in all these things there is nothing magical or insane (*in his omnibus nihil est magicum vel insanum*).[89]

Thus, astrological timing links at the core both of what we would call Roger's magical practices, namely, the *words* and *works* of wisdom. Bacon here completes the discussion raised above on the power of words. The analysis of the greatest power, such as that found in miracles, derives from [1] the utterance of words [2] contemplated with great and clear intention, desire and confidence, now combined with [3]

---

[87] As with the *res praesentes* just above.

[88] Nam Plinius docet septimo Naturalium, quod multi homines sunt talis complexionis, qui alios inficiunt et laedunt ex sua praesentia, et per visum et per verbum. Nam quia verbum ab interioribus membris naturalibus generatur et formatur ex cogitatione et sollicitudine, et delectatur homo in eo, et propriisimum est instrumentum animae rationalis, ideo maximam efficaciam habet inter omnia quae fiunt ab homine, praecipue cum ex intentione certa, desiderio magno, et vehementi confidentia profertur (398, 37–399, 8).

[89] Cuius signum est, quod omnia fere miracula quae facta sunt per sanctos a principio fiebant per virtutem verborum, unde in verbis summa est potestas, sicut explicavi. Et si huiusmodi multiplicatio speciei et verbi prolatio fiant in constellatione debita, necesse est quod operatio valida consequatur; et in his omnibus nihil est magicum vel insanum (399, 8–14).

propitious astrological timing. Roger also makes certain to reaffirm that this is neither magical nor insane. One would also think that Roger's analysis here would be highly relevant for understanding how prayer works.

On the basis of this rich and informative analysis, Roger returns to the term '*fascinatio*' itself:

> And if this be called '*fascinatio*', we can change the word (*nomen*, term), if we wish. But the thing (*res*) has [no] calumny among the wise. I have mentioned "fascination" [sc. before]; but here it is explained more fully. But magicians and evil old women (*vetulae maledictae*) do not consider fascination along these lines (*per has vias*). Nay rather, they assert that sudden qualitative changes (*alterationes subitae*) happen indifferently by any person whomsoever, when this is seen in the morning or evening, or spoken about; therefore their opinion is vain and reproached by the wise.[90]

If the term itself has been compromised, we can change it, Roger contends, precisely as Thomas had suggested with "fate" in my Chap. 5. The thing itself, the *res*, however, is fine, at least when practiced by the wise, and it contains nothing worthy of reproach. Bacon also emphasises that his analysis is quite different than the magicians' and witches' understanding of how '*fascinatio*' works, and that theirs should, therefore, be rejected. Unfortunately, he does not clearly explain to us what it entails.

8.12 Conclusion: Motivating the Pope to Action

After establishing this cumulative picture of the works (*opera,* talismans) and words (*verba*) of the wise—and developing his rich account of *fascinatio*—Bacon now shifts his approach and starts heating up his rhetoric in order to achieve his own ultimate aim (with his own passionate and focused intention) of terrifying the pope into immediate action, including, no doubt, to patronize him. Roger argues in effect that malicious terrorists—namely, the enemies of the Church—already possess these supremely powerful tools, which they will undoubtedly use as weapons of mass destruction. In the process, Roger paints a terrifying portrait of malevolent power:

> I do not write these things only for the considerations of wisdom (*pro consideratione sapientali*), but because of dangers (*pericula*) which do and will happen (*contingunt et contingent*) to Christians and to the Church of God by infidels (*per infideles*), and especially by Antichrist (*per Antichristum*), because he will use the power of wisdom (*potestas sapientiae*) and turn everything towards evil (*omnia convertet in malum*).[91]

[90] Et si vocetur fascinatio, nomen si volumus, possumus mutare. Sed res in nullo habet calumniam apud sapientes. Feci vero mentionem de fascinatione; sed hic uberius explicatur. Sed quia magici et vetulae maledictae non considerant fascinationem per has vias, immo asserunt alterationes subitas indifferenter fieri per homines quoscunque, cum videtur res in mane vel sero, vel loquuntur de illa, ideo vana est eorum sententia et a sapientibus reprobata (399, 14–22).

[91] Non solum pro consideratione sapientali haec scribo, sed propter pericula quae contingunt et contingent Christianis et ecclesiae Dei per infideles, et maxime per Antichristum, quia ipse utetur potestate sapientiae, et omnia convertet in malum (399, 23–27).

Here Antichrist returns to Roger's story as the archenemy of the Church and the faithful.[92]

Then Roger tells us precisely what means Antichrist will use, namely, the words and works of wisdom:

> And by starified words and works of this sort (*per huiusmodi* [1] *verba et* [2] *opera stellificanda*),[93] and made (*componenda*)[94] with [3] a great desire to do ill (*magno desiderio malignandi*) with [4] an extremely certain intention and passionate confidence (*cum intentione certissima et confidentia vehementi*), he [sc. Antichrist] brings bad fortune and fascinates (*infortunabit et infascinabit*) not only individuals, but cities and regions. And by this wondrous path (*per hanc viam magnificam*) he can do whatever he wants without war [sc. without having to fight for it],[95] and people will obey him just like beasts do, and he will make kingdoms and cities fight against each other for him, so that friends will destroy their friends, and thus he will do whatever he desires concerning the world.[96]

In Roger's portrait, then, Antichrist uses the very same works and words of wisdom, but focused instead with a passionately malevolent intention. Roger here makes an important distinction between means and ends. Taking perfectly fine means—namely, both the words and works of wisdom, which he had just insisted were neither magical nor insane, and indeed, result in miracles—Antichrist uses them with intensely evil intentions towards utterly maleficent ends. People who do this are '*malefici*' (evil-doers) by definition. The ultimate terrorist, Antichrist, thus possesses and will most certainly employ these powerful techniques towards maleficent ends. Based on this intelligence, Bacon clearly implies that the pope should retain him as the ecclesiastical equivalent of a national security advisor.

*

From terrifying generalities involving Antichrist to focus the pope's attention, Bacon then turns to more palpable current threats from the most sinister contemporary infidels: "And thus the Tartars [= Mongols] and Saracens [= Arabs] can do [sc. such things]."[97] Bacon begins with the Tartars, encouraging the pope to better

[92] Abate offers an insightful analysis of this material in his "Roger Bacon and the Rage of Antichrist."

[93] That is, made with astrological elections, and thus charged with celestial influences.

[94] As before with '*compositio*'.

[95] As above, in the first example of the power of the works of wisdom from the *Secretum secretorum*.

[96] Et per huiusmodi verba et opera stellificanda, et magno desiderio malignandi componenda cum intentione certissima et confidentia vehementi, ipse infortunabit et infascinabit non solum personas singulares, sed civitates et regiones. Et per hanc viam magnificam faciet sine bello quid volet, et obedient homines ei sicut bestiae, et faciet regna et civitates pugnare ad invicem pro se, ut amici destruant amicos suos, et sic de mundo faciet quod desiderabit (399, 27–34).

[97] "Et sic poterunt Tartari et Saraceni facere (399, 35)." North discusses the complexities of Roger's terminology and much else of interest in his "Roger Bacon and the Saracens."

understand the Church's enemies by means of what we would call "astrologico-magical" intelligence:

> For it is agreed that the Tartars have more leisure for the science of the stars (*astronomia*) than others do, for this reason, that, although there are wise astrologers (*astronomi sapientes*) in many nations, nevertheless, princes of their republic are not ruled (*reguntur*) except by them [sc. the wise astrologers]. But those who are among the Tartars as prelates are among us, are astrologers (*astronomi*).[98]

Bacon warns the pope that the Tartars take the advice of their astrologers very seriously indeed, and that their equivalent of priests are precisely astrologers. A rocket scientist is not required to see where this is going.

Bacon now provides a pointed and detailed contemporary example, and one that he wants Pope Clement to take seriously to heart:

> When the Lord King of France, Louis [IX] (*dominus rex Franciae Lodovicus*) sent brother William of the Franciscan Order to the Tartars, Mangu Cham, Emperor of the Tartars, living in the year of our Lord 1253 said to the Christians gathered before him, with the said brother (*frater*) present: "We have the law from God through our divines (*legem a Deo per divinos nostros*), and we do everything that they say. And you Christians have a law from God through your prophets, but you do not do it."[99] He said this because in the East (*in oriente*) there are bad Christians (*mali Christiani*), such as the Nestorians and many others, who do not live well according to Christ's law. And this Frater wrote to the lord king that if he had known a little about the stars (*si parum scivisset de astris*), he would have been received well by them (*ipse fuisset bene receptus apud eos*); but because he did not know the terms of astrology (*astronomia*), therefore they looked down upon him. And therefore, the Tartars proceed in everything by the path of *astronomia*, both in the prevision of the future and in the works of wisdom (*et in praevisione futurorum et in operibus sapientiae*).[100]

The Tartar priests represented here possess and actually use Bacon's greatest powers for both foreknowledge of the future as well as for making *opera sapientiae*, namely, talismans. Thus, they use astrology in both key respects: knowledge-based (*iudiciaria*) and operative (*operativa*). The obvious implication here is that Christian rulers—and the pope in particular—should do so as well, especially given how powerful these techniques are, with the further implication that that power comes primarily from astrological knowledge and practice. Not only would this be good for national security, but also in foreign relations, especially if Christian leaders

[98] Nam constat, quod Tartari plus vacant astronomiae quam alii, eo quod licet in multis nationibus sint astronomi sapientes, tamen principes reipublicae non reguntur nisi per eos. Sed illi qui sunt apud Tartaros sicut praelati apud nos, sunt astronomi (399, 35–400, 2).

[99] Bridges's n. 1 (400) is valuable.

[100] Mangu Cham, imperator Tartarorum existens anno Domini 1253, quando dominus rex Franciae Lodovicus misit fratrem Willielmum de ordine Minorum ad Tartaros, dixit Christianis congregatis coram eo, praesente dicto fratre; 'Nos habemus legem a Deo per divinos nostros, et facimus omnia quae ipsi dicunt. Et vos Christiani habetis legem a Deo per prophetas, sed non facitis eam.' Hoc dixit, quia in oriente sunt mali Christiani, ut Nestoriani et alii multi, qui non vivunt bene secundum legem Christi. Et ille frater scripsit domino regi, quod si parum scivisset de astris, ipse fuisset bene receptus apud eos; sed quia nescivit terminos astronomiae, ideo despexerunt eum. Et ideo Tartari procedunt in omnibus per viam astronomiae, et in praevisione futurorum et in operibus sapientiae (400, 3–16).

were to send ambassadors skilled in the science of the stars to Eastern nations. We should note that Bacon explicitly uses the phrase '*opus sapientiae*' here too. [101]

Bacon then offers proof by discussing the Tartars' military strength:

> The indication of this is obvious because—even though they are [1] a people (*gens*) having few and weak people, and [2] they neither eat nor drink what would strengthen their nature (*naturam confortet*), and [3] they are not strong in foot soldiers and [4] are unarmed, if we speak properly, except that [5] they have arrows for terrifying those whom they pursue while they are fleeing, [6] nor do they ever fight up close by the power of waging war, so that they are drawn up in formation army against army—they have already prostrated the entire latitude of the earth from the North to the East, and from the East to the South. Only two corners of Christendom (*terrae Chistianorum*) are lacking, namely Egypt and Africa. For if the Lord (*Dominus*) did not repress them and permit discord to be sown among them regularly, they would already have occupied the entire world by now. And therefore, it is fitting that through works of wisdom (*per opera sapientiae*) wondrous things come forth (*procedant magnifica*) by which they trample the world. For the said Frater in his book on the customs of the Tartars (*De moribus Tartarorum*) which he sent to the king, tells that 14,000 Tartars subdued the Sultan of Turkey (*devicerunt Soldanum Turkiae*), who had with him 200,000 horses without [sc. even counting the] foot soldiers. But they could not do this by the power of waging war (*per potestatem bellandi*), as is clear, and therefore they did it through the works of wisdom (*per opera sapientiae*), and especially through astrology (*per astronomiam*), by which they profess that they are ruled and led in everything (*qua profitentur se regi et duci in omnibus*).[102]

Bacon's proof is implicit: How else could the Tartars have taken over so much of the world—given their small and weak armies—in the face of the huge and powerful armies of their foes? Surely it must have been through these great powers, namely, the *opera sapientiae*. Thus, astrology *qua* knowledge and the *opera sapientiae qua* power and action have allowed the Tartars to rule most of the world with far fewer men, defeating even the Sultan of Turkey. This is similar to Aristotle's advice to Alexander that we explored earlier in this chapter, except that here it is in the hands

[101] Bacon took this account from the still extant report of William of Rubruck, which Bridges quotes in n. 1. The modern translation with a valuable introduction is very helpful; *The Mission of Friar William of Rubruck: His Journey to the Court of the Great Khan Möngke, 1253–55*, Peter Jackson (tr), London: Hakluyt Society, 1990. Abate is very useful on all of this; "Rage of Antichrist," 255–64.

[102] Cuius signum est evidens, quod cum sit gens habens parvos et debiles homines, et quasi non comedentes nec bibentes quod naturam confortet, et nihil pedibus valentes, et inermes si proprie loquamur, nisi quod habent sagittas ad terrorem quibus persequantur fugientes, nec unquam de prope confligunt potestate bellandi, ut acies contra acies ordinentur, eum iam totam latitudinem mundi prostraverunt ab aquilone ad orientem, et ab oriente usque ad meridiem, solum eis deficiunt duo anguli terrae Christianorum, scilicet Aegyptus et Africa. Nisi enim Dominus reprimeret eos et permitteret seminari discordias inter eos frequentes, iam totum mundum occupassent. Et ideo oportet quod per opera sapientiae procedant magnifica quibus mundum conculcant. Nam narrat dictus frater in libro suo de moribus Tartarorum quem Domino regi transmisit, quod quatuordecim millia Tartarorum devicerunt Soldanum Turkiae, qui habuit secum ducentos millia equitum sine peditibus. Sed hoc facere non poterant per potentiam bellandi, ut planum est, et ideo per opera sapientiae, et maxime per astronomiam, qua profitentur se regi et duci in omnibus (400, 16–401, 6).

of the truly bad guys, namely, the enemies of Christendom. Hence the seriousness of Bacon's warning to the pope.

Roger then turns to the Saracens, but much more briefly: "Likewise, the Saracens use astrology (*astronomia*) much, and the wise among them know how to make these works (*sciunt sapientes inter eos facere haec opera*)."[103] Bacon continues by discussing both:

> And it is to be greatly feared that the Tartars and Saracens remaining in their regions will send people to the Christians, by whom they will manufacture the works of astrology (*opera astronomiae conficiant*) for causing misfortune (*ad infortunandum*) and for placing discord among [sc. Christian] princes because the enemies of the Christians rely on this most of all, that they place among them wars and discord.[104]

In addition to outright conquests, these Tartars and Saracens are also very clever at sowing seeds of discord among Christian rulers by means of talismans. Thus, Bacon implies, Christians too need to possess these powers, both to defend against their enemies, but also to act positively toward the good.

*

Bacon continues in his efforts to terrify the pope into action with two particularly flagrant examples of how these powers could be used towards malevolent ends. In this case, Bacon even addresses the pope directly (and with a striking familiarity) in the second person singular. The first example concerns a sort of medieval pied piper, and took place at the turn of the thirteenth century:

> For many things of this sort [sc. using *opera*] have been done, although the stupid multitude (*multitudo stulta*) do not consider how they happen. Perhaps you have seen or heard for a certainty that the children (*pueri*) of the French kingdom at one point came together in an infinite multitude (*in infinita multitudine*) behind some awful man (*post quendam malignum hominem*), in such a way that they could not be held back either by their fathers, mothers or friends; and they were placed in ships and sold to the Saracens; and that was not 64 years ago.[105]

Evil men using the secret powers of wisdom can thus create havoc in Christendom in part by destroying the structure of the family.

Bacon himself was an eyewitness to the second example:

---

[103] Similiter Saraceni multum utuntur astronomia, et sciunt sapientes inter eos facere haec opera (401, 6–7).

[104] Et timendum est valde quod Tartari et Saraceni in suis regionibus commorantes ad Christianos mittant homines, quibus opera astronomiae conficiant ad infortunandum et ad ponendum discordias inter principes, quia ad hoc maxime nituntur inimici Christianorum, ut ponant inter eos bella et discordias (401, 8–12).

[105] "Nam multa facta sunt huiusmodi, licet multitudo stulta non consideret unde accident. Forsan vidistis aut audistis pro certo quod pueri de regno Franciae semel occurebant in infinita multitudine post quendam malignum hominem, ita quod nec a patribus nec a matribus nec amicis poterant detineri, et positi sunt in navibus et Saracenis venditi; et non sunt adhuc quatuor et sexaginta anni (401, 13–19)." Ca. 1203. Matthew Paris dates it to 1213, so 54 years before; n. 1, p. 401.

> Likewise in our times, the Master Pastor[106] disturbed all of Germany and France, and a multitude of men ran behind him, and he held grace (*gratiam habuit* [mass?]) before an entire crowd of lay people in contempt of the priest (*clerus*) and to the confusion of the Church. And he said to Queen Blanche that he would go to his son beyond the sea, deceiving the wisest woman (*sapientissimam mulierem decipiens*) with such words. The wise do not doubt that these were the spokesmen (*nuntii* [sc. ambassadors]) of the Tartars and Saracens, and that they had some works (*aliqua opera*, sc. magical objects) with which they fascinated the people (*fascinabant plebem*). And I saw him with my own eyes (*vidi cum oculis meis*) openly carrying something in his hand (*portare patenter in manu sua quiddam*) as if it were a holy thing (*tanquam esset res sacra*), as if he were a person carrying relics (*reliquiae*). And he went with bare feet, and there was around him a multitude of armed men (*multitudo armatorum*). They were so dispersed in the fields, nevertheless, that what he was carrying in his hand with great ostentation (*quod portabat in manu cum magna ostentatione*) could be seen with him by all those arriving (*ab omnibus occurrentibus potuit videri cum illo*).[107]

Although it is not described in detail, the object in his hand seems to have been a talisman, an *opus sapientiae*, which he treated and displayed as if it were a sacred object, even though it was bent toward evil ends.

Turning from these examples, Bacon unsurprisingly concludes his hortatory diatribe with Antichrist:

> But whatever the case is concerning the Tartars and Saracens, it is certain concerning Antichrist and his minions that they will do these things (*haec operabuntur*). And unless the Church takes action through holy councils for impeding and destroying works of this sort (*ad impediendum et destruendum opera huiusmodi*), it will be aggrieved intolerably (*aggravabitur intolerabiliter*) by the whipping of Christians (*flagellis Christianorum*). And all wise people (*ab omnibus sapientibus*) believe that we are not much removed from the times of Antichrist (*non sumus multum remoti a temporibus Antichristi*), just as is clear in the chapter about sects revolved into one by means of *astronomia*.[108]

For Bacon, the time of Antichrist is nigh, as he had shown in the discussion in *Opus Maius* IV (253–69) discussed in Chap. 6 above. And as with Pierre d'Ailly a century and a half later (to be discussed in Part IV below), Bacon believed strongly that one may astrologically derive profound insights into the timing of Antichrist's

[106] Abate calls him the Master of Hungary; "Rage of Antichrist," 124–25.

[107] Similiter in temporibus nostris magister Pastor totam Alemanniam et Franciam commovit, et cucurrit post eum multitudo hominum, et gratiam habuit coram toto vulgo laicorum in contemptu cleri et ecclesiae confusionem. Et dixit Dominae Blachiae, quod iret ad filium suum ultra mare, talibus verbis sapientissimam mulierem decipiens. Non dubitent sapientes, quin ipsi fuerunt nuntii Tartarorum aut Saracenorum, et quin habuerunt aliqua opera unde fascinabant plebem. Et vidi cum oculis meis portare patenter in manu sua quiddam tanquam esset res sacra, ac si homo deferret reliquias, et ivit nudis pedibus, et erat circa eum multitudo armatorum, ita tamen dispersa in campis, quod ab omnibus occurrentibus potuit videri cum illo quod portabat in manu cum magna ostentatione (401, 19–402, 4).

[108] Quicquid sit de Tartaris et Saracenis, certum est de Antichristo et suis, quod haec operabantur. Et nisi ecclesia occurrat per sancta consilia ad impediendum et destruendum opera huiusmodi, aggravabitur intolerabiliter flagellis Christianorum. Et creditur ab omnibus sapientibus quod non sumus multum remoti a temporibus Antichristi, sicut in capitulo de sectis per astronomiam in uno revolutis patet (402, 5–11).

advent.[109] Thus, according to Bacon's dire warnings, all wise men agree that Christians must act quickly to defend against these attacks by impeding and destroying the powerful *opera* of their enemies!

*

After all this fearmongering, Bacon returns to the positive benefits to be derived from this knowledge, as he moves towards the conclusion of this rich section of the *Opus minus*:

> Therefore, if Christians knew that these works (*opera*) were to be made under papal authority (*auctoritate papali*) for impeding bad things against Christians (*mala Christianorum*), it would be sufficiently laudible, and not only for repelling bad things (*propter mala repellenda*), but also for the promotion of certain useful things (*ad promotionem quorumcunque utilium*). And because [1] individual people, cities and regions, can be altered for the better (*possunt alterari in melius*) in accordance with what was said before, and [2] as life can be prolonged as much as is sufficient, and [3] all things can be looked after usefully, and [4] much greater things can be done than ought to be recommended in the present writing, not only among natural things (*in naturalibus*), but also in the moral sciences and the arts, as is made clear by Moses and Aristotle.[110]

Bad things can be repelled and useful things promoted on many levels of society, as we saw before. Here Bacon again alludes to the fact that he should not write too much about these powerful objects and the theories behind them. Therefore, the pope should invite him to Rome, Roger strongly implies, in order to get a fully detailed account.

Bacon then reiterates how these *opera* work:

> And these things especially can be done, when the species and power of the rational soul (*species et virtus animae rationalis*)—which is more worthy than the heavens (*dignior caelo*)—runs from the part of the one making [sc. an *opus*] (*currit a parte operantis*) with the power and species of the heavens (*cum virtute et specie coeli*), in such a way that the thought of a strong person (*fortis cogitatio*), and a burning desire (*ardens desiderium*), fixed intention (*certa intentio*) and full confidence (*plena confidentia*), and especially the sanctity of [sc. their] life (*sanctitas vitae*) are present. Because nature obeys the thoughts and feelings of the soul (*natura obedit cogitationibus et affectibus animae*), and especially holiness (*sanctitas*).[111]

[109] For a deeply influential study of many related issues, but with a very different prophetic style, see Marjorie Reeves, *The Influence of Prophecy in the Later Middle Ages: A Study in Joachimism*, Oxford: Clarendon, 1969.

[110] Si igitur Christiani scirent haec opera auctoritate papali facienda ad impedienda mala Christanorum, satis esset laudabile, et non solum propter mala repellenda, sed ad promotionem quorumcunque utilium. Et quia personae, et civitates, et regiones secundum praedicta possunt alterari in melius, et ut vita quantum sufficit prolongetur, et omnes res utiliter procurari, atque multo maiora fieri, quam praesenti scripturae debeant commodari, non solum in naturalibus, sed in moralibus scientiis et artibus, sicut patuit per Moysen et Aristotelem (402, 12–21).

[111] Et haec praecipue fieri possunt, quando cum virtute et specie coeli currit a parte operantis species et virtus animae rationalis, quae est dignior coelo, ita ut adsit fortis cogitatio, et ardens desiderium, certa intentio et plena confidentia, et maxime sanctitas vitae; et quia natura obedit cogitationibus et affectibus animae, et maxime sanctitati (402, 21–26).

Here we encounter Bacon's fundamental analysis again: the most powerful works are composed of focused and passionate attention and desire—derived from the soul—working together with the optimal celestial influences derived from nature. With such knowledge and its concomitant power, great good can be done for both individuals and the entire republic of the faithful. This is especially true for holy people, namely, those who embody a *sanctitas vitae*, among whom we may consider that Bacon includes himself.

Bacon then further clarifies this picture by describing the hierarchy of nature and man, which he supports with some authoritative examples:

> For in the eighth book of *De animalibus*, Avicenna gives an example concerning a vigorous hen (*de gallina strenua*), who defeated a wretched cock (*gallum unum miserum*), and from the glory of the victory a horn suddenly grew on its leg. And Avicenna says there that in this we can understand that nature obeys the thoughts of the sensitive soul (*natura obedit cogitationibus animae sensitivae*). And he gives a [sc. second] example in the fourth book of *De anima* about someone who, walking on a piece of wood placed on water, falls because he was thinking about [sc. having] an accident and lost hope. Nature obeys the thoughts (*cogitationes*) and feelings (*affectiones*) of the soul. Whence he says that heat does not come forth from heat, nor cold from cold, but only from the cogitation of the soul; and thus one becomes sick, and every qualitative change of its own body (*omnis alteratio proprii corporis*) takes place.[112]

Since the soul is more dignified than nature (whether celestial or terrestrial), nature "naturally" follows the soul's orientation, especially when it is focused with a passionate and clear intention and desire. Bacon teaches here that the mind and feelings can thus have powerful affects for good or ill on one's own body, thus indicating that this is also relevant for medicine, providing, among other things, its psychosomatic foundations. The mind thus acts "intransitively" on its own body.

But thoughts and feelings can also affect the bodies of others, Bacon argues, as he turns from intransitive internal (psychosomatic) responses to transitive effects and their influence on other people and things. We should know, however, that the intensity of the passion is by no means the only factor. Sanctity or holiness is as well, as Bacon had just mentioned, adding a new element to his analysis:

> And doubtless many things can be done to another body (*multa fieri possunt in corpore alieno*). Because there is a certain family of wolves (*quoddam genus luporum*), which make a person hoarse if it looks at him first, as is commonly held among authorities. And the rational soul can change (*immutare*) many things much more strongly due to the nobility of its species with the five conditions mentioned above, and especially with sanctity/holiness because sanctity is nobler than thought or desire (*quia nobilior est sanctitas quam cogitatio*

[112] Nam octavo de Animalibus Avicenna ponit exemplum de gallina strenua, quae vicit gallum unum miserum, et ex gloria victoriae statim crevit ei cornu in crure, et dicit ibi Avicenna, in hoc cognoscimus quod natura obedit cogitationibus animae sensitivae. Et ponit exemplum in quarto de Anima, de eo qui ligno posito super aquam ambulans cadit, quia cogitat de casu et desperans est, et natura obedit cogitationibus et affectibus animae; unde ait, quod provenit calor non ex calore, et frigus non ex frigore, sed ex sola animae cogitatione et sic fit infirmitas, et omnis alteratio proprii corporis (402, 27–36).

> *vel desiderium*), and therefore the matter of the world (*materia mundi*) obeys a holy soul (*anima sancta*) more than the other four conditions.[113]

As there is a hierarchy in nature such that the soul is more dignified than the stars, likewise there is a hierarchy in the functions of our own souls, whereby holiness is nobler than either thought or desire, powerful though they both are in themselves. Therefore, a holy soul can act more powerfully in the world.

Bacon then offers examples of the *sancti* and their miracles:

> And we see this fulfilled among the saints (*sancti*) who made miracles (*fecerunt miracula*), whom the elements of the world (*elementa mundi*) obeyed. And Avicenna teaches in the fourth book of *De anima* that the soul which is holy and cleansed of sin (*anima sancta et munda a peccatis*) can qualitatively change what is universal and the elements (*potest universale et elementa alterare*), and by its power, storms, rain, and all the qualitative changes of bodies in the world (*pluvia, tempestates, et omnes alterationes corporum mundi*) can be made. Moreover, it is true that the grace of God does much (*gratia Dei multum facit*), but the holy soul working together with grace makes grace (*anima sancta cooperatur gratiae gratum facienti*), so that a person is not saved by grace alone (*non salvetur homo per solam gratiam*), but it is fitting that the soul work together with such grace (*oportet animam cooperari gratiae tali*); therefore, it can work together much more strongly with the grace given to what is full of grace, which is the grace of miracles.[114]

Just as a material substrate co-ordinated with celestial virtue and a person's focused intention mutually increase their power, so it is the same when the soul works together with God's grace. Miracles happen in these circumstances, Bacon contends, as he rises to a religious crescendo.

Bacon concludes this extensive section with a strong appeal to keep this powerful information secret:

> But as with other wondrous things (*in aliis mirabilibus*), it is better to think than to write (*melius est cogitare quam scribere*), until such time as greater certainty is demanded by apostolic authority, and these things are possible. And if you and the successors of Your Blessedness wish, everything can be fulfilled. These are the roots of the treatise in summary that I have proposed to write about things of this sort; but due to impediments I was unable to write more.[115]

---

[113] Et proculdubio multa fieri possunt in corpore alieno; quia quoddam genus luporum est, quod reddit hominem raucum, si primo aspiciat eum, et vulgatum est apud auctores. Et multo fortius anima rationalis potest propter nobilitatem suae speciei multa immutare cum quinque conditionibus supradictis et maxime cum sanctitate, quia nobilior est sanctitas quam cogitatio vel desiderium, et ideo animae sanctae plus obediet materia mundi quam aliis quatuor conditionibus (402, 36–403, 6).

[114] Et hoc videmus adimpleri in sanctis qui fecerunt miracula, quibus elementa mundi obediebant. Et Avicenna quarto de Anima docet, quod anima sancta et munda a peccatis potest universale et elementa alterare, ut eius virtute fiant pluvia, tempestates, et omnes alterationes corporum mundi. Verum est autem quod gratia Dei multum facit, sed anima sancta cooperatur gratiae gratum facienti, ut non salvetur homo per solam gratiam, sed oportet animam cooperari gratiae tali; multo ergo fortius poterit cooperari gratiae gratis datae quae est gratia miraculorum (403, 6–16).

[115] Sed in aliis mirabilibus melius est cogitare quam scribere, donec apostolica auctoritate requiratur maior certitudo, et haec possibilia sunt. Et si vos et successores Vestrae Beatitudinis velint, poterunt omnia adimpleri. Hae sunt radices in summa tractatus, quam proposui de huiusmodi conscribere; sed propter impedimenta non potui plura scribere (403, 16–22).

Thus Roger completes this richly informative section of the *Opus minus*. I have now reconstructed the basic structures of Roger Bacon's in some respects normal, but in others, highly idiosyncratic understanding of astrology in relation to natural knowledge, theology/religion and magic within (primarily) the *Opus maius* and *minus*.[116]

8.13 Conclusion of Parts I–III

I have now closely examined works by Albertus Magnus (ca. 1200–1280), Thomas Aquinas (1225–74) and Roger Bacon (ca. 1214/20–1292)—and the deliberately anonymous *Speculum astronomiae* (ca. 1260s)—that were all written in contexts with strongly natural philosophical and/or theologico-religious dimensions. Essentially employing a shared astrologizing Aristotelian natural philosophical discourse with differing levels of depth, each of these texts has a fundamental astrological component, either with a pronounced practical orientation—as in the cases of Roger Bacon and the *Speculum astronomiae*—or—as with Albertus Magnus and Thomas Aquinas—with a strongly natural philosophical and theological orientation concerned, directly or indirectly, with establishing and securing astrology's scientific and theological foundations within the economy of nature, as a centrally-positioned and integrating feature of God's very deliberate and exceedingly wise creation.

Among other things, these texts also establish a strongly religious ideology for pursuing the study of astrology *qua* natural knowledge as an integral and integrating feature of nature as God's creation, and thereby as the best way to understand both nature and the divine—and in the process inspire a fervent awe and love for Him via His Creation. Astrology plays a central—starring!—role in this process, including for understanding God's providential governance of the world, as we saw in both Thomas Aquinas and the *Speculum astronomiae*. Thus, in Parts I and II, I have reconstructed astrology's natural philosophical and theological foundations.

*

Albertus Magnus's translation-commentaries on fundamental texts of Aristotle's natural philosophy, composed between 1250 and 1254, were all written in order to teach Aristotle in a sound and systematic manner to the students in the Dominican *studium* in Cologne, and elsewhere throughout the order. Among other things, they provided astrology's natural philosophical foundations to generations of students. Albert also participated in a deeply astrological disputation on fate at the papal curia in Anagni, Italy in 1256, which he probably conducted before Pope Alexander IV himself (1199 [or ca. 1185]-1261).

[116] In a comprehensive treatment of astrology and magic in Roger Bacon's works, it would be very worthwhile to integrate the analysis here with other texts by Roger, including *The Secret Works of Art and Nature* and more fully with the *Secretum secretorum*, as well as with his "experimental philosophy," as Hackett has tried to do. But this will have to wait for another occasion.

In 1266–67, Roger Bacon wrote his three famous *Opera* for Pope Clement IV (ca. 1190[–1200]–1268) that at least in part made their way into the papal curia and its great *studium* there, as David Lindberg has shown. In these detailed and influential works, Roger laid out the fundamental patterns of *his* highly mathematicized and deeply astrologizing understanding of nature—in which astrology once again played a central role, in close association with what Roger calls "true mathematics"—which has a great practical value for understanding both human (individual and collective) and religious affairs. In these works, Roger also employed a more apocalyptically-inflected rhetorical style than the other writers we have examined.

We also explored a range of texts by Thomas Aquinas that he composed between 1256 and 1259 (with the *De veritate*) and his last works on these subjects, the *Secunda Secundae* of his justly famous *Summa theologiae* (1271–72) along with several insightful consultations, including the *De operationibus occultis naturae* and *De iudiciis astrorum*. In the *Summa theologiae* and *De iudiciis astrorum*, Thomas argued strongly for astrology's legitimacy as a mode of understanding and predicting what will happen in the future. This took place within certain specific limitations, however: as long as certainty in predictions was not claimed, with its concomitant deterministic interpretation of nature that would thereby undermine both human free will and God's absolute power.

These are themes that both Albertus Magnus in *De fato* and Roger Bacon in *Opus maius* IV also took great pains to address, as did the author of the deliberately anonymous *Speculum astronomiae*, probably written in the 1260s and certainly by 1271–72. These texts, and especially Thomas's analyses in the *Summa theologiae* and the somewhat earlier *Summa contra gentiles*, continued to inform astrological debate throughout the sixteenth and seventeenth, and in some cases well into the eighteenth century, as we will see in greater detail in volume III. Astrology's scientific and theological foundations thus endured—more or less soundly, and in varying measure—for around 500 years in the premodern European map of knowledge.

To provide these solid and long-lasting foundations, astrology seems to have been exceedingly well integrated within both Aristotelian, Ptolemaic and Galenic natural knowledge and Roman Catholic theology during the thirteenth century, at least in the works of these important and influential authors, who seem themselves to have in large measure constructed this synthesis. Albertus Magnus seems to have been the most influential author on the natural philosophy side, with his admirably clear and learned edition of Aristotle's complete works, including various supplementary texts, that were composed primarily during the 1250s and '60s. In turn (and in large measure on this basis), Thomas Aquinas seems to have constructed astrology's theological foundations with his articulation of God's providential governance of the world—by means of the celestial bodies—in the third book of the deeply influential *Summa contra gentiles* composed between 1263 and 1265. In both respects, astrology would have been hard pressed to inspire more learned, influential and energetic champions.[117]

---

[117] I would be curious to know if this central astrologizing strain of Thomas Aquinas's thought was explored by Thomas's Neo-Thomist expositors in the 19th and 20th centuries. The critical

In this shared discourse informed in large measure by the twelfth and thirteenth-century translations of Aristotle, astrology, medicine and magic primarily from Arabic but also from Greek into Latin, astrological concepts and structures came increasingly to be employed towards natural philosophical and theological, but also towards what we would normally call magical ends. This discourse centrally embraced a well articulated geometrical-optical model of celestial influences, although neither Thomas Aquinas nor the anonymous author of the *Speculum astronomiae* seem to have described it in much detail, nor contributed to it directly at all. This is in sharp contrast to both Albertus Magnus and Roger Bacon, for whom it seems to have been utterly central to their understanding of astrologizing Aristotelian natural knowledge, in relation to both theologico-religious and magical themes. For all of these influential thinkers, astrology *qua* knowledge—*astronomia iudiciaria* in Roger's useful phrasing, as articulated in terms of a broadly influential astrologizing Aristotelian natural philosophy—was perfectly legitimate. Everyone seems to agree in this respect, as long as certainty in predictions and its concomitant necessity in nature were neither claimed nor implied.

*

Astrology and Aristotelian natural philosophy were thus explored in Part I, and astrology and theology/religion in Part II. What about astrology and magic in Part III? Here we found a strikingly different dynamic, especially between Thomas Aquinas and his esteemed teacher, Albertus Magnus, particularly in relation to talismans. Both Albert in the *De mineralibus* and the deliberately anonymous author of the *Speculum astronomiae* considered one type of talisman to be legitimate as long as it is "all-natural," and thus has no demonic taint as indicated (i.a.) by characters and/or *carmina*.

In both the *Opus maius* and *minus*, and his important edition of the *Secretum secretorum*, Roger Bacon also seems to embrace the use of talismans—which he like Albert and the Magister Speculi call '*imagines*'—but he utterly and vehemently rejects applying the term "magic" to these objects except as a term of abuse. He would certainly never use it as a term of art as Albert does, and later Giovanni Pico della Mirandola, as we will see in volume II. Roger's usage of the terms "magic" and "divination" parallels Thomas Aquinas's equally and consistently negative use of the term '*divinatio*' in the *Summa theologiae*. In addition, Roger employs the power of words among what we would call his magical interests, thus aligning himself much more closely to the content of al-Kindi's controversial and deeply influential *De radiis stellarum*, which in several respects is certainly one of—if not

reception of Thomas Litt's masterpiece, *Les corps célestes dans l'univers de saint Thomas d'Aquin* (1963), would make an interesting historiographical case study, especially since it appeared at the very same time as two other epoch-making if not downright revolutionary works of scholarship, namely, Thomas Kuhn's *Structure of Scientific Revolutions* (1962) and Frances Yates's *Giordano Bruno and the Hermetic Tradition* (1964).

the most—fundamental text for understanding the entire tradition reconstructed in Parts I–III.

On the other hand, Thomas Aquinas draws the line of legitimacy quite differently than Albert and the Magister Speculi do, and thus rejects astrology's operational or magical dimension altogether, at least in his definitive later works, especially *Summa theologiae* II.II.96.2 and *De operationibus occultis naturae*. For the mature later Thomas, there were no legitimate uses for talismans whatsoever, thus inspiring Weill-Parot's characterization of his position as "un rejet global." This strongly negative position seems to have been his normative view, and was certainly his unambiguous position at the end of his career. In the earlier *Summa contra gentiles* III.105, however, Thomas seems to have left a gaping loophole for legitimate all-natural talismans of the type defended by the authentic Albert in his *De mineralibus*, and in the *Speculum astronomiae*. Thomas seems to have closed this loophole completely in his later works.

Furthermore, both the authentic Albert and the *Speculum astronomiae* reject *carmina* and characters from their legitimate talismans, whereas Roger allows them and even explicitly integrates the soul's intention with the stellar rays to make the *opera sapientiae* even more powerful. In this, Roger follows al-Kindi more fully than Albert does, who really only embraces celestial influences, along with their optimal timing using elections, as well as the language of rays. In addition, Roger does not bother with the niceties associated with legitimate all-natural talismans. His primary concern, by contrast, is whether they are being used for good or evil purposes, that is, by the saints preferably, and certainly not by Antichrist and his malevolent and disruptive minions.

With respect to what we call magic—both the *opera* and *verba sapientiae*—Bacon thus seems much closer to al-Kindi than to both Albert and the *Speculum astronomiae*, and he is even further away from Thomas Aquinas, with his emphasis on both celestial influences (in terms of rays) *and* of the soul's intention, as well as his use of words, *carmina* and characters. Bacon's combination of "the words and works of the wise" seems to have a solidly full-spectrum Kindian pedigree.[118] This configuration of elements contrasts starkly with Albert, as we have seen, who prefers an all-natural approach to talismans articulated in terms of his geometrical-optical model of celestial influences, in which he embraces only natural celestial influences as mediated by images and figures, and heightened by astrological timing. In this, Albert's is similar to the position in the *Speculum astronomiae*, but with the natural philosophical foundations more fully articulated.

Nevertheless, these three—Roger, Albert and the *Speculum astronomiae*—all differ significantly from Thomas Aquinas's explicit no-talisman policy, rejecting them altogether as illicit, since for him they always and only have either implicit or explicit truck with demons. This seems to be the case at least in his last words on the

[118] But only up to a point. As Rosier has shown (and Weill-Parot re-emphasized), Roger does not embrace al-Kindi's ontological determinism, opting instead to defend and more solidly safeguard human free will. Weill-Parot usefully summarizes Rosier's argument at *Les "images astrologiques,"* 334–35. Rosier discusses this more fully at *La parole comme acte*, 223–27.

subject, the striking loophole in the earlier *Summa contra gentiles* notwithstanding. Thus, as Weill-Parot has argued, a spectrum of positions may be discerned.[119] Furthermore, as we saw, Thomas's complex and in part contradictory position had a rich and varied Nachleben. *Fata sua habent libelli*! I should also emphasize once again the fundamental fact that the representatives of all three thirteenth-century positions on the spectrum *all* embrace astrology as a legitimate mode of knowledge if practiced within its proper limitations.

Comparing astrology's configuration with magic in Roger Bacon to Albertus Magnus, Thomas Aquinas and the *Speculum astronomiae*, we can see that their intellectual profiles have significant points of overlap as well as profound divergences. The main similarity is that they all embrace astrology as an art useful for deriving *knowledge* about the past, the present, and especially the future, and they all embrace and espouse a similar version of astrologizing Aristotelian natural knowledge, although Albert and Roger articulate it much more fully than Thomas or the Magister Speculi do. Comparing the similarities concerning astrology in both respects with the differences concerning magic can give us a more precise sense of each individual's unique and characteristic intellectual physiognomy, at least in relation to these types of knowledge and practice.

We now have a more solid basis for characterizing each thinker's unique intellectual configuration, their personal knowledge-power profile, at least of what we would normally call their esoteric or occult interests, although I personally resist this terminology. Roger would have called them "secrets of art and nature," certainly not "magic." In addition, we have noticed tensions regarding talismanic practices both *within* Thomas's own writings over time and *between* him and his esteemed teacher, Albert (as well as with the *Speculum astronomiae*), and also in the terminology used with respect to magic and divination. This is so both within the thirteenth-century context, and in relation to our own conceptual typologies (e.g., natural vs. judicial astrology, occult and esoteric sciences and traditions, etc.).

*

When placed within their institutional and socio-political contexts, then, all three of these conceptual patterns (natural philosophical, theologico-religious and magical)—replete with their many harmonies and tensions—provide the fundamental structures for comparing these thirteenth-century configurations with those of the fifteenth—as I do in volume II—in order, ultimately, to compare and contrast the views of Marsilio Ficino and Giovanni Pico della Mirandola, and thus characterize their positions to see if there is indeed a "tournant Ficinien," as Nicolas Weill-Parot, D.P. Walker and Brian Copenhaver have all argued in their different ways, and whether we may profitably call this a properly Renaissance configuration, as I will argue we should. A well articulated and richly contoured understanding of astrology with its natural philosophical foundations in relation to both theology/religion and magic provides the proper conceptual and associated institutional structures for

[119] He lays out the spectrum at *Les "images astrologiques,"* 335.

understanding this entire knowledge-power dynamic, thereby enriching our interpretations of what we normally call science (or natural knowledge), theology/religion and magic in Medieval, Renaissance and Early Modern Europe.

As we have seen, Roger Bacon's distinction between *astronomia iudiciaria et operativa* is extremely valuable for characterizing both the knowledge and the operative (or magical) sides of astrology. We saw that *astronomia iudiciaria* in Roger, Albert, the *Speculum astronomiae* and Thomas all normatively refer to astrological interpretations differentiated into one or more of the four canonical types of astrological practice. Thus "judicial astrology," to use the modern rough equivalent, simply means interpretative knowledge-oriented astrology of all types. "Natural" astrology—or causality—on the other hand, is to be contrasted with "demonic" in analyzing the causal foundations for both predicting the future (in contrast to "divination" in Thomas and Roger) and for making and using legitimate talismans (as in the authentic Albert, the *Speculum astronomiae* and Roger, but not in Thomas, at least at the end). Natural causality thus provides the causal foundations for the legitimate practice of judicial or interpretive as well as of operative astrology, thereby offering a more crisply historicized conceptual structure that is very different from—and much more complex and refined than—that reflected in the normal modern natural vs. judicial astrology distinction.

I have thus established that there is no precise medieval equivalent to the natural vs. judicial astrology distinction of the sort often found in the scholarly literature. Rather, the main premodern distinction is between *astronomia iudiciaria et operativa.* We begin to see something like the modern distinction between judicial and natural astrology in the sixteenth century, as we will see in volume III, with Rule IX of the *Index of Prohibited Books* (1564), which relies in large measure on the more tolerant pole of Thomas Aquinas's construction in *Summa theologiae* II.II.92–95. And although we find Thomas centrally employing the occult (or "hidden") powers of nature as an important and characteristic part of his natural philosophical analysis to distinguish between legitimate (natural) and illegitimate (demonic) types of powers derived from the heavens, this is quite different from our normal understanding of the occult sciences or traditions. If our modern usage were sound, then natural philosophy should itself be characterized as an occult science, and Thomas Aquinas as an occultist, thereby rendering all such descriptions completely absurd.

*

Although this configuration is richly complex, the basic outlines seem clear: in the thirteenth century, astrology and its natural philosophical foundations were fundamentally understood to be a legitimate part of knowledge about nature and humanity, and one that was well integrated with both theological and medical discourses. Astrology integrated natural philosophy and theology within Aristotelian, Ptolemaic and Galenic patterns of mathematically-articulated natural knowledge as well as within influential theological constructs of the Roman Catholic Church. This was true for both the Dominicans and the Franciscans, the two great mendicant orders, as articulated in influential works of both scientific and theologico-religious natures by members of both orders. In my study, Roger Bacon primarily, but also Robert Grosseteste represent the Franciscans, and Albertus Magnus and Thomas Aquinas the Dominicans. Although there are large areas of overlap and agreement, there are also significant points of tension and contrast both terminologically and conceptually—and both within and between these two Orders—as we have seen.

Nevertheless, with the evidence explored here, I would be hard pressed to argue for a distinctively Dominican or Franciscan style of astrologizing Aristotelian natural knowledge, based on my analysis of a range of Albertus Magnus's, Thomas Aquinas's and Roger Bacon's works. The major discernable differences I have found are Roger's predilection for arguments with a strongly practical-astrological dimension as well as his distinctively apocalyptic content and tone. He also embraced the full spectrum of al-Kindi's magic with his *opera et verba sapientiae*. These are all strikingly different in both style and content from Albert's or Thomas's works examined here, although they too refer to the practical-astrological dimension from time to time—and sometimes even centrally—as with Albert's use of the structure of the horoscope to articulate the central concept of the "period of life" in Aristotle's *De generatione et corruptione*. With respect to the intricately wrought structures of astrologizing Aristotelian natural philosophy and its geometrical-optical model of celestial influences—using al-Kindi's rays—that they both employed at a very high level, however, Roger and Albert could hardly have been closer, as I argued in Chap. 2.

*

Now that I have broadly articulated some of astrology's central conceptual structures in the thirteenth century—and its configuration within natural knowledge (primarily mathematics, natural philosophy and medicine), and in relation to theology/religion and magic—we should now situate astrology within its primary institutional location, namely its teaching and practice in medieval and Renaissance European universities, ca. 1250–1500. I will focus here on articulating the characteristic premodern disciplinary configurations and their curricular instantiations. These central themes will then be traced out over the next 500 years, especially arguments concerning the legitimacy of astrology in relation to natural philosophy and mathematics, on the one hand, and theology/religion and magic, on the other. Thus, the conceptual structures reconstructed in Parts I–III, together with their disciplinary configurations and institutional foundations to be reconstructed in Part IV, will then set the stage for the explorations in volume II on Ficino and Pico—astrology in the Renaissance—and for discussing astrology's various circumstances in the 16th through 18th centuries in volume III, both continuities and transformations.

Part IV
Institutional, Socio-political and Cultural Structures: Universities, Cities and Courts (1300–1500)

Introduction: Astrology, Mathematics and Humanism

A historically significant occasion may be used to mark the beginning of the final phase of premodern science, at least with respect to mathematics and astronomy: Regiomontanus's famous series of lectures at the University of Padua in 1464 on the Arabic astronomer al-Farghani's *De scientia stellarum*. We no longer possess the text of the lectures themselves, only that of his inaugural oration for the course in praise of mathematics, which was written and delivered in Latin.[1]

Regiomontanus's oration reveals a high degree of humanistic learning.[2] In this respect (among others), he plays a pivotal early role in Paul Lawrence Rose's seminal and richly informative study, *The Italian Renaissance of Mathematics* (90–117), where Rose convincingly demolishes the misconception that mathematical studies were fundamentally inimical to humanist concerns.[3] In fact, Rose shows in great detail, here and in other works,[4] how crucially important humanistic text-critical philological skills were, both in the recovery of Greek mathematical learning, and in the subsequent development of these disciplines. This is particularly so in the numerous cases where humanistic textual skills were combined with advanced

[1] Noel M. Swerdlow, "Science and Humanism in the Renaissance: Regiomontanus's Oration on the Dignity and Utility of the Mathematical Sciences," in *World Changes: Thomas Kuhn and the Nature of Science*, P. Horwich (ed.), Cambridge, MA: MIT Press, 1993, 131–168, 132. I use a facsimile of the text first published by Johannes Schöner in 1537: Alfraganus, *Rudimenta astronomica*, Nuremburg: Petreius, ff a (beta) 1r-5r, reprinted in *Regiomontani Opera Collectanea*, F. Schmeidler (ed.), Osnabruck: O. Zeller, 1972.

[2] Swerdlow "Science and Humanism," *passim*.

[3] *The Italian Renaissance of Mathematics: Studies on Humanists and Mathematicians from Petrarch to Galileo*, Geneva: Droz, 1975. He states the problem most succinctly in a publication preliminary to the full-length study; "Humanist Culture and Renaissance Mathematics: The Italian Libraries of the *Quattrocento*," *Studies in the Renaissance* 20 (1973): 46–105, esp. 46–48.

[4] See especially Rose and Stillman Drake, "The Pseudo-Aristotelian 'Questions of Mechanics' in Renaissance Culture," *Studies in the Renaissance* 18 (1971): 65–104.

mathematical abilities—as with Federico Commandino most spectacularly[5]—and, in the special case of astronomy, with observational skills as well. Indeed, this triumvirate of scholarly and scientific abilities were developed to an extraordinarily high degree in Regiomontanus,[6] as they were also, later, in Kepler.[7]

Unfortunately, however, Rose replaced his exploded false dichotomy between mathematics and humanism with another. He strongly asserts at several points that it was not towards mathematics that the humanists were antagonistic, but, rather, towards "judicial astrology."[8] Ample evidence will be provided in this chapter and in what follows in volumes II and III to show that categorically opposing humanism to astrology is also profoundly mistaken,[9] as witness *inter alios* Regiomontanus himself, Marsilio Ficino,[10] Philipp Melanchthon,[11] Joachim Camerarius,[12] Tycho Brahe[13] and Kepler.[14]

*

[5] Rose, *Italian Renaissance of Mathematics*, 185–221.

[6] For Regiomontanus's skill in observation, see e.g. Grössing, *Humanistische Naturwissenschaft.*

[7] See Anthony Grafton's penetrating studies on this issue, which is a leitmotif in his collection of essays, *Defenders of the Text: The Traditions of Scholarship in an Age of Science, 1450–1800*, Cambridge, MA: Harvard University Press, 1991; see especially his introductory comments (1–22), and his essay on Kepler, "Humanism and Science in Rudolphine Prague: Kepler in Context," 178–203. Although Kepler did not himself have very good vision, he was able to rely on Tycho Brahe's extraordinarily accurate observational records over many years. I will discuss Kepler further in volume III.

[8] Rose states this position with great vigour; "Humanistic Culture and Renaissance Mathematics," 50: "What humanists from Petrarch to Pico detest almost unanimously was judicial astrology, not mathematical astronomy." On the following page, he refers to "a humanist distaste for astrology." In the late nineteenth century, Karl Hartfelder put forth the same view; *Philip Melanchthon als Praeceptor Germaniae*, Berlin: A. Hofmann, 1889, 196. Because of this misconception, Hartfelder strikingly characterizes Luther, in this respect, as more of a humanist than Melanchthon.

[9] A further false dichotomy between humanism and religion has also been laid to rest; see Charles Stinger, *Humanism and the Church Fathers: Ambrogio Traversari (1386–1439) and Christian Antiquity in the Italian Renaissance*, Albany: State University of New York Press, 1977, and John D'Amico, *Renaissance Humanism in Papal Rome: Humanists and Churchmen on the Eve of the Reformation*, Baltimore: Johns Hopkins University Press, 1983. The false dichotomy between astrology and scholastic Aristotelian natural philosophy has already been treated in Part I.

[10] See volume II, part 2.

[11] See Sachiko Kusukawa, *The Transformation of Natural Philosophy: The Case of Philip Melanchthon*, Cambridge: Cambridge University Press, 1995.

[12] In the opinion of contemporaries, he was second only to Erasmus as a humanist; see the introduction to *Joachim Camerarius (1500–1574)*, F. Baron (ed.), Munich: Fink, 1978, 7–9.

[13] For Tycho's humanism, see Victor E. Thoren, *The Lord of Uraniborg: A Biography of Tycho Brahe*, Cambridge: Cambridge University Press, 1990; for his astrology, see index s.v. "astrology," and my volume III.

[14] For Kepler as a humanist, see Grafton, "Humanism and Science in Rudolphine Prague"; for his astrology, see, e.g., Max Caspar, *Kepler*, Doris Hellman (ed and tr), New York: Dover, 1993, index s.v. "astrology"; Gérard Simon, *Kepler Astronome Astrologue*, Paris: Gallimard, 1979, and my volume III.

In Part IV, Chaps. 9 and 10, I will reconstruct astrology's curricular and disciplinary patterns—including the figure of the astrologer as university professor—and situate these patterns in their primary institutional location, the universities. Chap. 11 will more broadly illustrate the range of roles astrology could play in society, politics and culture, primarily in Renaissance Italy. In this way, I will bring my story up to the end of the fifteenth century. Shifting focus from Northern Europe to Northern Italy, I will thereby build a bridge between volumes I and II.[15]

[15]This structure (however imperfect) reflects the earlier origins of this material, now split between volumes I and II, as originally together in one volume. With the appearance of volume II, it will become clear that Part IV is not a bridge to nowhere.

Chapter 9
Disciplinary Configurations: Astrology and the Mathematical Disciplines

9.1 Introduction

Now that the central place of astrology within the medieval map of knowledge as the integrating link between natural philosophy and metaphysics, and its richly mathematical nature, have been reconstructed—as well as some features of its relationships with theology/religion and magic—it is time to situate these conceptual structures within their primary institutional locations, namely, their instruction and practice at the universities, with an emphasis on Padua, Bologna and Ferrara, where Giovanni Pico della Mirandola, Nicolaus Copernicus and many others studied and taught. These conceptual patterns and their institutional settings will also be brought up to the end of the fifteenth century, mainly through the work of Regiomontanus (1436–1476), a seminal early figure in the "Renaissance of mathematics." This chapter thus provides further essential components of the interpretive framework being constructed in volume I. It will be in relation to these disciplinary and curricular patterns and their institutional realization in the universities *ca.* 1250–1500 that we may directly approach the fundamental historical question of astrology's removal, during the seventeenth and eighteenth centuries, from its now securely established central position in the premodern map of knowledge.

9.2 Regiomontanus and the Mathematical Disciplines

In this section, I bring the conceptual patterns established in Parts I–III up to 1500—and begin to provide their institutional context—by examining Regiomontanus's inaugural oration for his course of lectures on mathematical astronomy at the University of Padua delivered in 1464. I focus here on the standard premodern configuration of the mathematical disciplines, the *quadrivium*, establishing its

H D. Rutkin, *Sapientia Astrologica: Astrology, Magic and Natural Knowledge, ca. 1250–1800*, Archimedes 55, https://doi.org/10.1007/978-3-030-10779-6_9

continuation into the second half of the fifteenth century in the work of Johannes Müller von Königsberg, better known as Regiomontanus, who lived from 1436 to 1476. The several interconnected facets of Regiomontanus's work—scholarly (humanistic), mathematical and observational—established a solid foundation for future developments in mathematics and astronomy.[1]

One of Regiomontanus's most important contributions, in which he combined all three requisite skills, was to make Ptolemy's foundational text in mathematical astronomy, the *Almagest*, accessible in a more compact yet accurate format in his *Epitome of the Almagest*.[2] While situating Regiomontanus within the intellectual context of two of the most important thirteenth-century natural philosophers with a mathematical bent, Albertus Magnus and Roger Bacon (as reconstructed in Part I)—to which Pietro d'Abano will soon be added—I will also indicate their Ptolemaic roots. Regiomontanus may best be understood as a pivotal figure in the Euclido-Ptolemaic quadrivial tradition who wished to reform its mathematical and astronomical foundations within an explicitly astrological context.

Furthermore, it must be emphasized that a technological factor of the greatest importance had recently arisen: Regiomontanus had the great good fortune to pursue his scholarly activities in the earliest period of printing. In fact, he has no small role in the history of Renaissance and early modern culture as the earliest publisher of astronomico-mathematical material.[3] I will conclude volume I with a discussion of Regiomontanus's relevant publications—and those of an important fellow printer of mathematical and astrological texts, Erhard Ratdolt.

---

[1] For Regiomontanus's epoch-making contributions to knowledge, see Ernst Zinner, *Leben und Wirken des Joh. Müller von Königsberg Gennant Regiomontanus*, 2nd ed., Osnabrück: O. Zeller, 1968. The English translation, *Regiomontanus: His Life and Works*, E. Brown (tr), Amsterdam: North Holland, 1990, should be used with caution; see Jerzy Dobrzycki's review, *Journal for the History of Astronomy* 23 (1992): 305–7. See also Rose, *The Italian Renaissance of Mathematics*; Helmuth Grössing, *Humanistische Naturwissenschaft*; Swerdlow, "Science and Humanism in the Renaissance," and James S. Byrne, "A Humanist History of Mathematics? Regiomontanus's Padua Oration in Context," *Journal of the History of Ideas* 67 (2006): 41–61.

[2] Swerdlow, "Science and Humanism," 131. The *Editio princeps* is *Epitoma in Almagestum Ptolemaei*, Venice: Johannes Hamman, Aug 31, 1496. It is conveniently reprinted in *Regiomontani Opera Collectanea*. Technical features are discussed in Noel M. Swerdlow and Otto Neugebauer, *Mathematical Astronomy in Copernicus's* De revolutionibus, 2 vols., New York: Springer, 1984; and less fully and technically but more accessibly by André Goddu, *Copernicus and the Aristotelian Tradition: Education, Reading and Philosophy in Copernicus's Path to Heliocentrism*, Leiden: Brill, 2010.

[3] See Rose, *Italian Renaissance of Mathematics*, and Swerdlow, "Science and Humanism."

9.3 Regiomontanus in Praise of Mathematics

Examining Regiomontanus's inaugural oration can help us articulate how the mathematical disciplines were configured in the second half of the fifteenth century.[4] The oration is presented in the expected humanistic form of an epideictic speech in praise of the subject of the lectures, in this case, mathematics.[5] What makes Regiomontanus's speech in particular of such interest to historians of science is that he described in detail to his learned audience at the University of Padua—and thence to us—the current state of the mathematical disciplines at his time, including important, recently rediscovered Greek mathematical manuscripts that had not yet been translated into Latin, for example, Diophantus's *Arithmetic*.[6] These details are well known and may be consulted in Rose's or Swerdlow's accounts. For our purposes, I will first paint with large brushstrokes to set out Regiomontanus's picture of the mathematical disciplines overall, focusing on our disciplines of central concern, the two-fold science of the stars: astronomy and astrology.

After an introductory section meant to capture the attention and good will of his audience (the *captatio benevolentiae*), Regiomontanus turns to his subject proper and begins to sketch out the mathematical disciplines. He discusses first geometry, then arithmetic in some detail, especially the history of these subjects—their great names and great books.[7] We will now follow Regiomontanus into the next relevant section:

> Thus far we have remained concerned with the primary sciences [sc. geometry and arithmetic], those, namely, that consider pure quantity. Now I proceed to the secondary sciences, wonderfully agreeable and very pleasing to know, which are concerned with quantity applied in some way. [...] Of such a kind are astronomy (*astronomia*), which treats movable magnitude (*magnitudo mobilis*); music, which considers the numbers of sounds (*numeri sonorum*); and optics (*perspectiva*), which treats rays of light (*linea radialis*). Moreover, there are in addition others less familiar, as for example the science of weights,

---

[4] Swerdlow ("Science and Humanism") provides a thorough sketch of the oration, of both its contents and background, to which the reader may be referred for further detail. The translations are from Swerdlow's soon to be published version, slightly modified, for the use of which I am grateful.

[5] For background on inaugural orations, see Charles Trinkaus, "A Humanist's Image of Humanism: The Inaugural Orations of Bartolommeo della Fonte," *Studies in the Renaissance* 7 (1960): 90–147. A full study of such texts in praise of mathematics from the 12th through 18th centuries would be of immense interest.

[6] "Diofanti autem tredecim libros subtilissimos nemo usquehac ex Graecis Latinos fecit, in quibus flos ipse totius Arithmeticae latet, ars videlicet rei et census, quam hodie vocant Algebram Arabico nomine (46)." See Swerdlow, "Science and Humanism," 133. See also Swerdlow's illustrated account in *Rome Reborn: The Vatican Library and Renaissance Culture*, A. Grafton (ed.), Washington, DC: Library of Congress, 1993; "The Recovery of the Ancient Sciences of Antiquity: Mathematics, Astronomy, Geography," 125–67.

[7] Swerdlow notes that this is much more a rhetorical use of history than a critical history of the subjects. We need not follow him in the details, interesting as they are; see "Science and Humanism," 140–1, and Rose, *Italian Renaissance of Mathematics*. The concern here is, rather, with the basic structural relationships of the mathematical disciplines.

> of aqueducts, of the proportion of speeds in motions, and others.[8] Among all these sciences, *astronomia*, like a pearl, far surpasses not only her sisters, that is, the other *scientiae mediae*,[9] but even geometry and arithmetic, the mothers of all [sc. mathematical] sciences. On account of its extraordinary antiquity, we have not satisfactorily discovered the origins of astronomy, so that you might justly think it eternal or born together with the world (4).[10]

Regiomontanus then launches into a myth-historical sketch of astronomy's origins, from Abraham to Moses, through Prometheus and Hercules, to Hipparchus and Ptolemy, whither we need not follow him.[11] What is important for our purposes is the way Regiomontanus configured the mathematical disciplines—first geometry and arithmetic, then astronomy and music, the traditional subjects of the *quadrivium*, with the addition of a few other mixed mathematical sciences (including *perspectiva*)—and that he placed the science of the stars (*astronomia*) as by far the most important.

After an extensive historico-biobibliographic sketch of astronomy, much like that for geometry and arithmetic, Regiomontanus provides much briefer sketches of music and *perspectiva* (where he mentions, among others, Roger Bacon),[12] and the other *scientiae mediae*. Finally, after discussing mathematics' usefulness for understanding Aristotle's natural philosophy, and an empassioned discussion of its

---

[8] Almost exactly the same list is found in Dominicus Gundisalvi in the twelfth century; see Guy Beaujouan, "Motives and Opportunities for Science in the Medieval Universities," in *Scientific Change: Historical Studies in the Intellectual, Social and Technical Conditions for Scientific Discovery and Technical Invention, from Antiquity to the Present*, A. Crombie (ed.), New York: Basic Books, 1963, 219–36, 220.

[9] That is, the mathematical disciplines connected with matter, Aristotle's mixed mathematical sciences, which treat existing corporeal entities mathematically, thus taking part in both mathematics and natural philosophy. For a penetrating discussion, see James Lennox, "Aristotle, Galileo, and 'Mixed Sciences'," in *Reinterpreting Galileo*, William A. Wallace (ed.), Washington, D.C: Catholic University of America Press, 1986, 29–51.

[10] "Stetimus hucusque circa disciplinas primarias, quae videlicet absolutam contemplantur quantitatem. Nunc ad secundarias descendo, quae circa quantitatem contracto quodam modo versantur, suaves mirum in modum et scitu iucundissimae. Quemadmodum enim diversarum rerum simplicium compagine grata quaedam surgit temperies, ita unamquanque harum promiscue obiecti sui consideratio intellectui desideratissimam efficit, quales sunt, Astronomia quae magnitudinem mobilem pertractat: Musica quae numeros sonorum contemplatur: Perspectiva quae lineam radialem contrectat. Accedunt insuper aliae minus usitatae, ut est scientia de ponderibus, de aquaeductibus, de proportione velocitatum in motibus, et caeterae. Inter omnes autem hasce disciplinas Astronomia instar margaritae non modo sorores suas, reliquas inquam scientias medias, verumetiam omnium disciplinarum matres Geometriam et Arithmeticam longe antecellit, cuius ortum prae vetustate nimia haud satis comperimus, ita ut aeternam aut mundo concreatam non inique putaveris (46)." In the text, I cite the pagination from the typescript of Swerdlow's translation. In the footnotes, I cite the Latin pagination from Schmeidler's facsimile.

[11] For a later example, see Nicholas Popper, "'Abraham, Planter of Mathematics': Histories of Mathematics and Astrology in Early Modern Europe," *Journal of the History of Ideas* 67 (2006): 87–106.

[12] I would like to know much more about the reading of Roger Bacon in the fifteenth century, especially in Italy. In addition to Regiomontanus here, Giovanni Pico della Mirandola also mentions him several times. For the reading of Bacon in England, see Amanda Power, "A Mirror for Every Age: The Reputation of Roger Bacon," *English Historical Review* 121 (2006): 657–92.

greater certitude,[13] Regiomontanus launches into a full folio of the purplest prose in praise of mathematics, in which he directly addresses the mathematical disciplines themselves:

> O constant companions of mortals, who will not rest until the world itself comes to an end! O divine goddesses of philosophers (*divina philosophorum numina*), worthy to be awarded the highest honors! O delightful instructors of students, who fear to undertake nothing in all the world! You (pl.) measure the depth of the earth, you ascend the heights of the heavens (9).[14]

Regiomontanus continues for quite a while in this vein, praising the value of the mathematical disciplines mainly for celestial measurements, then for their value in the three traditional higher faculties—medicine, law and theology—in rapid succession. In discussing medicine, he makes a brief pro-astrological reference among a series of rhetorical questions.[15]

Then, after mentioning Cardinal Bessarion (Regiomontanus's current patron), Georgius Gemistus Plethon (Bessarion's Platonist teacher), and Cardinal Nicholas of Cusa and his concern with squaring the circle,[16] Regiomontanus makes an unexpected transition (at least for a modern reader), launching into rapturous praise of astrology, about which he has thus far said little[17]:

[13] Here he abuses, in proper humanist fashion, the vain sophistries of the schoolmen. But this must not be interpreted to mean that Regiomontanus rejected all medieval thought. On the contrary, he speaks quite respectfully of some medieval Arabic and Latin contributions to the mathematical disciplines. In addition to Roger Bacon above, Regiomontanus refers respectfully to Alhazen's and Witelo's work on *perspectiva*, and, as we shall see, he speaks very respectfully of Albertus Magnus. Regiomontanus's tolerant attitude towards both Greek natural knowledge and later Arabic and Latin contributions contrasts sharply with Nicolò Leoniceno's rabidly pro-Hellenic position, as discussed in Chap. 4 of my dissertation; see Daniela Mugnai Carrara, *La biblioteca di Nicolò Leoniceno: tra Aristotele e Galeno, cultura e libri di un medico umanista*, Florence: Olschki, 1991, and now Hasse, *Success and Suppression*, on this entire theme of Greek vs. Arabic antagonism. For more on the debate concerning certainty in mathematics, primarily in the sixteenth century, see Paolo Mancosu, *The Philosophy of Mathematics and Mathematical Practice in the Seventeenth Century*, New York: Oxford University Press, 1996 (with further bibliography).

[14] O perpetuae mortalium comites, non prius cessaturae quam mundus ipse desinat. O divina philosophorum numina summis prosequenda honoribus. O suavissimae discipulorum doctrices, quae nihil prorsus in mundo aggredi formidatis. Vos terrae profunditatem dimetimini. Vos coelorum conscenditis fastigia (51).

[15] "Nonne influentiales qualitates signorum zodiaci ac alia Astronomica o domini Medici parente vestro Hippocrate suscepimus? quasi nemo bonus medicus evadere possit nisi prius Astrologiae studuerit. Longe a maioribus vestris degenerare, sed reprimam me." Here Regiomontanus, speaking at the University of Padua after all, places himself directly into the Paduan medically-oriented astrological tradition, as we will see in more detail below.

[16] Regiomontanus was quite critical of Nicholas's work on the subject. In fact, Regiomontanus adressed his criticism to Nicholas's great friend, Paolo Toscanelli; see Eugenio Garin, "Paolo Toscanelli," in his collection of essays, *Portraits from the Quattrocento*, V. A. and E. Velen (trs), New York: Harper and Row, 1972, 118–136, 130. Toscanelli studied at Padua from 1417 to 1424, where he began his lifelong friendship with (later cardinal) Nicholas of Cusa. The translations in this volume should be treated with caution.

[17] Except for the medical reference just above, and a brief notice close to the beginning: "Accedit interea, quod potissimi facio momenti, spectatissimus Mathematicarum splendor, divinae demum

> I would declare openly before all how much [Leon] Battista Alberti, Paulo [Toscanelli] of Florence, Giovanni Bianchini, and many other illustrious men venerate your honor today,[18] except that the order of lecturing calls me especially to the excellences of *astronomia*. Just as by the common term (*generali vocabulo*) '*astronomia*' philosophers are accustomed to express both the theory of motions and the foreseeing of effects, so we shall in a general way acclaim the excellence of both parts together (10).[19]

Regiomontanus here begins to praise the science of the stars (*astronomia*), which he immediately distinguishes into two parts: the investigation of motions (*motuum speculatio*), which we would call "astronomy," and the foreseeing of effects (*effectuum providentia*), which we would call "astrology." He notes that philosophers refer to them by the same term, *astronomia*, and that he will now praise them together. This intimate linking of astronomy and astrology is central to our concerns.

Regiomontanus continues (I will be highly selective):

> Therefore I call upon you, divine goddess of astrology (*divinum Astrologiae numen*). I should be pleased if you would give assistance to your praises, if you would come forth to demonstrate your immeasurable benefits to mortals. You are without doubt the most trustworthy messenger of immortal God (*fidelissima immortalis Dei nuncia*). By interpreting their secrets (*secretis suis interpretandis*), you disclose the law for the sake of which the Almighty ordained to create the heavens, upon which he everywhere impressed glittering stars as monuments of the future. Through this angelic science we are brought near to immortal God no less than through the other arts we are set apart from wild beasts (10).[20]

> I could enumerate very many teachers of the judicial part (*pars iudicialis*), both ancient and modern, Indians, Greeks, Arabs and Latins, so that it would become evident by how many and how illustrious men of genius the philosophy of celestial influences (*philosophia influentiarum coelestium*) was discovered, enlarged and confirmed. I could show by the most certain reasons (*certissimis* [...] *rationibus*) how necessary experience of the stars (*astrorum peritia*) is to physicians, how useful to professors of canon and civil law, etc (11–12).[21]

Astrologiae decus clarissimum, quae laudi suae demonstrandae praeconem me hodiernum constituere, quod amator earum singularis semper extiterim (43)."

[18] Recall that Regiomontanus is still directly adressing the mathematical disciplines *en masse*.

[19] Quantum Baptista de Albertis, Paulus Florentinus, Ioannes de Blanchinis, aliique multi viri clarissimi decus vestrum hodie revereantur, palam omnibus recitarem, nisi ordo dicendorum ad laudes Astronomiae singulariter me vocaret, cuius quemadmodum generali vocabulo tam motuum speculationem quam effectuum providentiam exprimere solent philosophi, ita ambarum partium excellentiam summatim conclamabimus (51).

[20] "Te igitur divinum Astrologiae numen appello, tuis velim aspires praeconiis, beneficia tua immensa mortalibus demonstratura venias. Tu es procul dubio fidelissima immortalis Dei nuncia, quae secretis suis interpretandis legem praebes, cuius gratia coelos constituere decrevit omnipotens, quibus passim ignes sidereos, monimenta futurorum impressit. Per hanc disciplinam angelicam non minus immortali deo propinqui reddimur, quam per caeteras artes a beluis segregamur (51–52)." These themes are similar to those Galileo and Kepler used in the rhetorically elaborate dedicatory letters to their respective patrons in their epoch-making astronomical works, *Sidereus Nuncius* and *Astronomia Nova.* See my "Celestial Offerings."

[21] Possem enumerare plurimos tam veteres quam modernos partis iudicialis professores Indos, Graecos, Arabes ac Latinos, ut apertum fieret, quam multis et quam claris ingeniis philosophia influentiarum coelestium inventa, adaucta et confirmata extiterit. Possem certissimis ostendere

As interesting as it would be, there is no need to continue with Regiomontanus for the remaining two folios of his rapturous praise of astrology.[22]

We should notice, first of all, the unmistakably religious overtones of Regiomontanus's rhetoric: astrology is divine, the most faithful messenger of immortal God. She brings us closer to God who had decided to create the heavens in the first place precisely to allow divine law to be revealed through the science of the stars. The second point is that astrology is intimately linked with astronomy, and both are configured within the mathematical disciplines, with astrology considering the effects of celestial motions. The third is that Regiomontanus calls the interpretive astrological part the *pars iudicialis* as we also saw in the *Speculum astronomiae* and Roger Bacon, and as we will soon see in Pietro d'Abano.

Furthermore, Regiomontanus's statement that astrology is necessary for physicians explicitly places him within the integrated tradition of scientific instruction at the University of Padua that was enriched by Pietro d'Abano in the early fourteenth century, as discussed further below. In his oration, Regiomontanus names Pietro d'Abano, Giovanni Dondi dell'Orologio, Antonius de Monte Ulmi and Georg Peurbach, four of his distinguished predecessors who taught the science of the stars at Padua.[23] The purpose here has been to establish the unchanged structural continuity of (1) astronomy and astrology as the sister sciences of the stars, and (2) their configuration among the mathematical disciplines in the very figure responsible, in large measure personally, for inaugurating a new stage in the history of mathematics and astronomy. The premodern configuration of the mathematical disciplines has now been brought up to the second half of the fifteenth century.[24]

---

rationibus, quam necessaria sit medicis Astrorum peritia, quam utilis Canonum atque Legum professoribus, etc. (53).

[22]Anyone who could claim after reading this oration that Regiomontanus was luke-warm to negative about astrology (as Rose, "Humanist Culture and Renaissance Mathematics," cited above) would surely have the onus on them to demonstrate their point. A simple statement that Regiomontanus rejects astrology, as Rose claims, is hardly sufficient proof; see also Thorndike, *HMES*, V, 377 (n. 200).

[23]Favaro discusses them in "I Lettori di matematiche nella università di Padova dal principio del secolo XIV alla fine del XVI," in *Memorie e Documenti per la Storia della Università di Padova*, I, Padua 1922, 1–70, 7–40. For Pietro and Antonio da Montulmi, see also Fabrizio Bonolì and Daniela Piliarvu, *I Lettori di astronomia presso lo studio di Bologna dal XII al XX secolo*, Bologna: CLUEB, 2001, 57–59 and 77–79. My sincere thanks to Professor Bonolì for a copy of this book on my splendid stay in Bologna, for which I am most grateful to Professors Giuliano Pancaldi and Marta Cavazza. For more on Peurbach's practice of astrology at King Ladislaus's court, see Shank, "Academic Consulting," 264.

[24]Additional evidence is offered later in this chapter to further establish these characteristic structures.

9.4 Ancient Ptolemaic Roots

The ideas Regiomontanus expresses here have roots reaching back much further than the thirteenth century, however, as he himself points out. I will now sketch in central ideas of the most fundamental and influential figure at the root of this tradition of scientific astrology, namely, Ptolemy himself. Arguably the greatest developer and exponent of the mathematical disciplines in Antiquity (along with Euclid and Archimedes), Ptolemy had already established the integral relationship between astronomy and astrology in no uncertain terms in the second century CE, and in its canonical form. He characterized their relationship in the first chapter of his classic astrological treatise, the *Tetrabiblos*[25]:

> Of the means of prediction through the science of the stars (*'astronomia'* [Gk]/*scientie stellarum*), O Syrus, two are the most important and valid. One, which is first both in order and in effectiveness, is that whereby we apprehend the patterns of the movements of sun, moon, and stars in relation to each other and to the earth, as they occur at different times. The second is that in which, by means of the natural character (*'dia tes physikes'* [Gk]/*propter figuras* [...] *naturales*) of these celestial patterns themselves, we investigate the changes which they bring about in that which they surround.[26]

The first "means of prediction through the science of the stars" is thus what we would call "astronomy," which is concerned with the movements of the sun, moon and planets, in relation to each other and to the earth at different times. The second "means of prediction through the science of the stars," which is completely dependent on the first for its primary data, is concerned with the physical influences of those celestial bodies on the world below, what we would call "astrology." It should be noted, as I have emphasized in this characterization, that Ptolemy considers both astronomy and astrology to be the two most important members of the overarching category: what can be predicted through the science of the stars (*'to di' astronomias prognostikon telos'* [Gk]/*pronosticabilis scientie stellarum profectio*).

Ptolemy's extensive writings were profoundly influential in all of the mathematical fields in which he wrote, whether astronomy, astrology, geography, optics or

[25] I use Frank Robbins's Loeb volume, with introduction, Greek text and English translation; Ptolemy, *Tetrabiblos*, Cambridge, MA: Harvard University Press, 1940. For a revised Greek text, see now, *Claudii Ptolemaei opera quae extant omnia,* vol. III, 1, *Apotelesmatika*, W. Hübner (ed.), Stuttgart: Teubner, 1998.

[26] I use Robbins's translation, with some minor modifications for clarity; Ptolemy, *Tetrabiblos*, 3. Here is the Latin translation by Plato of Tivoli from Ratdolt's 1484 Venice edition in David Juste's transcription, for the use of which I am grateful: "Rerum, Iesure, in quibus est pronosticabilis scientie stellarum profectio, magnas et precipuas duas esse deprehendimus. Quarum altera, que precedit et est fortior, est scientia Solis et Lune necnon v stellarum erraticarum figuras demonstrans, quas suorum motuum causa et unius ad aliam eorumque ad terram collatione contingere manifestum est. Altera vero est scientia qua explanantur mutationes et opera que accidunt et complentur propter figuras circuitus earum naturales eis in rebus quas circumdant."

harmonics—the most significant branches of mixed mathematics for our purposes. This is true equally for later antiquity,[27] Arabic science,[28] and the Latin Middle Ages, especially after the extraordinary translating activities of the twelfth-century renaissance.[29] There was quite a Ptolemy industry in the Renaissance as well, the full story of which remains to be told.[30]

In brief, Plato of Tivoli's and Aegidius de Thebaldis's twelfth- and thirteenth-century Latin translations of the *Tetrabiblos* from the Arabic were published several times in the late-fifteenth and early-sixteenth centuries, holding sway until 1535, when Johannes Petreius published Joachim Camerarius's first edition of the entire

---

[27] See (e.g.) Sylvia Fazzo, "Alessandro d'Afrodisia e Tolomeo: Aristotelismo e astrologia fra il II e il III secolo d.C.," *Rivista di Storia della Filosofia* 4 (1988): 627–649. Porphyry, Plotinus's student and editor, wrote commentaries on the *Tetrabiblos* and the harmonics; Theon of Alexandria wrote a commentary on the *Almagest;* Proclus also on the *Tetrabiblos*. A thorough study of Ptolemy's influence in later antiquity would be of great interest. See also Peter Adamson, "Plotinus and Astrology" (*Oxford Studies in Ancient Philosophy* 35 [2008]: 265–291) that I will discuss in greater detail in volume II in relation to Ficino.

[28] See (e.g.) Fuat Sezgin, *Geschichte des Arabischen Schrifttums, Band VII: Astrologie-Meteorologie und Verwandtes. Bis ca. 430 H*, Leiden: Brill, 1979; and Gutas, *Greek Thought, Arabic Culture.*

[29] See (e.g.) Haskins, *Studies in the History of Medieval Science*; d'Alverny, "Translations and Translators"; and several essays by Charles Burnett now conveniently collected in his *Arabic into Latin in the Middle Ages: The Translators and the Intellectual and Social Context.*

[30] See (e.g.) Francis J. Carmody, *Arabic Astronomical and Astrological Sciences in Latin Translation,* on the translations. The "new Carmody" being brought up to date by David Juste and Charles Burnett is eagerly anticipated. See also Baron, *Joachim Camerarius*, on Camerarius's editorial activities on the Greek text of Ptolemy's *Tetrabiblos*; Patrick Gautier Dalché, "The Reception of Ptolemy's *Geography* (End of the Fourteenth to Beginning of the Sixteenth Century)," in *The History of Cartography, Vol. 3: Cartography in the European Renaissance*, 2 vols., Chicago: University of Chicago Press, 2007, I: 285–364; Anthony Grafton on Cardano's commentary on Ptolemy's *Tetrabiblos*; *Cardano's Cosmos*, ch. 8, "Classical Astrology Restored"; Bruce Stephenson on Ptolemy's music theory (*The Music of the Heavens: Kepler's Harmonic Astronomy*, Princeton: Princeton University Press, 1994); Clarke, *Giovanni Antonio Magini (1555–1617) and Late Renaissance Astrology*, 41–3, on the *Tetrabiblos*; and Lemay, "Late Medieval Astrological School," 344 ff. on the *Centiloquium*. Unfortunately, there is as yet no full study of Ptolemy in the *Catalogus Translationum et Commentariorum: Medieval and Renaissance Latin Translations and Commentaries*, Washington, D.C: Catholic University of America, 1960-; this is a significant desideratum. I am also beginning to contribute to the better understanding of Ptolemy's Nachleben, ca. 1250–1800, throughout the several volumes of this monograph, and in three free-standing essays: "The Use and Abuse of Ptolemy's *Tetrabiblos* in Renaissance and Early Modern Europe"; "Teaching Astrology in the sixteenth Century," and my "*Optimus Malorum*: Giovanni Pico della Mirandola's Complex and Highly Interested Use of Ptolemy in the *Disputationes adversus astrologiam divinatricem* (1496), A Preliminary Survey," forthcoming. I am particularly excited about the scholarly results that will undoubtedly appear from the "Ptolemaeus Arabus et Latinus" project under the direction of David Juste, Dag Nikolaus Hasse and Benno van Dalen that has been funded by the Bavarian Academy of Sciences. See a description of their project with several invited essays (including by Alexander Jones, Charles Burnett and Noel Swerdlow) in *Akademie Aktuell* (the official journal of the Bavarian Academy of Science and Humanities) 46 (2013). See also now Hasse, *Success and Suppression*, Chap. 6 that has much valuable information on the use of Ptolemy during the Middle Ages, the Renaissance and the early modern period.

Greek text with a new Latin translation of the first two books (and a summary of books III and IV). The first complete Latin translation from the Greek came in 1548 from Antonius Gogova, who took over whole cloth Camerarius's translation of books I and II, adding his own fresh translation of books III and IV. The second complete Latin translation was accomplished in 1553 by Philipp Melanchthon, who translated the entire work anew based on Camerarius's revised edition of the Greek text.[31]

I will limit myself here to indicating this Ptolemaic pattern in the *Speculum astronomiae*, and in the works of Pietro d'Abano, a seminal figure in adapting the thirteenth-century astrologizing Aristotelianism reconstructed in Part I to a medical end, and thereby solidly establishing its characteristic features in the medically-oriented Italian universities.[32]

9.5 Thirteenth Century Patterns: *Speculum astronomiae*

I will now briefly review this Ptolemaic structure in the *Speculum astronomiae,* which Regiomontanus in his oration explicitly attributed to Albertus Magnus.[33] As we will recall, the *Speculum astronomiae* provided a conceptual and bibliographic introduction to the science of the stars as well as a defense of astrology against mainly theological objections, primarily perceived threats to human free will and to God's omnipotence. After the overall introduction establishing the *Speculum* as a

[31] The textual history of the *Tetrabiblos* will also be discussed more fully in volume III. I use the useful but incomplete and in part inaccurate information in Robbins's introduction (xii–xiii) as the basis for this paragraph. The full bibliography for these editions is found in the introduction to Hübner's edition, LII–LIII. I also ignore the rich history of the *Centiloquium* entirely, which I hope will be a significant part of the *CTC* entry. Jean-Patrice Boudet is working on the textual and cultural history of the *Centiloquium* as is Maria Mavroudi.

[32] In addition to the *Speculum astronomiae*, whoever its author, Pietro d'Abano was also strongly influenced by undoubtedly authentic works of Albertus Magnus. See Luigi Olivieri's richly detailed philological analysis, which treats this issue among many others; *Pietro d'Abano e il pensiero neolatino: filosofia, scienza e ricerca dell'Aristotele Greco tra i secoli XIII e XIV*, Padua: Antenore, 1988. See also, Fabio Seller, *Scientia astrorum: la fondazione epistemologica dell'astrologia in Pietro d'Abano*, Naples: Giannini, 2009.

[33] Regiomontanus refers to Albert twice, first in discussing astronomy, where he explicitly refers to the *Speculum astronomiae* (47): "Item Geber Hispalensis Gerardo quodam Cremonensi traductus, quem Albertus Magnus in speculo Astronomiae correctorem Ptolemaei vocare non formidat [...]." Soon after (48), Regiomontanus speaks in glowing terms about Albert—*Albertus noster Magnus*—in treating illustrious earlier German students of astronomy. Albert comes just before Peurbach. As Paravicini Bagliani shows, the *Speculum astronomiae* was increasingly attributed to Albert during the fourteenth century and beyond, as we will also see in volumes II and III; *Le* Speculum astronomiae.

defense of legitimate astrology, the anonymous author turns to his task, beginning Chap. 1 thus:

> There are two great wisdoms (*magnae sapientiae*) and each is understood by the name '*astronomia*'. The first of these deals with the science of the shape (*figura*) of the first heaven; and with the quality of its motion (*motus*) about the poles of the equator of day [sc. and night] and of the heavens (*caeli*) placed under it (209).[34]

The basic features of mathematical astronomy are then outlined in some detail (Chap. 1), and a solid bibliography provided for its study (Chap. 2).

Chapter 3 turns to the other science of the stars:

> The second great wisdom, which is likewise called '*astronomia*', is the science of the judgments of the stars (*scientia iudiciorum astrorum*), which is the link between natural philosophy and metaphysics (219, 221).[35]

In these passages, the *Speculum astronomiae* presents the very same fundamental structure for the science of the stars as did Ptolemy before and Regiomontanus afterwards: what we would call "astronomy," concerned with the structures and motions of the heavens, is intimately linked with its sister science of the stars (also called *astronomia*),[36] what we would call "astrology." The latter is concerned with understanding and predicting the influences of the celestial motions on the earth and its inhabitants. It is called the *pars iudicialis* in Regiomontanus, the *scientia iudiciorum astrorum* in the *Speculum astronomiae*, and *astronomia iudiciaria* in Roger Bacon, as we saw above.

Considering disciplinary configurations alone, we could stop discussing the *Speculum astronomiae* here. Nevertheless, we should also briefly recall the rest of Chap. 3 because of its religious orientation. Similar ideas were just discussed in Regiomontanus, where he calls astrology the divine messenger of immortal God, and refers to astrology as an angelic science. In the *Speculum astronomiae*, a text Regiomontanus mentions favorably in his oration, astrology's association with theologico-religious themes was hardly a rhetorical flourish, as we saw. If the author was actually Albert, a philosopher and theologian who taught in the more theologically-oriented universities of Northern Europe, namely, Paris and Cologne, it would not be surprising for him to emphasize the theological benefits of studying astrology. For the *Speculum*'s author, astrology is simply the best way to learn about the divine by means of His creation. This theologically oriented use of astrology is

---

[34] Duae sunt magnae sapientiae et utraque nomine astronomiae censetur. Quarum prima est in scientia figurae caeli primi et qualitate motus eius super polos aequatoris diei et caelorum sub eo positorum (I.2–5)[.].

[35] Secunda magna sapientia, quae similiter astronomia dicitur, est scientia iudiciorum astrorum, quae est ligamentum naturalis philosophiae et metaphysicae (III.2–3).

[36] Or *astrologia*; the same applies whether in Latin or Greek. Both terms are used indifferently for the most part, as I discussed in the overall introduction. Very occasionally, however, they are distinguished terminologically as in our modern usage (although sometimes the terms used are opposite to ours) as we will see below with Pietro d'Abano.

to be compared with its medical orientation in the Italian universities, to which we will now turn.[37]

9.6 Pietro d'Abano: Astrology and Medicine

This precise configuration of the science of the stars is also found in another extraordinarily influential teacher and writer, Pietro d'Abano (*ca.* 1250-ca. 1316), who contributed significantly to an integrated scientific curriculum for the study of medicine at the University of Padua in the early fourteenth century.[38] As we saw, Regiomontanus mentioned Pietro favorably 150 years later as he situated himself within the very same astrologically-inflected medical tradition at the University of Padua. Pietro expressed this configuration clearly in his extremely influential *Conciliator controversiarum, quae inter philosophos et medicos versantur* (*Reconciler of the Controversies that Take Place between Philosophers and Physicians*), which was originally written in 1303, while he was at the University Paris, and completed in Padua in 1310.[39] Pietro treated this configuration in *Differentia* X, in discussing whether the science of the stars can help a physician heal a sick person. I discuss other features of this revealing chapter just below.

Pietro discussed the structure of the science of the stars early in his response to seven objections *quod non* (15H^{2}1–16D^{1}9)[40]:

[37] We will, of course, remember in this context the analysis of astrology's relation to theology/religion in Albertus Magnus, Thomas Aquinas and Roger Bacon in Part II above.

[38] For basic orientation, see Thorndike, *HMES*, II, 874–947; Nancy G. Siraisi, *Arts and Sciences at Padua*; Eugenia Paschetto, *Pietro d'Abano, medico et filosofo*, Florence: Nuovedizioni E. Vallecchi, 1984, and Graziella Federici Vescovini, *Il 'Lucidator dubitabilium astronomiae' di Pietro d'Abano: opere scientifiche inedite*, Padua: Programma e 1 + 1 Editori, 1988. See also the recent monograph on his astronomy and astrology by Fabio Seller; *Scientia astrorum*, and *Médicine, astrologie et magie entre Moyen Âge et Renaissance: autour de Pietro d'Abano*, J.-P. Boudet, F. Collard and N. Weill-Parot (eds.), Florence: SISMEL—Galluzzo, 2013. Vescovini makes a strong claim about Pietro's importance ("I programmi degli insegnamenti del collegio di medicina, filosofia e astrologia dello statuto dell'universitaria di Bologna del 1405," in *Roma, magistra mundi: itineraria culturae medievalis. Melanges offerts au Pere L.E. Boyle*, J. Hamesse (ed), Louvain: La Neuve, 1998, I, 193–223, 204): "Pietro d'Abano può essere considerato il fondatore del sapere medico astrologico del Medioevo latino a livello teorico-pratico fino alla fine del Rinascimento, chiamato da tutti il 'Conciliatore' dei medici e dei filosofi, il *praeceptor meus* di Michele Savonarola. In altri termini può essere considerato il caposcuola di una medicina astrologica a cui si richiamerà espressamente anche Michele, che aveva studiato a Padova e con accenti diversi anche da Marsilio Ficino nel suo *De triplici vita*." On Michele Savonarola, see now, *Michele Savonarola: Medicina e cultura di corte*, Florence: SISMEL—Galluzzo, 2011.

[39] I discuss Pietro's biography and bibliography more fully in Chap. 10. I cite the *Conciliator* from the facsimile of the Venice: Giunta edition of 1565; Pietro d'Abano, *Conciliator*, E. Riondato and L. Olivieri (eds.), Padua: Antenore, 1985.

[40] The pages in this edition are divided into two columns with a set of letters running down the center (A-D or E-H which correspond to recto and verso respectively). I add a superscript number

> Moreover, *astronomia*, or *astrologia* is the law or *ratio* considering in general celestial figures and motions in themselves and in their effects. [...][41] The two-fold part of the science of the stars (*duplex astrologiae pars*) is also touched on in this description: one is what is called the science of the quantities of figures, arrangement and motions of bodies (*scientia* [...] *quantitatum figurarum, ordinationis, et motuum corporum*), or the science of the whole: and this is through its first distinctions. But through its second [sc. distinctions], as in its effects (*in suis effectibus*), and what remains, the other part touches, which is called the science of judgments (*scientia iudiciorum*), which is two-fold.[42]

Pietro then immediately describes the twofold nature of the science of judgments, what we call astrology. He follows the description from *Speculum astronomiae* Chap. 4 precisely:

> One part is introductory to judgments, the other is called practical (*exercitativa*),[43] which, again, is separated into four parts, of which one is on revolutions, the second on nativities and their revolutions, the third on interrogations, the fourth on elections, with which the science of talismans (*imaginum* [...] *scientia*) is associated. Moreover, the part on revolutions has three parts, of which one is on the 120 conjunctions; another is on the revolution of the year considered according to the entry of the sun into the beginning of Aries, or as received from the preceding conjunction or opposition; the third is on the change of times (*tempora* = seasons), as of rains, and this sort of thing, as I have shown in the first *differentia* of the *Lucidator*.[44]

(1 or 2) to indicate whether the text at issue is in the left or right column respectively; I have also numbered the lines in each section. So 16E^{1}1 refers to the first line of section E in the left column of page 16.

[41] Then he explicitly argues *against* a crisp terminological distinction between these two parts of the science of the stars: "Quidam vero assignantes differentiam appellare volunt astronomiam scientiam de motibus: astrologiam autem iudiciorum scientiam: quod non oportet: fere enim idem videntur importare: nam astronomia ab astro: et nomos quod est lex dicitur: astrologia vero ab astro, et logos, quod est ratio, vel sermo (16B^{2}2–7)." Because both terms have the same etymology, Pietro argues, they should also have the same conceptual referents.

[42] Est autem astronomia, seu astrologia lex vel ratio figuras celestes, et motus in se, et in suis effectibus universaliter considerans (16A^{2}18–16B^{2}2). [...] Tangitur quoque in hac descriptione duplex astrologiae pars: una quidem quae scientia dicitur quantitatum figurarum, ordinationis, et motuum corporum, vel scientia totius: et hoc per eius primas differentias: per secundas vero, ut in suis effectibus, et reliqua tangitur altera pars, quae dicitur scientia iudiciorum: quae duplex existit (16B^{2}7–12)[.].

[43] Pietro's discussion of *exercitiva* falls entirely into Roger Bacon's *astronomia iudiciaria*, except where elections treat of *imagines*, which Pietro mentions explicitly here. He treat talismans elsewhere, and magic explicitly in the *Lucidator*. For more discussion, see Weill-Parot, *Les "images astrologiques,"* 500–31.

[44] "Una quidem introductiva ad iudicia: altera exercitativa appellatur: quae et iterum in quatuor separatur partes, quarum una est de revolutionibus: secunda de nativitatibus, et earum revolutionibus: tertia de interrogationibus: quarta de electionibus, cui et imaginum supponitur scientia. Quae autem de revolutionibus tres habet partes, quarum una est de. 120. coniunctionibus: alia de revolutione anni secundum introitum solis in caput arietis considerata, vel ex coniunctione, seu oppositione solis praecedente accepta: tertia vero de mutatione temporum, ut pluviarum, et huius, seu lucidatoris differentia monstra vi prima (16B^{2}12–16C^{2}4, my transcription)." These types of practical astrology were discussed more fully in the excursus to the overall introduction, and will be illustrated in practice in Chap. 11 below.

Pietro d'Abano's configuration of the twofold science of the stars—with one part, what we would call "astronomy," treating celestial configurations and motions, and the other, what we call "astrology," treating their effects on the earth—coheres perfectly with the Ptolemaic tradition of the *Tetrabiblos* as precisely reflected in the thirteenth century in the *Speculum astronomiae*, and more generally in the works of Roger Bacon and Albertus Magnus explored in Parts I–III.

*

We can now see more clearly how Regiomontanus's oration fits squarely into the premodern configuration of the mathematical disciplines (with theological overtones and medical ramifications)—and that it has deep roots extending back to the most influential investigators into nature of late antique Greek, and medieval Arabic and Latin learning—even as it points forward to the rediscovery of important Greek mathematical texts, and the subsequent profound transformations in our view of the universe and our place in it.

In fact, Regiomontanus's inaugural oration can be read as a manifesto for scientific astrology, articulating the belief that astrology provides unique insight into both God Himself and into nature as God's creation. The *Speculum astronomiae* seems to have provided the first full articulation of this ideology for the Latin West in the later thirteenth century, which Pietro d'Abano adapted at the beginning of the fourteenth and integrated into an influential system of scientific education at the medically-oriented University of Padua. We are fortunate to possess a detailed curricular expression of this tradition in the Bologna statutes of 1405 to be discussed in Chap. 10. Regiomontanus articulated this same ideology brilliantly in his 1464 Paduan oration as he modernized it with the rich palette of humanistic learning—sharpening its focus and turning attention to the reform of its mathematical and astronomical foundations—but by no means neglecting its deeply astrological ends.

*

Before turning to the Bologna statutes, however, we should continue exploring the *Conciliator* to get a fuller sense of Pietro d'Abano's views on astrology's natural philosophical foundations. Although there are many other passages of interest in Pietro d'Abano's works, I will focus here on another section of *Differentia* X, where he directly addresses the primary question of the entire *differentia*: "Whether an actual physician (*quis medicus existens*) can contribute to the health of a sick person (*possit conferre in salutem aegroti*) through knowledge of the science of the stars (*per scientiam astronomiae*) or not." Pietro addresses this question directly in the third section ($16E^1 1$–$17F^2 17$), where he treats in order five central issues: (1) that the celestial bodies act on these things below (*quod corpora coelestia in haec agunt inferiora*); (2) how many celestial bodies there are and in what ways they act (*quot, et quibus modis*); (3) that an actual physician-astrologer can have this knowledge (*quod medicus existens astrologus huius potest cognitionem habere*); (4) that the science of the stars can contribute to the health of a sick person (*quod confert astrologia ad salutem aegroti*); and (5) in what ways (*quibus utique modis*, 16E^{1}1–5).

The structure of these questions breaks down into two main sections: (1–2) which treat the cosmological and natural philosophical framework, demonstrating, first, *that* the celestial bodies act on the world below, which he then differentiates (2) by describing how many types of celestial influences there are and in what ways they act. The second section (3–5) takes this physico-cosmological (ontological) structure as the basis for assessing essential features of the physician's relationship to this framework. First, he treats the fundamental epistemological issue (3): can the physician have knowledge of this cosmological structure. Answering in the affirmative, Pietro argues that the physician can help a sick person through this knowledge (4), and in what ways (5). Pietro is very careful in his argument structure as well as his language, as we will see.

I will only treat the first part here in order to bring out the structure and basic features of Pietro's argument. This section is worthy of more detailed investigation, both in itself and with respect to its influence. The first two statements provide the natural philosophical foundations for the more specifically medical arguments to follow. Pietro first establishes that the celestial bodies act on things below:

> [A] The first of this [sc. set of five issues] is shown because this lower world (*iste mundus et inferior*) is necessarily continuous with the higher movements (*continuus est superioribus lationibus*): so that all of its power is ruled from there (*omnis virtus eius gubernetur inde*).[45]
>
> [B] The change of earthly things with respect to this and that [sc. individual] is by the change of the celestial bodies (*mundanorum mutatio ad hoc et ad illud, corporum coelestium mutatione est*, 16E^{1}9–10)[.]
>
> [C] And therefore one who intends to investigate the causes of things (*causae rerum*) should first contemplate the heavenly bodies (*coelestia*).[46]

The beginning of Pietro's first argument is significant in several respects: first, in addition to embracing the structure of practical astrology in the *Speculum astronomiae* (as mentioned just above), here Pietro takes over the precise phrasing of another fundamental statement there. In *Speculum astronomiae* 3, to conclude his extensive introductory sentence, the author states: '*quid desideratius concionatori quam habere mediam scientiam, quae doceat nos qualiter mundanorum ad hoc et ad illud mutatio caelestium fiat corporum mutatione* (my underline)?' This *media scientia* is, of course, the science of the stars.

Secondly, Pietro's statement in (C) that the *coelestia* are the *causae rerum* is strikingly similar to passages in Roger Bacon's *Opus maius* IV (and elsewhere),

[45] Pietro supports this claim with a reference to the first book of Aristotle's *Meteorologica*, and also to the astrologers; in particular, the beginning of the *Centiloquium*: "Primum huius ostenditur, quia ex necessitate iste mundus et inferior continuus est superioribus lationibus: ut omnis virtus eius gubernetur inde. Tactum prius Metheo. 1. quod etiam dixere astrologi. unde principio centiloquii (16E^{1}5–9)."

[46] Et ideo causas rerum rimaturus primo coelestia contemplare (16E^{1}10–11).

particularly where Roger articulates his views on the nature of mathematics, by which he means primarily astronomy and astrology, and how one must know astrology in order to fully understand the causes of things, which are, after all, the *coelestia*, as we saw in Chap. 2. Finally, as we will see in volume II, the first of the astrologers' strongest arguments that Pico attacks in *Disputationes* Book III is precisely Pietro's first argument here, that all changes in the lower world are dependent on those of the heavens.

*

Although there is more of interest in Pietro's first response—including the magnetic analogy he used for understanding the type of dependence inferior things have on superiors[47]—we should now turn to his second response where he describes how many kinds of celestial influences there are and in what ways they work. Once again, I will be very selective. He begins thus:

> With respect to the second [sc. of the five issues], one should know that the *superiora* seem to impress themselves on things below (*in haec inferiora imprimere*) in three ways, of which one is universal, another particular, and the third median.[48]

Pietro then treats the three in turn, although not following the order just articulated.

He begins thus: "The universal [sc. mode] is perfected by motion (*motus*) and light (*lux*), for superior bodies act universally on inferiors in these two ways (*De caelo*, II)."[49] Pietro then describes these modes of action in turn, first motion (16G^{1}8–18), then light (16G^{1}18–H^{1}13), whither we shall not follow him, beyond noting that in treating light, he argues that one can see this most readily with respect to the planetary aspects, mainly conjunctions, oppositions and trines.

Pietro then treats the median mode: "The median *modus operandi* is what acts by means of their [sc. the planets] complexion [i.e. the four qualities]."[50] He then discusses the traditional attribution of a set of two of the four primary qualities to each planet, for which he gives an extensive analysis, whither, once again, we shall not follow him, beyond noting that in both the first and second modes, Pietro reveals himself as belonging fully to the tradition of astrologizing Aristotelian natural philosophy reconstructed in Part I.

Finally Pietro treats the particular mode, which I will discuss in greater detail:

> But the particular mode is a certain influence of the stars (*quaedam influxus stellarum*) imparted to each individual somewhat differently than to another by virtue of their own nativity or its revolution. And therefore we see that Saturn and Mars, although they are unfortunate *simpliciter*, and they [sc. can] take years away from the *alcocoden*, nevertheless,

---

[47] 16E^{1}12 ff.

[48] Propter secundum huius sciendum, quod tribus modis videntur superiora in haec inferiora imprimere: quorum unus est universalis, alius particularis, tertius. q. medius (16G^{1}3–6).

[49] Universalis vero motu perficitur, et luce: corpora namque superiora his duobus modis universaliter in haec inferiora agunt. unde. 2. caeli et mundi (16G^{1}6–8).

[50] Modus vero operandi medius est, quo per eorum complexionem operantur: nam quaedam earum dicuntur frigidae et siccae, ut Saturnus [...] (16H^{1}13–15).

> when they themselves *are* the *alcocoden*, that is, the givers of the forms of life (*datores formarum vitae*),[51] they preserve their minor, median or major years in accordance with their good fortune [...]. And therefore every planet exists fortunately for one person and unfortunately for another according to the diversity of their nativity and revolution (*secundum diversitatem nativitatis, et revolutionis*, my emphasis).[52]

In this third mode, then, celestial influences act in a particular manner which is strongly conditioned by the particular nature of the individual's nativity—the particular planetary configuration at their birth—and its annual revolutions—two types of practical astrology that Pietro had already identified. Although he does not say so explicitly, the configuration at birth represented in a horoscope imprints a person's root complexion, as we saw with Albertus Magnus and Roger Bacon in Part I.

As we will see, although Pico embraces Pietro d'Abano's analysis of the first, universal mode of planetary action, namely, in terms of motion and light, he takes pains in *Disputationes* III to attack the other two modes, median and particular. In treating the remaining three issues (3–5)—and in their fleshing-out in the rest of the *Conciliator*—we can see Pietro adapting central patterns of the astrologizing Aristotelian natural philosophy reconstructed in Part I towards a primarily medical end.[53]

9.7 Conclusion

The consistency of these conceptual and disciplinary patterns from the 13th through the 15th centuries and their roots in Antiquity have now been established. We will further solidify our understanding of these patterns by reconstructing their curricular instantiations and institutional foundations at the universities. Turning from conceptual and disciplinary configurations considered in themselves, we will now explore how they were embedded within their primary knowledge-centered institutional location, the universities (Chap. 10). In Chap. 11, we will focus on how this knowledge was put into practice in cities and courts.

---

[51] For more on the *alcocoden*, see Vitali, *Lexicon*, 18–19; Hasse, *Success and Suppression*, on the length of life (262–65), and Hand, "Uses of Military Astrology," Appendix II s.v. Azzolini discusses the use of the alcocoden in practice in the context of other predictive prorogators in interpreting Galeazzo Maria Sforza's nativitiy; *The Duke and the Stars*, 109.

[52] Modus vero particularis est quidam influxus stellaris unicuique individuo fere differenter ab alio propria nativitate, aut revolutione eiusdem inditus. Et ideo videmus Saturnum, et Martem: licet simpliciter sint infortunae, et habeant annos auferre ab alcocoden: tamen quando fuerint alcocoden, id est datores formarum vitae, annos suos minores medios seu maiores praestabant secundum eorum fortunationem [...]. Et ideo omnis planeta secundum diversitatem nativitatis, et revolutionis uni existit fortuna: alii vero infortuna ($16F^2 19$–$G^2 10$).

[53] Pietro d'Abano will be treated further in Chap. 10 in relation to the University of Padua.

Chapter 10
Institutional Foundations: The Universities

> *From the thirteenth century to the fifteenth, the idea of the quadrivium gradually became unbalanced in the universities, to the advantage of astronomy. At Bologna, Salamanca, Cracow and elsewhere, mathematics became an auxiliary to the study of judicial astrology. Scientific texts are very often listed in manuscripts in the following order: algorism, the sphere and the computus (all three by Sacrobosco), the quadrant attributed to Robertus Anglicus, the astrolabe of Messahalla, Theorica planetarum Gerardi. This corpus occurs so often that one is forced to recognize in it the frame-work of the teaching of the exact sciences in the fourteenth century. Such classes were given to a limited number of pupils who were supposed to have some practical knowledge of calculation and of using the astrolabe, an instrument of great pedagogical value. The drawing up of horoscopes had, at least, the merit of compelling one to do practical exercises. However poor an opinion we may have of judicial astrology, we must admit that the care given to the use and to the improvement of astronomical tables is one of the most interesting manifestations of the medieval scientific mind.*

Guy Beaujouan, "Motives and Opportunities for Science in the Medieval Universities," 223; see Lemay, "The Teaching of Astronomy in Medieval Universities, Principally at Paris in the Fourteenth Century," *Manuscripta* 1976, 197–217, for a critique thereof. In his treatment of the teaching of mathematics at Padua from Pietro d'Abano to Galileo, Antonio Favaro makes an important related point, as we will see below; "I lettori," 14–16. For more information on this teaching corpus, see Olaf Pedersen's many related studies, including, "The 'Corpus Astronomicum' and the Traditions of Medieval Latin Astronomy," *Studia Copernicana* 13 (1975): 57–96. On the 'Quadrivium' at the medieval universities, see John D. North, "The Quadrivium," in *A History of the University in Europe*, 4 vols., Volume I: *Universities in the Middle Ages*, Hilde de Ridder-Symoens (ed.), Cambridge: Cambridge University Press, 1994, 337–59.

H D. Rutkin, *Sapientia Astrologica: Astrology, Magic and Natural Knowledge, ca. 1250–1800*, Archimedes 55, https://doi.org/10.1007/978-3-030-10779-6_10

10.1 Introduction

The broad framework and diverse contents of the astrologizing Aristotelian type of natural knowledge reconstructed so far were also studied and taught within the premier educational institution, the universities, and thereby passed down as "normal science" from generation to generation during the entire premodern period from the thirteenth into the seventeenth century, and sometimes beyond.[1] Although the evidence is piecemeal, the teaching of astrology seems to have had its institutional foundation in the medieval universities in three distinct scientific disciplines, where it was taught in different respects:[2] [1] Technical features of astrological theory and practice—including how to use tables and astrolabes in constructing horoscopes—were taught in the mathematics course, along with arithmetic, geometry, music and astronomy (and sometimes *perspectiva*).[3] [2] Astrology's natural philosophical foundations were studied in the natural philosophy course in teaching core Aristotelian texts.[4] [3] After prerequisites in mathematics and natural philosophy, aspects of astrological theory and practice directly related to medicine were taught in the medical course.[5] All three were taught in the combined faculties of arts and

---

[1] For the history of the universities, a unique contribution of medieval Europe to world culture, one must begin with F. Hastings Rashdall, *The Universities of Europe in the Middle Ages*, F. M. Powicke and A.B. Emden (eds.), new edition, 3 vols, Oxford: Oxford University Press, 1936. See also Olaf Pedersen, *The First Universities:* Studium generale *and the Origins of University Education in Europe*, R. North (tr), Cambridge: Cambridge University Press, 1997. For the Italian universities in particular, see Grendler, *The Universities of the Italian Renaissance*. For many relevant studies, see the journal, *History of Universities*.

[2] The institutional foundations for the teaching, study and practice of astrology in the premodern universities needs full-length synthetic treatment, for which this is a preliminary study. I intend to explore this major topic in further research. Additional evidence for the sixteenth through eighteenth centuries is presented in volume III. My profound thanks to Domenico Bertoloni Meli for gently nudging me into this realization of astrology's three-fold location in university curricula.

[3] For an informative general sketch for both arts and medicine, see Lemay, "The Teaching of Astronomy," which is mainly for the fourteenth and fifteenth centuries, and has a very useful collection of evidence. The rich material he indicates deserves to be treated in much greater depth. For a detailed study of the teaching of mathematics-astronomy-astrology at the University of Ingolstadt in the fifteenth and sixteenth centuries, see Christoph Schöner, *Mathematik und Astronomie an der Universität Ingolstadt im 15. und 16. Jahrhundert*, Berlin: Speyer and Peters, 1994. My thanks to Thony Christie for this reference.

[4] Albertus Magnus's commentaries on Aristotle clearly illustrate this point, as I established in Part I. Albert's works were quite influential in the Renaissance as well. Mahoney has shown this for Padua in the fifteenth and sixteenth centuries; "Albert the Great and the *Studio Patavino*." Johannes Stöffler, Philip Melanchthon's esteemed teacher at the University of Tübingen, was also deeply influenced by Albert's natural philosophy, as I will discuss in volume III. Fuller study of Albert's influence into the seventeenth century and beyond would be very interesting, as also would a systematic study of astrological discussions in Aristotle commentaries by other scholars in the thirteenth through seventeenth centuries. I will provide further evidence in volume III.

[5] For a useful brief orientation, see Siraisi, *Medieval and Early Renaissance Medicine*, 67 ff. See also Cornelius O'Boyle, *Medieval Prognosis and Astrology: A Working Edition of* Aggregationes de crisi et criticis diebus, Cambridge: Wellcome, 1991; Monica Azzolini, "Reading Health in the

medicine found in Italy, most influentially in Bologna[6] and Padua,[7] but also (among many others) at Ferrara,[8] Pavia,[9] and Pisa,[10] and outside Italy at Paris, Cologne, Cracow and elsewhere.[11]

Stars: Politics and Medical Astrology in Renaissance Milan," in *Horoscopes and Public Spheres*, 183–205, and *Astro-Medicine: Astrology and Medicine, East and West*, A. Akasoy, C. Burnett and R. Yoeli-Tlalim (eds.), Florence: SISMEL—Galluzzo, 2008. This topic is discussed in more detail below. Much more work needs to be done on astrology and medicine—including the interesting topic of physiognomy's relationship to both, for which see the valuable essays collected in Agrimi, Ingeniosa scientia nature.

[6] See Beaujouan, "Motives and Opportunities," 232 ff., and Lemay, "The Teaching of Astronomy," 199, and "The Late Medieval Astrological School," 339 ff. For a useful discussion of the teaching of astrology at Bologna, see Siraisi, *Taddeo Alderotti and his Pupils*, esp. 139–45. See now also Bonolì and Piliarvu, *I lettori*, for much useful information.

[7] See esp. Siraisi, *Arts and Sciences at Padua*, and Favaro, "I Lettori."

[8] See Graziella Federici Vescovini, "L'astrologia all'università di Ferrara nel Quattrocento," in *La rinascita del sapere: libri e maestri dello studio Ferrarese*, Patrizia Castelli (ed.), Venice: Marsilio, 1991, 293–306.

[9] Azzolini usefully reconstructs the teaching of astrology at Pavia in Chap. 1 of *The Duke and the Stars*, "The Science of the Stars: Learning Astrology at the University of Pavia," 22–64.

[10] I discuss astrology at Pisa in the sixteenth century extensively in volume III. For now, see my "Teaching Astrology in the sixteenth Century," and my "Galileo Astrologer."

[11] See Lemay, "The Teaching of Astronomy," especially 199 ff. on the establishment of a college of astrology and medicine in the 1360s; and Boudet, *Le Recueil des plus Celebres Astrologues*, vol. I., and now his "A 'College of Astrology and Medicine"? Charles V, Gervais Chrétien, and the Scientific Manuscripts of Maître Gervais's College," *Studies in History and Philosophy of Biological and Biomedical Sciences* 41 (2010): 99–108. For the situation at Montpellier, see Luke E. Demaitre, *Doctor Bernard de Gordon: Professor and Practitioner*, Toronto: Pontifical Institute of Mediaeval Studies, 1980; and Lemay, "The Teaching of Astronomy," 207. Boudet states that astrology was better integrated within the curriculum in Italy than in France; *Le Recueil*, I.238. See also Lemay, "The Late Medieval Astrological School," which mainly treats the University of Cracow, but also those of Prague, Leipzig and Erfurt. For more on Cracow (and as influenced by Cologne), see Goddu, *Copernicus and the Aristotelian Tradition*. See also Carey, *Courting Disaster*, for England; Grössing, *Humanistische Naturwissenschaft*, for Vienna in the fifteenth century; and Beaujouan, "Motives and Opportunities," 220–221. John D. North's account of astrology at the University of Oxford in the fourteenth century in *Chaucer's Universe* is very useful. Now also see Hasse, who provides much useful evidence; *Success and Suppression*, 17–27, and esp. for astrology, 23–26, with an emphasis on Bologna, Ferrara and Cracow. In addition, he treats medicine and philosophy as well, with an emphasis in all three sections on the Arabic authors in Latin translation that were studied and taught at university. I will treat astrology in Italy as normative in what follows, but one should also be aware of local and national differences—and differences over time—that should all be taken into account in a comprehensive study.

10.2 The 1405 Statutes: Curricular Patterns at the University of Bologna (Rubric 78)

A revealing source for understanding both how astrology was configured within the disciplines of natural knowledge and how that configuration was institutionalized in the universities is the 1405 statutes for one of the premier universities of the day, and one of the oldest, the University of Bologna.[12] The curricular patterns are clearly articulated in rubric 78, where the multi-year three-fold course of study in arts and medicine is spelled out in detail.[13] The prescribed multi-year curriculum for all three tracks—[1] mathematics-astronomy-astrology (the *quadrivium*), [2] natural philosophy, and [3] medicine—is clearly described. These patterns are utterly central to my larger argument.

Beginning with the mathematical disciplines, the 4-year course—called '*astrologia*' and leading up to the science of the stars—provides the necessary foundations as follows. This precious evidence is important (and brief) enough to treat in full. In the first year, the student began with the prerequisites in arithmetic (the *Algorismus*, on integers and fractions) and the first book of Euclid's geometry, with Campanus of Novara's thirteenth-century commentary. He also began studying two essential astronomical tools: one practical, the Alphonsine tables with its canons; the other, the *Theorica planetarum*, on planetary theory. In the first year of the mathematics-astronomy-astrology course, then, quadrivial works in elementary arithmetic, geometry and mathematical astronomy were treated.[14]

---

[12] For the early history of Bologna, see, e.g. Rashdall, *Universities*, and Pearl Kibre, *Scholarly Privileges in the Middle Ages: The Rights, Privileges and Immunities of Scholars and Universities at Bologna, Padua, Paris and Oxford*, Cambridge, MA: Medieval Academy of America, 1962. Siraisi, *Taddeo Alderotti*, provides a useful summary with bibliography. For a helpful in-depth treatment of the relevant statutes, see Vescovini, "I programmi." All the statutes are printed in Carlo Malagola, *Statuti delle Università e dei college dello studio bolognese*, Bologna: N. Zanichelli, 1888. See also Grendler, *Universities*, 3–40, for its origins and development, and 408 ff. for the statutes. Grendler treats terminological issues, especially concerning the name of the professorship of astrology, astronomy and/or mathematics, 415–29. He uses the changes in terminology to argue for a change in disciplinary identity—primarily with respect to the rise of "mathematics" and the fall of astrology—but not altogether successfully. At the very least, he would need to treat the case of Giovanni Antonio Magini, who taught mathematics, astronomy and astrology at Bologna from 1588–1617. I treat this issue in some depth in volume III in establishing significant curricular and disciplinary continuities.

[13] Siraisi, *Medieval and Early Renaissance Medicine,* 71, states that these statutes "probably formalize arrangements made in the late thirteenth century"; see also her *Taddeo Alderotti*, 96 ff. Vescovini ("I programmi," 211) conveniently reproduces the Latin of the mathematics-astronomy-astrology curriculum. Thorndike translates the relevant statutes and arranges them usefully; *University Records and Life in the Middle Ages*, translated with introduction and notes by Lynn Thorndike, New York: Columbia University Press, 1944, 279–82. See also Grendler's annotated translation; *Universities*, 410–11, and Hasse's description; *Success and Suppression*, 23–24. Both Hasse and Azzolini emphasize the amount and importance of Arabic material in the curriculum for the teaching of astrology; *The Duke and the Stars*, 33 and 39 ff.

[14] "In astrologia, in primo anno, primo legantur algorismi de minutis et integris, quibus lectis, legatur primus geometrie Euclidis cum commento Campani. Quo lecto, legantur Tabule Alfonsi cum

In the second year, the student read the *Sphere* of Sacrobosco, that phenomenally successful thirteenth-century textbook of elementary cosmography, both celestial and terrestrial. He continued studying Euclid's geometry with the second book. He also studied two astronomical works: John de Lineriis's canons on the Alphonsine tables, and "Messahalah" on the astrolabe,[15] another essential tool in the astrologer's mathematico-astronomical toolkit.[16]

In the third year, however—that is, at the beginning of the second half of the 4-year course—the students began their study of astrology proper with Alcabitius's influential introductory textbook, the *Liber introductorius*.[17] They also studied [ps.-] Ptolemy's *Centiloquium* with [ps.-]Haly's commentary,[18] a fundamental astrological text for medical practitioners.[19] To complete the third year, the student also worked on the third book of Euclid's geometry, and a treatise on another astronomical instrument, the quadrant.[20] Finally, in the fourth year, the student moved on to more advanced works in both astrology and astronomy—Ptolemy's *Tetrabiblos* (all) and *Almagest* (Book III)—as well as a more medically-oriented text, the *De urina non visa* (*On urine unseen*).[21]

---

canonibus. Quibus lectis legantur theorica planetarum." It is clear from the way the rubric is composed that each text is to be read in order. See (e.g.) Grendler and *DSB* for more information on the texts, their authors, textual tradition and influence.

[15] Although normally attributed to Messahalah, this text was actually a compilation from an Arabic treatise on the astrolabe by Ibn as-Shafar; see Hasse, *Success and Suppression*, 23 and 390.

[16] "In secundo anno, primo legitur tractatus de sphera, quo lecto, legetur secundus geometrie Euclidis, quo lecto legantur Canones super tabulis de Lineriis. Quibus lectis, legatur tractatus astrolabii Mes[sa]chale." Part I of North's *Chaucer's Universe* will give the interested reader a very good sense of what was studied in the mathematics course towards an astronomico-astrological end in the fourteenth century.

[17] For Alcabitius, see the recent edition by Burnett and Yamamoto: *Al-Qabisi (Alcabitius): The Introduction to Astrology*. For the teaching of Alcabitius and much other valuable information, see Rüdiger Arnzen, "Vergessene Pflichtlektüre: Al-Qabisis astrologische Lehrschrift im Europäischen Mittelalter," *Zeitschrift für Geschichte der Arabisch-Islamischen Wissenschaften* 13 (1999): 93–128. His study of the European reception of the *Liber Introductorius* from the twelfth through the seventeenth centuries—including translations into French, German, English and Spanish—is particularly interesting (97–107). He also briefly discusses its use in university lectures (103–4). Hasse also has a section on the printing history of Alcabitius's *Liber Introductorius*, which he characterizes as "the most often printed work of Arabic origin"; *Success and Suppression*, 12–13 and 328–30.

[18] Hasse discusses this pseudo-Haly at *Success and Suppression*, 374–75; the true commentator on the *Centiloquium* is Ahmad ibn Yusuf.

[19] See Lemay, "The Teaching of Astronomy," 206–9, and "The Late Medieval Astrological School," 344–49. For the arguments establishing the *Centiloquium's* pseudonymity, see Franz Böll, *Studien über Claudius Ptolemäus: Ein Beitrag zur Geschichte der Griechischen Philosophie und Astrologie*, Leipzig: Teubner, 1894, 180–81. Pico accepted the *Centiloquium's* authenticity in the *Disputations* (1496), but Cardano strongly questioned it in his commentary on Ptolemy's *Tetrabiblos* (1554), to be discussed in volume III. A proper study of the *Centiloquium* in all of its relevant culture-historical and textual respects would be extraordinarily interesting.

[20] In tertio anno, primo legatur Alkabicius, quo lecto, legatur Centiloquium Ptolomei cum commento Haly. Quo lecto, legatur tertius geometrie, quo lecto, legatur tractatus quadrantis.

[21] "In quarto anno, primo legatur Quadripartitus totus quo lecto, legatur liber de urina non visa. Quo lecto legatur dictio tertia Almagesti." A text on astrological interrogations for medical purposes, *De*

After teaching this 4-year sequence in the prescribed order, the professor was then required to start over again from the beginning. The heavy handed "legalese" of the statutes comes out clearly here: "When the said years are completed and the said books completed at the said end, a circuit should be made and a return to the reading (*lectura*) of the first year, after that, to the reading of the second year, and thus in order."[22] The lecturer's statutory requirements for teaching are not left in any doubt whatsoever.

But astrology was not only taught in the mathematics curriculum. It was also taught, in a different respect, in the natural philosophy course, as we saw in Part I, which is reflected but not spelled out explicitly in the Bologna statutes. This three-year course required, among others, the core texts of Aristotelian natural philosophy, including but hardly limited to those which provided astrology's natural philosophical foundations: the *Physics*, *De caelo*, *De generatione et corruptione* and *Meteorology*, during the first 2 years of the course.[23]

Finally, in relation to the 3-year course in natural philosophy and the 4-year course in mathematics-astronomy-astrology, there was also the 4-year medical course proper.[24] This focused, as one would expect, on medical texts in their own right, but it also included texts with a strongly astrological character, specifically Galen's *De diebus decretoriis*, which was studied during each of the first 3 years of

urina non visa by Guillelmus Anglicus is also found in the specialized astrologico-medical manuscript compiled by Bartolomeo Valdizocco, possibly in Padua (1467–1470), discussed below. Although the *De urina non visa* is medical in aim, it is purely astrological in content. See now the critical edition by Laurence Moulinier-Brogi, *Guillaume l'Anglais, le frondeur de l'uroscopie médiévale (XIIIe siècle): Édition commentée et traduction du* De urina non visa, Geneva: Librarie Droz, 2011. My thanks to David Juste for this reference and his clarifications.

[22] Dictis annis completis et completis dictis libris in dicto termino, fiat circulus et redeatur ad lecturam primi anni, postea ad lecturam secundi anni et sic per ordinem.

[23] Thorndike, *University Records,* 279; Vescovini, "I programmi degli insegnamenti," 200–202. See also Grendler, *Universities*, 279; his useful chapter on natural philosophy is at 267–313. A thorough and systematic treatement of the astrological dimension of the Aristotelian commentaries actually taught at medieval and Renaissance universities in addition to those by Albertus Magnus would be of great interest, and would potentially provide more evidence for my argument. For a major step in this direction that analyzes astrologizing natural philosophy in the sixteenth century in the Coimbra commentaries on Aristotle, see Carolino, "*Agant corpora coelestia in sublunarem mundum an non?*"

[24] These three curricula are treated under the same rubric (78) in the 1405 arts and medicine statutes. It is unclear, however, what their order of study would have been, concurrent or *seriatim*; see Siraisi, *Medieval and Early Renaissance Medicine*, 70 ff., esp 72. We know that study in arts was propaedeutic to medicine, and that sometimes students began their study of medicine with a strong arts preparation, but sometimes not. On the relationship of natural philosophy and medicine more generally, see Schmitt, "Aristotle among the Physicians"; and esp. Siraisi, *Taddeo Alderotti*, 147–236. Grendler treats the medical curriculum at *Universities*, 314–52. There is also much useful and relevant information in Angus Clarke's unpublished Warburg dissertation on Giovanni Antonio Magini, and in Hasse, *Success and Suppression*.

the course.[25] Medical training, as we will see in more detail below, was one of the two primary ends toward which the arts and medicine faculties of the Italian universities were oriented, along with natural philosophy itself.[26]

10.3 The University Astrologer: Job Description (Rubric 60)

The figure of the astrologer at the University of Bologna comes into sharper relief if we examine rubric 60 of the 1405 statutes and begin to embed him more deeply into the broader university culture as well as into the wider society.[27] In addition to mandating the curriculum, the 1405 statutes also required the professor of astrology at Bologna to make a public annual prognostication (*iudicium anni*), to participate in formal public disputations, and to provide astrological services—free, and in a timely manner—for the scholars of the university.[28]

[25] See Thorndike, *University Records*, 280–81. In the first year, book I was taught by the ordinary professor, book II by the extraordinary professor. In the second year, book III was taught by the ordinary professor, book I by the extraordinary. In the third year, book II was taught by the ordinary, and book III by the extraordinary; thus there were two concurrent cycles of teaching. See also Siraisi, *Taddeo Alderotti*, 107. For a rich treatment of critical days, see dell'Anna, *Dies critici*. This tradition continued well into the seventeenth century, where we can see that Giovanni Antonio Magini (1555–1617), Andrea Argoli (1570–1657) and Placido Titi (1603–68)—professors of mathematics at Bologna, Padua and Pavia respectively—all wrote separate textbooks on critical days, as we will see in volume III. For more on Galen's *On Critical Days*, its Greek text and its translations into both Arabic and Latin, see numerous studies by Glen M. Cooper, including his "Approaches to the Critical Days in Late Medieval and Renaissance Thinkers," *Early Science and Medicine* 18 (2013): 536–65, with much additional bibliography.

[26] Favaro, "I lettori," 5–6, discusses the close connection between astrology and medicine at Padua from the fourtheenth into the eighteenth century; it is discussed further below. Vescovini's 1991 comment is worth quoting here; "L'astrologia all'università di Ferrara nel Quattrocento," 296: "Tuttavia la storia dei rapporti tra medicina e astrologia tra Medioevo e Rinascimento resta quasi tutta ancora da scrivere." David A. Lines argues persuasively that natural philosophy was increasingly becoming an end in itself, especially in the sixteenth century. See especially his "Natural Philosophy in Renaissance Italy: The University of Bologna and the Beginnings of Specialization," in *Science and Universities of Early Modern Europe: Teaching, Specialization, Professionalization*, a special issue of *Early Science and Medicine* 6 (2001): 267–323.

[27] It is unclear how far the highly developed situation in Bologna may be generalized to apply either to other Italian universities or elsewhere. Regardless, this is an important piece of evidence to help build up a broader picture at the University of Bologna. For another attempt, see Casali, *Le spie del cielo*, ch. 1, "Professione astrologo," 5–34. A systematic comparison of astrologers and their practices at different universities and courts over the entire period and throughout Europe would be tremendously interesting.

[28] Most of my information for this section comes from Albano Sorbelli, "Il 'Tacuinus' dell'università di Bologna e le sue prime edizioni," *Gutenburg Jahrbuch*, 1938: 109–14; Alberto Serra-Zanetti, "I Pronostici di Girolamo Manfredi," in *Studi riminesi e bibliografici in onore di Carlo Lucchesi*, Faenza, 1952, 193–213; Franz Hammer, "Astrologie und Buchdruck im 15. Jahrhundert," in *Colligere fragmenta: Festschrift Alban Dold*, B. Fischer and V. Fiala (eds.), Beuron in Hohenzellern:

Rubric 60 states clearly that the professor of astrology was required to compose a judgment for the year (an annual prognostication) and to make it publically accessible in a standardized manner:

> Again, they have established, ordered and affirmed that the doctor chosen or to be chosen by the said university for lecturing on the science of the stars (*ad legendum in astrologia*) shall be required [1] to give judgments (*iudicia* [sc. astrological interpretations]) for free to the scholars of the said university within one month after they have been requested,[29] also [2] one time only (*singulariter*) to post a judgment of the year (*iudicium anni* [sc. an annual prognostication]) in writing at the office of the general Bidells (*ponere ad stationem generalium Bidellorum*), and also he is required [3] to lecture at specified times (*legere secundum puncta*), save only on feast days and vacations, with a penalty, for each in turn in each of the said cases, of twenty Bolognese soldi.[30]

The professor of astrology is thus required to give individual interpretations (*iudicia*) for his colleagues (and perhaps for the students as well), a general forecast for the year (*iudicium anni*) to be posted publically (that is, "published" as it was done prior to the printing press's arrival in Italy),[31] and to teach (lecture, read) at set times, as we will also see in the rolls (rotuli) discussed below.

There are over 100 extant prognostications in manuscripts, including 43 for Munich and Paris alone (see *CCAL* I-II, index under "pronostication").[32] There are also extant manuscript prognostications by Bologna professors of astrology, e.g. Stefano da Faenza, Giovanni Fondi, Matteo da Brescia, Martinus de Polonia, Georgius de Russia and Girolamo Manfredi.[33] Many printed annual prognostications also survive from later in the fifteenth century, some of which I will discuss

Beuroner Kunstverlag, 1952, 281–87; Curt F. Bühler, *The University and the Press in Fifteenth-Century Bologna*, Notre Dame: Mediaeval Institute, 1958; and Robert S. Westman, "Copernicus and the Prognosticators," *Universitas* 5 (1993): 1–5, and, of course, the text of Rubric 60 itself. Disappointingly, Westman adds little of value in his longer treatment in *The Copernican Question*. Tur's "À l'entrée du soleil en Bélier" complements and in many respects supersedes all of these studies for issues related to annual astrological prognostications.

[29] One would like to know what sorts of requests were made and the practices undertaken; and, if records were kept of such consultations, whether any survive. I do not know of any.

[30] Sorbelli, "Il 'Tacuinus'," 109. The text comes from Malagola, *Statuti*, 264: "Item statuerunt, ordinaverunt et firmaverunt quod doctor electus vel eligendus per dictam Universitatem ad legendum in astrologia, teneatur iudicia dare gratis scolaribus dicte Universitatis infra unum mensem postquam fuerint postulata, et etiam singulariter judicium anni in scriptis ponere ad stationem generalium Bidellorum, et etiam teneatur legere secundum puncta, ea servando solum diebus festivis et vacationum, pena pro qualibet vice in quolibet dictorum casuum, viginti solidorum bon."

[31] Tur discusses the situation in Bologna in particular as well as the other principle sites of production; "À l'entrée du soleil en bélier," chapter 7.3.

[32] In volume II of his thesis, Tur discusses in detail 86 extant prognostications composed and/or promulgated between 1405 and 1484; "À l'entrée du soleil en Bélier."

[33] My thanks to David Juste for this information. See also Bonolì and Piliarvu, *I Lettori*, for the Bolognese professors, and now Alexandre Tur's "À l'entrée du soleil en Bélier." His thesis includes a valuable prosopography with up-to-date bibliography of all the known prognosticators, at least one of whose prognostications we possess.

just below.[34] We should also note that the subject is explicitly called '*astrologia*' here (*ad legendum in astrologia*), as it was also in rubric 78, where the 4-year curriculum was stipulated.

*

Alexandre Tur has recently discovered valuable evidence revealing that there was a university-wide ceremony (probably annual) at the University of Bologna to promulgate the university's statutorily-mandated annual astrological judgment.[35] This uniquely detailed evidence[36] comes from the prologue to the manuscript of Stefano da Faenza annual prognostication for 1426, which is written entirely in indirect speech except for the prologue:

> Since every person naturally desires to know, and the human mind longs to hear novelties (*novitates*), hence it is that I, together with Johannes Custodis, resolved to dispatch this present judgment (*iudicium*) to you most deserving lords, [sc. a judgment] made (*practicatum*) by a most excellent doctor of arts and promulgated publically in the Bolognese schools, about which [sc. I have written] below. In making it public (*publicatio*), the entire community (*tota universitas*) of all the faculties of this most famous and bountiful *studium* were arranged presently, listening to and most diligently considering the future events coming forth from the influence of the superior [sc. bodies] (*futuros eventus ex superiorum* [*?*] *influentia evenientes*), since inferior things are ruled by these superiors (*cum enim inferiora regantur ex ipsis superioribus*).[37]

Tur claims that these ceremonies of promulgation constituted one of the most important events in the academic year for the students in astrology.[38]

*

[34] On these in particular, see Jonathan Green, *Printing and Prophecy* and "Printing the Future," and Robin B. Barnes, "Astrology and Popular Print in Germany, C. 1470–1520," in *Books Have Their Own Destiny: Essays in Honor of Robert V. Schnucker*, R. Barnes, R. Kolb and P.L. Presley (eds.), Kirksville, MO: Thomas Jefferson University Press, 1998, 17–26 primarily for the situation in Germany (and all with further bibliography), and *I pronostici* on Domenico Maria da Novara's annual prognostications between 1484 and 1504.

[35] Tur claims that these ceremonies of promulgation also took place in Cracow, most likely in Louvain and Leipzig, and probably elsewhere; "À l'entrée du soleil en Bélier," 249–50.

[36] Tur provides two pieces of comparatively minor evidence as well: that Stefano da Faenza's annual prognostication of 1422 at Bologna was '*publice pronuntia[tum]*', and that Jean Spierinck's for 1464 at Louvain was '*prenuntiata a Universitati Louvaniense*'; "À l'entrée du soleil en Bélier," 250.

[37] "Cum enim omnis homo naturaliter scire desiderat, et mens humana audire novitates affectatur, hinc est quod ego, una cum Johanne Custodis, disposui vestris dignissimis dominationibus transmittere hoc presens judicium praticatum per excellentissimum artium doctorem ac in scolis Bononiensibus publice promulgatum, de quo infra, in cujus publicatione tota universitas hujus almi ac preclarissimi studii omnium facultatum fuit praesentialiter constituta auscultans ac diligentissime considerans futuros eventus ex superiorum [?] influentia evenientes, cum enim inferiora regantur ex ipsis superioribus (f. 168r; Tur's question mark)." Tur, "À l'entrée du soleil en Bélier," 251.

[38] "À l'entrée du soleil en Bélier," 250.

Rubric 60 also required the professor of astrology to publically dispute at least two *questiones* on the science of the stars (*astrologia*) every year, and to participate in an astrological *quodlibet*, both of which were also "published":

> Again, that the doctor chosen for the salary in *astrologia* be required and obligated [1a] to dispute in each year two questions (*disputare duas questiones*) in *astrologia* and [1b] to make a determination on them within eight days from the day of the said disputation (*disputatio*), and also [2a] let him be required to dispute *de quolibet in astrologia* at least one time (*semel adminus*), and [2b] to make a determination on the said *quodlibet*, and further, [3] to post the said questions and the said *quodlibet* in writing at the office (*statio*)[39] and to do it in good handwriting (*de bona littera*) and on good parchment sheets (*in bonis cartis membraneis*), not erased (*non abrasis*), and in the manner of the larger mode (*ad formam modi maioris*) within fifteen days after the determination. And let the said questions stay at the office continually (*continue stent in statione*), and from these a copy may be had.[40]

As far as I know, no trace of these disputations exist. If we could find some, however, they might well shed interesting light on the subject of this study.

*

The requirement to make an annual prognostication is also clear from the official university rolls, where, for example, under the academic year 1475–76, under the notice: "morning lecture on *astronomia*," are the names D[ominus] M[agister] Hieronymus de Manfredis and D. M. Aurelius de Roma, which are immediately followed by: "With this, that he make a *tacuinum* in good form (*bona forma*) according to the form of the roll, otherwise let him not have his salary."[41] The situation is clear: no prognostication (*tacuinum*), no salary! This is standard for the rotuli.[42] Westman points to Domenico Maria da Novara's almanacs with their prognostica-

[39] Sc. of the Bidells, understood as above.

[40] Sorbelli, "Il 'Tacuinus'," 110: "Item quod doctor electus ad salarium astrologie teneatur et debeat quolibet anno disputare duas questiones in astrologia et eas determinare infra octo dies a die dicte disputationis et etiam teneatur disputare de quolibet in astrologia semel adminus, et dictum quodlibet determinare, et supra, et dictas questiones et dictum quodlibet in scriptis ad stationem ponere et dare de bona littera et in bonis cartis membranis, non abrasis ad formam modi maioris infra quindecim dies post determinationem. Et dicte questiones continue stent in statione, ut de eis copia habeatur." Grendler does not treat this required feature of the professor of astrology's responsibilities, although he discusses others; *Universities*, 409–12. I hope to research further these public astrological disputations and their literary Nachlaß.

[41] Cum hoc, quod faciat Tacuinum bona forma secundum formam rotuli, aliter non habeat salarium.

[42] Sorbelli, "Il 'Tacuinus,'" 110. In the Bolognese rotuli, these annual prognostications are either called '*iudicium*' or '*tacuinum*', often both together; in the fifteenth century '*pronosticon*' or '*pronosticatio*' was sometimes used to indicate both. In Germany, '*practica*' became the most common term, especially for vernacular prognostications. According to Sorbelli, the *iudicium* originally referred to prognostications concerning the major events of the year, whereas the *tacuinum* was specifically concerned with health issues. Requiring the astrology professor by statute to make an annual prognostication seems to have been peculiar to the University of Bologna. Aurelius is discussed in Bonolì and Piliarvu, *I lettori*, 115. I discuss Manfredi below. Tur, Green and Kremer further discuss this distinctive and sometimes ambiguous and confusing terminology.

tions over a 20 year period, but does not analyze or describe them in detail, even though these are in fact his only extant writings. Fortunately, they have now all been transcribed and published, and thus may now be properly studied.[43]

10.4 Structure of Annual Prognostications (*Iudicium anni*)

Now that we have a clearer picture of the professor of astrology's role at the university, both with respect to teaching and his other functions, we can get a fuller sense of his impact on society by focusing on one of these roles, namely, his writing of an annual prognostication or judgment (interpretation) for the year, which falls under the astrological practice of general/universal astrology or revolutions.[44]

The annual prognostications in Bologna followed a standard but flexible pattern: (1) A dedication, whether to the signore, the pope, or another powerful figure; (2) an *exordium*, sometimes containing praise of the ruling family (or the papacy), sometimes a celebration of the city, sometimes a defense of astrology, etc; (3) astronomical phenomena, usually limited to eclipses of the luminaries and to comets; and (4) astrological phenomena, including the major planetary aspects and their influence on natural and political matters.[45]

Franz Hammer describes the structure of master Georg Drohobicz's (a.k.a. Georgius of Russia's) prognostication for 1483/84.[46] Its ten chapters have the following headings:[47] (1) [General] On the ruler (*dominus*) of the year and its general and seasonal disposition. (2) [Religious] On the disposition of the faith and sects. (3) [Political, 3–6] On the state of the pope and his cardinals. (4) On the state of the emperor, kingdoms and important (*famosi*) cities. (5) On wars and battles (*De bellis et guerris*) and a judgment (*iudicium*) on two lunar eclipses. (6) On the end of the war between the Venetians and the Ferrarese. (7) [Agriculture] On the fertility and famine of things in general (*in generali*) and in particular (*in speciali*). (8) [Public

[43] *I pronostici di Domenico Maria da Novara.*

[44] We will see in Chap. 11 how seriously these very public prognostications could be taken. Tur discussus their influence in the political realm; "À l'entrée du soleil en bélier," 282–95.

[45] Sorbelli, "Il 'Tacuinus,'" 111. Annual prognostications later became a central feature of annual almanacs, an extraordinarily successful printed genre from the beginning of printing throughout the eighteenth century and beyond. For Italy, see Elide Casali, *Le spie del cielo: Oroscopi, lunari e almanacchi nell'Italia moderna*, Turin: Einaudi, 2003, who also discusses the professor of astrology at Bologna; for England, see Bernard Capp, *English Almanacs 1500–1800*. For Germany, see Green, *Printing and Prophecy* and "Printing the Future." He focuses on the period from 1450 to 1620. Almanacs and annual prognostications will be discussed further in Chap. 11 below and in volume III.

[46] Hammer, "Astrologie und Buchdruck," 282–83 (GW 9060). See also Pessina Longo, *Georgius de Russia, rettore a Bologna nel XV secolo* (Bologna: CLUEB, 1988), where the prognostication printed in facsimile is mostly illegible, and Bonolì and Piliarvu, *I lettori*, 115–17. Tur discusses Georgius in the prosopography; "À l'entrée du soleil en bélier," 521–22.

[47] I note the general themes in brackets.

health] On the state of the plague and death. (9) [Human beings] On the state of men subject to individual planets.[48] (10) [General again] On the disposition of the months and days of the entire year.[49]

This prognostication thus touched on many issues of public concern, including religion, politics, war, agriculture, fertility, public health (plague and death), the general effects on people ruled by different planets, and the nature of each day and month of the year.[50] It was dedicated to pope Sixtus IV (r. 1471–84, whom we will meet again below) with a 15 line poem, four lines of which Hammer quotes (283):

> Omnia celesti mortalia lege fruuntur:
> Quis neget ergo astris inferiora regi?
> Non cogunt, licet ipsa regant, si forte minantur:
> Obverti possunt hae ratione minae.

> All mortal things delight in celestial law:/ Who would deny, therefore, that things below (*inferiora*) are ruled by the stars?/ They do not compel (*non cogunt*), although they do rule, if perchance they threaten:/ But these threats can be opposed by reason (*ratio*).

In this quatrain, we can see the same significant larger themes that were treated in just the same way in the thirteenth century: there is celestial law and it rules the world below, but it does not compel, that is, there is no necessity and the human will may act freely.

[48] We will find a resonance with this theme in Francis Bacon's proposals for astrological reform discussed in volume III.

[49] "1. De domino anni et eius generali et quartarum dispositione. 2. De dispositione fidei et sectarum. 3. De statu Summi Pontificis et suorum Cardinalium. 4. De statu Imperatoris, regnorum et urbium famosarum. 5. De bellis et guerris et iudicio duarum eclipsium lunae. 6. De exitu seu fine belli Venetorum et Ferrariensium. 7. De fertilitate et caristia rerum in generali et speciali. 8. De statu pestis et mortalitatis. 9. De statu hominum singulis planetis subjectorum. 10. De dispositione mensium et dierum totius anni." In volume II of his thesis, "À l'entrée du soleil en Bélier," Tur provides the particular detailed structure ("plan") for each of the prognostications he analyses between 1405 and 1484. He discusses them analytically in Chap. 4.

[50] In the introductory material to Ratdolt's 1492 edition of Regiomontanus's *Almanach* for 15 years (1491–1506) on how to use an ephemeris (*Usum ephemeridis cuiuslibet breviter exponemus*, a2r), Ratdolt prints some practical astrological information on how to make the prognostication in a '*tacuinum*' (*Prognostica in tacuinos ex prolixis in compendium redacta*, a3v) that he seems to have written himself. He begins with a discussion of the revolution of the ascendents of the years of the world and the months of the year along with the new and full moons (*Quam vim revolutiones ascendentium annorum mundi ac mensium anni in caeli figura habeant quam solis cum luna coitus et plenilunium*, a3v–a4v) and the planetary aspects. Then he treats how to predict the weather (*aeris mutationes*, a4v–5ar); how to make elections for the beginnings of actions (*Electiones in initiis operum*, 5ar–6ar), for bleeding (*Pro deiectione sanguinis electio*, 6ar), for taking medications (*Electio pro farmacia*, 6av), and for planting seeds (*Electiones in seminum iactu: arborum et vinearum cultu*, 6av). He finishes this practical introduction with a special section on al-Kindi on weather prediction, in which he uses the language of rays (*De mutatione aeris secundum Alkindum*, 6av–8ar). It ends with a brief explicit that he printed this introduction in 1492 (*Explicitum est hoc opus anno Christi domini 1492*, 8ar). Tur discusses all of these features in volume I of his thesis; "À l'entrée du soleil en Bélier." Ephemerides in the sixteenth and seventeenth centuries continued this trend and increasingly added more useful introductory matter, as we will see in volume III.

We will see that annual prognostications in Italy were deeply political through the end of the fifteenth and into the sixteenth century. This seems to have changed profoundly from the mid-sixteenth century onwards as Elide Casali has argued, and as I will discuss in relation to the *Index of Prohibited Books* in volume III. The political dimension of astrological prognostications was often controversial, as we will see, especially in Chap. 11.

10.5 Publication of Annual Prognostications at Bologna

Bologna seems to have been the major site for publishing annual prognostications in Quattrocento Italy, as opposed to Venice, which was the major site for publishing astrological texts.[51] Using the evidence gathered in Curt Bühler, *University and the Press in Fifteenth-Century Bologna*, and as supplemented by the information in the International Short Title Catalogue (ISTC), one may begin to see how regularly these prognostications were published, as well as who wrote them.[52] I list them chronologically:[53]

1475: Girolamo Manfredi[54] (not before Feb 14, 1475 [ISTC im00193300]).
1476: Girolamo Manfredi (before Mar. 1476 [ISTC im00193500]).
1477: Girolamo Manfredi.
1478: Girolamo Manfredi (Jan 1478).

[51] The only theoretical astrological works published in Bologna from the beginning of printing to 1500 were Alcabitius, *Liber isagogicus* (4.A.1. Vurster, 1473), and Michael Scot on the *Sphere* (41.A.1). Manilius, *Astronomicon* (March 20, 1474, by Ugo Ruggeri, 6.A.1.), and Ficino's *De vita* (36.C.1) were also published there as was Pico's anti-astrological *Disputations* as a part of his posthumous *Opera omnia*. These citations are from Bühler, *The University and the Press*, with his numeration. The famous 1477 edition of Ptolemy's *Geography* was published at Bologna (9.A.3), with the assistance of several distinguished mathematician-astronomer-astrologers: Girolamo Manfredi, Pietro Buono Avogaro, Galeotto Marzio da Narni and Cola Montana as well as Filippo Beroaldo, a distinguished humanist; Serra-Zanetti, "I Pronostici di Girolamo Manfredi," 201. There is a facsimile of this edition: *Cosmographia: Bologna, 1477*, Amsterdam: Theatrum Orbis Terrarum, 1964. For Manfredi's participation in this edition, see Duranti, *Mai sotto Saturno*, 53–65. Azzolini also discusses it in the context of the murder of Galeazzo Maria Sforza; *The Duke and the Stars*.

[52] Some were published in Latin and some in the vernacular; those published at Bologna were often republished and translated elsewhere.

[53] Bühler lists them by publisher. Rather, he does so where he can, since 27 of the 29 dated but unassigned books are prognostications. I discuss some of these figures and their prognostications in volume III.

[54] He also published a medical work, *Liber de homine*, on July 1, 1474 (ISTC im00191000); the ISTC listing is under 'Manfredis, Hieronymus de'. See Girolamo Manfredi, *Liber de homine: Il Perché*, with useful introductory essays by F. Foresti, A.L. Trombetti Budriesi and A.M. Nada Patrone, Bologna: Luigi Parma, 1988. See now Duranti, *Mai sotto Saturno*, and Bonolì and Piliarvu, *I Lettori*, 111–12. I discuss Manfredi further just below. Manfredi also had his prognostications published in places other than Bologna, for example in Rome and Louvain. I put ISTC numbers for those I could find online.

1479: Girolamo Manfredi; Marcus Scribanarius.[55]

1480: Girolamo Manfredi (ISTC im00193730); Nicolaus de Insula Marie (ISTC in00110300).

1481: Girolamo Manfredi (after 2 Jan 1481 [ISTC im00193820]).

1482: Girolamo Manfredi (after 22 Feb 1482 [ISTC im00195000]); Scribanarius (not in ISTC); Franciscus Guasconus (ISTC ig00539050).

1483: Scribanarius (In Italian, after 20 Feb [ISTC is00338650]);[56] Paul of Middleburg.[57]

1484: Paul of Middelburg (after Dec 31, 1484).

1485: Scribanarius (before Mar. 1485 [ISTC is00339300]).

1486: Carolus Susena (ISTC is00870900).

1487: Domenico Maria da Novara (GW 8661 [ISTC id00307150]).[58]

1489: Domenico Maria da Novara (Jan [ISTC id00307230]).

1490: Girolamo Manfredi (in Italian, after 13 Dec 1489 [ISTC im00195500]).

1491: Girolamo Manfredi (in Italian, after 12 Feb 1491 [ISTC im00195600]); Baldinus de Baldinis (before March 1 [GW 3206]).

1492: Girolamo Manfredi (in Italian, after 23 Feb 1492 [ISTC im00195700]); Domenico Maria da Novara (ISTC id00307300).

1493: Girolamo Manfredi; Pietro Buono Avogaro (Jan 1493 and after Jan 25);[59] Antonio Arquato (Dec 6); Scribanarius (after 1 Jan 1493 [ISTC is00339400]; there is also one in Italian for the same year also after 1 Jan [ISTC is00339500]).

[1494][60]

1495: Scribanarius (ISTC is00339600).

1496:[61] Pietro Buono Avogaro (Jan 1496 [ISTC ia00057400]); Domenico Maria da Novara (Jan).

---

[55] See Tur's prosopography; "À l'entrée du soleil en Bélier," 541.

[56] Although Scribinarius has other prognostications before 1483 listed in ISTC, none are for Bologna.

[57] For more on Paul of Middleburg (1445–1533), who became Bishop of Fossombrone, see Vanden Broecke, *Limits of Influence*, 61–65, and Stephan Heilen, "Paul of Middelburg's *Prognosticum* for the Years 1484 to 1504," in *From Masha'allah to Kepler*, 231–78. See also Tur's prosopography; "À l'entrée du soleil en Bélier," 556.

[58] Copernicus's mentor, he was professor of astrology at Ferrara briefly before teaching for many years at Bologna. See Bonolì and Piliarvu, *I lettori*, 118–21, and now *I pronostici*. I discuss his career below. ISTC lists him under Dominicus Maria de Novaria.

[59] He was professor at Ferrara from 1467–1506. He is no. 17 in the list of professors in *La riniscita da sapere*. It is listed under the Latin form of his name, Petrus Bonus Advogarius, in ISTC: ISTC ia00057087. There is also an Italian version from the same year: ISTC ia00057088. He has several published elsewhere for other years as well as those mentioned here for Bologna, including Milan and Ferrara.

[60] There is a *Defensio pronostici eiusdem* of June 3, 1494 attributed to Scipio de Manfredis.

[61] Although Bühler cites a prognostication for Girolamo Manfredi for 1496 (UN.A.21), as Serra-Zanetti rightly points out (correcting Hellman), this could hardly be the case, since Manfredi died in 1493; "I pronostici di Girolamo Manfredi," 203 (n. 18). Perhaps his name was used even though he was already dead to sell more copies, as was so often the case later in England; see Capp, *English Almanacs*, and Curry, *Prophecy and Power*.

1498: Pietro Buono Avogaro (ISTC ia00057620); Pietromellara (Jan);[62] Scribanarius (after 1 Jan 1498 [ISTC is00339700]).
1499: Domenico Maria da Novara; Scribanarius (Feb 1); Johannes Antonius Scandolariis (after 25 Nov 1498 [ISTC is00304200]).
1500: Domenico Maria da Novara (ISTC id00307700); Hieronymus Catinellus (after 17 Dec 1499 [ISTC ic00287000]); Pietramellara.
1501: Domenico Maria da Novara (Jan).

Although many were, the prognostications published at Bologna were not only composed by professors of astrology there. For example, Pietro Buono Avogaro taught at the University of Ferrara.

10.6 Girolamo Manfredi (1425–92)

Girolamo Manfredi, whom we just encountered and whom we will meet again in Chap. 11, was a major figure in fifteenth-century prognosticatory culture. Born into a distinguished Bolognese family, he received his doctorate in philosophy and medicine there in 1455. Manfredi taught several disciplines at the University of Bologna during his long career: logic from 1455–57; philosophy (extraordinary) during 1458–60 and 1462–65; philosophy (ordinary) and medicine during 1465–66; medicine alone, 1466–69; medicine and *astronomia*, 1469–71; medicine, 1471–74; *astronomia*, 1474–83; medicine, 1483–86; and *astronomia* alone from 1486 until his death in 1492.[63] As his career pattern shows, Manfredi went back and forth between teaching astrology and medicine, sometimes teaching both in the same year.[64]

Author of the best-selling medical text, *Liber de homine* or *Il perche*, Manfredi also authored a famous aphorism that clearly indicates the importance of astrology for medicine: "Although medicine in itself is a complete body of knowledge

[62] This prognostication stems from the same press that published Pico's *Opera omnia* with the *Editio princeps* of the *Disputationes adversus astrologiam divinatricem*, namely, Benedictus Hectoris. For Giacomo Pietramellara, see Bonolì and Piliarvu, *I Lettori*, 122–3.

[63] Serra-Zanetti, "I pronostici di Girolamo Manfredi," 195 (n. 4), gathered from the essential work by U. Dallari, *I rotuli dei lettori legisti e artisti dello studio Bolognese* [...], Bologna: Merlani, 1888–1924, I, 43 ff. Serra-Zanetti also provides a full bibliographical conspectus of Manfredi's prognostications (207–13). See also Bonolì and Piliarvu, *I Lettori*, 101–2, and Ferdinando Gabotto, "L'astrologia nel Quattrocento in rapporto colla civiltà: osservazioni e documenti storici," *Rivista di filosofia scientifica*, 1889, 373–413, 401–3. See now Duranti, *Mai sotto Saturno*; for his career in particular, 42–53, and Tur, "À l'entrée du soleil en Bélier," 163 and 526.

[64] This complex career trajectory challenges Westman's suggestion of a more linear path from mathematics to natural philosophy to medicine; "Astronomer's Role." It is disappointing that Westman did not develop this interesting line of research in his *Copernican Question*. For the fullest treatment of career trajectories that I know, see David Lines, "Natural Philosophy in Renaissance Italy," and his *Aristotle's Ethics in the Italian Renaissance (ca. 1300–1650): The Universities and the Problem of Moral Eduction*, Leiden: Brill, 2002.

(*scientia* [...] *perfecta*), nevertheless, a physician (*medicus*) in his work (i.e. in practice) is not complete without astrology (*sine astrologia*)."[65] Manfredi's career overlapped with another famous professor of astronomy/astrology at Bologna, Domenico Maria da Novara, who taught there from 1483–1504.[66]

Manfredi was apparently quite well known in his day. According to his contemporaries, especially Giovanni Garzoni (whom we will meet again just below), Manfredi was the most learned contemporary astrologer. In addition to possessing honors and glory, he also become very wealthy.[67] He even had the dubious distinction of being singled out by Giovanni Pico della Mirandola in the *Disputationes* (II.9), where Pico mentions the public annual prognostication Manfredi composed in the year he died (composed in 1492 for the year 1493)—contemporaneous with Pico's composition of the *Disputationes*—and a prognostication Manfredi made concerning the husband of Pico's sister Lucrezia, Pino Ordelaffi, condottiere and signore of Forli, both of which I will discuss in volume II.

In addition to this minor notoriety, 20 years earlier Manfredi also excited the heated displeasure of Galeazzo Maria Sforza, duke of Milan (r. 1466–76), in 1474. Duke Galeazzo construed Manfredi's intentions in his prognostication for that year to refer to the duke's own imminent death, and to troubled times ahead for Milan. The duke communicated his displeasure effectively to Manfredi through formal and informal diplomatic channels, as we will see in detail in Chap. 11.[68] Manfredi, however, was not alone. The duke also took action against Pietro Buono Avogaro, professor of astrology at Ferrara, and a certain Marsiglio of Bologna.

Finally, Jerome Torrella, a spanish physician who lived and studied in Italy for 20 years in the late Quattrocento tells us that Manfredi gave him the recipe for an astrological talisman, which he prints in his *Opus praeclarum de imaginibus astrologicis* (250–53). Manfredi also wrote a plague treatise in 1480 with a significant astrological dimension. I discuss this in volume II in relation to Marsilio Ficino's own contemporary plague treatise.[69]

---

[65] Bonolì and Piliarvu, *I Lettori*, 111: "[Q]uamvis medicina de se scientia sit perfecta: medicus tamen in opere suo sine astrologia non est perfectus." For Manfredi's career and writings as a physician, see Duranti, *Mai sotto Saturno*, section III: "Il Medico," 67–138, which also treats the *Centiloquium*.

[66] Serra-Zanetti, "I pronostici di Girolamo Manfredi," 202, and Bonolì and Piliarvu, *I Lettori*, 118–21. For Domenico Maria's biography, career and writings, see now the first chapter to *I pronostici*, "Domenico Maria Ferrarese da Novara, Maestro di Copernico," 1–34, by Fabrizio Bonolì.

[67] Serra-Zanetti, "I pronostici di Girolamo Manfredi," 196.

[68] Serra-Zanetti does not mention this prognostication in his catalogue ("I pronostici di Girolamo Manfredi," 207); the first in his catalogue is for 1475. Bonolì and Piliarvu note that Manfredi began compiling prognostications in 1469; *I lettori*, 111.

[69] Girolamo Manfredi, *Tractato de la pestilentia = Tractatus de peste*, T. Duranti (ed.), Bologna: CLUEB, 2008, and Teodoro Katinis, *Medicina e filosofia in Marsilio Ficino: il Consilio contro la pestilenza*, Rome: Edizioni di Storia e Letteratura, 2007.

10.7 Astrologers at Italian Universities

Now that we have a better sense of astrology's integrating position in the arts and medicine curriculum and some of the socio-professional roles performed by university professors of astrology, I would like to further develop the picture of astrology at the Italian universities from ca.1300 to ca.1500—roughly from Pietro d'Abano to Regiomontanus and his immediate influence—by gathering together some results of relevant research. The evidence will be of different types, including archival lists of *rotuli*, the structure and content of extant manuscripts, inventory lists from personal libraries, and publication lists of early printed books. I focus on those Italian universities where both Pico and Copernicus happened to study, namely, Bologna, Padua and Ferrara. A remarkably consistent picture will emerge.[70] Bologna and Padua were both originally independent university towns in vibrant cities that were both absorbed by larger cities and/or territorial entities in the normal Italian way.[71]

10.8 University of Bologna

In their valuable recent study, *I lettori di astronomia presso lo studio di Bologna dal XII al XX secolo*, Fabrizio Bonolì and Daniela Piliarvu provide as complete a list as possible of lecturers in mathematics-astronomy-astrology at the University of Bologna, relying primarily on the official annual rolls (*rotuli*) that list lecturers in the various disciplines.[72] The first extant list is for the academic year 1370/71, but they are quite intermittent until 1438, from which point most exist until they cease being composed in the academic year 1799/1800. They also list a special lectureship (*Lecture Universitatis*) for a young, poor and foreign (i.e. non-Bolognese) scholar under 25 years of age that was won in a public disputation and held for only 1 year.[73]

With respect to terminology, Bonolì and Piliarvu mention both *astrologia* and *astronomia* in their overall introduction to indicate the indifferent usage of both terms in the roduli to refer to the chair in question.[74] This issue will return in volume

[70] There were many other astrologers at that time in Bologna who did not teach at the university; among these are Galeotto Marzio da Narni and perhaps a Marsilio of Bologna. We also know of Giovanni Garzoni, who was well-known in the city, Giovanni Pasio and Giovanni Bianchini. Some of these figures will be discussed in greater detail below. What follows is, once again, a highly selective sketch, not a complete and systematic study. It may, however, provide a basis for further research.

[71] Lauro Martines, *Power and Imagination: City-States in Renaissance Italy*, 2nd ed., Baltimore: Johns Hopkins University Press, 1988. Padua was swallowed up by Venice in 1405, and Bologna by the Papal States in 1506.

[72] Duranti also has two chapters on the University of Bologna and its faculty of arts and medicine; *Mai sotto Saturno*, 27–42.

[73] Bonolì and Piliarvu, *I lettori*, 263–65. See also, Grendler, *Universities*, 146.

[74] Sono così emersi altri studiosi, non menzionati nei Rotuli, che, contemporaneamente ai Lettori ufficiali, tenevano lezioni di Astronomia o di Astrologia, come talora venne indifferentemente indicata la cattedra astronomica (21).

III when I address and attempt to account for the change in the lectureship's terminology over the fourteenth to sixteenth centuries from *astrologia* to *astronomia*, and ultimately to *matematica*, which has been interpreted as indicating a downgrading and then rejection or suppression of astrology either altogether or as a separate discipline.[75] For now, however, I will simply indicate the terminology in use in the rotuli and try to determine if the lecturer taught, wrote about or practiced astrology, as well as the overall pattern of their career trajectory.

The first evidence for the teaching of *astronomia* at Bologna comes from 1297 in a manuscript of Bartolomeo da Parma,[76] but Bonolì and Piliarvu think it was already taught there from the beginning of the thirteenth century. They point to the presence of Guido Bonatti (d. 1296/7) in 1233 and of Gherardo Sabbionetta. For them, astrology was taught at the university in order to satisfy the needs of the ruling classes, both lay and ecclesiastic; for example, Guido da Montefeltro's relationship with Guido Bonatti, or the horoscopes compiled by Bartolomeo da Parma (38).

In 1303, the commune of Bologna appointed an official astrologer (*astrologus*), Giovanni di Luni (57), and in 1334, astrology became an officially salaried subject by the municipality (55). They also increased the number of lecturers in astrology and astronomy at the university (55–6). Giovanni Fondi (sic),[77] who taught for 40 years at the university, was the commune's official astrologer, thus directly linking the university to the town and its concerns. 1475 saw Aurelio Romano—whom we met before in the company of Girolamo Manfredi—compile the first officially required annual astronomical-astrological *Taccuino* or *Iudicium* mentioned in the rotuli (115).[78]

I would now like to mention some of the more noteworthy fifteenth century professors of astrology. Pietro Giovanetti (?–1443) was a physician and an expert astrologer, who took his degree in arts and medicine at Bologna in 1383. In addition to teaching *astrologia* in 1404/5 (and possibly 1413/14), he also taught medicine in 1410/11 and again from 1419–34, during which time he also taught *astrologia* on feast days in 1429/30 and 1432/33. In 1438 he transferred to Siena, where he taught medicine for 2 years, called by Aeneas Sylvius Piccolomini, the future Pope Pius II. Giovanetti returned to Bologna in 1439 and held the first chair of medicine there until 1443, the year of his death (96–7).

Stefano da Faenza (fl. 1407–28) was a professor of various disciplines and was ultimately named Archdoctor. Between 1407 and 1428, he seems to have taught *astrologia* and *astronomia* (there are some gaps in the rotuli), sometimes

[75] I discuss Thorndike's, Grendler's, and Bonolì and Piliarvu's positions in Volume III. This evidence is also relevant in attempting to account for changing disciplinary patterns in the movement toward the modern configuration of the mathematical disciplines. This first section summarizes Bonolì and Piliarvu; page references in parentheses are to their study.

[76] See also the essays collected in *Seventh Centenary of the Teaching of Astronomy in Bologna 1297–1997*, Pierluigi Battistini, et al. (eds.), Bologna: CLUEB, 2001.

[77] This is not a typo for Giovanni *D*ondi (dall'Orologio), whom I will briefly discuss in the section on Padua. Tur treats Fondi in his prosopography; "À l'entrée du soleil en Bélier," 399–400.

[78] I wonder if this is the Aurelius C. discussed in Tur prosopography; "À l'entrée du soleil en Bélier," 516–18.

supplemented by metaphysics (1415–17) and sometimes by practical medicine (1425/26 [100]).[79]

Giovanni Fondi (fl. 1428–73), whom we just met, taught for around 40 years and was considered one of the preeminent astrologer-astronomers in Bologna. He took his degree in arts and medicine there in 1428, and began teaching *astrologia* the following academic year. He also lectured on *astrologia* and *astronomia* for the commune of Bologna in 1435 and '37, from which two manuscripts exist, and in 1451 he wrote a defense of astrology against Nicole Oresme and others. His name is listed indifferently in the rotuli as teaching *astronomia* and *astrologia*. A manuscript of his *Tacuinus astronomico-medicus*, namely, his annual prognostication for 1435 (dated 7 Feb. 1435) exists in the University Library at Bologna (104–6).[80]

Matteo da Brescia (fl. 1462–79) was university lecturer in *astronomia* in 1462/63, and a proper lecturer in *astronomia* in the morning from 1471 to '74. He seems to have been the only university lecturer to enjoy this increase in status. From 1474 to '78, Matteo taught moral philosophy.[81] In 1479, he made an astrological prediction addressed to the doge in Venice (114). Once again we can see that the professor of astrology often taught a fairly wide range of subjects (including medicine, metaphysics and moral philosophy) and that he had a broader impact on society, primarily by means of annual prognostications.

10.9 Giovanni Garzoni (1419–1505)

To focus on the second half of the fifteenth century, I will discuss Giovanni Garzoni, professsor of medicine at the University of Bologna from 1466–1505.[82] In a short but informative article, Pearl Kibre brought attention to the broad range of Garzoni's library, both to his own works and those he owned—literary and scientific—focusing on his astrological writings. Garzoni taught practical medicine at the University of Bologna during his entire extensive career there, lecturing on Avicenna's *Canon*, book three, *fens* nine through twelve, which deal with the individual diseases affecting the human body.[83]

[79] Tur discusses him in the prosopography; "À l'entrée du soleil en Bélier," 369–70.

[80] MS Bologna, Biblioteca Universitaria, 1 IV (2), fol. 1–10r. My thanks to David Juste for the correct shelf mark.

[81] For the teaching of moral philosophy with a wealth of detailed information, see the fundamental researches of David A. Lines, *Aristotle's* Ethics *in the Italian Renaissance (ca. 1300–1650)*; for Matteo da Brescia in particular, see 412.

[82] I derive most of the information in this section from Pearl Kibre, "Giovanni Garzoni of Bologna (1419–1505), Professor of Medicine and Defender of Astrology," *Isis* 58 (1967): 504–14. See also the *DBI* article by Roberto Ridolfi (vol. 52 [1999], 438–40), and the recent edition of his letters cited below. Page references in this section refer to Kibre's article.

[83] Kibre, "Giovanni Garzoni," 505. For the teaching of Avicenna's *Canon* at the Italian universities, see Nancy G. Siraisi, *Avicenna in Renaissance Italy: The Canon and Medical Teaching in Italian Universities after 1500*, Princeton: Princeton University Press, 1987.

Kibre adduces no direct evidence that Garzoni ever taught astrology at the university, nor do Bonolì and Piliarvu mention him as such. There are, however, multiple indications of his profound (and practical) interest in the subject, beginning with three horoscopes in an autograph manuscript, two in his own hand, all dated to 1474. These occur in a three-volume medical collection that was apparently related directly to Garzoni's lectures on Avicenna. Two of the three horoscopes are medical interrogations, whereby a medical practitioner *cum* astrologer attempts to gain insight into the nature of a medical problem by casting a horoscope for the moment a question was asked.[84]

In addition to these explicit medical references to astrology, Kibre also draws attention to the astrological books in Garzoni's library:

> [W]e may take note of the several works pertaining to astrology, the study of which was deemed essential for a physician. From the evidence of the works formerly in his possession, he was acquainted with such well-known Arabic astrologers and astronomers as Alcabitius, Alfraganus, and Messahala, whose writings had been translated into Latin in the course of the twelfth and thirteenth centuries. He also had tracts on the astrolabe, the *Algorismus*, the *Sphere* of John of Sacrobosco, and works by Gerard of Cremona, Giovanni Campanus, and a calendar by an otherwise unknown Friar Fucus of Ferrara.[85]

Garzoni also had literary and humanistic interests, into which fall the last two references to astrology. First, he praised astrology highly in his work on the history of Bologna:

> No other art or science could be compared with astrology, Garzoni asserted, reaffirming that the great merit and significance of astrology lay in the use men could make of it to predict such happenings as the death of princes, political revolutions, wars, pestilence, high prices and famine. And in providing warnings astrologers might point out the means by which the effects of such predicted happenings might be avoided or mitigated (510).

In addition to this clear reference to the contents of annual prognostications, Garzoni also discussed the dependence on astrology of philosophers, poets and physicians as well as geographers, navigators, farmers and military men "all of whom, according to Garzoni, cultivated astrology with the greatest diligence (510)."

Finally, there are five[86] extant orations in praise of astrology:[87] "although Garzoni himself appears to have written no systematic treatise on astrology, he did, as the collection reveals, set forth his views on the subject in several addresses or orations (508)[.]" Kibre describes their contents briefly (508–10). It would be worthwhile to

---

[84] Kibre, "Giovanni Garzoni," 511. She describes the medical contents of the manuscript at 511–12.

[85] Kibre, "Giovanni Garzoni," 507–8; see also Thorndike, "Notes upon Some Medieval Astronomical, Astrological and Mathematical Manuscripts at Florence, Milan, Bologna and Venice," *Isis* 50 (1959): 43–45.

[86] One is a duplicate.

[87] Actually, the first one Kibre describes (508) seems to be opposed to an improper sort of astrology, which Garzoni classifies with the magic arts and divination. Unfortunately, her description does not give enough information to get at Garzoni's criticism.

compare their contents in detail with Regiomontanus's nearly contemporaneous (1464) inaugural oration at Padua as well as with earlier examples, including Roger Bacon's *De laudibus mathematicae*, and later ones like Tycho Brahe's. Like Regiomontanus and Bacon, Garzoni explicitly argues for the utility of astrology, supporting it with historical examples, and "[h]e had no doubt that it belonged with the liberal arts, as was indeed true for the curriculum of the University of Arts and Medicine at Bologna (509)." Furthermore, among his historical examples, Garzoni mentions Albertus Magnus, whose work he claims to have read.

Unfortunately, in such a brief survey, Kibre can only mention the high points, but without any significant depth of analysis. For example, Garzoni seems to be defending astrology against specific attacks, but she mentions no names in her account, nor are any of the orations dated any closer than to the fifteenth century. Further, she does not at all discuss the circumstances of the orations: when were they given; in what sort of context: academic, civic, courtly or other; before whom, etc? These orations in particular are worthy of further research, both in themselves and for reconstructing in richer detail the context of astrology at the University of Bologna in the second half of the fifteenth century.[88] Finally, in addition to Garzoni's obvious and extensive interest in astrology, there is one further point of interest for our purposes: When the young Girolamo Savonarola was at the Dominican studium at Bologna from 1476 to '79, he sought out Garzoni's counsel. Their correspondence concerns Girolamo asking Garzoni's advice on improving his preaching![89]

10.10 Domenico Maria da Novara (1454–1504)

I would now like to briefly mention Domenico Maria da Novara, professor of mathematics at Bologna and Copernicus's master. In a tantalizingly brief preliminary study, "Copernicus and the Prognosticators: The Bologna Period, 1496–1500," Robert S. Westman makes some intriguing suggestions concerning Copernicus's relationship to the astrologico-prognosticatory culture in Bologna during his time there, 1496–1500.[90] He associates Copernicus closely with Domenico Maria da Novara's work as a prognosticator. He also presents evidence that Copernicus was informed in his concern to accurately order the planets by

[88] This was precisely Kibre's point, to draw attention to this important material, all of which exists in manuscript in the libraries of the University of Bologna, some of which I examined *in situ*.

[89] Weinstein, *Savonarola and Florence: Prophecy and Patriotism in the Renaissance*, Princeton: Princeton University Press, 1970, 83. Weinstein does not date the epistolary exchange. The letters are printed in *The Letters of Giovanni Garzoni: Bolognese Humanist and Physician, 1419–1505*, L. R. Lind (ed.), Atlanta: Scholars Press, 1992. Lind dates them generally to 1476–79, Savonarola's years in Bologna. Lind's edition would be vastly more user friendly with a detailed *index rerum*.

[90] *Universitas* 5 (1993): 1–5.

specific arguments made in Pico's *Disputationes,* which had just then (1496) issued from the press at Bologna. Both of these interesting topics would benefit from fuller treatment, and Westman himself has done this to some extent in his magnum opus, *The Copernican Question.*[91]

From 1483 to 1504, Dominico Maria da Novara was listed in the rotuli as the ordinary professor of *astronomia* in the morning who was to make a *iudicium* and a *taccuinum.*[92] Many of his prognostications from 1484–1504 still exist; most are extant in the Biblioteca Columbina in Seville.[93] They have now all been transcribed and published in the extremely valuable volume, *I pronostici di Domenico Maria da Novara.* According to contemporaries, Domenico Maria was one of the most important teachers in the faculty of arts, and a major figure among fifteenth century students of nature. We do not know when or where he graduated in arts and medicine, but we do know that before teaching at Bologna, he taught at Ferrara, Rome and possibly Perugia.

His only writings to survive are the prognostications, whose structure is the same as the other *iudicia* compiled in the fifteenth century: there is an introduction, followed by chapters concerning important astronomical phenomena. Then there are predictions on the conditions of citizens and rulers, disease and war, and on cities, such as Bologna, Venice, Florence and Pisa. Finally, there is a table reporting new and full moons. Of course, Domenico Maria owes his posthumous fame to being Copernicus's master, but Copernicus never once mentions him. In fact, their relationship is only recorded in a much later letter by Georg Joachim Rheticus (118–21).[94]

*

I will conclude this section on astrology at the University of Bologna with a brief quotation from Girolamo Manfredi, this time from the prologue to his final annual *Iudicium* (1493):

> Astrology (*l'astrologia*) is not concerned with supernatural or artificial matters, but only with natural things (*cose naturale*) [...]. Thus, it pertains to the astrologer only to inquire naturally (*naturalmente*) into things down here (*quazò* = *qua giù*) and their causes (*casone* = *cagione*) by means of celestial influences (*per l'influxi celesti*).[95]

---

[91] Andrè Goddu discuss Domenico Maria as well in relation to Copernicus; *Copernicus and Aristotelian Tradition*, 187–193.

[92] [A]d lecturam Astronomiae de mane diebus continuis et ordinariis, faciat judicium et tachuinum.

[93] Westman refers to them, but unfortunately does not discuss them in detail. To me this is one of the most disappointing features of his *Copernican Question.* Tur discusses the importance of Fernand Colomb's collection of annual prognostications for our knowledge of this important literary genre; "À l'entrée du soleil en Bélier," 243–46.

[94] See also Lino Sighinolfi, *Domenico Maria Novara e Niccolò Copernico allo Studio di Bologna*, Modena: Ferraguti, 1920. Goddu too is very useful on this; *Copernicus and Aristotelian Tradition*, 151–52 and 184.

[95] "L'astrologia non è de cose sopranaturale nè de le artificiale, ma solo delle cose naturale [...]. Solo doncha pertene a l'astrologo inquirere naturalmente le cose che quazò si volgono e loro casone per l'influxi celesti." This is cited in Gianfranco Fioravanti, "Pico e l'ambiente Ferrarese,"

At the end of the fifteenth century, then, and in a great university center, we find an explict and very public allusion by a famous professor there to astrology's role in the causal understanding of nature by means of celestial influences.

10.11 University of Padua

Astrology was also taught at the University of Padua, which was influenced in various ways by the University of Bologna,[96] including that it was originally formed in the early thirteenth century by an exodus of students and teachers from Bologna.[97] For reconstructing the curricular patterns of astrological instruction at Padua, however, we do not have the same fullness of sources as we do for Bologna.[98] Nevertheless, based on his extensive archival research, Antonio Favaro has provided as solid a foundation as we are likely to ever have—given the state of the evidence—for reconstructing the teaching of mathematics, astronomy and astrology at the University of Padua from its early fourteenth-century beginnings to the time of

in *Giovanni Pico della Mirandola: Convegno Internazionale di Studi nel Cinquecentesimo Anniversario della Morte (1494–1994)*, Gian Carlo Garfagnini (ed.), 2 vols., Florence: Olschki, 1997, 157–72, 166. I discuss Fiorovanti's valuable study below.

[96] Grendler treats mathematics at Padua, but not the problem concerning the lack of primary sources; *Universities*, 416–19. The first mathematician Grendler mentions is Biagio Pelicani da Parma (416).

[97] See Siraisi, *Arts and Sciences*, chapter one, "The *Studium* of Padua," 15–31, where she conveniently summarizes the relevant evidence and bibliography. The traditional date for the founding of the university is 1222, which is just around the time that Albertus Magnus was a student there, and was inducted into the Dominican Order: "[I]t is certain that Albert joined the Dominican Order when he was a student at Padua, receiving the habit from Jordan of Saxony in 1223." James Weisheipl, "The Life and Works of St. Albert the Great," in *Albertus Magnus and the Sciences*, 13–51, 19.

[98] Siraisi (*Arts and Sciences*, 151) discusses the sources for reconstructing the curriculum of the medical course for the period before 1350, when her study ends. The sources she discusses, however, are relevant for more than the medical course, on the one hand, and go through 1465, on the other. The third source she discusses is most relevant for our concerns and is already familiar: "Thirdly, the statutes of the University of Arts and Medicine of Bologna compiled in 1405 include a list of required medical works, a number of which are of early date and may have been in continuous use at Bologna from the thirteenth century. Since, as has already been demonstrated, the Bolognese medical school definitely influenced that of Padua in the thirteenth and fourteenth centuries, it is quite probable that the texts mentioned were also in use in Padua."

Galileo primarily,[99] but beyond as well.[100] One may thus trace the teaching of astrology at Padua as a part of the curriculum in arts and medicine, a pattern solidly established by Pietro d'Abano who taught philosophy, astrology and medicine there from 1306-ca.1315.[101]

[99] His two main relevant works are the chapter, "Le matematiche nello studio di Padova prima di Galileo," in *Galileo Galilei e lo studio di Padova,* Florence: Le Monnier, 1883, 100–136, and "I lettori." This may now be supplemented for the period 1490–1500 by Paolo Sambin, "Professori di astronomia e matematica a Padova nell'ultimo decennio del Quattrocento," *Quaderni per la Storia dell'Università di Padova* 7 (1974): 59–67. Sambin characterizes Favaro as a "robusto e infaticabile storico dell'Universita di Padova (59)." Favaro's general comments on the place of astrology in the history of science are profound and worth quoting in full ("I lettori," 14–16): "Il frammischiare continuo che fa Pietro d'Abano delle cose cosmografiche e fisiche colle astrologiche, non diminuisce in modo alcuno agli occhi nostri i pregi delle opere di lui, e quando qualche scrittore anche reputatissimo ci parla della storia dell'astrologia ponendovi parallela 'la storia della imbecillità umana,' noi non sappiamo invero come da critici assennati possa pretendersi di giudicare con critieri odierni gli uomini e le cose d'altri tempi.

Noi confessiamo candidamente che, il leggere presso la quasi totalità degli scrittori e dei biografi espressioni di sprezzo verso autori astronomici del medio-evo e dei primi tempi del rinascimento, perchè non seppero mantenersi immuni dagli errori della astrologia giudiziaria, vivamente ci accora, come di una gravissima ingiustizia commessa verso uomini egregi, che subivano necessariamente la influenza dell'epoca nella quale vivevano, e forse in gran parte non erano del tutto in buona fede e probabilmente erano costretti a seguire la corrente ed a farsi interpreti delle pretese influenze degli astri per compiacere principi e monarche, dei quali la storia ci addita taluni crudeli ed efferati, null'altro temere quanto una contraria e fatale congiunzione di stelle.

In verità, la postuma affettazione di sprezzo e le energiche manifestazioni ai riguardi della astrologia giudiziaria e degli uomini che più o meno se ne occuparono, averebbero una ragione, qualoro cosifatti studi avessero per loro stessi costituito un ostacolo ai progressi della astronomia, ma ciò non è, e chi con animo spassionato si fa ad interrogare la storia, apprende facilmente di quanto la scienza astronomica vada debetrice agli studi astrologici. *Verumtamen*, scrive il Keplero, *ut in arborem fibris anni, sic in tota divinissimae artis compositione lineamenta quaedam apparent ortus huius, ut matrem et nutricem astrologiam abnegare non possit astronomia filia et alumna.*"

[100] Favaro states that the tradition of studying mathematics and astronomy in relation to medicine continued into the eighteenth century; "I lettori," 6: "Noi avremo motivo di vedere più innanzi quanto stretto si conservasse anche in seguito, nello Studio di Padova, il nesso fra medicina da un lato ed astronomia e matematica dall'altro, anzi ancora fino alla seconda metà del decimottavo secolo, quando cioè era fatto obligo agli scolari medici di frequentare le lezioni di astronomia, geografia e meteore e veniva lasciate in loro arbitrio di iscriversi alla scuola di chimica od a quella di analisi, a quella di matematica e navigazione od a quella di malattie delle donne e dei bambini." In volume III, I discuss some of the later developments at Padua in the figure of Giuseppe Toaldo (1719–97), including his proposals for astrological reform.

[101] Pietro refers to himself in his will and in other official documents as: "Providus et discretus vir Magister Petrus filius quondam domini Constanti de Abano de Contracta Sante Lucie de Padua, Artis, Medicine, Philosophie, et Astrologie Professor." Favaro, "I lettori," 7–8. Much of Siraisi's useful reconstruction of the teaching at Padua of mathematics (particularly astrology), natural philosophy and medicine during the fourteenth century (*Arts and Sciences*, 81–89) is based on a close reading of Pietro d'Abano's various works. The most relevant section for our purposes is her discussion of his specifically astrological writings (albeit with their ubiquitous medical orientation), to which I referred in Chap. 9.

Before returning to Pietro, however, I will first present Ferdinando Gabotto's succinct treatment of astrology at Padua, both at court and university.[102] First, the court:

> Astrology had also been cultivated in Padua for a long time. The chronicler Rolandino tells that Ezzelino II, the monk, and his wife, Adelaide, were great admirers of it. Another old Brescian chronicler, Iacopo Malvezzi, tells likewise how their son, the famous Ezzelino III, was surrounded by an entire cohort of astrologers. He records the names of Riprandino da Verona, Paolo da Brescia, Salione da Padova, and, primarily, Guido Bonatti; other sources, with a few variant names, confirm this important notice.

Gabotto then turns to the university: Only in the fourteenth century do we find true professors of this material in Guglielmo di Montorso, and in the extremely famous Pietro d'Abano. Because most of the *rotuli* of the professors no longer exist, this prevents us from following the development of the chair of astrology at Padua as we can for Bologna. Nevertheless, in the fifteenth century we have the names of Prosdocimo de Beldomandi, Cando di Cando, Giovanni di Camposampiero (*theologiae magister*), Antonio di Padova (*artium et medicinae doctor*), Georg Peurbach (who was earlier a professor at Bologna), Giovanni Battista Capuano di Manfredonia, Federigo Crisogono, and especially Johannes Müller von Königsberg, the famous Regiomontanus.[103]

10.12 Pietro d'Abano

Although it seems likely that astrology was taught in Padua during the thirteenth century,[104] our first solid evidence comes with Pietro d'Abano, who appears to have inaugurated a new stage in the Italian university tradition of closely integrating astrological study with both natural philosophy and medicine.[105] For Pietro, a deep knowledge of astrology was considered absolutely essential for the study and practice of medicine.[106] In this respect, he was squarely in the tradition of the influential pseudo-Hippocratic *Astrologia*, which he himself translated as *De medicorum astrologia*.[107] As we saw in Chap. 9, Pietro was also deeply influenced by what came

[102] "L'astrologia nel Quattrocento in rapporto colla civiltà," 386–87. I will discuss this pioneering article further in Chap. 11.

[103] Some of these will be treated further below.

[104] As Siraisi plausibly suggests, based mainly on a reading of passages in the chronicler Rolandino of Padua and the fact that Guido Bonatti was an astrologer in the employ of the signore, Ezzelino; *Arts and Sciences*, 78–81. See also Paolo Marangon, *Alle origini dell'Aristotelismo padovano (sec. XII–XIII)*, Padua: Antenore, 1977.

[105] Favaro, "I lettori," 7: "La serie dei professori che nello Studio di Padova insegnarono indubbiamente la astrologia, si inaugura con uno dei più bei nomi nella storia della scienza, con Pietro d'Abano[.]" He discusses Pietro's teaching at Padua, 7–14. Pietro also inaugurated a new stage in the study of Aristotle; see Olivieri, *Pietro d'Abano e il pensiero neolatino*.

[106] Siraisi, *Art and Sciences*, 83.

[107] Siraisi quotes another passage from Pietro; *Arts and Sciences*, 83 (n. 88): "'Medicus qui astrologiam ignorat nullus de se in eius manus ponere' (*Conciliator, Diff.* 1, fol 3r). The pseudo-Hippo-

to be called the *Speculum astronomiae,* providing exactly the same structure for practical astrology in *Differentia* X of the *Conciliator* as the *Speculum astronomiae* did in Chaps. 7–10.[108] Pietro also refers explicitly to Albertus Magnus's authentic and deeply astrologizing *De natura locorum.*[109]

Much of Pietro's life remains obscure. In fact, most of the information we have comes from his own writings, since there is little documentary evidence.[110] We know that he was born in Abano near Padua around 1250 (23), that he studied Greek in Constantinople at some point,[111] that he completed his *Compilatio physionomiae*

---

cratic writer was even more emphatic: 'Dixit Ypocras qui fuit medicus et magister optimus, cuiusmodo medicus qui astronomiam ignorat, nullus homo debet commitere se in manus illius qui non est medicus perfectus' (*Astronomia Hypocratis*[...] fol. 90r)." For more information on this text, see Pearl Kibre, "'Astronomia' or 'Astrologia Ypocratis,'" in her *Studies in Medieval Science: Alchemy, Astrology, Mathematics and Medicine*, London: Hambledon Press, 1984, 133–56 (originally published, 1978), and her *Hippocrates Latinus: Repertorium of Hippocratic Writings in the Latin Middle Ages*, revised ed., New York: Fordham University Press, 1985.

108 Zambelli, Speculum astronomiae *and its Enigma* (116, with notes), discusses the *Speculum*'s influence on Pietro d'Abano. *Differentia* X of the *Conciliator* was discussed more fully in Chap. 9.

109 Siraisi states that Pietro frequently cited or incorporated material from Albert's *De natura locorum*; *Arts and Sciences*, 119–20. Mahoney discusses Albert's influence on Pietro d'Abano, Jacopo Dondi and Biagio Pelicani da Parma—and later on, Gaetano da Thiene, Nicoletto Vernia, Agostino Nifo, Pietro Pomponazzi and Marcantonio Zimara—all of whom were professors at Padua; "Albertus and the *Studio Patavino*," 542. For his influence on Pietro, see also Martin Grabmann, *Mittelalterliches Geistesleben,* II: 290, 395–400, 407–8. See also Siraisi, *Arts and Sciences*, 117–25, 141–2. Mahoney's conclusion to his opening section is worth quoting (542): "While these allusions to Albert in Pietro d'Abano, Jacopo Dondi and Biagio Pelicani da Parma demonstrate that he [Albertus] was studied and cited during the fourteenth century, only in the next century would the wide range of his writings be known to philosophers at Padua and have a striking influence on the discussions of major philosophical issues." For Albert's influence on Biagio Pelicani da Parma, see also Vescovini, "Su alcune testimonianze dell'influenza di Alberto Magno come 'metafisico', scienziato e 'astrologo' nella filosofia padovana del cadere del secolo XIV: Angelo di Fossombrone e Biagio Pelacani da Parma," in *Albert der Grosse: Seine Zeit, Sein Werk, Seine Wirkung*, A. Zimmerman (ed.), Berlin: de Gruyter, 1981, 155–176, section II, "Biagio Pelicani e Alberto Magno come scienzato e 'astrologo'," 167–76. I hope to focus on Pietro d'Abano in future research, especially to investigate the extent to which Pietro adapted the astrologizing Aristotelian natural philosophical tradition reconstructed in Part I from its more explicitly theological orientation at the University of Paris (where Pietro studied and taught) for the much more natural philosophical and medically oriented University of Padua.

110 In this section I follow Vescovini's account; *Il 'Lucidator'*, "La vita," 21–30, to which the page numbers in parentheses refer. Her opening sentence sets the tone: "La scarsità delle fonti, di documenti precisi e attendibili sulle vicende della vita di Pietro d'Abano, unitamente all'abondanza di leggende e di racconti fabulosi sulla sua attività e sulle sue disavventure con le autorità ecclesiastiche, hanno grandemente contribuito a ricostruzioni contradittorie, se non fallaci, della sua biografia come del suo pensiero (21)." See also Paschetto, *Pietro d'Abano, medico e filosofo*; Paolo Marangon, *Ad cognitionem scientiae festinare: gli studi nell'università di Padova nei secoli XIII e XIV*, Tiziana Pesenti (ed.), Trieste: LINT, 1997; Bruno Nardi, *Saggi sul Aristotelismo padovano dal secolo XIV al XVI*, Florence: Sansoni, 1958; and Thorndike's still useful account; *HMES* II, 874–946.

111 The dates vary from 1270 to 1285–90 (24).

at Paris in 1295 (25), and that he completed his revision of the *Conciliator* in 1310 at the University of Padua, where he had been invited to teach in 1306 (27). He died there ca. 1315–16 during his last Inquisitorial trial, which began in 1315 (28 ff). Many consequent myths and legends have arisen, thus further obscuring our understanding of his life.

Fortunately, we possess several of his works, which Vescovini divides into three categories (31–4): (1) original treatises, (2) commentaries on or reworkings of texts, and (3) translations. In the first group there are two sorts of texts: (A) works more specifically psychological or medical, including the *Compilatio physionomiae* and the *Conciliator*— both of which contain substantial astrological sections—and (B) works dealing specifically with the science of the stars: the *Lucidator dubitabilium astronomiae*, *De motu octavae spherae,* and *De imaginibus (Astrolabium).* In group two, we find his treatment of the *Problemata* attributed to Aristotle; the *Additiones* to Mesue, the last part of Mesue's *Universal Canons on the Consolation of Medical Simples and the Correction of their Actions* (*De consolatione medicinarum simplicium et correctione operationum earum Canones universales*), which had been completed by Pietro;[112] and the *De natura medica* of Dioscurides.

Finally, in the translations we find philosophical, medical and astrological works: the *Problems* of Alexander of Aphrodisias; several of Galen's medical works, in particular the second book of the *De complexionibus*; and the astrologico-medical work by Pseudo-Hippocrates, *De medicorum astrologia.* He also translated and/or revised the existing translations of astronomical and astrological works by Abraham ibn Ezra: *Principium sapientie*, *Liber rationum*, *Liber nativitatum et revolutionum earum*, *De electionibus*, *De interrogationibus*, *Tractatus particulares* and *Liber luminarium*; only the last one deals with astrological medicine.[113]

I would like to end this brief discussion of Pietro d'Abano with an extensive quotation from Siraisi's *Arts and Sciences at Padua*:

> The astrology and astronomy known to Rolandino [of Padua] were simple indeed compared to the highly sophisticated knowledge of those topics found at Padua in the early years of the fourteenth century. In Peter of Abano, the College of Arts and Medicine had a professor of astrology of more than merely local reknown and of immense learning. In his writings, Peter devoted more attention to astrology than to any other subject except medicine. His two major compilations, the *Conciliator* and the *Expositio Problematum Aristotelis* are replete with astrological references and interpretations; he was, as noted, the author of three exclusively astrological and astronomical works, the *Imagines*, the *Lucidator*, and *De motu octave sphere* and the translator of a group of astrological treatises by Abraham ibn Ezra, or Avenezra and of an astrological work attributed to Hippocrates. [...][114]

---

[112] Hasse discusses Mesue and the printing history of his texts; *Success and Suppression*, 391–96.

[113] My thanks to David Juste for sharpening up this list. Vescovini also discusses several works spuriously attributed to him; *Il 'Lucidator'*, 35–6. For more on Abraham Ibn Ezra, see many fine studies by Shlomo Sela, including *Abraham Ibn Ezra and the Rise of Medieval Hebrew Science*, Leiden: Brill, 2003, and *Abraham Ibn Ezra on Elections, Interrogations and Medical Astrology*, Leiden: Brill, 2011.

[114] For Pietro's astronomy and astrology, see Vescovini's edition with a valuable introduction and annotations of his *Lucidator*, and Fabio Seller's recent *Scientia astrorum*.

> Peter vigorously defended the validity of astrology as a science; of the surviving portion of his principal astrological treatise, the *Lucidator*, more than a third is devoted to a lengthy disquisition proving the value of the study of the stars. [...] Elsewhere, approaching the matter from a practical professional standpoint, Peter remarked that no one would willingly entrust himself to a physician who was ignorant of astrology, a sentiment that echoes the words of the supposedly Hippocratic treatise on astrological medicine translated by Peter. He firmly rejected the criticisms of astrological medicine brought by Albubatir and even those of Averroes, who had apparently claimed that an immense number of possible astrological conjunctions and the complexity of the calculations involved made it impossible to determine what, if any, effect the stars had on human health. If the most distinguished and influential professor of the College of Arts and Medicine at Padua held views such as those just outlined, it is hardly open to doubt that the study of the stars played an important part in the Paduan arts and medical curriculum (81–4).

With Pietro d'Abano, then, astrology was firmly established at the University of Padua, a tradition that continued well into the seventeenth century and beyond, as we will see in volume III.

*

There were, of course, other astrologers teaching at Padua between Pietro d'Abano's death and Valdizocco's compilation to be discussed just below. Among others, Jacopo and Giovanni Dondi dall'Orologio,[115] Biagio Pelacani da Parma,[116] Antonio de Montolmo and Regiomontanus spring readily to mind. The interested reader should consult, among others, Favaro, Bonolì and Piliarvu, Weill-Parot and the relevant *DBI* and *DSB* articles.

10.13 Valdizocco's Astrological Manuscript

To further develop the picture of astrological study at the University of Padua and to bring it up towards the end of the fifteenth century, I will now discuss an extensive manuscript of astrological texts, some medically-oriented, compiled from 1467 to '70—possibly at Padua—by one of the city's first typographers, Bartolomeo Valdizocco, as studied by Tiziana Pesenti Marangon.[117] The manuscript comprises

[115] Giovanni's famous '*orologio*' seems to have also had at least some astrological functions, as we can see in part II, Chap. 2 of his *Tractatus astrarii*. Here, in a description of how to determine the rising degree and the four cardines, Giovanni explicitly mentions that one can thus make a horoscope (*figura*) with all the 12 house cusps (310, 66–77). Likewise, after discussing the location of the other six planets (in addition to the sun), he says that it will be very easy then to discover the planetary aspects (314, 149–52). I use the text from Emmanuel Poulle's critical edition; Giovanni Dondi dall'Orologio, *Tractatus astrarii,* edition critique et traduction de la version A par E. Poulle, Geneva: Droz, 2003.

[116] See Vescovini, *Astrologia e Scienza*, which is a study of Biagio.

[117] "La miscellanea astrologica del prototipografo padovano Bartolomeo Valdizocco e la diffusione dei testi astrologici e medici tra i lettori padovani del '400," *Quaderni per la storia dell'università*

31 texts[118]—mainly from medieval Latin authorities and translations from the Arabic, but with some contemporaries as well[119]—only one of which appears in the basic astrological curriculum as articulated in the 1405 Bologna statutes, namely, the *De urina non visa.*

Pesenti Marangon claims that this manuscript demonstrates the continuity of the astrological medical tradition at Padua articulated by Pietro d'Abano through the second half of the fifteenth century. The manuscript was compiled in 1467–70, soon after Regiomontanus gave his course on al-Farghani there with the extensive inaugural rhapsody in praise of astrology. She further claims that Giovanni Pico della Mirandola himself actually owned this manuscript. Regardless of her argument's validity about Pico's ownership[120]—and if she is correct about the manuscript's provenance—the manuscript would then reflect a serious interest in astrology and medical astrology at Padua soon before Pico studied there (1480–82), and in a manuscript written in the same manner as those intended for a university audience.[121] As Pesenti Marangon rightly emphasizes, the depth of interest is revealed by both the serious nature of the texts and tables collected, as well as by the learned interpretive apparatus of marginal glosses that indicate extensive further reading.[122]

---

di Padova 11 (1978): 87–106. Valdizocco also appears in Martin Lowry, *The World of Aldus Manutius: Business and Scholarship in Renaissance Venice*, Ithaca: Cornell University Press, 1979. Monica Azzolini also discusses this manuscript in relation to her reconstruction of a *corpus astrologicum* on analogy with the medieval *corpus astronomicum*; *The Duke and the Stars,* 40 for Valdizocco, 39–50 for the *corpus astrologicum.* Pesenti Marangon argues for the Paduan and university context of this manuscript, but David Juste is not persuaded. In particular, she did not notice that the tables of houses on ff. 8r–13v are for Bologna and Ferrara, not Padua.

[118] Pesenti Marangon provides incipits, explicits and some other information, including basic bibliography; "La miscellanea astrologica," 88–93. David Juste has examined the manuscript. Since he does not plan to publish this description in a later volume of *CCAL,* he has kindly allowed me to publish his description of the manuscript as an appendix to this chapter. It is worthwhile to do so because Pesenti Marangon often made up her own titles and misidentified several works.

[119] Pesenti Marangon, "La miscellanea astrologica," 93: "La scelta dei testi rivela un cultore delle scienze astrologiche non occasionale, ne mosso da curiosità pratiche. Con scarse concessioni alle effemeridi tanto diffuse in quel periodo il Valdizocco organizza i suoi interessi per la medicina astrologica e per l'astrologia giudiziaria intorno ad autori ed opere fondamentali: tra gli arabi Haly, Alkindi, Messahala, Alcabizio, Albumasar e l'ebreo Abraham ibn Ezra, tra i medievali due classici della filosofia naturale, Guglielmo Anglico e Arnaldo da Villanova, e tra gli astrologi Guido Bonatti, vissuto anche alla corte di Ezzelino, Niccolo Paganica, uno dei più celebri medici astrologici del '300, e due recenti glorie padovane: il Beldomandi e Niccolo Conti, contemporaneo del Valdizocco e 'astrologus per ea tempora celeberrimus."

[120] I discuss this just below.

[121] Pesenti Marangon, "La miscellanea astrologica," 88: "In scrittura cancelleresca ed usando per rubriche, paraffi ed iniziali l'inchiostro rosso, come nei manoscritti destinati allo Studio[.]"

[122] It is worth quoting Pesenti Marangon at some length here; "La miscellanea astrologica," 93–94: "Un fitto apparato di glosse marginali concorre a qualificare in modo ancora più convincente la natura teoretica e il serio approfondimento degli interessi dello scrittore, che per interpretare i suoi 'auctores' si serve di un'ulteriore bibliografia, altrettanto solida. Le *Regule ad sciendum de diversis ad egrotantes spectantibus*, tratte sopratutto dal Bonatti, sono illustrate con citazioni di passi del *Liber astronomicus*, l'opera fondamentale del celebre astrologo, del gia ricordato *Liber novem iudicium*, del *De diebus criticis* di Avicenna, del *Centiloquium* pseudo-tolemaico, del *Liber de*

But the history of the manuscript's ownership brings this significant collection of texts and glosses much more directly into Pico's immediate context when we find out that the manuscript was owned by Cardinal Domenico Grimani (1461–1523), whom we know bought Pico's library in 1498.[123] Although it is plausible that Pico owned this manuscript (as Pesenti Marangon argues), it has no explicit indications of ownership.[124] Nevertheless, she claims that Pico did in fact own it on secondary grounds, namely, that entries in both of the extant manuscript catalogues of Pico's library refer, with a *certa sicurezza,* to the Valdizocco manuscript.[125] Without additional confirming evidence, however, her claim is far too strong, since it rests solely on two extremely general descriptions in both library catalogues. In the first, the entry reads: *Liber astronomie teorice in papiro sine n(omine)*; in the second: *Diversa in astrologia.*[126] To be sure, these two entries certainly *could* refer to our manuscript, but without any further information, there is little reason to think that they could not also just as easily apply to any other astrological miscellany.[127]

Lacking explicit indication of Pico's ownership, then, only a close reading of the manuscript (both texts and glosses) against Pico's *Disputations* could reveal whether Pico used it, which use would then make his ownership much more likely. Pesenti Marangon does not attempt this in her article, nor can I do so here. Certainly the fact that Cardinal Grimani owned it makes it more likely that Pico also had, since Pico's library became the basis of Grimani's own famous collection. But since Grimani's library came to contain 15,000 volumes as compared with Pico's 1190, this (once again) significantly reduces the likelihood.[128] At the very least, the presence of this manuscript at Padua—or either Bologna or Ferrara—in the second half of the fifteenth century provides further

iudiciis astrorum di Abenragel e di 'libri geomantice iudiciorum' non meglio identificabili. Per chiarire il senso di taluni passi del *De urina non visa* di Guglielmo Anglico, il Valdizocco si serve di un'altra opera di Abraham ibn Ezra, il *De interrogationibus* ed ancora del *Liber de iudiciis* di Abenragel, mentre il *Centiloquium* ritorna nelle note agli *Aforismi* di Caciaguerra, e nel *De hora conceptionis et nativitatis*, il testo più fittamente glossato[.] [...] Gli autori che il Valdizocco cita più di frequente: Albumasar, ibn Ezra e il *Centiloquium*, sono particolarmente cari alla tradizione padovana." I left out the references to folio citations. This manuscript and its glosses deserve further study.

[123] See Martin J. C. Lowry, "Two Great Venetian Libraries in the Age of Aldus Manutius," *Bulletin of the John Rylands University Library of Manchester* 57 (1974): 128–66.

[124] Nor do any of the extant manuscripts that Grimani owned have explicit indication of Pico's former ownership; Marangon, "La miscellanea astrologica," 104.

[125] Pesenti Marangon, "La miscellanea astrologica," 105.

[126] Pesenti Marangon, "La miscellanea astrologica," 105: "In entrambi i cataloghi la miscellanea del Valdizocco e riconoscibile con una certa sicurezza: nel primo nel *Libro astronomie teorice in papiro sine n(omine)*, nel secondo nei *Diversa in astrologia* che la Kibre fa corrispondere al titolo precedente." Marangon also provides the reference in Kibre (266, nº 1116).

[127] In fact, it seems unlikely that a manuscript entitled (or described as) *Libro astronomie teorice* would refer to astrological texts. Rather, it would much more likely refer to texts treating astronomical *theorica*, such as the *Theorica planetarum.*

[128] On the extent of Grimani's library, see, Lowry, "Two Great Venetian Libraries," 147.

evidence of the serious interest in astrology toward a medical end in the Northern Italian university milieu.[129]

10.14 University of Ferrara

Let us now journey from Padua to Ferrara. Gianfranco Fioravanti sets the stage perfectly for a discussion of astrology's place at the University of Ferrara:

> Ferrara's cultural ambient possessed another characteristic throughout the 15th century: the continual and pervasive presence of astrology, at the university, at court and in society. If Battista Guarino taught there for forty years, Pietro Bono Avogaro did the same (from 1467–1506), dedicating a long series of annual prognostications to Ercole d'Este. Next to his are the names of Antonio Arquato, Domenico Maria Novara, Giovan Battista Piasio, as regards the university; Giovanni Bianchini, Pellegrino Prisciani for the ducal court.[130]

We will encounter some of these characters in Chap. 11.

The organization of studies at the University of Ferrara was modelled on the Universities of Bologna and Padua.[131] The integrated instruction in astrology, natural philosophy and medicine articulated by Pietro d'Abano at Padua seems to have provided the model for Ferrara as well. In fact, an important fifteenth-century professor of medicine at Ferrara, Michele Savonarola (Girolamo's grandfather),[132] explicitly refers to Pietro d'Abano as his teacher (*praeceptor meus*), but he means this metaphorically, of course, since Pietro had died over half

---

[129] Pesenti Marangon also adduces other relevant, mainly manuscript evidence to put Valdizocco's collection within its proper learned context: first, the traditionally learned contexts of medicine and law; then the less traditional, artisans and merchants; "La miscellanea astrologica," 93–103. Boudet, *Le Receuil de plus celebres astrologues*, II, 235 (n. 25) describes a work published by Valdizocco at Padua in 1474, an influential prognostication: "Le *Pronosticum super Antechristi adventu judeorumque messie*, publié à Padoue en avril 1474 par B. de Valdezochio, eut une influence perceptible, dans tout l'Occident, pendant plus d'un demi-siècle." For more on astrology at the University of Padua, see Giancarlo Zanier, "Ricerche sull'occultismo a Padova nel. Sec. XV," in *Scienza e filosofia all'università di Padova nel Quattrocento*, A. Poppi (ed.), Trieste: LINT, 1983, 345–72.

[130] Fioravanti, "Pico e l'ambiente ferrarese," 163. There is also information on astrology and astronomy at Ferrara in Noel M. Swerdlow, "Regiomontanus on the Critical Problems of Astronomy," in *Nature, Experiment and the Sciences*, T. H. Levere and W. R. Shea (eds.), Dordrecht: Kluwer, 1990, 165–95; he focuses primarily on Bianchini.

[131] Vescovini, "L'astrologia all'università di Ferrara nel Quattrocento," 293: "Come è noto l'ordinamento didattico dell'Università di Ferrara era ai suoi inizi, ancora quello dei secoli precedenti, sul modello di Bologna (l'*Alma Mater*) e della vicina Padova." The University of Ferrara was only solidly established in the middle of the fifteenth century.

[132] The Italian "nipote" is ambiguous. My thanks to Stefano Dall'Aglio for clearing up this confusion for me.

a century before Michele was born.[133] As at Bologna, the course in astrology at Ferrara was for 4 years.[134]

Introducing the debate over astrology at Ferrara that ensued after the publication of Pico's *Disputationes*,[135] Fioravanti first places it in the context of astrology's place within Aristotelian natural philosophy, in relation to the *De caelo*, *De generatione et corruptione* and *Meteorologica*. He focuses especially on the role of the *Meteorologica*, discussing intriguing evidence concerning its study at Ferrara in the 1490s which seems to bear directly on Pico's attack.[136]

*

I would like to conclude this all too brief section on astrology at the University of Ferrara with an interesting text from Pietro Buono Avogaro's annual prognostication for 1477, part of the preface dedicated to Ercole d'Este, duke of Ferrara:

> If everything were diligently gathered by reason (*ratio*), and if you were to blend [1] the positive and negative [sc. influences] of the stars, and if you were to join [2] the force of places (*locorum vis*) and [3] the powers (*potestates*) of signs and their parts [= degrees] in a balanced comparison, you will be able to trace out (*designare*) the entire person (*homo*) from his first to his most recent day, and his entire life and substance, with the easy unfolding of an explanation.[137]

[133] Vescovini, "L'astrologia all'università di Ferrara," 296: "Combattuta da alcuni indirizzi del sapere cristiano medievale, sopratutto nelle Facoltà teologiche, questa impostazione sarà invece accolta e rielaborata in modo organico ed unitario da quel medico, astrologo e filosofo padovano che può essere considerato il fondatore del sapere medico astrologico del Medievo latino fino alla fine del Rinascimento: Pietro d'Abano (c.1254–c.1315), il Conciliatore, *praeceptor meus,* come lo chiamerà Michele Savonarola. Pietro d'Abano è il caposcuola di una medicina astrologica a cui si richiamera espressamente anche Michele." In this light, it is not surprising that Niccolò Leoniceno takes Pietro as one of the major focuses of his attacks; see William F. Edwards, "Niccolò Leoniceno and the Origins of Humanist Discussions of Method," in *Philosophy and Humanism: Renaissance Essays in Honor of Paul Oskar Kristeller*, E. P. Mahoney (ed.), Leiden: Brill, 1976, 283–305. I hope to develop this line of research. See also T. Pesenti Marangon, "Michele Savonarola a Padova: L'ambiente, le opere, la cultura medica," *Quaderni per la Storia dell'Università di Padova* 9–10 (1976–77): 88–102, and now *Michele Savonarola: Medicina e cultura di corte.*

[134] Vescovini, "L'astrologia all'università di Ferrara," 296: "Nel curricolo dell'insegnamento dell'Università di Ferrara sappiamo che l'astrologia era insegnata da un lettore per un corso di quattro anni e che la domenica si insegnava l'astronomia."

[135] Fioravanti mentions one of the earliest responses, that of Pietro Buono Avogaro in the prologue to his annual prognostication, the *Judicium* of 1497; "Pico e l'ambiente Ferrarese," 166: "Sapiano questi tali impugnatori de la scientia de le stelle [...] che Aristotele nel primo de la Methaura e nel terzo De generatione con firme ratione conclude questo inferiore mondo per moto, lume et influentia essere dal cielo gubernato." We shall note that Pietro Buono here defends astrology with two of Aristotle's "charters of scientific astrology" identified by Lemay.

[136] Fioravanti, "Pico e l'ambiente ferrarese," 166: "In effetti i commenti ai *Meteorologica* sono per i filosofi naturali, a partire da Alberto Magno fino al Pomponazzi ed oltre, il luogo deputato a parlare, e se è il caso discutere, dell'astrologia." This too deserves fuller study.

[137] "Si omnia fuerint diligenter ratione collecta et si benivolam stellarum malivolamque mixturas, locorum vim et signorum et partium potestates equata comparatione sociaveris, totum hominem a primo usque ad novissimum diem et omnem eius vitam ac substantiam [...] facili definitionis

This professor of astrology at the University of Ferrara, where he taught for 40 years, certainly made major claims for the applications of his art. The conditional nature of the protases present the conditions of complete knowledge, which, for Pietro Buono, astrology could give in theory, a claim not normally made in the texts we have examined so far. I will also note that the importance of place as a causal factor is mentioned here.

10.15 Conclusions

We should also note in this context what appears to be a reasonably well established career trajectory from teaching mathematics-astronomy-astrology—or, alternately, natural philosophy—while a student of medicine, as a stepping stone on the path to the much more prestigious and lucrative position of professor of medicine, where one could do very well indeed as a university professor and also in private practice.[138] Thus, the study of mathematics-astronomy-astrology was not only a necessary conceptual background for the practice of medicine, it also provided career opportunities while pursuing one's training in a higher faculty.[139] According to Lines, these career patterns seem to have changed abruptly around the turn of the sixteenth century, as I will discuss in volume III.

*

We now have a much better sense of astrology's disciplinary configurations with mathematics, natural philosophy and medicine within the Aristotelian-Ptolemaic-Galenic map of knowledge, and its characteristic curricular instantiations and institutional locations in the finest premodern universities, primarily in Italy (but also elsewhere), through ca. 1500. With this background, we may now explore astrology's roots in and influences on society.

---

explicatione poteris designare." Fioravanti, "Pico e l'ambiente ferrarese," 163. The elipses are supplied by Fioravanti.

[138] For preliminary sketches of this career pattern, see, for the mathematical orientation, Siraisi, *Arts and Sciences*, 146; Lemay, "The Late Medieval Astrological School," n. 16; Westman, "Astronomer's Role," 116–21; Sachiko Kusukawa, "*Aspectio divinorum operum*: Melanchthon and Astrology for Lutheran Medics," in *Medicine and the Reformation*, O. P. Grell and A. Cunningham (eds.), London: Routledge, 1993, 33–56, *passim*. For natural philosophy, see now David Lines, "Natural Philosophy in Renaissance Italy," and Siraisi, *Medieval and Early Renaissance Medicine*, 69. For evidence which complicates this picture, see Girolamo Manfredi's career trajectory (above) where he goes back and forth in his teaching between medicine and astrology. This important subject requires further detailed study along the lines indicated by David Lines.

[139] Westman "Astronomer's Role," 117–18. See also the extensive essay by Nancy Siraisi and Anthony Grafton, "Between the Election and My Hopes: Girolamo Cardano and Medical Astrology," which challenges many accepted, rather simplistic views on the relationship between astrology and medicine by providing an in-depth treatment of a major figure deeply experienced in both; *Secrets of Nature*, 69–131.

To sum up, there were three primary disciplinary configurations and institutional locations for the teaching and study of astrology at the finest premodern universities: in the mathematical, the natural-philosophical, and the medical courses. These disciplinary patterns were strikingly similar at three major Italian universities—Bologna, Padua and Ferrara—and may thus be generalized for other universities, especially in Italy.[140] With this reconstruction, we are now in a much better position to understand how deeply rooted astrology was in the premodern map of knowledge, ca. 1250–1500, and thus evaluate how seriously astrology was taken by a significant percentage of the intellectual elite of Northern Italy from the beginning of the fourteenth through the end of the fifteenth centuries, and by extension throughout Europe, although this too needs to be more fully developed on the basis of detailed localized studies, such as those in Germany by Mentgen, France by Boudet, and England by Carey. These patterns will also provide the primary structures in relation to which we may measure continuities and transformations ca. 1450–1800 in volumes II and III. Now I will bring the story to life by placing astrology within the passionate pageant that life was in late Quattrocento Italy.

Appendix: David Juste's Description of Valdizocco's Astrological Manuscript[141]

*Vicenza, Biblioteca Civica Bertoliana, 208 (132)

- c. XV^2 (1467, cf. f. 90v; 1470, cf. f. 256r); or. : Northern Italy, copied by Bartolomeo Valdizocco of Padua (cf. f. 90v et 256r); prov.: cardinal Grimani de San Marco, Venice (f. IIIv).
- Paper, III+267 f., one hand. Astrology throughout, except for a table of adjoined material (IIv–IIIr); tables of houses, with canons, for the « fifth clima » (1r–7v), for Ferrara and Bologna (8r–13v) and for the « seventh clima » (14r–19v); Nicolaus de Comitibus, *De triplici motu octave spere* (258r–265v). Blank: Ir–IIr, IIIv (except for the ex-libris Grimani), 59v–60v, 84v, 256v–257v.

→ G. Mazzatinti, *Inventari dei manoscritti delle biblioteche d'Italia*, II, Forli, 1892, pp. 26–27 (n°132); T. Pesenti Marangon, « La miscellanea astrologica del prototipografo padovano Bartolomeo Valdizocco e la diffusione dei testi astrologici e medici tra i lettori padovani del '400 », in *Quaderni di Storia dell'Università di Padova*, 11, 1978, pp. 87–106; N.G. Marchioli, L. Granata, e.a., *I manoscritti medievali di Vicenza e provincia*, Firenze, 2007, pp. 59–60 (selective description), with reproduction of f. 213r, plate 97; Moulinier-Brogi 2011, pp. 238–240.

[140] A systematic study of all the evidence we have for the teaching of astrology in all three disciplines throughout all of Europe ca. 1250–1800 would be very valuable, and would surely sharpen and nuance the picture offered here. The basic structures, however, seem perfectly clear.

[141] I have translated his French descriptions into English. For a revised version with corrections, see his online entry (https://ptolemaeus.badw.de/ms/516).

20r–20v	Table « Pro almutaz » for each degree (20r) of Aries, Leo and Sagittarius, and (20v) of Taurus, Virgo and Capricorn. The other signs are missing.
21r–33r	« In siemdo (!) utrum eger liberetur aut non ab illa egritudine. Si tibi facta fuerit questio per ipsum infirmum aut per allium (!) — qui habet plures dignitates in domo illa. Finis » (with glosses citing numerous authorities f. 21r–22v, 28v et 33r).
33r–42r	Guiellmus Anglicus, *De urina non visa* [Azzolini 2013, 40].
42v	Table (12x7): « Tabulla signorum planetarum in corpore humano » (zodiacal melothesia for each planet in the 12 signs).
43r–43v	« Pro siemdo (!) intencionem querentis. Dixit Alkimdus (!): Quando volueris sire (!) intentionem querentis numera gradus qui sunt… Capitulum de tempore mortis interrogantis interrogacionibus (?) absollute quando moriatur (?). Si fuerit interrogatio absolluta et non determinari — si mobilia dies, si communia menses. Finis ».
43v–50r	Caciaguerra de Faventia, *Aphorismi* (« Incipiunt anforismi Cazaguerra medici de Faventia. Maxime considerabis si significator sexte domus sit in allico (!) gradu azemona — sed alliis (!) medicinis et curationibus cura secure etc. Finis »).
50r–52r	« Hic infra possite (!) sunt quedam regulle extravagantes et utiles in sientia (!) medicine extracte de libris sapientum. Si queratur fissicus (!) ire ad infirmum et ipse iverit (?) Luna separando se a coniunctione cum Sole — ut predictum est iudicandum erit. Finis ».
52r–53v	« De fortitudinibus planetarum. Fortitudines planetarum sunt hec ut sint in asspectibus bonorum planetarum — ab eodem loco. Et hec est regulla generallis. Finis ».
53v–54r	« De divissione (!) dierum et dominacione humorum. Dies naturalis dividitur in quatuor partes — et commendatum in hoc. Finis ».
54v	« Aries, Cancer, Libra, Capricornus sunt signa mobilia — et eorum particeps est Luna et est fleumatica. Finis ».
55r	« De planetis quando exaltantur et quando deprimuntur. Saturnus in Capricorno et in Aquario — (Cauda) deprimitur in Gemini. Finis ».
55v–56r	Two astrological tables: « Anni minores, medii, maiores » of the planets, and « Orbis luminis planetarum ».
56r	« De vemtis (!). Pars orientallis continet in se istos ventos — ut dicit Aristotilles in libro Metaurorum ».
56v–59r	« Incipit opuscullum Gerbis in medicinis et in omnibus. Et scito quod in omnibus interrogacionibus iudicia tribus modis currunt. Primo ut sias (!) pro quo venit ad te interrogator vel de quo interrogat — et sapientia qui sit omnia antequam fiant. Finis ».
59r	Incomplete horoscope.
61r–77v	Nicolaus de Paganica, *Compendium medicinalis astrologie* [Azzolini 2013, 40].
78r–84r	« Hic est liber quem collegit magister Romanus ex dictis philosophorum, astrologorum, medicorum ad cuiusdam infirmitatem per viam astrologie et vocatur liber pronosticationis mortis et vite. Capitulum primum. ationne (!) quidem sue compossitionis (!) cum hec inferiora alterationi — capitis passionnem susstinnet (!), et sic et de alliis domibus » (this text differs from Magister Romanus, *De XVI impedimentis in astronomia*).

85r–90v	« Incipiunt mansiones Lune secundum sententiam famosissimi viri magistri Perdocimi de Beldomandis [Prosdocimo de Beldomandi] de Padua. ichil prestancius in humano genere apud quemlibet — cavendum est tunc a cuiuslibet operis initio. Et sic est finis huius brevissimi tractatulli de elleccionibus secundum situm Lune in suis 28 mansionibus per Prosdocimum de Beldomandis Patavinum ab ellecionibus Indorum anno domini 1413 in castro Montagnane in Paduani dictrictus. Deo gracias. Amen. Finis. Finis per me Bartholameum de Valdezochio de Padua anno domini 1467, die 7 Februarii » [Azzolini 2013, 40].
91r–95r	Arnald of Villanova, *Introductorium ad iudicia astrologie quantum pertinet ad medicinam.*
95r–103v	Abraham Avenezra, *Liber luminarium* (tr. Pietro d'Abano).
103v–111r	Robert Grosseteste, *De impressionibus aeris* (attr. Haly: « Incipit libellus Ali de impressionibus — Libellus Haly de impressionibus finit »).
111v–115r	Alkindi (?), *Saturnus in Ariete sub radiis...* [Bos/Burnett, 458].
115r–120r	Messahallah, *Epistola de rebus eclipsium.*
120r–126v	Messahallah, *De mercibus.*
126v–136r	*Capitula Almansoris* [Azzolini 2013, 40].
136v–142v	Hermes, *Centiloquium* [*TCEM*, 32; Azzolini 2013, 40].
143r–143v	Hermes, *Liber de sex rerum principiis* (extract, § 325–339: « De tonitruis secundum Hermetem (title in margin). n quocumque signo sit tonitruum sive in die sive in nocte verum erit — homines infirmabuntur nec tamen multi morientur ») [*TCEM*, 21].
143v–148v	« Capitula Zaellis utillia in interrogacionibus. Cum coniuncta fuerit Luna alicui planete et perfecerit coniunctionem...; (144v) Capitulum Zaellis quando malus planeta significat bonum et prosperitatem. alus planeta cum fuerit orientallis...; (145v) Capitullum Zaellis de iudiciis. Quando planeta bonus est impeditus et debillis...; (147v) Capitulum Zaellis de malis planetis quando significant magnum malum et impedimentum. Mali significant difficultatem...; (148r) Capitulum Zaellis de significacione Lune et domini (?) ascendentis. Luna cum fuerit cursu vacua et nulli planetarum iuncta — erit minus forte eorum impedimentum ».
148v–154r	Bethen, *Centiloquium.*
154r–157v	« Sol-Luna: Hec dies vallet ad agendum negocia regna ad loquendum senibus ducibus et principibus...; (157r) Luna cum fuerit cum Capite vel Cauda in uno signo est dies cavenda in omnibus operibus — vel sextilli asspectu intellige. Finis » (elections after the aspects of the Moon with the other planets).
157v–164v	Gergis, *De significatione septem planetarum in domibus* (attr. Alcabitius: « Incipit liber Alcabicii de significatione planetarum in domibus 12. Capitulum primum de significacione Solis. Sol cum fuerit in ascendente significat principatum et sublimitatem — Explicit liber Alchabicii de significacionibus septem planetarum et Capitis et Caude Draconis in 12 domibus, a Ioanne Isspanno (!) interpretatus de Hebreo in Latinum »).
164v–168r	Alcabitius, *Tractatus de coniunctionibus planetarum* (« Incipit liber Alkabicii in revolucione annorum et in coniunctionibus planetarum in 12 signibus (!) a Ioanne Isspanensis translatus de Hebreo in Latinum. Si in quolibet anno que res care aut viles seu mediocres — Explicit liber Alcabicii in revolutione annorum mundi et de significacione coniunctionis planetarum in unoquoque signorum, interpretatus a Ioanne Isspanensi »).
168r–213r	Albumasar, *De revolutionibus annorum mundi* [Azzolini 2013, 40].
213r–256r	Ps.-Ptolemy, *Centiloquium* (« ... Finis. B V scripsit 1470, die 21 Septembris »).

266r	« Racio spere Pitagore quam Apulegius descripsit de qua re consullere volueris — in inferiorum numeris nigris morietur » (sphere of Apuleius/Pythagoras, with numerical alphabet and figure).
266v–267v	« Capitulum ad sciendum per artem geomancie utrum eger liberetur ab infirmitate quam habet vel non. Hec est ars consimilli sexdecim iudicum geomantice iudicacium de vita vel morte infirmum… (267r) Si quis te interrogaverit de aliqua re tunc tantum considera signum quod tunc ellevatur ab oriente — Si Pissis in omnia prosperabitur » (the final paragraph corresponds to the *Epistola Argafalau*, 4).

Chapter 11
Astrology in Society, Politics and Culture

As we have now seen in detail, the astrologizing natural knowledge reconstructed in Parts I–III may usefully be characterized as fundamentally Aristotelian natural philosophy within an integrated Ptolemaic cosmographic framework, comprised of mathematical astronomy integrated with mathematical geography, and employing a geometrical-optical model of celestial influences. What we call "astrology"—one of the two sister sciences of the stars (along with astronomy)—was studied, taught and practised within this mathematically articulated scientific framework, and often towards a medical end. As interesting as all this is—and as essential as it is for understanding major continuities and transformations in the history of science—it is only one of the many facets of astrology's historical interest, namely, astrology as a historian of science or philosophy would approach it. Historians of politics or art, for example, are primarily interested in other aspects of the subject.

In treating the history of astrology as part of the history of science—as well as more generally within the history of culture—it is thus extremely important to construct an explicit interpretive framework, precisely because astrology ramified so broadly and deeply within vast regions of premodern culture. Aristotle with Ptolemy along with their Arabic and Latin progeny provided the fundamental patterns of the scientific framework—the basic patterns of analysis and interpretation of the cosmos, nature and humankind—as reconstructed in Parts I–III, and located in its disciplinary and institutional contexts so far in Part IV. But there are other socio-political and cultural contexts in which these ideas were practised and perpetuated at various levels of society—from private astrological consultations with political leaders, for example, Regiomontanus's consulting work for cardinals, popes, and the king of Hungary,[1] and Kepler's, later, for the Holy Roman Emperor[2]—to published annual

[1] See Zinner, *Leben und Wirken*, and Shank, "Academic Consulting."

[2] See Barbara Bauer, "Die Rolle des Hofastrologen und Hofmathematicus als Fürstlicher Berater"; Katrin Bauer (no relation), "Johannes Kepler between two Emperors," in *Astrologers and their Clients in Medieval and Early Modern Europe*, and Caspar, *Kepler*. We saw Roger Bacon seeking

H D. Rutkin, *Sapientia Astrologica: Astrology, Magic and Natural Knowledge, ca. 1250–1800*, Archimedes 55, https://doi.org/10.1007/978-3-030-10779-6_11

prognostications, circulated far and wide that were a contracted part of a university professor of astrology's professional responsibilities, at least at the University of Bologna.[3] This chapter will explore a broad range of astrology's manifestations in society, politics and culture.[4]

11.1 Annual Prognostications and Astrological History: Great Conjunctions

Proferred in many cities throughout Europe, annual prognostications also articulated an astrological framework for understanding the broader patterns of history, namely, the theory of great conjunctions, which were given renewed credence in the work of Cardinal Pierre d'Ailly (1350–1420).[5] The basic patterns of historical astrology have already been described, so I will be very brief here.[6] It is an astronomical fact that the two outermost planets of the premodern planetary system, Jupiter and Saturn, conjoin every 20 years, that is, they meet each other at the same

---

such a role in the thirteenth century with both a pope and a king. We will see more examples in volume III.

[3] See Thorndike, *HMES*, IV, 455 ff. He provides examples of those at Bologna, Padua, Ferrara, and elsewhere. As we saw in Chap. 10, quite a few were published from the very beginning of printing. I explore one highly dramatic example below. A full study of these prognostications in their social and political contexts would be very interesting. A useful step in this direction for Italy is Elide Casali, *Le spie del cielo.* Capp's *English Almanacs* is the most successful. For an excellent orientation and splendid evocation, see Grafton, *Cardano's Cosmos*, ch. 3, "The Prognosticator." For the fifteenth century, Tur's MA thesis supercedes all of these studies; "À l'entrée du soleil en Bélier."

[4] In this chapter, I will piece my story together primarily from revealing scholarship in various scholarly fields. I will focus on Italy simply because it is the area I know best, and it will also provide the proper socio-political and cultural context for volume II. A comprehensive comparative treatment for all of Europe ca. 1100–1800 would be of great interest.

[5] Laura Smoller usefully discusses Roger Bacon's profound influence on d'Ailly's views of astrology, and, in particular, great conjunctions; *History, Prophecy and the Stars*, 52–57.

[6] See Smoller, *History, Prophecy and the Stars*, 20–22, and Garin, *Astrology of Life*, 14–25. The relationship of astrologically-based prognostications to other types of prophecy is yet another huge area for further research. Westman ("Copernicus and the Prognosticators") goes too far, I think, in distancing the annual prognostications from astrology. Although the astrologers did not usually provide foundational theorizing in their prognostications, they often used astrological argumentation to support their claims, as we will see below. There were other sorts of prognostications, however, which are more of a religio-prophetic than an astrological cast, for example, those of Johannes Lichtenberger, for whom see Dietrich Kurze, *Johannes Lichtenberger: Eine Studie zur Geschichte der Prophetie und Astrologie*, Lübeck: Matthiesen, 1960; Green, *Printing and Prophecy*, 39–55, and Wiebke Deimann, "Astrology in an Age of Transition: Johannes Lichtenberger and his Clients," *Astrologers and their Clients*, 83–104, and later Paracelsus. For an orientation to a major type of prophecy, see Reeves, *Influence of Prophecy.* See also Richard W. Southern, "Aspects of the European Tradition of Historical Writing: 3. History as Prophecy," *Royal Historical Association, Transactions*, fifth ser, 22 (1972): 159–80; and for the sixteenth century, Robin Barnes, *Prophecy and Gnosis: Apocalypticism in the Wake of the Lutheran Reformation*, Stanford: Stanford University Press, 1988. Once again, close attention to terminology and usage (both theirs and ours) well repays the effort.

degree of longitude on the zodiac. Or at least very close to the same degree, for a conjunction need not be exact to have an astrologically significant effect, since each planet has its own accepted sphere of influence, as discussed in my Excursus to the overall introduction. This conjunction is called a "great" conjunction *simpliciter*. These conjunctions would then continue to occur in a strikingly triangular pattern within the bounds of the same elemental triplicity for 240 years (i.e., twelve great conjunctions), at which time the conjunction would then switch to the next triplicity, say, from earth to air; this would mark a "greater" conjunction.[7] Finally, after 960 years, the conjunctions would return to the initial point of departure, thus marking a "greatest" conjunction (see Image 6.1).[8]

A great deal of ingenuity was expended during the Middle Ages and beyond to fit the major events of history to this model.[9] In particular, major transformations in politics and religion were associated with the greater and greatest conjunctions. This view of history was introduced by Messahalah in the late eighth and early ninth centuries; it was greatly promoted soon after by Albumasar.[10] Such an analysis was used to great effect by Cardinal d'Ailly to interpret the cataclysmic events of the Avignonese papacy (the so-called Babylonian captivity, 1309–1378), and the Great Schism (1378–1414), for whose resolution he was in good measure personally responsible, thus indicating an important area where astrology had a significant impact on society.

*

There were two distinct periods in d'Ailly's views about the Great Schism within the Church, which began September 20, 1378.[11] At first he considered the Schism as

[7] We will recall that there are four triplicities corresponding to the four elements, each embracing three signs of the zodiac: the "watery" triplicity is composed of the three "water" signs: Cancer, Scorpio and Pisces; the "fiery" triplicity is composed of the three "fire" signs: Aries, Leo and Sagitarrius, etc. See North, "Astrology and the Fortunes of Churches," and Germana Ernst, "From the Watery Trigon to the Fiery Trigon: Celestial Signs, Prophecies and History," in Astrologi Hallucinati," 265–80.

[8] The illustration in Kepler's *Harmonices mundi* is a clear visual expression of this view; *The Harmony of the World*, E. J. Aiton, A. M. Duncan and J. V. Field (trs), Philadelphia: American Philosophical Society, 1997.

[9] See Friedrich von Bezold, "Astrologische Geschichtskonstruktion im Mittelalter," in his *Aus Mittelalter und Renaissance: Kulturgeschichtliche Studien*, Munich, 1918, 165–95, and now Hasse's valuable study of great conjunctions in *Success and Suppression*, 272–89. We saw Roger Bacon's attempts in Chap. 6.

[10] See Smoller, *History, Prophecy and the Stars*, and David Pingree, "Astrology," in *The Cambridge History of Arabic Literature: Religion, Learning and Science in the 'Abbasid Period*, M.J.L. Young, J.D. Latham, and R.B. Sarjeant (eds.), Cambridge: Cambridge University Press, 1990, 290–300, 294 and 297 ff., and his *From Astral Omens to Astrology*, and Gutas, *Greek Thought, Arabic Culture*. For Albumasar's fundamental text on this subject, see Abu Ma 'Sar, *On Historical Astrology*.

[11] The following account is based on Smoller, *History, Prophecy and the Stars*, 95–111, to which the page numbers in parentheses refer.

the immediate preamble to the imminent arrival of Antichrist and apocalypse. His main sources for this view—which he held through ca. 1400, when he thought it would actually take place—were mainly latter-day prophets, primarily Hildegard of Bingen, Joachim of Fiore and Arnald of Villanova, but also the Book of Revelations. Moreover, during this period, "he learned of astrological attempts to calculate the time of apocalypse, although in these years he rejected any such prognostication (101)." His ideas continued to develop:

> In subsequent years, however, d'Ailly revised his early interpretation of events. He began to hope that human efforts, with God's assistance, could end the Schism. With the church healed and reformed, he believed, God would withdraw the torments he had prepared for the earth, and the apocalypse would be postponed. D'Ailly's later examinations of the apocalypse bear another striking difference from his early works. Whereas in the 1380s he had denied astrology's ability to predict the End, he now looked to the stars to forecast the arrival of Antichrist. Astrological calculations confirmed his new hopeful interpretation of the Schism, for they put the fiend's advent in the distant future (102).

Thus d'Ailly came to believe that a reform of the church could postpone the apocalypse, and that insight into its timing could be gleaned through astrology.

Traces of this view occur as early as 1403, but it was in a series of works dated to 1410–14 that d'Ailly came to his mature understanding. These views were deeply influenced by astrological theory, including the theory of great conjunctions, which he learned, in particular, from Roger Bacon's writings. In the *Concordantia astronomie cum hystorica narratione* (finished May 10, 1414), "he openly offered an astrological conjecture for the timing of the apocalypse (104)."

> Having introduced the topic of Antichrist's reign, d'Ailly went on to examine this question by means of astrology, following the theories outlined by Roger Bacon. D'Ailly noted a number of astrological phenomena pointing to the importance of 1789. A greatest conjunction of Saturn and Jupiter, of the sort that happens once every 960 years, would occur in the year 1692. After this conjunction, another astrologically significant period would end with the completion of ten trips through the zodiac by the planet Saturn in 1789. Further, from the years 1764 to 1789, the eighth sphere would "stand still." Based on these observations, d'Ailly concluded, "if the world shall last until that time, which only God knows, then there will be many great and marvellous alterations and changes in the world, and chiefly with respect to laws and sects [i.e. religions]." Whence, he continued, we can conclude with due probability that around that time will arrive Antichrist and "his damnable law or sect."
>
> Unlike so many of d'Ailly's astrological writings, this passage does not appear to be a borrowing from another author. Rather, the cardinal himself seems to have derived the date of 1789 from his own list of greatest conjunctions and revolutions of Saturn. His choice of date rested on sound theory. He incorporated in his prediction the three most important signifiers of religious change according to Albumasar (Abu Ma'shar): the greatest conjunctions, occurring once every 960 years; Saturn's completion of ten revolutions, every 300 years; and the period of the eighth sphere's *accessus* and *recessus*. D'Ailly must have been struck by the coincidence of the last two of these three celestial phenomena in 1789. Albumasar had taught that great changes were likely when the eighth sphere's completion of its access and recess went along with a shift in signs for Saturn (105–6).[12]

[12] It was precisely this apparent subordination of religion to the heavenly motions in historical astrology that Pico attacked with his characteristic vehemence in *Disputationes*, Book V, where he attacked d'Ailly's theories explicitly and in detail. We should note that d'Ailly's relevant works

We can see that Cardinal d'Ailly took Roger Bacon's views about astrology's utility for religion very seriously indeed, and that he put them to good use. As it turns out, choosing 1789 as a year of profound political and religious change is not entirely without its merits, especially in France!

Annual prognostications are to be understood in this broader context, especially since 1484 was also considered a year of profound transformation, with a great conjunction in the fixed water sign Scorpio.[13] Likewise, the great conjunction of 1524 in the mutable water sign Pisces provoked tremendous anticipation concerning a second flood of biblical proportions.[14] Another area where great conjunctions were used to understand tremendous changes in the world arose in relation to two profound and apparently apocalyptic public health crises, namely, the Black Death of 1347–8 (and after) and the Great Pox, the arrival of syphilis in Europe ca. 1492. Both were understood to be caused by great conjunctions of the three superior planets: Mars, Jupiter and Saturn.[15]

---

received their *Editiones principes* in ca. 1480–82 and 1490, soon before Pico wrote. (1) *Imago mundi et tractatus alii*, Louvain: Johannes de Westfalia, about 1480–82 (ISTC ia00477000); (2) *Concordantia astronomiae cum theologia*, ed. Johannes Angeli, Augsburg: Ratdolt, 2 Jan 1490 (ISTC ia00471000). A full discussion of their influence, especially in the fifteenth and sixteenth centuries, would be of great interest. Smoller, unfortunately, does not discuss this major issue beyond the case of Christopher Columbus.

[13] Anthony Grafton evokes this splendidly in "Giovanni Pico della Mirandola: Trials and Triumphs of an Omnivore," in his *Commerce with the Classics: Ancient Books and Their Renaissance Readers*, Ann Arbor: University of Michigan Press, 1997, 93–134; see also Weinstein, *Savonarola and Florence*, and Aby Warburg's classic treatment, where he discusses in detail the efforts to date Martin Luther's birth to 1484; "Pagan Antique Prophecy in Words and Images in the Age of Luther," in his *The Renewal of Pagan Antiquity*, D. Britt (tr), Los Angeles: Getty Center for the History of Art and the Humanities, 1999, 597–697. See also my "Teaching Astrology in the sixteenth Century: Giuliano Ristori and Filippo Fantoni on Pseudo-Prophets and Other Effects of Great Conjunctions." Michael J.B. Allen discusses Ficino's views of 1484 as marking the beginnings of a Golden Age; *Synoptic Art: Marsilio Ficino on the History of Platonic Interpretation*, Florence: Olschki, 1998. See also Claudia Rousseau, "Cosimo I de Medici and Astrology: The Symbolism of Prophecy," PhD Thesis, Columbia University, 1983, and Janet Cox Rearick, *Dynasty and Destiny in Medici Art: Pontormo, Leo X, and the Two Cosimos*, Princeton: Princeton University Press, 1984. Paul of Middelburg's prognostication for 1484 was very influential in this context, especially after it had been adapted by Johannes Lichtenberger. On Paul and his prognostication, see Heilen, "Paul of Middelburg's *Prognosticum* for the Years 1484–1504."

[14] There has been a good deal of scholarly literature on the 1524 conjunction; see especially, "*Astrologi Hallucinati*"; Talkenberger, *Sintflut*, and Niccoli, *Prophecy and People.* Of course, the epoch-making religious and political events of 1517 and their aftermath, including the skillful use of the press, played a major role as well. See now also Barnes, *Astrology and Reformation.*

[15] For astrological analyses of the plague, see (e.g.) Chase, "Fevers, Poisons, and Apostemes"; and Campbell, *Black Death and Men of Learning*. For astrology and the Great Pox (i.e. the French disease according to the Italians, or the Italian disease according to the French!), see Arrizabalaga, Henderson and French, *The Great Pox.* I treat astrological analyses of plagues more fully and specifically in relation to Girolamo Manfredi and Marsilio Ficino concerning the pandemic of 1478–79 (with *spiritus* and talismans) in volume II. For a revised interpretation of the Black Death, see Samuel K. Cohn, *The Black Death Transformed: Disease and Culture in Early Renaissance Europe*, London: Arnold, 2002.

11.2 Astrology, Humanism and Chivalry in Courtly Culture

Moreover, in addition to annual prognostications and great conjunctions, astrology had other equally significant social, political and cultural expressions, mainly in the secular (princely and republican) and ecclesiastical courts: in Italy, to be sure, but also throughout Europe.[16] Astrology played major roles in the political life of the Italian courts, especially in the closely interconnected courts of Northern Italy, including Ferrara, Bologna, Mantua, Urbino, Rimini, Florence and Milan, as well as Naples and Rome to the south.[17] These are also the major courts where humanism took root and flourished: Guarino Veronese and his son Battista established their school at Ferrara, and Vittorino da Feltre established his in Mantua. Later, Angelo Poliziano taught at the studio in Florence, to name only the most illustrious.[18] Further, at the very same time, chivalric knightly culture on the Anglo-French-Burgundian model was also in full flower in Italy.[19] A splendid pictorial

[16] See e.g. Carey, *Courting Disaster*, for England; Boudet's extensive treatment of astrology at the French royal court, *Le* Recueil des plus celebres astrologues, II, 240–291, ch. III, B: "Astrologie et politique: le cas du royaume de France"; and Gerd Mentgen, *Astrologie und Öffentlichkeit*, part two, for Germany. The cultural dimension is also significant for fully assessing continuities and transformations concerning astrology's range of influential historical roles.

[17] Boudet, *Le* Recueil des plus celebres astrologues, II, 238: "La prépondérance de l'Italie et de l'Empire en matière astrologique au XVe siècle s'explique à la fois par des raisons intellectuelles, institutionelles et politiques. La critique de l'aristotélisme y a été menée plus loin qu'ailleurs et l'integration de l'étude de l'astronomie-astrologie à l'enseignement universitaire semble y avoir été beaucoup plus complète qu'en France. Le foisonnement intellectuel et éditorial, la sympathie évidente de certains papes comme Sixte IV et Alexandre VI et de multiples princes—Médicis, Visconti, Sforza, Este, rois aragonais de Naples, Habsbourg, sans parler de Mathias Corvin—, les rivalités multiples entre les cités de la péninsule et les villes impériales ont abouti à une sorte de surenchères astrologique, dans laquelle les professeurs d'arts, d'astronomie, de philosophie et de médicine d'un grand nombre d'universitès ont joué les premiers rôles." See also Caroti, *L'astrologia in Italia*, for all of Italy, and Signorini for Mantua, Azzolini for Milan, and Rousseau and Cox-Rearick for Florence.

[18] See (e.g.) Anthony Grafton and Lisa Jardine, *From Humanism to the Humanities; Education and the Liberal Arts in Fifteenth- and Sixteenth-Century Europe*, Cambridge, MA: Harvard University Press, 1986; W.H. Woodward, *Vittorino da Feltre and Other Humanist Educators*, Cambridge: Cambridge University Press, 1897, rep. N.Y., 1963; and Peter Godman, *From Poliziano to Machiavelli: Florentine Humanism in the High Renaissance*, Princeton: Princeton University Press, 1998.

[19] See (e.g.) Thomas Tuohy on Ferrara (*Herculean Ferrara: Ercole d'Este, 1471–1505, and the Invention of a Ducal Capital,* Cambridge: Cambridge University Press, 1996); Joanna Woods-Marsden on Mantua (*The Gonzaga of Mantua and Pisanello's Arthurian Frescoes,* Princeton: Princeton University Press, 1988); and Lucia Ricciardi on Florence (*Col senno, col tesoro e colla lancia: riti e giochi cavallereschi nella Firenze del magnifico Lorenzo*, Florence: Le Lettere, 1992). A passage from Ricciardi is appropriate here: "Molti studiosi hanno insistito—e oggi tornano a insistere—sulla 'rottura' fra medioevo ed età moderna, magari assumendo l'umanesimo fiorentino come svolta nella quale l'Occidente ha intrapreso la strada di un mondo rinovato. Nelle sue linee di fondo questo e un attegiamento e schematico e sorpassato: al quale del resto si è tentato di rispondere con altri schemi ormai invecchiati anch'essi, ma capaci di darci l'illusione di comprendere un processo storico che, altrimenti, ci disorienterebbe per la sua complessità. Alla luce di

representation is the duke of Urbino, Federico da Montefeltro, kneeling in prayer, dressed in a full suit of knightly armor.[20] Federico also had one of the premier libraries of the day, which included an excellent collection of astrological texts.[21] Likewise, astrology played an important part in ecclesiastical court life, although this part of the subject does not seem to have attracted as much scholarly attention, at least for the late Quattrocento.[22]

Recent scholarship indicates clearly that chivalry, humanism and astrology—each in their individual and interrelated domains: academic, political and cultural—

questi schemi di comodo il Quattrocento fiorentino e apparso come il momento de trapasso dal medioevo all'età moderna: tempo di letterati e di artisti di genio, di riscoperta dell'antico, di decollo dell'economia e della mentalità 'borghese', esso si sarebbe scrollato di dosso quella civiltà 'gotica' che altrimenti sarebbe perdurata a lungo non solo oltralpe ma anche nelle corti principesche della penisola.

Se guardiamo pero con attenzione al clima culturale dell'eta laurenziana, scopriamo qualcosa che gli studiosi ben conoscono ma che il resto di un pubblico magari colto e amante delle letture storiche ignora. Scopriamo un turbinar di cavalli, un volteggiar di cavalieri, un rutilar di insegne araldiche: un medioevo 'gotico' nel cuore della Firenze umanistica (8)." For more on chivalry in Florence, see also Richard C. Trexler, *Public Life in Renaissance Florence*, Ithaca, NY: Cornell University Press, 1980.

[20] See Alison Cole, *Virtue and Magnificence: Art of the Italian Renaissance Courts*, New York: Abrams, 1995, 86 (pl. 63). Another striking image is of Galeazzo Maria Sforza ca. 1477 in full armor kneeling in prayer with knights in the field behind him; see Alison Wright, "A Portrait for the Sforza Visit," in *Lorenzo the Magnificent: Culture and Politics*, M. Mallett and N. Mann (eds.), London: Warburg Institute, 1996, 65–92, 85, n. 56 (fig. 5). Wright usefully discusses the contemporary chivalric associations and relations between Milan and Florence, esp. 68–75. For a magnificent painting of Federico reading a book while dressed in armor along with his son, see the beautifully illustrated *Federico da Montefeltro and his Library,* M. Simonetta (ed.), Vatican City: Y. Press, 2007. The painting, "Double Portrait of Federico da Montefeltro and his Son Guidobaldo," is reproduced and discussed at 102–109.

[21] See Cecil H. Clough, "The Library of the Dukes of Urbino," *Librarium. Revue de la Société des Bibliophiles Suisses* IX (1966): 101–105, 192, which is now conveniently reprinted in his *The Duchy of Urbino in the Renaissance,* London: Variorum, 1981, with the same pagination. The collection is now at the Biblioteca Apostolica Vaticana; C. Stornajolo, *Codices Urbinates Latini,* 3 vols, Rome: Vatican Press, 1902–21 and ibid., *Codices Urbinates Graeci*, Rome: Vatican Press, 1895. See also Patrizia Castelli, "Gli astri e i Montefeltro," *Res Publica Litterarum: Studies in the Classical Tradition* 6 (1983): 75–89. Guidobaldo was one of Vespasiano da Bisticci's preferred customers (M. Giussepe and B. Cagni, *Vespasiano da Bisticci e il suo epistolario*, Rome: Edizioni di Storia e Letteratura, 1969), and a correspondent of Ficino (Enrico Gamba, "Astrologi alla corte dei Montefeltro a Urbino e degli Sforza a Pesaro," *Rivista della societa pesarese di studi storici* 1: 1991, 75–83, 76; my thanks to Domenico Bertoloni Meli for an offprint of this article).

[22] One should begin, of course, with Thorndike's indices in the relevant volumes of *HMES*. The various papal courts have received more attention in this respect than cardinals' courts have. There is relevant information in David S. Chamber's book on Cardinal Francesco Gonzaga; *A Renaissance Cardinal and His Worldly Goods: The Will and Inventory of Francesco Gonzaga (1444–1483)*, London: Warburg Institute, 1992. Explicitly following Pico, Paolo Cortesi rejected astrology from a cardinal's court in his 1510, *De cardinalatu*; see, John F. d'Amico, "*Contra divinationem*: Paolo Cortesi's Attack on Astrology," in his *Roman and German Humanism, 1450–1550*, P. Grendler (ed.), Aldershot: Variorum, 1993, 281–291 (originally published, 1985). Astrological features of Leo X's papacy have received more attention, especially by art historians; see (eg) Cox-Rearick's and Rousseau's relevant works already cited.

provided defining elements of these Renaissance courts.[23] Ferrara and Florence are two major courts where all three of these components were fully developed and, fortunately, most fully studied. Lorenzo de' Medici's coming of age *giostra* in 1469—and his brother Giuliano's of 1475—illustrate this configuration splendidly. The *giostra* itself is, after all, a joust, one of the most characteristic of chivalric activities.[24] Furthermore, one of the major poets for the 1475 occasion was, of course, Angelo Poliziano, one of the premier Renaissance humanists.[25] Finally, Lorenzo's helmet for his *giostra*, as depicted in a contemporary engraving (1470), carries a representation of his sun sign, Capricorn, and the ruling planet, Mars, of his ascendent, Scorpio.[26]

No simple patterns emerge from this rich cultural nexus. Sometimes diametrically opposing tendencies arose within the same groups (as with humanist views of astrology); more often, however, there are rich variegations of color and texture, with complex patterns individuated by their particular combinations of elements.[27] Thus, following Aby Warburg's lead, for astrology's cultural history to be written effectively, it must be acutely attuned to the details of time and place, so that a nuanced appreciation of each cultural moment—including the rich variety of individual views (with their own continuities and transformations, and developments over time)—may emerge, and only then be compared with others and set into their own properly differentiated contexts.[28]

[23] Mary Hollingsworth clearly articulates the constellation of chivalry and humanism in these courts; *Patronage in Renaissance Italy: From 1400 to the Early Sixteenth Century*, London: John Murray, 1994, 157–9.

[24] See Tuohy, *Herculean Ferrara*, ch. 8.

[25] *Stanze per la giostra: Orfeo, rime: con un'appendice di prose volgari*, Novara: Istituto Geografico De Agostino, 1969. Another major poet, Luigi Pulci, author of the chivalric epic, *Morgante*, commemorated Lorenzo's giostra.

[26] The image is reproduced in Gioia Mori, *Arte e astrologia*, Florence: Giunti, 1987, 26. More generally, see Ricciardi, *Riti e giochi cavallereschi*, 166–74, and Paola Ventrone, *Le tems revient: feste e spettacoli nella Firenze di Lorenzo il Magnifico*, Cinisello Balsamo: Silvana, 1992.

[27] Garin (*Astrology in the Renaissance*, 24–5) expresses the multiplicity well: "Astrology and religion, astrology and politics, astrology and propoganda, but also astrology and medicine, and astrology and science: a philosophy of history, a conception of reality, a fatalistic naturalism and an astral cult—astrology was all this and more." One must use the English translation with caution; see Michael J. B. Allen's review, *Renaissance Quarterly* 36 (1983): 577–80.

[28] For insightful analyses of Aby Warburg and his method, see in particular Carlo Ginzburg, *Clues, Myths and the Historical Method*, J. and A. C. Tedeschi (trs), Baltimore: Johns Hopkins University Press, 1989, ch. 1, and Ernst Gombrich, *Aby Warburg: An Intellectual Biography*, 2nd ed., Chicago: University of Chicago Press, 1986. At long last, there is now a splendid English translation of the idiosyncratic German of all of Warburg's works by David Britt, *The Renewal of Pagan Antiquity*, with an extensive interpretive introduction by Kurt W. Forster.

11.3 Astrology in Visual and Literary Culture

One of the most attractive and revealing areas where astrology was woven deeply into this courtly cultural fabric is the realm of artistic expression. As everyone knows, the art produced in and for these courts is one of the greatest contributions of the Italian Renaissance (however defined) to world civilization.[29] As it turns out, a significant percentage of that art is informed, in various ways, with astrological imagery: (*inter alia*), in the early fourteenth century, the frescoes and ceiling paintings in the Sala della Ragione in Padua;[30] in the fifteenth century, the Palazzo Schifanoia in Ferrara;[31] in the early sixteenth century, the Villa Farnesina in Rome,[32] and certain Medici buildings in Florence and thereabouts;[33] and, in the seventeenth century, the Palazzo Barberini in Rome.[34] Dieter Blume's monograph, *Regenten des*

---

[29] Alison Cole, *Virtue and Magnificence*, provides a fine orientation for the Northern Italian courts. Loren Partridge does the same for Rome in the same series; *The Art of Renaissance Rome, 1400–1600*, New York: Abrams, 1996.

[30] Pietro d'Abano's works provided the encyclopedic program that Giotto painted; see Dieter Blume, *Regenten des Himmels: Astrologische Bilder in Mittelalter und Renaissance,* Berlin: Akademie Verlag GmBH, 2000, 70–85; Mori, *Arte e astrologia*, and Seznec, *Survival of the Pagan Gods*.

[31] See (e.g.) Warburg, "Italian Art and International Astrology," and Marco Bertozzi, *La tirannia degli astri: gli affreschi astrologici di Palazzo Schifanoia*, Livorno: Sillabe, 1999.

[32] Mary Quinlan-McGrath, "The Astrological Vault of the Villa Farnesina: Agostino Chigi's Rising Sign," *Journal of the Warburg and Courtauld Institutes*, 47 (1984): 91–105, and her "The Villa Farnesina: Time-Telling Conventions and Renaissance Astrological Practice," *Journal of the Warburg and Courtauld Institutes*, 58 (1995): 52–71.

[33] Rousseau, "Cosimo I de Medici and Astrology," and Cox Rearick, *Dynasty and Destiny in Medici Art*.

[34] John Beldon Scott, *Images of Nepotism: The Painted Ceilings of Palazzo Barberini*, Princeton: Princeton University Press, 1991. On this topic in general, see the splendidly illustrated *Arte e Astrologia* by Gioia Mori. A quotation from her introduction is apt here (6): "L'arte, specchio di cultura, riflette questa ramificata penetrazione della sapienza astrologica nel pensiero occidentale, con una proliferazione di immagini collocate nelle chiese, nei palazzi di corte, in quelli pontifici.

Dal Medioevo al Rinascimento venne realizzata in Italia una notevole serie di cicli astrologici, con caraterristiche tipologiche differenziate: dalle strutture a calendario (con l'associazione dei segni zodiacali a ogni mese dell'anno) alle esposizioni degli influssi dei Pianeti e dei segni sulle attività dell'uomo; dalle mappe celesti, alle raffigurazioni delle origini mitologiche delle costellazioni.

La complessità maggiore si riscontra nei cicli in cui è raffigurato un oroscopo: si tratta della particolare situazione del cielo in corrispondenza di eventi importanti o al momento della nascita (la "genitura") del committente.

I potenti del Rinascimento, come gli imperatori romani, affermavano l'inevitabilità della loro gloria attraverso il tema della propria genitura. Cosimo il Vecchio fece affrescare un oroscopo nella Sagrestia Vecchia di San Lorenzo; e dopo di lui tutta la famiglia de' Medici si interesso di astrologia e affido una lunga serie di cicli a grandi artisti: da Luca della Robbia a Vasari, al Bronzino.

Il potente banchiere senese Agostino Chigi fece realizzare il proprio oroscopo alla Farnesina da Baldassare Peruzzi, affidando poi a Rafaello la decorazione con tema astrologico della propria cappella funebre. Lo stesso Rafaello dipinse, su richiesta di Giulio II, la situazione del cielo al momento della sua elezione al papato, nella Stanza della Segnatura in Vaticano.

Himmels: Astrologische Bilder in Mittelalter und Renaissance is a valuable synthetic contribution to our knowledge.[35]

There are several types of artistic evidence: from large-scale painting and fresco cycles to elaborate deluxe manuscript miniatures (like the exquisitely illuminated *De sphera* in the Biblioteca Estense and in many Books of Hours),[36] as well as mass produced wood-block prints and engravings used to illustrate calendars, almanacs, broadsheets, books and pamphlets.[37] There are two main types of painting and fresco, mainly ceiling cycles. The first is general and encyclopedic, with visual representations of the planets, the signs they rule, and the sorts of earthly activities they inform, for example, the Sala della Ragione in Padua (a sort of visual astrological *summa*) and the Palazzo Schifanoia in Ferrara.[38] These images are also encountered in miniatures and in mass produced ephemeral literature.[39]

The second major type of astrological imagery is much more particular, illustrating the celestial configuration at the moment of major events, for example, the election or coronation of a pope, or for the moment of a powerful person's birth, their nativity. The Sagrestia Vecchia in Florence, which commemorates the Council of Florence, is an example of the former.[40] The Villa Farnesina in Rome, in which Baldassare Peruzzi exquisitely represents the heavens at the birth of his patron and its owner, the Sienese banker and modern day Maecenas, Agostino Chigi, is a splendid example of the latter.[41]

Le decorazioni degli appartamenti vaticani riflettono in modo inequivocabile la complessa realtà culturale dell'epoca: negli appartamenti di Alessandro VI Borgia, Pinturicchio illustro gli influssi dei Pianetti sulle attività e il carratere dei loro "figli"; nella Sala dei Pontefici, Giovanni da Udine e Perin del Vaga dipinsero su commissione di Leone X Medici le immagini dei papi accanto a quelle delle costellazioni[.]"

[35] See also his article, "Picturing the Stars: Astrological Imagery in the Latin West, 1100–1550," in *A Companion to Astrology in the Renaissance*, B. Dooley (ed.), Leiden: Brill, 333–98.

[36] See (e.g.) Mori, *Arte e astrologia*, and the splendidly illustrated, *Astrologia: arte e cultura in età rinascimentale—Art and Culture in the Renaissance*, D. Bini (ed.), Modena: Bulino, 1996. A bilingual text is provided, but the English would have greatly benefited from proofreading by a native English speaker.

[37] Warburg uses a broad range of visual evidence in his essays; see *The Renewal of Pagan Antiquity*, *passim*. Casali, *Le spie del cielo*, and Caroti, *L'astrologia in Italia*, use a number of such illustrations.

[38] Mori, "Arte e astrologia," 18–21, 27–31.

[39] The prints of the so-called Baccio Baldini from a calendar ca. 1464 also fall into this category; see Warburg, "Italian Art and International Astrology"; Ventrone, *Le Tems Revient*, and Marguerite Brown, "Visioning the Planets in Fifteenth Century Florentine Printmaking," *Visual Pursuits: Prints from the British Museum (Uncategorized)*, posted June 26, 2014 (http://visualpursuits.org/2014/06/26/visioning-the-planets-in-15th-century-florentine-printmaking/).

[40] Mori, *Arte e astrologia*, 24, states that this is the first such example. Another example is that of the coronation of Julius II by Raphael; Mori, *Arte e astrologia*, 34. John Beldon Scott (*Images of Nepotism*) has argued that Pietro da Cortona's fresco in the Palazzo Barberini, a seventeenth-century example, represents the moment of the election of Maffeo Barberini to the papacy as Urban VIII.

[41] See esp. Mary Quinlan-McGrath, "The Astrological Vault of the Villa Farnesina," and "The Villa Farnesina: Time-Telling Conventions and Renaissance Astrological Practice." For a different

*

There has also been excellent scholarship on some of astrology's places in contemporary literary culture, especially in the fields of poetry,[42] historiography,[43] critical philology,[44] and philosophy.[45] The literary evidence is varied, coming from personal and official correspondence, but also from poetry and historiography, and publications of a more ephemeral nature. This evidence also reveals that there are several kinds of astrological praxis with strong political reverberations. First are the annual prognostications themselves. Second is the use of election horoscopes to determine the best time for the beginning of significant endeavors, such as military operations,[46]

---

analyis, see Kristin Lippincott, "Two Astrological Ceilings Reconsidered: The *Sala di Galatea* in the Villa Farnesina and the *Sala dei Mappamundo* at Caprarola," *JWCI* 53 (1990): 185–96. For a broad and variegated portrait of the rich culture in Rome at the very end of the fifteenth and the first decades of the sixteenth centuries, and Chigi's central role in it, see Ingrid D. Rowland, *The Culture of the High Renaissance: Ancients and Moderns in Sixteenth-Century Rome*, Cambridge: Cambridge University Press, 1998. At the beginning of her useful bibliography, Gioia Mori (*Arte e Astrologia*, 66) notes: "Non esiste un testo in cui siano catalogati tutti i cicli astrologici italiani: sono dunque qui citati i contributi più importanti relativi ad alcuni cicli." Dieter Blume's *Regenten des Himmels* has gone a long way towards fulfilling this desideratum.

42 See e.g. Benedetto Soldati's classic study, *La poesia astrologica nel 400,* Florence: Le Lettere, 1986, with a new introduction by Cesare Vasoli (originally published 1906); Lorenzo Buonincontri, *De rebus naturalibus et divinis*, Stephan Heilen (ed.), Stuttgart: Teubner, 1999, with introduction and philological commentary, and Matteo Soranzo, *Poetry and Identity in Quattrocento Naples*, Farnham: Ashgate, 2014, on Giovanni Gioviano Pontano.

43 In general, see Bezold, "Astrologische Geschichtskunstruktion." On the use of historical astrology by the Villani in Florence, see Nicolai Rubinstein, "Some Ideas on Municipal Progress and Decline in the Italy of the Communes," in *Fritz Saxl 1890–1948: A Volume of Memorial Essays from His Friends in England*, D. J. Gordon (ed.), London: T. Nelson, 1957, 165–183, 178–83, and Lewis Green, *Chronicle into History: An Essay on the Interpretation of History in Florentine Fourteenth-Century Chronicles*, Cambridge: Cambridge University Press, 1972. For Giovanni Villani's use of astrology relating to personal character, see Viktor Stegemann, "Giovanni Villanis historische Charakterbilder und die astrologischen Texte der planetarischen Anthropologie: Ein Beitrag zur geistigen Physiognomie der Frührenaissance," in *Lebenskräfte in der abendländischen Geistesgeschichte: Festschrift Walter Goetz*, B. Bischoff et al. (eds.), Marburg: Simons, 1948, 125–98; for an earlier period, on Rolandino of Padua, see Siraisi, *Arts and Sciences at Padua.*

44 For Pico's critical philological work on the history of astrology, see in particular Grafton, "Giovanni Pico della Mirandola: Trials and Triumphs." For a later period, concerning Joseph Scaliger's work on Manilius, see Chap. 7 of Grafton's *Joseph Scaliger* I, "Scaliger's Manilius: From Philology to Cultural History," 180–226. For Kepler, see Grafton, "Humanism and Science in Rudolphine Prague."

45 For many of the astrological features of Marsilio Ficino's work on Plato and the Neoplatonists, see, in particular, Michael J. B. Allen's many erudite studies, including, *Nuptial Arithmetic: Marsilio Ficino's Commentary on the Fatal Number in Book VIII of Plato's* Republic, Berkeley: University of California Press, 1994, and *Synoptic Art.* See also the valuable but problematic introduction to Kaske and Clark's edition and translation of Ficino's *De vita.* I develop this picture more fully in volume II.

46 See Eugenio Casanova, "L'astrologia e la consegna del bastone al capitano generale della repubblica fiorentina," *Archivio Storico Italiano*, ser. 5, VIII, 1891, 134–44, and Trexler, *Public Life in*

or the coronation of a pope (or king).[47] Election horoscopes were also cast to time the beginning of significant building projects, including the placement of their cornerstones, for example, Julius II's rebuilding of St. Peters.[48]

As we will see, other types of astrological practice could also have significant political implications, for example, the medical interrogation to be discussed presently. Much of our information on these practices comes from official correspondence found in various Italian archives.[49] I will treat some of these features of the astrological landscape in what follows, where I will articulate in greater detail some of the contexts necessary for understanding astrology's broadly encompassing and deeply rooted locations in society and culture.[50]

11.4 Knowledge and Power: Astrology, Politics and Culture in Late-Quattrocento Northern Italian Courts

As we can now see more clearly, practical astrology offered four primary ways of acting in and on society by means of its four canonical practices: general astrology or revolutions, nativities, interrogations and elections. All of the astrologer's training and experience was put into practice here, often in very public high-stakes intensely-competitive circumstances. We are now entering into a world in many ways vastly different from our own.[51] The rest of this chapter gathers a selection of

---

Renaissance Florence. The astrological timing for the ritual handing of the baton of command to a new captain general of the Florentine army is discussed in detail below.

[47] Julius II apparently postponed his coronation several times on the advice of his astrologers; see Mori, *Arte e astrologia*, 34: "L'elezione al soglio pontificio di Giulio II avvenne il 31 ottobre 1503, poche ore dopo il tramonto, e la sua incoronazione fu più volte rimandata perche, secondo gli astrologi di corte, il momento non era propizio."

[48] See Mary Quinlan-McGrath, "The Foundation Horoscope(s) for St. Peter's Basilica, Rome, 1506: Choosing a Time, Changing a *Storia*," *Isis* 92 (2001): 716–41. I will explore other examples in volume III, including that for the Fortezza da Basso in Florence.

[49] For a valuable and up-to-date guide to these materials, astrological and otherwise, see now *Italian Renaissance Diplomacy: A Sourcebook*, Monica Azzolini and Isabella Lazzarini (eds.), Turnhout: Brepols, 2017. Azzolini uses a broad range of this evidence to good effect in her The *Duke and the Stars, passim.*

[50] I rely almost entirely in what follows on the valuable researches of a broad range of scholars.

[51] But not entirely! Richard Trexler's anthropologically sophisticated *Public Life in Renaissance Florence* is an excellent entry point into the characteristic features of this world. That Nancy Reagan regularly consulted an astrologer during her husband's years as president of the United States during the 1980s, however, indicates important points of continuity; Joan Quigley, *What does Joan say?* In her own words: "I was responsible for timing all press conferences, most speeches, the State of the Union addresses, the takeoffs and landings of Air Force One. I picked the time of Ronald Reagan's debate with [Jimmy] Carter and the two debates with Walter Mondale; all extended trips abroad as well as the shorter trips and 1-day excursions, the announcement that Reagan would run for a second term, briefings for all the summits except Moscow, although I selected the time to begin the Moscow trip. I timed congressional arm-twisting, the second Inaugural Oath of Office, the announcement of Anthony Kennedy's Supreme Court nomination. I

relevant evidence—most of which is drawn from the intensely localized and detailed archivally-based researches characteristic of Italian historical scholarship of the late nineteenth and early twentieth centuries—and places it within the broader comparative framework of the Northern Italian courts and Rome, all of which belonged to the same cultural and political cosmos. What follows is a panorama—a series of vignettes, portraits in a landscape—all meant to give a fuller sense of the many roles astrology could play in society, politics and culture, and, hence, why Pico and other opponents were so passionately devoted to eradicating it entirely from the cultural, political and religious landscape, however unsuccessfully.[52]

11.5 Astrology and the Papacy

To appreciate how astrology was configured within Quattrocento Italian political life at the very highest echelons, we should begin with a scene taken from the strikingly interesting archival evidence gathered in Ferdinando Gabotto's richly detailed, "L'astrologia nel Quattrocento in rapporto colla civiltà: osservazioni e documenti storici."[53] In it, Gabotto transcribed a tremendously interesting letter of July 20, 1492 from the astrologer, Ambrogio Varese da Rosate, to Ludovico il Moro, duke of Milan (1452–1508).[54] Ambrogio's letter answers a question put to him by the duke

delayed President Reagan's first operation for cancer from July 10, 1985, to July 13, and chose the time for Nancy's mastectomy (12)." These are all forms of astrological elections; I will mention one more of her examples below. The archive of her astrological activities is now at the Hoover Institution of Stanford University, where I examined them *in situ*.

[52] This is merely a sketch. There is much rich material here for microhistorical case studies, which may then be woven together into a vividly detailed cultural tapestry. I intend to develop this material further. For now, see the recent studies on Mantua by Signorini, and on Milan by Azzolini. Serra-Zanetti's point is apt; "I Pronostici di Girolamo Manfredi," 194: "Nello stesso tempo l'astrologia giudiziaria, o divinatrice, penetrò profondamente nella vita intellettuale, politica e sociale: governanti, signori, principi e condottieri manifestarono una constante fiducia—a volte assoluta, a volte interessata—nella potenza profetica dell'astrologia, tennero in sommo pregio gli astrologi, li ricercarono per averli loro servizio e li colmarono di ricchezze e di onori. In un periodo turbato da lotte, da ambiziosi disegni di grandezza e di dominio, da vasti intrighi, e contrassegnato da fortunosi trionfi individuali, l'astrologo rappresentò spesso uno strumento importante nel gioco politico, poichè serviva non solo a suggerire consigli e rimedi per evitare le insidie della sorte e prevenire i pericoli imminenti, ma anche a ordire trame a danno dei nemici e a confondere gli avversari."

[53] *Rivista di filosofia scientifica* 8 (1889): 377–413. The numbers in parentheses in my text are to this article.

[54] See also Thorndike, *HMES*, IV, 434. Ambrogio will reappear later in this section. On Ambrogio and other astrologers in the Milanese context, see Monica Azzolini, "Anatomy of a Dispute: Leonardo, Pacioli and Scientific Courtly Entertainment in Renaissance Milan," *Early Science and Medicine* 9 (2004): 115–35. For this letter, see her "Reading Health in the Stars," 183–85, where she also retranscribes part of it, and her full scale study of astrology in late Quattrocento Milan, *The Duke and the Stars*.

concerning the health of the ailing pope, Innocent VIII (1432–92). The pope himself seems to have made Ludovico the middleman between himself and the astrologer.[55]

Ambrogio performed the astrological service of a medical interrogation, of the sort we saw theoretical texts for in Chap. 10 in the roughly contemporary astrologico-medical miscellany compiled by Valdizocco, ca. 1470. To gain insight into the pope's illness, the astrologer would normally cast a horoscope for the time the medical question was asked. It is worthwhile to examine this letter closely, for it opens a particularly clear window onto a critical role astrology could play at the highest levels of Quattrocento political culture. It also shows how different astrological practices could relate; in this case, nativities (or their lack), annual revolutions and interrogations (medical or otherwise).[56]

In the first section, Ambrogio informs Ludovico that he has constructed a figure of the heavens for the time the letter with the original question was sent—not the time at which he actually received it—and that he has diligently examined the heavens at that time for their influences.[57] But since Ambrogio does not know Innocent VIII's natal horoscope, he must use the general signifier (*el signatore generale*) for the pope in the annual revolution, which, according to some astrologers, is the sun with Mars, and, according to others, is Mercury with Jupiter and the sun.[58]

Having provided this background information, Ambrogio now turns to the judgment: the pope will surely die![59] He adduces three pieces of astrological evidence for this claim that he had gathered in both [1] comparing the horoscope of the interrogation with the horoscope of the revolution of the year of the world (*in la revolutione de li anni del mundo*) or annual revolution,[60] and [2] in his reading of the annual revolution horoscope itself.[61] Ambrogio then treats the next logical question:

---

[55] "Ill.mo et Ex.mo S.re, per satisfare a la domanda del Pontefice quale l'Ex. V. per una sua mi fece… (382)." At least this is the way the letter represents it. Azzolini, however, offers a different interpretation, as we will see below.

[56] As we will recall, this was also the case with interrogations, nativities and elections in *Speculum astronomiae* 7–10.

[57] Ill.mo et Ex.mo S.re, per satisfare a la domanda del Pontefice quale l'Ex. V. per una sua mi fece, et fumi presentata heri, circa 21 hora, da le quale in qua cun più diligentia et studio ho saputo, ho considerato et revoltato il sito de li Corpi celesti in Celo, et loro influxo a l'hora de la domanda, che fu l'altr'heri a due hore de nocte quando gionsano le littere, et insuma examinato la interrogatione sua (382).

[58] "Perche m'e ignota la nativitate d'epso Pontefice, rivoltando insieme ancora el signatore generale del pastore et principe della fede chrystiana in la revolutione de li anni del mundo, quali secundo alcuni Astrologi sono el sole cum Marte, et secondo altri Mercurio cum Jove et il Sole (382)." The signifier for princes in an annual revolution for 1474 is the sun, as we will see just below.

[59] This sort of prediction is usually what was prohibited by law when astrology (or certain of its practices) were proscribed. I discuss the colorful and dramatic case of Orazio Morandi in the 1620s in volume III.

[60] That is, the second type of revolution from the *Speculum astronomiae*.

[61] "Insuma ritrovo, si havendo respecto a l'hora de la interrogatione de la Ex.ma S. V., (1) in la quale Marte fu significatore cum Jove, per essere Marte in el Caso suo, et Jove ancora sotto li ragi del Sole, la luna sotto la terra, coniuncta per generale aspecto al Sole signore de la Casa de la infirmitade et Jove signore de lo assendente adusto, et fra 25 di coniuncto cum Marte in la interrogatione Signore de la casa de la Morte, *epso Pontefice dovere morire*. (2) Il medesimo [sc. outcome,

if the pope is going to die, when, in fact, will this happen? This information is derived solely from the present interrogation, providing two possible (and very close) time frames, either within 22 days, because of the conjunction of Mars with Jupiter, or within 15, because of the square of the moon with Venus, the lord of the house of death in the interrogation figure. At any rate, there is no way he could live until the end of August.[62]

To conclude the letter and his judgment, Ambrogio informs Duke Ludovico that Innocent's successor will be very well disposed toward him (once again for solid astrological reasons), but, because he, Ambrogio, has not yet had enough time to fully study the matter, he cannot give a full interpretation on the nature and condition of Innocent's successor.[63] This would also, of course, be an invitation for Duke Ludovico to consult Ambrogio again on this most important matter.

Needless to say, solid information on the forthcoming death of a pope would have been extremely valuable information for anyone involved in contemporary politics at any level, whether cardinal, prince or courtier, and especially for a less than fully legitimate ruler of a major Renaissance city-state, as was the case with Ludovico.[64] Here we have a type of practical astrology, a medical interrogation, with profound political ramifications. As it turns out, Innocent died 3 days after the query was put, which would certainly not have hurt Ambrogio's reputation as some-

---

death] si denotta per la interrogatione per essere stata Venere sua significatrice a l'hora de la revolutione combusta et giuncta cum lo significatore de la Morte, quale in lo presente anno che denotava la morte. (3) Il medesimo ritrovo per la revolutione de li anni del mondo, quale incomezo a di 10 di Marzo, in la quale fu el signatore del Pastore de la fede chrystiana insieme cum Venere damnati per adustione del sole in la casa de Infirmitade, per la qual cosa si denotta la morte del pastore de la fede chrystiana (382–83; my numbering and emphasis)." I provide the extensive quotations throughout this chapter mainly for the language and the astrological details, and because some of the articles are not easily accessible.

[62] "Il quando mo debbia morire ritrovo per la presente interrogatione, che debbe seguire la morte aut fra 22 di che sara ad 10 o 11 d'augusto per la coniunctione de Marte cum Jove in la domanda aut fra 15 di per il quarto aspecto de la Luna ad Venere signora in la domanda de la Casa de la Morte in modo che per via de la domanda mi si demonstra non possa pervenire insino al fine d'augusto, che cosi el nostro signore Dio per la sua clementia permetta et conceda si per il bene publico come per il privato (383)." The astrologer in this case becomes the interpreter intermediary between God's will and man's knowledge. Astrologers still used this sort of rhetoric into the seventeenth century, as I have shown in the case of Galileo; "Celestial Offerings: Astrological Motifs in the Dedicatory Letters of Galileo's *Sidereus Nuncius* and Kepler's *Astronomia Nova*," in *Secrets of Nature*, 133–72.

[63] Quanto ad la intelligentia et amicitia quale ha ad seguire cum lo successore del pontefice, respondo che la Ill.[ma] S. V. sara bene amata, et gli parturira comodo, pero che partendose la luna ultimamente da Jove, fu recepta dal Sole da s.[le] aspecto de la casa propria del Sole, che denotta l'excellentia v. dal successore essere amata come uno amico o vero parente quale alogiasse in casa un altro per farli conzo et servitio. E questo e quanto in questo poco tempo ho potuto cavare per la virtude del Celo et corpi celesti in queste cose inferiore. De quale natura et quale conditione habia ad essere questo successore per la brevitate del tempo non ho ancora potuto comprendere: non perderò tempo per considerare et intendere questa parte, et trovato habia alcuna cosa, subito daro adviso a la Illma S. V., a la quale humiliter me ricomando. Mediolani, 20 julij 1492. Servulus Ambrosius Varisius de Rosate (383).

[64] For a splendidly clear discussion of the extremely complex political situation, see Azzolini, *The Duke and the Stars*.

one able to access insider information accurately. We are left, then, with the striking image of Innocent VIII, the pope who both sponsored the notorious and profoundly influential *Malleus maleficarum* and condemned Pico's 900 *Conclusiones*, consulting, on his deathbed, a famous professional astrologer for insight into the nature and resolution of his malady.[65]

*

But Innocent VIII was neither the first nor the last pope to have an interest in astrology. In the thirteenth century, Albertus Magnus disputed his deeply astrological *De fato* before Pope Alexander IV and his curia, and Roger Bacon addressed his three profoundly astrological *Opera* to Clement IV. There were Pius II and Alexander VI in the 15th; Julius II, Leo X, and Paul III in the 16th; and Urban VIII in the seventeenth century, to name only a few popes with significant interest in astrology who spring readily to mind.[66] The cardinals' courts, as for any princes of the time, would also have been deeply influenced by astrology, otherwise Paolo Cortesi would have felt no need to strongly advocate its removal in his treatise, *De cardinalatu* (1510), on how a cardinal should run his household.[67]

I will briefly mention one fifteenth-century pope here, namely, the most humanist of all humanist popes, Aeneas Sylvius Piccolomini (1405–64), who reigned as Pope Pius II from 1458 until 1464. Although Thorndike does not treat Pius at much length in his chapter, "Humanism and Science" (*HMES* IV, 393–4), the description of his interest in astrology is revealing and worth quoting *in toto*:

> The humanist Aeneas Sylvius who later became pope Pius II felt that some knowledge of astrology was essential for a ruler. In his educational treatise he affirms that "a prince must not be ignorant of astronomy, which unfolds the skies and by that means interprets the secrets of Heaven to mortal men."[68] Aeneas's faith in astrology is further attested by such a passage in his letters as that to Piero da Noceto from Basel on May 21, 1437, in which he describes the stormy session of May seventh when contradictory resolutions were proposed

[65] Azzolini rejects Gabotto's suggestion that the pope himself requested the interrogation; *The Duke and the Stars*, 191–99, and her "Reading Health in the Stars," n. 1. For what actually happened at the death of a pope, see Agostino Paravicini Bagliani's provocative *The Pope's Body*, D. Peterson (tr.), Chicago: University of Chicago Press, 2000. He treats Innocent VIII's death in some detail, 293 and n. 34. This theme of the pope's death will recur below in volume III in Galileo's milieu.

[66] A full study of astrology and the papal courts ca. 1250–1800 would be of great interest. In the meantime, as on so many other topics, the relevant volumes of Thorndike, *HMES*, provide an essential point of departure.

[67] Such a study would also be of great interest for cardinals' households. For example, one should look into Regiomontanus's role in Cardinal Bessarion's household. For Cortesi, see d'Amico, "*Contra divinationem*: Paolo Cortesi's Attack on Astrology." We saw earlier that Cardinal Pierre d'Ailly had a strong personal interest in astrology. Apparently Cardinal Nicholas of Cusa did as well, as we can see from the books in his library; J. Marx, *Verzeichnis der Handschrift-Sammlung des Hospitals zu Cues*, Trier: Selbstverlag des Hospitals, 1905. Kepler was so impressed with the astrological holdings in Cardinal Nicholas's library that he offered to buy them for his own library. For Cardinal Ascanio Sforza and astrology, see Azzolini, *The Duke and the Stars*.

[68] There is a translation of Aeneas Sylvius's educational treatise, *De liberorum educatione*, in Woodward, *Vittorino da Feltre and Other Humanist Educators*, 136–58.

> as to the place, Avignon or Florence, where representatives of the Greek and Roman churches should meet. He gloomily ascribes the fact that things are not going well at the council to the discordant influence of the planet Mars. But this redounds to the advantage of Florence, which city is said to have been founded under that planet. Aeneas is further impressed by the fact that the events to which he alludes occured on Mars's day or Tuesday, that all the stars agreed in causing dissension, and that certain of the astrologers affirm that it was under Mars that the great schism began which was healed at Constance, and also that Mohammedanism originated, with Jupiter in the tail of the Scorpion. He hopes that these are fables and dreams rather than the presages of a true seer, but he cannot shut his eyes to the fact that two conflicting decrees have been published in a single session, and that attempts have been made to depose the papal presidents at the council (393).

It is clear from this letter that Pius II, one of the premier humanist clerics of the mid-Quattrocento, used astrology here to analyze and interpret significant public events, much as Cardinal Pierre d'Ailly had also recently done.[69] Both acted within the very same intensely competitive struggle for political and spiritual power between popes and councils, as the Renaissance popes continued to develop the absolutist monarchical ideology of their medieval forebears.[70]

11.6 Astrology in Milan

Moving from ecclesiastical to secular courts, astrology played many and varied roles in Renaissance Milan, one of the most powerful and important Italian city-states.[71] By articulating some of astrology's varied textures within the fabric of Renaissance society, the Milanese material sets the stage splendidly for describing other courts, especially Ferrara and Florence. The mainly archival evidence provided by Gabotto and Azzolini helps us reconstruct in rich detail several of astrology's interrelated places within the political culture of the day, thus evoking a vivid impression of political processes and tensions, and, more generally, of how

[69] For more on Pius II, see Shank, "Academic Consulting," 260–62 with respect to his time in Vienna, esp. n. 53. Concerning Alexander VI, the Borgia pope, Thorndike does not go into depth; he does, however, indicate that several annual prognostications were dedicated to him; *HMES*, IV, 461, etc. For Alexander's broader cultural interests, see Rowland, *The Culture of the High Renaissance*. For the specifically Egyptianizing elements, see Brian A. Curran, *The Egyptian Renaissance: The Afterlife of Ancient Egypt in Early Modern Italy*, Chicago: University of Chicago Press, 2007, and his "De sacrorum litterarum Aegyptiorum interpretatione." Reticence and Hubris in Hieroglyphic Studies of the Renaissance: Piero Valeriano and Annius of Viterbo," *Memoirs of the American Academy in Rome*, 43/44 (1998/1999), 139–182, 167ff.

[70] On the medieval papacy, see (i.a.) Geoffrey Barraclough, *The Medieval Papacy*, New York: Norton, 1968; on the Renaissance papacy, see (i.a.) Charles L. Stinger, *The Renaissance in Rome*, Bloomington, IN: Indiana University Press, 1985.

[71] For a much fuller treatment than I can offer here, and with a focus on astrology's roles in politics and medicine at Milan ca. 1450–1500, see now Monica Azzolini's marvelously clear, learned and lively, *The Duke and the Stars*. In addition, I have learned a great deal from her insightful esssays on astrology, medicine and politics in fifteenth century Milan, several of which I have read in pre-publication drafts, for which I heartily thank her.

centrally astrology was woven into this public political fabric at the heart of the knowledge-power nexus.

Gabotto focuses most fully on Galeazzo Maria Sforza's relation to astrology. A vivid picture emerges of the dramatic roles astrology could play at the very center of one of the most powerful Renaissance courts, especially during his troubled rule (1466–76).[72] The first evidence Gabotto presents derives from the time when Galeazzo Maria became the sole hereditary prince. He obtained a natal horoscope with interpretation from Raffaele da Vimercato, which exists in a manuscript in the Biblioteca Trivulziana with the title: "Liber iudiciorum in nativitate Comitis Galeazzi Mariae Vicecomitis Ligurum futuri Ducis dignanter electi." It was composed precisely on the 8th hour of Tuesday, June 2, 1461.[73]

On succeeding his father Francesco as duke in 1466, Galeazzo Maria continued his relationship with the same astrologer, Raffaele da Vimercato, who also made a nativity for his son, Gian Galeazzo, as appears in a letter dated September 2, 1474 from Vimercato himself to the duke.[74] We shall examine this interesting letter, but only after first describing a splendidly well documented incident of July 1474, which provides lively and detailed information on the exactly contemporaneous place of astrology within the culture of Galeazzo Maria's court.

Before turning to this incident, however, I should note that Raffaele da Vimercato was not Galeazzo Maria's only astrologer. There was also Francesco Montano, whom he invited to Milan in a letter of March 11, 1472, with the substantial gift of 50 gold Milanese ducats for comfortable transport.[75] Another letter of Feb 12, 1472 mentions the same Montano as dead. The February letter describes memorable features of his astrological practice, mainly predicting the deaths of important persons, including popes and cardinals. His predictions were often accurate: for the previous pope, Paul II, and for himself. Galeazzo Maria wanted his own complete copy of the predictions for himself.[76]

---

[72] For a detailed treatment of this court, see Gregory Lubkin, *A Renaissance Court: Milan under Galeazzo Maria Sforza*, Berkeley: University of California Press, 1994. Azzolini treats astrology in Galeazzo Maria's court in *The Duke and the Stars*, Chap. 3, "Astrology is Destiny: Galeazzo Maria Sforza and the Political Uses of Astrology," 100–34.

[73] Gabotto does not describe its contents; "L'astrologia nel Quattrocento," 398. Azzolini discusses this horoscope in depth; *The Duke and the Stars*, 103–14.

[74] "L'astrologia nel Quattrocento," 398–99. Gabotto identifies the letter in the archive thus: "Nell'archivio di Stato di Milano: *Astrologi*." Many of the documents cited in his article were found under this heading, but they have since been reorganized. Azzolini, who followed up many of Gabotto's leads, discusses this; *The Duke and the Stars*, 8–9. For her discussions of Vimercato, see her index, s.v.

[75] "L'astrologia nel Quattrocento," 399.

[76] "L'astrologia nel Quattrocento," 400: "Intendemo essere passato de la presente vita uno Francesco Montagna, quale se delectava de fare iudicii et pronostro la morte del quandam papa Paolo, como credemo sapiati, et che a la morte sua se gli e trovato sotto la testa certo iudicio per lo quale s'e pronosticata la morte propria, pronostica la morte del presente pontifice et d'alcuni cardinali, fa mentione de la cometa, et dice anchora alcune cose de li facti nostri. Volimo vediati di haverne copia integra et ne la mandati piu presto ve sia possibile et a la lettera vostra, nella quale sara incluso dicto iudicio, farite una croce sopra el soprascripto." There is obviously a problem with the

*

Gabotto prints an extraordinary series of letters, mainly from the duke of Milan, Galeazzo Maria Sforza, all dated between July 14 and 28, 1474. These letters illustrate splendidly how vital a role astrological prognostications could play in the public political dynamics of the day, and how seriously this was taken by the main political figures in the intensely competitive world of premodern Italian politics. I will treat these letters in chronological order.[77]

First, there is the matter of three annual prognostications of the sort described above, which were promulgated by professors of astrology at two major centers of astrological study, practice and literary production, namely, the universities of Bologna and Ferrara, two of whom I discussed above: Pietro Buono Avogaro and Girolamo Manfredi. They soon reached the duke in Milan.[78] Duke Galeazzo found elements of these prognostications deeply disturbing because, in his somewhat paranoid interpretation, they predicted evils for himself and his lands, and would thereby stir up the populace who got wind of them to unrest. The duke fully intended to stop this sort of troublesome, very public prognosticating. These letters open a window into the heated high-stakes dynamics of this process. I begin with a set of four letters written between July 14 and 18, 1474. The first two are more general, one to the secretary of the Milanese embassy in Bologna, Gerardo Cerruti,[79] the other to another of the duke's agents, Sacramoro da Rimini.

In the letter of July 14, Galeazzo informs Sacramoro at Ferrara how badly the populace (*li populi*) reacts when there are negative prognostications: it really disturbs them when they hear about the deaths of kings, wars and famine; it makes

---

dates, even if it were the custom in Milan to begin the new year at the Spring equinox (which it was not; see Lubkin, *Renaissance Court*, 69). If so, February would have been at the end of the same year in which the duke of Milan invited Francesco there. There are apparently other problems with the dating as well, in that their earlier editor, Cantu, dated them both to the reign of Ludovico il Moro.

[77] Gabotto's order for presenting the letters obscures their interrelationship. I would like to record here my sincere appreciation to Massimo Scalabrini of Indiana University, who helped me to grasp the obscurities of language in these letters. Azzolini also discusses this episode, but more briefly in her *Duke and the Stars*, 129–32, with much valuable analysis, and with some of the texts retranscribed according to modern criteria. Refer to her treatment for further information and bibliography on the figures mentioned here and their historico-political context. Duranti also discusses this episode in his study of Girolamo Manfredi; *Mai sotto Saturno*, 11–15.

[78] Gabotto, "L'astrologia nel Quattrocento," 403: "Cosi che niuna meraviglia se avendo egli fatto il 'giudizio,' come portava l'ufficio suo, se ne sparsero subito copie per tutta Italia e una capito, con altri due giudizi di un Marsilio da Bologna e di Pietro Bono, nelle mani di quel gran cultore di astrologia ch'era Galeazzo Maria Sforza." The universities were closely controlled by the local signore; see Peter Denley, "*Signore* and *Studio*: Lorenzo in a Comparative Context," in *Lorenzo the Magnificent: Culture and Politics,* 203–16, with further bibliography.

[79] Duranti has also edited Cerruti's diplomatic correspondence for precisely the time he was in Bologna with a rich introduction, *Il carteggio di Gerardo Cerruti, oratore sforzesco a Bologna (1470–74),* 2 vols., Bologna: CLUEB, 2007. The relevant letters between July 14 and 28, 1474 are on pp. 472–83.

trouble.[80] For this reason, Duke Galeazzo wants his agent there to persuade the Ferrarese authorities to stop the astrologers prognosticating about particular rulers. They should only make general claims because particulars are dangerous (and religion prevents such superstition).[81] Then Galeazzo Maria gets down to the important business at hand. He says that the Ferrarese judgment speaks well of the duke of Burgundy (Ferrara's ally) and badly of the king of France (Milan's ally) because the astrologer knows that it will please his lord, Duke Ercole d'Este.[82] Galeazzo ends the letter by letting his agent know that any correspondence on this obviously important matter should be adressed directly to himself or to his trusted secretary, Cicco Simonetta, and to no one else![83] The letter of July 15th to Cerruti at Bologna is much shorter. Duke Galeazzo wants to know about the astrologers there who made annual judgments, and whether they said anything specific about him or Milan.[84] He should inform the astrologers either not to mention his affairs, or only to say good things.[85]

[80] "Voi sapete de quanta importantia sia quando in li populi si divulga qualche sinistra opinione de li mali eminenti et che possono intervenire, como fanno spesse volte questi astrologi temerarij et legeri, li quali mettendo suo studio in divinare et fantasticare de le cose occullte, reservate solo in arbitrio de Idio, stultamente predicono la morte di principi, guerre et carestie, et descendono usque ad individua in signare ad chi credono debia toccare la sorte; et licet li homini de bon sentimento et gravi poco prestino fede ad simile pronosticatione, tamen el vulgo pur li presta orechie et istano cun li animi suspesi, et fanse spesso pensieri che generano scandolo in li Stati et principati (404)." I quote so much of the original here to give a fuller taste of the pungent language; also, Gabotto's article is not easily accessable, and Azzolini does not quote all of the Italian.

[81] Nostro parere seria che la S.[ta] de n. S.[re] faci excommunicare tutti et singuli Astrologi et Mathematici, li quali presumeranno in li loro iudicii nominare ne specificare alcuno principe ne S.[or], ne farne alcuna mentione tacita vel expressa; ma solamente li sia permesso dire de le cose universale, perche le particularitate possono mettere turbatione et sonno pericolose, et la Religione et fede christiana prohibisce queste tale superstitione (404).

[82] Pero la inhibitione sotto le censura ecclesiastiche sera utile et honestissima, et voi non li mancate de sollicitudine che se faci, maxime che, oltra contra el precepto divino simile prenosticare, el fanno etiam cum vitio seguitando l'appetito di soi S.[ri] per adularli et gratificarseli. et si soa S.[ta] vole bene vedere la verita, se faci dare el iuditio de quello de Ferrara, che dice bene del duca di Borgogna perche sa che piace al suo S.[re] Duca Hercule, et dice male del re de Franza et anco soa S.[ta], et el simile costume seguitano gli altri astrologi (404).

[83] "Volemo che quante lettere vostre accadra scriverne sopra tale materia, tutte le facciate drizare in manibus nostris proprijs, aut Ioannis Iacobi Simonettae, en non de altri (404)." For more on Simonetta, see Lubkin, *A Renaissance Court*, and Azzolini, *The Duke and the Stars*. For the structures of diplomacy in general, see Garrett Mattingly, *Renaissance Diplomacy*, Boston: Houghton Mifflin, 1955.

[84] Dux Medilani. Gerardo. Volemo che dextramente sapii da quelli Astrologi ly, de li quali ne hai mandato el iudicio de questo anno, si epsi in particularita hanno veduto qualche pronosticatione de noi et si implicite per loro iudicij ne hanno voluto sentire alcuna cosa che pertenghi ad noi et al nostro Stato (403).

[85] "Et fate dire el tutto in scripto et bene expresso; pregandoli che de qui innanze non voglino piu fare mentione di facti nostri per alcuno modo, et si pur ne faranno mentione, la facino in bene et commodo nostro, et non in contrario. Cassani, xv Julij MCCCCLXXIIII. Jo. Jacobus S[imonetta]." On back: "Egregio viro Gerardo Cerruto, Secretario n[ost]ro car[issi]mo Bononie. Cito (403)."

But the duke of Milan was not content simply to vent his anger and frustration. At the same time, Galeazzo also wrote other letters to his ambassadorial agents in Ferrara and Bologna with specific official instructions regarding how to communicate the seriousness of his concerns to the local authorities: the duke of Ferrara, Ercole d'Este, and the signore of Bologna, Giovanni II Bentivoglio. Both letters are written to the same Giovanni Battista da Cotignola, the Milanese ambassador to Ferrara. The instructions are dated July 18, 1474:

> Go to Duke Ercole and tell him that master Pietro Buono [Avogaro], an astrologer in your principality, published an annual judgment for all of Italy. Our natural philosophers, deeply learned in astrology (*questi nostri physici peritissimi de Astrologia*), received it. Having read the details (*particularita*) of this judgment (*iudicio*) closely, they declared that the said master Pietro Buono infers openly that this year we [the royal we] will encounter danger with respect to our own life and also our state, as you can understand from selections of the judgment appended below. This is undeniable because he mentions other leaders and cities, naming them explicitly, but he made no explicit mention of us, about whom he only wrote implicitly.[86]

"We believe it happened without your consent," the secretary is to inform Duke Ercole:

> We want you to tell that astrologer from now on not to discuss us either implicitly or explicitly, so that he will make no more conjectures about our affairs in his judgments. And when he does the contrary [...], we will show him how much it displeases us for him to prognosticate about us and our allies, as he did concerning the most Christian king of France, when he spoke with obvious zeal (*cum manifesta passione*), speaking badly of his majesty and well of his enemy.[87]

After this warning for the astrologer, Duke Galeazzo turns to warn Duke Ercole directly: And ensure the duke that we can return the favor, for we also have skilled

[86] "Cassani/die XVIIJ Julij/1474 (*). Instructio data d. Io. Bap. de Cotignola famri. M. Zoan Bapta. Voi andarete da lo Ill.[mo] Duca Hercule et per nostra parte li exponerete che avendo M[ro] Petro Bono astrologo de la S[ria] sua divulgato el suo iudicio de questo anno per tutta Italia, e acapitato in le mano di questi nostri physici peritissimi de Astrologia, li quali, lecto epso iudicio et examinato bene la particularita che sonno in quello, ne hanno dechiarato ch'el dicto M[ro] Petro Bono apertamente inferisce che noi in questo anno havemo ad incurrerre in pericolo della vita, et anche del Stato, como per le parole tolte da li capitoli de epso iudicio et notate qua de soto se po comprendere. Et questo non se po negare, perche habiando lui facto mentione de caduno S[re], Potentato, Repu., Regnante et Communita, nominandoli expressamente, de noy, che ha omesso farne parola, implicite scrive quanto li e parso (405)." (*) With respect to the date, Gabotto adds this note (405, n.1): "Cosi nell'originale. Ma probabilmente deve leggersi 17, perche il documento seguente, che ha precisamente questa seconda data, ha un 'anche' che presuppone scritto precedemente il documento presente."

[87] Et benche crediamo questo sia proceduto senza consentimento de la S[ria] sua, perche como quella che ne ama, non lo haveria potuto, tamen volemo la pregate che voglia admonir dicto astrolog che da qui inante, nec explicite, nec implicite, parle de noi; ita mai in li soi iudicii se posse fare conjectura sopra li facti nostri; che quando lui faci el contrario(?), et nunc volemo ne excusiate cum la S[ria] sua, che ad dicto astrologo demonstreremo quanto ne despiacia che lui voglie prenosticare de noi, ne anche de nostri adherenti et confederati, como ello ha facto del christ.mo S[re] Re de Franza, dove parla cum manifesta passione, dicendo male de la M[ta] et de lo adversario bene (405)[.]

and learned astrologers, but this does not seem to be the proper duty of a royal prince (*sia officio alieno da Reale Principe*).[88]

To complete the instructions, Galeazzo Maria instructs his agent:

> [A]ffirm our good intentions to the duke, but beg his pardon in advance if the aforementioned astrologer does not cease prognosticating about our concerns. We will see to it that he be made unhappy (*mal contento*); he should think twice if he is prudent. And communicate the same thing directly to the astrologer.[89]

Gabotto also printed the letters written for the astrologers to be discussed just below.

Then follow excerpts from the offending prognostication. We will recognize their structure from the discussion in Chap. 10. The extracts come from three sections, of which I will discuss two. The second is on the state of princes (*De statu principum*), where it says that "a lord, unless I am mistaken, will die suddenly from Jupiter in the house of illness (*in domo langoris*), with a diametrically [sc. opposite] ray (*radiatio*) aspecting Saturn."[90] The third section on the state of prominent men (*De statu hominum*) states that kings, emperors and lords are all under the sun's rule,[91] who for the most part will be exalted by the sun illuminating the royal house. Nevertheless, one lord will die suddenly from Saturn pouring forth rays (*radii*) from a squared aspect to the sun, and with Jupiter, the lord of the midheaven, lying in an angle of weakness. This lord will also live in great terror, and the end of his rule will occur in the fourth month. Likewise, his wife will die. In addition to the immediate astrological circumstances, we should also notice the language of rays, precisely as we saw in Part I.

The language of the prognostication is colorful and terrifying:

> In this matter, one lord (*dominus*) will experience fears and terror, and he will gather great wealth of which he will consume a great part, and he will disturb hateful *fratres* or the same men with sedition and strife, and the end of his power will appear suddenly in the fourth month, in which, thirsting after blood (*cruorem sitiens*), he will take a lethal potion (*laetale sumet poculum*) on a long slow journey with great anxiety and harm.[92]

---

[88] [E]t certificate la S^{ria} sua che quando noi volessemo usare simile arte, havemo chi anche saperia astrologare non cum manco doctrina et rasone che se facino li altri, ma ne pare che sia officio alieno da Reale Principe (405).

[89] Farete adonqa bene intendere el tutto al p^{to}. Duca, et excusatene cum epso che, non abstenendose dicto astrologo dal prenosticare di facti nostri, cercaremo de farlo mal contento, et pur cosi deve dubitare, se elle e prudente. Et el medesimo direte ad epso astrologo (405).

[90] Ex secundo capitulo de Statu principum. Et unus dominus, nisi fallor, morietur repente, ex Jove in domo langoris diametra radiatione Saturnum respiciente (405).

[91] This is an example of the planetary rulership of major political figures, as we also saw in the medical interrogation for Innocent VIII above.

[92] Ex capitulo tertio de Statu hominum. Sub sole sunt Reges, imperatores et domini, qui ut plurimum hoc anno exaltabuntur ex sole domum Regiam illustrante. Unus tamen dominus repente morietur ex Saturno radios quadrature ad Solem diffondente Jove domino medij celi confracto et in angulo infirmitatis iacente. Hac etiam in re unus Dominus metus sentiet et terrorem, et es magnum congregabit, ex quo magnam consumet partem et fratres odiosos vel eosdem seditione perturbat (sic) et rixa et exitus imperii sui repente parabitur mense quarto, in quo cruorem sitiens, laetale sumet poculum et ex itinere tardo longa anxietate et damno. Et infra. Et unius principis uxor ad

The identity of this ruler would then be the 64,000 ducat question! And if the other major political figures had been explicitly named (as Duke Galeazzo claimed), it would not take a genius to interpret the prognostication by a process of elimination. If such a judgment were to become widely known, we can easily imagine the threat to public order. As it turns out, Galeazzo Maria's paranoia was hardly unfounded. He was brutally murdered relatively soon after in 1476.[93] We can thus see how a major Renaissance leader thought annual prognostications could effect public opinion.

*

The other letter from Cicco Simonetta to the same Giovanni Battista da Cotignola, dated July 17, 1474, concerns Bologna: Also go to Bologna and let the signore know what master Girolamo Manfredi and master Marsiglio wrote in this year's judgment, which they cannot deny.[94] These Bolognese astrologers apparently used the same strategy as Pietro Buono, mentioning all the other Italian princes, thus necessarily implying (in Galeazzo Maria's view, at any rate) that the certain great prince in Italy who will die is himself.[95] The rest of the letter covers much the same ground as the one to Ferrara,[96] including excerpts, once again, from the offending prognostications, each in turn.[97] Once again, we can see the structure of annual prognostications referred to here.

---

ultimum deveniet finem ex Venere tempore alti Solstitij adusta in imo celi cum Sole existente (405–6).

[93] Azzolini discusses all of this material in relation to and as in part informing the circumstances surrounding Galeazzo Maria's actual murder.

[94] Dux Mediolani, etc. M. Joan Baptista. Voi andarete anche ad Bologna et farete intendere ad quello M[co] Regimento quello che implicite M[ro] Hieronymo di Manfredi et M[ro] Marsilio in loro iuditij de questo anno de noi haniano scripto; el che non possono negare (406).

[95] Could this have been a well co-ordinated negative public relations effort orchestrated by his enemies, and thus an early instance in which the media has played a less than savory role in politics? This is in part the tenor of Azzolini's analysis. We will see a later example with Orazio Morandi in volume III.

[96] Perche havendo facto particulare mentione de tutti li altri principi de Italia, ex necessario conseguenti vene ad inferirse de noi dove epsi parlano de quodam principe magno in Italia qui morietur, etc. Et de nostra parte precarete le loro M[tie] che admoniscano epsi Astrologi che da qui inante mai piu in li iuditij loro implicite nec explicite facino mentione ne parola de noi, aut di nostri amici, confederati et colligati che se possa coniecturare sopra li facti nostri. Perche quando facessero el contrario, ex nunc volemo essere excusati, che gli faremo demonstrare che ad noi despiacera summamente el iudicare loro, et forse anche cosi ne li faremo pentire, ecc. Et el medesimo direte alli prenominati Astrologi, partecipando pero el tutto cum Gerardo Cerruto nostro Secretario la, cum lo quale anderete ad visitare el M[ro] nostro Comp.[re] d. Zoanne, et communicarete pur con epso el tutto (406).

[97] Excerpta ex iudicio M[ri] Marsilij

Ex çapitulo de ijs que ad ventos spectant.

Nec qui illos malivolentia prosequitur, miseriarum calamitatumque immunis est.

De ijs que ad Marchionem Mantue pertinent.

Is praeterea, quem aeque ac fratrem in eo amore complectitur, de vita exibit. Difficile est dictu quam dolori ea mors illi sit futura. Nec potentior deerit, qui principi terrorem injciat.

While Cotignola went from Ferrara to Bologna to deal with this matter in an official diplomatic manner, he also sent Lorenzo Belleto to Ferrara with a letter for Pietro Buono himself, dated July 16, 1474:

> To Master Pietro, the Ferrarese astrologer. You made judgments about others' but not about your own imminent danger because the duke of Milan has ordered that you be cut into pieces (*tagliare a pezi*) […]! Pay attention and you will see a certain Albanian named Zorzo turn up, short, with a dark face, and Johannes de Lucoli, large, with a red face; he has long hair and a slight limp. Pay attention! I am not speaking idly.[98]

The letters to Marsiglio and Manfredi in Bologna were similarly menacing.[99]

Apparently this combination of aboveboard diplomacy and underhanded threats was effective, at least at Bologna; nothing regarding the situation in Ferrara was known to Gabotto, nor now to Azzolini. Having contacted the two astrologers, Cerruti was able to write the duke on July 20, 1474 that the astrologers in question claim that they were not in fact referring to Galeazzo Maria in either their public judgments or secretly: "One of them [sc. Marsiglio], who made his judgment in the volgare, swore to me on his soul that he had not seen your revolution for that year." The other, Manfredi, "swore, and showed me how he made the calculations. I saw the conclusions," each was better than the other. "I had him set it out and make a

---

Que ad Principem Constantium Pisaurensem spectant.

Quem incredibili [li] benivolentia prosequitur, is de hac vita migrabit.

De ijs quae ad diversa hominum genera pertinent.

Rex vero aut maximus princeps de vita exibit, huius morte bella suscipientur asperrima et populi excruciabuntur.

Ex iudicio M[ri] Hieronymi.

De capitulo Imperatoris, Regum et Principem (sic).

In Italia non video certitudinem de alicuius principis morte excepto uno, de quo infra mentionem faciemus.

Unus vel rex vel dominus periculum magnum patietur ex parte animalium brutorum, vel ex parte inimicorum occultorum, seu alicuius tractatus vel capture. Ideo principes et magnates notent hoc verbum.

Et de quanto exequirete, ne farete la relatione in scripto. Cassani, XVIJ Julij 1474. Cichus (406–7).

[98] Ferrariensi Astrologo, ecc.,/M[ro] Petro Bono. Voy astrologati e fati iudicio d'altri e non sapeti astrologare ne fare iudicio de periculi vostri imminente, perche il Duca di Milano ha mandato li per farve tagliare a pezi e tutta via ne manda de li altri per fare questo, che, non potendolo uno, venghi facto all'altro; e azo credati ve dicha el vero, se fate ponere mente ad le Bollete et al le Porte, trovareti che tra li altri ve capitara uno Zorzo Albanese di piccola statura et homo scuro in faza, et l'altro Iohanne de Lucoli, grande, rubicono, cum li capelli longhi di colore castano, et va uno pocho zoppo. State advertente, che non ve parlo senza casone (407).

[99] The letter to Manfredi warns him to expect the arrival of a terrifying "Philippo de Monte Cridolfo, homo nero de faza a ha una ferita suso la massalla destra, l'altro se chiama Anselmo Vainacher Alamanno, biancho de faza cum li capelli longhi et bianchi, e ha uno ochio pyo basso de l'altro; et amby duy sonne de non troppo grande statura (407–8)." Azzolini informs us that Pietro Buono's prognostication is still extant and is in the Milanese archives; *The Duke and the Stars,* 130.

copy, which is included here. For the future, the prognostications will only happen as you desire."[100]

Put into this new situation, Manfredi naturally gave an excellent judgment; in fact, he now became convinced of the very opposite of what he had previously affirmed! It was much too dangerous to act otherwise. Thus every prosperity was announced for the duke, for whom Manfredi wrote the extant revised prognostication for the year 1474, based on the revolution of the duke's nativity:[101] "The celestial configuration of the present revolution is without a doubt the best, and it signifies an augmentation in the state, in dignities, and in reputation, etc."[102]

From a fragment of Giovanni Battista da Cotignola's letter, the signore of Bologna, Giovanni II Bentivoglio, also appears to have responded satisfactorily. And Marsilio of Bologna made an excuse even greater than Manfredi's, claiming that he had never seen the duke's nativity, and that he had not made a judgment on matters pertaining to him. Interpreting every suspect phrase of the judgment in a positive light, he declared finally that, after becoming aware of the duke's anger, he now wished to see his nativity, so that "when he could, he would say great things,

---

[100] Ill[me] Princeps D[ne] mi sing[me]. Li astrologi de qui non fano mentione di vostra Ill[ma] S. nelli iudicij che dano in pubblico. Che non lo faciano secretamente, o per loro piacere, o ad requisitione di qualche amico, questo non se gli puo tuorre, ma se guardano bene di farlo che persona el sapia, per via de la quale ne potesse venir notitia alla V. Cel[ne]. Io so uno di questi, et anche mi non ho curato di intender s'el se fa, perche, vedendo questo V. Ex[a] l'ha a sdegno, venire non me po in animo de investigare liberamente. Hora visto quel che V. Subl[ta] mi commette in questa materia, ma ne so[no] con bel modo ritrovato con li duoj de qui che hano indicato. L'uno, cioe quello che gli ha dato opera per vulgare, mi jura per animam suam ch'el non ha visto la revolutione vostra de questo anno. L'altro, che e M[ro] Hieronymo di Manfredi, se ne volse ascondere; et pur assicurato, in fine me ha confessato et mostro como ne fece il calculo et vide le conclusioni, ma non le mise in forma, contento assai de haverne visto lo effecto. Qualunque (sic) elle piu tosto grate et felici che altramente. Io me l'ho facto distendere et toltone copia, la quale sara qui inclusa. Per l'avenire, quanto al publicare, non ne sera se non come commette et desidera la V. Cel[ne]. Alla quale divote me recomando. Bon. 20 Julij 1474. Dev[mus] servulus Gerardus (408).

[101] Thus a *revolutio nati*, not a *revolutio anni*, which would be for an annual prognostication in general.

[102] D. M[ri] Hieronymi de Manfredi, pro anno 1474. Revolutio nativitatis ex[mi] Dni Ducis Mediolani die 14[ma] Ian. pti dedit initium 7° gradu piscium ascendente. Applicatio Mediolani ad Taurum. Venus (o Domus?) diu Mediolani in medio coeli coniuncta Jovi cum receptione.

Optima haec praesentis revolutionis procul dubio extitit constellatio, et augmentum in statu, dignitatibus ac reputatione significat. Reverentiam quoque a subditis on. dit. Quid enim melius ac in revolutionibus annorum felicius quam Dni Mediolani in domo regia Jovi commistum in receptione iudicandum esse? Hic igitur annus ad exaltationes et dignitates pro hoc Ill[mo] principe est dispositus.

Predicta bona constellatio fortunam in negocijs suis largitur et apud alios principes et Dominos convenientiam et domum suam in statu ac dominio felicitatem ac prosperam fortunam immittit.

Nec in statu suo felicissimo, nec in persona est per hoc anno timor nec periculum, nisi bonum et fortuna atque salus.

Si aliquo in negocio arduo hoc anno intromiserit, fortuna ei succedet.

Mars in 6[a] retrogradus in servis immittit infortunium et forte lites et garras ac eorum quasdam acutas egritudines vel vulnera. Hoc est quod peius videro de hac revolutione.

Observet iste Ill[mus] princeps finem huius revolutionis propter Martis combustionem. Et erit finis decembris cum prima medietate Januarii (408–9).

about which the heavens exalt his majesty."[103] Personally, there will be no fear or danger, only good fortune and safety!

The last act of this episode is a letter, dated July 28, 1474, from the astrologer, Marsiglio of Bologna, to Galeazzo Maria. He tells the duke that his agent interrogated him about certain features of the prognostication, which he will now report. In his judgment, he referred to an enemy of the Venetians who would die or come into a great calamity. In his defense, he names some of the Serenissima's enemies: the emperor, the king of France, the king of Hungary, the Turk and others (conspicuously not naming his correspondent), but which one of them will die, he does not know.[104] Then he assures the duke that he need not be concerned about either his own death or any detriment to his state until he is 42, but even then it may be avoided.[105] The remainder of the letter is given over to supplications.[106] Gabotto comments: "And in this manner the little comedy (*comediola*) is finished, which, among other things, is new proof for us of the extraordinary influence exercised by astrology in the entire life of the 15th century."[107] Here we see the heated political use of annual prognostications derived from three different kinds of astrological practice: annual revolutions (*revolutio anni* or *annorum mundi*), the nativities of rulers, and their revolutions (*revolutio nati*).[108]

---

[103] [Q]uando havesse licentia, diria cose grande, de la quale i cieli esaltano la S. S^ria^ (409).

[104] Ill^me^ Princeps et Ex^me^ Domine, D. mi sing^me^. Per Misser Johanne Baptista da Codignola sono stato examinato sopra certi capituli dil mio iudicio, et fra li altri quello voglio dire nel capitulo di Venetiani dove dico uno inimico suo doverire morire o vero pervenire in calamitade. Et come ho dicto a lui, dico etiam a V. Illma S. che Venetiani hano di molti inimici, come el imperatore, il Re di Francia, il Re di Hungaria, el Turco o de li altri assai che non scio, et di uno di loro voglio dire. Non tamen scio qual sia. Fra V. Illma S. e loro non intendo che sia inimicicia, saltem che si sappia, ma piu presto ligame d'amicicia per la ligha [...] (409).

[105] Dico etiam et per conclusione a V. Illma S. che quella non tema, ne habia alcuna paura di Morte o vero detrimento nel stato suo fino a la eta di anni quarantadue. Et anco in quel tempo, savendo V. S. el periculo, facilmente se porra guardare et evitarlo. Quia sapiens dominabit astrus (sic). Affirmo etiamche secondo la demostration di cieli V. S. vivera piu lungamente che non fece il signore vostro patre (409).

[106] Ho durato questa fatica con credere di fare piu presto cosa grata a V. Illma S. che altramente. Si tamen forsi io havesse (sic) errato o aliqualiter offeso la mente di quella, supplico che se digni darmi perdono et attribuire questo mio errore a grande affectione et carita et devotione ch'io li porto; promettendo a la prefata V. S. di mai piu non fare iudicio ne revolutione sua nisi per suo commandamento. Et cusi e mia intentione di non fare mai piu iudicio universale, se io non sero astrecto da questi miei Signori, e quando loro me astrenzano, sempre ne sara exceptuata V. Illma S., a la quale umilmente me recomando. Date Bononie, XXVIII Iulij 1474. Servus Marsilius de Bononia (409–10).

[107] E a questo modo finiva la comediola, che per altro a noi e nuova prova dell'influsso straordinario esercitato dall'astrologia sulla vita tutta del Quattrocento (410).

[108] This should be compared with Roger Bacon's discussion of how studying a ruler's nativity and its revolution could offer insight into a city's political dynamics, as discussed in Chap. 4 above. We can thus begin to see in practice just how valuable insider access to this information could be, and thus why collecting genitures of powerful and otherwise famous people became so popular in the sixteenth century and beyond.

*

The immediate crisis of July 1474 ended with the above letter. Gabotto also transcribes three related letters, although he does not treat them as such. Let us now return to the letter mentioned above, dated soon after the incident just recounted (September 2, 1474), from the duke's astrologer, Raffaele da Vimercato, to Galeazzo Maria concerning his son, Gian Galeazzo, the count of Pavia. After an extensive excuse for why this information arrived so late,[109] Raffaele gives Galeazzo Maria the summary of a judgment he had made concerning his son.[110] He also mentions interpreting a revolution for the horoscope of Galeazzo Maria's coronation.

The last two letters, dated a year later, November 4 and 5, 1475, are also worthy of note. Both were written towards the same end, to solicit judgments from astrologers on the qualities of the times in general (including the weather), and on the individual days of the coming year (*fare uno iudicio de qualitate temporum et singulorum dierum anni futuri*), in other words, to commission select astrologers to make revolutions of the year. In the first letter (November 4, 1475), Duke Galeazzo requests, through his agent Guido, the astrological services of the

[109] "Illmo et ex.^{mo} Signore, Signore, mando a V. subl.^{ta} quelo summario sopra li juditij del inclito conte de Pavia, v. primogenito de la excelsa consorte, quale summario, ne le feste del natale prosimo passato, habiendolo portato nel Castello qua a Milano con intentione poi de presentarvelo mi steso, fu a me referto per parte de la V. S. restase quela volta, et quamvix molte altre da poy, mediante la gratia de vra excelentia, intrase et dimorase longi tempi in saleta, niente de mancho, maye non sape exquisire el tempo commodo a fare questa mia offerta, sempre dubitando mi reportarne titulo d'esser temerario et imprudente. Hora vedendo la cosa deduta mosto in longo, non ho volluto pyu diferire, e perdoname vostra celsitudine, se in propria persona non l'o portato: certo non he facto per scarsitate ne di faticha, ne di tempo, che veramente ogni mia gran faccenda reputo nulla comparandola a le cose da mi esser fate quale spetano a V. S.; ma son restato per paura d'essere insolente, sentendo mi V. S. haver altre occupatione, sperando mi anchora che V. Ex.^{tia} non dubita punto ch'io sia promptissimo sempre a meter la roba, eciam che la fusse grandissima, li figlioli, e la vita propria a li servitij soy (398–99)." For more on the unfortunate Gian Galeazzo and medical astrology, see Azzolini, *The Duke and the Stars,* Chap. 4.

[110] Non c'e dietim meter in scito nove cose per satisfare a li desideri de V. S., perche, oltra quelo ha veduto V. subl.^{ta} tam sopra la nativitate vostra quam sopra la nativitate del conte de Pavia, restano molte altre bele speculatione da considerare circa lla intronizatione de V. S. e s' el primo principe, qual sera da po' V. celsitudine de' essere de la casa vostra, e qual de casa vostra, silicet o figlio o fratelo o chi altro, e, se figlio, quale o magiore o minore. Simelmente se pos cognoscere del stato de V. celsit.^{ne}, s'el ha a crescere o manchare, e cosi de la vita se po cognoscere como per la nativitate propria. Similmente, facendo revolutione de dita intronizatione, annuatim potremo intendere se debeno venire gente d'arme ne la patria a fargli guerra e de qual parte hano a venire, e molte altre cose, quale numerare sarebe longo dire. Unde per cognoscere tali inditij dispenso tempo asay in servitio de V. celsit.^{ne}, credendo mi fargli cosa grata; che se altramente pensase o cognoscese, farebe voto non tochar may ne libro ne pena per meter in scrito cosa alcuna de simille facultate, ma pyo tosto tendareve ad altri studij molto penosi e de gran stima, ben che questo sia penosissimo, maxime piacendo a tanto principe como V. S., a la quale sempre con le genogie in terra humilmente mi racomando. Valete, mi princeps invictissime. Data Mediolani, die sabati secundo septembris 1474./E. J. D. V./Servulus fidelissimus/Raphael de Vicomercato (398–99).

Dominican friar, Johannes of Viterbo, none other than the notorious forger, Annius of Viterbo.[111]

The second letter, written the following day, is closely related; it is the response to an exactly similar request for the same services as reported by Simonetta.[112] This time the requests for astrological judgments were made to master Raffaele da Vimercato, and to master Nicolo di Arzago. The astrological services are the same, but the astrologers are not to know about each other, and the matter is to remain secret. This could be in order to get an "objective judgment," namely, one would suspect, where the two judgments agree, but also, and perhaps more importantly, to keep the duke's knowledge for himself alone. Master Raffaele will deliver his in a month. Master Nicolo is also pleased to provide his services to the duke, but he would prefer, instead of discussing the nature of every single day, to provide selected elections of important days. Simonetta calls them all "judgments" (*iudicii*).

*

To end his rich article, Gabotto treats astrological practice during the reign of Ludovico il Moro (1483–99), during whose regency and dukedom astrology enjoyed even greater favor in Milan.[113] According to Gabotto, this prince was given over to

---

[111] "Dux Mediolani, etc. Dilectissime noster. Volemo che, recevuta questa, parlati con magistro frate Johanne da Viterbio de l'ordene di predicatori, in quella nostra inclyta Citta, quale intendemo essere doctissimo astronomo, et lo exortarete strectamente da nostra parte che omnino voglia fare uno iudicio de qualitate temporum et singulorum dierum anni futuri, facendolo con omne diligentia et doctrina che per divina gratia, longo studio et grande experientia ha de simile faculta acquistata. Et facto che l'habia, con sue et vostre lettere mel manderete. Galiate, die iiij Novembris 1474. Cichus. Magnifico Militi Guidoni Vicecomiti/Genuae vicegubernatori nostro dilectissimo (401)." In addition to Curran, "De sacrarum litterarum interpretatione" and "Ancient Egypt and Egyptian Antquities," see also Anthony Grafton, "Traditions of Invention and Inventions of Tradition in Renaissance Italy: Annius of Viterbo," in *Defenders of the Text*, 76–103, and *Forgers and Critics: Creativity and Duplicity in Western Scholarship*, Princeton: Princeton University Press, 1990. For the astrological side of his career, see now Monica Azzolini, "Annius of Viterbo Astrologer: Predicting the Death of Ferrante of Aragon," *Bruniana e Campanelliana* 14 (2008): 619–32. She also discusses Annius in *The Duke and the Stars*.

[112] Illmo S.^{re} mio. Ho recevuto la lettera de la ex.^{tia} vra per la quale me scrive debia dire ad m.^{ro} Raphaele da Vimercato et m.^{ro} Nicolo de Arsago phisico che caduno de loro faccia uno iudicio de qualitate temporum et singularitate dierum anni futuri, et che l'uno non sapia de l'altro, et debiano tenere questa cosa secreta. Et per exequire quanto e dicto, ho havuto da me dicti phisici l'uno separato da l'altro, et factoli la commissione como me commette vra ex.^{tia}. Esso M.^{ro} Raphaele ha tolto voluntere el carico de farlo, perche ha caro, per fare cosa grata a la ex.^{tia}, che quella lo adoperi; et domandandoli in quanto tempo lo havera facto, me ha resposto lo fara in uno mese. Dicto M.^{ro} Nicolo similmente dice lo fara voluntere, ma perche vra S.^{ria} scrive se debia fare dicto iudicio de qualitate temporum et singulorum dierum, luy voria esser meglio chiarito da V. S., cioe se dove dice de qualitate singulorum dierum vole dir altro, cioe de electione dierum; sicche V. S. lo po chiarire, se li piace, de questa parte. Et como serano facti dicti iudicij, me li portarano sigillati, et io li mandero ad V. S., a la quale me recommando. Mediolani, v novembris 1475./J. D. V./F. Servitor Johannes Symoneta (399).

[113] Azzolini discusses astrology in depth during Ludovico's long reign, during all of which Leonardo da Vinci was in Milan; *The Duke and the Stars*, ch. 5, "The Viper and the Eagle: The Rise and Fall of Astrology under Ludovico Sforza," 167–209. For the scarce evidence concerning

astrology more than anyone else. During his reign, he called Giovan Lazzaro Sicleri, Giovanni Otto (a German), Annibale Belenco and Marsilio Cremasci to teach astrology at the University of Pavia.[114] But Ambrogio Varese da Rosate, whom we met earlier performing a medical interrogation for Pope Innocent VIII, was Ludovico's preferred astrologer. We know that he taught for a long time at Milan and Pavia; that he became a senator of Ludovico's secret council, a feudatory of Rosate and Corticella, a citizen of Novara, and also a count; that he cured Lodovico of diverse maladies; and that Ambrogio was also cherished by Maximillian, Ludovico's son, who ruled after the French invasion (1494) at the time of the holy league. We also know that on September 17, 1494, Ambrogio was sent by his *signore* to care for Charles VIII, king of France, at Asti; King Charles was infected, as Ambrogio recognized, with smallpox.[115]

We get other information on Ambrogio Varese from the famous Venetian diarist, Marin Sanudo, who tells us that in January 1497:

> The duke wrote that 300 Venetian men of arms needed to come to his aid, that they were in Lodi, and that they did not pass Adda because he wished that they should pass at the moment determined astrologically (*per punto de astrologia*)[116] by master Ambrogio da Rosate.[117]

Sanudo also tells us that in March 1498: "on the 17th day, Saturday, at hour 17, the duke of Milan entered Genoa, which hour he got from master Ambrogio da Rosate, from whose counsel he never parted."[118] Here we see Duke Ludovico relying on his astrologer for information regarding the conduct of war as well as on the best moment for a ducal entry; both are forms of astrological elections. Gabotto also discusses a letter, dated May 23, 1499, which he believes refers to Ambrogio, in which Ludovico (although unnamed) informs his correspondent, Augustino Tomentio, that the Holy Roman Emperor was not displeased by what our astrologer (*nostro astrologo*) told him about his majesty's current situation.[119] We now have, I

Leonardo's own views on astrology, see Graziella Federici Vescovini, "Note di commento a alcuni passi del *Libro di pittura*: 'L'astrologia che nulla fa senza la prospettiva'," in *Leonardo e Pico: analogie, contatti, confronti*, F. Frosini (ed.), Florence: Olschki, 2005, 99–129. Azzolini discusses this issue briefly; *The Duke and the Stars*, 22–23. She notes that among Leonardo's books were an unidentified work by Albumasar, and a vernacular translation of Alcabitius's *Liber Introductorius*.

[114] "L'astrologia nel Quattrocento," 410. Gabotto provides no dates for when they taught at Pavia.

[115] "L'astrologia nel Quattrocento," 411. For more on Ambrogio, see Azzolini, *The Duke and the Stars*, index, s.v.

[116] This is the same phraseology used in the Florentine records discussed by Casanova below, where the selection of times for holding the ceremony of the baton were also astrologically determined.

[117] "L'Astrologia nel Quattrocento," 411: "[I]l ducha scrisse che li trecento homeni d'arme (*veneziani*) doveano andar in suo aiuto, zonti che fossena a Lodi, che non passasseno Adda, perche per punto de astrologia del suo maestro Ambrosio da Rosato voleva passasseno (*Diarii*, I, 475)."

[118] Ibid., 411–12: "[A] di 17, sabato, a hore 17, el ducha de Milam era entrato in Zenoa, la quale hora have dal suo maistro Ambrosio de Rosato, dal consejo dil qual mai non di parte (*Diarii*, I, 910)."

[119] Poiche a la Cesarea Maesta non e dispiaciuto quello li significassimo esserne dicto dal nostro astrologo supra le presente occorrentie de sua M.ta, l'havemo de novo ricercato supra quelle cose et epso ne ha posto in scripto che vederay per lo incluso exemplo, quale ti mandamo a cio glilo faci

trust, a much fuller sense of the significant roles astrology could play in the political life of a major Italian city-state in the late Quattrocento.[120]

11.7 Astrology in Ferrara

Astrology also had a high cultural profile in another Renaissance court: Ferrara. Fortunately this aspect of Ferrarese culture has been explored to some extent, so I will be brief.[121] One must begin, of course, with the obligatory reference to Marchese Leonello d'Este choosing the color scheme of his daily wardrobe in accordance with each day of the week's planetary rulership. Much more interesting, however, is a letter—discovered by Aby Warburg during his researches on the Palazzo Schifanoia—from Pellegrino Prisciani (ca. 1435–1518), a courtier and humanist functionary during Borso and Ercole d'Este's rule,[122] to Leonora of Aragon, duchess of Ferrara, dated October 26, 1487.[123]

The duchess had recently communicated to Pellegrino the existence of critical matters requiring divine assistance, but the nature of these matters were not revealed in the letter. He wrote back to tell her that a particularly propitious celestial configuration was about to occur, which modern and older authorities both acknowledge; it

vedere cun piacere suo, cun dirli che a nuy he in molto cose pronosticato el vero et che pregamo N. S. Dio li facia succedere el medesmo a ley. Et benche sua M.ta no li presti piu fede che la facia, non dovera pero dispiacerli intendere ch'el dica bene de ley (412).

[120] For more readily accessable material on astrology in Milan and a much fuller and up-to-date treatment, see Azzolini, *The Duke and the Stars*.

[121] See e.g. Warburg, "Italian Art and International Astrology," and Bertozzi, *La tirannia degli astri*, on the Palazzo Schifanoia; Rotondo, "Pellegrino Prisciani," on Pellegrino Prisciani; and Cesare Vasoli, "L'astrologia a Ferrara tra la metà del Quattrocento e la metà del Cinquecento," in *Il Rinascimento nelle corti padane: Società e cultura*, Bari: De Donato, 1977, 469–567, and his "Gli astri e la corte (L'astrologia a Ferrara nell'età ariostesca)," in his *La cultura delle corti*, Florence: Il Portolano, 1980, 129–158. For the broader culture of the court itself, see Werner L. Gundersheimer, *Ferrara: The Style of a Renaissance Despotism*, Princeton: Princeton University Press, 1973, and Tuohy, *Herculean Ferrara.*

[122] Rotondo, "Pellegrino Prisciani," 70–1. Prisciani was a professor of notarial arts at the University of Ferrara in 1455. At the court of Ercole d'Este he played important roles, as did his father, Prisciano Prisciani, at Borso's court: He was ambassador to Venice in 1481, '85, '89, '91-'92 and '96-'98; in 1483–84 he was podestà of Reggio; and in 1501 he was ambassador to Rome. Also, in 1488 he was the ducal archivist and librarian. In addition to his work on the Palazzo Schifanoia, he was a painter and architect in his own right, and he wrote a history of Ferrara (*Historiae Ferrariae*) and an astronomical treatise (*Ortopasca*, 1508). His career should be compared with Giovanni Gioviano Pontano, the astrological poet and humanistically trained court functionary in Naples; see (i.a.) Jerry H. Bentley, *Politics and Culture in Renaissance Naples*, Princeton: Princeton University Press, 1987, and works by Matteo Soranzo.

[123] I use the text printed in Warburg, *Renewal of Pagan Antiquity*, 586–88 (tr. 588–90); it is printed as an appendix to the English translation of "Italian Art and International Astrology." This letter is worthy of fuller treatment. Pellegrino wrote a very similar letter to Isabella d'Este on August 15, 1509; see Rotondo, "Pellegrino Prisciani," 79.

occurs infrequently and is greatly anticipated.[124] When the head of the dragon (*caput draconis* [sc. the north node of the moon]) conjoins Jupiter in the midheaven with the moon approaching them, and a prayer is offered to God, it is much more likely to be successful.[125] Some people think, he continues, that the heavenly situation should also be sculpted on silver or some other metal (*in questo tempo far sculpire in argento o alcuno metallo la situatione del cielo in quello tempo*), but he thinks it unecessary (*non mi parere necessario*). Rather, she should utter a particular prayer explicitly written for the occasion, which is also contained in the letter.[126] Pellegrino then instructs her to go into her proper place for prayer and devotion, a private chapel of some sort, on the second of November (on Venus's day, thus Friday), at the 24th hour and three quarters. Then she should kneel and begin reciting the prayer, which he has written out.[127] She should repeat the prayer (*oratione*) continuously until the clock strikes one.[128] Within a few days, he strongly assures her (*tenga per fermo*), she will see the results of her prayer.[129]

Here we have another splendid example of the place of astrology at court, but of a different sort than we have encountered before, namely, an astrologer prescribing the most propitious time for prayer, a type of election. Prisciani shies away here from the more controversial talismanic practice of inscribing a figure of the heavens on metal, opting for verbal means alone. We will recall, of course, that we saw both discussed together by Roger Bacon in Chap. 8 as the '*opera et verba sapientiae*', the "works and words of the wise"; we also saw both in al-Kindi's *De radiis stellarum*.[130] Prisciani recommends using the power of words, in this case called '*orationes*', at an astrologically propitious time, namely, when Jupiter in the midheaven is conjoined with the north node of the moon (*caput draconis*), and the moon is approaching them.[131]

---

[124] "Nel tempo qua di sopto annotato: corre quella constellatione de cui non tanto li doctori moderni: ma li antiqui ancora: fano festa: et la qual da mi da molto anni in qua: come credo ancora da molti altri: è stato cum grandissimo desiderio expectata (586)." The lines of the text are unnumbered.

[125] Pellegrino supports this claim with three authorities: Almansor, Pietro d'Abano and Albumasar.

[126] "El perche Illma Madama mia alcuni qualche volte soleno in questo tempo far sculpire in argento o alcuno metallo la situatione del cielo in quello tempo: per non mi parere necessario: piu presto ho ordinato certe parole molto al proposito previe ala Oratione (587)[.]" Prisciani refers here to both the words and works of the wise, choosing the former over the latter as more fitting.

[127] The prayer in itself deserves further study.

[128] Vostra Ill.[ma] Sig.[ia] adonche: a dui di de novembre proximo futuro che sera de Venere di: la sira sonate le vintiquatro hore et tri quarti posta in sua bona devotione et loco apto: ingenochiata incomenciara la Oratione sua dicendo [...]. Et stagi cusi reiterando la Oratione insino chel sonera una hora di nocte (587)[.]

[129] Et tenga per fermo che non possaran troppo giorni vedera per efecto haver consequito la adimantada gratia (587).

[130] We will also see both together in Ficino's *De vita*, which was published two years after this letter was composed. I discuss *De vita* in detail in volume II.

[131] There is much relevant material on '*orationes*' discussed in Skemer, *Binding Words*.

We find ourselves, once again, in the very tricky—and highly contested—terrain where astrology dwells between magic and religion, and at the higher levels of political society. In this case, however, this domestic procedure would most likely have far less public political impact than a public annual prognostication could, depending, of course, on the nature of Duchess Leonora's mysterious concerns.[132] Furthermore, in addition to citing Albumasar and Pietro d'Abano as his authorities, Pellegrino also refers to Manilius at the end of the letter while indicating the particular significance of the conjunction's occurrence in Aquarius.

These sources are revealing because they are also the three main sources Prisciani used to design the astrological frescoes in the Palazzo Schifanoia 20 years earlier.[133] With respect to the well-known frescoes themselves, I will only mention their structure here. They are composed of three levels, each representing a different level of reality: Above is the representation of a planetary god, an Olympian god, the one who rules the month in question. Second is a series of three images representing the decans for each sign, each of which rules ten degrees of the sign. Finally, on the lowest level are representative images of life on earth, including, for the July fresco, the recent marriage of Galeotto Pico della Mirandola (Giovanni's eldest brother) to Maria Bianca d'Este.[134]

I would like to conclude this section by returning to a theme treated briefly above that will recur with a much higher profile in volume II. In the letter to Leonora, Pellegrino also indicates that the heavens are animated. The first section of the prayer he wrote for her begins with an invocation to God that describes His creation of the heavens. The language is very interesting, both with respect to the animation of the heavens, but also in relation to the material reconstructed in Parts I and II:[135]

> Omnipotent and eternal God, who created everything visible and invisible from nothing (*de nihilo*), and arranged the heavens themselves (*celos ipsos*) in such a wondrous order (*miro ordine*), and decorated them so wondrously (*mirabiliter*) with wandering and fixed stars, bestowing them in addition with rays (*radii*), lights (*lumina*), motions (*motus*), power (*potestas*) and that force (*vis*) which pleased You, and which (sc. *celos*) You animated (*ani-*

[132] Another very interesting area in which astrology played a major role was in the domestic sphere and associated political dimensions of family life, including determining propitious times for weddings and exploring whether potential marriage partners were astrologically well suited to each other. Azzolini discusses much evidence about such matters in fifteenth century Milan and related cities in her *The Duke and the Stars*, passim.

[133] Rotondo, "Pellegrino Prisciani," 71: "Nel 1469–70 fa [Pellegrino] parte del gruppo di esperti incaricati di soprintendere all'esecuzione degli affreschi di Palazzo Schifanoia."

[134] This is a large-scale painting of the encyclopedic type described above. The frescoes are very large and make a powerful impression. I should also note that this marriage made the duke of Ferrara, Ercole d'Este, Giovanni Pico's brother in law. So was the signore of Bologna, Giovanni II Bentivoglio, whose sister Pico's other brother, Antonio Maria, had married. Both of Pico's brothers and their father were all condottiere-princes of Mirandola and Concordia. I discuss this revealing biographical and socio-professional information in greater detail in Chap. 4 of my dissertation and in volume II.

[135] It should also be compared with Pico's creation story in the *Oratio*, also to be discussed in volume II.

masti) with separated intelligences (*intelligentiis seperatis*) and Your holy angels (*angelis sanctis tuis*).[136]

We find this same view in an earlier letter of Prisciani's to Paolo Antonio Trotti in Milan (1474), where he was Ercole d'Este's ducal secretary. Prisciani there describes the stars' roles as mediators between God and man, with God as the first cause and the heavens—*divinis animata mentibus*—as secondary causes.[137] Astrology was thus alive and well in fifteenth century Ferrara, as it was also in Florence, whither we shall now venture.

11.8 Astrology in Florence

As Anthony Grafton reminds us, Florence's intellectuals were haunted by astrology.[138] Concerning the fourteenth-century background, astrological views were woven deeply into two extremely influential works: Dante's *Divine Comedy*,[139] and the city chronicles by the Villani.[140] In his *Public Life in Renaissance Florence*, Richard Trexler provides further evidence of astrology's places in Florentine public and private life in the fourteenth century. For example, military forces needed to cross the liminal space of the city walls at astrologically propitious times. Matteo

---

[136] "Omnipotens et Eterne Deus qui de nihilo cuncta visibilia et invisibilia creasti: et celos ipsos tam miro ordine collocasti: errantibus et fixis stellis sic mirabiliter decorasti: radios insuper: lumina: motus: potestatem: et vim eam illis tribuens: quam tibi libuit: et quos intelligentiis separatis et angelis sanctis tuis animasti (587)[.]" We should note that these heavens are thus animated with intelligences and angels, but not explicitly with souls.

[137] Rotondo, "Pellegrino Prisciani," 79–80. In a contemporary text, Marsilio Ficino also holds the view that the heavens are animated in his *De vita* of 1489. Pico, however, seems to hold this view in his earlier period (1486–89), but explicitly does not do so in the later *Disputationes* (1493–94). Neither, emphatically, does the author of the *Speculum astronomiae,* as we saw in Chap. 1. This shift from an animated to a non-animated universe provides evidence for a major transformation of Pico's views, as I discuss in Chaps. 5 and 6 of my dissertation. I discuss this feature of Ficino's and Pico's thought in detail in volume II.

[138] "Giovanni Pico della Mirandola: Trials and Triumphs of an Omnivore," 113. Sebastiano Gentile gathers some very interesting evidence in his informative introduction to the first volume of Ficino's letters; *Marsilio Ficino, Lettere I: Epistolarum familiarum liber I*, Florence: Olschki, 1990, XXXVI-XLII. Some of this information is discussed in Chap. 5 of my dissertation. I discuss this more fully in volume II and the overall conclusion to volume I below in relation to that extraordinary year 1484. Once again, since there is a significant amount of scholarship concerning astrology at Florence, this treatment will be very selective. Grafton's just mentioned chapter and several of Michael J. B. Allen's works (cited below) are excellent places to begin.

[139] See Richard Kay, *Dante's Christian Astrology*, Philadelphia: University of Pennsylvania Press, 1994, and his "L'astrologia di Dante," in *Dante e la scienza,* P. Boyde and V. Russo (eds.), Ravenna: Longo, 1995, 119–132. Albertus Magnus was Dante's main scientific authority. For Dante's astronomy, see (e.g.) Alison Cornish, *Reading Dante's Stars*, New Haven: Yale University Press, 2000.

[140] See e.g. Nicolai Rubinstein, "Some Ideas on Municipal Progress," 178–83. Their views on historical cycles were informed by great conjunction theory.

Villani records that in 1362, the Florentine troops broke rank in order to make it through the walls by an astrologically determined time.[141] Likewise, the timing of the vitally important civic ritual of the scrutiny was determined by astrologers: "In 1385, for example, the times at which the commissions executed the different steps of the scrutiny were determined by a clerical astrologer hired by the commune."[142] Both of these are further examples of elections, the determination of optimal timing by astrological means. With respect to private life, the moment of birth was carefully recorded by the parents so that they could accurately construct the child's nativity, at least for a male child: "When a male child was born, parents carefully recorded the day, the hour, and the minute, for this moment's astrological character could reveal the child's future."[143]

For the fifteenth century, Trexler discusses the astrological timing of Lorenzo de' Medici's departure from Florence on his way to attend Sixtus IV's coronation at Rome:[144]

> [I]n September 1471 he [Lorenzo] assumed the role of communal ambassador for the first time, traveling to Rome with other emissaries to congratulate Pope Sixtus IV on his coronation. We do not know how the skeptical Romans judged him, but the Florentines thought the Magnificent, as he was now called, impressive indeed. Observing the conventions of his elders, Lorenzo left Florence for Rome when the stars were right and with 'many magnificences,' as one eyewitness reported.[145]

Even Lorenzo de' Medici's public actions were informed by astrological timing. We will see other examples below.

*

In this next section, I rely heavily on Eugenio Casanova's strikingly interesting article, "L'astrologia e la consegna del bastone al capitano generale della repubblica fiorentina."[146] He presents vivid archival evidence for reconstructing an utterly central role astrology had in Florentine public life during the second half of the Quattrocento. Although the evidence is limited, it is rich and detailed, and it features several of the main characters in late Quattrocento political and cultural history: Sigismondo Malatesta, Ercole d'Este and Lorenzo de' Medici himself.

Among the ceremonies celebrated in Florence at the time of the republic, one of particular interest is that in which the highest magistrate of the republic solemnly

---

[141] Trexler, *Public Life in Renaissance Florence*, 48 and n. 9

[142] Trexler, *Public Life in Renaissance Florence*, 334 and n. 7.

[143] Trexler, *Public Life*, 73 and n. 134: "A nativity astrologer first asked for information on the time of birth; Filarete, *Treatise*, 22. Typical are two futures read by the important astrologer Alessio: "Due responsi astrologici dell'anno 1382 resi da maestro Allesio da Firenze concernante due pistoiesi," ed. A. Chiappelli, *Bolletino Storico Pistoiese* XXIV (1922), 133–38. The same astrologer was used by Datini; Mazzei, I, 187."

[144] Trexler, *Public Life in Renaissance Florence*, 439.

[145] Trexler, *Public Life in Renaissance Florence*, 439 and n. 110: "Dissesi che parte in quella hora per punto d'astrologia"; ser Giusto, ff. 99rv. Trexler's main treatment of astrology occurs at 77–80.

[146] *Archivio Storico Italiano* 7 (1891): 134–43.

consigns the spruce baton and lilied banner, signs of command, to the captain general of Florence's troops.[147] There is evidence of the order taken in this ceremony, of the retinue, and of the speeches ("discorsi") in the *Libro cerimoniale della repubblica fiorentina* by the heralds Francesco Filarete and Angelo Manfido, which covers the years 1450–1522.[148] Casanova's essay focuses elsewhere, however—on the brief archival notices he discovered concerning the precise moment chosen for this solemn occasion, and on the practices followed for observing the propitious moment. These notices are found in a few documents at the Archivio di Stato in Florence, which Casanova published for the first time in his 1891 article.

After having selected and hired the captain general of the republic, the magistrates for military matters commissioned their astrologer and others to astrologically discover the most favorable moment for transferring the citizens' loyalty to the office (not the person) of the chosen captain general. This was obviously a matter of great importance in the public life of any city-state, but especially for Florence with its republican traditions. In Florence, the Dieci di Balia affirmed the antiquity of this procedure in a letter of September 21, 1478.[149]

We should accept this assertion of antiquity, Casanova argues, even if the documents which record these customs only come from the second half of the fifteenth century because similar customs were also found to be performed normally at other solemn occasions. He provides two examples. The first custom is used to determine the proper moment to begin the foundation of a fortress by laying down its first stone, for example, in the fortress of S. Giovanni in Florence.[150] The second example Casanova provides is for determining the proper moment for a military assault.[151] In the war of Pisa in 1404, for example, the Florentines assaulted Vice Pisano after

---

[147] For a more detailed description of the bastone itself, see Trexler's discussion just below.

[148] See Trexler's edition with an informative introduction; *The Libro Cerimoniale of the Florentine Republic by Francesco Filarete and Angelo Manfidi: Introduction and Text*, Geneva: Droz, 1978.

[149] For the Dieci di Balia, an extraordinary committee formed during wartime, see Nicolai Rubinstein, *The Government of Florence under the Medici (1434 to 1494)*, second ed., Oxford: Clarendon Press, 1997; Michael Mallett, "The Florentine *Otto di Pratica* and the Beginning of the War of Ferrara," in *Florence and Italy: Renaissance Studies in Honour of Nicolai Rubenstein*, P. Denley and C. Elam (eds.), London: Westfield College, 1988, 3–12; and Patrizia Salvadori, "I signori di Mirandola, Firenze e i Medici," in *Pico, Poliziano e l'umanesimo di fine Quattrocento: Biblioteca Medicea Laurenziana*, Paolo Viti (ed), Florence: Olschki, 1994, 287–302, 298.

[150] Casanova, "Consegna del bastone," 135 (2). See also John Hale's splendid description, "The End of Florentine Liberty: The Fortezza di Basso," in his *Renaissance War Studies*, London: The Hambledon Press, 31–62, 49–50. Francesco Guicciardini provides curious notices about these matters in a letter written from Bologna to his brother, Luigi, dated July 21, 1534. Casanova provides the archival reference (135, n. 2). See also *I Guicciardini e le scienze occulte: L'oroscopo di Francesco Guicciardini, lettere di alchimia, astrologia e cabala a Luigi Guicciardini*, R. Castagnola (ed.), Florence: Olschki, 1990. This interesting material will be discussed more fully in volume III.

[151] Casanova, "Consegna del bastone," 135, n. 3.

having gotten the moment (*avere il punto*) from the astrologer.[152] These are both, of course, further examples of astrological elections.[153]

*

Astrological skills were required anywhere the ceremony of the baton took place and by whatever magistrate it was determined. Casanova presents three examples culled from Filarete's manual: [1] Describing a ceremony celebrated for giving the baton to Costanzo Sforza, signore of Pesaro, on Oct 4, 1481, Filarete wrote: "On day 4, hour 17 1/3 on Thursday, he took the baton." Casanova adds that even if he does not say so, it is certain that the hour was indicated by the astrologer.[154] [2] We also know that in Florence, at the *ringhiera* of the Palazzo del Priori, "Count Niccola Orsino, count of Pitigliano, took the baton on June 24, with the election of the astrologer at the 18 1/2 hour, in the year 1485."[155] [3] On June 1, 1498, to Paolo Vitelli "is given the baton at the 14 1/2 hour with the celestial configuration (*constellatione*) watched for by the astrologer."[156]

In order to get a fuller picture of the ceremony of the baton, I quote Trexler's description in full:[157]

> The *presa di bastone*, as the pompous ceremony was called through which the *condottieri* became leaders of Florence's army, reached back to those times when the republican commune had depended entirely on mercenary forces and foreign generals for the city's military welfare. Typically the ceremony had involved the solemn entry of the marshal into the city with a small retinue of his troops. This procession led to the *ringhiera* or platform in front of the Palazzo della Signoria, where the marshal was received by the government of the city, heard his praises sung by orators, and received symbols of authority from and through the commune. The most important of these was the *bastone*, described by the communal master of ceremonies in 1515 as a "thin-veined, iron-gray spruce a yard and a half long." From an earlier description we know that it was "without ornament."

---

152 As we saw, this same idiom is used for the same purpose in the Milanese material recently discussed.

153 For much more on military astrology, see Hand, "The Use of Military Astrology in Late Medieval Italy."

154 Casanova, "Consegna del bastone," 135, n. 6: "Nel margine della c. 15 t° [Filarete, *Cerimoniale*], descrivendo la cerimonia celebrata il 4 ottobre 1481 sulla ringhiera medesima per dare il bastone a Costanzo Sforza, signore di Pesaro, l'araldo scrisse: 'Adi 4, hore 17 1/3 in giovedi, prese el bastone.'" For astrology at the Sforza court in Pesaro, of Costanza and later Giovanni Sforza, see Gamba, "Astrologi alla corte dei Montefeltro a Urbino e degli Sforza a Pesaro," 1: 80–2, for a discussion of Camillo Leonardi, who served them both.

155 Casanova, "Consegna del Bastone," 135 and n. 5; Filarete, *Cerimoniale,* c. 17 t°: "In Firenze sulla ringhiera del Palazzo dei Priori il 'Conte Niccola Orsino, conte di Pitigliano, prese il bastone adi 24 di giugno, con electione dello astrolago a hore 18 1/2, anno 1485.'" Here the election itself is mentioned explicitly.

156 Casanova, 135 and n. 6; Filarete, *Cerimoniale,* c. 31 t°: "e il 1° giugno 1498 a Paolo Vitelli 'diessi el bastone a hore 14 1/2 con costellatione pervigiliata da l'astrologo.'"

157 Richard Trexler and Mary Elizabeth Lewis, "Two Captains and Three Kings: New Light on the Medici Chapel," in his *Church and Community 1200–1600: Studies in the History of Florence and New Spain*, Rome: Edizioni di Storia e Letteratura, 1987, 169–244 (originally published, 1981).

> The portentiousness of the ceremony, surrounded by astrological lore, became evident at the moment of bestowal. The Standard Bearer of Justice handed the truncheon over to the new captain at the correct stellar moment. If an unexpected "bad disposition of the heavens" arose, the *presa* might be postponed.[158] Favorable omens at the moment of bestowal, on the contrary, were a cause of joy: in one case at that very moment a wind came up which blew the captain's flag in the direction of the city to be attacked, and this was said to embolden the captain. Once received, the *bastone* along with the other symbols had to be solemnly but punctually carried out of the city along astrologically determined routes, through similarly determined gates, by a certain astrological deadline.[159] This was true even if the captain had to consult further with the government about strategy. The captain deposited the symbols in a church outside the walls and then reentered in private status without these symbols of authority.
>
> Such astrological procedures reflected evident political preoccupations. Nothing should suggest that the new captain had signorial rights over the city. This concern was all the more pressing in a republican city because signorial regimes in Italy bestowed similar batons to grant civil authority as well as military powers abroad. Florentine history itself showed more than one case in which the mercenary captain had wanted to interpret the *presa di bastone* as conferring the "blood jurisdiction" of princely Europe (177–79).

This vivid description thus indicates how deeply rooted astrology was in a major Florentine ritual of public power.

*

Casanova presents two detailed examples that bring these practices to life by placing them within their intensely vital circumstances. In the first example, from 1453, the time for bestowing the *bastone* was chosen while the Florentine army was already gathered in the field after having confronted the enemy, since the captain general needed to be formally invested by the commissioners of the republic. In that case, after a period of neglect, the astrologers recalled the magistrates to the ancient practice on the morning of September 29, 1453, after terrible and frightening earthquakes (*terremoti terribilissimi et spaventevoli*) struck Florence in the middle of the previous night. Recognizing astral influences, the astrologers reminded the "signori priori" of the ancient custom. Thus they delayed the ceremony of the baton, which had already been arranged, until the time became propitious. One is strongly reminded here of ancient Roman augural practices.[160] Casanova prints this letter as

[158] An example is examined below.

[159] Trexler refers here to: *Croniche di Giovanni, Matteo e Filippo Villani*, 2 vols. (Trieste, 1858), M. Villani, bk. XI, cap. 51 (1363).

[160] For a rich introduction to the subject, see Jerzy Linderski, "The Augural Law," *Aufstieg und Niedergang der Römischen Welt*, W. Haase (ed.), II, 16. 3, Berlin: de Gruyter, 1986, 2146–2312. Lily Ross Taylor brings their use in political elections to life in her Sather Lecture, *Party Politics in the Age of Caesar*, Berkeley: University of California Press, 1949.

document I.[161] It is usually assumed that the Paolo mentioned in this letter is Paolo Toscanelli, but, as Jane Jervis points out, the evidence is not very strong.[162]

In *Public Life in Renaissance Florence,* Trexler discusses the related issue of omens in Florence in his section on astrological timing:

> Omens cross almost every page of Florentine history, making secular events sacred, formalizing and concluding most chapters of the civic past. Machiavelli did not really understand why "no great events ever occur in any city or country that have not been predicted by soothsayers, revelations, or by portents and other celestial signs." But he suggested that certain philosophers were right in explaining that "the air is peopled with spirits, who by their superior intelligence foresee future events, and out of pity for mankind warn them by such signs, so that they may prepare against the coming evils." Whatever the explanation of this curious fact, its importance to the historian was indisputable: "Be that as it may, however, the truth of the fact exists, that these portents are invariably followed by the most remarkable events." The great political writer simply repeated a truism of the age.[163] [...] "I am a dead man!" Lorenzo the Magnificent is reported to have exclaimed on being told that the collapse of part of the cathedral's cupola in 1492 had sent boulders careening northward toward his palace (80–1).[164]

As an unusual natural phenomenon, the earthquake of 1453 should be understood in this context.

At the magistrates' request, the astrologers responded conscientiously. They studied the paths of the stars and, having found the propitious moment with their calculations, they gave a minute and precise notice for the magistrate to record. The astrologers urged the magistrate not to depart in any manner from the precepts

---

[161] Casanova, 136. Document I is printed at 143: (R. Archivio di Stato in Firenze. —Riformagioni. —Signori. Carteggio: Missive, Registri, I Cancelleria n.º 38, a c. 147 t.º). "Commissariis campi. Spectabiles concives car.[mi] Noi vi scrivemo hiersera quanto fu di bisogno rispondendo a la vostra ultima de' di 26 et dipoi non abbiamo vostra lettera, ne a noi occorre dirvi altro di nuovo. Questa solo vi facciamo per darvi notitia come questa mattina sentendosi da m.º Paolo [Toscanelli] et da altri la commissione che v'abbiamo data di dare il bastone al m.[co] s. messer Sigismondo ci anno ricordato che per insino adi diecj d'ottobre non c' e alcun punto da fare tale acto, ma tutti sono punti contrarii. Et annoci confortato a fare differire il dare del bastone insino a quel tempo. Et cosi ci contentiamo che si facci. Et in questo mezo faremo vedere con piu diligentia quale et quando sara il punto buono et darenvene notitia accio che lo facciate allora. Et nondimeno e nostra intentione che la sua M.[tia] abbia l'obedientia et il governo et sia Cap.º generale et cosi exerciti quello uficio come se avesse avuto il bastone. Et pero sarete con la Sua M.[tia] et con quelle parole et modi che vi parranno migliori li darete notitio di questo che vi scriviamo et faretelo rimanere contento che soprasegga al dare di decto bastone ceme e detto; non obstante che a voi et allui abbiamo per altre scripto che voi glele diate a sua posta; avisandovi che la nocte passata dopo le cinque hore ci cominciorono terremoti terribilissimi et spaventevoli, quali non furono mai piu uditi per huomo che ci sia, con guastamento di molte case et edifici. Donde e proceduto che c'e stato ricordato questa cattiva dispositione de' Cieli che corre al presente et debba correre anchora alchun di [...]. Dat. Florentie die xxviiij[a] septembris 1453 hora 17."

[162] Jane L. Jervis, *Cometary Theory in Fifteenth-Century Europe*, Wroclaw: Ossolineum, 1985, 45–46.

[163] See also Ottavia Niccoli, *Prophecy and People in Renaissance Italy*, L. Cochrane (tr), Princeton: Princeton University Press, 1990, for a slightly later period.

[164] For a suggestive but not fully developed argument that astrology deeply informed Machiavelli's world view, see Parel, *Machiavelli's Cosmos*. I discuss and critique Parel's argument in volume III.

because the work of the captain, to whom the command of the army was given with the standards, will not fail on their part.[165] Thus, in the wake of the ominous earthquake, the magistrates capitulated to the exhortations of the astrologers on September 29, 1453, countermanding a previous order already given for performing the ceremony of the baton, which had already been received by the authorities in the field. After being informed of Sigismondo Pandolfo Malatesta's election as captain general and their participation at his appointment, these same authorities had already sent instructions the day before (September 28) for performing the ceremony of the baton.[166] In these first orders there is no mention of astrological timing.[167] Thus, it is clear that before the ominous occurrence of the earthquake they felt no concern to interrogate the stars first before performing the ceremony.

But that very night the terrible earthquake struck and the astrologers warned the priors that they should follow ancient custom and consult an astrologer for the ceremony's proper timing, so they consulted master Paolo and his colleagues. He gave an order to delay the passing of the baton until the tenth of October, on which day alone the proper moment for taking this action would occur. But the new order arrived too late. As soon as they received the letter of September 28, the officials tried to execute the orders contained in it, as they notified the Signori. In their return letter of Oct 3, 1453, however, they expressed little concern at not observing the revised precepts. In fact, they had derived good omens from Sigismundo's promises

[165] Casanova, "Consegna del Bastone," 136.

[166] Casanova, "Consegna del Bastone," 138, n.1: (R. Arch. di Stato in Firenze. —Riformagioni. —Signori Cart. Miss. Registri, I Cancelleria: n.º 38, a c. 144 t.º-145) "Commissariis campi [...]. Et a cio che per effecto si dimostri la fede che noi habbiamo sempre avuta nella persona del s. Sigismondo, vi diciamo che noi siamo contenti, come ne richiedete, che lui sia Capitano generale del nostro exercito et cosi possa comandare con quella auctorita et con quelle preeminentie che a tanta dignita si conviene et si come ne'capitoli facti con lui gli fu promesso. Et pero significherete alla sua M. quanto si dice; gli darete in titolo et in segno di tale auctorita nel nome dell'altissimo Dio a sua poste il bastone, confortandolo et pregandolo che queste nostre cose posate sopra le spalle della sua sapientia vogla con sperientia mostrare a questo popolo quello che moltissime volte la sua M. et per sue lettere et per sua parte ci e stato significato... Dat. Flor. die xxvij septembris 1453, hor. 3 noctis." —And, in the same place, a c. 145 t.º: "[...] Come per l'ultima scrivemmo a'nostri Commissarii, noi vi diamo et concediamo et confermiamo quella auctorita de l'essere governatore et capitano del nostro exercito et commettemo loro che in segno di cio nel nome dello eterno Dio vi dessono il bastone, si come pe'capitoli fermati fra la V. M. et il nostro Comune vi fu promesso [...]."

[167] Casanova, 138 and n. 2: (R. Arch. di Stato in Firenze. —Riformagioni. —Signori Cart. Miss. Registri, I Cancelleria: n.º 38, a c. 147) "Circa la parte della obedientia messer Sigismondo, [...] gl'abbiamo dato et diamo la obedientia di governo et titolo di Capitano generale et cosi voglamo ch'egl'abbi in segno di quello il bastone [...]. Et cosi per queste vi commettiamo che con 'l nome di Dio glelo diate a ogni sua volunta, acio che niente manchi che cognosca essere utile al buono governo delle nostre genti et a conseguire il fine disiderato delle imprese disegnate. Et cosi con quelle parole et modi che vi parranno convenienti gli direte et farete [...]. Dat. Flor. die xxviij septembris 1453, hor. 4."

and vivacity, from the field of battle, and from the successful beginning of the undertaking, "which is a good sign and augur" (*che è il segno et augurio buono*).[168]

The second example places astrological timing at the very heart of Lorenzo de' Medici's Florence at one of its darkest moments, September 1478, soon after the murder of his brother Giuliano and the unsuccessful attempt on Lorenzo's own life.[169] There exists a precious document, the letter which the Dieci di Balia wrote to Lorenzo de' Medici (who had been elected *sindaco*, September 26, 1478), for handing over the baton of command to the captain general of the Florentine army, Ercole d'Este, duke of Ferrara:

> [T]hey sent him the same note of the astrologer, so that he could diligently observe the determinate moment, which is the 16th hour and a fifth; and they exhort him to attend strictly to their instructions, since it is of the greatest importance that the ceremony take place at that precise time and not before because all the moments preceding that one are unpropitious. But if he is unable to observe that moment with the recommended diligence on account of finding oneself in an open region and thus deprived of the means of measuring time exactly, rather than running the risk of anticipating the hour indicated by even a minute, let it wholly pass by, always taking care to give a time to the beginning of the ceremony as close as possible to the earlier time.[170]

[168] Casanova, "Consegna del Bastone," 139, n.3: (R. Arch. di Stato in Florence. —Riform. —Sig. Cart. Miss., I Canc. n.° 38, a c. 150 t.°) "Commissariis campi. Karissimi nostri. Rispondendo alle vostre ultime lettere che sono de' di xxx del passato et de' di 2 del presente, verremo solamente agli effecti. Et prima vi diciamo che non obstante che noi v'avessimo scripto che voi soprasedessi di dare il bastone al m.co s. Sigismondo, nondimanco, poi che gliel'avete dato, ne siamo contenti, poi che la sua S.a se ne contenta; et insieme con la sua M. et con voi ci rallegriamo del conforto che scrivete che per la sua S.a et per cotesto campo se n'e preso, sperando che sara stato in buon'ora et in buono punto, et maxime veduto il felice principio, ch'e il segno et agurio buono come scrivete; et cosi piaccia a Dio farlo succedere di bene in meglio [...]. Dat. Flor., die iij octobris 1453, hor. 2."

[169] For a lively and learned account of the famous Pazzi conspiracy, see Lauro Martines, *April Blood: Florence and the Plot against the Medici*, Oxford: Oxford University Press, 2003.

[170] Document II (144): (Dieci di Balia. Carteggio: Missive, Registri, n.° 7, a c. 90 bis). "Die xxvj sept. 1478. Laurentio de Medicis. Mag.ce collega come.e noster. Mandianti in questa la poliza di mano della astrolago nella quale e segnato il punto che si debba dare il bastone a cotesto ill.mo Cap.° Siamo stati confortati a mandartela perche piu a punto possiate observare; perche e stimato d'importantia gravissima, perche dopo quelle 16 hore, che sono segnate, et maximamente nel punto disegnato il Cielo promette ogna cosa felice; ma innanzi a quelle hore 16 ogni cosa e in contrario. Non ci pare per cosa alcuna che se pretermetta alcuna diligentia che v'acostiate piu a quello punto che e possibile. E difficile et specialmente costi, secondo crediamo, misurarlo cosi ad unguem; ma vuolsi in ogni modo passare le 16 hore et mettere ogni industria possibile per acostarsi il più che si puo a quello punto disegnato." Joan Quigley makes a similar point; *What does Joan Say?*: "In addition to birth charts [sc. nativities], there are charts [sc. horoscopes] for the beginning of events [namely, elections]. When a meeting takes place, an important document is signed, a journey begins, the astrologer can warn you about avoidable dangers and tell you how something potentially good can be made even better yet. He can predict whether a venture will end in failure

Astrology thus played significant roles in Florentine public life during the fourteenth and fifteenth centuries.[171] The last mention of astrology in Lorenzo's Florence (in this chapter, at any rate) takes us to the end of his reign, to the carnival festivities of 1490–91, when Lorenzo himself designed the carnival pageant around the theme of the seven planets, for which he wrote the poetry himself.[172]

*

As we can now see much more clearly, astrology played significant roles in the knowledge-power dynamics of Quattrocento Italy and elsewhere on both highly public and more private stages. This brief series of microhistorical vignettes has indicated some of the broad range of the roles that astrology actually played in this highly-dynamic, brutally-passionate and intensely-competitive culture into whose highest reaches Giovanni Pico della Mirandola (1463–94) was born, lived and passionately participated. We are now in a better position to engage with central features of his writings, and those of his extremely influential contemporary, Marsilio Ficino (1433–99). Now that astrology has been placed into its institutional and socio-political contexts and brought up to the end of the fifteenth century—and we have shifted geographical focus from Northern Europe to Northern Italy—we may now return in volume II to the intensive exploration of conceptual patterns in the works of two of the most famous figures in the historiography of Renaissance thought.

---

or success. In some cases, such charts must be very exact because of technical factors. For example, the announcement of Anthony Kennedy's nomination to the Supreme Court was timed with a stopwatch, not only to the minute but to the exact second of time. In other cases, being off a few minutes is not of great importance (15)."

[171] We will see in volumes II and III that it continued to do so throughout the sixteenth century and beyond.

[172] This ceremonial performance—and its political significance—are discussed in Paola Ventrone's well-illustrated, *Le tems revient*, and in her essay, "Lorenzo's *Politica Festiva*," in *Lorenzo the Magnificent: Culture and Politics*, 105–116; Mario Martelli, "Una vacanza letteraria di Lorenzo: Il carnevale del 1490," in his *Studi Laurenziani*, Florence: Olschki, 1965, 37–49, and Vescovini, "Lorenzo il Magnifico e l'astronomia a Firenze," *Il Ponte* 48, 11 (1992): 77–106, 93 ff. Other features of astrology's varied presence in Lorenzo's Florence are treated in Chaps. 5 and 6 of my dissertation, and in volume II. I also discuss features of astrology in 16th and seventeenth century Florence, especially in the courts of Dukes Cosimo I and II de' Medici, in volume III.

Overall Conclusion to Volume I

The *Annus Mirabilis* of 1484: Towards "Renaissance" Astrology and Magic

Introduction

Astrologizing *Aristotelian* natural philosophy held the day virtually unchallenged in the Middle Ages from the mid-13th throughout most of the fifteenth century, but that situation changed dramatically in the early years of the 1480s when Marsilio Ficino published his philosophical masterpiece, the *Theologia Platonica,* in 1482, and his epoch-making and extraordinarily influential translation of all of Plato's works from Greek into Latin in 1484, to correspond, in fact, with the much anticipated Great Conjunction in Scorpio of that year.[1] Nevertheless, the medieval structures reconstructed in volume I provide the touchstone, the structures against which to measure the range of continuities and transformations in the Renaissance, the Reformation and early modern Europe. In this conclusion to volume I, I will first briefly resume these medieval structures in curricular-disciplinary and conceptual respects, and draw some final conclusions. Then I will address an epoch-making technological invention—the printing press with movable type—and its ever increasing importance to astrology (and vice versa) from the second half of the fifteenth century onwards.

[1] In general, and with much detail, see James Hankins, *Plato in the Italian Renaissance*. Of course, Ficino also translated and published works by several Neoplatonists in the 1480s and '90s, including Plotinus, Iamblichus and Proclus, and he composed and published numerous of his own commentaries on Plato's dialogues. Ficino's *Theologia Platonica* is now easily accessible with a marvellous English translation in the I Tatti Renaissance Library; *Platonic Theology*, Latin text edited by James Hankins with William Bowen; English translation by Michael J.B. Allen, 6 vols., Cambridge, MA: Harvard University Press, 2001–2006.

H D. Rutkin, *Sapientia Astrologica: Astrology, Magic and Natural Knowledge, ca. 1250–1800*, Archimedes 55, https://doi.org/10.1007/978-3-030-10779-6

[1] Medieval Structures

As I have now shown in some depth and detail—and using a broad range of evidence—the disciplinary and curricular patterns characteristic of premodern Aristotelian-Ptolemaic-Galenic natural knowledge as taught at European universities ca. 1250-ca. 1500 provide the primary structures for developing sharper and more accurate analytic tools. With these tools I have constructed a well-articulated, flexible, and historically- and conceptually-sound interpretive framework to use in more sharply focusing and ultimately reframing the larger historical question of how astrology came to be removed from its previously central conceptual and institutional locations in medieval, Renaissance and early modern natural knowledge during the seventeenth and eighteenth centuries. It is also relevant for articulating both the significant differences *and* the deep-structural continuities between medieval and Renaissance astrology and magic (as I do in volume II), and in relation to Copernicus's education and life's work, and to those of other major Scientific Revolution figures, including Galileo Galilei, Johannes Kepler, Francis Bacon and Isaac Newton (to be treated in volume III) as well as for many other issues and figures not mentioned here.

In approaching these larger questions, we must first accurately understand astrology's status and locations in the medieval map of knowledge ca. 1250 to 1500, which eventually came to be so greatly transformed. In volume I, I have focused primarily on two scientific disciplines: mathematics and natural philosophy, although I have also taken the third, medicine, into account, albeit not as fully. Reconstructing characteristic features of these three bodies of knowledge—in all of which astrology was significantly configured—is a major component of my interpretive framework.

As we have seen, in the mathematics curriculum, astrology as a theoretical and practical doctrine was taught as a central feature of the "science of the stars" along with mathematical astronomy, but only after the preliminary study of arithmetic and geometry. This integral configuration of astrology with astronomy differs greatly, of course, from our modern disciplinary structures, where astrology is no longer considered a legitimate part of mathematics, nor is it taught as such in modern universities.

Furthermore, astrology's foundations in nature were taught in the natural philosophy course by reading core texts of Aristotle, most notably, the *De caelo* and *De generatione et corruptione*. Her foundations in medicine, in their turn, were taught in the medical course with core texts by Galen, including the *De diebus decretoriis*. To establish my interpretive framework, I have explored in detail the characteristic structures of these conceptual and curricular patterns by reconstructing their understanding in the thirteenth century, and their further development and institutionalization during the fourteenth and fifteenth centuries. The basic curricular structures for astrology's configuration within mathematics, natural philosophy and medicine were clearly articulated in the 1405 statutes for the University of Bologna, and were strikingly consistent until well into the seventeenth century, and in some places beyond.

*

I will now briefly recall the central conceptual structures of what I have come to call an "astrologizing Aristotelian" system of natural knowledge. As taught in Italian and other universities in the fourteenth and fifteenth centuries (and beyond), three main aspects of the topic received the most attention: natural philosophy, cosmography, and a geometrical-optical model of planetary influences. According to Aristotle's natural philosophy, the movements of the heavens, especially the sun, were utterly fundamental to life on earth. In *De generatione et corruptione* II.9–10, Aristotle argued that the sun in its annual motion around the ecliptic, as the universal efficient cause, was ontologically prior to and necessary for generation and corruption. Therefore, in processes of generation (human and otherwise), the sun as efficient cause was required along with the male, who provided the formal cause (the specific form) in the seed, and the female, who provided the material cause in the womb. In his commentary on this passage, Albertus Magnus expanded the realm of the efficient cause to also include the rest of the planets. In Book II of *De caelo,* Aristotle discussed celestial influences, which he limited to motion and light. The luminaries' and planets' motions and light both provided the basis for their mathematization. Albert also emphasized that these Aristotelian texts provided the natural philosophical/scientific foundations for astrological practice.

These Aristotelian natural philosophical structures were fitted into a fundamentally Ptolemaic cosmographic framework, composed of mathematical astronomy calibrated with mathematical geography. This cosmographic framework allowed the planetary motions to be uniquely determined for any time at any place by means of the horizon. This is important because place was essential for analyzing the role of celestial influences in generation, as both Albertus Magnus and Roger Bacon made abundantly clear.

Once the planets' motions were mapped in this way, and related to each place on earth, their influences could then be analyzed using the other primary feature of this system, a geometrical-optical model of planetary influences. These irradiating influences were thought to act omnidirectionally and in straight lines, on the model of the planets' other mode of influence, light. The angular relationship of the planets to each other—the planetary aspects—and their collective relationship to each place on earth could then be fully articulated. When the different qualitative natures of each planet—in themselves and as modified by each sign of the zodiac—were taken into account, and the variation in effect from their varying angular relationships (both to themselves and to each place on earth), the result was an integrated and highly mathematicized natural philosophy of richness and sophistication. As we saw in Parts II and III, this astrologizing Aristotelian system of natural knowledge also provided the primary conceptual framework for analyzing astrology's complex and sometimes controversial relationship to theology/religion and to magic, thus offering a range of fundamental conceptual, curricular and disciplinary structures for assessing both continuities and transformations that should all be taken into account in constructing a comprehensive framework.

Establishing and properly historicizing this framework serves two fundamental and interrelated purposes in my study. First, it articulates astrology's primary conceptual and institutional structures within the medieval map of knowledge (ca. 1250–1500), as I have done in volume I. Secondly, it provides a means to analyze in detail astrology's gradual removal from this central position and other relevant questions to be explored in volumes II and III. The interpretive framework presented here is intended to orient future research in this utterly central yet incompletely understood part of the history of science and culture. It may also be used to grind more accurate conceptual lenses through which to examine the premodern material in greater depth and in sharper detail, as I argued in the methodological section of the overall introduction. I hope that volume I exemplifies the success of this endeavor.

*

We have now seen that astrology had solid conceptual and/or institutional foundations in two major areas that we would call science and religion: it had both conceptual and institutional foundations in the study and teaching of mathematics, natural philosophy and medicine (as reconstructed in Parts I and IV), and conceptual foundations in theology (as explored in Part II).[2] Astrological theory and practice were thus supported, maintained and at least sometimes defended by means of these well-established structures of [1] astrologizing Aristotelian natural knowledge, as I have analyzed primarily in the works of Albertus Magnus and Roger Bacon, and of [2] Roman Catholic theology, as I have argued with Thomas Aquinas's conception of divine providence. Astrology was thus in fundamental ways subordinated to and dependent upon—or, perhaps better, subalternated to—both disciplines for both its practical and epistemological legitimacy.

Magic, however, presents a different aspect when viewed through the focus of *imagines astronomicae* or talismans. In this configuration, magic did not provide astrology's foundations in any respect. Quite the contrary! For the most part, magic was itself dependent on practical astrology (especially elections) and its mathematical, natural philosophical and theological foundations (despite Thomas's ambivalent protestations), which thus provided magic's practical and ontological foundations—at least for the making and use of talismans—as well as whatever legitimacy it might, from time to time, have enjoyed. This strong linkage between astrology and magic then seems to have provided the fundamental pivot in the Enlightenment when astrology finally lost its mathematical, natural philosophical and theological foundations within the map of legitimate knowledge, as science and religion were themselves both profoundly transformed under the increasing and multifold pressures of incipient modernity.

[2] I hope to develop theology's institutional foundations more fully in volume III.

As it turns out, when astrology's foundations in nature and theology were undermined and ultimately dissolved during the seventeenth and eighteenth centuries, she seems to have survived this epoch-making rupture by linking up much more closely with magic and the other so-called occult or esoteric sciences/traditions/currents, but this is an Enlightenment story of appropriation, reconfiguration and rejection to be explored in detail in volume III. In this epoch-making adaptation, astrology found a new culturo-evolutionary niche within the rapidly transforming map of knowledge and culture in which to survive her transition into the modern world, where she is currently alive and well, and active in many parts of the world—both East and West—although, at least in the West, lacking in most of her former conceptual and institutional legitimacy.

*

In my reconstruction, a nested and strikingly architectonic structure seems to emerge with theoretical and practical astrology at the center. To its left (as it were) astrology is subalternated to and both conceptually and institutionally dependent upon and integrated with mathematics, natural philosophy and medicine in the first instance. Then, in a second nesting, this astrologizing Aristotelian system of natural knowledge—the combination of theoretical and practical astrology as integrated within and grounded upon its mathematical, natural philosophical and medical foundations—is in its turn conceptually and institutionally subalternated to the main science, art and/or discipline in medieval Catholic Europe, namely, theology, where it is integrated within and grounded upon, among other things, Thomas Aquinas's analysis of Divine Providence. Astrologizing Aristotelian natural knowledge is thus revealed to be nothing less than theology's handmaiden, as we might have surmised. In the other direction, this highly-articulated nested architectonic structure supports, grounds and legitimates the one operative part of astrology found within a subset of astrological elections, namely, all-natural talismans as powered by celestial energies alone. This richly and complexly integrated medieval scientifico-theologico-practical system was ultimately passed on to Renaissance and early modern Europe.

I have found that astrology was subalternated to both natural philosophy (more closely) and theology (more distantly), both of which provided her with essential conceptual and institutional foundations within the medieval map of knowledge. Thus established, astrology itself—as grounded in and legitimized by astrologizing Aristotelian natural knowledge (which it itself helped to articulate, as we have seen)—could then provide the relevant foundations in both theory and practice for one of her own more controversial parts, namely, the design, manufacture and use of talismans (*imagines astronomicae*) as a part of elections, along with the other canonical branches of practical astrology.

These are the normative medieval conceptual and institutional structures against which Renaissance, Reformation and Early Modern continuities and transformations should be measured. At least, this is how I view the situation. Others may disagree. Regardless, I hope that I have persuaded the curious reader that the subject is interesting and worthy of further investigation, and that my interpretive framework may be helpful in this endeavor. The next step will be to explore Renaissance continuities and transformations in volume II, to which we will soon turn. First, however, we should briefly explore certain features of astrology's relationship to printing.

[2] The Early Printing of Mathematical, Astronomical and Astrological Texts: Regiomontanus and Ratdolt

To complete volume I, we should now examine another facet of astrology's role within an important feature of the late fifteenth century intellectual and institutional context, namely, the new epoch-making means of disseminating information, the printing press, which, in addition to printing annual prognostications and almanacs,[3] also supplied books to be read at the universities. Regiomontanus played a central role in this story as the first publisher of scientific texts. I will here examine his publication list, and that of one of his protégés, Erhard Ratdolt, who published for a far longer time than Regiomontanus himself.[4] As it turns out—and as we would expect—books by both Regiomontanus and Ratdolt formed part of the astronomico-astrological contents of Giovanni Pico della Mirandola's library.[5] Furthermore, their publications fit neatly into the curricular patterns examined in Part IV, and they served to make astrological materials much more broadly and economically accessible.

After six productive years in Hungary, Regiomontanus returned to Nuremberg in 1471 in order to reform astronomy at its theoretical, mathematical and observational roots, and to publish the most important works in the closely related fields of mathematics, mathematical astronomy and astrology.[6] His press was active from 1472–75, publishing nine works in total, including Peurbach's *Novae theoricae planetarum* (1472 [ISTC ip01134000]), Manilius's ancient astrological poem, *Astronomicon* (ISTC im00202000), and a few works of his own: the *Disputationes contra Cremonensia in planetarum theoricas deliramenta*, criticizing the work of Gerard

[3] There is much valuable information on the earliest printing of almanacs and annual prognostications in Jonathan Green, *Printing and Prophecy*, and his "Printing the Future: The Origin and Development of the *Practica Teütsch* to 1620," *Archiv für Geschichte des Buchwesens* 67 (2012): 1–18, which supercedes the former; and in Richard L. Kremer, "Incunable Almanacs and *Practica* as Practical Knowledge Produced in Trading Zones," in *The Structures of Practical Knowledge*, Matteo Valleriani (ed.), Dordrecht: Springer, 2017, 333–69.

[4] For an assessment of Ratdolt's significance, see Hammer, "Astrologie und Buchdruck," 286, and the bibliography below.

[5] I treat this in detail in Chap. 3 of my dissertation.

[6] This has been discussed in detail by Zinner, *Leben und Wirken*, 163–236; Rose, *Italian Renaissance of Mathematics*, 90–117; and Swerdlow, "Science and Humanism in the Renaissance." I will thus be very selective here. For more on the printing of astrological (and medical and philosophical) texts from the beginning of printing through 1700 with a focus on Latin translations from the Arabic, see Hasse, *Success and Suppression*, 7–17, and his valuable catalogue, 317–410.

Sabbioneta of Cremona (13th c. [ISTC ir00104000]),[7] two calendars (Latin [ISTC ir00092000] and German [ISTC ir00100300]), and his ephemerides for 1475–1506 (1474 [ISTC ir00104500]). In 1474 he also published a prospectus of books that he hoped to publish (ISTC ir00091800), but his early death in 1476 made the completion of his plans impossible. Fortunately, others were willing and able to pick up the torch.

I will briefly describe his publication program.[8] There were two parts, books by other authors and those he wrote (or hoped to write) himself. Of works by others authors, all are concerned with mathematics, astronomy and astrology. In addition to Peurbach's *Theoricae novae* and Manilius's *Astronomicon*, he listed works on arithmetic[9] and geometry,[10] conic sections,[11] spherical geometry,[12] pneumatics[13] and mechanics,[14] and an ancient astronomical poem, Hyginus's *Astronomia.*

Of most interest for our purposes are the works he intended to publish in astronomy, perspective and astrology. First and foremost, he intended to publish Ptolemy's major works: new translations of the *Almagest*,[15] *Tetrabiblos*, and *Centiloquium*,[16] as well as editions of his works on optics and harmonics. Furthermore, he hoped to produce a new translation of the *Geography* with the assistance of Theodore of Gaza and Paolo Toscanelli, as well as a large commentary on it.[17]

Of works on astrology proper, in addition to the works of Manilius and Ptolemy just mentioned, Regiomontanus also hoped to publish editions of what remained of Firmicus Maternus's writings, and the works of Leopold of Austria, Antonio de Montulmo, and other astrological prognosticators worthy of note.[18] In optics, in addition to Ptolemy, he hoped to publish Witelo's great *Perspectiva*, a work Regiomontanus called an *opus ingens ac nobile*.

---

[7] Michael H. Shank has intensively researched this work. His critical edition of this text is forthcoming in the collection, "Science et Savoirs," Paris: Les Belles Lettres. Among other things, he has focused on Regiomontanus's and later Ratdolt's innovative and influential techniques for printing geometrical diagrams; see his "The Geometrical Diagrams in Regiomontanus's Edition of his own *Disputationes* (c. 1475): Background, Production and Diffusion," *Journal for the History of Astronomy* 43 (2012): 27–55. See also Renzo Baldasso's insightful 2007 Columbia PhD thesis, "Illustrating the Book of Nature in the Renaissance: Drawing, Painting and Printing Geometrical Diagrams."

[8] Zinner published a facsimile in *Leben und Wirken*, pl. 26.

[9] The *Quadripartitum numerorum* and Jordanus's *Elementa arithmetica.*

[10] Euclid's *Elements* and the works of Archimedes.

[11] Apollonius's *Conica* and Serenus's *Cylindrica.*

[12] A new edition of Menelaus's *Spherica*, and a new translation of Theodosius's.

[13] Hero's *Pneumatics (Inventa spiritalia).*

[14] Aristotle's *Problemata mechanica.*

[15] He also wanted to publish the commentary by Theon of Alexandria.

[16] Regiomontanus considered this an authentic work of Ptolemy's.

[17] *Commentaria magna in Cosmographiam Ptolomei.*

[18] We will recall from Chap. 9 that Regiomontanus mentioned Antonio de Montulmo in the inaugural oration as one of his predecessors at the University of Padua.

Most of the astronomical works Regiomontanus hoped to publish, on the other hand, were those he wrote himself.[19] Most involve critical treatments of his immediate predecessors, for example, Jacobus Angelus's translation of Ptolemy's *Geography*, and a defense of Theon of Alexandria's commentary on Ptolemy's *Almagest* against George of Trebizond. In addition to the ephemerides mentioned above, his most important and influential works were those on mathematical astronomy, his *Epitome of the Almagest*, and the tables he constructed: *On Directions*, and *On the Motion of the First Sphere*. He also wrote influential treatises on triangles, directions, dividing the astrological houses and calendar reform.

*

Although Regiomontanus's premature death in 1476 sounded the death knell for his press, it hardly did so for his publication program. I will examine here the printed production of Erhard Ratdolt (ca. 1460–1527), who had probably worked for Regiomontanus.[20] He published in Venice from 1476–86, and at Augsburg from 1486–1516. Although strongly influenced by Regiomontanus and his works, Ratdolt's publications were oriented somewhat differently, namely, more towards theoretical astrology (in our sense) than mathematical astronomy. His earliest publication—while Ratdolt was associated with Bernhard Maler and Peter Löslein—was Regiomontanus's *Calendar* for 1476 (ISTC ir00093000),[21] published just after Regiomontanus died. The only other related works published by this threesome were geographical: the *De situ orbis* of Dionysus, and Pomponius Mela's *Cosmographia* (both 1477). Ratdolt and Maler alone published another edition of Regiomontanus's German *Calendar* for 1478 (ISTC ir00100500). After these texts, Ratdolt published on his own.

From 1478–1480, Ratdolt only published one book of interest to us, the *Fasciculus temporum*,[22] a work on chronology, along with other works on law,

[19] Of those not by him, besides Ptolemy's *Almagest*, he only lists Proclus's *Sufformationes astronomicae*.

[20] See Gilbert R. Redgrave, *Erhard Ratdolt and His Work at Venice*, London: Chiswick Press, 1894 (repr. 1899), 3 ff., for the arguments, which without other evidence seem convincing. See now also Renzo Baldasso, "La stampa dell'Editio Princeps degli Elementi di Euclide (Venezia, Erhard Ratdolt, 1482)," in *The Books of Venice/Il libro veneziano*, L. Pon and C. Kallendorf (eds.), New Castle, Delaware: Oak Knoll Press, 2009, 61–100. For further bibliography, see *Der Buchdruck im 15. Jahrhundert, Eine Bibliographie*, S. Corsten and R.W. Fuchs (eds.), Stuttgart: A. Hiersemann, 1988, I, 653–54, the all too brief article in the *Neue Deutsche Biographie*, s.v., and the studies by Shank and Baldasso cited above. Jonathan Green also discusses Ratdolt; *Printing and Prophecy*, 135–36. A full culture-historical study of Ratdolt in context would be of great interest. For the astrological textbooks discussed here, and these and their later publication history throughout the sixteenth century, see Lerch, Scientia astrologiae.

[21] Redgrave notes two editions for 1476; they are nos. 1 and 2 in his catalogue. For the Venetian works, I rely mainly on Redgrave, *Erhard Ratdolt*. Much further information may be derived for both Ratdolt and Regiomontanus from the British Museum's online Incunabula Short Title Catalogue: The International Database of fifteenth century European Printing (http://data.cerl.org/istc/_search).

[22] Nov 24, 1480; he published another edition in Dec 21, 1481, and again in 1483(?), '84 and '85.

grammar and logic. In 1481, however, one begins to note a striking change in what issued from Ratdolt's press, beginning with an edition of Regiomontanus's *Ephemerides* for 1482–1506 (ISTC ir00105000). This is followed in rapid succession by Alcabitius's *Liber isagogicus* (Jan 16, 1482 [ISTC ia00362000]),[23] Abraham ibn Ezra's *De luminaribus et diebus criticis* (Feb 7, 1482),[24] Euclid's *Elements,* (May 25, 1482),[25] and an edition of Sacrabosco's *Sphaera mundi* (July 6, 1482 [ISTC ij00405000]),[26] which included Sacrobosco's *De sphaera,* Regiomontanus's *Disputationes contra cremonensia in planetarum theoricas deliramenta*, and Peurbach's *Theoricae novae planetarum.* These are followed by his first solo edition of Mela's *Cosmographia* (July 18, 1482), and of Regiomontanus's *Calendar* (Aug 9, 1482 [ISTC ir00094000]).[27] His first edition of Hyginus's *Poeticon astronomicon liber* appeared Oct 14, 1482,[28] which we will recall from Regiomontanus's publishing prospectus. Before 1482 he also published another edition of Regiomontanus's *Ephemerides*, with a letter and nine couplets addressed by Ratdolt to Duke Federigo da Montefeltro of Urbino, who died in 1482.[29]

After a break of 6 months during which Ratdolt began printing religious works (later his bread and butter), he again started publishing works relevant for our purposes, beginning with an edition of the Alphonsine Tables, with canons by John of Saxony (July 4, 1483 [ISTC ia00534000]), and an edition of Eusebius's *Chronicon* (Sep 13, 1483). In 1484, he published Ptolemy's *Tetrabiblos* and *Centiloquium* with Haly's commentary (Jan 15, 1484 [ISTC ip01088000])—which, without Haly's commentary, we also found on Regiomontanus's list. 1484 also saw Pietro Borgo's *Arithmetica mercantile* (Aug 2, 1484) and Abraham ibn Ezra's *De nativitatibus* (Dec 24, 1484).[30]

His last relevant publications before returning to Augsburg came out in 1485: Haly filius Abenragel's *Liber in iudiciis astrorum* (July 4, 1485 [ISTC ih00004000])[31]; an *Opusculum repertorii pronosticon in mutationes aeris* (before Nov 4, 1485

[23] Tr. Johannes Hispalensis; an enlarged edition with a commentary by John of Saxony appeared in 1485. See Hasse, *Success and Suppression*, 12–13 and 328–30.

[24] Only the 1485 edition is in ISTC.

[25] Campanus of Novara's edition of Adelard of Bath's and Gerard of Cremona's translation.

[26] Another edition appeared in 1485.

[27] Ratdolt published Regiomontanus's *Calendar* also for 1483, '85 and '89.

[28] Eds. Jacobus Sentinus and Johannes Lucilius Santritter. Another edition came out in 1485. There is also a German edition of 1491.

[29] He published another edition in 1484 (ISTC ir00107000). There is no ISTC entry for the 1482 edition, which I have examined at the Vatican Library.

[30] The 1484 edition is not in ISTC, but the 1485 edition is with exactly the same date of publication. Perhaps the 1484 edition is a bibliographical ghost. For more on Abraham ibn Ezra's works overall, including his *De nativitatibus* in both the Hebrew and Latin versions, see Shlomo Sela, *Abraham ibn Ezra and the Rise of Hebrew Science*. For the *De nativitatibus* in particular, see ibid., *Abraham ibn Ezra on Nativities and Continuous Horoscopy: A Parallel Hebrew-English Critical Edition of the Book of Nativities and the Book of Revolution*, (Abraham ibn Ezra's Astrological Writings, Volume 4), Leiden: Brill, 2014.

[31] Ed. Bartholomaeus de Alten.

[ISTC if00191300]), which includes works on meteorological prognostications and Pietro d'Abano's Latin translation of Hippocrates's book on medical astrology (*De medicorum astrologia*)[32]; Abraham ibn Ezra's *De nativitatibus* with Henry Bate of Maline's *Magistralis compositio astrolabii* (Dec 24, 1485 [ISTC ia00009200]); and a work on chiromancy, this time published by Ratdolt.[33]

Ratdolt returned to Augsburg in 1486, where he continued to practice his trade until 1516. In addition to the increasing load of theologically related works, he also printed[34]:

Boethius,	*De institutione arithmetica*, May 20, 1488 (ISTC ib00828000).
Albumasar,	*Flores astrologiae*, tr. Johannes Hispalensis, Nov 18, 1488 (ISTC ia00356000) and Sep 14, 1495 (ISTC ia00357000).[35]
Johannes Angelus,	*Astrolabium*, Oct 6 or Nov 27, 1488 (ISTC ia00711000).
Leopold, Duke of Austria,	*Compilatio de astrorum scientia*, Jan 9, 1489 (ISTC il00185000).
Albumasar,	*Introductorium in astronomiam*, tr. Hermannus Dalmata, Feb 7, 1489 (ISTC ia00359000).
Albumasar,	*De magnis coniunctionibus*, tr. Johannes Hispalensis, ed. Johannes Angelus, Mar 31, 1489 (ISTC ia00360000).
Pierre d'Ailly,	*Concordantia astronomiae cum theologia*, ed. Johannes Angelus, Jan 2, 1490 (ISTC ia00471000).
Johannes Regiomontanus,	*Tabulae directionum et profectionum. Tabella sinus recti*, ed. Johannes Angelus, Jan 2, 1490 (ISTC ir00112000).
Guido Bonatus,	*Decem tractatus astronomiae*, ed. Johannes Angelus, Mar 26, 1491 (ISTC ib00845000).[36]

We can see from this brief discussion that Ratdolt continued publishing in Regiomontanus's tradition, but with a much greater emphasis on theoretical astrological texts. His publications in theoretical mathematical astronomy are meager indeed, and his publications on their mathematical foundations nonexistent. Furthermore, wholly unlike Regiomontanus, Ratdolt published a significant number

[32] Perhaps this was the same work Regiomontanus hoped to publish.

[33] An edition of a work on chiromancy had come out at Padua in 1484 with diagrams by Ratdolt.

[34] For the Augsburg works, I also rely on *Bibliothecae Apostolicae Vaticanae Incunabula,* William J. Sheehan (ed.), 4 vols., Vatican City: La Biblioteca, 1997, which goes through 1500.

[35] For the printing history of the three works mentioned here by Albumasar, see Hasse, *Success and Suppression*, 326–28.

[36] According to Sheehan's catalogue, there are no new works printed by Ratdolt concerning astrology from 1491–1500, only another edition of Regiomontanus's *Calendarium* for 1496. ISTC only adds a German almanac for 1492.

of works by Arabic authors. It is unfortunate that Ratdolt did not write more dedicatory letters (I know of one)[37] or append informative prefatory material to his editions explaining their circumstances; fortunately his colophons are dated.[38] One would particularly like to know if the great conjunction of 1484 was one of the contributing factors to the change in his publishing direction.

As we will see in volume III, the Nuremberg printer Johannes Petreius (1497–1550), who published (*inter alia*) Copernicus's *De revolutionibus orbium coelestium* (1543) and several astrological writings by Girolamo Cardano, continued this tradition of astrologically-oriented scientific publishing. In this way, the publication of astrological works was put on a solid foundation and they were thus made more affordable and accessible. In addition to the publication of annual prognostications discussed in Part IV, the incipient but increasingly prolific publishing of annual astrological almanacs with prognostications in Latin and various vernaculars took off throughout Europe with the advent of printing.[39]

[3] 1484

To complete volume I and set the tone for volume II, we should return to Italy and briefly discuss the momentous Great Conjunction of Saturn and Jupiter in Scorpio that took place in 1484. That tumultuous year saw the death of Pope Sixtus IV (August 12), which brought Giovanni Battista Cibo (1432–92) to the papal throne as Innocent VIII (August 29).[40] The year before, Louis XI had died and Charles VIII was crowned king of France; his subsequent reign was of great consequence for Italy.[41] This year also witnessed the first publication of Ficino's Latin translation of all of Plato's *Dialogues* from the Greek,[42] as well as the dramatic presence in Rome

[37] In fact, Baldasso says that it is the first dedicatory letter ever printed by a publisher; "La stampa dell'Editio Princeps," 63.

[38] I reviewed most of his editions *in situ* at the Biblioteca Apostolica Vaticana.

[39] In addition to the works by Jonathan Green cited above, see also Capp, *English Almanacs, 1500–1800*, Casali, *Le spie del cielo*, and Barnes, *Astrology and Reformation.*

[40] At the very beginning of his reign, Innocent promulgated a papal bull against witchcraft (the *Summis desiderantes affectibus*), and he commissioned the infamous *Malleus maleficarum* from the Dominican inquisitors, Heinrich Kramer and Jacob Sprenger, which came to be published in 1486. For both, see Henricus Institoris and Jacobus Sprenger, *Malleus Maleficarum.* I will discuss the *Malleus maleficarum*'s discussion of talismans in volume II.

[41] See (i.a.) *The French Descent into Renaissance Italy, 1494–5: Antecedents and Effects*, David Abulafia (ed.), Aldershot: Variorum, 1995. Azzolini discusses various features of the "descent" in *The Duke and the Stars.*

[42] Apparently Ficino rushed his Plato translation into print so that it would be published during the Great Conjunction, perhaps in order to help usher in a new Golden Age; see Paul O. Kristeller, "The First Printed Editions of Plato's Works and the Date of Its Publication (1484)," in *Science and History: Studies in Honor of Edward Rosen*, Ossolineum, 1978, 25–35, and James Hankins, *Plato in the Renaissance*, I, 303. For a modern rhapsody on the significance of 1484 more generally, see Ioan P. Couliano, *Eros and Magic in the Renaissance*, M. Cook (tr), Chicago: University of Chicago Press, 1987, 184–91.

of Giovanni Mercurio da Corregio, robed in black satin, presiding unmolested at Easter as the Hermetic figure, Pimander.[43]

To focus on Florence, Cristoforo Landino's 1481 commentary on Dante refers to the impending Great Conjunction of 1484 as the return of a Golden Age:

> And it is certain (*certo*)[44] that in the year 1484, on the 25th of November at the 13th hour and 41st minute of that day will be the conjunction of Saturn and Jupiter in Scorpio, and the 5th degree of Libra will be on the ascendant, which shows a transformation (*mutatione*) in religion. And because Jupiter is stronger than Saturn, it signifies that such a transformation will be positive. Since no religion can be truer than ours, I have a firm hope (*ferma speranza*) that the *Res Publica Christiana* will return to the finest life and government, in a form that we can truly say: *Iam redit et virgo, redeunt Saturnia regna*.[45]

It was also claimed that prophets and pseudo-prophets were born (or arose) at this time of religious transformation. After the fact, these were identified as Martin Luther by some[46] and as Girolamo Savonarola by others.[47] With these medieval structures firmly in mind, let us now turn to the Renaissance and explore the relevant works of two of its most famous and influential representatives, namely, Marsilio Ficino and Giovanni Pico della Mirandola.

---

[43] Donald Weinstein, *Savonarola and Florence*, 199–200; see also Grafton, "Giovanni Pico della Mirandola: Trials and Triumphs," 94–95. Mirella Brini (in her contribution on Ludovico Lazzarelli to *Testi Umanistici su l'Ermetismo*, E. Garin, M. Brini, C. Vasoli and P. Zambelli (eds), Rome: Fratelli Bocca, 1955) wants to associate Pico more closely with both Lazzarelli and Giovanni Mercurio (30), but the evidence she presents is inconclusive at best.

[44] This use of '*certo*' in the context of an astrological prognostication is striking, considering Thomas Aquinas's analysis of divination in relation to astrology in *Summa theologiae* II.II.92–95 discussed in Part II. Realistically, though, the '*certo*' most likely refers to the astronomical dimension.

[45] Cristoforo Landino, *Comento* [...] *sopra la comedia di Dante Alighieri poeta* (1481), as printed in Ficino, *Lettere*, Vol. I, XLI (n. 75): "Et certo nell'anno MCCCCLXXXIIII nel dì vigesimo quinto di novembre et a hore XIII et minuti XLI di tale dì, sarà la coniunctione di Saturno et di Iove nello Scorpione, nell'ascendente del quinto grado della Libra, la quale dimostra mutatione di religione. Et perchè Iove prevale a Saturno, significa che tale mutatione sarà in meglio. Il perchè non potendo essere religione alchuna più vera che la nostra, ho ferma speranza che la repubblica christiana si ridurrà a optima vita et governo, in forma che poteremo veramente dire: 'Iam redit et virgo, redeunt Saturnia regna.'" On Landino and his teaching in the Florentine studio, see (i.a.) Simona Foà's *DBI* article: 63 (2004).

[46] For a rich discussion, see Aby Warburg, "Pagan-Antique Prophecy in Words and Images in the Age of Luther" (originally published 1920), in his *Renewal of Pagan Antiquity*, 597–697, 760–75.

[47] For example, in his teaching commentary of the 1540s on Ptolemy's *Tetrabiblos* at the University of Pisa, Giuliano Ristori (1492–1556) identified this pseudo-prophet as Savonarola. On the other hand, Filippo Fantoni (ca. 1530–91) who taught this same course at Pisa in the 1580s—and who did so on the basis of Ristori's lectures—rejected this interpretation and crossed out Savonarola's name ad loc. I treat this material more fully in my "Giuliano Ristori and Filippo Fantoni on Pseudo-Prophets, Great Conjunctions and Other Astrological Effects." I will also treat Ristori and Fantoni in volume III.

Bibliography

Primary Sources

Albumasar

Liber introductorii majoris. ad scientiam iudiciorum astrorum, 9 vols., Richard Lemay (ed.). Naples: Istituto Universitario Orientale, 1995.

Abu Ma'shar on. Historical Astrology: The Book of Religions and Dynasties (On the Great Conjunctions), Keiji Yamamoto and Charles S.F. Burnett (ed.), 2 vols. (Leiden, 2000).

Albertus Magnus

De mineralibus in. *Opera omnia*, E. Borgnet (ed.), 38 vols., Paris: Vives, 1890–99, vol. 5.

Physica Pars I,. *Opera omnia* 4.1, Paul Hossfeld (ed.), Monasterii Westfalorum: Aschendorff, 1987. (The *Editio Coloniensis*)

Physica Pars. II, *Opera omnia* 4.2, Paul Hossfeld (ed.), 1993.

De caelo et mundo, Opera Omnia, 5.1, Paul Hossfeld (ed.), 1971.

De generatione. et corruptione, *Opera omnia* 5.2, Paul Hossfeld (ed.), 1980, 107–219.

Liber de causis proprietatum. elementorum, *Opera omnia*, 5.2, Paul Hossfeld (ed.), 1980, 49–104.

De fato, *Opera omnia* 17.1, Paul Simon (ed.), 1975.

On the Causes. of the Properties of the Elements (Liber de causis proprietatum elementorum), translated from the Latin with an introduction and notes by Irven M. Resnick. Milwaukee: Marquette University Press, 2010.

H D. Rutkin, *Sapientia Astrologica: Astrology, Magic and Natural Knowledge, ca. 1250–1800*, Archimedes 55, https://doi.org/10.1007/978-3-030-10779-6

Alcabitius

Al-Qabisi. (Alcabitius): The Introduction to Astrology: Editions of the Arabic and Latin Texts and an English Translation, Charles S. F. Burnett and Keiji Yamamoto (ed and Trans.). London: Warburg Institute, 2004.

Al-Kindi

d'Alverny, Marie-Thérèse, and Françoise Hudry, "Al-Kindi: *De Radiis*," *Archives d'Histoire Doctrinale et Litteraire du Moyen Age* 41 (1975): 139–259.

The Philosophical. Works of Al-Kindi, Peter Adamson and Peter E. Pormann (ed and Trans.). Oxford: Oxford University Press, 2012.

Ammonius and Boethius

Ammonius on Aristotle. On interpretation, David Blank (Trans.), with *Boethius on Aristotle* On interpretation 9, Norman Kretzmann (Trans.), with essays by Richard Sorabji, Norman Kretzmann and Mario Mignucci, London: Duckworth, 1998.

Andrea Argoli

Argoli, Andrea, *Exactissimae coelestium motuum ephemerides* [...], Padua: P. Frambotti, 1648.

Aristotle

Latin

Aristoteles Latinus. IX.1, *De generatione et corruptione, Translatio Vetus*, Joanna Judycka (ed.). Leiden: Brill, 1986.

Greek

Aristotle On Coming-To-Be and Passing-Away (De generatione et corruptione): *A Revised Text with Introduction and Commentary,* H. H. Joachim (ed.). Oxford: Clarendon Press, 1922.

Arnald of Villanova

Arnaldi de Villanova, Aphorismi De gradibus, Michael McVaugh (ed.). Granada: Seminarium Historiae Medicae Granatensis, 1975.

Roger Bacon

The Opus maius. *of Roger Bacon*, J.H. Bridges (ed.), 3 vols., Oxford: Clarendon, 1897–1900.

The Opus Majus. of Roger Bacon, Robert Belle Burke (Trans.), 2 vols., Philadelphia: University of Pennsylvania Press, 1928.

Opus tertium, in *Fr. Rogeri Bacon Opera quaedam hactenus inedita*, J.S. Brewer (ed.), vol. I in the series *Rerum Britannicarum Medii Aevi Scriptores*, London: Longman, 1859.

Part of the. Opus Tertium *of Roger Bacon*, A. G. Little (ed.). Aberdeen: The University Press, 1912.

Roger Bacon. and the Origins of Perspectiva *in the Middle Ages: A Critical Edition and English Translation of Bacon's* Perspectiva, *with Introduction and Notes*, David C. Lindberg (ed and Trans.). Oxford: Clarendon, 1996.

Roger Bacon's. Philosophy of Nature: A Critical Edition with English Translation, Introduction and Notes of De multiplicatione specierum *and* De speculis comburentibus, David C. Lindberg (ed and Trans.). Oxford: Clarendon, 1983.

*Filosofia, scenza, teologia dall'*Opus maius, Valeria Sorge and Fabio Sellar (ed and Trans.). Rome: Armando, 2010.

Secretum secretorum. cum glossis et notulis [...] *fratris Rogeri*, Robert Steele (ed.). Oxford: Clarendon, 1920.

Lorenzo Buonincontri

De rebus naturalibus. et divinis, Stephan Heilen (ed.). Stuttgart: Teubner, 1999.

Campanus of Novara

Campanus of Novara. and Medieval Planetary Theory: Theorica planetarum, Francis S. Benjamin, Jr. and G.J. Toomer (ed and Trans.). Madison: University of Wisconsin Press, 1971.

Gerardo Cerruti

Il carteggio di Gerardo. Cerruti, oratore sforzesco a Bologna (1470–74), Tommaso Duranti (ed.), 2 vols., Bologna: CLUEB, 2007.

Cicero and Boethius

Cicero, *On Fate.* (*De fato*) and Boethius, *The Consolation of Philosophy* (*Philosophiae consolationis*), Robert W. Sharples (ed and Trans.). Westminster: Aris and Phillips, 1991.

John Dee

John Dee on. Astronomy: Propaedeumata Aphoristica *(1558 &1668), Latin and English,* Wayne Shumaker and John Heilbron (ed.). Berkeley: University of California Press, 1978.

Domenico Maria da Novara

I pronostici di Domenico. Maria da Novara, Fabrizio Bonolì, Giuseppe Bezza, Salvo de Meis and Cinzia Colavita (ed.). Florence: Olschki, 2012.

Giovanni Dondi dall'Orologio

Tractatus astrarii, edition critique et traduction de la version A par E. Poulle, Geneva: Droz, 2003.

Marsilio Ficino

Lettere I: Epistolarum. familiarum liber I, Sebastiano Gentile (ed.). Florence: Olschki, 1990.

_____, *Platonic Theology*, Latin text edited by James Hankins with William Bowen; English translation by Michael J.B. Allen, 6 vols., Cambridge, MA: Harvard University Press, 2001–2006.

_____, *Three Books. on Life*, Carol V. Kaske and John R. Clark (ed and Trans.). Binghamton, N.Y.: Medieval and Renaissance Texts and Studies, 1989.

Francesco Filarete and Angelo Manfidi

The Libro Cerimoniale. of the Florentine Republic by Francesco Filarete and Angelo Manfidi: Introduction and Text, Richard Trexler (ed.). Geneva: Droz, 1978.

Galileo Galilei

Le opere di Galileo. Galilei, Appendice Volume III, Testi, Germana Ernst (ed.). Florence: Giunti, 2017, 103–93. (Galileo's Astrological Manuscript 81.)

Giovanni Garzoni

The Letters of Giovanni. Garzoni: Bolognese Humanist and Physician, 1419–1505, L.R. Lind (ed.). Atlanta: Scholars Press, 1992.

Robert Grosseteste

Ludwig Baur, "Die Philosophischen Werke des Robert Grosseteste," *Beiträge zur Geschichte der Philosophie des Mittelalters*, ix (1912).

Hugh of Saint Victor

The Didascalicon. *of Hugh of St. Victor: A Medieval Guide to the Arts*, Jerome Taylor (Trans.). New York: Columbia University Press, 1961.

Abraham Ibn Ezra

Abraham Ibn Ezra. on Elections, Interrogations and Medical Astrology, Shlomo Sela (ed and Trans.). Leiden: Brill, 2011.

———. *on Nativities and Continuous Horoscopy: A Parallel Hebrew-English Critical Edition of the Book of Nativities and the Book of Revolution*, Shlomo Sela (ed and Trans.). Leiden: Brill, 2014.

Henricus Institoris and Jacobus Sprenger

Malleus Maleficarum, Christopher S. Mackay (ed and Trans.), Latin text and English translation, 2 vols., Cambridge: Cambridge University Press, 2006.

Johannes Kepler

The Harmony. of the World, E. J. Aiton, A. M. Duncan and J. V. Field (Trans.), Philadelphia: American Philosophical Society, 1997.

Girolamo Manfredi

Liber de homine: Il Perché, with introductory essays by F. Foresti, A.L. Trombetti Budriesi and A.M. Nada Patrone, Bologna: Luigi Parma, 1988.

Tractato de la pestilentia. = *Tractatus de peste*, T. Duranti (ed.). Bologna: CLUEB, 2008.

Isaac Newton

Certain Philosophical. Questions: Newton's Trinity Notebook, J.E. McGuire and Martin Tamny (ed.). Cambridge: Cambridge University Press, 1983.

Pseudo-Ovid

Pseudo-Ovidius. De vetula: *Untersuchungen und Text*, Paul Klopsch (ed.). Leiden: Brill, 1967.

The Pseudo-Ovidian. De vetula, Dorothy Mae Robathan (ed.). Amsterdam: Hakkert, 1969.

John Pecham

John Pecham and. the Science of Optics: Perspectiva Communis, David C. Lindberg (ed.). Madison: University of Wisconsin Press, 1970.

Pietro d'Abano

Conciliator, E. Riondato. and L. Olivieri (ed.). Padua: Antenore, 1985 (facsimile of the Venice: Giunta edition of 1565).

Il 'Lucidator dubitabilium. astronomiae' di Pietro d'Abano: opere scientifiche inedite, Graziella Federici Vescovini (ed.). Padua: Programma e 1 + 1 Editori, 1988.

Angelo Poliziano

Stanze per la giostra: Orfeo, rime: con un'appendice di prose volgari. Novara: Istituto Geografico De Agostino, 1969.

Regiomontanus

Alfraganus, *Rudimenta. astronomica*, Nuremburg: Petreius, 1537, ff a (beta) 1r-5r, reprinted in facsimile in *Regiomontani Opera Collectanea*, F. Schmeidler (ed.). Osnabruck: O. Zeller, 1972. (Regiomontanus's inaugural oration for his course on Alfraganus at the University of Padua in 1464.)

Almanach magistri. Johannis de Monte Regio ad annos xv accuratissime calculata. Augsburg: Ratdolt, 1492.

Richard of Wallingford

Richard of Wallingford: An Edition of His Writings, 3 vols., John D. North (ed.). Oxford: Clarendon Press, 1976.

Sacrobosco

The Sphere of Sacrobosco. and its Commentators, Lynn Thorndike (ed and Trans.). Chicago: University of Chicago Press, 1949.

Girolamo Savonarola

Contro gli astrologi, Claudio Gigante (ed.). Rome: Salerno Editrice, 2000.

Michael Scot

Glen M Edwards,. "The *Liber Introductorius* of Michael Scot," PhD thesis, University of Southern California, 1978.

Simon de Phares

Recueil des plus. celebres astrologues de Simon de Phares, J. P. Boudet (ed.), 2 vols., Paris: Champion, 1997–99.

Thomas Aquinas

De operationibus. occultis naturae, Opera omnia 43, Rome: De propaganda fide, 1976, 183–86. (The Leonine edition)
De iudiciis astrorum, Opera omnia 43, 201.
De veritate, Opera omnia 22.1/2, 1970.
Summa theologiae, Opera omnia 9, 1897.
Summa theologia*: Latin text and English Translation*, 61 vols., London: Blackfriars, 1964–80.
On the Truth of. the Catholic Faith (Summa Contra Gentiles), Anton C. Pegis, James F. Anderson, Vernon J. Bourke, and Charles J. O'Neil (Trans.), 5 vols., New York: Doubleday, 1955–57; reprinted as *Summa contra gentiles,* Notre Dame, IN: University of Notre Dame Press, 1975. (III.82 and 83 are in volume 3:1; III.84-105 are in 3:2; both originally published 1956.). E-text with some revision at dhspriory.org/thomas/ContraGentiles.htm.

Placido Titi

Tocco di paragone, Giuseppe Bezza (ed.). Milan: Nuovi Orizzonti, 1992.

Jérôme Torrella (Hieronymus Torrella)

Opus praeclarum. de imaginibus astrologicis, Nicolas Weill-Parot (ed.). Florence: SISMEL–Galluzzo, 2008.

William of Rubruck

The Mission of Friar. William of Rubruck: His Journey to the Court of the Great Khan Möngke, 1253–55, Peter Jackson (Trans.), London: Hakluyt Society, 1990.

Secondary Sources

Abate, Mark T. 2000. *Roger Bacon and the Rage of Antichrist: The Apocalypse of a Thirteenth Century Natural Philosopher*, PhD thesis. Boston University.
Adamson, Peter. 2007. *Al-Kindi.* Oxford: Oxford University Press.
———. 2008. Plotinus and Astrology. *Oxford Studies in Ancient Philosophy* 35: 265–291.
———. 2015. Neoplatonism: The Last Ten Years. *The International Journal of the Platonic Tradition* 9: 205–220.
Adelard of Bath: An English Scientist and Arabist of the Early 12th Century, Charles S.F. Burnett (ed.). London: Warburg Institute, 1987.
Agrimi, Jole. 2002. *Ingeniosa scientia nature: Studi sulla fisiognomica medievale*. Florence: SISMEL—Galluzzo.

Agrimi, Jole, and Chiara Crisciani. 1992. Immagini e ruoli della vetula tra sapere medico e antropologia religiosa (secoli XIII-XV). In *Poteri carismatici e informali: chiesa e società medioevali*, ed. A. Vauchez and A. Paravicini Bagliani, 224–261. Palermo: Sellerio.

———. 1993. Savoir médical et anthropologie religieuse: Les représentations et les fonctions de la vetula (XIIIe-XVe siècle). *Annales ESC* 5: 1281–1308.

Albertus Magnus and the Sciences: Commemorative Essays 1980, James A. Weisheipl (ed.). Toronto: Pontifical Institute, 1980.

Allen, Michael J.B. 1994. *Nuptial Arithmetic: Marsilio Ficino's Commentary on the Fatal Number in Book VIII of Plato's Republic*. Berkeley: University of California Press.

———. 1998. *Synoptic Art: Marsilio Ficino on the History of Platonic Interpretation*. Florence: Olschki.

Anzulewicz, Henryk. 1998. Perspektive und Raumvorstellung in den Frühwerken des Albertus Magnus. *Miscellanea Mediaevalia* 25: 249–286.

———. 1999. Der Einfluss der Gestirne auf die sublunare Welt und die menschliche Willensfreiheit nach Albertus Magnus. In *Actes de la Vème Conférence Annuelle de la SEAC, Gdansk 1997*, ed. T. Mikocki, 263–277. Warsaw: Institute of Archaeology.

———. 2001. *Fatum*: Das Phänomen des Schicksals und die Freiheit des Menschen nach Albertus Magnus. In *Nach der Verurteilung von 1277: Philosophie und Theologie an der Universität von Paris im letzten Viertel des 13. Jahrhunderts. Studien und Texte*, ed. J.A. Aertsen, K. Emery Jr., and A. Speer, 507–534. Berlin: De Gruyter.

———. 2011. Magie im Verständnis Alberts des Großen. In *Mots médiévaux offerts à Ruedi Imbach*, ed. I. Atucha et al., 419–431. Porto: Fédération Internationale des Instituts d'Études Médiévales.

Arnzen, Rüdiger. 1999. Vergessene Pflichtlektüre: Al-Qabisis astrologische Lehrschrift im Europäischen Mittelalter. *Zeitschrift für Geschichte der Arabisch-Islamischen Wissenschaften* 13: 93–128.

Arrizabalaga, Jon, John Henderson, and Roger French. 1997. *The Great Pox: The French Disease in Renaissance Europe*. New Haven: Yale University Press.

Astro-Medicine: Astrology and Medicine, East and West, A. Akasoy, C. Burnett and R. Yoeli-Tlalim (ed.). Florence: SISMEL—Galluzzo, 2008.

Astrologers and their Clients in Medieval and Early Modern Europe, Wiebke Deimann and David Juste (ed.). Cologne: Böhlau Verlag, 2015.

"Astrologi Hallucinati"*: Stars and the End of the World in Luther's Time*, Paola Zambelli (ed.). Berlin: de Gruyter, 1986.

Astrologia: arte e cultura in età rinascimentale—Art and Culture in the Renaissance, D. Bini (ed.). Modena: Bulino, 1996.

Astrology, Science and Society: Historical Essays, Patrick Curry (ed.). Woodbridge: Boydell, 1987.

Astronomy before the Telescope, Christopher Walker (ed.). New York: St. Martins Press, 1997.

Avalos, Ana Cecilia. 2007. *As Above, So Below: Astrology and the Inquisition in 17th-Century New Spain*, PhD thesis. European University Institute.

Azzolini, Monica. 2004. Anatomy of a Dispute: Leonardo, Pacioli and Scientific Courtly Entertainment in Renaissance Milan. *Early Science and Medicine* 9: 115–135.

———. Reading Health in the Stars: Politics and Medical Astrology in Renaissance Milan. In *Horoscopes and Public Spheres*, 183–205.

———. 2008. Annius of Viterbo Astrologer: Predicting the Death of Ferrante of Aragon. *Bruniana e Campanelliana* 14: 619–632.

———. 2013. *The Duke and the Stars: Astrology and Politics in Renaissance Milan*. Cambridge, MA: Harvard University Press.

Bagchi, David. 2016. Printing, Propoganda, and Public Opinion in the Age of Martin Luther. *Oxford Research Encyclopedia of Religion*, Aug. 2016 (online publication date), 1–31.

Bailey, Michael D. 2013. *Fearful Spirits, Reasoned Follies: The Boundaries of Superstition in Late Medieval Europe*. Ithaca: Cornell University Press.

Baldasso, Renzo. 2007. *Illustrating the Book of Nature in the Renaissance: Drawing, Painting and Printing Geometrical Diagrams*, PhD thesis. Columbia University.

———. 2009. La stampa dell'Editio Princeps degli Elementi di Euclide (Venezia, Erhard Ratdolt, 1482). In *The Books of Venice/Il libro veneziano*, ed. L. Pon and C. Kallendorf, 61–100. New Castle: Oak Knoll Press.

Barnes, Robin B. 1988. *Prophecy and Gnosis: Apocalypticism in the Wake of the Lutheran Reformation*. Stanford: Stanford University Press.

———. 1998. Astrology and Popular Print in Germany, c. 1470–1520. In *Books Have Their Own Destiny: Essays in Honor of Robert V. Schnucker*, ed. R. Barnes, R. Kolb, and P.L. Presley, 17–26. Kirksville: Thomas Jefferson University Press.

———. 2016. *Astrology and Reformation*. New York: Oxford University Press.

Barraclough, Geoffrey. 1968. *The Medieval Papacy*. New York: Norton.

Barton, Tamsyn S. 1994a. *Knowledge and Power: Astrology, Physiognomics, and Medicine under the Roman Empire*. Ann Arbor: University of Michigan Press.

———. 1994b. *Ancient Astrology*. London: Routledge.

Bauer, Barbara. 1989. Die Rolle des Hofastrologen und Hofmathematicus als fürstlicher Berater. In *Höfischer Humanismus*, ed. A. Buck, 93–117. Wienheim: VCH.

Bauer, Katrin. Johannes Kepler between two Emperors. In *Astrologers and their Clients in Medieval and Early Modern Europe*, 205–219.

Beaujouan, Guy. 1963. Motives and Opportunities for Science in the Medieval Universities. In *Scientific Change: Historical Studies in the Intellectual, Social and Technical Conditions for Scientific Discovery and Technical Invention, from Antiquity to the Present*, ed. A. Crombie, 219–236. New York: Basic Books.

Bentley, Jerry H. 1987. *Politics and Culture in Renaissance Naples*. Princeton: Princeton University Press.

Berschin, Walter. 1988. *Greek Letters and the Latin Middle Ages: From Jerome to Nicholas of Cusa*. Revised and expanded edition, Trans. J.C. Frakes. Washington, DC: Catholic University of America Press.

Bertozzi, Marco. 1999. *La tirannia degli astri: gli affreschi astrologici di Palazzo Schifanoia*. Livorno: Sillabe.

Bezza, Giuseppe. 1995. *Arcana mundi: Antologia del pensiero astrologico antico*. Vol. 2. Milan: Rizzoli.

Bibliothecae Apostolicae Vaticanae Incunabula, William J. Sheehan (ed.), 4 vols. Vatican City: La Biblioteca, 1997.

Biologie, Logique et Metaphysique Chez Aristote, Daniel Devereux and Pierre Pellegrin (ed.). Paris: Editions du CNRS, 1990.

Blume, Dieter. 2000. *Regenten des Himmels: Astrologische Bilder in Mittelalter und Renaissance*. Berlin: Akademie Verlag GmBH.

———. Picturing the Stars: Astrological Imagery in the Latin West, 1100–1550. In *A Companion to Astrology in the Renaissance*, 333–398.

Böll, Franz. 1894. *Studien über Claudius Ptolemäus: Ein Beitrag zur Geschichte der Griechischen Philosophie und Astrologie*. Leipzig: Teubner.

Bono, James J. 1981. *The Languages of Life: Jean Fernel (1497–1558) and Spiritus in Pre-Harveian Thought,* PhD thesis. Harvard University.

———. 1984. Medical Spirits and the Medieval Language of Life. *Traditio* 40: 91–130.

Bonolì, Fabrizio, and Daniela Piliarvu. 2001. *I Lettori di astronomia presso lo studio di Bologna dal XII al XX secolo*. Bologna: CLUEB.

Boockmann, Friederike. 2005. Die Horoskopsammlung von Johannes Kepler. In *Miscellanea Kepleriana: Festschrift für Volker Bialas*, ed. F. Boockmann, D. di Liscia, and H. Kothmann, 183–203. Augsburg: Rauner.

Boudet, Jean-Patrice. 2006. *Entre science et nigromance: astrologie, divination et magie dans l'Occident médiéval, XIIe-XVe siècle*. Paris: Publications de la Sorbonne.

———. 2008. Un traité de magie astrale Arabo-Latin: Le *Liber de imaginibus* du Pseudo-Ptolémée. In *Natura, scienze e società medievali: Studi in onore di Agostino Paravicini Bagliani*, ed. Claudio Leonardi and Francesco Santi, 17–35. Florence: SISMEL—Galluzzo.

———. 2010. A 'College of Astrology and Medicine'? Charles V, Gervais Chrétien, and the Scientific Manuscripts of Maître Gervais's College. *Studies in History and Philosophy of Biological and Biomedical Sciences* 41: 99–108.

Bowden, Mary Ellen. 1974. *Scientific Revolution in Astrology: The English Reformers, 1558–1686*, PhD thesis. Yale University.

Brind'Amour, Pierre. 1993. *Nostradamus astrophile: Les astres et l'astrologie dans la vie et l'oeuvre de Nostradamus*. Ottawa: Presses de l'Universite d'Ottawa.

Brosseder, Claudia. 2004. *Im Bann der Sterne: Caspar Peucer, Philipp Melanchthon und andere Wittenberger Astrologen*. Berlin: Akademie Verlag.

Brown, Marguerite. 2014. Visioning the Planets in Fifteenth Century Florentine Printmaking, *Visual Pursuits: Prints from the British Museum (Uncategorized)*, posted June 26, 2014 (http://visualpursuits.org/2014/06/26/visioning-the-planets-in-15th-century-florentine-printmaking/).

Bühler, Curt F. 1958. *The University and the Press in Fifteenth-Century Bologna*. Notre Dame: Mediaeval Institute.

Burnett, Charles S.F. Adelard, Ergaphalau and the Science of the Stars. In *Magic and Divination*, 133–145. (originally published, 1987).

———. The Earliest Chiromancy in the West. In *Magic and Divination*, 189–195. (originally published 1987).

———. Chiromancy: Supplement. The Principal Latin Texts on Chiromancy Extant in the Middle Ages. In *Magic and Divination*, 1–29.

———. 1990. A New Source for Dominicus Gundissalinus's Account of the Science of the Stars. *Annals of Science* 47: 361–374.

———. 1996a. Astrology. In *Medieval Latin: An Introduction and Bibliographical Guide*, ed. F.A.C. Mantello and A.G. Rigg, 369–382. Washington, D.C: Catholic University of America Press.

———. 1996b. *Magic and Divination in the Middle Ages: Texts and Techniques in the Islamic and Christian Worlds*. Aldershot: Ashgate.

———. Talismans: Magic as Science? Necromancy among the Seven Liberal Arts. In *Magic and Divination in the Middle Ages*, 1–13.

———. 2009a. *Arabic into Latin in the Middle Ages: The Translators and their Intellectual and Social Context*. Farnham: Ashgate/Variorum.

———. 2009b. The Theory and Practice of Powerful Words in Medieval Magical Texts. In *The Word in Medieval Logic, Theology and Philosophy*, ed. Tetsuro Shimizu and Charles Burnett. Turnhout: Brepols.

Byrne, James S. 2006. A Humanist History of Mathematics? Regiomontanus's Padua Oration in Context. *Journal of the History of Ideas* 67: 41–61.

Cadden, Joan. Charles V, Nicole Oresme, and Christine de Pizan: Unities and Uses of Knowledge in 14th-Century France. In *Texts and Contexts in Ancient and Medieval Science*, 208–244.

Cameron, Euan. 2010. *Enchanted Europe: Superstition, Reason, and Religion, 1250–1750*. Oxford: Oxford University Press.

Campbell, Anna M. 1931. *The Black Death and Men of Learning*. New York: Columbia University Press.

Cambridge History of Late Antique and Early Medieval Philosophy, A.H. Armstrong (ed.). Cambridge: Cambridge University Press, 1970.

Cambridge History of Later Medieval Philosophy, N. Kretzmann, A. Kenny, and J. Pinborg (ed.). Cambridge: Cambridge University Press, 1982.

Cambridge History of Medieval Philosophy, Robert Pasnau (ed.), 2 vols. Cambridge: Cambridge University Press, 2010.

Campion, Nicholas. 2009. *History of Western Astrology*. Vol. 2. London: Continuum.

Capp, Bernard. 1979. *English Almanacs, 1500–1800: Astrology and the Popular Press*. Ithaca: Cornell University Press.

Carey, Hilary M. 1992. *Courting Disaster: Astrology at the English Court and University in the Later Middle Ages*. London: Macmillan.

Carmody, Francis J. 1956. *Arabic Astronomical and Astrological Sciences in Latin Translation, A Critical Bibliography*. Berkeley: University of California Press.

———. 1960. *The Astronomical Works of Thabit b. Qurra*. Berkeley: University of California Press.

Carolino, Luis Miguel Nunes. 2000. *Agant corpora coelestia in sublunarem mundum an non?* Ciencia, Astrologia e Sociedade em Portugal (1593–1755), PhD thesis. Universidade de Évora.

Caroti, Stefano. 1983. *L'astrologia in Italia: profezie, oroscopi e segreti celesti, dagli zodiaci romani alla tradizione islamica, dalle corti rinascimentali alle scuole moderne: storia, documenti, personaggi*. Rome: Newton Compton.

Carrara, Daniela Mugnai. 1991. *La biblioteca di Nicolò Leoniceno: tra Aristotele e Galeno, cultura e libri di un medico umanista*. Florence: Olschki.

Casali, Elide. 2003. *Le spie del cielo: Oroscopi, lunari e almanacchi nell'Italia moderna*. Turin: Einaudi.

Casanova, Eugenio. 1891. L'astrologia e la consegna del bastone al capitano generale della repubblica fiorentina. *Archivio Storico Italiano* (ser. 5) VIII: 134–44.

Caspar, Max. 1993. *Kepler*, (Doris Hellman ed and Trans.). New York: Dover.

Cassirer, Ernst. 1964. *The Individual and the Cosmos in Renaissance Philosophy*, Trans. Mario Domandi. New York: Harper and Row, (originally published 1927).

Castagnola, Raffaella. 1989. Un oroscopo per Cosimo I. *Rinascimento* 29: 125–189.

Castelli, Patrizia. 1983. Gli astri e i Montefeltro. *Res Publica Litterarum: Studies in the Classical Tradition* 6: 75–89.

Catalogus Codicum Astrologorum Latinorum (CCAL). Vol. 1: *Les manuscrits astrologiques latins conservés à la Bayerische Staatsbibliothek de Munich*, David Juste (ed.), (*Documents, Études et Répertoires* 81). Paris: CNRS Éditions, 2011.

——— (CCAL). Vol. II: *Les manuscrits astrologiques latins conservés à la Bibliothèque nationale de France à Paris*, David Juste (ed.) (*Documents, Études et Répertoires* 84). Paris: CNRS Éditions, 2015.

Catholic Church and Modern Science: Documents from the Archives of the Roman Congregation of the Holy Office and the Index, Ugo Baldini and Leen Spruit (ed.); Vol. 1: 16th Century Documents (4 vols.). Rome: Libreria Editrice Vaticana, 2009.

Chabas, Jose, and Bernard R. Goldstein. 2012. *A Survey of European Astronomical Tables in the Late Middle Ages*. Leiden: Brill.

Chamber, David S. 1992. *A Renaissance Cardinal and His Worldly Goods: The Will and Inventory of Francesco Gonzaga (1444–1483)*. London: Warburg Institute.

Chase, Melissa P. 1985. Fevers, Poisons and Apostemes: Authority and Experience in Montpellier Plague Treatises. *Annals of the New York Academy of Sciences* 441: 153–170.

Clark, Stuart. 1997. *Thinking with Demons: The Idea of Witchcraft in Early Modern Europe*. New York: Oxford University Press.

Clarke, Angus G. 1985. *Giovanni Antonio Magini (1555–1617) and Late Renaissance Astrology*, PhD thesis. University of London.

Clough, Cecil H. 1981. The Library of the Dukes of Urbino, *Librarium. Revue de la Société des Bibliophiles Suisses* IX (1966): 101–105, 192, reprinted in his *The Duchy of Urbino in the Renaissance*. London: Variorum.

Clulee, Nicholas. 1977. Astrology, Magic and Optics: Facets of John Dee's Early Natural Philosophy. *Renaissance Quarterly* 30: 632–680.

Cohn, Samuel K. 2002. *The Black Death Transformed: Disease and Culture in Early Renaissance Europe*. London: Arnold.

Cole, Alison. 1995. *Virtue and Magnificence: Art of the Italian Renaissance Courts*. New York: Abrams.

Collins, David J. 2010. Albertus, *Magnus* or *Magus?* Magic, Natural Philosophy, and Religious Reform in the Late Middle Ages. *Renaissance Quarterly* 63: 1–44.

A Companion to Albert the Great: Theology, Philosophy, and the Sciences, Irven M. Resnick (ed.). Leiden: Brill, 2013.

A Companion to Astrology in the Renaissance, Brendan Dooley (ed.). Leiden: Brill, 2014.

Conrad, Lawrence I., Michael Neve, Vivian Nutton, Roy Porter, and Andrew Wear. 1995. *The Western Medical Tradition: 800 B.C.–1800 A.D.* Cambridge: Cambridge University Press.

Cooper, Glen M. 2013. Approaches to the Critical Days in Late Medieval and Renaissance Thinkers. *Early Science and Medicine* 18: 536–565.

Copenhaver, Brian P. 1984. Scholastic Philosophy and Renaissance Magic in the *De Vita* of Marsilio Ficino. *Renaissance Quarterly* 37: 523–554.

Cordonier, Valérie. 2011. Sauver le Dieu du Philosophe: Albert le Grand, Thomas d'Aquin, Guillaume de Moerbeke et l'invention du *Liber de bona fortuna* comme alternative autorisée à l'interprétation averroïste de la théorie aristotélicienne de la providence divine. In *Christian Readings of Aristotle from the Middle Ages to the Renaissance*, ed. Luca Bianchi, 65–114. Turnhout: Brepols.

———. 2014. La doctrine Aristotélicienne de la Providence selon Thomas d'Aquin. In *Fate, Providence and Moral Responsibility in Ancient, Medieval and Early Modern Thought. Studies in Honour of Carlos Steel*, ed. P. D'Hoine and G. van Riel, 495–515. Leuven: Peeters.

Cornelius, Geoffrey. 2003. *The Moment of Astrology: Origins in Divination.* Revised and expanded second edition. Bournemouth: The Wessex Astrologer, (originally published 1994).

———. Interpreting Interpretations: The Aphorism in the Practice of the Renaissance Astrologers. In *From Masha'Allah to Kepler*, 101–121.

Cornish, Alison. 2000. *Reading Dante's Stars.* New Haven: Yale University Press.

Couliano, Ioan P. 1987. *Eros and Magic in the Renaissance*, Trans. M. Cook. Chicago: University of Chicago Press.

Rearick, Cox. 1984. *Janet, Dynasty and Destiny in Medici Art: Pontormo, Leo X, and the Two Cosimos.* Princeton University Press: Princeton.

Curran, Brian A. 1998/1999. "De sacrorum litterarum Aegyptiorum interpretatione." Reticence and Hubris in Hieroglyphic Studies of the Renaissance: Piero Valeriano and Annius of Viterbo. *Memoirs of the American Academy in Rome* 43/44: 139–182.

———. 2007. *The Egyptian Renaissance: The Afterlife of Ancient Egypt in Early Modern Italy.* Chicago: University of Chicago Press.

Curry, Patrick. 1989. *Prophecy and Power: Astrology in Early Modern England.* Cambridge: Polity.

Curry, Patrick, and Roy G. Willis. 2004. *Astrology, Science and Culture: Pulling Down the Moon.* Oxford: Berg.

Dalché, P. Gautier. 2007. The Reception of Ptolemy's *Geography* (End of the Fourteenth to Beginning of the Sixteenth Century), In *The History of Cartography, Vol. 3: Cartography in the European Renaissance*, 2 vols. Chicago: University of Chicago Press, I: 285–364.

———. 2011. Vers une *perfecta locorum doctrina*: Lieu et espace géographique selon Roger Bacon. In *Représentations et conceptions de l'espace dans la culture médiévale. Repräsentationsformen und Konzeptionen des Raums in der Kultur des Mittelalters*, ed. T. Suarez-Nani and M. Rohde, 9–43. Berlin: De Gruyter.

Dallari, U. 1888–1924. *I rotuli dei lettori legisti e artisti dello studio Bolognese [...].* Merlani: Bologna.

D'Alverny, Marie-Thérèse. 1982. Translations and Translators. In *Renaissance and Renewal in the 12th Century*, ed. R.L. Benson, G. Constable, and C. Lanham, 421–462. Cambridge, MA: Harvard University Press.

D'Amico, John. 1983. *Renaissance Humanism in Papal Rome: Humanists and Churchmen on the Eve of the Reformation.* Baltimore: Johns Hopkins University Press.

———. 1993. *Contra divinationem*: Paolo Cortesi's Attack on Astrology. In *Roman and German Humanism, 1450–1550*, ed. P. Grendler, 281–291. Aldershot: Variorum. (originally published, 1985).

Dane, Joseph A. 2003. *The Myth of Print Culture: Essays on Evidence, Textuality, and Bibliographical Method*. Toronto: University of Toronto Press.

Dear, Peter. 2009. *Revolutionizing the Sciences: European Knowledge and its Ambitions, 1500–1700*. 2nd ed. Princeton University Press: Princeton.

Deimann, Wiebke. Astrology in an Age of Transition: Johannes Lichtenberger and his Clients. In *Astrologers and their Clients*, 83–104.

Delaurenti, Béatrice. 2007. *La Pouissance des mots, <<Virtus verborum>>: Débats doctrinaux sur le pouvoir des incantations au Moyen Age*. Paris: Editions du Cerf.

De Libera, Alain. 1991. *Penser au Moyen Age*. Paris: Éditions du Seuil.

Dell'Anna, Giuseppe. 1999. *Dies critici: la teoria della ciclicità delle patologie nel XIV secolo*. Galatina (Lecce): Congedo.

Demaitre, Luke E. 1980. *Doctor Bernard de Gordon: Professor and Practitioner*. Toronto: Pontifical Institute of Mediaeval Studies.

Denley, Peter. *Signore* and *Studio*: Lorenzo in a Comparative Context. In *Lorenzo the Magnificent: Culture and Politics*, 203–216.

Der Buchdruck im 15. Jahrhundert, Eine Bibliographie, S. Corsten and R.W. Fuchs (ed.). Stuttgart: A. Hiersemann, 1988, I, 653–54.

de Romilly, Jacqueline. 1975. *Magic and Rhetoric in Ancient Greece*. Cambridge, MA: Harvard University Press.

Divination: Perspectives for a New Millenium, Patrick Curry (ed.). Burlington: Ashgate, 2010.

Dobbs, Betty Jo Teeter. 1975. *The Foundations of Newton's Alchemy or 'The Hunting of the Greene Lyon'*. Cambridge: Cambridge University Press.

Drevillon, Herve. 1996. *Lire et ecrire l'avenir: l'astrologie dans la France du Grand Siecle, 1610–1715*. Seyssel: Champ Vallon.

Duranti, Tommaso. 2008. *Mai sotto Saturno: Girolamo Manfredi, medico e astrologo*. Bologna: CLUEB.

Eade, J.C. 1984. *The Forgotten Sky: A Guide to Astrology in English Literature*. Oxford: Clarendon.

Easton, Stewart C. 1952. *Roger Bacon and His Search for a Universal Science*. New York: Columbia University Press.

Edwards, William F. 1976. Niccolò Leoniceno and the Origins of Humanist Discussions of Method. In *Philosophy and Humanism: Renaissance Essays in Honor of Paul Oskar Kristeller*, ed. E.P. Mahoney, 283–305. Leiden: Brill.

Eisenstein, Elizabeth L. 1979. *The Printing Press as an Agent of Change: Communications and Cultural Transformations in Early-Modern Europe*. Cambridge: Cambridge University Press.

Ernst, Germana. From the Watery Trigon to the Fiery Trigon: Celestial Signs, Prophecies and History. In *Astrologi Hallucinati*, 265–280.

Fanger, Clare. 1999. Things Done Wisely by a Wise Enchanter: Negotiating the Power of Words in the Thirteenth Century. *Esoterica* I: 97–132.

Faracovi, Ornella Pompeo. 1999. *Gli oroscopi di Cristo*. Venice: Marsilio.

Favaro, Antonio. 1883. Le matematiche nello studio di Padova prima di Galileo. In *Galileo Galilei e lo studio di Padova*, 100–136. Florence: Le Monnier.

———. 1922. I Lettori di matematiche nella università di Padova dal principio del secolo XIV alla fine del XVI. In *Memorie e Documenti per la Storia della Università di Padova*, I, Padua, 1–70.

Fazzo, Sylvia. 1988. Alessandro d'Afrodisia e Tolomeo: Aristotelismo e astrologia fra il II e il III secolo d.C. *Rivista di Storia della Filosofia* 4: 627–649.

Federico da Montefeltro and his Library, M. Simonetta (ed.). Vatican City: Y. Press, 2007.

Feke, Jacqueline. 2018. *Ptolemy's Philosophy: Mathematics as a Way of Life*. Princeton: Princeton University Press.

Feola, Vittoria. 2014. Antiquarianism, Astrology, and the Press in William Lilly's Network(?). In *Antiquarianism and Science in Early Modern Urban Networks*, ed. V. Feola, 186–203. Paris: Blanchard.

Feyerabend, Paul. 1978. *Science in a Free Society*. London: NLB.

Fichte, Joerg O. 1996. Providentia—Fatum—Fortuna. *Das Mittelalter* I: 5–20.

Fioravanti, Gianfranco. 1997. Pico e l'ambiente Ferrarese. In *Giovanni Pico della Mirandola: Convegno Internazionale di Studi nel Cinquecentesimo Anniversario della Morte (1494–1994)*, ed. Gian Carlo Garfagnini, vol. 2, 157–172. Florence: Olschki.

Flint, Valerie I.J. 1991. *The Rise of Magic in Early Medieval Europe*. Princeton: Princeton University Press.

Forgotten Stars: Rediscovering Manilius's Astronomica, K. Volk (ed.). Oxford: Oxford University Press, 2011.

Franklin, James. 2001. *The Science of Conjecture: Evidence and Probability before Pascal*. Baltimore: Johns Hopkins University Press.

The French Descent into Renaissance Italy, 1494–5: Antecedents and Effects, David Abulafia (ed.). Aldershot: Variorum, 1995.

French, Roger. 1994. Astrology in Medical Practice. In *Practical Medicine from Salerno to the Black Death*, ed. L. Garcia-Ballestar, R. French, J. Arrizabalaga, and A. Cunningham, 30–59. Cambridge: Cambridge University Press.

Freudenthal, Gad. 2000. Providence, Astrology, and Celestial Influences on the Sublunar World in Shem-Tov ibn Falaquera's *De'ot ha-Filosofim*. In *The Medieval Hebrew Encyclopedias of Science and Philosophy*, ed. Steven Harvey, 335–370. Dordrecht: Kluwer.

———. 2002. The Medieval Astrologization of Aristotle's Biology: Averroes on the Role of the Celestial Bodies in the Generation of Animate Beings. *Arabic Sciences and Philosophy* 12: 111–137.

———. 2009. The Astrologization of the Aristotelian Cosmos: Celestial Influences on the Sublunar World in Aristotle, Alexander of Aphrodisias, and Averroes. In *New Perspectives on Aristotle's De caelo*, ed. Alan C. Bowen and Christian Wildberg, 239–281. Leiden: Brill.

From Masha'allah to Kepler: Theory and Practice in Medieval and Renaissance Astrology, Charles Burnett and Dorian Gieseler Greenbaum (ed.). Ceredigion, Wales: Sophia Centre Press, 2015.

Furth, Montgomery. 1988. *Substance, Form and Psyche: An Aristotelian Metaphysics*. Cambridge: Cambridge University Press.

Gabotto, Ferdinando. 1889. L'astrologia nel Quattrocento in rapporto colla civiltà: osservazioni e documenti storici. *Rivista di filosofia scientifica*: 373–413.

Gamba, Enrico. 1991. Astrologi alla corte dei Montefeltro a Urbino e degli Sforza a Pesaro. *Rivista della societa pesarese di studi storici* 1: 75–83.

Garin, Eugenio. 1972. "Paolo Toscanelli," In *Portraits from the Quattrocento*, Trans. V.A. and E. Velen, 118–136. New York: Harper and Row.

———. 1976. *Lo zodiaco della vita: la polemica sull'astrologia dal Trecento al Cinquecento*. Rome: Laterza.

Geertz, Clifford. 1973. Thick Description: Toward an Interpretive Theory of Culture. In *The Interpretation of Cultures: Selected Essays*, 3–30. New York: Basic Books.

Gelber, Hester. Providence. In *The Cambridge History of Medieval Philosophy*, 761–772.

Geneva, Ann. 1995. *Astrology and the 17th Century Mind: William Lilly and the Language of the Stars*. Manchester: Manchester University Press.

Geomancy and Other Forms of Divination, Alessandro Palazzo and Irene Zavaterro (ed.), (Micrologus Library, 87). Florence: SISMEL—Galluzzo, 2017.

Ginzburg, Carlo. 1989. *Clues, Myths and the Historical Method,* Trans. J. and A.C. Tedeschi. Baltimore: Johns Hopkins University Press.

Giussepe, M., and B. Cagni. 1969. *Vespasiano da Bisticci e il suo epistolario*. Rome: Edizioni di Storia e Letteratura.

Goddu, André. 2010. *Copernicus and the Aristotelian Tradition: Education, Reading and Philosophy in Copernicus's Path to Heliocentrism*. Leiden: Brill.

Godman, Peter. 1998. *From Poliziano to Machiavelli: Florentine Humanism in the High Renaissance*. Princeton: Princeton University Press.

Gombrich, Ernst. 1986. *Aby Warburg: An Intellectual Biography*. 2nd ed. Chicago: University of Chicago Press.

Grabmann, Martin. 1936. Der Einfluss Alberts der Grossen auf das Mittelalterliche Geistesleben. In *Mittelalterliches Geistesleben: Abhandlungen zur Geschichte der Scholastik und Mystik*, 3 vols, 1926–56; vol 2. Munich: M. Hueber.

Grafton, Anthony. 1980. The Importance of Being Printed. *Journal of Interdisciplinary History* 11: 265–286.

Grafton, Anthony, and Joseph Scaliger. 1983. *A Study in the History of Classical Scholarship, vol. I: Textual Criticism and Exegesis*. Oxford: Clarendon.

Grafton, Anthony. 1990. *Forgers and Critics: Creativity and Duplicity in Western Scholarship*. Princeton: Princeton University Press.

———. 1991. *Defenders of the Text: The Traditions of Scholarship in an Age of Science, 1450–1800*. Cambridge, MA: Harvard University Press.

———. Humanism and Science in Rudolphine Prague: Kepler in Context. In *Defenders of the Text*, 178–203.

———. Traditions of Invention and Inventions of Tradition in Renaissance Italy: Annius of Viterbo. In *Defenders of the Text*, 76–103.

———. 1997. Giovanni Pico della Mirandola: Trials and Triumphs of an Omnivore. In *Commerce with the Classics: Ancient Books and Their Renaissance Readers*, 93–134. Ann Arbor: University of Michigan Press.

———. 1999. *Cardano's Cosmos: The Worlds and Works of a Renaissance Astrologer*. Cambridge, MA: Harvard University Press.

———. 2000a. Starry Messengers: Recent Work in the History of Western Astrology. *Perspectives on Science* 8: 70–83.

———. 2000b. Geniture Collections: Origins and Use of a Genre. In *Books and the Sciences in History*, ed. M. Frasca-Spada and N. Jardine, 49–68. Cambridge: Cambridge University Press.

Grafton, Anthony, and Lisa Jardine. 1986. *From Humanism to the Humanities; Education and the Liberal Arts in Fifteenth- and Sixteenth-Century Europe*. Cambridge, MA: Harvard University Press.

Grant, Edward. 1987. Medieval and Renaissance Scholastic Conceptions of the Influence of the Celestial Region on the Terrestrial. *Journal of Medieval and Renaissance Studies* 17: 1–23.

———. 1994. *Planets, Stars, and Orbs: The Medieval Cosmos, 1200–1687*. Cambridge: Cambridge University Press.

Green, Jonathan. 2011. *Printing and Prophecy: Prognostication and Media Change 1450–1550*. Ann Arbor: The University of Michigan Press.

———. 2012. Printing the Future: The Origin and Development of the *Practica Teütsch* to 1620. *Archiv für Geschichte des Buchwesens* 67: 1–18.

Green, Lewis. 1972. *Chronicle into History: An Essay on the Interpretation of History in Florentine Fourteenth-Century Chronicles*. Cambridge: Cambridge University Press.

Green, Steven J. 2014. *Disclosure and Discretion in Roman Astrology: Manilius and his Augustan Contemporaries*. Oxford: Oxford University Press.

Gregory, Tullio. 1992. Astrologia e teologia nella cultura medievale. In *Mundana Sapientia: Forme di conoscenza nella cultura medievale*, 291–328. Rome: Edizioni di Storia e Letteratura.

Grendler, Paul F. 2002. *The Universities of the Italian Renaissance*. Baltimore: Johns Hopkins University Press.

Grössing, Helmuth. 1983. *Humanistische Naturwissenschaft: Zur Geschichte der Wiener Mathematischen Schulen des 15. und 16. Jahrhunderts*. Baden-Baden: Valentin Koerner.

Gundersheimer, Werner L. 1973. *Ferrara: The Style of a Renaissance Despotism*. Princeton: Princeton University Press.

Gutas, Dmitri, and Greek Thought. 1998. *Arabic Culture: The Graeco-Arabic Translation Movement in Baghdad and Early Abbasid Society (2nd-4th/8th–10th Centuries).* London: Routledge.
Hackett, Jeremiah M.G. The Attitude of Roger Bacon to the *Scientia* of Albertus Magnus. In *Albertus Magnus and the Sciences*, 53–72.
———. Roger Bacon on Astronomy-Astrology: The Sources of the *Scientia Experimentalis.* In *Roger Bacon and the Sciences*, 175–198.
———. Roger Bacon on *Scientia Experimentalis.* In *Roger Bacon and the Sciences*, 277–316.
———. 2000. Aristotle, *Astrologia,* and Controversy at the University of Paris (1266–1274). In *Learning Institutionalized: Teaching in the Medieval University*, ed. John van Engen, 69–110. Notre Dame: University of Notre Dame Press.
Halleux, Robert. 1982. Albert le Grand et l'alchimie. *Revue des Sciences Philosophiques et Théologiques* 66: 57–80.
Hammer, Franz. 1952. Astrologie und Buchdruck im 15. Jahrhundert. In *Colligere fragmenta: Festschrift Alban Dold*, ed. B. Fischer and V. Fiala, 281–287. Beuron in Hohenzellern: Beuroner Kunstverlag.
Hand, Robert S. 2014. *The Use of Military Astrology in Late Medieval Italy: The Textual Evidence*, PhD thesis. The Catholic University of America.
Hanegraaff, Wouter. 1996. *New Age Religion and Western Culture: Esotericism in the Mirror of Secular Thought.* Leiden: Brill.
———. 2012. *Esotericism and the Academy: Rejected Knowledge in Western Culture.* Cambridge: Cambridge University Press.
Hankins, James. 1991. *Plato in the Italian Renaissance.* Vol. 2. Brill: Leiden.
———. 2007. Ficino, Avicenna and the Occult Powers of the Rational Soul. In *Tra antica sapienza e filosofia naturale: La magia nell'Europa moderna*, ed. F. Meroi, 3 vols. Florence: Olschki, I: 35–52.
Harmening, Dieter. 1979. *Superstitio: Überlieferungs- und theoriegeschichtliche Untersuchung zur kirchlich-theologischen Aberglaubensliteratur des Mittelalters.* Berlin: E. Schmidt.
Hartfelder, Karl. 1889. *Philip Melanchthon als Praeceptor Germaniae.* Berlin: A. Hofmann.
Harvey, E. 1975. *Ruth, The Inward Wits: Psychological Theory in the Middle Ages and the Renaissance.* London: Warburg Institute.
Haskins, Charles H. 1927. *Studies in the History of Mediaeval Science.* 2nd ed. Cambridge, MA: Harvard University Press. (first edition, 1924).
Hasse, Dag Nikolaus. 2016. *Success and Suppression: Arabic Science and Philosophy in the Renaissance.* Cambridge, MA: Harvard University Press.
Hayton, Darin. 2015. *The Crown and the Cosmos: Astrology and the Politics of Maximilian I.* Pittsburgh: University of Pittsburgh Press.
Heilen, Stephan. Ptolemy's Doctrine of the Terms and its Reception. In *Ptolemy in Perspective*, 45–93.
———. Paul of Middelburg's *Prognosticum* for the Years 1484 to 1504. In *From Masha'allah to Kepler*, 231–278.
Hexter, J.H. 1979. The Historical Method of Christopher Hill. In *On Historians: Reappraisals of Some of the Masters of Modern History*, 227–251. Cambridge, MA: Harvard University Press.
Hissette, Roland. 1980. Albert le Grand et l'expression *Diluvium ignis. Bulletin de philosophie médiévale* 22: 78–81.
Hollingsworth, Mary. 1994. *Patronage in Renaissance Italy: From 1400 to the Early Sixteenth Century.* London: John Murray.
Horoscopes and Public Spheres: Essays on the History of Astrology, Günther Oestmann, H Darrel Rutkin and Kocku von Stuckrad (ed.). Berlin: Walter de Gruyter, 2005.
Howe, Ellic. 1968. *Astrology: A Recent History including the Untold Story of its Role in World War II.* New York: Walker and Company.

Hub, Berthold. 2016. Aristotle's 'Bloody Mirror' and Natural Science in Medieval and Early Modern Europe. In *The Mirror in Medieval and Early Modern Culture: Specular Reflections*, ed. Nancy M. Frelick, 31–71. Turnhout: Brepols.

I Guicciardini e le scienze occulte: L'oroscopo di Francesco Guicciardini, lettere di alchimia, astrologia e cabala a Luigi Guicciardini, R. Castagnola (ed.). Florence: Olschki, 1990.

Iliffe, Rob. 2017. *Priest of Nature: The Religious Worlds of Isaac Newton*. Oxford: Oxford University Press.

Imponderable: The Archive of Tony Oursler, Ann Wehr (ed.). Zurich: JPS/Ringier, 2016.

Individuation in Scholasticism: The Later Middle Ages and the Counter-Reformation (1150–1650), Jorge J.E. Gracia (ed.). Albany: State University of New York Press, 1994.

Jervis, Jane L. 1985. *Cometary Theory in Fifteenth-Century Europe*, 45–46. Wroclaw: Ossolineum.

Joachim Camerarius (1500–1574), F. Baron (ed.). Munich: Fink, 1978.

Johns, Adrian. 1998. *The Nature of the Book: Print and Knowledge in the Making*. Chicago: University of Chicago Press.

Jolly, Karen. Medieval Magic: Definitions, Beliefs, Practices. In *Witchcraft and Magic in Europe: The Middle Ages*, 3–71.

Juste, David. 2000. Astrologie et philosophie naturelle chez Roger Bacon. *Villiers* 15: 10–18.

Kahm, Nicholas. 2013. Divine Providence in Aquinas's Commentaries on Aristotle's *Physics* and *Metaphysics*, and its Relevance to the Question of Evolution and Creation. *American Catholic Philosophical Quarterly* 87: 637–656.

Katinis, Teodoro. 2007. *Medicina e filosofia in Marsilio Ficino: il Consilio contro la pestilenza*. Rome: Edizioni di Storia e Letteratura.

Kay, Richard. 1994. *Dante's Christian Astrology*. Philadelphia: University of Pennsylvania Press.

———. 1995. L'astrologia di Dante. In *Dante e la scienza*, ed. P. Boyde and V. Russo, 119–132. Ravenna: Longo.

Kennedy, Edward S. 1998. The Astrological Houses as Defined by Islamic Astronomers. In *Astronomy and Astrology in the Medieval Islamic World*, 535–578. Ashgate: Aldershot. (originally published 1996).

Kibre, Pearl. 1962. *Scholarly Privileges in the Middle Ages: The Rights, Privileges and Immunities of Scholars and Universities at Bologna, Padua, Paris and Oxford*. Cambridge, MA: Medieval Academy of America.

———. 1967. Giovanni Garzoni of Bologna (1419–1505), Professor of Medicine and Defender of Astrology. *Isis* 58: 504–514.

———. 1984. 'Astronomia' or 'Astrologia Ypocratis,'. In *Studies in Medieval Science: Alchemy, Astrology, Mathematics and Medicine*, 133–156. London: Hambledon Press. (originally published, 1978).

———. 1985. *Hippocrates Latinus: Repertorium of Hippocratic Writings in the Latin Middle Ages*, revised ed. New York: Fordham University Press.

Kieckhefer, Richard. 1989. *Magic in the Middle Ages*. Cambridge: Cambridge University Press.

———. 1998. *Forbidden Rites: A Necromancer's Manual of the Fifteenth Century*. University Park: The Pennsylvania State University Press.

Klaassen, Frank. 2013. *The Transformations of Magic: Illicit Learned Magic in the Later Middle Ages and Renaissance*. University Park: The Pennsylvania State University Press.

Klibansky, Raymond, Erwin Panofsky, and Fritz Saxl. 1964. *Saturn and Melancholy: Studies in the History of Natural Philosophy, Religion and Art*. London: Nelson. repr. 1979 (Kraus).

Klibansky, Raymond, Erwin Panofsky and Fritz Saxl. 1992. *Saturn und Melancholie: Studien zur Geschichte der Naturphilosophie und Medizin, der Religion und der Kunst*, Trans. Christa Buschendorf. Frankfurt am Main: Suhrkamp.

Klingshirn, William E. 2003. Isidore of Seville's Taxonomy of Magicians and Diviners. *Traditio* 58: 59–90.

———. 2005. Divination and the Disciplines of Knowledge according to Augustine. In *Augustine and the Disciplines: From Cassiciacum to Confessions*, ed. K. Pollmann and M. Vessey, 113–140. Oxford: Oxford University Press.

Koyré, Alexandre. 1978. *Galileo Studies,* Trans. J. Mepham. Atlantic Highlands: Humanities Press, (originally published 1939).

Kremer, Richard L. 2017. Incunable Almanacs and *Practica* as Practical Knowledge Produced in Trading Zones. In *The Structures of Practical Knowledge*, ed. Matteo Valleriani, 333–369. Dordrecht: Springer.

Kristeller, Paul Oskar. 1943. (repr. Gloucester, MA: Peter Smith, 1964). *The Philosophy of Marsilio Ficino*, 146–170. New York: Columbia University Press.

———. 1956. The School of Salerno: Its Development and Its Contribution to the History of Learning. In *Studies in Renaissance Thought and Letters*, 495–551. Rome: Edizioni di Storia e Letteratura.

———. 1992. Thomism and Italian Thought of the Renaissance. In *Medieval Aspects of Renaissance Learning: Three Essays by Paul Oskar Kristeller,* (ed and Trans., E.P. Mahoney), 2nd ed., 27–91. New York: Columbia University Press, (originally published, 1974).

———. 1978. The First Printed Editions of Plato's Works and the Date of Its Publication (1484). In *Science and History: Studies in Honor of Edward Rosen*, 25–35. Wroclaw: Ossolineum.

Kuhn, Thomas S. 2012. *Structure of Scientific Revolutions*. 4th ed. Chicago: University of Chicago Press. (first published, 1962).

Kurze, Dietrich. 1960. *Johannes Lichtenberger: Eine Studie zur Geschichte der Prophetie und Astrologie*. Lübeck: Matthiesen.

Kusukawa, Sachiko. 1993. *Aspectio divinorum operum*: Melanchthon and Astrology for Lutheran Medics. In *Medicine and the Reformation*, ed. O.P. Grell and A. Cunningham, 33–56. London: Routledge.

———. 1995. *The Transformation of Natural Philosophy: The Case of Philip Melanchthon*. Cambridge: Cambridge University Press.

Labrousse, Elisabeth. 1974. *L'entrée de Saturne au Lion (L'eclipse de soleil du 12 Aout 1654)*. The Hague: Nijhoff.

Lang, Helen S. 1992. *Aristotle's Physics and its Medieval Varieties*. Albany: SUNY Press.

Lanuza Navarro, Tayra. 2005. *Astrologia, Ciencia y Sociedad en la España de los Austrias,* Phd thesis. Universitat de Valencia.

Leff, Gordon. 1968. *Paris and Oxford in the 13th and 14th Centuries*. New York: Wiley.

Lemay, Richard J. 1962. *Abu Ma'shar and Latin Aristotelianism in the 12th Century: The Recovery of Aristotle's Natural Philosophy through Arabic Astrology*. Beirut: Publications of the Faculty of Arts and Sciences.

———. The True Place of Astrology in Medieval Science and Philosophy: Towards a Definition. In *Astrology, Science and Society*, 57–73.

———. Roger Bacon's Attitude Toward the Latin Translations and Translators of the Twelfth and Thirteenth Centuries. In *Roger Bacon and the Sciences*, 25–47.

Lennox, James. 1986. Aristotle, Galileo, and 'Mixed Sciences'. In *Reinterpreting Galileo*, ed. William A. Wallace, 29–51. Washington, D.C: Catholic University of America Press.

Lerch, Andreas. 2015. *Scientia astrologiae: Der Diskurs über die Wissenschaftlichkeit der Astrologie und die lateinischen Lehrbücher 1470–1610*. Leipzig: Akademischen Verlagsanstalt.

Lindberg, David C. 1971. Lines of Influence in Thirteenth-Century Optics: Bacon, Witelo, and Pecham. *Speculum* 46: 66–83.

———. 1976. *Theories of Vision from al-Kindi to Kepler*. Chicago: University of Chicago Press.

———. 1982. On the Applicability of Mathematics to Nature: Roger Bacon and his Predecessors. *The British Journal for the History of Science* 15: 3–25.

———. 1987. Roger Bacon and the Origins of *Perspectiva* in the West. In *Mathematics and its Applications to Science and Natural Philosophy in the Middle Ages: Essays in Honor of Marshall Clagett*, ed. Edward Grant and John E. Murdoch, 249–268. Cambridge: Cambridge University Press.

———. 2007. *The Beginnings of Western Science: The European Scientific Tradition in Philosophical, Religious, and Institutional Context, 600 B.C. to A.D. 1450*. 2nd ed. Chicago: University of Chicago Press. (first edition, 1992).

Linderski, Jerzy. 1986. The Augural Law. In *Aufstieg und Niedergang der Römischen Welt*, ed. W. Haase, II, 16. 3, 2146–2312. Berlin: de Gruyter.

Lines, David A. 2001. Natural Philosophy in Renaissance Italy: The University of Bologna and the Beginnings of Specialization. In *Science and Universities of Early Modern Europe: Teaching, Specialization, Professionalization*, ed. David Lines, a special issue of *Early Science and Medicine* 6: 267–323.

———. 2002. *Aristotle's Ethics in the Italian Renaissance (ca. 1300–1650): The Universities and the Problem of Moral Eduction*. Brill: Leiden.

Lippincott, Kristin. 1990. Two Astrological Ceilings Reconsidered: The *Sala di Galatea* in the Villa Farnesina and the *Sala dei Mappamundo* at Caprarola. *Journal of the Warburg and Courtauld Institutes* 53: 185–196.

Lipton, Joshua. 1978. *The Rational Evaluation of Astrology in the Period of the Arabo-Latin Translations, ca. 1126–1187*, PhD thesis. UCLA.

Litt, Thomas. 1963. *Les corps celestes dans l'univers de Saint Thomas d'Aquin*. Louvain: Publications Universitaires.

Lodge, David. 1975. *Changing Places: A Tale of Two Campuses*. London: Secker and Warburg.

Lohr, Charles H. 1988. *Latin Aristotle Commentaries*. Florence: Olschki.

Longo, Pessina. 1988. *Georgius de Russia, rettore a Bologna nel XV secolo*. Bologna: CLUEB.

Lorenzo the Magnificent: Culture and Politics, M. Mallett and N. Mann (ed.). London: Warburg Institute, 1996.

Lowry, Martin J.C. 1974. Two Great Venetian Libraries in the Age of Aldus Manutius. *Bulletin of the John Rylands University Library of Manchester* 57: 128–166.

———. 1979. *The World of Aldus Manutius: Business and Scholarship in Renaissance Venice*. Ithaca: Cornell University Press.

Lubkin, Gregory. 1994. *A Renaissance Court: Milan under Galeazzo Maria Sforza*. Berkeley: University of California Press.

Lucentini, Paolo. 2007. L'Ermetismo magico nel secolo XIII. In *Platonismo, ermetismo, eresia nel medioevo*, 264–324. Louvain-la-Neuve: Fédération Internationale des Instituts d'Études Médiévales. (originally published, 2000).

Magdalino, Paul. 2006. *L'Orthodoxie des astrologues: La science entre le dogme et la divination à Byzance (VIIe-XIVe siècle)*. Paris: Lethielleux.

Mahoney, Edward P. Albert the Great and the *Studio Patavino* in the Late Fifteenth and Early Sixteenth Centuries. In *Albertus Magnus and the Sciences*, 537–563.

Maier, Anneliese. 1949. *Die Vorläufer Galileis im 14.Jahrhundert: Studien zur Naturphilosophie der Spätscholastik*. Rome: Edizioni di Storia e Letteratura. (second ed. 1966; 1977).

———. 1951. *Zwei Grundprobleme der scholastischen Naturphilosophie. Das Problem der intensiven Grösse. Die Impetustheorie*. Rome: Edizioni di Storia e Letteratura. (second ed; third ed. 1968).

Malagola, Carlo. 1888. *Statuti delle Università e dei college dello studio bolognese*. Bologna: N. Zanichelli.

Mallett, Michael. 1988. The Florentine *Otto di Pratica* and the Beginning of the War of Ferrara. In *Florence and Italy: Renaissance Studies in Honour of Nicolai Rubenstein*, ed. P. Denley and C. Elam, 3–12. London: Westfield College.

Mancosu, Paolo. 1996. *The Philosophy of Mathematics and Mathematical Practice in the Seventeenth Century*. New York: Oxford University Press.

Mantik, Schicksal und Freiheit im Mittelalter, Loris Sturlese (ed.). Cologne: Böhlau Verlag, 2011.

Marangon, Paolo. 1977. *Alle origini dell'Aristotelismo padovano (sec. XII–XIII)*. Antenore: Padua.

———. 1997. In *Ad cognitionem scientiae festinare: gli studi nell'università di Padova nei secoli XIII e XIV*, ed. Tiziana Pesenti. Trieste: LINT.

Marangon, Tiziana Pesenti. 1976–77. Michele Savonarola a Padova: L'ambiente, le opere, la cultura medica. *Quaderni per la Storia dell'Università di Padova* 9–10: 88–102.

———. 1978. La miscellanea astrologica del prototipografo padovano Bartolomeo Valdizocco e la diffusione dei testi astrologici e medici tra i lettori padovani del '400. *Quaderni per la Storia dell'Università di Padova* 11: 87–106.
Marenbon, John. 1987. *Later Medieval Philosophy (1150–1350): An Introduction*. London: Routledge.
Martelli, Mario. 1965. Una vacanza letteraria di Lorenzo: Il carnevale del 1490. In *Studi Laurenziani*, 37–49. Florence: Olschki.
Martin, Craig. 2011. *Renaissance Meteorology: Pomponazzi to Descartes*. Baltimore: Johns Hopkins University Press.
Martines, Lauro. 1988. *Power and Imagination: City-States in Renaissance Italy*. 2nd ed. Baltimore: Johns Hopkins University Press.
———. 2003. *April Blood: Florence and the Plot against the Medici*. Oxford: Oxford University Press.
Martinich, Aloysius P. 2008. *The Philosophy of Language*. fifth ed. New York: Oxford University Press.
Marx, J. 1905. *Verzeichnis der Handschrift-Sammlung des Hospitals zu Cues*. Selbstverlag des Hospitals: Trier.
Mattingly, Garrett. 1955. *Renaissance Diplomacy*. Boston: Houghton Mifflin.
Maurer, Wilhelm. 1967–69. *Der Junge Melanchthon zwischen Humanismus und Reformation*. Vol. 2. Göttingen: Vandenhoeck and Ruprecht.
Mavroudi, Maria. 2015. Translations from Greek into Latin and Arabic during the Middle Ages: Searching for the Classical Tradition. *Speculum* 90: 28–59.
McAllister, Joseph Bernard. 1939. *The Letter of Saint Thomas Aquinas De operationibus occultis naturae ad quemdam militem ultramontanum: A Dissertation*. Washington, D.C: Catholic University of America Press.
Médicine, astrologie et magie entre Moyen Âge et Renaissance: autour de Pietro d'Abano, J.-P. Boudet, F. Collard and N. Weill-Parot (ed.). Florence: SISMEL—Galluzzo, 2013.
Medieval Prognosis and Astrology: A Working Edition of Aggregationes de crisi et criticis diebus, Cornelius O'Boyle (ed.). Cambridge: Wellcome Unit, 1991.
Mentgen, Gerd. 2005. *Astrologie und Öffentlichkeit im Mittelalter*. Stuttgart: Hiersmann.
Michele Savonarola: Medicina e cultura di corte, Chiara Crisciani and Gabriella Zuccolin (ed.). Florence: SISMEL—Galluzzo, 2011.
Molland, George. Roger Bacon's Knowledge of Mathematics. In *Roger Bacon and the Sciences*, 151–174.
———. Roger Bacon's *De laudibus mathematicae*: A Preliminary Study. In *Texts and Contexts*, 68–83.
Mori, Gioia. 1987. *Arte e astrologia*. Florence: Giunti.
Moulinier-Brogi, Laurence. 2011. *Guillaume l'Anglais, le frondeur de l'uroscopie médiévale (XIIIe siècle): Édition commentée et traduction du De urina non visa*. Geneva: Librarie Droz.
Murdoch, John E. 1991. Pierre Duhem and the History of Late Medieval Science and Philosophy in the Latin West. In *Gli studi di filosofia medievale fra Otto e Novecento*, ed. R. Imbach and A. Maierù, 253–302. Rome: Edizioni di Storia e Letteratura.
Murdoch, John, and Edith D. Sylla. 1978. The Science of Motion. In *Science in the Middle Ages*, ed. David C. Lindberg, 206–264. Chicago: University of Chicago Press.
Nardi, Bruno. 1958. *Saggi sul Aristotelismo padovano dal secolo XIV al XVI*. Florence: Sansoni.
Nella luce degli astri: l'astrologia nella cultura del Rinascimento, Ornella Pompeo Faracovi (ed.). Sarzana, La Spezia: Agorà, 2004.
Nello specchio del cielo: Giovanni Pico della Mirandola e le Disputationes contro l'astrologia divinatoria, Marco Bertozzi (ed.). Florence: Olschki, 2008.
Neugebauer, Otto. 1969. *The Exact Sciences in Antiquity*. 2nd ed. New York: Dover. (originally published, 1957).
———. 1975. *A History of Ancient Mathematical Astronomy*. Berlin: Springer.

Neugebauer, Otto, and H.B. Van Hoesen. 1959. *Greek Horoscopes*. Philadelphia: American Philosophical Society.

Newman, William R. 1991. *The Summa Perfectionis of Pseudo-Geber: A Critical Edition, Translation and Study*. Leiden: Brill. (based on his 1986 Harvard PhD thesis).

———. 2004. *Promethean Ambitions: Alchemy and the Quest to Perfect Nature*. Chicago: University of Chicago Press.

———. 2009. Brian Vickers on Alchemy and the Occult: A Response. *Perspectives on Science* 17: 482–506.

———. 2019. *Newton the Alchemist: Science, Enigma, and the Quest for Nature's "Secret Fire,"*. Princeton: Princeton University Press.

Newman, William R., and Anthony Grafton. Introduction: The Problematic Status of Astrology and Alchemy in Premodern Europe. In *Secrets of Nature*, 1–37.

Niccoli, Ottavia. 1990. *Prophecy and People in Renaissance Ital,* Trans. L. Cochrane. Princeton: Princeton University Press.

North, John D. 1980. Astrology and the Fortunes of Churches. *Centaurus* 24: 181–211. reprinted in his *Stars, Minds and Fate*, 59–90.

———. 1986a. *Horoscopes and History*. London: Warburg Institute.

———. 1988. *Chaucer's Universe*. Oxford: Clarendon Press.

———. 1986b. Celestial Influence—The Major Premiss of Astrology. In *"Astrologi Hallucinati": Stars and the End of the World in Luther's Time*, ed. Paola Zambelli, 45–100. Berlin: de Gruyter. reprinted in his *Stars, Minds and Fate*, 243–98.

———. 1989. *Stars, Minds and Fate: Essays in Ancient and Medieval Cosmology*. London: Hambledon.

———. 1994. The Quadrivium. In *A History of the University in Europe*, 4 vols., Volume I: *Universities in the Middle Ages*, ed. Hilde de Ridder-Symoens, 337–59. Cambridge: Cambridge University Press.

———. 1999. Roger Bacon and the Saracens. In *Filosofia e scienza classica, arabo-latina medievale e l'eta moderna*, ed. Graziella Federici Vescovini, 129–160. Turnhout: Brepols.

———. 2013. Astronomy and Astrology. In *The Cambridge History of Science, Vol 2: Medieval Science*, ed. David C. Lindberg and Michael H. Shank, 456–484. Cambridge: Cambridge University Press.

Nothaft, C., and E. Philipp. 2012. *Dating the Passion: The Life of Jesus and the Emergence of Scientific Chronology (200–1600)*. Brill: Leiden.

———. 2016. *Vanitas vanitatum et super omnia vanitas*: The Astronomer Heinrich Selder and a Newly Discovered 14th-Century Critique of Astrology. *Erudition and the Republic of Letters* 1: 261–304.

Nutton, Vivian. 2013. *Ancient Medicine*. 2nd ed. Oxford: Routledge.

Occult and Scientific Mentalities in the Renaissance, Brian Vickers (ed.). Cambridge: Cambridge University Press, 1984.

Oestmann, Günther. 2004. *Heinrich Rantzau und die Astrologie: Ein Beitrag zur Kulturgeschichte des 16. Jahrhunderts*. Braunschweig: Braunschweig-druck GmbH.

Olivieri, Luigi. 1988. *Pietro d'Abano e il pensiero neolatino: filosofia, scienza e ricerca dell'Aristotele Greco tra i secoli XIII e XIV*. Padua: Antenore.

Omodeo, Pietro Daniel. 2014. *Copernicus in the Cultural Debates of the Renaissance*. Leiden: Brill.

Osler, Margaret J. 2010. *Reconfiguring the World: Nature, God, and Human Understanding from the Middle Ages to Early Modern Europe*. Baltimore: Johns Hopkins University Press.

Page, Sophie. 2013. *Magic in the Cloister: Pious Motives, Illicit Interests, and Occult Approaches to the Medieval Universe*. University Park: The Pennsylvania State University Press.

Palazzo, Alessandro. 2008. The Scientific Significance of Fate and Celestial Influences in Some Mature Works by Albert the Great: *De fato, De somno et vigilia, De intellectu et intelligibili, Mineralia*, In Per perscrutationem philosophicam: *Neue Perspektiven der mittelalterlichen*

Forschung, Loris Sturlese zum 60. Geburtstag gewidmet, ed. A. Beccarisi, R. Imbach and P. Porro. Hamburg: Felix Meiner Verlag, 55–78.
———. 2009. Albert the Great's Doctrine of Fascination in the Context of his Philosophical System. In *Via Alberti: Texte—Quellen—Interpretationen*, ed. L. Honnefelder, H. Möhle, and S. Bullido del Barrio, 135–215. Münster: Aschendorff Verlag.
———. Albert the Great's Doctrine of Fate. In *Mantik, Schicksal und Freiheit im Mittelalter*, 65–95.
Paravicini Bagliani, Agostino. 1991. *Medicina e scienze della natura alla corte dei papi nel Duecento*. Centro Italiano di studi sull'alto Medioevo: Spoleto.
———. 2000. *The Pope's Body*, Trans. D. Peterson. Chicago: University of Chicago Press.
———. 2001. *Le Speculum Astronomiae, une énigme? Enquête sur les manuscrits*. Florence: SISMEL—Galluzzo.
Parel, Anthony. 1992. *The Machiavellian Cosmos*. New Haven: Yale University Press.
Partridge, Loren. 1996. *The Art of Renaissance Rome, 1400–1600*. New York: Abrams.
Paschetto, Eugenia. 1984. *Pietro d'Abano, medico et filosofo*. Florence: Nuovedizioni E. Vallecchi.
Pedersen, Olaf. 1975. The 'Corpus Astronomicum' and the Traditions of Medieval Latin Astronomy. *Studia Copernicana* 13: 57–96.
———. 1997. *The First Universities:* Studium generale *and the Origins of University Education in Europe,* Trans. R. North. Cambridge: Cambridge University Press.
Peters, Edward. 1978. *The Magician, The Witch and the Law*. Philadelphia: University of Pennsylvania Press.
———. The Medieval Church and State on Superstition, Magic and Witchcraft: From Augustine to the Sixteenth Century. In *Witchcraft and Magic in Europe: The Middle Ages*, 173–245.
Philosophical Issues in Aristotle's Biology, Allan Gotthelf and James G. Lennox (ed.). Cambridge: Cambridge University Press, 1987,
Pingree, David. 1990. Astrology. In *The Cambridge History of Arabic Literature: Religion, Learning and Science in the 'Abbasid Period*, ed. M.J.L. Young, J.D. Latham, and R.B. Sarjeant, 290–300. Cambridge: Cambridge University Press.
———. 1997. *From Astral Omens to Astrology: From Babylon to Binaker*. Rome: Istituto Italiano per l'Africa et l'Oriente.
Popper, Nicholas. 2006. 'Abraham, Planter of Mathematics': Histories of Mathematics and Astrology in Early Modern Europe. *Journal of the History of Ideas* 67: 87–106.
Porro, Pasquale. 2012. *Lex necessitatis vel contingentiae*: Necessità, contingenza e provvidenza nell'universo di Tommaso d'Aquino. *Revue des Sciences Philosophiques et Théologiques* 96: 401–450.
Power, Amanda. 2006. A Mirror for Every Age: The Reputation of Roger Bacon. *English Historical Review* 121: 657–692.
———. 2013. *Roger Bacon and the Defense of Christendom*. Cambridge: Cambridge University Press.
Price, Betsey Barker. The Physical Astronomy and Astrology of Albertus Magnus. In *Albertus Magnus and the Sciences*, 155–185.
Principe, Lawrence. 2011. *The Scientific Revolution: A Very Short Introduction*. Oxford: Oxford University Press.
Ptolemy in Perspective: Use and Criticism of his Work from Antiquity to the Nineteenth Century, Alexander Jones (ed.). Dordrecht: Springer, 2010.
Quigley, Joan. 1990. *"What Does Joan Say?": My Seven Years as White House Astrologer to Nancy and Ronald Reagan*. New York: Birch Lane Press.
Quinlan-McGrath, Mary. 1984. The Astrological Vault of the Villa Farnesina: Agostino Chigi's Rising Sign. *Journal of the Warburg and Courtauld Institutes* 47: 91–105.
———. 1995. The Villa Farnesina: Time-Telling Conventions and Renaissance Astrological Practice. *Journal of the Warburg and Courtauld Institutes* 58: 52–71.
———. 2001. The Foundation Horoscope(s) for St. Peter's Basilica, Rome, 1506: Choosing a Time, Changing a *Storia. Isis* 92: 716–741.

———. 2013. *Influences: Art, Optics and Astrology in the Italian Renaissance*. Chicago: University of Chicago Press.

Rashdall, F. 1936. Hastings, *The Universities of Europe in the Middle Ages*, ed. F.M. Powicke and A.B. Emden, new edition, 3 vols. Oxford: Oxford University Press.

Redgrave, Gilbert R. 1894. *Erhard Ratdolt and His Work at Venice*. London: Chiswick Press. (repr. 1899).

Reeves, Marjorie. 1969. *The Influence of Prophecy in the Later Middle Ages: A Study in Joachimism*. Oxford: Clarendon.

Ricciardi, Lucia. 1992. *Col senno, col tesoro e colla lancia: riti e giochi cavallereschi nella Firenze del magnifico Lorenzo*. Florence: Le Lettere.

Ridder-Patrick, Jane. 2012. *Astrology in Early Modern Scotland ca. 1560–1726,* PhD thesis. University of Edinburgh.

Riddle, John M., and James Mulholland. Albert on Stones and Minerals. In *Albertus Magnus and the Sciences*, 203–234.

Robichaud, Denis J.-J. 2018. *Plato's Persona: Marsilio Ficino, Renaissance Humanism, and Platonic Traditions*. Philadelphia: University of Pennsylvania Press.

Rochberg, Francesca. 1998. *Babylonian Horoscopes*. Philadelphia: American Philosophical Society.

———. 2004. *The Heavenly Writing: Divination, Horoscopy, and Astronomy in Mesopotamian Culture*. Cambridge: Cambridge University Press.

———. 2016. *Before Nature: Cuneiform Knowledge and the History of Science*. Chicago: University of Chicago Press.

Rodriguez Arribas, Josefina. 2010. *Testimonies* in Medieval Astrology: Finding Degrees of Certitude in Astrological Judgments. In *Doxa: Études sur les formes et la construction de la croyance*, ed. Pascale Hummel, 115–133. Paris: Philologicum.

Roger Bacon and the Sciences: Commemorative Essays, J. Hackett (ed.). Leiden: Brill, 1997.

Roger Bacon Essays, A.G. Little (ed.). Oxford: Clarendon Press, 1914.

Rose, Paul Lawrence. 1973. Humanist Culture and Renaissance Mathematics: The Italian Libraries of the *Quattrocento*. *Studies in the Renaissance* 20: 46–105.

———. 1975. *The Italian Renaissance of Mathematics: Studies on Humanists and Mathematicians from Petrarch to Galileo*. Geneva: Droz.

Rose, Paul Lawrence, and Stillman Drake. 1971. The Pseudo-Aristotelian 'Questions of Mechanics' in Renaissance Culture. *Studies in the Renaissance* 18: 65–104.

Rousseau, Claudia. 1983. *Cosimo I de Medici and Astrology: The Symbolism of Prophecy*, PhD Thesis. Columbia University, .

Rosier, Irène. 1994. *La parole comme acte: sur la grammaire et la sémantique au XIIIe siècle*. Paris: Vrin.

Rowland, Ingrid D. 1998. *The Culture of the High Renaissance: Ancients and Moderns in Sixteenth-Century Rome*. Cambridge: Cambridge University Press.

Rubinstein, Nicolai. 1997. *The Government of Florence under the Medici (1434 to 1494)*. 2nd ed. Oxford: Clarendon Press.

———. 1957. Some Ideas on Municipal Progress and Decline in the Italy of the Communes. In *Fritz Saxl 1890–1948: A Volume of Memorial Essays from His Friends in England*, ed. D.J. Gordon, 165–183. London: T. Nelson.

Rutkin, H. Darrel. Celestial Offerings: Astrological Motifs in the Dedicatory Letters of Galileo's *Sidereus Nuncius* and Kepler's *Astronomia Nova*. In *Secrets of Nature*, 133–172.

Rutkin, H Darrel. 2002. Astrology, Natural Philosophy and the History of Science, c. 1250–1700: Studies Toward an Interpretation of Giovanni Pico della Mirandola's *Disputationes adversus astrologiam divinatricem,* PhD thesis. Indiana University.

Rutkin, H. Darrel. 2005. Galileo Astrologer: Astrology and Mathematical Practice in the Late-Sixteenth and Early-Seventeenth Centuries. *GALILAEANA: Journal of Galilean Studies* 2: 107–143.

———. 2006. Astrology. In *The Cambridge History of Science, Vol. 3: Early Modern Science*, ed. Lorraine Daston and Katharine Park, 541–561. Cambridge: Cambridge University Press.
———. The Use and Abuse of Ptolemy's *Tetrabiblos* in Renaissance and Early Modern Europe: Two Case Studies (Giovanni Pico della Mirandola and Filippo Fantoni). In *Ptolemy in Perspective*, 135–149.
———. Astrology and Magic. In *A Companion to Albert the Great*, 451–505.
———. 2013. Astrological Timing and Architectural Sites: The Deborah Loeb Brice *Loggiato* in One of its Historical Contexts. In *Renaissance Studies in Honor of Joseph Connors*, ed. Mächtelt Israels and Louis Waldman, 133–137. Milan: Officina Libraria.
———. 2015. Understanding the History of Astrology Accurately: Methodological Reflections on Terminology and Anachronism. *Philosophical Readings* 7: 42–54. (a special issue on astrology edited by Donato Verardi).
———. Teaching Astrology in the 16th Century: Giuliano Ristori and Filippo Fantoni on Pseudo-Prophets and Other Effects of Great Conjunctions. In *From Masha'allah to Kepler*, 353–406.
———. Astrology, Politics and Power in 16th-century Florence: Giuliano Ristori's Extensive Judgment on Cosimo I's Nativity (1537). In *Astrologers and their Clients in Medieval and Early Modern Europe*, 139–150.
———. 2017. An Idealized Astrological Courtier at a 13th-Century Papal and Royal Court: The Case of Roger Bacon. In *De Frédéric II à Rodolphe II: Astrologie, divination et magie dans les cours (XIIIe-XVIIe siècle)*, Micrologus Library 85, ed. Jean-Patrice Boudet, Martine Ostorero, and Agostino Paravicini Bagliani, 53–69. Firenze: Sismel edizioni del Galluzzo.
———. 2018. How to Accurately Account for Astrology's Marginalization in the History of Science and Culture: The Central Importance of an Interpretive Framework. *Early Science and Medicine* 23: 217–243. (a special issue on the marginalization of astrology edited by Rienk Vermij and Hiro Hirai).
Ryan, Michael A. 2011. *A Kingdom of Stargazers: Astrology and Authority in the Late Medieval Crown of Aragon*. Ithaca: Cornell University Press.
Saif, Liana. 2015. *The Arabic Influences on Early Modern Occult Philosophy*. Basingstoke: Palgrave Macmillan.
Salvadori, Patrizia. 1994. I signori di Mirandola, Firenze e i Medici. In *Pico, Poliziano e l'umanesimo di fine Quattrocento: Biblioteca Medicea Laurenziana*, ed. Paolo Viti, 287–302. Florence: Olschki.
Sambin, Paolo. 1974. Professori di astronomia e matematica a Padova nell'ultimo decennio del Quattrocento. *Quaderni per la Storia dell'Università di Padova* 7: 59–67.
Sarton, George. 1959. *Ancient Science and Modern Civilization*. New York: Harper.
Schabel, Christopher D. 2000. *Theology at Paris, 1316–1345: Peter Aureol and the Problem of Divine Foreknowledge and Future Contingents*. Aldershot: Ashgate.
Schabel, Christopher D., and Fritz S. Pederson. 2014. Miraculous, Natural, or Jewish Conspiracy?: Pierre Ceffons' Question on the Black Death, with Astrological Predictions by Gersonides and Jean de Murs/ Firman de Beauval. *Recherches de Théologie et Philosophie Médiévales* 81: 137–179.
Schmitt, Charles B. 1983a. *John Case and Aristotelianism in Renaissance England*. Kingston: McGill-Queens University Press.
———. 1983b. *Aristotle and the Renaissance*. Cambridge, MA: Harvard University Press.
———. 1985. Aristotle among the Physicians. In *The Medical Renaissance of the 16th Century*, ed. A. Wear, R.K. French, and I.M. Lonie, 1–15. Cambridge: Cambridge University Press.
Schmitt, Charles B., and Dilwyn Knox. 1985. *Pseudo-Aristoteles Latinus: A Guide to Latin Works Falsely Attributed to Aristotle before 1500*. London: Warburg Institute.
Schöner, Christoph. 1994. *Mathematik und Astronomie an der Universität Ingolstadt im 15. und 16. Jahrhundert*. Berlin: Speyer and Peters.
Scott, John Beldon. 1991. *Images of Nepotism: The Painted Ceilings of Palazzo Barberini*. Princeton: Princeton University Press.

Secrets of Nature: Astrology and Alchemy in Early Modern Europe, William R. Newman and Anthony Grafton (ed.). Cambridge, MA: MIT Press, 2001.

Seeing with Different Eyes: Essays on Astrology and Divination, Patrick Curry and Angela Voss (ed.). Newcastle: Cambridge Scholars Publishing, 2007.

Sela, Shlomo. 2003. *Abraham Ibn Ezra and the Rise of Medieval Hebrew Science*. Leiden: Brill.

Seller, Fabio. 2009. *Scientia astrorum: la fondazione epistemologica dell'astrologia in Pietro d'Abano*. Naples: Giannini.

Serra-Zanetti, Alberto. 1952. I Pronostici di Girolamo Manfredi. In *Studi riminesi e bibliografici in onore di Carlo Lucchesi*, 193–213. Faenza: Fratelli Legov.

Seventh Centenary of the Teaching of Astronomy in Bologna 1297–1997, Pierluigi Battistini, et al. (ed.). Bologna: CLUEB, 2001.

Sezgin, Fuat. 1979. *Geschichte des Arabischen Schrifttums, Band VII: Astrologie-Meteorologie und Verwandtes. Bis ca. 430 H.* Brill: Leiden.

Seznec, Jean. 1953. *The Survival of the Pagan Gods: The Mythological Tradition and its Place in Renaissance Humanism and Art*, Trans. Barbara F. Sessions. New York: Pantheon.

Shank, Michael H. Academic Consulting in 15th-Century Vienna: The Case of Astrology. In *Texts and Contexts in Ancient and Medieval Science*, 245–270.

———. 2012. The Geometrical Diagrams in Regiomontanus's Edition of his own *Disputationes* (c. 1475): Background, Production and Diffusion. *Journal for the History of Astronomy* 43: 27–55.

Shanley Brian, J.O.P. 2012. Thomas Aquinas on Demonstrating God's Providence. In *The Science of Being as Being: Metaphysical Investigations*, ed. Gregory T. Doolan, 221–242. Washington, D.C: Catholic University of America Press.

Sharples, Robert. 1982. Alexander of Aphrodisias on Divine Providence: Two Problems. *The Classical Quarterly* 32: 198–211.

———. 2009. Fate, Prescience and Free Will. In *Cambridge Companion to Boethius*, ed. John Marenbon, 207–227. Cambridge: Cambridge University Press.

Sighinolfi, Lino. 1920. *Domenico Maria Novara e Niccolò Copernico allo Studio di Bologna*. Modena: Ferraguti.

Signorini, Rodolfo. 2007. *Fortuna dell'astrologia a Mantova: arte, letteratura, carte d'archivio*. Mantua: Sometti.

Simon, Gérard. 1979. *Kepler Astronome Astrologue*. Paris: Gallimard.

Siraisi, Nancy G. 1973. *Arts and Sciences at Padua: The Studium of Padua before 1350*. Toronto: Pontifical Institute of Medieval Studies.

———. 1981. *Taddeo Alderotti and His Pupils: Two Generations of Italian Medical Learning*. Princeton: Princeton University Press.

———. 1987. *Avicenna in Renaissance Italy: The Canon and Medical Teaching in Italian Universities after 1500*. Princeton: Princeton University Press.

———. 1990. *Medieval and Early Renaissance Medicine: An Introduction to Knowledge and Practice*. Chicago: University of Chicago Press.

Siraisi, Nancy, and Anthony Grafton. Between the Election and My Hopes: Girolamo Cardano and Medical Astrology. *Secrets of Nature*: 69–131.

Skemer, Don. 2006. *Binding Words: Textual Amulets in the Middle Ages*. University Park: The Pennsylvania State University Press.

Smoller, Laura. 1994. *History, Prophecy, and the Stars: The Christian Astrology of Pierre D'Ailly, 1350–1420*. Princeton: Princeton University Press.

———. 2007. Astrology and the Sybils: John of Legnano's *De adventu Christi* and the Natural Theology of the Middle Ages. *Science in Context* 20: 423–450.

Soldati, Benedetto. 1986. *La poesia astrologica nel 400*. Florence: Le Lettere. with a new introduction by Cesare Vasoli (originally published 1906).

Solmsen, Friedrich. 1960. *Aristotle's System of the Natural World: A Comparison with his Predecessors*. Ithaca: Cornell University Press.

Soranzo, Matteo. 2014. *Poetry and Identity in Quattrocento Naples*. Farnham: Ashgate.

Sorbelli, Albano. 1938. Il 'Tacuinus' dell'università di Bologna e le sue prime edizioni. *Gutenburg Jahrbuch* 13: 109–114.

Southern, Richard W. 1972. Aspects of the European Tradition of Historical Writing: 3. History as Prophecy. *Royal Historical Association, Transactions* (fifth ser.) 22: 159–80.

———. 1992. *Robert Grosseteste: The Growth of an English Mind in Medieval Europe*. 2nd ed. Oxford: Clarendon Press.

Stegemann, Viktor. 1948. Giovanni Villanis historische Charakterbilder und die astrologischen Texte der planetarischen Anthropologie: Ein Beitrag zur geistigen Physiognomie der Frührenaissance. In *Lebenskräfte in der abendländischen Geistesgeschichte: Festschrift Walter Goetz*, ed. B. Bischoff et al., 125–198. Marburg: Simons.

Stephenson, Bruce. 1994. *The Music of the Heavens: Kepler's Harmonic Astronomy*. Princeton: Princeton University Press.

Stinger, Charles. 1977. *Humanism and the Church Fathers: Ambrogio Traversari (1386–1439) and Christian Antiquity in the Italian Renaissance*. Albany: State University of New York Press.

———. 1985. *The Renaissance in Rome*. Bloomington: Indiana University Press.

Stolzenberg, Daniel. 2013. *Egyptian Oedipus: Athanasius Kircher and the Secrets of Antiquity*. Chicago: The University of Chicago Press.

Stornajolo, C. 1902–21. *Codices Urbinates Latini*. Vol. 3. Rome: Vatican Press.

———. 1895. *Codices Urbinates Graeci*. Rome: Vatican Press.

Studi per le "sorti": Gioco, immagini, poesia oracolare a Venezia nel Cinquecento, Paolo Procaccioli (ed.). Treviso: Edizione Fondazione Benetton, 2007.

Sturlese, Loris. 1980. Saints et magiciens: Albert le Grand en face d'Hermes Trismégiste. *Archives de Philosophie* 43: 615–634.

———. Thomas von Aquin und die Mantik. In *Mantik, Schicksal und Freiheit*, 97–107.

Swerdlow, Noel M. 1990. Regiomontanus on the Critical Problems of Astronomy. In *Nature, Experiment and the Sciences*, ed. T.H. Levere and W.R. Shea, 165–195. Dordrecht: Kluwer.

———. 1993a. Science and Humanism in the Renaissance: Regiomontanus's Oration on the Dignity and Utility of the Mathematical Sciences. In *World Changes: Thomas Kuhn and the Nature of Science*, ed. P. Horwich, 131–168. Cambridge, MA: MIT Press.

———. 1993b. The Recovery of the Ancient Sciences of Antiquity: Mathematics, Astronomy, Geography. In *Rome Reborn: The Vatican Library and Renaissance Culture*, ed. A. Grafton, 125–167. Washington, DC: Library of Congress.

Swerdlow, Noel M., and Otto Neugebauer. 1984. *Mathematical Astronomy in Copernicus's De revolutionibus*. Vol. 2. New York: Springer.

Sylla, Edith D. 1971. Medieval Quantification of Qualities: The 'Merton School'. *Archive for History of Exact Sciences* 8: 9–39.

———. 1973. Medieval Concepts of the Latitude of Forms: The Oxford Calculators. *Archives d'Histoire Doctrinale et Litteraire du Moyen Age* 40: 223–283.

Tachau, Katherine H. 1988. *Vision and Certitude in the Age of Ockham: Optics, Epistemology, and the Foundations of Semantics, 1250–1345*. Leiden: Brill.

———. 1997. *Et maxime visus, cuius species venit ad stellas et ad quem species stellarum veniunt*: *Perspectiva* and *Astrologia* in Late Medieval Thought. In *Micrologus 5: Natura, scienze e società medevali*, 201–224. Florence: SISMEL—Galluzzo.

Talkenberger, Heike. 1990. *Sintflut: Prophetie und Zeitgeschehen in Texten und Holzschnitten astrologischer Flugschriften*. Tübingen: Niemeyer.

Tarnas, Richard. 2006. *Cosmos and Psyche: Intimations of a New World View*. New York: Viking.

———. 1995. *Prometheus the Awakener: An Essay on the Archetypal Meaning of the Planet Uranus*. Woodstock: Spring Publications.

Taylor, Lily Ross. 1949. *Party Politics in the Age of Caesar*. Berkeley: University of California Press.

Tester, S. 1987. *Jim, A History of Western Astrology*. Woodbridge: Boydell Press.

Testi Umanistici su l'Ermetismo, E. Garin, M. Brini, C. Vasoli and P. Zambelli (ed.). Rome: Fratelli Bocca, 1955.

Texts and Contexts in Ancient and Medieval Science, Edith Sylla and Michael McVaugh (ed.). Leiden: Brill, 1997.

Thomas, Keith. 1971. *Religion and the Decline of Magic*. New York: Scribner.

Thoren, Victor E. 1990. *The Lord of Uraniborg: A Biography of Tycho Brahe*. Cambridge: Cambridge University Press.

Thorndike, Lynn. 1923–58. *History of Magic and Experimental Science*. Vol. 8. New York: Columbia University Press.

———. 1944. *University Records and Life in the Middle Ages*. New York: Columbia University Press.

———. 1955. The True Place of Astrology in the History of Science. *Isis* 46: 273–278.

———. 1959. Notes upon Some Medieval Astronomical, Astrological and Mathematical Manuscripts at Florence, Milan, Bologna and Venice. *Isis* 50: 43–45.

———. 1965. *Michael Scot*. London: Nelson.

Tilmann, Sister Jean Paul, O.P. 1971. *An Appraisal of the Geographical Works of Albertus Magnus and His Contribution to Geographical Thought*, Michigan Geographical Publication No. 4. Ann Arbor: Dept. of Geography, University of Michigan.

Torrell, Jean-Pierre. 1996. *Saint Thomas Aquinas*, Vol. 1, *The Person and His Work*, Trans. Robert Royal. Washington, D.C.: Catholic University of America Press.

Travaglia, Pinella. 1999. *Magic, Causality and Intentionality: The Doctrine of Rays in al-Kindi*. Florence: SISMEL—Galluzzo.

Trexler, Richard C. 1980. *Public Life in Renaissance Florence*. Ithaca: Cornell University Press.

Trexler, Richard C., and Mary Elizabeth Lewis. 1987. Two Captains and Three Kings: New Light on the Medici Chapel. In *Church and Community 1200–1600: Studies in the History of Florence and New Spain*, ed. Trexler, 169–244. Rome: Edizioni di Storia e Letteratura. (originally published, 1981).

Trinkaus, Charles. 1960. A Humanist's Image of Humanism: The Inaugural Orations of Bartolommeo della Fonte. *Studies in the Renaissance* 7: 90–147.

Tuohy, Thomas. 1996. *Herculean Ferrara: Ercole d'Este, 1471–1505, and the Invention of a Ducal Capital*. Cambridge: Cambridge University Press.

Tur, Alexandre. 2014. À l'entrée du soleil en Bélier: Les prédictions astrologiques annuelles latines dans l'Europe du XVème siècle (1405–1484), 3 vols. Thèse pour le diplome d'archiviste paléographe, École Nationale des Chartes.

Broecke, Vanden. 2003. *Steven, The Limits of Influence: Pico, Louvain, and the Crisis of Renaissance Astrology*. Brill: Louvain.

Van Hasselt, Marie-Cécile. 1997. "Les livres des sortes en Italie de 1482 à 1551: L'imaginaire astrologique, les systemes de causalité et le marge de liberté, accordée à l'individu," PhD thesis, 2 vols., Université de la Sorbonne Nouvelle (Paris III).

Vasoli, Cesare. 1977. L'astrologia a Ferrara tra la metà del Quattrocento e la metà del Cinquecento. In *Il Rinascimento nelle corti padane: Società e cultura*, 469–567. Bari: De Donato.

———. 1980. Gli astri e la corte (L'astrologia a Ferrara nell'età ariostesca). In *La cultura delle corti*, 129–158. Florence: Il Portolano.

Ventrone, Paola. 1992. *Le tems revient: feste e spettacoli nella Firenze di Lorenzo il Magnifico*. Cinisello Balsamo: Silvana.

———. Lorenzo's *Politica Festiva*. In *Lorenzo the Magnificent: Culture and Politics*, 105–116.

Vescovini, Graziella Federici. 1979. *Astrologia e scienza: La crisi dell'aristotelismo sul cadere del Trecento e Biagio Pelacani da Parma*. Florence: Vallecchi.

———. 1981. Su alcune testimonianze dell'influenza di Alberto Magno come 'metafisico', scienziato e 'astrologo' nella filosofia padovana del cadere del secolo XIV: Angelo di Fossombrone e Biagio Pelacani da Parma. In *Albert der Grosse: Seine Zeit, Sein Werk, Seine Wirkung*, ed. A. Zimmerman, 155–176. Berlin: de Gruyter.

———. 1983. L'influenza di Alberto Magno come metafisico, scienziato e astrologo sul pensiero di Angelo da Fossombrone e Biagio Pelicani. In *"Arti" e filosofia nel secolo XIV: Studi sulla tradizione Aristotelica e i "moderni,"*, 77–100. Florence: Vallecchi.

———. 1991. L'astrologia all'università di Ferrara nel Quattrocento. In *La rinascita del sapere: libri e maestri dello studio Ferrarese*, ed. Patrizia Castelli, 293–306. Venice: Marsilio.

———. 1992. Lorenzo il Magnifico e l'astronomia a Firenze. *Il Ponte* 48 (11): 77–106.

———. 1998. I programmi degli insegnamenti del collegio di medicina, filosofia e astrologia dello statuto dell'universitaria di Bologna del 1405. In *Roma, magistra mundi: itineraria culturae medievalis. Melanges offerts au Pere L.E. Boyle*, ed. J. Hamesse. Louvain-la-Neuve: Fédération des Instituts d'Etudes Médiévales, I: 193–223.

———. 2005. Note di commento a alcuni passi del *Libro di pittura*: 'L'astrologia che nulla fa senza la prospettiva'. In *Leonardo e Pico: analogie, contatti, confronti*, ed. F. Frosini, 99–129. Florence: Olschki.

———. 2008. *Medioevo Magico. La magia tra religione e scienza nei secoli XIII e XIV*. Torino: UTET.

Vitali, Gerolamo. 2003. *Lexicon Mathematicum Astronomicum Geometricum* (anastatic reprint of the 1668 Paris edition), edited by Giuseppe Bezza with a preface by Ornella Pompeo Faracovi (Medium Coeli: Testi Filosofici e Scientifici dei Secoli XVI e XVII, II). La Spezia: Agora Edizioni.

Vodraska, Stanley Luis, 1969. Pseudo-Aristotle *De causis proprietatum et elementorum:* Critical Study and Edition, PhD thesis. University of London.

Volk, Katharina. 2009. *Manilius and His Intellectual Background*. Oxford: Oxford University Press.

Von Bezold, Friedrich. 1918. Astrologische Geschichtskonstruktion im Mittelalter. In *Aus Mittelalter und Renaissance: Kulturgeschichtliche Studien*, 165–195. Munich: R. Oldenbourg.

Von Stuckrad, Kocku. 2003. *Geschichte der Astrologie: Von den Anfängen bis zur Gegenwart*. Beck: Munich.

Walker, Daniel P. 1958. *Spiritual and Demonic Magic from Ficino to Campanella*. London: Warburg Institute.

Wallace, William A. 1972–74. *Causality and Scientific Explanation*. Vol. 2. Ann Arbor: University of Michigan Press.

Wallis, Richard T. 1995. *Neoplatonism*, second ed. with a foreword and updated bibliography by Lloyd P. Gerson. London: Duckworth, (originally published, 1972).

Warburg, Aby. 1999. Pagan Antique Prophecy in Words and Images in the Age of Luther. In *The Renewal of Pagan Antiquity*, Trans. D. Britt, 597–697. Los Angeles: Getty Center for the History of Art and the Humanities.

Weill-Parot, Nicolas. 2002a. *Les "images astrologiques" au Moyen Age et à la Renaissance: Spéculations intellectuelles et pratiques magiques (XIIe-XVe siècle)*. Paris: Champion.

———. 2002b. Astral Magic and Intellectual Changes (Twelfth-Fifteenth Centuries): 'Astrological Images' and the Concept of 'Addressative' Magic. In *The Metamorphosis of Magic from Late Antiquity to the Early Modern Period*, ed. Jan Bremmer and Jan R. Veenstra, 167–187. Louvain: Peeters.

Weinstein. 1970. *Savonarola and Florence: Prophecy and Patriotism in the Renaissance*. Princeton: Princeton University Press.

Weisheipl, James. The Life and Works of St. Albert the Great. In *Albertus Magnus and the Sciences*, 13–51.

Westfall, Richard S. Newton and Alchemy. In *Occult and Scientific Mentalities*, 315–335.

Westman, Robert S. 1980. The Astronomer's Role in the Sixteenth Century: A Preliminary Study. *History of Science* 18: 105–147.

———. 1993. Copernicus and the Prognosticators: The Bologna Period, 1496–1500. *Universitas* 5: 1–5.

———. 2011. *Copernican Question: Prognostication, Skepticism, and Celestial Order*. Berkeley: University of California Press.

Whitfield, Peter. 2001. *Astrology: A History*. New York: Abrams.

Wiesner, Hillary Suzanne, 1993. *The Cosmology of al-Kindi*, Phd thesis. Harvard University.

Willard, Charity C. 1984. *Christine de Pizan: Her Life and Works*. New York: Cannon.

Williams, Steven J. 1994. Roger Bacon and his Edition of the Pseudo-Aristotelian *Secretum secretorum*. *Speculum* 69: 57–73.

———. 1995. Defining the Corpus Aristotelicum. *Journal of the Warburg and Courtauld Institutes* 58: 29–51.

———. 2003. *The Secret of Secrets: The Scholarly Career of a Pseudo-Aristotelian Text in the Latin Middle Ages*. Ann Arbor: The University of Michigan Press.

Witchcraft and Magic in Europe: The Middle Ages, Bengt Ankarloo and Stuart Clark (ed.). Philadelphia: University of Pennsylvania Press, 2002.

Woods-Marsden, Joanna. 1988. *The Gonzaga of Mantua and Pisanello's Arthurian Frescoes*. Princeton: Princeton University Press.

Woodward, David, and Herbert M. Howe. Roger Bacon on Geography and Cartography. In *Roger Bacon and the Sciences*, 199–222.

Woodward, W.H. 1897. *Vittorino da Feltre and Other Humanist Educators*. Cambridge: Cambridge University Press. (rep. N.Y., 1963).

Wootton, David. 2015. *The Invention of Science: A New History of the Scientific Revolution*. New York: HarperCollins.

Wright, Alison. A Portrait for the Sforza Visit. In *Lorenzo the Magnificent: Culture and Politics*, 65–92.

Yates, Frances A. 1964. *Giordano Bruno and the Hermetic Tradition*. Chicago: University of Chicago Press.

———. 1979. *The Occult Philosophy in the Elizabethan Age*. London: Routledge.

Zambelli, Paola. 1992. *The Speculum Astronomiae and its Enigma: Astrology, Theology and Science in Albertus Magnus and his Contemporaries*. Dordrecht: Kluwer.

Zanier, Giancarlo. 1983. Ricerche sull'occultismo a Padova nel. Sec. XV. In *Scienza e filosofia all'università di Padova nel Quattrocento*, ed. A. Poppi, 345–372. Trieste: LINT.

Zika, Charles. 2007. *The Appearance of Witchcraft: Print and Visual Culture in Sixteenth-Century Europe*. Abingdon: Routledge.

Zinner, Ernst. 1968. *Leben und Wirken des Joh. Müller von Königsberg Gennant Regiomontanus*. 2nd ed. O. Zeller: Osnabrück.

———. 1990. *Regiomontanus: His Life and Works,* Trans. E. Brown. Amsterdam: North Holland.

Index

H D. Rutkin, *Sapientia Astrologica: Astrology, Magic and Natural Knowledge, ca. 1250–1800*, Archimedes 55, https://doi.org/10.1007/978-3-030-10779-6

B

C

G